VECTOR DIFFERENTIAL OPERATIONS

$$\nabla \Phi = \bar{a}_x \frac{\partial \Phi}{\partial x} + \bar{a}_y \frac{\partial \Phi}{\partial y} + \bar{a}_z \frac{\partial \Phi}{\partial z}$$

$$\nabla \cdot \bar{D} = \frac{\partial D_x}{\partial x} + \frac{\partial D_y}{\partial y} + \frac{\partial D_z}{\partial z}$$

$$\nabla \times \bar{H} = \bar{a}_x \left(\frac{\partial H_z}{\partial y} - \frac{\partial H_y}{\partial z} \right) + \bar{a}_y \left(\frac{\partial H_x}{\partial z} - \frac{\partial H_z}{\partial x} \right) + \bar{a}_z \left(\frac{\partial H_y}{\partial x} - \frac{\partial H_x}{\partial y} \right)$$

$$\nabla^2 \Phi = \frac{\partial^2 \Phi}{\partial x^2} + \frac{\partial^2 \Phi}{\partial y^2} + \frac{\partial^2 \Phi}{\partial z^2}$$

$$\nabla^2 \bar{A} = \bar{a}_x \nabla^2 A_x + \bar{a}_y \nabla^2 A_y + \bar{a}_z \nabla^2 A_z$$

$$\nabla \Phi = \bar{a}_r \frac{\partial \Phi}{\partial r} + \bar{a}_\phi \frac{1}{r} \frac{\partial \Phi}{\partial \phi} + \bar{a}_z \frac{\partial \Phi}{\partial z}$$

$$\nabla \cdot \bar{D} = \frac{1}{r} \frac{\partial}{\partial r} (r D_r) + \frac{1}{r} \frac{\partial D_\phi}{\partial \phi} + \frac{\partial D_z}{\partial z}$$

$$\nabla \times \bar{H} = \bar{a}_r \left[\frac{1}{r} \frac{\partial H_z}{\partial \phi} - \frac{\partial H_\phi}{\partial z} \right] + \bar{a}_\phi \left[\frac{\partial H_r}{\partial z} - \frac{\partial H_z}{\partial r} \right] + \bar{a}_z \left[\frac{1}{r} \frac{\partial (r H_\phi)}{\partial r} - \frac{1}{r} \frac{\partial H_r}{\partial \phi} \right]$$

$$\nabla^2 \Phi = \frac{1}{r} \frac{\partial}{\partial r} \left(r \frac{\partial \Phi}{\partial r} \right) + \frac{1}{r^2} \frac{\partial^2 \Phi}{\partial \phi^2} + \frac{\partial^2 \Phi}{\partial z^2}$$

$$\nabla^2 \bar{A} = \bar{a}_r \left(\nabla^2 A_r - \frac{2}{r^2} \frac{\partial A_\phi}{\partial \phi} - \frac{A_r}{r^2} \right) + \bar{a}_\phi \left(\nabla^2 A_\phi + \frac{2}{r^2} \frac{\partial A_r}{\partial \phi} - \frac{A_\phi}{r^2} \right) + \bar{a}_z (\nabla^2 A_z)$$

$$\nabla \Phi = \bar{a}_r \frac{\partial \Phi}{\partial r} + \bar{a}_\theta \frac{1}{r} \frac{\partial \Phi}{\partial \theta} + \frac{\bar{a}_\phi}{r \sin \theta} \frac{\partial \Phi}{\partial \phi}$$

$$\nabla \cdot \bar{D} = \frac{1}{r^2} \frac{\partial}{\partial r} (r^2 D_r) + \frac{1}{r \sin \theta} \frac{\partial}{\partial \theta} (\sin \theta D_\theta) + \frac{1}{r \sin \theta} \frac{\partial D_\phi}{\partial \phi}$$

$$\nabla \times \bar{H} = \frac{\bar{a}_r}{r \sin \theta} \left[\frac{\partial}{\partial \theta} (H_\phi \sin \theta) - \frac{\partial H_\theta}{\partial \phi} \right]$$

$$+ \frac{\bar{a}_\theta}{r} \left[\frac{1}{\sin \theta} \frac{\partial H_r}{\partial \phi} - \frac{\partial}{\partial r} (r H_\phi) \right] + \frac{\bar{a}_\phi}{r} \left[\frac{\partial}{\partial r} (r H_\theta) - \frac{\partial H_r}{\partial \theta} \right]$$

$$\nabla^2 \Phi = \frac{1}{r^2} \frac{\partial}{\partial r} \left(r^2 \frac{\partial \Phi}{\partial r} \right) + \frac{1}{r^2 \sin \theta} \frac{\partial}{\partial \theta} \left(\sin \theta \frac{\partial \Phi}{\partial \theta} \right) + \frac{1}{r^2 \sin^2 \theta} \frac{\partial^2 \Phi}{\partial \phi^2}$$

$$\nabla^2 \bar{A} = \bar{a}_r \left[\nabla^2 A_r - \frac{2}{r^2} \left(A_r + \cot \theta A_\theta + \csc \theta \frac{\partial A_\phi}{\partial \phi} + \frac{\partial A_\theta}{\partial \theta} \right) \right]$$

$$+ \bar{a}_\theta \left[\nabla^2 A_\theta - \frac{1}{r^2} \left(\csc^2 \theta A_\theta - 2 \frac{\partial A_r}{\partial \theta} + 2 \cot \theta \csc \theta \frac{\partial A_\phi}{\partial \phi} \right) \right]$$

$$+ \bar{a}_\phi \left[\nabla^2 A_\phi - \frac{1}{r^2} \left(\csc^2 \theta A_\phi - 2 \csc \theta \frac{\partial A_r}{\partial \phi} - 2 \cot \theta \csc \theta \frac{\partial A_\theta}{\partial \phi} \right) \right]$$

VECTOR FORMULAS

$$\nabla(\Phi + \psi) = \nabla\Phi + \nabla\psi$$

$$\nabla \cdot (\bar{A} + \bar{B}) = \nabla \cdot \bar{A} + \nabla \cdot \bar{B}$$

$$\nabla \times (\bar{A} + \bar{B}) = \nabla \times \bar{A} + \nabla \times \bar{B}$$

$$\nabla(\Phi\psi) = \Phi \nabla\psi + \psi \nabla\Phi$$

$$\nabla \cdot (\psi\bar{A}) = \bar{A} \cdot \nabla\psi + \psi\nabla \cdot \bar{A}$$

$$\nabla \cdot (\bar{A} \times \bar{B}) = \bar{B} \cdot \nabla \times \bar{A} - \bar{A} \cdot \nabla \times \bar{B}$$

$$\nabla \times (\Phi\bar{A}) = \nabla\Phi \times \bar{A} + \Phi\nabla \times \bar{A}$$

$$\nabla \times (\bar{A} \times \bar{B}) = \bar{A}\nabla \cdot \bar{B} - \bar{B}\nabla \cdot \bar{A} + (\bar{B} \cdot \nabla)\bar{A} - (\bar{A} \cdot \nabla)\bar{B}$$

$$\nabla \cdot \nabla\Phi = \nabla^2\Phi$$

$$\nabla \cdot \nabla \times \bar{A} = 0$$

$$\nabla \times \nabla\Phi = 0$$

$$\nabla \times \nabla \times \bar{A} = \nabla(\nabla \cdot \bar{A}) - \nabla^2\bar{A}$$

$$\nabla(\bar{A} \cdot \bar{B}) = (\bar{A} \cdot \nabla)\bar{B} + (\bar{B} \cdot \nabla)\bar{A} + \bar{A} \times (\nabla \times \bar{B}) + \bar{B}(\nabla \times \bar{A})$$

$$\bar{A} \cdot \bar{B} \times \bar{C} = \bar{B} \cdot \bar{C} \times \bar{A} = \bar{C} \cdot \bar{A} \times \bar{B}$$

$$\bar{A} \times (\bar{B} \times \bar{C}) = \bar{B}(\bar{A} \cdot \bar{C}) - \bar{C}(\bar{A} \cdot \bar{B})$$

Stokes's Theorem $\int_S \nabla \times \bar{A} \cdot d\bar{S} = \oint_l \bar{A} \cdot d\bar{l}$ Open surface S bounded by line l

Divergence Theorem $\int_V \nabla \cdot \bar{A}\, dV = \oint_S \bar{A} \cdot d\bar{S}$ Volume V bounded by surface S

PHYSICAL CONSTANTS

Vacuum permittivity	$\epsilon_0 = 8.854 \times 10^{-12}$ farads/meter
Vacuum permeability	$\mu_0 = 4\pi \times 10^{-7}$ henrys/meter
Electron charge magnitude*	$e = 1.60210 \times 10^{-19}$ coulombs
Electron rest mass*	$m_0 = 9.1091 \times 10^{-31}$ kilograms
Electron charge-to-mass ratio*	$= 1.758796 = 10^{11}$ coulombs/kilogram
Proton rest mass*	$= 1.67252 \times 10^{-27}$ kilograms
Speed of light in vacuum*	$c = 2.997925 \times 10^8$ meters/second
Conductivity of copper†	$= 5.800 \times 10^7$ mhos/meter

* Values recommended by National Academy of Sciences—National Research Council Committee on Fundamental Constants in 1963.
† Bureau of Standards value in *Handbook of Chemistry and Physics*, Chemical Rubber Publishing Co. (1962–1963) Cleveland, Ohio.

FIELDS AND WAVES IN COMMUNICATION ELECTRONICS

FIELDS AND WAVES IN COMMUNICATION ELECTRONICS

SECOND EDITION

SIMON RAMO
TRW Inc.

JOHN R. WHINNERY
University of California, Berkeley

THEODORE VAN DUZER
University of California, Berkeley

JOHN WILEY & SONS
NEW YORK   CHICHESTER   BRISBANE   TORONTO   SINGAPORE

Library of Congress Cataloging in Publication Data:

Ramo, Simon.
  Fields and waves in communication electronics.

  Includes indexes.
  1. Electromagnetic fields.   2. Electromagnetic waves.
3. Telecommunication.   I. Whinnery, John R.   II. Van
Duzer, Theodore.   III. Title.
QC665.E4R36   1984   537   84-7355
ISBN 0-471-87130-3

Printed in the United States of America
10 9

To our wives

Virginia
Patricia
Janice

PREFACE

This book is an intermediate-level text on electromagnetic fields and waves. It represents a substantial revision of the first edition, which in turn built upon an earlier volume by two of the authors.* It assumes an introductory course in field concepts, which can be the lower-division physics course in many colleges or universities, and a background in calculus. Material on vectors, differential equations, Fourier analysis, and complex notation for sinusoids is included in a form suitable for review or a first introduction. We have given such introductions wherever the material is to be used and have related them to real problems in fields and waves.

Although the fundamentals of electromagnetics have not changed since this book's first edition, new applications have appeared, and emphases have consequently changed. The use of coherent optics in communications and information processing, for example, has become well established. The importance of field interactions in a variety of materials continues to grow. Miniaturization and integration of devices at all frequencies have introduced new interaction problems and new types of wave-guiding devices. Thus, we have placed additional stress on each of these subjects.

We have substantially reorganized this text in response to the suggestions of many of its users. The main feature of this reorganization is that simple wave concepts are introduced as soon as possible after the review of static fields so that

---

* S. Ramo and J. R. Whinnery, *Fields and Waves in Modern Radio*, John Wiley & Sons (first edition, 1944; second edition, 1953). The first edition was prepared with the assistance of the General Electric Company when the authors were employed in its laboratories.

a first-term student will encounter dynamic as well as static concepts. Thus, material on product solutions and conformal transformations is presented later on (in Chapter 7) when these techniques are needed for wave-guiding problems. This delay permits the user to immediately apply the methods to wave as well as static problems. However, once the student has covered the material of the first three chapters, he or she is provided with considerable flexibility in using this text, especially with respect to the techniques of analysis (Chapter 7) and the electromagnetics of circuits (Chapter 4). Similarly, the later chapters on materials, optics, and microwave networks can be used in various orders.

In addition to making suggestions on content and order, users of the earlier edition requested more examples. We have therefore added new examples and have clearly designated existing ones. There was less uniformity of opinion on whether problems should be grouped at the end of a chapter or distributed throughout, as was done previously. Users indicated a slight preference for the former arrangement; hence, we have grouped problems by chapters, and have identified related sections with problem numbering. In addition, we have added many new problems. These problems are not, of course, related solely to a single section; they are designed to integrate previous material as well.

The text has benefited from the generous help of our students and colleagues at Berkeley, users in other universities, and the thoughtful reviews of the manuscript. We are deeply appreciative. We particularly thank Professors D. J. Angelakos and C. K. Birdsall for their advice in the development of this text. We gratefully acknowledge the work of Bettye Fuller and Lillian On in carefully and skillfully preparing the manuscript.

Canoga Park, California   Simon Ramo
Berkeley, California   John R. Whinnery
                                                             Theodore Van Duzer

CONTENTS

CHAPTER 1 STATIONARY ELECTRIC FIELDS   1

1.1 Introduction   1
     BASIC LAWS AND CONCEPTS OF ELECTROSTATICS   2
1.2 Force Between Electric Charges; The Concept of Electric Field   2
1.3 The Concept of Electric Flux and Flux Density; Gauss's Law   6
1.4 Examples of the Use of Gauss's Law   9
1.5 Surface and Volume Integrals; Gauss's Law in Vector Form   13
1.6 Tubes of Flux; Plotting of Field Lines   15
1.7 Energy Considerations; Conservative Property of Electrostatic
     Fields   17
1.8 Electrostatic Potential: Equipotentials   20
1.9 Capacitance   27
     DIFFERENTIAL FORMS OF ELECTROSTATIC LAWS   28
1.10 Gradient   28
1.11 The Divergence of an Electrostatic Field   30
1.12 Laplace's and Poisson's Equations   35
1.13 Static Fields Arising from Steady Currents   37
1.14 Boundary Conditions in Electrostatics   38
1.15 Direct Integration of Laplace's Equation: Field Between Coaxial
     Cylinders with Two Dielectrics   42
1.16 Direct Integration of Poisson's Equation: The pn Semiconductor
     Junction   44
1.17 Uniqueness of Solutions   47

ix

SPECIAL TECHNIQUES FOR ELECTROSTATIC
PROBLEMS   48

1.18   The Use of Images   48
1.19   Graphical Field Mapping   53
1.20   Examples of Information Obtained from Field Maps   56
      ENERGY IN FIELDS   58
1.21   Energy of an Electrostatic System   58
Problems   61

CHAPTER 2   STATIONARY MAGNETIC FIELDS   **68**

2.1   Introduction   68
      STATIC MAGNETIC FIELD LAWS AND CONCEPTS   70
2.2   Concept of a Magnetic Field   70
2.3   Ampère's Law   71
2.4   The Line Integral of Magnetic Field   75
2.5   Inductance from Flux Linkages; External Inductance   79
      DIFFERENTIAL FORMS FOR MAGNETOSTATICS AND
THE USE OF POTENTIAL   82
2.6   The Curl of a Vector Field   82
2.7   Curl of Magnetic Field   86
2.8   Relation Between Differential and Integral Forms of the Field
Equations   88
2.9   Vector Magnetic Potential   91
2.10   Distant Field of Current Loop: Magnetic Dipole   94
2.11   Divergence of Magnetic Flux Density   96
2.12   Differential Equation for Vector Magnetic Potential   96
2.13   Scalar Magnetic Potential for Current-Free Regions   98
2.14   Boundary Conditions for Static Magnetic Fields   99
2.15   Materials with Permanent Magnetization   100
      MAGNETIC FIELD ENERGY   104
2.16   Energy of a Static Magnetic Field   104
2.17   Inductance from Energy Storage; Internal Inductance   106
Problems   107

CHAPTER 3   MAXWELL'S EQUATIONS   **111**

3.1   Introduction   111
      LARGE-SCALE AND DIFFERENTIAL FORMS OF
MAXWELL'S EQUATIONS   113

Contents

3.2   Voltages Induced by Changing Magnetic Fields   113
3.3   Faraday's Law for a Moving System   116
3.4   Continuity of Charge and the Concept of Displacement Current   119
3.5   Physical Pictures of Displacement Current   120
3.6   Maxwell's Equations in Differential Equation Form   123
3.7   Maxwell's Equations in Large-Scale Form   126
3.8   Maxwell's Equations for the Time-Periodic Case   127
       EXAMPLES OF USE OF MAXWELL'S EQUATIONS   130
3.9   Maxwell's Equations and Plane Waves   130
3.10   Uniform Plane Waves with Steady-State Sinusoids   133
3.11   The Wave Equation in Three Dimensions   135
3.12   Power Flow in Electromagnetic Fields—Poynting's Theorem   137
3.13   Poynting's Theorem for Phasors   141
3.14   Continuity Conditions for ac Fields at a Boundary; Uniqueness of
       Solutions   143
3.15   Boundary Conditions at a Perfect Conductor for ac Fields   146
3.16   Penetration of Electromagnetic Fields into a Good Conductor   147
3.17   Internal Impedance of a Plane Conductor   151
3.18   Power Loss in a Plane Conductor   154
       POTENTIALS FOR TIME-VARYING FIELDS   156
3.19   A Possible Set of Potentials for Time-Varying Fields   156
3.20   The Retarded Potentials as Integrals over Charges and Currents   158
3.21   The Retarded Potentials for the Time-Periodic Case   160
Problems   161

CHAPTER 4   THE ELECTROMAGNETICS OF CIRCUITS   168

4.1   Introduction
       THE IDEALIZATIONS OF CLASSICAL CIRCUIT THEORY   169
4.2   Kirchhoff's Voltage Law   169
4.3   Kirchhoff's Current Law and Multimesh Circuits   174
       SKIN EFFECT IN PRACTICAL CONDUCTORS   178
4.4   Distribution of Time-Varying Currents in Conductors of Circular
       Cross Section   178
4.5   Impedance of Round Wires   180
       CALCULATION OF CIRCUIT ELEMENTS   184
4.6   Self-Inductance Calculations   184
4.7   Mutual Inductance   187
4.8   Inductance of Practical Coils   191
4.9   Self- and Mutual Capacitance   193

CIRCUITS THAT ARE NOT SMALL COMPARED WITH
WAVELENGTH   196

4.10   Distributed Effects and Retardation   196
4.11   Circuit Formulation Through the Retarded Potentials   198
4.12   Circuits with Radiation   203
Problems   206

CHAPTER 5   TRANSMISSION LINES   210

5.1   Introduction   210
       TIME AND SPACE DEPENDENCE OF SIGNALS ON
       IDEAL TRANSMISSION LINES   211
5.2   Voltage and Current Variations Along an Ideal Transmission
       Line   211
5.3   Relation of Field and Circuit Analysis of Transmission Lines 215
5.4   Reflection and Transmission at a Resistive Discontinuity   216
       SINUSOIDAL WAVES ON IDEAL TRANSMISSION
       LINES WITH DISCONTINUITIES   223
5.5   Reflection and Transmission Coefficients and Impedance and
       Admittance Transformations for Sinusoidal Voltages   223
5.6   Standing-Wave Ratio   226
5.7   The Smith Transmission-Line Chart   229
5.8   Purely Standing Wave on an Ideal Line   238
       NONIDEAL TRANSMISSION LINES   241
5.9   Power Loss and Attenuation in Low-Loss Lines   241
5.10   Input Impedance and Quality Factor for Resonant Transmission
       Lines   243
5.11   Transmission Lines with General Forms of Distributed
       Impedances; $\omega-\beta$ Diagram   247
5.12   Group and Energy Velocities   254
5.13   Backward Waves   257
5.14   Transmission Lines Used with Pulses   258
5.15   Nonuniform Transmission Lines   261
Problems   263

CHAPTER 6   PLANE WAVE PROPAGATION AND
REFLECTION   270

6.1   Introduction   270
       PLANE WAVE PROPAGATION   27

6.2	Uniform Plane Waves in a Perfect Dielectric	271
6.3	Polarization of Plane Waves	276
6.4	Waves in Imperfect Dielectrics and Conductors	279
	PLANE WAVES NORMALLY INCIDENT ON DISCONTINUITIES	283
6.5	Reflection of Normally Incident Plane Waves from Perfect Conductors	283
6.6	Transmission-Line Analogy of Wave Propagation: The Impedance Concept	285
6.7	Normal Incidence on a Dielectric	288
6.8	Reflection Problems with Several Dielectrics	291
	PLANE WAVES OBLIQUELY INCIDENT ON DISCONTINUITIES	296
6.9	Incidence at Any Angle on Perfect Conductors	296
6.10	Phase Velocity and Impedance for Waves at Oblique Incidence	299
6.11	Incidence at Any Angle on Dielectrics	302
6.12	Total Reflection	306
6.13	Polarizing or Brewster Angle	308
6.14	Multiple Dielectric Boundaries with Oblique Incidence	309
Problems		312

CHAPTER 7 TWO- AND THREE-DIMENSIONAL BOUNDARY VALUE PROBLEMS

		317
7.1	Introduction	317
	THE BASIC DIFFERENTIAL EQUATIONS AND NUMERICAL METHODS	318
7.2	Roles of Helmholtz, Laplace, and Poisson Equations	319
7.3	Numerical Solution of the Laplace, Poisson, and Helmholtz Equations	320
	METHOD OF CONFORMAL TRANSFORMATION	326
7.4	Method of Conformal Transformation and Introduction to Complex-Function Theory	326
7.5	Properties of Analytic Functions of Complex Variables	328
7.6	Conformal Mapping for Laplace's Equation	331
7.7	The Schwarz Transformation for General Polygons	340
7.8	Conformal Mapping for Wave Problems	343
	PRODUCT-SOLUTION METHOD	346
7.9	Laplace's Equation in Rectangular Coordinates	346

7.10   Static Field Described by a Single Rectangular Harmonic   349
7.11   Fourier Series and Integral   351
7.12   Series of Rectangular Harmonics for Two- and
       Three-Dimensional Static Fields   355
7.13   Cylindrical Harmonics for Static Fields   360
7.14   Bessel Functions   364
7.15   Bessel Function Zeros and Formulas   369
7.16   Expansion of a Function as a Series of Bessel Functions   371
7.17   Fields Described by Cylindrical Harmonics   373
7.18   Spherical Harmonics   375
7.19   Product Solutions for the Helmholtz Equation in Rectangular
       Coordinates   382
7.20   Product Solutions for the Helmholtz Equation in Cylindrical
       Coordinates   383
Problems   384

CHAPTER 8   WAVEGUIDES WITH CYLINDRICAL CONDUCTING BOUNDARIES

                                                                     392

8.1   Introduction   392
       GENERAL FORMULATION FOR GUIDED WAVES   393
8.2   Basic Equations and Wave Types for Uniform Systems   393
       CYLINDRICAL WAVEGUIDES OF VARIOUS CROSS
       SECTIONS   396
8.3   Waves Guided by Perfectly Conducting Parallel Plates   396
8.4   Guided Waves Between Parallel Planes as Superposition of
       Plane Waves   402
8.5   Parallel-Plane Guiding System with Losses   405
8.6   Stripline and Microstrip Transmission Systems   407
8.7   Rectangular Waveguides   411
8.8   The TE_{10} Wave in a Rectangular Guide   418
8.9   Circular Waveguides   422
8.10   Higher Order Modes on Coaxial Lines   428
8.11   Excitation and Reception of Waves in Guides   430
       GENERAL PROPERTIES OF GUIDED WAVES   434
8.12   General Properties of TEM Waves on Multiconductor Lines   434
8.13   General Properties of TM Waves in Cylindrical Conducting
       Guides of Arbitrary Cross Section   438
8.14   General Properties of TE Waves in Cylindrical Conducting
       Guides of Arbitrary Cross Section   442

8.15   Waves Below and Near Cutoff   444
8.16   Dispersion of Signals Along Transmission Lines and Waveguides 446
Problems   449

CHAPTER 9   SPECIAL WAVEGUIDE TYPES   **456**

9.1   Introduction   456
9.2   Dielectric Slab Guides   457
9.3   Parallel-Plane Radial Transmission Lines   460
9.4   Circumferential Modes in Radial Lines; Sectoral Horns   464
9.5   Duality; Propagation Between Inclined Planes   466
9.6   Waves Guided by Conical Systems   467
9.7   Ridge Waveguide   469
9.8   The Idealized Helix and Other Slow-Wave Structures   471
9.9   Surface Guiding   474
9.10   Periodic Structures and Spatial Harmonics   477
Problems   482

CHAPTER 10   RESONANT CAVITIES   **486**

10.1   Introduction   486
10.2   Elemental Concepts of Cavity Resonators   487
        RESONATORS OF SIMPLE SHAPE   489
10.3   Fields of Simple Rectangular Resonator   489
10.4   Energy Storage, Losses, and Q of Simple Resonator   491
10.5   Other Modes in the Rectangular Resonator   492
10.6   Circular Cylindrical Resonator   495
10.7   Wave Solutions in Spherical Coordinates   498
10.8   Spherical Resonators   502
        SMALL-GAP CAVITIES AND COUPLING   504
10.9   Small-Gap Cavities   504
10.10   Coupling to Cavities   507
10.11   Cavity Q and Other Figures of Merit   509
10.12   Cavity Perturbations   512
10.13   Dielectric Resonators   515
Problems   520

CHAPTER 11   MICROWAVE NETWORKS   523

11.1   Introduction   523
11.2   The Network Formulation   525
11.3   Conditions for Reciprocity   528
        TWO-PORT WAVEGUIDE JUNCTIONS   530
11.4   Equivalent Circuits for a Two Port   530
11.5   Determination of Circuit Parameters by Measurement   532
11.6   Scattering and Transmission Coefficients   535
11.7   Scattering Coefficients by Measurement   538
11.8   Cascaded Two Ports   538
11.9   Examples of Microwave and Optical Filters   542
        N-PORT WAVEGUIDE JUNCTIONS   547
11.10   Circuit and *S*-Parameter Representation of *N* Ports   547
11.11   Directional Couplers and Hybrid Networks   551
        FREQUENCY CHARACTERISTICS OF WAVEGUIDE
        NETWORKS   555
11.12   Properties of a One-Port Impedance   555
11.13   Equivalent Circuits Showing Frequency Characteristics of One
        Ports   558
11.14   Examples of Cavity Equivalent Circuits   562
11.15   Circuits Giving Frequency Characteristics of *N* Ports   566
        JUNCTION PARAMETERS BY ANALYSIS   567
11.16   Quasistatic and Other Methods of Junction Analysis   567
Problems   572

CHAPTER 12   RADIATION   577

12.1   Introduction   577
12.2   Some Types of Practical Radiating Systems   579
        FIELD AND POWER CALCULATIONS WITH
        CURRENTS ASSUMED ON THE ANTENNA   582
12.3   Electric and Magnetic Dipole Radiators   582
12.4   Systemization of Calculation of Radiating Fields and Power
        from Currents on an Antenna   586
12.5   Long Straight Wire Antenna; Half-Wave Dipole   589
12.6   Radiation Patterns and Antenna Gain   592
12.7   Radiation Resistance   595
12.8   Antennas Above Earth or Conducting Plane   596
12.9   Traveling Wave on a Straight Wire   598

12.10	V and Rhombic Antennas	600
12.11	Methods of Feeding Wire Antennas	604
	RADIATION FROM FIELDS OVER AN APERTURE	607
12.12	Fields as Sources of Radiation	607
12.13	Plane Wave Sources	610
12.14	Examples of Radiating Apertures Excited by Plane Waves	612
12.15	Electromagnetic Horns	617
12.16	Resonant Slot Antenna	619
12.17	Lenses for Directing Radiation	621
	ARRAYS OF ELEMENTS	623
12.18	Radiation Intensity with Superposition of Effects	623
12.19	Linear Arrays	627
12.20	Radiation from Diffraction Gratings	631
12.21	Polynomial Formulation of Arrays and Limitations on Directivity	633
12.22	Yagi–Uda Arrays	635
12.23	Frequency-Independent Antennas: Logarithmically Periodic Arrays	637
	FIELD ANALYSIS OF ANTENNAS	640
12.24	The Antenna as a Boundary-Value Problem	640
12.25	Direct Calculation of Input Impedance for Wire Antennas	647
12.26	Mutual Impedance Between Thin Dipoles	650
	RECEIVING ANTENNAS AND RECIPROCITY	652
12.27	A Transmitting–Receiving System	652
12.28	Reciprocity Relations	655
12.29	Equivalent Circuit of the Receiving Antenna	658
	Problems	659

CHAPTER 13 ELECTROMAGNETIC PROPERTIES OF MATERIALS

		666
13.1	Introduction	666
	LINEAR ISOTROPIC MEDIA	667
13.2	Characteristics of Dielectrics	667
13.3	Imperfect Conductors and Semiconductors	672
13.4	Perfect Conductors and Superconductors	676
13.5	Diamagnetic and Paramagnetic Responses	679
	NONLINEAR ISOTROPIC MEDIA	680
13.6	Materials with Residual Magnetization	680
13.7	Nonlinear Optics	685

ANISOTROPIC MEDIA 689

13.8   Representation of Anisotropic Dielectric Crystals 689
13.9   Plane Wave Propagation in Anisotropic Crystals 691
13.10   Plane Wave Propagation in Uniaxial Crystals 695
13.11   Electro-Optic Effects 698
13.12   Permittivity of a Stationary Plasma in a Magnetic Field 703
13.13   *TEM* Waves on a Plasma with Infinite Magnetic Field 706
13.14   Space-Charge Waves on a Moving Plasma with Infinite
        Magnetic Field 707
13.15   *TEM* Waves on a Stationary Plasma in a Finite Magnetic Field 709
13.16   Faraday Rotation 713
13.17   Permeability Matrix for Ferrites 715
13.18   *TEM* Wave Propagation in Ferrites 719
13.19   Ferrite Devices 721

Problems 726

CHAPTER 14   OPTICS 732

14.1   Introduction 732
RAY OR GEOMETRICAL OPTICS 733
14.2   Geometrical Optics Through Applications of Laws of Reflection
       and Refraction 733
14.3   Geometrical Optics as Limiting Case of Wave Optics 739
14.4   Rays in Inhomogeneous Media 741
14.5   Paraxial Ray Optics—Ray Matrices 746
14.6   Guiding of Rays by a Periodic Lens Systems or in Spherical
       Mirror Resonators 749
DIELECTRIC OPTICAL WAVEGUIDES 752
14.7   Dielectric Guides of Planar Form 752
14.8   Dielectric Guides of Rectangular Form 756
14.9   Dielectric Guides of Circular Cross Section 759
14.10   Weakly Guiding Fibers: Dispersion 762
14.11   Propagation of Gaussian Beams in Graded-Index Fibers 765
GAUSSIAN BEAMS IN SPACE AND IN OPTICAL
RESONATORS 768
14.12   Propagation of Gaussian Beams in a Homogeneous Medium 768
14.13   Transformation of Gaussian Beams by Ray Matrix 771
14.14   Gaussian Modes in Optical Resonators 774
14.15   Stability and Resonant Frequencies of Optical-Resonator Modes 779

BASIS FOR OPTICAL INFORMATION PROCESSING 781

14.16 Fourier Transforming Properties of Lenses 781

Problems 784

Appendix 1 Conversion Factors Between Systems of Units 789
Appendix 2 Coordinate Systems and Vector Relations 791
Appendix 3 Sketch of the Derivation of Magnetic Field Laws 797
Appendix 4 Complex Phasors as Used in Electrical Circuits 800

Index 805

FIELDS AND WAVES IN COMMUNICATION ELECTRONICS

1

STATIONARY ELECTRIC FIELDS

1.1 Introduction

Electric fields have their sources in electric charges—electrons and ions. Nearly all real electric fields vary to some extent with time, but for many problems the time variation is slow and the field may be considered stationary in time (static) over the interval of interest. For still other important cases (called *quasistatic*) the spatial distribution is nearly the same as for static fields even though the actual fields may vary rapidly with time. Because of the number of important cases in each of these classes, and because static field concepts are simple and thus good for reviewing the vector operations needed to describe fields quantitatively, we start with static electric fields in this chapter and static magnetic fields in the next. The student approaching the problem in this way must remember that these are special cases, and that the interactions between time-varying fields can give rise to other phenomena—most notably the wave phenomena to be studied later in the text.

Before beginning the quantitative development of this chapter, let us comment briefly on a few applications to illustrate the kinds of problems that arise in electrostatics or quasistatics. Electron and ion guns are good examples of electrostatic problems where the distribution of fields is of great importance in the design. Electrode shapes are designed to accelerate particles from a source and focus them into a beam of desired size and velocity. Electron guns are used in cathode-ray oscilloscopes, television tubes, in the microwave traveling wave tubes of radar and satellite communication systems, in electron microscopes, and for electron-beam lithography used for precision definition of integrated-circuit device features.

Many electronic circuit elements may have quite rapidly varying currents and voltages and yet at any instant have fields that are well represented by those calculated from static-field equations. This is generally true when the elements are small in comparison with wavelength. The passive capacitive, inductive, and resistive elements are thus commonly analyzed by such *quasistatic* laws, but so also are the semiconductor diodes and transistors which constitute the active elements of electronic circuits.

Transmission lines, including the strip line used in microwave and millimeter-wave integrated circuits even for frequencies well above 10 GHz, have properties that can be calculated using the laws for static fields. This is far from being a static problem, but we will see later in the text that for systems having no structural variations along one axis (along the transmission line) the fields in the transverse plane satisfy, exactly or nearly exactly, static field laws.

There are many other examples of application of knowledge of static field laws. The electrostatic precipitators used to remove dust and other solid particles from air, xerography, and power switches and transmission systems (which must be designed to avoid dielectric breakdown) all utilize static-field concepts. Electric fields generated by the human body are especially interesting examples. Thus the fields that are detected by electroencephalography (fields of the brain) and electrocardiography (fields of the heart) are of sufficiently low frequency to be distributed in the body in the same way that static fields would be.

In all the examples mentioned, the general problem is that of finding the distribution of fields produced by given sources in a specified medium with defined boundaries on the region of interest. Our approach will be to start with a simple experimental law (Coulomb's law) and then transform it into other forms which may be more general or more useful for certain classes of problems.

Most readers will have met this material before in physics courses or introductory electromagnetics courses, so the approach will be that of review with the purpose of deepening physical understanding, and improving familiarity with the needed vector algebra before turning to the more difficult time-varying problems.

BASIC LAWS AND CONCEPTS OF ELECTROSTATICS

1.2   Force Between Electric Charges; The Concept of Electric Field

It was known from ancient times that electrified bodies exert forces upon one another. The effect was quantified by Charles A. Coulomb through brilliant experiments using a torsion balance.[1] His experiments showed that like charges repel one another whereas opposite charges attract; that the force is proportional to the product of charge magnitudes; that it is inversely proportional to the square of the distance between charges; and that force acts along the line joining charges. Coulomb's experiments were done in air, but later generalizations show that force also depends upon the medium in which charges are placed. Thus force magnitude may be written

$$f = K \frac{q_1 q_2}{\varepsilon r^2} \tag{1}$$

where q_1 and q_2 are charge strengths, r is the distance between charges, ε is a constant representing the effect of the medium, and K is a constant depending upon units. Direction information is included by writing force as a vector $\mathbf{f}$ (denoted here as boldface) and defining a vector $\hat{\mathbf{r}}$ of unit length pointing from one charge directly away from the other,

$$\mathbf{f} = K \frac{q_1 q_2}{\varepsilon r^2} \hat{\mathbf{r}} \tag{2}$$

Various systems of units have been used, but that to be used in this text is the International System (SI for the equivalent in French) introduced by Giorgi in 1901. This is a meter-kilogram-second (mks) system, but the great advantage is that electric quantities are in the units actually measured: coulombs, volts, amperes, etc. Conversion factors to the classical systems still used in many references are given in Appendix 1. Thus in the SI system, force in (2) is in newtons (kg-m/s^2), q in coulombs, and r in meters. The constant K is chosen as $1/4\pi$ and the value of ε for vacuum is found from experiment to be

$$\varepsilon_0 = 8.854 \times 10^{-12} \approx \frac{1}{36\pi} \times 10^{-9} \frac{F}{m} \tag{3}$$

For other materials,

$$\varepsilon = \varepsilon_r \varepsilon_0 \tag{4}$$

where ε_r is the *relative permittivity* or *dielectric constant* of the material and is the value usually tabulated in handbooks. Here we are considering materials for which ε is a scalar independent of strength and direction of the force and of

---

[1] An excellent description of Coulomb's experiments is given in R. S. Elliott, *Electromagnetics*, McGraw-Hill, New York, 1966. For a detailed account of the history of this and other aspects of electromagnetics, see E. T. Whittaker, *A History of the Theories of Aether and Electricity*, Philosophical Library, New York, 1952; or P. F. Mottelay, *Biographical History of Electricity and Magnetism*, Charles Griffin & Co., London, 1922.

position. More general media will be considered in Chapter 13. Thus in these units Coulomb's law is written

$$\mathbf{f} = \frac{q_1 q_2}{4\pi\varepsilon r^2} \hat{\mathbf{r}} \tag{5}$$

Generalizing from the example of two charges, we deduce that a charge placed in the vicinity of a system of charges will also experience a force. This might be found by adding vectorially the component forces from the individual charges of the system, but it is convenient at this time to introduce the concept of an *electric field* as the force per unit charge for each point of the region influenced by charges. We may define this by introducing a test charge Δq at the point of definition, small enough not to disturb the charge distribution to be studied. Electric field $\mathbf{E}$ is then

$$\mathbf{E} = \frac{\mathbf{f}}{\Delta q} \tag{6}$$

where $\mathbf{f}$ is the force acting on the infinitesimal test charge Δq.

The electric field arising from a point charge q in a homogeneous dielectric is then given by the force law (5):

$$\mathbf{E} = \frac{q}{4\pi\varepsilon r^2} \hat{\mathbf{r}} \tag{7}$$

Since $\hat{\mathbf{r}}$ is the unit vector directed from the point in a direction away from the charge, the electric field vector is seen to point away from positive charges and toward negative charges as seen in the lower half of Fig. 1.2a. The units of electric field magnitude in the SI system are in volts per meter, as may be found by substituting units in (7):

$$E = \frac{\text{coulombs meter}}{\text{farads (meter)}^2} = \frac{\text{volts (V)}}{\text{meter (m)}}$$

We can see from the form of (7) that the total electric field for a system of point charges may be found by adding vectorially the fields from the individual charges, as is illustrated at point P of Fig. 1.2a for the charges q and $-q$. In this manner the electric field vector could be found for any point in the vicinity of the two charges. An *electric field line* is defined as a line drawn tangent to the electric field vector at each point in space. If the vector is constructed for enough points of the region, the electric field lines can be drawn in roughly by following the direction of the vectors as illustrated in the top half of Fig. 1.2a. Easier methods of constructing the electric field will be studied in later sections, but the present method, although laborious, demonstrates clearly the meaning of the electric field lines.

For fields produced by continuous distributions of charges, the superposition is by integration of field contributions from the differential elements of charge. For

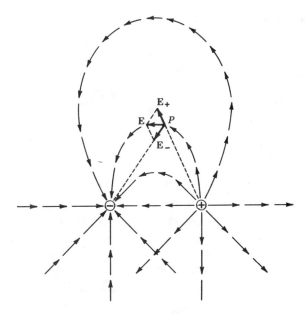

Fig. 1.2a Electric fields around two opposite charges. Lower half of figure shows the separate fields E_+ and E_- of the two charges. Upper half shows the vector sum of E_+ and E_-. Construction of E is shown at one point P.

volume distributions, the elemental charge dq is $\rho \, dV$ where ρ is charge per unit volume (C/m³) and dV is the element of volume. For surface distributions, a surface density ρ_s (C/m²) is used with elemental surface dS. For filamentary distributions, a linear density ρ_l (C/m) is used with elemental length dl. An example of a continuous distribution follows.

_____ **Example 1.2** _____
Field of a Ring of Charge

Let us calculate the electric field at points on the z axis for a ring of positive charge of radius a located in free space concentric with and perpendicular to the z axis, as shown in Fig. 1.2b. The charge ρ_l along the ring is specified in units of coulombs per meter so the charge in a differential length is $\rho_l \, dl$. The electric field of $\rho_l \, dl$ is designated by $\mathbf{dE}$ in Fig. 1.2b and is given by (7) with $r^2 = a^2 + z^2$. The component along the z axis is $dE \cos \theta$ where $\cos \theta = z/(a^2 + z^2)^{1/2}$. Note that, by symmetry, the component perpendicular to the axis is canceled by that of the charge element on the opposite side of the ring. The total field at points on the

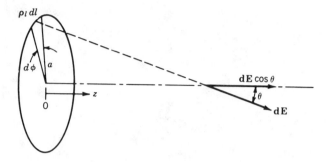

Fig. 1.2*b* Electric field of a ring of charge.

axis is directed along the axis and is the integral of the differential axial components. Taking $dl = a\,d\phi$ we have

$$E = \int_0^{2\pi} \frac{\rho_l az\,d\phi}{4\pi\varepsilon(a^2 + z^2)^{3/2}} = \frac{\rho_l az}{2\varepsilon(a^2 + z^2)^{3/2}} \tag{8}$$

1.3   The Concept of Electric Flux and Flux Density; Gauss's Law

It is convenient in handling electric field problems to introduce another vector more directly related to charges than is electric field **E**. If we define

$$\mathbf{D} = \varepsilon\mathbf{E} \tag{1}$$

we notice from Eq. 1.2(7) that **D** about a point charge is radial and independent of the material. Moreover, if we multiply the radial component D_r by the area of a sphere of radius r, we obtain

$$4\pi r^2 D_r = q \tag{2}$$

We thus have a quantity exactly equal to the charge (in coulombs) so that it may be thought of as the *flux* arising from that charge. **D** may then be thought of as the *electric flux density* (C/m²). For historical reasons it is also known as the *displacement vector* or *electric induction*.

It is easy to show (Prob. 1.3c) that for an arbitrarily shaped closed surface as in Fig. 1.3a, the normal component of **D** integrated over the surface surrounding a point charge also gives q. The analogy is that of fluid flow in which the fluid passing surface S of Fig. 1.3b in a given time is the same as that passing plane $S_\perp$ perpendicular to the flow. So fluid flow rate out of a region does not depend

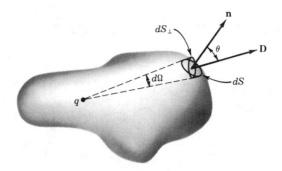

Fig. 1.3a Charge q and arbitrary surrounding surface.

upon the shape of the surface used to monitor it, so long as all surfaces enclose the same source. By superposition, the result can be extended to a system of point charges or a continuous distribution of charges, leading to the conclusion

$$\textit{electric flux flowing out of a closed surface} = \textit{charge enclosed} \qquad (3)$$

This is Gauss's law, and although argued here from Coulomb's law for simple media, is found to apply to more general media. It is thus a most general and important law. Before illustrating its usefulness, let us look more carefully at the role of the medium.

A simplified picture showing why the force between charges depends upon the presence of matter is illustrated in Fig. 1.3c. The electron clouds and the nuclei of the atoms experience oppositely directed forces as a result of the presence of the isolated charges. Thus the atoms are distorted or *polarized*. There is a shift of the centre of symmetry of the electron cloud with respect to the nucleus in each atom as indicated schematically in Fig. 1.3c. Similar distortions can occur in molecules, and an equivalent situation arises in some materials where naturally polarized molecules have a tendency to be aligned in the presence of free charges. The directions of the polarization are such for most materials that the equivalent

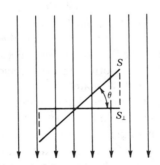

Fig. 1.3b Flow of a fluid through surfaces S and $S_\perp$.

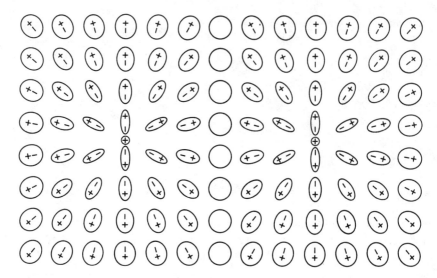

Fig. 1.3c Polarization of the atoms of a dielectric by a pair of equal positive charges.

charge pairs in the atoms or molecules tend to counteract the forces between the two isolated charges. The magnitude and the direction of the polarization depend upon the nature of the material.

The above qualitative picture of polarization introduced by a dielectric may be quantified by giving a more fundamental definition than (1) between **D** and **E**:

$$\mathbf{D} = \varepsilon_0 \mathbf{E} + \mathbf{P} \tag{4}$$

The first term gives the contribution that would exist if the electric field at that point were in free space and the second measures the effect of the polarization of the material (as in Fig. 1.3c) and is called the *electric polarization*.

The polarization produced by the electric field in a material depends upon material properties. If the properties do not depend upon position, the material is said to be *homogeneous*. Most field problems are solved assuming homogeneity; inhomogeneous media, exemplified by the earth's atmosphere, are more difficult to analyze. If the response of the material is the same for all directions of the electric field vector, it is called *isotropic*. Special techniques are required for handling a field problem in an anisotropic medium such as an ionized gas with an applied steady magnetic field. A material is called *linear* if the ratio of the response **P** to the field **E** is independent of amplitude. Nonlinearities are generally not present except for high-amplitude fields. It is also possible that the character of a material may be *time variable*, imposed, for example, by passing a sound wave through it. Throughout most of this text, the media will be considered to be homogeneous, isotropic, linear, and time invariant. Exceptions will be studied in the final chapters.

For isotropic, linear material the polarization is proportional to the field intensity and we can write the linear relation

$$\mathbf{P} = \varepsilon_0 \chi_e \mathbf{E} \tag{5}$$

where the constant χ_e is called the *electric susceptibility*. Then (4) becomes the same as (1),

$$\mathbf{D} = \varepsilon_0(1 + \chi_e)\mathbf{E} = \varepsilon\mathbf{E} \tag{6}$$

and the relative permittivity, defined in Eq. 1.2(4) is

$$\varepsilon_r = \frac{\varepsilon}{\varepsilon_0} = 1 + \chi_e \tag{7}$$

1.4   Examples of the Use of Gauss's Law

The simple but important examples to be discussed in this section show that Gauss's law can be used to find field strength in a very easy way for problems with certain kinds of symmetry and given charges. The symmetry gives the direction of the electric field directly and ensures that the flux is uniformly distributed. Knowledge of the charge gives the total flux. Symmetry is then used to get flux density $\mathbf{D}$ and hence $\mathbf{E} = \mathbf{D}/\varepsilon$.

_____ Example 1.4a _____
Field in a Planar Semiconductor Depletion Layer

For the first example, we consider a one-dimensional situation where a metal is in intimate (atomic) contact with a semiconductor. We assume that some of the typically valence 4 (e.g., silicon) atoms have been replaced by "dopant" atoms of valence 5 (e.g., phosphorous). The one extra electron in each atom is not needed for atomic bonding and becomes free to move about in the semiconductor. Upon making the metal contact, it is found that the free electrons are forced away from the surface for a distance d. The region $0 \leq x \leq d$ is called a *depletion region* because it is depleted of the free electrons. Since the dopant atoms were neutral before losing their extra electrons, they are positively charged when the region is depleted. This can be modeled as in Fig. 1.4a. In the region $x > d$ the donors are assumed to be completely compensated by free electrons and it is therefore charge-free. (The abrupt change from compensated to uncompensated behavior at $x = d$ is a commonly used idealization.) By symmetry, the flux is x-directed only. The surface used in application of Gauss's law consists of two infinite parallel planes, one at $x < d$ and one at $x = d$. This approximation is made because the transverse dimensions of the contact are assumed to be much larger than d. With no applied fields, there is no average movement of the electrons in

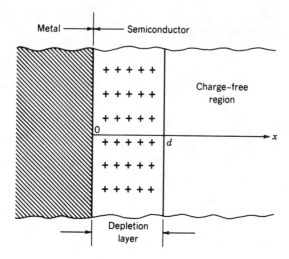

Fig. 1.4a Model of a metal–semiconductor contact.

the compensated region $x \geq d$ so **E** must be zero there. Thus **D** is also zero in that region and all the flux from the charged dopant atoms must terminate on the metal contact. Therefore, Gauss's law gives

$$D_x(x) = N_D e(x - d) \tag{1}$$

where N_D and e are the volume density of donor ions and the charge per donor (electronic charge), respectively. Then the x component of electric field is

$$E_x = \frac{N_D e(x - d)}{\varepsilon} \tag{2}$$

It should be clear that the simplicity of solution depended upon the symmetry of the system, that is, that there were no variations in y and z.

---------------------------------- **Example 1.4b** ----------------------------------
Field About a Line Charge or Between Coaxial Cylinders

We introduced the line charge in ring form in Ex. 1.2. Let us now find the field **E** produced by a straight, infinitely long, line of uniformly distributed charge. The radius of the line is negligibly small and can be thought of as the two-dimensional equivalent of a point charge. Practically, a long thin charged wire is a good approximation. The symmetry of this problem reveals that the force on a test charge, and hence the electric field, can only be radial. Moreover, this electric field will not vary with angle about the line charge, nor with distance along it. If the strength of the radial electric field is desired at distance r from the line charge, Gauss's law may be applied to an imaginary cylindrical surface of radius r and

Fig. 1.4b Coaxial line.

any length l. Since the electric field (and hence the electric flux density **D**) is radial, there is no normal component at the ends of the cylinder and hence no flux flow through them. However, **D** is exactly normal to the cylindrical part of the surface, and does not vary with either angle or distance along the axis, so that the flux out is the surface area $2\pi rl$ multiplied by the electric flux density D_r. The charge enclosed is the length l multiplied by the charge per unit length q_l. By Gauss's law, flux out equals the charge enclosed:

$$2\pi rlD_r = lq_l$$

If the dielectric surrounding the wire has constant ε,

$$E_r = \frac{D_r}{\varepsilon} = \frac{q_l}{2\pi\varepsilon r} \tag{3}$$

Hence, the electric field about the line charge has been obtained by the use of Gauss's law and the special symmetry of the problem.

The same symmetry applies to the coaxial transmission line formed of two coaxial conducting cylinders of radii a and b with dielectric ε between them (Fig. 1.4b). Hence the result (3) applies for radius r between a and b. We will use this result to find the capacitance in Sec. 1.9.

_____ **Example 1.4c** _____
Field Between Concentric Spherical Electrodes
with Two Dielectrics

Figure 1.4c shows a structure formed of two conducting spheres of radii a and c, with one dielectric ε_1 extending from $r = a$ to $r = b$, and a second, ε_2, from $r = b$ to $r = c$. This problem has spherical symmetry about the center, which implies that the electric field will be radial, and independent of the angular direction

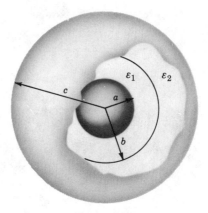

Fig. 1.4c Spherical electrodes separated by two layers of dielectric materials.

about the sphere. If the charge on the inner sphere is Q and that on the outer sphere is $-Q$, the charge enclosed by an imaginary spherical surface of radius r selected anywhere between the two conductors is only that charge Q on the inner sphere. The flux passing through it is the surface $4\pi r^2$ multiplied by the radial component of flux density D_r. Hence, using Gauss's law,

$$D_r = \frac{Q}{4\pi r^2} \tag{4}$$

The equation for the flux density is the same for either dielectric, since the flux passes from the positive charge on the center conductor continuously to the negative charge on the outer conductor. The electric field has a different value in the two regions, however, since in each dielectric D and E are related by the corresponding permittivity:

$$E_r = \frac{Q}{4\pi\varepsilon_1 r^2} \qquad a < r < b \tag{5}$$

$$E_r = \frac{Q}{4\pi\varepsilon_2 r^2} \qquad b < r < c \tag{6}$$

The radial flux density is continuous at the dielectric discontinuity at $r = b$, but the radial electric field is discontinuous there.

_____ **Example 1.4d** _____
Fields of a Spherical Region of Uniform Charge Density

Consider a region of uniform charge density ρ extending from $r = 0$ to $r = a$. As in the preceding example, Gauss's law can be written as (4) where, in this case,

$Q = \frac{4}{3}\pi r^3 \rho$ for $r \le a$ and $Q = \frac{4}{3}\pi a^3 \rho$ for $r \ge a$. Then the flux densities for the two regions are

$$D_r = \frac{r}{3}\rho \qquad r \le a \tag{7}$$

$$D_r = \frac{a^3}{3r^2}\rho \qquad r \ge a \tag{8}$$

1.5   Surface and Volume Integrals; Gauss's Law in Vector Form

Gauss's law, given in words by Eq. 1.3(3), may be written

$$\oint_S D \cos \theta \, dS = q \tag{1}$$

The symbol $\oint_S$ denotes the integral over a surface and of course cannot be performed until the actual surface is specified. It is in general a double integral. The circle on the integral sign is used if the surface is closed.

The surface integral can also be written in a still more compact form if vector notation is employed. Define the unit vector normal to the surface under consideration, for any given point on the surface, as $\hat{n}$. Then replace $D \cos \theta$ by $\mathbf{D} \cdot \hat{n}$. This particular product of the two vectors $\mathbf{D}$ and $\hat{n}$ denoted by the dot between the two is known as the *dot product* of two vectors, or the *scalar product*, since it results by definition in a scalar quantity equal to the product of the two vector magnitudes and the cosine of the angle between them. Also the combination $\hat{n} \, dS$ is frequently abbreviated further by writing it $\mathbf{dS}$. Thus the elemental vector $\mathbf{dS}$, representing the element of surface in magnitude and orientation, has a magnitude equal to the magnitude of the element dS under consideration, and the direction of the outward normal to the surface at that point. The surface integral in (1) may then be written in any of the equivalent forms,

$$\oint_S D \cos \theta \, dS = \oint_S \mathbf{D} \cdot \hat{n} \, dS = \oint_S \mathbf{D} \cdot \mathbf{dS} \tag{2}$$

All of these say that the normal component of the vector $\mathbf{D}$ is to be integrated over the general closed surface S.

If the charge inside the region is given as a density of charge per unit volume in coulombs per cubic meter for each point of the region, the total charge inside the region must be obtained by integrating this density over the volume of the region. This is analogous to the process of finding the total mass inside a region when the variable mass density is given for each point of a region. This process may also be denoted by a general integral. The symbol $\int_V$ is used to denote this, and, as with the surface integral, the particular volume and the variation of density over that volume must be specified before the integration can be performed. In the general case, it is performed as a triple integral.

Gauss's law may then be written in this notation:

$$\oint_S \mathbf{D} \cdot \mathbf{dS} = \int_V \rho \, dV \tag{3}$$

Although the above may at first appear cryptic, familiarity with the notation will immediately reveal that the left side is the net electric flux out of the region and the right side is the charge within the region.

Example 1.5
Round Beam of Uniform Charge Density

Consider the circular cylinder of uniform charge density ρ and infinite length shown in Fig. 1.5. A region of integration is taken in the form of a prism of square cross section. We will demonstrate the validity of (3), utilizing the fact that $\mathbf{D}$ has only a radial component. The right side of (3) is

$$\int_V \rho \, dV = \int_0^l dz \int_{-b/2}^{b/2} dy \int_{-b/2}^{b/2} \rho \, dx = lb^2\rho \tag{4}$$

At any radius, D_r can be found as in Sec. 1.4 to be

$$D_r = \frac{(\pi r^2)\rho}{2\pi r} = \frac{r\rho}{2} \tag{5}$$

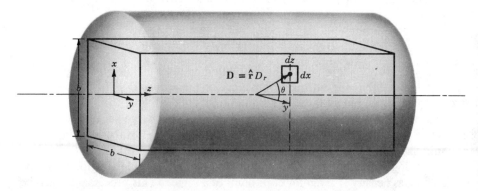

Fig. 1.5 Square cylindrical region of integration in a circular cylinder of free charge.

To do the surface integration on the left side of (3), we notice that $\mathbf{D} \cdot \mathbf{dS} = D_r\, dx\, dz \cos \theta$, that $\cos \theta = b/2r$, and that the four sides make equal contributions to the integral. Thus

$$\oint_S \mathbf{D} \cdot \mathbf{dS} = 4 \int_0^l dz \int_{-b/2}^{b/2} D_r \cos \theta\, dx = lb^2\rho \qquad (6)$$

and from (6) and (4) we see that (3) is satisfied. Problem 1.5b contains a similar situation, but is somewhat complicated by having the charge density dependent upon radius.

1.6   Tubes of Flux; Plotting of Field Lines

For isotropic media, the electric field $\mathbf{E}$ is in the same direction as flux density $\mathbf{D}$. A charge-free region bounded by $\mathbf{E}$ or $\mathbf{D}$ lines must then have the same flux flowing through it for all selected cross sections, since no flux can flow through the sides parallel with $\mathbf{D}$, and Gauss's law will show the conservation of flux for this source-free region. Such a region, called a *flux tube* is illustrated in Fig. 1.6a. Surface S_3 follows the direction of $\mathbf{D}$, so there is no flow through S_3. That flowing in S_1 must then come out S_2 if there are no internal charges. These tubes are analogous to the flow tubes in the fluid analogy used in Sec. 1.3.

Fig. 1.6a Tube of flux.

The concept of flux tubes is especially useful in making maps of the fields, and will be utilized later (Sec. 1.19) in a useful graphical field-mapping technique. We show in the following example how it may be used to obtain field lines in the vicinity of parallel line charges.

_____ **Example 1.6** _____
Flux Tubes and Field Lines About Parallel Lines
of Opposite Charge

To show how field lines may be found by constructing flux tubes, we use, as an example, two infinitely long, parallel lines of opposite charge. It is obvious from

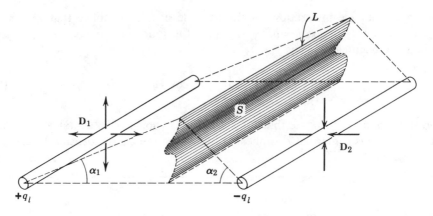

Fig. 1.6b Construction of flux tubes about line charges.

symmetry that the plane in which the two charge lines lie will contain **D** lines and hence can be a boundary of a flux tube. We introduce the *flux function*

$$\psi = \int_S \mathbf{D} \cdot \mathbf{dS} \tag{1}$$

which measures the flux crossing some chosen surface. For example, suppose S is the cross-hatched surface in Fig. 1.6b. First, let the angles α_1 and α_2 shrink to zero so S also vanishes. Then the flux function will be zero along the plane of the lines of charge. (The surface S also could be chosen in some other way, making the flux function different from zero on that plane. The resulting additive constant is arbitrary, so we take it to be zero.) We get other flux-tube boundaries by taking nonzero values of the angles α_1 and α_2. First we derive an expression for the flux passing between the $\psi = 0$ plane and line L in Fig. 1.6b. Then paths will be formed along which L may be moved while keeping the same flux between it and the $\psi = 0$ surface. Moving L along such a path therefore generates a surface which is the boundary of a flux tube of infinite length parallel to L. The flux may be divided into the part from the positive line charge and the part from the negative line charge, since the effects are superposable. The flux from the positive line goes out radially so that the amount (per unit length) crossing S is $q_l(\alpha_1/2\pi)$. The flux passing radially inward toward the negative line charge through S adds directly to that of the positive line charge and has the magnitude $q_l(\alpha_2/2\pi)$. The total flux per unit length crossing S is

$$\psi = \frac{q_l}{2\pi}(\alpha_1 + \alpha_2) \tag{2}$$

The surface generated by moving L in such a way as to keep ψ constant is a circular cylinder that passes through the charge lines with its axis in the plane normal to the $\psi = 0$ line midway between the line charges. Figure 1.6c shows

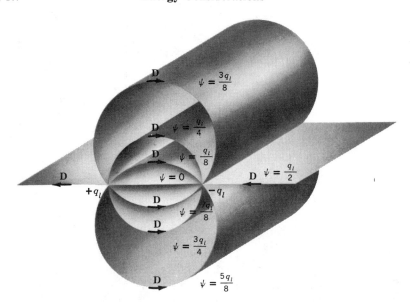

Fig. 1.6c Tubes of flux between line charges.

several flux tubes with the values of ψ indicating the amount of flux between the $\psi = 0$ surface and the one being considered, as α_2 is increased from zero to 2π. Note that, as a path is taken around one of the lines, the flux function goes from 0 to q_l; the total flux per unit length coming from a line charge is q_l. It is clear from this example that the flux function ψ is not single valued since it continues to increase as α_1 or α_2 increases; more flux lines are crossed as motion about the line charge continues. We must therefore limit α_1 and α_2 to the range 0 to 2π to ensure unique values for ψ.

Since the boundaries of the flux tubes lie along **D** vectors and **D** = ε**E**, they also lie along **E** vectors. Thus, by plotting flux tubes, we find the directions of the electric field vectors surrounding the charges. There is a given amount of flux in each tube so the flux density **D** and, therefore, also **E** become large where the cross section of the tube becomes small.

1.7 Energy Considerations; Conservative Property of Electrostatic Fields

Since a charge placed in the vicinity of other charges experiences a force, movement of the charge represents energy exchange. Calculation of this requires integration of force components over the path (line integrals). It will be found

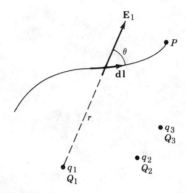

Fig. 1.7a Integration path for force on test charge.

that the electrostatic system is conservative in that no net energy is exchanged if a test charge is moved about a closed path, returning to its initial position.

Consider the force on a small charge Δq moved from infinity to a point P in the vicinity of a system of charges: q_1 at Q_1, q_2 at Q_2, q_3 at Q_3, and so on, Fig. 1.7a. The force at any point along its path would cause the particle to accelerate and move out of the region if unconstrained. A force equal to the negative of that from surrounding charges must then be applied in order to bring Δq from infinity to its final position.

Consider, first, the work integral arising from q_1. The differential work done on the system is the negative of the force component in the direction of the path, multiplied by differential path length:

$$dU_1 = -\mathbf{F}_1 \cdot \mathbf{dl}$$

Or, using the definition of the scalar product, the angle θ as defined in Fig. 1.7a and the force as stated above, we write the line integral

$$U_1 = -\int_\infty^{PQ_1} \mathbf{F}_1 \cdot \mathbf{dl} = -\int_\infty^{PQ_1} \frac{\Delta qq_1 \cos\theta\, dl}{4\pi\varepsilon r^2} \qquad (1)$$

where r is the distance from q_1 to the differential path element $\mathbf{dl}$ at each point in the integration. Since $dl \cos\theta$ is dr, the integral is simply

$$U_1 = -\int_\infty^{PQ_1} \frac{\Delta qq_1\, dr}{4\pi\varepsilon r^2}$$

and similarly for contributions from other charges, so that the total work integral is

$$U = -\int_\infty^{PQ_1} \frac{\Delta qq_1}{4\pi\varepsilon r^2}\, dr - \int_\infty^{PQ_2} \frac{\Delta qq_2}{4\pi\varepsilon r^2}\, dr - \int_\infty^{PQ_3} \frac{\Delta qq_3}{4\pi\varepsilon r^2}\, dr - \cdots$$

Note that there is no component of the work arising from the test charge acting upon itself. Integrating,

$$U = \frac{\Delta q q_1}{4\pi\varepsilon PQ_1} + \frac{\Delta q q_2}{4\pi\varepsilon PQ_2} + \frac{\Delta q q_3}{4\pi\varepsilon PQ_3} + \cdots \tag{2}$$

Equation (2) shows that the work done is only a function of final positions and not of the path of the charge. This conclusion leads to another: if a charge is taken around any closed path, no net work is done. Mathematically this is written as the closed line integral

$$\oint \mathbf{E} \cdot \mathbf{dl} = 0 \tag{3}$$

This general integral signifies that the component of electric field in the direction of the path is to be multiplied by the element of distance along the path, and the sum taken by integration as one moves about the path. The circle through the integral sign signifies that a closed path is to be considered. As with the designation for a general surface or volume integral, the actual line integration cannot be performed until there is a specification of a particular path and the variation of **E** about that path.

In the study of magnetic fields and time-varying electric fields, we shall find corresponding line integrals which are not zero.

_____ **Example 1.7** _____
Demonstration of Conservative Property

To illustrate the conservative property and the use of line integrals, let us take the line integral (3) around the somewhat arbitrary path through a uniform sphere of charge density ρ shown in Fig. 1.7b. The path is chosen, for simplicity, to lie in the $x = 0$ plane. The integrand $\mathbf{E} \cdot \mathbf{dl}$ involves electric field components E_y and E_z. The radial electric field $E_r = \rho r/3\varepsilon_0$ is found from Eq. 1.4(7). The components are $E_y = E_r(y/r)$ and $E_z = E_r(z/r)$ so $E_y = \rho y/3\varepsilon_0$ and $E_z = \rho z/3\varepsilon_0$. The integral (3) becomes

$$\oint \mathbf{E} \cdot \mathbf{dl} = \int_{c_1}^{c_2} E_z \, dz + \int_{b_1}^{b_2} E_y \, dy + \int_{c_2}^{c_1} E_z \, dz + \int_{b_2}^{b_1} E_y \, dy$$

$$= \frac{\rho}{3\varepsilon_0} \left\{ \frac{z^2}{2}\Big|_{c_1}^{c_2} + \frac{y^2}{2}\Big|_{b_1}^{b_2} + \frac{z^2}{2}\Big|_{c_2}^{c_1} + \frac{y^2}{2}\Big|_{b_2}^{b_1} \right\} = 0$$

The general conservative property of electrostatic fields is thus illustrated in this example.

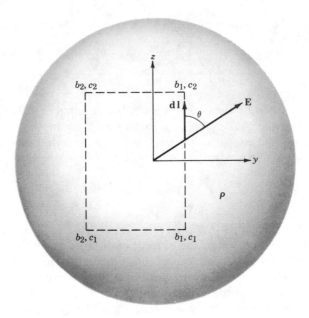

Fig. 1.7b Path of integration (broken line) through electric field of sphere of uniform charge to show conservative property.

1.8   Electrostatic Potential: Equipotentials

The energy considerations of the preceding section lead directly to an extremely useful concept for electrostatics—that of potential. The electrostatic potential function is defined as the work done per unit charge. Here we start generally and define a potential difference between points 1 and 2 as the work done on a unit test charge in moving from P_1 to P_2.

$$\Phi_{P_2} - \Phi_{P_1} = -\int_{P_1}^{P_2} \mathbf{E} \cdot \mathbf{dl} \tag{1}$$

The conclusion of the preceding section that the work in moving around any closed path is zero shows that this potential function is single valued; that is, corresponding to each point of the field there is only one value of potential.

Only a difference of potential has been defined. The potential of any point can be arbitrarily fixed, and then the potentials of all other points in the field can be found by application of the definition to give potential differences between all points and the reference. This reference is quite arbitrary since the potential differences alone have significance. For example, in certain cases it may be convenient to, define the potential at infinity as zero and then find the corresponding potentials of all points in the field; for the determination of the field between two conductors, it will be more convenient to select the potential of one of these as zero.

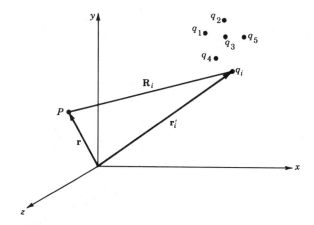

Fig. 1.8a Potential of $q_1, q_2, q_3, ..., q_i$ is found at point P.

If the potential at infinity is taken as zero, it is evident that the potential at the point P in the system of charges is given by U of Eq. 1.7(2) divided by Δq, so

$$\Phi = \frac{q_1}{4\pi\varepsilon PQ_1} + \frac{q_2}{4\pi\varepsilon PQ_2} + \frac{q_3}{4\pi\varepsilon PQ_3} + \cdots \tag{2}$$

This may be written in a more versatile form as

$$\Phi(\mathbf{r}) = \sum_{i=1}^{n} \frac{q_i}{4\pi\varepsilon R_i} \tag{3}$$

where R_i is the distance of the ith charge at $\mathbf{r}'_i$ from the point of observation at $\mathbf{r}$, as seen in Fig. 1.8a.

$$R_i = |\mathbf{r} - \mathbf{r}'_i| = [(x - x'_i)^2 + (y - y'_i)^2 + (z - z'_i)^2]^{1/2} \tag{4}$$

Here x, y, and z are the rectangular coordinates of the point of observation and x'_i, y'_i, z'_i are the rectangular coordinates of the ith charge.[2] Generalizing to the case of continuously varying charge density

$$\Phi(\mathbf{r}) = \int_V \frac{\rho(\mathbf{r}') \, dV'}{4\pi\varepsilon R} \tag{5}$$

The $\rho(\mathbf{r}')$ is charge density at point (x', y', z'), and the integral signifies that a summation should be made similar to that of (2) but continuous over all space. If the reference for potential zero is not at infinity, a constant must be added of such value that potential is zero at the desired reference position.

$$\Phi(\mathbf{r}) = \int_V \frac{\rho(\mathbf{r}') \, dV'}{4\pi\varepsilon R} + C \tag{6}$$

[2] Throughout the text we use primed coordinates to designate the location of sources and unprimed coordinates for the point at which their fields are to be calculated.

It should be kept in mind that (2)–(6) were derived assuming that the charges are located in an infinite, homogeneous, isotropic medium. If conductors or dielectric discontinuities are present, differential equations for the potential (to be given shortly) are used for each region.

We will see in Sec. 1.10 how the electric field **E** can be found simply from $\Phi(\mathbf{r})$. It is usually easier to find the potential by the scalar operations in (3) and (5) and,

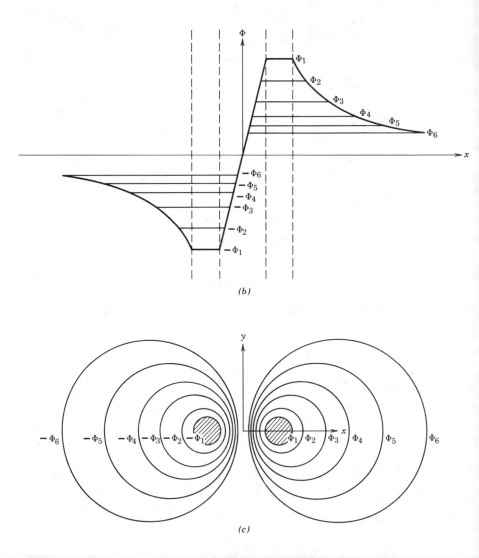

Fig. 1.8 (*b*) Plot of a two-dimensional potential distribution using the third dimension to show the potential. (*c*) Equipotentials for the same potential system as in (*b*) plotted onto the *x–y* plane.

from it, the field **E** than to do the vector summations discussed in Sec. 1.2. Such convenience in electrostatic calculations is one reason for introducing this potential.

In any electrostatic field, there exist surfaces on which the potential is a constant, so-called equipotential surfaces. Since the potential is single valued, surfaces for different values of potential do not intersect. The pictorial representation of more than one such surface in a three-dimensional field distribution is quite difficult. For fields that have no variation in one dimension and are therefore called two-dimensional fields, the third dimension can be used to represent the potential. Figure 1.8*b* shows such a representation for the potential around a pair of infinitely long, parallel wires at potentials Φ_1 and $-\Phi_1$. The height of the surface at any point is the value of the potential. Notice that lines of constant height or constant potential can be drawn. These equipotentials can be projected onto the *x*–*y* plane as in Fig. 1.8*c*. In such a representation, the equipotentials look like the contour lines of a topographic map and, in fact, measure potential energy of a unit charge relative to a selected zero-potential point just as contours measure potential energy relative to some reference altitude, often sea level. It should be kept in mind that these lines are actually traces in the *x*–*y* plane of three-dimensional cylindrical equipotential surfaces.

We will discuss the boundary conditions on conductors in some detail in Sec. 1.14. At this point it is sufficient to say that the electric fields inside of a metallic conductor can be considered to be zero in electrostatic systems. Therefore, (1) shows that the conductor is an equipotential region.

Example 1.8a
Potentials Around a Line Charge and
Between Coaxial Cylinders

As an example of the relations between potential and electric field, consider first the problem of the line charge used as an example in Sec. 1.4, with electric field given by Eq. 1.4(3). By (1) we integrate this from some radius r_0 chosen as the reference of zero potential to radius r:

$$\Phi = -\int_{r_0}^{r} E_r \, dr = -\int_{r_0}^{r} \frac{q_l \, dr}{2\pi\varepsilon r} = -\frac{q_l}{2\pi\varepsilon} \ln\left(\frac{r}{r_0}\right) \tag{7}$$

Or this expression for potential about a line charge may be written

$$\Phi = -\frac{q_l}{2\pi\varepsilon} \ln r + C \tag{8}$$

Note that it is not desirable to select infinity as the reference of zero potential for the line charge, for then by (7) the potential at any finite point would be infinite. As in (6) the constant is added to shift the position of the zero potential.

In a similar manner, the potential difference between the coaxial cylinders of Fig. 1.4b may be found:

$$\Phi_a - \Phi_b = -\int_b^a \frac{q_l\, dr}{2\pi\varepsilon r} = \frac{q_l}{2\pi\varepsilon}\ln\!\left(\frac{b}{a}\right) \tag{9}$$

_____ **Example 1.8b** _____
Potential Outside a Spherically Symmetric Charge

We saw in Eq. 1.4(4) that the flux density outside a spherically symmetric charge Q is $D_r = Q/4\pi r^2$. Using $\mathbf{E} = \mathbf{D}/\varepsilon_0$ and taking the reference potential to be zero at infinity, we see that the potential outside the charge Q is the negative of the integral of $\hat{\mathbf{r}}E_r \cdot \hat{\mathbf{r}}\, dr$ from infinity to radius r

$$\Phi(r) = -\int_\infty^r \frac{Q\, dr_1}{4\pi\varepsilon_0 r_1^2} = \frac{Q}{4\pi\varepsilon_0 r} \tag{10}$$

_____ **Example 1.8c** _____
Potential of a Uniform Distribution of Charge
Having Spherical Symmetry

Consider a volume of charge density ρ that extends from $r = 0$ to $r = a$. Taking $\Phi = 0$ at $r = \infty$, the potential outside a is given by (10) with $Q = \frac{4}{3}\pi a^3 \rho$ so

$$\Phi(r) = \frac{a^3\rho}{3\varepsilon_0 r} \qquad r \geq a \tag{11}$$

In particular, at $r = a$

$$\Phi(a) = \frac{a^2\rho}{3\varepsilon_0} \tag{12}$$

Then to get the potential at a point where $r \leq a$ we must add to (12) the integral of the electric field from a to r. The electric field is given as $E_r = \rho r/3\varepsilon_0$ (Ex. 1.7) and the integral is

$$\Phi(r) - \Phi(a) = -\int_a^r \frac{\rho r_1}{3\varepsilon_0}\, dr_1 = \frac{\rho}{6\varepsilon_0}(a^2 - r^2) \tag{13}$$

So the potential at a radius r inside the charge region is

$$\Phi(r) = \frac{\rho}{6\varepsilon_0}(3a^2 - r^2) \qquad r \leq a \tag{14}$$

_____ **Example 1.8d** _____
Electric Dipole

A particularly important set of charges forms the _electric dipole_ which consists of two closely spaced point charges of opposite sign.

Assume two charges, having opposite signs to be spaced by a distance 2δ as shown in Fig. 1.8d. The potential at some point a distance r from the origin displaced by an angle θ from the line passing from the negative to positive charge can be written as the sum of the potentials of the individual charges,

$$\Phi = \frac{q}{4\pi\varepsilon}\left(\frac{1}{r_+} - \frac{1}{r_-}\right) \tag{15}$$

Using the law of cosines, we have

$$r_+^2 = r^2 + \delta^2 - 2r\delta \cos\theta$$

and similarly for r_-. For $\delta \ll r$,

$$r_+ \approx r - \delta \cos\theta$$
$$r_- \approx r + \delta \cos\theta$$

Substituting in (15) and again using the restriction $\delta \ll r$, one obtains

$$\Phi \approx \frac{2\delta q \cos\theta}{4\pi\varepsilon r^2} \tag{16}$$

We define an _electric dipole moment_ **p** of a pair of equal charges as the product of the charge and the separation. The direction of vector **p** is from the negative charge to the positive one. Thus (16) may be written as

$$\Phi \approx \frac{\mathbf{p}\cdot\hat{\mathbf{r}}}{4\pi\varepsilon r^2} \tag{17}$$

where $\hat{\mathbf{r}}$ is a unit vector directed outward toward the point of observation. It is seen that the dipole potential decreases as $1/r^2$ with increasing distance from the origin, whereas the potential of the single charge decreases only as the first power. The increased rate of decay of the potential is to be expected as a result of the partial cancellation of the potentials of opposite sign. Equipotential lines are shown plotted on a plane passing through the dipole in Fig. 1.8e. The values shown are relative. The equipotentials are surfaces of revolution generated by rotating the lines in Fig. 1.8e about the dipole axis.

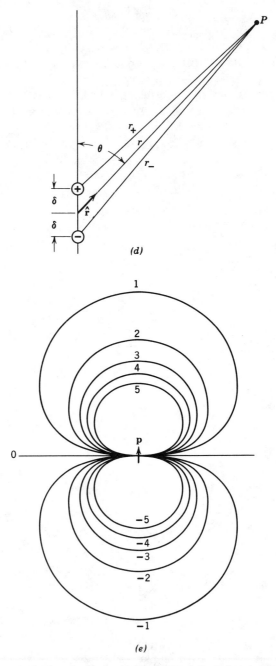

Fig. 1.8 (*d*) Electric dipole. (*e*) Equipotentials of electric dipole.

1.9 Capacitance

The capacitance between two electrodes, widely used in circuit calculations, is a measure of the charge Q on each electrode per volt of potential difference $\Phi_a - \Phi_b$ between them:

$$C = \frac{Q}{\Phi_a - \Phi_b} \tag{1}$$

For capacitance systems having two electrodes, the excess negative charge on one equals the deficiency of negative charge on the other.

Consider first the parallel plates in Fig. 1.9a. The separation is small compared with the width. The result is that charges accumulate mainly on the most closely separated surfaces. We shall idealize the structure as a portion of infinitely wide plates (Fig. 1.9b) and thereby neglect the *fringing fields* that do not pass straight from one plate to the other. Such idealizations of real situations are extremely useful, but their limitations must always be remembered. The flux density $\mathbf{D}$ is found from Gauss's law to equal the surface charge density ρ_s on each plate. Since the field $\mathbf{E} = \mathbf{D}/\varepsilon$ is uniform, the potential difference $\Phi_a - \Phi_b$ is, by Eq. 1.8(1)

$$\Phi_a - \Phi_b = \frac{\rho_s d}{\varepsilon} \tag{2}$$

The total charge on each plate of area A is $\rho_s A$ so (1) and (2) give the familiar expression

$$C = \frac{\varepsilon A}{d} \quad \text{F} \tag{3}$$

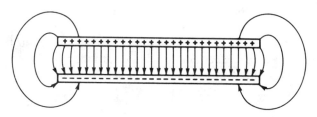

Fig. 1.9a Parallel-plane capacitor with fringing fields.

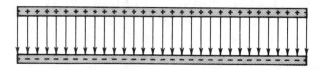

Fig. 1.9b Idealization of the capacitor in Fig. 1.9a.

In practice, (3) is modified by the fringing fields, which are increasingly important as the ratio of plate spacing to area is increased.

Next consider a capacitor made of coaxial, circular cylindrical electrodes. We assume that fields are only radial and neglect any fringing at the ends if it is of finite length. The charge on each conductor is distributed uniformly in this idealization, as required by symmetry, with the total charge per unit length being q_l. The potential difference found from the field produced by this charge is given by Eq. 1.8(9). The capacitance per unit length is thus

$$C = \frac{2\pi\varepsilon}{\ln(b/a)} \quad \text{F/m} \tag{4}$$

where b and a are the radii of larger and smaller conductors, respectively.

Finally, consider two concentric spherical conductors of radii a and b, with $b > a$, separated by a dielectric ε. Using symmetry, Gauss's law and $\mathbf{E} = \mathbf{D}/\varepsilon$, it is clear that at any radius

$$E_r = \frac{Q}{4\pi\varepsilon r^2} \tag{5}$$

where Q is the charge on the inner conductor (equal in magnitude and opposite in sign to the charge on the inside of the outer sphere). Integration of (5) between spheres gives $\Phi_a - \Phi_b$ and this, substituted into (1) yields

$$C = \frac{4\pi\varepsilon}{(1/a) - (1/b)} = \frac{4\pi\varepsilon ab}{b - a} \quad \text{F} \tag{6}$$

The flux tubes in these three highly symmetric structures are very simple, being bounded by parallel surfaces in the first example and by cylindrically or spherically radial surfaces in the last two. In Sec. 1.19 we will see a way of finding capacitance graphically for two-dimensional structures of arbitrary shape in which the flux tubes have more complex shapes.

Capacitance of an isolated electrode is sometimes calculated; in that case, the flux from the charge on the electrode terminates at infinity and the potential on the electrode is taken with respect to an assumed zero at infinity. More extensive considerations of capacitance are found in Sec. 4.9.

DIFFERENTIAL FORMS OF ELECTROSTATIC LAWS

1.10 Gradient

We have looked at several laws of electrostatics in macroscopic forms. It is also useful to have their equivalents in differential forms. Let us start with the relation between electric field and potential. If the definition of potential difference is applied to two points a distance **dl** apart,

$$d\Phi = -\mathbf{E} \cdot \mathbf{dl} \tag{1}$$

where **dl** may be written in terms of its components and the defined unit vectors:

$$\mathbf{dl} = \hat{\mathbf{x}}\,dx + \hat{\mathbf{y}}\,dy + \hat{\mathbf{z}}\,dz \tag{2}$$

We expand the dot product,

$$d\Phi = -(E_x\,dx + E_y\,dy + E_z\,dz)$$

Since Φ is a function of x, y, and z, the total differential may also be written

$$d\Phi = \frac{\partial\Phi}{\partial x}\,dx + \frac{\partial\Phi}{\partial y}\,dy + \frac{\partial\Phi}{\partial z}\,dz$$

From a comparison of the two expressions,

$$E_x = -\frac{\partial\Phi}{\partial x}, \qquad E_y = -\frac{\partial\Phi}{\partial y}, \qquad E_z = -\frac{\partial\Phi}{\partial z} \tag{3}$$

so

$$\mathbf{E} = -\left(\hat{\mathbf{x}}\frac{\partial\Phi}{\partial x} + \hat{\mathbf{y}}\frac{\partial\Phi}{\partial y} + \hat{\mathbf{z}}\frac{\partial\Phi}{\partial z}\right) \tag{4}$$

or

$$\mathbf{E} = -\,\text{grad}\,\Phi \tag{5}$$

where grad Φ, an abbreviation of the *gradient* of Φ, is a vector showing the direction and magnitude of the maximum spatial variation of the scalar function Φ, at a point in space.

Substituting back in (1), we have

$$d\Phi = (\text{grad}\,\Phi) \cdot \mathbf{dl} \tag{6}$$

Thus the change in Φ is given by the scalar product of the gradient and the vector **dl**, so that, for a given element of length **dl**, the maximum value of $d\Phi$ is obtained when that element is oriented to coincide with the direction of the gradient vector. From (6) it is also clear that grad Φ is perpendicular to the equipotentials because $d\Phi = 0$ for **dl** along an equipotential.

The analogy between electrostatic potential and gravitational potential energy discussed in Sec. 1.8 is useful for understanding the gradient. It is easy to see in Fig. 1.8b that the direction of maximum rate of change of potential is perpendicular to the equipotentials (which are at constant heights on the potential hill).

If we define a vector operator $\mathbf{\nabla}$ (pronounced del)

$$\mathbf{\nabla} \triangleq \hat{\mathbf{x}}\frac{\partial}{\partial x} + \hat{\mathbf{y}}\frac{\partial}{\partial y} + \hat{\mathbf{z}}\frac{\partial}{\partial z} \tag{7}$$

then grad Φ may be written as $\nabla\Phi$ if the operation is interpreted as

$$\nabla\Phi = \hat{x}\frac{\partial\Phi}{\partial x} + \hat{y}\frac{\partial\Phi}{\partial y} + \hat{z}\frac{\partial\Phi}{\partial z} \qquad (8)$$

and

$$\mathbf{E} = -\text{grad }\Phi \triangleq -\nabla\Phi \qquad (9)$$

The gradient operator in circular cylindrical and spherical coordinates is given on the inside back cover.

Example 1.10
Electric Field of a Dipole

As an example of the use of the gradient operator, we will find an expression for the field around an electric dipole. The potential for a dipole is given in Sec. 1.8 in spherical coordinates so the spherical form of the gradient operator is selected from the inside of the back cover.

$$\nabla\Phi = \hat{r}\frac{\partial\Phi}{\partial r} + \hat{\theta}\frac{1}{r}\frac{\partial\Phi}{\partial\theta} + \frac{\hat{\phi}}{r\sin\theta}\frac{\partial\Phi}{\partial\phi}$$

Substituting Eq. 1.8(16) and noting the independence of ϕ, we get

$$\nabla\Phi = -\hat{r}\frac{\delta q\cos\theta}{\pi\varepsilon r^3} - \hat{\theta}\frac{\delta q\sin\theta}{2\pi\varepsilon r^3}$$

Then from (9) the electric field is

$$\mathbf{E} = \frac{\delta q}{\pi\varepsilon r^3}\left(\hat{r}\cos\theta + \hat{\theta}\frac{\sin\theta}{2}\right) \qquad (10)$$

1.11 The Divergence of an Electrostatic Field

The second differential form we shall consider is that of Gauss's law. Equation 1.5(3) may be divided by the volume element ΔV and the limit taken:

$$\lim_{\Delta V\to 0}\frac{\oint_S \mathbf{D}\cdot d\mathbf{S}}{\Delta V} = \lim_{\Delta V\to 0}\frac{\int_V \rho\, dV}{\Delta V} \qquad (1)$$

The right side is, by inspection, merely ρ. The left side is the outward electric flux per unit volume. This will be defined as the *divergence* of flux density, abbreviated div $\mathbf{D}$. Then

$$\text{div }\mathbf{D} = \rho \qquad (2)$$

This is a good place to comment on the size scale implicit in our treatment of fields and their sources; the comments also apply to the central set of relations, Maxwell's equations, toward which we are building. In reality, charge is not infinitely divisible—the smallest unit is the electron. Thus, the limit of ΔV in (1) must actually be some small volume which is still large enough to contain many electrons in order to average out the granularity. For our relations to be useful, the limit volume must also be much smaller than important dimensions in the system. For example, in order to neglect charge granularity, the thickness of the depletion layer in the semiconductor in Ex. 1.4a should be much greater than the linear dimensions of the limit volume which, in turn, should be much greater than the average spacing of the dopant atoms. Similarly the permittivity ε is an average representation of atomic or molecular polarization effects such as that shown in Fig. 1.3c. Therefore, when we refer to the field at a point, we mean that the field is an average over a volume small compared with the system being analyzed but large enough to contain many atoms. Analyses can also be made of the fields on a smaller scale, such as inside an atom, but in that case an average permittivity cannot be used. In this book, we will concentrate on situations where ρ, ε_r, and other quantities are averages over small volumes. There are thin films of current practical interest which are only a few atoms thick, and semiconductor devices such as that of Ex. 1.4a where these conditions are not well satisfied so that results from calculations using ρ and ε as defined must be used with caution.

Now let us return to seeking an understanding of (2). Consider the infinitesimal volume as a rectangular parallelepiped of dimensions Δx, Δy, Δz as shown in Fig. 1.11a. To compute the amount of flux leaving such a volume element as compared with that entering it, note that the flux passing through any face of the parallelepiped can differ from that which passes through the opposite face only if the flux density perpendicular to those faces varies from one face to the other. If the distance between the two faces is small, then to a first approximation the difference in any vector function on the two faces will simply be the rate of change

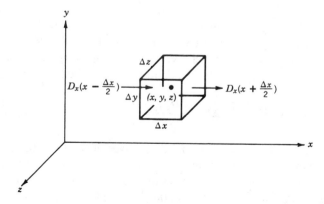

Fig. 1.11a Volume element used in div $\mathbf{D}$ derivation.

of the function with distance times the distance between faces. According to the basis of calculus, this is exactly correct when we pass to the limit, since the higher order differentials are then zero.

If the vector **D** at the center x, y, z has a component $D_x(x)$, then

$$D_x\left(x + \frac{\Delta x}{2}\right) \cong D_x(x) + \frac{\Delta x}{2}\frac{\partial D_x(x)}{\partial x}$$

$$D_x\left(x - \frac{\Delta x}{2}\right) \cong D_x(x) - \frac{\Delta x}{2}\frac{\partial D_x(x)}{\partial x}$$

(3)

In this functional notation, the arguments in parentheses show the points for evaluating the function D_x. When not included, the point (x, y, z) will be understood. The flux flowing out the right face is $\Delta y\,\Delta z D_x(x + \Delta x/2)$, and that flowing in the left face is $\Delta y\,\Delta z D_x(x - \Delta x/2)$, leaving a net flow out of $\Delta x\,\Delta y\,\Delta z(\partial D_x/\partial x)$, and similarly for the y and z directions. Thus the net flux flow out of the parallelepiped is

$$\Delta x\,\Delta y\,\Delta z\frac{\partial D_x}{\partial x} + \Delta x\,\Delta y\,\Delta z\frac{\partial D_y}{\partial y} + \Delta x\,\Delta y\,\Delta z\frac{\partial D_z}{\partial z}$$

By Gauss's law, this must equal $\rho\,\Delta x\,\Delta y\,\Delta z$. So, in the limit,

$$\frac{\partial D_x}{\partial x} + \frac{\partial D_y}{\partial y} + \frac{\partial D_z}{\partial z} = \rho$$

(4)

An expression for div **D** in rectangular coordinates is obtained by comparing (2) and (4):

$$\text{div } \mathbf{D} = \frac{\partial D_x}{\partial x} + \frac{\partial D_y}{\partial y} + \frac{\partial D_z}{\partial z}$$

(5)

If we make use of the vector operator ∇ defined by Eq. 1.10(7) in (4), then (5) indicates that div **D** can correctly be written as $\nabla \cdot \mathbf{D}$. It should be remembered that ∇ is not a true vector but rather a vector operator. It has meaning only when it is operating on another quantity in a defined manner. Summarizing:

$$\nabla \cdot \mathbf{D} \triangleq \text{div } \mathbf{D} = \frac{\partial D_x}{\partial x} + \frac{\partial D_y}{\partial y} + \frac{\partial D_z}{\partial z} = \rho$$

(6)

The divergence is made up of spatial derivatives of the field, so (6) is a partial differential equation deduced from the previous laws for comparatively large systems. It will be so important, that we should become accustomed to looking at it as an expression for Gauss's law applied to a point in space. The physical significance of the divergence must be clear. It is, as defined, a description of the

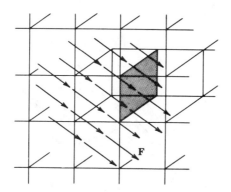

Fig. 1.11b Solid divided into subvolumes to illustrate the divergence theorem.

manner in which a field varies at a point. It is the amount of flux per unit volume emerging from an infinitesimal volume at a point. With this picture in mind, (6) seems a logical extension of Gauss's law. In fact (6) can be converted back to the large-scale form of Gauss's law through the *divergence theorem*, which states that the volume integral of the divergence of any vector $\mathbf{F}$ throughout a volume is equal to the surface integral of that vector flowing out of the surrounding surface,

$$\int_V \nabla \cdot \mathbf{F} \, dV = \oint_S \mathbf{F} \cdot \mathbf{dS} \tag{7}$$

Although not a proof, this is made plausible by considering Fig. 1.11b. The divergence multiplied by volume element for each elemental cell is the net surface integral out of that cell. When summed by integration, all internal contributions cancel since flow out of one cell goes into another, and only the external surface contribution remains. Applications of (7) to (6) with $\mathbf{F} = \mathbf{D}$ gives

$$\oint_S \mathbf{D} \cdot \mathbf{dS} = \int_V \nabla \cdot \mathbf{D} \, dV = \int_V \rho \, dV \tag{8}$$

which is the original Gauss's law.

It will be useful to have expressions for the divergence and other operations involving ∇ in other coordinates systems for simpler treatment of problems having corresponding symmetries. Let us, as an example, develop here the divergence of $\mathbf{D}$ in spherical coordinates. We use the left side of (1) as the definition of div $\mathbf{D}$ and apply it to the differential volume shown in Fig. 1.11c. We will find first the net radial outward flux from the volume. Both the radial component D_r and the element of area $r^2 \, d\theta \sin \theta \, d\phi$ change as we move from r to $r + dr$. Thus the net flux flow out the top over that in at the bottom is

$$d\psi_r = (r + dr)^2 \sin \theta \, d\theta \, d\phi \left(D_r + \frac{\partial D_r}{\partial r} \, dr \right) - r^2 \sin \theta \, d\theta \, d\phi D_r$$

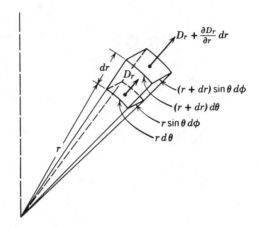

Fig. 1.11c Element of volume in spherical coordinates.

To first-order differentials, this leaves

$$d\psi_r = r^2 \sin\theta \, d\theta \, d\phi \frac{\partial D_r}{\partial r} dr + 2r \, dr \sin\theta \, d\theta \, d\phi D_r$$

$$= \sin\theta \, dr \, d\theta \, d\phi \frac{\partial}{\partial r}(r^2 D_r)$$

Similarly for the θ and ϕ directions:

$$d\psi_\theta = d\theta \frac{\partial}{\partial\theta}(D_\theta r \sin\theta \, d\phi \, dr) = r \, dr \, d\theta \, d\phi \frac{\partial}{\partial\theta}(\sin\theta D_\theta)$$

$$d\psi_\phi = d\phi \frac{\partial}{\partial\phi}(D_\phi r \, d\theta \, dr) = r \, dr \, d\theta \, d\phi \frac{\partial D_\phi}{\partial\phi}$$

The divergence is then the total $d\psi$ divided by the element of volume

$$\mathbf{\nabla \cdot D} = \frac{d\psi_r + d\psi_\theta + d\psi_\phi}{r^2 \sin\theta \, dr \, d\theta \, d\phi}$$

$$\mathbf{\nabla \cdot D} = \frac{1}{r^2}\frac{\partial}{\partial r}(r^2 D_r) + \frac{1}{r \sin\theta}\frac{\partial}{\partial\theta}(\sin\theta D_\theta) + \frac{1}{r \sin\theta}\frac{\partial D_\phi}{\partial\phi} \tag{9}$$

_____ **Example 1.11** _____
Uniform Sphere of Charge

We will show an example of the application of (6) using a sphere of charge of uniform density ρ having a radius a with divergence written in spherical coordinates. Since the spherical symmetry of the charge region leads to $D_\theta = D_\phi = 0$, the last two terms of (9) vanish. We use the value of $D_r = r\rho/3$ from Eq. 1.4(7) for the region inside the charge sphere and obtain

$$\mathbf{V} \cdot \mathbf{D} = \frac{1}{r^2} \frac{\partial}{\partial r} \left[(r^2)\left(\frac{\rho r}{3}\right) \right] = \rho \tag{10}$$

For the region outside the sphere we use $D_r(r) = a^3\rho/3r^2$, from Eq. 1.4(8) to show that

$$\mathbf{V} \cdot \mathbf{D} = \frac{1}{r^2} \frac{\partial}{\partial r} \left[(r^2)\left(\frac{a^3\rho}{3r^2}\right) \right] = 0 \tag{11}$$

This example shows that divergence is zero outside the charge region but equal to charge density within it. The same result is obtained if one uses divergence expressed in rectangular coordinates, or other coordinates not so natural to the symmetry (Prob. 1.11d).

1.12   Laplace's and Poisson's Equations

The differential relations of the two preceding sections allow us to derive an important differential equation for potential. Differential equations can be applied to more general problems than those solved by symmetry in the first part of the chapter and it is often convenient to work with potential as the dependent variable. This is because potential is a scalar and because the specified boundary conditions are often given in terms of potentials.

If the permittivity ε is constant throughout the region, the substitution of E from Eq. 1.10(5) in Eq. 1.11(6) with $\mathbf{D} = \varepsilon\mathbf{E}$ yields

$$\text{div(grad } \Phi) = \mathbf{V} \cdot \mathbf{V}\Phi = -\frac{\rho}{\varepsilon}$$

But, from the equations for gradient and divergence in rectangular coordinates [Eqs. 1.10(7) and 1.11(6)],

$$\mathbf{V} \cdot \mathbf{V}\Phi = \frac{\partial^2\Phi}{\partial x^2} + \frac{\partial^2\Phi}{\partial y^2} + \frac{\partial^2\Phi}{\partial z^2} \tag{1}$$

so that

$$\frac{\partial^2\Phi}{\partial x^2} + \frac{\partial^2\Phi}{\partial y^2} + \frac{\partial^2\Phi}{\partial z^2} = -\frac{\rho}{\varepsilon} \tag{2}$$

This is a differential equation relating potential variation at any point to the charge density at that point, and is known as *Poisson's equation*. It is often written

$$\nabla^2 \Phi = -\frac{\rho}{\varepsilon} \tag{3}$$

where $\nabla^2 \Phi$ (del squared of Φ) is known as the *Laplacian* of Φ.

$$\nabla^2 \Phi \triangleq \nabla \cdot \nabla \Phi = \text{div}(\text{grad } \Phi) \tag{4}$$

In the special case of a charge-free region, Poisson's equation reduces to

$$\frac{\partial^2 \Phi}{\partial x^2} + \frac{\partial^2 \Phi}{\partial y^2} + \frac{\partial^2 \Phi}{\partial z^2} = 0$$

or

$$\nabla^2 \Phi = 0 \tag{5}$$

which is known as *Laplace's equation*. Although illustrated in its rectangular coordinate form, ∇^2 can be expressed in cylindrical or spherical coordinates through the relations given on the inside back cover.

Any number of possible configurations of potential surfaces will satisfy the requirements of (3) and (5). All are called solutions to these equations. It is necessary to know the conditions existing around the boundary of the region in order to select the particular solution which applies to a given problem. We will see in Sec. 1.17 a proof of the uniqueness of a function that satisfies both the differential equation and the boundary conditions.

Quantities other than potential can also be shown to satisfy Laplace's and Poisson's equations, both in other branches of physics and in other parts of electromagnetic field theory. For examples, the magnitudes of the rectangular components of **E** and the component E_z in cylindrical coordinates also satisfy Laplace's equation.

A number of methods exist for solution of two- and three-dimensional problems with Laplace's or Poisson's equations. The separation of variables technique is a very general method for solving two- and three-dimensional problems for a large variety of partial differential equations including the two of interest here. Conformal transformations of complex variables yield many useful two-dimensional solutions of Laplace's equation. Increasingly important are numerical methods using digital computers. All these methods are elaborated in Chapter 7. Examples for this chapter, after discussion of boundary conditions, will be limited to one-dimensional examples for which the differential equations may be directly integrated. These show clearly the role of boundary and continuity conditions in the solutions.

1.13   Static Fields Arising from Steady Currents

Stationary currents arising from dc potentials applied to conductors are not static because the charges producing the currents are in motion, but the resulting steady-state fields are independent of time. Quite apart from the question of designation, there is a close relationship to laws and techniques for the fields arising from purely static charges.

We consider ohmic conductors for which current density is proportional to electric field **E** through conductivity[3] σ siemens per meter (S/m)

$$\mathbf{J} = \sigma\mathbf{E} \tag{1}$$

Since the electric field is independent of time, it is derivable from a scalar potential as in Sec. 1.10, so

$$\mathbf{J} = -\sigma\nabla\Phi \tag{2}$$

For a stationary current, continuity requires that the net flow out of any closed region be zero since there cannot be a buildup or decay of charge within the region in the steady state,

$$\oint_S \mathbf{J} \cdot \mathbf{dS} = 0 \tag{3}$$

or in differential form,

$$\nabla \cdot \mathbf{J} = 0 \tag{4}$$

Substitution of (2) in (4), with σ taken as constant, yields

$$\nabla \cdot \nabla\Phi \triangleq \nabla^2\Phi = 0 \tag{5}$$

Thus potential satisfies Laplace's equation as in other static field problems (Sec. 1.12). In addition to the boundary condition corresponding to the applied potentials, there is a constraint at the boundaries between conductors and insulators since there can be no current flow across such boundaries. Referring to (2), this requires that

$$\frac{\partial\Phi}{\partial n} = 0 \tag{6}$$

across metal–insulator boundaries, where n denotes the normal to such boundaries. Continuity relations between different conductors will be considered in the following section.

The SI unit for conductivity is siemens per meter (S/m) but older literature sometimes uses mho as the conductivity unit.

1.14   Boundary Conditions in Electrostatics

Most practical field problems involve systems containing more than one kind of material. We have seen some examples of boundaries between various regions in the examples of earlier sections, but now we need to develop these systematically in order to utilize the differential equations of the preceding sections.

Let us consider the relations between normal flux density components across an arbitrary boundary by using the integral form of Gauss's law. Consider an imaginary pillbox bisected as shown in Fig. 1.14a by the interface between regions 1 and 2. The thickness of the pillbox is considered to be small enough that the net flux out the sides vanishes in comparison with that out the flat faces. If we assume the existence of net surface charge ρ_s on the boundary, the total flux out of the box must equal $\rho_s \, \Delta S$. By Gauss's law,

$$D_{n1} \, \Delta S - D_{n2} \, \Delta S = \rho_s \, \Delta S$$

or

$$D_{n1} - D_{n2} = \rho_s \tag{1}$$

where ΔS is small enough to consider D_n and ρ_s to be uniform.

A second relation may be found by taking a line integral about a closed path of length Δl on one side of the boundary and returning on the other side as indicated in Fig. 1.14a. The sides normal to the boundary are considered to be small enough that their net contributions to the integral vanish in comparison

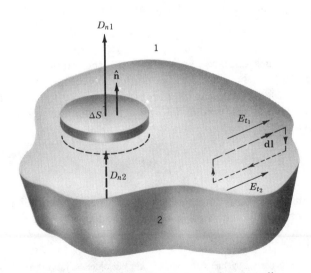

Fig. 1.14a Boundary between two different media.

with those of the sides parallel to the surface. By Eq. 1.7(3) any closed line integral of electrostatic field must be zero:

$$\oint \mathbf{E} \cdot \mathbf{dl} = E_{t1}\,\Delta l - E_{t2}\,\Delta l = 0$$

or

$$E_{t1} = E_{t2} \tag{2}$$

The subscript t denotes components tangential to the surface. The length of the tangential sides of the loop is small enough to take E_t as a constant over the length. Since the integral of $\mathbf{E}$ across the boundary is negligibly small,

$$\Phi_1 = \Phi_2 \tag{3}$$

across the boundary. Equations (1) and (2), or (1) and (3), form a complete set of boundary conditions for the solution of electrostatic field problems.

Consider an interface between two dielectrics with no charge on the surface. From (1) and Eq. 1.3(1),

$$\varepsilon_1 E_{n1} = \varepsilon_2 E_{n2} \tag{4}$$

It is clear that the normal component of $\mathbf{E}$ changes across the boundary, whereas the tangential component, according to (2), is unmodified. Therefore the direction of the resultant $\mathbf{E}$ must change across such a boundary except where either E_n or E_t are zero. Suppose that at some point at a boundary between two dielectrics the electric field in region 1 makes an angle θ_1 with the normal to the boundary. Thus, as seen in Fig. 1.14b,

$$\theta_1 = \tan^{-1}\frac{E_{t1}}{E_{n1}} \tag{5}$$

and

$$\theta_2 = \tan^{-1}\frac{E_{t2}}{E_{n2}} \tag{6}$$

Then using (2) and (4) we see that

$$\theta_2 = \tan^{-1}\!\left(\frac{\varepsilon_2}{\varepsilon_1}\tan\theta_1\right) \tag{7}$$

Let us now consider the boundary properties of conductors, as exemplified by a piece of semiconductor with metallic contacts. The currents flowing in both the semiconductor and the metallic contacts are related to the fields by the respective conductivities so potential drops occur in them, as developed in Sec. 1.13.

At the interface of two contiguous conductors, the normal component of current density is continuous across the boundary, because otherwise there would

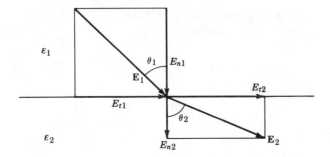

Fig. 1.14b Vector relations among electric field components at a point on a boundary between dielectrics ($\varepsilon_2 > \varepsilon_1$).

be a continual buildup of charges there. The normal component of electric field is therefore discontinuous and given by

$$E_{n1} = \frac{\sigma_2}{\sigma_1} E_{n2} \tag{8}$$

The argument used in dielectric problems for tangential components of electric field also applies in the case of a discontinuity of conductivity so the tangential electric field is continuous across the boundary.

In problems with metallic electrodes on a less conductive material such as a semiconductor, one normally assumes that the conductivity of the metallic electrode is so high compared with that of the other material that negligible potential drops occur in the metal. The electrodes are assumed to be perfect conductors with equipotential surfaces. In this text we normally assume for dc problems that the electrodes in either electrostatic or dc conduction problems are equipotentials and therefore the tangential electric fields at their surfaces are zero.

_____ **Example 1.14** _____
Boundary Conditions in a dc Conduction Problem

The structure in Fig. 1.14c illustrates some of the points we have made about boundary conditions. We assume that a potential difference is provided between the two metallic electrodes by an outside source. The electrodes are considered to be perfect conductors and, therefore, equipotentials. The space directly between them is filled with a layer of conductive material having conductivity σ and permittivity ε. The space surrounding the conductive system is filled with a dielectric having permittivity ε_1 and $\sigma = 0$.

The normal component of the current density at the conductor–dielectric interface must be zero since the current in the dielectric is zero. This follows

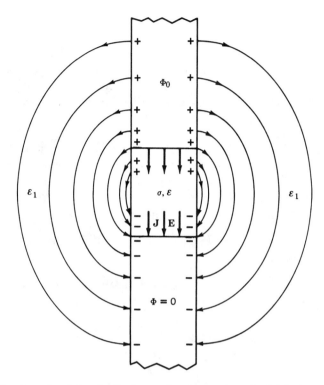

Fig. 1.14c Structure involving both dc conduction and static fields in a dielectric to illustrate boundary conditions.

formally from (8) using zero conductivity for the dielectric. Therefore, since the electric field is J/σ the normal component of electric field inside the conductor must be zero. Then the electric fields inside the conductor at the boundaries with the dielectric are wholly tangential. Since the electrodes are assumed to be equipotentials, the tangential field there is zero and the field lines are perpendicular to the electrode surfaces.

It is clear that since there is an electric field in the conductive medium and it has a permittivity, there is also a flux density following the same paths as the current density. The flux must terminate on charges so there will be charges on the electrode-to-conductor boundary.

The electric field in the dielectric region terminates on charges both on the perfectly conducting electrodes and on the sides of the other conductive medium. On the boundary of the imperfect conductor there is a tangential component and also surface charges so the field makes an oblique angle with the conductor surface.

1.15   Direct Integration of Laplace's Equation: Field Between Coaxial Cylinders with Two Dielectrics

The first simple example of Laplace's equation we will take is one of finding the potential distribution between two coaxial conducting cylinders of radii a and c (Fig. 1.15), with a dielectric of constant ε_1 filling the region between a and b, and a second dielectric of constant ε_2 filling the region between b and c. The inner conductor is at potential zero, and the outer at potential V_0. Because of the symmetry of the problem, the solution could be readily obtained by using Gauss's law as in the example of Sec. 1.4, but the primary purpose here is to demonstrate several processes in the solution by differential equations.

The geometrical form suggests that the Laplacian $\nabla^2\Phi$ be expressed in cylindrical coordinates (see inside back cover), giving for Laplace's equation

$$\nabla^2\Phi = \frac{1}{r}\frac{\partial}{\partial r}\left(r\frac{\partial\Phi}{\partial r}\right) + \frac{1}{r^2}\frac{\partial^2\Phi}{\partial\phi^2} + \frac{\partial^2\Phi}{\partial z^2} = 0 \tag{1}$$

It will be assumed that there is no variation in the axial (z) direction, and the cylindrical symmetry eliminates variations with angle ϕ. Equation (1) then reduces to

$$\frac{1}{r}\frac{d}{dr}\left(r\frac{d\Phi}{dr}\right) = 0 \tag{2}$$

Note that in (2) the derivative is written as a total derivative, since there is now only one independent variable in the problem. Equation (2) may be integrated directly:

$$r\frac{d\Phi}{dr} = C_1 \tag{3}$$

Integrating again, we have

$$\Phi_1 = C_1\ln r + C_2 \tag{4}$$

This has been labeled Φ_1 because we will consider that the result of (4) is applicable to the first dielectric region ($a < r < b$). The same differential equation with the same symmetry applies to the second dielectric region, so the same form of solution applies there also, but the arbitrary constants may be different. So, for the potential in region 2 ($b < r < c$), let us write

$$\Phi_2 = C_3\ln r + C_4 \tag{5}$$

The boundary conditions at the two conductors are:

(a) $\Phi_1 = 0$ at $r = a$

(b) $\Phi_2 = V_0$ at $r = c$

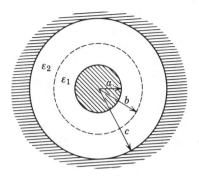

Fig. 1.15 Coaxial cylinders with two dielectrics.

In addition, there are continuity conditions at the boundary between the two dielectric media. The potential and the normal component of electric flux density must be continuous across this charge-free boundary (Sec. 1.14):

(c) $\Phi_1 = \Phi_2$ at $r = b$

(d) $D_{r1} = D_{r2}$ at $r = b$, or $\varepsilon_1 \dfrac{d\Phi_1}{dr} = \varepsilon_2 \dfrac{d\Phi_2}{dr}$ there.

The application of condition (a) to (4) yields

$$C_2 = -C_1 \ln a \tag{6}$$

The application of (b) to (5) yields

$$C_4 = V_0 - C_3 \ln c \tag{7}$$

Condition (c), applied to (4) and (5) gives

$$C_1 \ln b + C_2 = C_3 \ln b + C_4 \tag{8}$$

And condition (d), applied to (4) and (5) gives

$$\varepsilon_1 C_1 = \varepsilon_2 C_3 \tag{9}$$

Any one of the constants, as C_1, may be obtained by eliminating between the four equations, (6) to (9):

$$C_1 = \frac{V_0}{\ln(b/a) - (\varepsilon_1/\varepsilon_2)\ln(b/c)} \tag{10}$$

The remaining constants, C_2, C_3, and C_4, may be obtained from (6), (9), and (7), respectively. The results are substituted in (4) and (5) to give the potential distribution in the two dielectric regions:

$$\Phi_1 = \frac{V_0 \ln(r/a)}{\ln(b/a) + (\varepsilon_1/\varepsilon_2)\ln(c/b)} \qquad a < r < b \tag{11}$$

$$\Phi_2 = V_0\left[1 - \frac{(\varepsilon_1/\varepsilon_2)\ln(c/r)}{\ln(b/a) + (\varepsilon_1/\varepsilon_2)\ln(c/b)}\right] \qquad b < r < c \tag{12}$$

It can be checked that these distributions do satisfy Laplace's equation and the boundary and continuity conditions of the problem. Only in such simple problems as this will it be possible to obtain solutions of the differential equation by direct integration, but the method of applying boundary and continuity conditions to the solutions, however obtained, is well demonstrated by the example.

1.16   Direct Integration of Poisson's Equation: The *pn* Semiconductor Junction

The *pn* semiconductor junction is an important practical example in which properties can be found by direct integration of Poisson's equation. Figure 1.16a shows a simplified *pn* junction. The basic semiconductor is typically a valence 4 material such as silicon (or a compound semiconductor such as gallium arsenide that behaves much the same). The *n* region of the figure has been "doped" with valence 5 impurity atoms such as phosphorous (donors), which although electrically neutral in themselves, have more electrons than needed for bonding with adjacent silicon atoms and so contribute electrons which can move relatively freely about the material. The *p* region of the figure has valance 3 impurities such as boron (acceptors) which have fewer electrons than needed for bonding with adjacent silicon atoms. These too are electrically neutral in themselves, but leave *holes* that move from atom to atom with electric fields or other forces much like small positive charges. Although the transition between *p* and *n* regions must be over some finite region, we assume an idealized model in which it is abrupt—a step discontinuity.

When the junction is formed, the excess electrons in the *n*-type region at first diffuse into the *p*-type side. The holes diffuse to the *n*-type side. The electrons flowing into the *p*-type side fill the vacancies in the acceptor bonds causing them to become negatively charged. (Remember that they were originally electrically neutral.) Likewise the holes moving into the *n*-type side are filled by the excess electrons there. The result is a zone near the junction in which there is a net negative charge density in a region on the *p*-type side and a net positive charge density on the *n*-type side. The density on the *n*-type side is eN_D since each of the donor atoms has been stripped of one electron. The density on the *p*-type side is $-eN_A$ since each acceptor atom has one additional electron. The widths of the zones stabilize when the potential arising in the charge regions is sufficient to prevent further diffusion. Outside the charged regions, the semiconductors are neutral. No fields exist there in the equilibrium situation that we shall examine; if a field did exist it would cause motion of the charges and violate the assumption of equilibrium. The regions of charge are shown (not to scale) in Fig. 1.16b.

One can deduce the form of the gradient of potential $d\Phi/dx = -E_x = -D_x/\varepsilon$ from Gauss's law as was done in Ex. 1.4a for the metal–semiconductor contact. We have just argued that E_x is zero outside the charge regions $(x < -d_p$ and

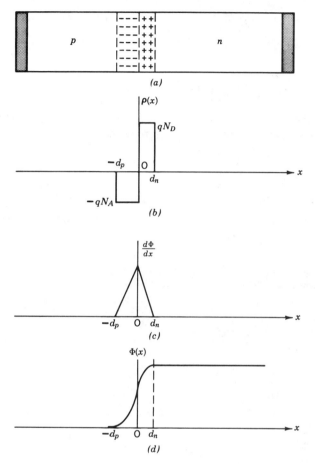

Fig. 1.16 (a) pn diode showing regions of uncompensated charge. (b) Charge density in the diode. (c) Potential gradient. (d) Potential.

$x > d_n$). All the flux from the positive charges in $0 < x < d_n$ must therefore end on the negative charges in $-d_p < x < 0$. This determines the relation between d_n and d_p in terms of the given values of N_a and N_b. The flux density D_x is negative and its magnitude increases linearly from $x = -d_p$ to $x = 0$ since the charge density is taken to be constant. It then falls linearly to zero between $x = 0$ and $x = d_n$. The gradient therefore takes the form shown in Fig. 1.16c. The potential is the integral of the linearly varying gradient so it has the square-law form in Fig. 1.16d.

Now let us directly integrate Poisson's equation to get the complete analytic forms. The boundary conditions on the integration are that the gradient is zero at $x = -d_p$ and at d_n. The potential may be taken arbitrarily to have its zero at $x = -d_p$. Specializing Poisson's equation Eq. 1.12(3) to one dimension and substituting for charge density, the value $-eN_A$ for the region $-d_p < x < 0$, we have

$$\frac{d^2\Phi}{dx^2} = \frac{eN_A}{\varepsilon} \tag{1}$$

The permittivity ε is not appreciably affected by the dopant, at least for low-frequency considerations. Integrating (1) from $x = -d_p$ to an arbitrary $x = 0$ making use of the zero boundary condition on the gradient at $x = -d_p$ gives

$$\left.\frac{d\Phi}{dx}\right|_x = \frac{eN_A}{\varepsilon}(d_p + x) \tag{2}$$

Integrating a second time taking $\Phi(-d_p) = 0$ gives the potential for $-d_p \le x \le 0$ as

$$\Phi(x) = \frac{eN_A}{2\varepsilon}(x + d_p)^2 \tag{3}$$

The gradient and potential evaluated at $x = 0$ are

$$\left.\frac{d\Phi}{dx}\right|_0 = \frac{eN_A}{\varepsilon}d_p \tag{4}$$

$$\Phi(0) = \frac{eN_A}{2\varepsilon}d_p^2 \tag{5}$$

These constitute the boundary conditions for the integration in the region $0 \le x \le d_n$. Poisson's equation for this region differs from (1) in the choice of charge density:

$$\frac{d^2\Phi}{dx^2} = -\frac{eN_D}{\varepsilon} \tag{6}$$

Integrating (6) with the boundary condition (4) gives

$$\left.\frac{d\Phi}{dx}\right|_x = -\frac{eN_D}{\varepsilon}x + \frac{eN_A}{\varepsilon}d_p \tag{7}$$

Then using the condition that the total positive charge must equal the total negative charge and therefore that $N_D = N_A(d_p/d_n)$, (7) can be put in the form

$$\left.\frac{d\Phi}{dx}\right|_x = \frac{eN_A d_p}{\varepsilon}\left(1 - \frac{x}{d_n}\right) \tag{8}$$

which is seen to be zero at $x = d_n$, as expected from Gauss's law as discussed above. Integrating (8) with the boundary condition (5), we find

$$\Phi(x) = \frac{eN_A d_p^2}{2\varepsilon}\left(1 + 2\frac{x}{d_p} - \frac{x^2}{d_n d_p}\right) \tag{9}$$

The maximum value of the potential is reached at $x = d_n$:

$$\Delta\Phi = \Phi(d_n) = \frac{eN_A d_p^2}{2\varepsilon}\left(1 + \frac{N_A}{N_D}\right) \tag{10}$$

Note that distance d_p in (10) is not immediately known. Since the potential barrier arises to stop the diffusion of the charge carriers, it is expected that diffusion must also be considered. Diffusion theory reveals that the height of the potential barrier (10) is

$$\Delta\Phi = \frac{k_B T}{e} \ln\left[\frac{N_A N_D}{n_i^2}\right] \tag{11}$$

where k_B is Boltzmann's constant, N_A and N_D are acceptor and donor doping densities, respectively, and n_i is the electron density of intrinsic (undoped) silicon, which is about 1.5×10^{10} electrons/cm^3 at $T = 300$ K. Once $\Delta\Phi$ is calculated, d_p can be found from (10) and all quantities in the field and potential expressions are then known.

1.17   Uniqueness of Solutions

It can be shown that the potentials governed either by the Laplace or Poisson equations in regions with given potentials on the boundaries are unique. With normal derivatives of potential (or equivalently, charges) specified on the boundaries, the potential is unique to within an additive constant. Here we will prove the theorem for a charge-free region with potential specified on the boundary. The proofs of the other parts of the theorem are left as problems.

The usual way to demonstrate uniqueness of a quantity is first to assume the contrary and then show this assumption to be false. Imagine two possible solutions, Φ_1 and Φ_2. Since they must both reduce to the given potential along the boundary,

$$\Phi_1 - \Phi_2 = 0 \tag{1}$$

along the boundary surface. Since they are both solutions to Laplace's equation,

$$\nabla^2\Phi_1 = 0 \quad \text{and} \quad \nabla^2\Phi_2 = 0$$

or

$$\nabla^2(\Phi_1 - \Phi_2) = 0 \tag{2}$$

throughout the entire region.

In the divergence theorem, Eq. 1.11(7), **F** may be any continuous vector quantity. In particular, let it be the quantity

$$(\Phi_1 - \Phi_2)\nabla(\Phi_1 - \Phi_2)$$

Then

$$\int_V \mathbf{V} \cdot [(\Phi_1 - \Phi_2)\mathbf{V}(\Phi_1 - \Phi_2)]\, dV = \oint_S [(\Phi_1 - \Phi_2)\mathbf{V}(\Phi_1 - \Phi_2)] \cdot \mathbf{dS}$$

From the vector identity

$$\mathrm{div}(\psi \mathbf{A}) = \psi\, \mathrm{div}\, \mathbf{A} + \mathbf{A} \cdot \mathrm{grad}\, \psi$$

the equation may be expanded to

$$\int_V (\Phi_1 - \Phi_2)\mathbf{V}^2(\Phi_1 - \Phi_2)\, dV + \int_V [\mathbf{V}(\Phi_1 - \Phi_2)]^2\, dV$$
$$= \oint_S (\Phi_1 - \Phi_2)\mathbf{V}(\Phi_1 - \Phi_2) \cdot \mathbf{dS}$$

The first integral must be zero by (2); the last integral must be zero, since (1) holds over the boundary surface. There remains

$$\int_V [\mathbf{V}(\Phi_1 - \Phi_2)]^2\, dV = 0 \tag{3}$$

The gradient of a real scalar is real. Thus its square can only be positive or zero. If its integral is to be zero, the gradient itself must be zero:

$$\mathbf{V}(\Phi_1 - \Phi_2) = 0 \tag{4}$$

or

$$(\Phi_1 - \Phi_2) = \text{constant} \tag{5}$$

This constant must apply even to the boundary, where we know that (1) is true. The constant is then zero, and $\Phi_1 - \Phi_2$ is everywhere zero, which means that Φ_1 and Φ_2 are identical potential distributions. Hence the proof of uniqueness: Laplace's equation can have only one solution which satisfies the boundary conditions of the given region. If by any sort of conniving we find a solution to a field problem that fits all boundary conditions and satisfies Laplace's equation, we may be sure it is the only one.

SPECIAL TECHNIQUES FOR ELECTROSTATIC PROBLEMS

1.18　The Use of Images

The so-called method of images is a way of finding the fields produced by charges in the presence of dielectric[4] or conducting boundaries with certain symmetries.

---

[4] W. R. Smythe, *Static and Dynamic Electricity*, McGraw-Hill, New York, 3rd ed., 1969.

Here we concentrate on the more common situations, those with conducting boundaries. (But see Prob. 1.18d.)

_____ Example 1.18a _____
Point Image in a Plane

The simplest case is that of a point charge near a grounded[5] conducting plane (Fig. 1.18a). Boundary conditions require that the potential along the plane be zero. The requirement is met if in place of the conducting plane an equal and opposite image charge is placed at $x = -d$. Potential at any point P is then given by

$$\Phi = \frac{1}{4\pi\varepsilon}\left(\frac{q}{r} - \frac{q}{r'}\right)$$

$$= \frac{q}{4\pi\varepsilon}\{[(x - d)^2 + y^2 + z^2]^{-1/2} - [(x + d)^2 + y^2 + z^2]^{-1/2}\} \qquad (1)$$

This reduces to the required zero potential along the plane $x = 0$, so that (1) gives the potential for any point to the right of the plane. The expression of course does not apply for $x < 0$, for inside the conductor the potential must be everywhere zero.

If the plane is at a potential other than zero, the value of this constant potential is simply added to (1) to give the expression for potential at any point for $x > 0$.

The charge density on the surface of the conducting plane must equal the normal flux density at that point. This is easily found by using

$$\rho_s = D_x = \varepsilon E_x = -\varepsilon\left.\frac{\partial\Phi}{\partial x}\right|_{x=0} \qquad (2)$$

Substituting (1) in (2) and performing the indicated differentiation gives

$$\rho_s = -\frac{qd}{2\pi}(d^2 + y^2 + z^2)^{-3/2} \qquad (3)$$

Analysis of (3) shows that the surface charge density has its peak value at $y = z = 0$ with circular contours of equal charge density centered about that point. The density decreases monotonically to zero as y and/or z goes to infinity. One application of the image method is in studying the extraction of electrons from a metallic surface as in the metal–semiconductor surface shown in Fig. 1.4a.

_____

[5] The use of the term "grounded" implies a source for the charge that builds up on the plane.

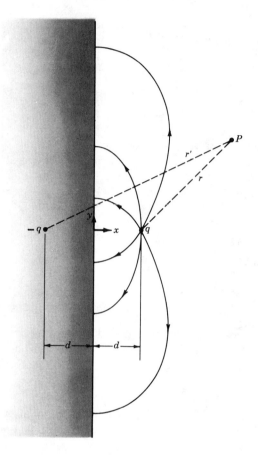

Fig. 1.18a Image of a point charge in a conducting plane. The field lines shown are the same for either the charge q with the conductor or the charge q with its image $-q$, without the conductor.

_____ **Example 1.18b** _____
Image of a Line Charge in a Plane

If there is a line charge of strength q_l C/m parallel to a conducting plane and distance d from it, we proceed as above, placing an image line charge of strength $-q_l$ at $x = -d$. The potential at any point $x > 0$ is then

$$\Phi = -\frac{q_l}{2\pi\varepsilon}\ln\left(\frac{r}{r'}\right) = \frac{q_l}{4\pi\varepsilon}\ln\left[\frac{(x+d)^2 + y^2}{(x-d)^2 + y^2}\right] \tag{4}$$

_____ **Example 1.18c** _____
Image of a Line Charge in a Cylinder

For a line charge of strength q_l parallel to the axis of a conducting circular cylinder, and at radius r from the axis, the image line charge of strength $-q_l$ is placed at radius $r' = a^2/r$, where a is the radius of the cylinder (Fig. 1.18b). The combination of the two line charges can be shown to produce a constant potential along the given cylinder of radius a. Potential outside the cylinder may be computed from the original line charge and its image. (Add q_l on axis if cylinder is uncharged.) If the original line charge is within a hollow cylinder a, the rule for finding the image is the same, and potential inside may be computed from the line charges.

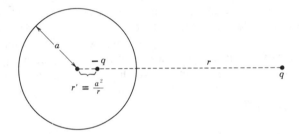

Fig. 1.18b Image of line charge q_l in a parallel conducting cylinder.

_____ **Example 1.18d** _____
Image of a Point Charge in a Sphere

For a point charge q placed distance r from the center of a conducting sphere of radius a, the image is a point charge of value $(-qa/r)$ placed at a distance (a^2/r) from the center (Fig. 1.18c). This combination is found to give zero potential

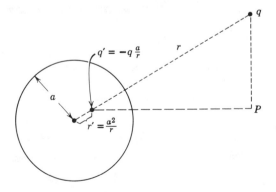

Fig. 1.18c Image of a point charge in a sphere.

52   **Stationary Electric Fields**

along the spherical surface of radius *a*, and may be used to compute potential at any point *P* outside of radius *a*. (Or, if the original charge is inside, the image is outside, and the pair may be used to compute potential inside.)

_____ **Example 1.18e** _____
Multiple Imagings

For a charge in the vicinity of the intersection of two conducting planes, as *q* in the region of *AOB* of Fig. 1.18*d*, there might be a temptation to use only one image in each plane, as 1 and 2 of Fig. 1.18*d*. Although +*q* at *Q* and −*q* at 1 alone would give constant potential as required along *OA*, and +*q* at *Q* and −*q* at 2 alone would give constant potential along *OB*, the three charges together would give constant potential along neither *OA* nor *OB*. It is necessary to image these images in turn, repeating until further images coincide, or until all further images are too far distant from the region to influence the potential. It is possible to satisfy exactly the required conditions with a finite number of images only if the angle *AOB* is an exact submultiple of 180 degrees, as in the 45-degree case illustrated by Fig. 1.18*d*.

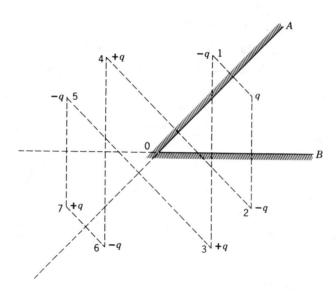

Fig. 1.18d Multiple images of a point or line charge between intersecting planes.

1.19   Graphical Field Mapping

Mainly to help develop some intuition about the distribution of potential, we will present a graphical technique that is easy to understand and implement for two-dimensional distributions. Such two-dimensional distributions exist around electrodes with shapes that are independent of one coordinate. An example is a pair of infinitely long parallel wires. Whereas no electrodes are truly of infinite length, the assumption of two-dimensionality is often a good approximation. We find the distribution in one cross-sectional plane and it is the same for every cross section. In addition to providing approximate potential distributions, the graphical method can give quick, useful values for certain properties of electrodes such as capacitance, as we will see in the next section. Complicated electrode shapes can be handled. Computer programs for finding potential distributions are convenient today, so that method is chosen where detailed information on potentials and fields is needed (see Chapter 7).

In the graphical method, the known potential difference between electrodes of the boundary is divided into a number of smaller potential divisions, and the equipotential lines (traces of the equipotential surfaces in the cross-sectional plane) are sketched by guess throughout the plot. The electric field lines are also sketched by guess, and these guesses are improved by means of certain properties of the field. As was shown in Sec. 1.10, equipotentials and electric field lines intersect at right angles; a correct field map must have this property. Moreover, if the difference of potential between adjacent equipotentials, and the amount of electric flux between adjacent field lines, are made to be constant throughout the plot, the side ratios for all the little "curvilinear rectangles" formed by the intersection of the equipotentials and the orthogonal field lines must be the same throughout the plot, as will be shown in the following. For convenience in estimating by eye how well this property is satisfied, the side ratio is usually chosen as unity, so that the plot is divided into small "curvilinear squares". This property is demonstrated by the simple plot of the field between coaxial cylinders (Fig. 1.19a). In this, the field lines are radial and the equipotentials are circles with the spacing between adjacent equipotentials proportional to radius.

To demonstrate the side ratio property, consider one of the curvilinear rectangles from a general plot, as in Fig. 1.19b. If Δn is the distance between two adjacent equipotentials, and Δs the distance between two adjacent field lines, the magnitude of electric field, assuming a small square, is approximately $\Delta\Phi/\Delta n$. The electric flux flowing along a flux tube bounded by the two adjacent field lines, for a unit length perpendicular to the page is then

$$\Delta\psi = D\,\Delta s = \varepsilon E\,\Delta s = \frac{\varepsilon\,\Delta\Phi\,\Delta s}{\Delta n}$$

or

$$\frac{\Delta s}{\Delta n} = \frac{\Delta\psi}{\varepsilon\,\Delta\Phi} \tag{1}$$

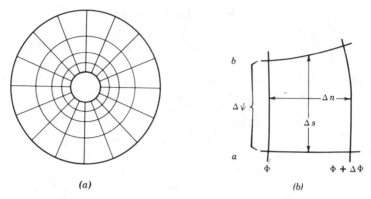

(a)   (b)

Fig. 1.19 (a) Map of field between coaxial conducting cylinders. (b) Curvilinear rectangle for graphical field mapping.

So, if the flux per tube $\Delta\psi$, the potential difference per division $\Delta\Phi$, and the permittivity ε are constant throughout the plot, the side ratio $\Delta s/\Delta n$ must also be constant, as stated above.

We saw in Sec. 1.14 that no potential variations can exist inside electrodes in an electrostatic field so their surfaces are equipotentials. Thus, the electric field lines meet the electrodes at right angles.

In applying the principles to the sketching of fields, each person will soon develop his or her own rules of procedure. In beginning the process, some schedule such as the following will be helpful.

1. Plan on making a number of rough sketches, taking only a minute or so apiece, before starting any plot to be made with care. The use of transparent paper over the basic boundary will speed up this preliminary sketching.
2. Divide the known potential difference between electrodes into an equal number of divisions, say four or eight to begin with.
3. Begin the sketch of equipotentials in the region where the field is known best, as for example in some region where it approaches a uniform field. Extend the equipotentials according to your best guess throughout the plot. Note that they should tend to hug acute angles of the conducting boundary, and be spread out in the vicinity of obtuse angles of the boundary.
4. Draw in the orthogonal set of field lines. As these are started, they should form curvilinear squares, but as they are extended, the condition of orthogonality should be kept paramount, even though this will result in some rectangles with ratios other than unity.
5. Look at the regions with poor side ratios, and try to see what was wrong with the first guess of equipotentials. Correct them and repeat the procedure until reasonable curvilinear squares exist throughout the plot.
6. In regions of low field intensity, there will be large figures, often of five or six sides. To judge the correctness of the plot in this region, these large units

should be subdivided. The subdivisions should be started back away from the region needing subdivision, and each time a flux tube is divided in half, the potential divisions in this region must be divided by the same factor.

There is little more that can be said in words except that the technique can be learned only by the study of some given plots, and by trying the technique on some examples. Some plots are given in Figs. 1.19c and 1.19d. Figure 1.19c shows the field between a plane conductor at potential zero, and a stepped plane at potential V_0, with a step ratio of $\frac{1}{2}$. Figure 1.19d shows a plot that was made to show the field about a particular two-conductor transmission line whose conductors were so shaped as to make difficult an exact mathematical solution.

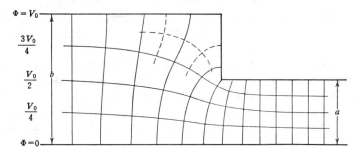

Fig. 1.19c Map of fields between a plane and stepped conductor.

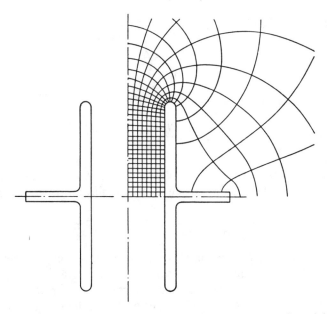

Fig. 1.19d Map of fields between transmission-line conductors of special shape.

1.20   Examples of Information Obtained from Field Maps

Field maps can, in principle, give fairly accurate values for electric fields if enough time is spent to make the map highly subdivided and carefully constructed. As an example, assume for the structure in Fig. 1.19c that the potential V_0 is 100 V and the spacing b is 1 cm. Then the uniform field at left side is 10^4 V/m and that at the right end is 2×10^4 V/m. One can estimate the field at the maximum point at the corner by seeing that the $V_0/4$ division extends over a distance of about 0.057 cm. This gives a maximum field intensity $E = 25/(5.7 \times 10^{-4})$ V/m $= 4.37 \times 10^4$ V/m. Because of the large steps, this is only a rough value, but it may nevertheless be useful in a preliminary design. More accurate values, if needed, can then be obtained by computer solution. There are certain integral properties of electrode systems, as seen below, which are not very sensitive to the details of the fields and they can easily be found with reasonable accuracy using graphical field maps.

Let us see how to calculate capacitance per unit length. By Gauss's law, the charge induced on a conductor is equal to the flux ending there. This is the number of flux tubes N_f multiplied by the flux per tube. The potential difference between conductors is the number of potential divisions N_p multiplied by the potential difference per division. So, for a two-conductor system, the capacitance per unit length is

$$C = \frac{Q}{\Phi_2 - \Phi_1} = \frac{N_f \, \Delta\psi}{N_p \, \Delta\Phi}$$

The ratio $\Delta\psi/\Delta\Phi$ can be obtained from Eq. 1.19(1):

$$C = \frac{N_f}{N_p}\left(\frac{\varepsilon \, \Delta s}{\Delta n}\right) \tag{1}$$

And, for a small squares plot with $\Delta s/\Delta n$ equal to unity,

$$C = \varepsilon \frac{N_f}{N_p} \quad \text{F/m} \tag{2}$$

For example, in the transmission line plot of Fig. 1.19d, there are 16 potential divisions and 66 flux tubes, so the capacitance, assuming air dielectric, is

$$C \approx \frac{10^{-9}}{36\pi} \times \frac{66}{16} = 36.5 \times 10^{-12} \text{ F/m} \tag{3}$$

This same technique can be used to find the conductance between two electrodes placed in a homogeneous, isotropic conductive material. The conductivity of the electrode materials must be much greater than that of the surrounding region to ensure that, when current flows, there is a negligible drop of voltage in the electrodes and they can be considered to be equipotential regions. When a potential difference is applied, one electrode becomes charged positively and the other negatively. The potentials and electric field are related in the same way as for the case where there is no conductivity (Sec. 1.13). As discussed in Ex. 1.14, there is still a flux density $\mathbf{D} = \varepsilon\mathbf{E}$ required by Gauss's law. Now, however, there is also a current density $\mathbf{J} = \sigma\mathbf{E}$ where σ is conductivity. This flow of current is parallel with the electric flux. Referring back to Eq. 1.19(1), we can see that current tubes replace the flux tubes of the dielectric problem. The current in a current tube is

$$\Delta J = J \, \Delta s = \sigma E \, \Delta s = \frac{\sigma \, \Delta\Phi \, \Delta s}{\Delta n} \tag{4}$$

so

$$\frac{\Delta s}{\Delta n} = \frac{\Delta J}{\sigma \, \Delta\Phi} \tag{5}$$

The conductance per unit length between two electrodes is defined as

$$G = \frac{I}{\Phi_2 - \Phi_1} \quad \text{S/m} \tag{6}$$

and the total current I per unit length is $N_f \, \Delta J$ since the current flows in the same tubes as the flux. Following the same analysis as in (1) and (2) above, we find

$$G = \sigma \frac{N_f}{N_p} \quad \text{S/m} \tag{7}$$

From (2) and (7) we see the useful conclusion that the capacitance between a pair of electrodes is related to the conductance between the same electrodes by

$$G = \frac{\sigma}{\varepsilon} C \tag{8}$$

This can be of use, for example, in transmission-line problems in giving the conductance per unit length between conductors when the capacitance is known.

Before digital computer methods for solving field problems became so readily available, the analogy seen above between the field distributions in conducting and dielectric media formed the basis for an important means for determining fields in dielectric systems. In such systems measurement of fields is not possible but resistive analogs with the same electrode forms do permit such measurement.

It should be pointed out that what has been said above applies to the situation where the conductive medium fills all space. If only a part of the region surrounding the electrodes is conductive as shown in Fig. 1.20, the graphical method can still be applied but with small changes. Note that $\mathbf{J} = \sigma\mathbf{E}$, and current does not flow in the nonconductive region. As pointed out in Sec. 1.14, the normal component of $\mathbf{E}$ inside the conductive region must vanish at the boundary with the dielectric region. Using this condition and the fact that $\mathbf{E}$ is

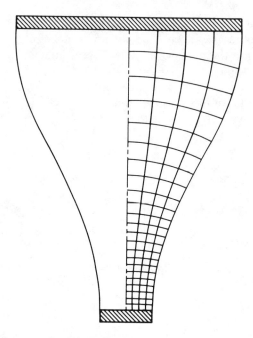

Fig. 1.20 Field map for a conductive medium partially filling the space between two highly conductive electrodes.

perpendicular to the equipotential electrode surfaces, the field in the conductor can be mapped. Of course, in calculating the conductance we note that current tubes exist only inside the conducting region.

Suppose, for example, the conductive materials between the electrodes in Fig. 1.20 is silicon with the common value of conductivity $\sigma = 100\ \text{S/m}$. Then

$$G = 100 \times \tfrac{8}{23} = 35\ \text{S/m}$$

ENERGY IN FIELDS

1.21   Energy of an Electrostatic System

The aim of this section is to derive an expression for electrostatic energy in terms of field quantities. The result we will obtain can be shown to be true in general; however, for simplicity, it is shown here for charges in an unbounded region.

The work required to move a charge in the vicinity of a system of charges was discussed in Sec. 1.7. The work done must appear as energy stored in the system, and consequently the potential energy of a system of charges may be computed from the magnitudes and positions of the charges. To do this, let us consider bringing the charges from infinity to their positions in space. No force is required to bring the first charge in since no electric field acts on the charge. When the second charge q_2 is brought to a position separated from the location of q_1 by a distance R_{12}, an energy

$$U_{12} = \frac{q_1 q_2}{4\pi\varepsilon R_{12}} \tag{1}$$

is expended, as was shown in Sec. 1.7. When the third charge is brought from infinity, it experiences the fields of q_1 and q_2 and an energy of

$$U_{13} + U_{23} = \frac{q_1 q_3}{4\pi\varepsilon R_{13}} + \frac{q_2 q_3}{4\pi\varepsilon R_{23}} \tag{2}$$

is expended. The total energy expended to assemble these three charges is the sum of (1) and (2).

In summing over the three charges, we may write

$$U = \frac{1}{2} \sum_{i=1}^{3} q_i \sum_{j=1}^{3} \frac{q_j}{4\pi\varepsilon R_{ij}}, \qquad i \neq j$$

With the factor $\frac{1}{2}$, this yields the sum of (1) and (2), since by convention i and j are summed over all the particles, and each contribution to energy enters twice. In physical terms, the factor of 2 would result from assuming all other charges in position when finding the energy of the ith charge. The term $i = j$ has been excluded since the self-energy of the point charge (i.e., the electron, ion, etc.) does not affect the energy of the field. Its contribution to the total system energy does not depend upon the relative positions of the charges. For n charges, the direct extension gives

$$U_E = \frac{1}{2} \sum_{i=1}^{n} q_i \sum_{j=1}^{n} \frac{q_i}{4\pi\varepsilon R_{ij}}, \qquad i \neq j \tag{3}$$

where the subscript E indicates energy stored in electric charges and fields. By use of Eq. 1.8(3) for potential, this becomes

$$U_E = \frac{1}{2} \sum_{i=1}^{n} q_i \Phi_i \tag{4}$$

Extending (4) to a system with continuously varying charge density ρ per unit volume, we have

$$U_E = \frac{1}{2} \int_V \rho \Phi \, dV \tag{5}$$

The charge density ρ may be replaced by the divergence of **D** by Eq. 1.11(2):

$$U_E = \frac{1}{2} \int_V (\nabla \cdot \mathbf{D}) \Phi \, dV$$

Using the vector equivalence of Prob. 1.11a,

$$U_E = \frac{1}{2} \int_V \mathbf{V} \cdot (\Phi \mathbf{D}) \, dV - \frac{1}{2} \int_V \mathbf{D} \cdot (\nabla \Phi) \, dV$$

The first volume integral may be replaced by the surface integral of $\Phi \mathbf{D}$ over the closed surface surrounding the region, by the divergence theorem [Eq. 1.11(7)]. But, if the region is to contain all fields, the surface should be taken at infinity. Since Φ dies off at least as fast as $1/r$ at infinity, D at least as fast as $1/r^2$, and area only increases as r^2, this surface integral approaches zero as the surface approaches infinity.

$$\int_V \mathbf{V} \cdot (\Phi \mathbf{D}) \, dV = \oint_{S_\infty} \Phi \mathbf{D} \cdot d\mathbf{S} = 0$$

Then there remains

$$U_E = -\frac{1}{2} \int_V \mathbf{D} \cdot (\nabla \Phi) \, dV = \frac{1}{2} \int_V \mathbf{D} \cdot \mathbf{E} \, dV \qquad (6)$$

This result seems to say that the energy is actually in the electric field, each element of volume dV appearing to contain the amount of energy

$$dU_E = \tfrac{1}{2}\mathbf{D} \cdot \mathbf{E} \, dV \qquad (7)$$

The right answer is obtained if this *energy density* picture is used. Actually, we know only that the total energy stored in the system will be correctly computed by the total integral in (6).

The derivation of (6) was based on a system of charges in an unbounded homogeneous region. The same result can be shown if there are surface charges on conductors in the region. With both volume and surface charges present, (5) becomes

$$U_E = \frac{1}{2} \int_V \rho \Phi \, dV + \frac{1}{2} \int_S \rho_s \Phi \, dS \qquad (8)$$

The proof that this leads to (6) is left to Prob. 1.21e. Note that in this equation, and the special case of (5), Φ is defined with its reference at infinity, since the last term of (3) is identified as Φ. The advantage of (6) is that it is independent of the reference for potential.

_____ **Example 1.21** _____
Energy Stored in a Capacitor

It is interesting to check these results against a familiar case. Consider a parallel-plate capacitor of capacitance C and a voltage V between the plates. The energy is known from circuit theory to be $\tfrac{1}{2}CV^2$, which is commonly obtained by integrating the product of instantaneous current and instantaneous voltage over the time of charging. The result may also be obtained by integrating the energy distribution in the field throughout the volume between plates according to (6). For plates of area A closely spaced so that the end effects may be neglected, the magnitude of field at every point in the dielectric is $E = V/d$ (d = distance between plates) and $D = \varepsilon V/d$. Stored energy U_E given by (6) becomes simply

$$U_E = \frac{1}{2}(\text{volume})(DE) = \frac{1}{2}(Ad)\left(\frac{\varepsilon V}{d}\right)\left(\frac{V}{d}\right) \tag{9}$$

This can be put in terms of capacitance using Eq. 1.9(3)

$$U_E = \frac{1}{2}\left(\frac{\varepsilon A}{d}\right)V^2 = \frac{1}{2}CV^2 \tag{10}$$

PROBLEMS

1.2a (i) Compute the force between two charges of 1 C, each placed 1 m apart in vacuum. (ii) The esu unit of charge (statcoulomb) is defined as one that gives a force of 1 dyne when placed 1 cm from a like charge in vacuum. Use this fact to check the conversion between statcoulombs and coulombs given in Appendix 1.

1.2b Calculate the ratio of the electrostatic force of repulsion between two electrons to the gravitational force of attraction, assuming that Newton's law of gravitation holds. The electron's charge is 1.602×10^{-19} C, its mass is 9.11×10^{-31} kg, and the gravitational constant K is 6.67×10^{-11} N-m^2/kg^2.

1.2c In his experiment performed in 1785, Coulomb suspended a horizontal rod from its center by a filament with which he could apply a torque to the rod. On one end of the rod was a charged pith ball. In the plane in which the rod could rotate was placed another, similarly charged, pith ball at the same radius. By turning the top of the filament he applied successively larger torques to the rod with the amount of torque proportional to the angle turned at the top. With the angle at the top set to 36 degrees, the angle between the two pith balls was also 36 degrees. Raising the angle at the top to 144 degrees decreased the angular separation of the pith balls to 18 degrees. A further increase of the angle at the top to 575.5 degrees decreased the angular separation of the balls to 8.5 degrees. Determine the maximum difference between his measurements and the inverse-square law. (For more details, see R. S. Elliott, Ref. 1.)

1.2d Construct the electric field vector for several points in the x–y plane for like charges q at $(d/2, 0, 0)$ and $(-d/2, 0, 0)$, and draw in roughly a few electric field lines.

1.2e* Repeat Prob. 1.2d for charges of $2q$ and $-q$ at $(d/2, 0, 0)$ and $(-d/2, 0, 0)$, respectively. Note any special points.

---

* The asterisk on problems denotes those longer or harder than the average; a double asterisk will denote unusually difficult or lengthy problems.

1.2f Calculate the electric field at points along the axis perpendicular to the center of a disk of charge of radius a located in free space. The charge on the disk is a *surface charge* ρ_s C/m² uniform over the disk.

1.3a Show by symmetry arguments and the results of Sec. 1.3 that there is no electric field at any point inside a spherical shell of uniform surface charge.

1.3b If Eq. 1.3(1) were used to define permittivity, with **D** in C/m² and **E** in volts per meter, show that ε has the units stated, farads per meter.

1.3c Show that the integral of normal flux density over a general closed surface as in Fig. 1.3a with charge q inside gives q. *Hint:* relate surface element to element of solid angle.

1.3d Calculate that electric flux emanating from a point charge q and passing through a mathematical plane disk of radius a located a distance d from the charge. The charge lies on the axis of the disk. Show that in the limit where $a/d \rightarrow \infty$, the flux through the disk becomes $q/2$.

1.4a A coaxial transmission line has an inner conducting cylinder of radius a, and an outer conducting cylinder of radius c. Charge q_l per unit length is uniformly distributed over the inner conductor and $-q_l$ over the outer. If dielectric ε_1 extends from $r = a$ to $r = b$ and dielectric ε_2 from $r = b$ to $r = c$, find the electric field for $r < a$, for $a < r < b$, for $b < r < c$, and for $r > c$. Take the conducting cylinders as infinitesimally thin. Sketch the variation of D and E with radius.

1.4b A long cylindrical beam of electrons of radius a moving with velocity $v_z = v_0[1 - (r/a)^2]$ m/s has a charge-density radial variation $\rho = \rho_0[1 + (r/a)^2]$ C/m³. Find the electric field in terms of the axial velocity v_0 and the total beam current I_0 and sketch its variation with radius.

1.4c Derive the expression for the field about a line charge, Eq. 1.4(3), from the field of a point charge.

1.4d* A sphere of charge of radius a has uniform density ρ_0 except for a spherical cavity of zero charge with radius b, centered at $x = d$, $y = 0$, $z = 0$, where $d < a$ and $b < a - d$. Find electric field along the x axis from $-\infty < x < \infty$. *Hint:* Use superposition.

1.4e* As in Prob. 1.4d, but now find an expression for electric field for a general point inside the cavity, showing that the field in the cavity is constant.

1.5a A point charge q is located at the origin of coordinates. Express the electric field vector in its réctangular coordinate components, and evaluate the surface integral for S chosen as the surface of a cube of sides $2a$ centered on the charge.

1.5b Perform the integrations in Eq. 1.5(3) for an infinitely long circular cylindrical ion beam with $\rho = \rho_0[1 + (r/a)^2]$ using the square prism shown in Fig. 1.5 and plane ends at $z = 0$ and l orthogonal to the axis.

1.5c If **A**, **B**, and **C** are vectors, show that

$$\mathbf{B} \cdot \mathbf{A} = \mathbf{A} \cdot \mathbf{B}$$

$$(\mathbf{A} + \mathbf{B}) + \mathbf{C} = \mathbf{A} + (\mathbf{B} + \mathbf{C})$$

$$\mathbf{A} \cdot (\mathbf{B} + \mathbf{C}) = \mathbf{A} \cdot \mathbf{B} + \mathbf{A} \cdot \mathbf{C}$$

1.5d Vector **A** makes angles α_1, β_1, γ_1 with the x, y, and z axes respectively, and **B** makes angles α_2, β_2, γ_2 with the axes. If θ is the angle between the vectors, make use of the scalar product **A** · **B** to show that

$$\cos \theta = \cos \alpha_1 \cos \alpha_2 + \cos \beta_1 \cos \beta_2 + \cos \gamma_1 \cos \gamma_2$$

1.6a Show how the flux function may be used to plot the field from *point* charges q and $-q$ distance d apart. *Hint:* Make use of solid angles and relate these to angle θ from the axis joining charges.

1.6b Plot the field from like charges q distance d apart (Prob. 1.2d) by making use of the flux function.

1.6c Plot the field of charges $2q$ and $-q$ distance d apart (Prob. 1.2e) by use of flux.

1.7a Evaluate $\oint \mathbf{F} \cdot \mathbf{dl}$ for vectors $\mathbf{F} = \hat{\mathbf{x}}xy$ and $\mathbf{F} = \hat{\mathbf{x}}y - \hat{\mathbf{y}}x$ about a rectangular path from $(0, 1)$ to $(1, 1)$ to $(1, 2)$ to $(0, 2)$ and back to $(0, 1)$. Repeat for a triangular path from $(0, 0)$ to $(0, 1)$ to $(1, 1)$ back to $(0, 0)$. Are either or both nonconservative?

1.7b A point charge q is located at the origin of a system of rectangular coordinates. Evaluate $\int \mathbf{E} \cdot \mathbf{dl}$ in the x–y plane first along the x axis from $x = 1$ to $x = 2$, and next along a rectangular path as follows: along a straight line from the point $(1, 0)$ on the x axis to the point $(1, \frac{1}{2})$; along a straight line from $(1, \frac{1}{2})$ to $(2, \frac{1}{2})$; along a straight line from $(2, \frac{1}{2})$ to $(2, 0)$.

1.8a A circular insulating disk of radius a is charged with a uniform surface density of charge ρ_s C/m². Find an expression for electrostatic potential Φ at a point on the axis distance z from the disk.

1.8b* A charge of surface density ρ_s is spread uniformly over a spherical surface of radius a. Find the potential for $r < a$ and for $r > a$ by integrating contributions from the differential elements of charge. Check the results by making use of Gauss's law and the symmetry of the problem.

1.8c Check the result Eq. 1.8(8) for the potential about a line charge by integrating contributions from the differential elements of charge. Note that the problem is one of handling properly the infinite limits.

1.8d Assume a slab of charge having infinite transverse extent, a finite thickness d, and charge density ρ_0 C/m³. Using Gauss's law and Eq. 1.8(1), find the dependence of the potential difference across the sheet on the thickness d.

1.8e Consider two parallel sheets of charge having equal surface charge densities but with opposite sign. The sheets are both of infinite transverse dimension and are spaced by a distance d. Using Gauss's law and Eq. 1.8(1), find the electric fields between and outside the sheets and find the dependence of potential difference between the pair of sheets on the spacing. (This is called a *dipole layer*.)

1.8f In a system of infinite transverse dimension, a sheet of charge of ρ_s C/m² lies between, and parallel to, two conducting electrodes at zero potential spaced by distance d. Find the distribution of electric field and potential between the electrodes for arbitrary location of the charge sheet. Sketch the results for the cases where the sheet is (i) in the center and (ii) at position $d/4$.

1.8g Show that all the equipotential surfaces for two parallel line charges of opposite sign are cylinders whose traces in the perpendicular plane are circles as shown in Fig. 1.8c.

1.8h A linear quadrupole is formed by two pairs of equal and opposite charges located along a line such that $+q$ lies at $+\delta$, $-2q$ at the origin, and $+q$ at $-\delta$. Find an approximate expression for the potential at large distances from the origin. Plot an equipotential line.

1.8i Show that the magnitude of the torque on a dipole in an electric field is the product of the magnitude of the dipole moment and the magnitude of the field component perpendicular to the dipole.

1.9 Find the capacitance of the system of two concentric spherical electrodes containing two different dielectric used as Ex. 1.4c.

1.10a Given a scalar function $M = \sin \alpha x \cos \beta y \sinh \gamma z$, find the gradient of M.

1.10b For two point charges q and $-q$ at $(d/2, 0, 0)$ and $(-d/2, 0, 0)$, respectively, find the potential for any point (x, y, z) and from this derive the electric field. Check the result by adding vectorially the electric field from the individual charges.

1.10c Three positive charges of equal magnitude q are located at the corners of an equilateral triangle. Find the potential at the center of the triangle and the force on one of the charges.

1.10d For two line charges q_l and $-q_l$ at $(d/2, 0)$ and $(-d/2, 0)$, respectively, find the potential for any point (x, y) and from this derive the electric field.

1.10e Find the electric field along the axis for Prob. 1.8a.

1.10f** Find the expression for potential outside the large sphere of Prob. 1.4d. Also find the electric field for that region as well as for the region outside the small sphere but inside the large sphere.

1.10g* Find E_x and E_y in the void in the sphere of charge in Prob. 1.4d by first finding the potential. The zeros for the potentials of both large and small spheres should be at infinity.

1.11a Utilize the rectangular coordinate form to prove the vector equivalences

$$\mathbf{V}(\psi \Phi) = \psi \mathbf{V} \Phi + \Phi \mathbf{V} \psi$$

$$\mathbf{V} \cdot (\psi \mathbf{A}) = \psi \mathbf{V} \cdot \mathbf{A} + \mathbf{A} \cdot \mathbf{V} \psi$$

where ψ and Φ are any scalar functions and $\mathbf{A}$ is any vector function of space.

1.11b Evaluate $\mathbf{V} \cdot \mathbf{F}$, where $\mathbf{F} = \hat{x}x^3 + \hat{y}xyz + \hat{z}yz^2$.

1.11c Derive the expression for divergence in the circular cylindrical coordinate system.

1.11d Evaluate the divergence of $\mathbf{D}$ in Exs. 1.4a, 1.4b, and 1.4c and compare with the known charge densities. Evaluate $\mathbf{V} \cdot \mathbf{D}$ for Ex. 1.11 using rectangular coordinates.

1.11e Give a vector $\mathbf{F} = \hat{x}x$, evaluate $\oint_s \mathbf{F} \cdot d\mathbf{S}$ for S taken as the surface of a cube of sides $2a$ centered about the origin. Then evaluate the volume integral of $\mathbf{V} \cdot \mathbf{F}$ for this cube and show that the two results are equivalent, as they should be by the divergence theorem.

1.11f The width of the depletion region at a metal-semiconductor contact (Ex. 1.4a) can be calculated using the relation $d^2 = (2\varepsilon \Phi_B/eN)$, where Φ_B is the barrier potential, e is electron charge, and N is the density of dopant ions. Calculate d for ion densities of $10^{16} \, \text{cm}^{-3}$, $10^{18} \, \text{cm}^{-3}$, and $10^{20} \, \text{cm}^{-3}$ assuming a barrier potential of 0.6 V. Comment on the applicability in this calculation of the concept of smoothed-out charge as assumed in using Poisson's equation. Take $\epsilon_r = 11.7$.

1.12a Find the gradient and Laplacian of a scalar field varying as $1/r$ in two dimensions and in three dimensions.

1.12b Find the electric field and charge density as functions of x, y, and z if potential is expressed as

$$\Phi = C \sin \alpha x \sin \beta y \, e^{\gamma z} \quad \text{where} \quad \gamma = \sqrt{\alpha^2 + \beta^2}$$

1.12c Find the electric field and charge density as functions of x for a space-charge-limited, parallel-plane diode with potential variation given by $\Phi = V_0(x/d)^{4/3}$.

1.12d Argue from Laplace's equation that relative extrema of the electrostatic potential cannot exist and hence that a charge placed in an electrostatic field cannot be in stable equilibrium (Earnshaw's theorem).

1.12e The potential around a perpendicular intersection of the straight edges of two large perfectly conducting planes, where the line of intersection is taken to be the axis of a cylindrical coordinate system, can be shown to be expressible as

$$\Phi = Ar^{2/3} \sin \tfrac{2}{3}\phi$$

Show that this function satisfies Laplace's equation in cylindrical coordinates and satisfies a zero-potential boundary condition on the planes.

1.13 Which of the following may represent steady currents: $\mathbf{J} = \hat{x}x + \hat{y}y$, $\mathbf{J} = (\hat{x}x + \hat{y}y)(x^2 + y^2)^{-1}$? Sketch the form of the two vector fields.

1.14 Sketch field and current lines of Fig. 1.14c if the perfectly conducting electrode region and dielectric region are interchanged, checking all continuity conditions.

1.15a Obtain by means of Laplace's equation the potential distribution between two concentric spherical conductors separated by a single dielectric. The inner conductor of radius a is at potential V_0, and the outer conductor of radius b is at potential zero.

1.15b Obtain by means of Laplace's equation the potential distribution between two concentric spherical conductors with two dielectrics as in Ex. 1.4c.

1.15c Two coaxial cylindrical conductors of radii a and b are at potentials zero and V_0, respectively. There are two dielectrics between the conductors, with the plane through the axis being the dividing surface. That is, dielectric ε_1 extends from $\phi = 0$ to $\phi = \pi$, and ε_2 extends from $\phi = \pi$ to $\phi = 2\pi$. Obtain the potential distribution from Laplace's equation.

1.15d Obtain the electrostatic capacitances for the two conductor systems described in Sec. 1.15 and in Probs. 1.15a, b, and c.

1.16a Assume the charge-density profile shown in Fig. 1.16b with $N_A = 10^{16}\,\mathrm{cm}^{-3}$ and $N_D = 10^{19}\,\mathrm{cm}^{-3}$, $T = 300$ K, and $\varepsilon_r = 12$. Find the height of the potential barrier and the width of the space-charge region $d_n + d_p$. Determine maximum value of electric field.

1.16b Calculate $\Phi(x)$ for the metal–semiconductor junction in Ex. 1.4a by integrating Poisson's equation. Call total barrier height Φ_i in this case. Find the width of the space-charge region d, assuming N_D constant.

1.17a Prove that, if charge density ρ is given throughout a volume, any solution of Poisson's equation 1.12(3) must be the only possible solution provided potential is specified on a surface surrounding the region.

1.17b Show that the potential in a charge-free region is uniquely determined, except for an arbitrary additive constant, by specification of the normal derivatives of potential on the bounding surfaces.

1.18a Prove that the line charge and its image as described for a conducting cylinder in Ex. 1.18c will give constant potential along a cylindrical surface at radius a in the absence of the conducting cylinder.

1.18b Prove that the point charge and its image as described for the spherical conductor in Ex. 1.18d gives zero potential along a spherical surface at radius a in the absence of the conducting sphere.

1.18c A circularly cylindrical electron beam of radius a and uniform charge density ρ passes near a conducting plane that is parallel to the axis of the beam and distance s from it. Find the electric field acting to disperse the beam for the edge near the plane and for the edge farthest from the plane.

1.18d* For a point charge q lying in a dielectric ε_1 distance $x = d$ from the plane boundary between ε_1 and a second dielectric ε_2, the given charge plus an image charge $q(\varepsilon_1 - \varepsilon_2)/(\varepsilon_1 + \varepsilon_2)$ placed at $x = -d$ with all space filled by a dielectric ε_1 may be used to compute the potential for any point $x > 0$. To find the potential for a point $x < 0$, a single charge of value $2q\varepsilon_2/(\varepsilon_1 + \varepsilon_2)$ is placed at the position of q with all space filled by dielectric ε_2. Show that these images satisfy the required continuity relations at a dielectric boundary.

1.18e Find and plot the surface charge density induced on the conducting plane as a function of y, when a line charge q_l lying parallel to the z axis is at $x = d$ above the plane.

1.18f Discuss the applicability of the image concept for the case of a line charge parallel to and in the vicinity of the intersection of two conducting planes with an angle $AOB = 270$ degrees. (See Fig. 1.18d.)

1.18g Find the potential at all points outside a conducting sphere of radius a held at potential Φ_0 when a point charge q is located a distance d from the center of the sphere $(a < d)$.

1.19a Map fields between an infinite plane conductor at potential zero and a second conductor at potential V_0, as in Fig. 1.19c but for step ratios a/b of $\frac{1}{4}$ and $\frac{3}{4}$.

1.19b Map fields between an infinite flat plane and a cylindrical conductor parallel to the plane. The conductor has diameter d, and its axis is at height h above the plane. Take $d/h = 1, \frac{1}{4}$.

1.19c The outer conductor of a two-conductor transmission line is a rectangular tube of sides $3a$ and $5a$. The inner conductor is a circular cylinder of radius a, with axis coincident with the central axis of the rectangular cylinder. Sketch equipotentials and field lines for the region between conductors, assuming a potential difference V_0 between conductors.

1.19d Two infinite parallel conducting planes defined by $y = a$ and $y = -a$ are at potential zero. A semi-infinite conducting plane of negligible thickness at $y = 0$ and extending from $x = 0$ to $x = \infty$ is at potential V_0. Sketch a graphical field map for the region between conductors.

1.19e In Prob. 1.12e take A to be 35 and r to be measured in centimeters and make a plot of the 100-V equipotential. Construct a graphical field map between the zero and 100 V equipotentials showing the 25, 50, and 75 V equipotentials. Then sketch in the same equipotentials found from the formula given in Prob. 1.12e to evaluate your field map. Find the radial distance from the corner where the gradient exceeds the breakdown field in air, 30 kV/cm. What does this suggest about the shape the corner should have to avoid breakdown?

1.20a Assume that Fig. 1.19c is full scale, and that V_0 is 1000 V. Find the approximate direction of the minimum and maximum electric fields in the figure. Plot a curve of electric field magnitude along the bottom plane as a function of distance along this plane, and a curve showing surface charge density induced on this plane as a function of distance.

1.20b Calculate the capacitance per unit length from your plots for Probs. 1.19b and 1.19c.

1.20c Find the conductance per unit length of the transmission line of Prob. 1.19c assuming the medium between conductors to have a uniform conductivity σ.

1.20d Describe the simplest way to use resistance paper to determine the capacitance per wire between a grid of parallel round wires and an electrode lying parallel to the grid.

Assume the grid to be infinitely long and wide. Defend all decisions made in the design of the analog.

1.20e For a certain real electronic device, two electrodes consisting of thin metal films in complex shapes are to be deposited near each other on the plane surface of a dielectric sheet of nonuniform thickness. It is desired to know the capacitance between the electrodes. Evaluate the suggestion that the electrodes of the same size and shape be deposited on a conductive sheet with the same thickness nonuniformity as the subject dielectric and the conductance measured in order to deduce the capacitance.

1.21a For a given potential difference V_0 between conductors of a coaxial capacitor, evaluate the stored energy in the electrostatic field per unit length. By equating this to $\frac{1}{2}CV^2$, evaluate the capacitance per unit length.

1.21b The energy required to increase the separation of a parallel-plate capacitor by a distance dx is equal to the increase of energy stored. Find the force acting between the plates per unit cross-sectional area assuming constant charge on the plates.

1.21c Discuss in more detail the exclusion of the self-energy term in Eq. 1.21(3), and explain why the problem disappeared in going to continuous distributions, as in Eq. 1.21(5).

1.21d Show the equality of the energies found using Eqs. 1.21(5) and (6) for a spherical volume of charge of radius a and charge density ρ C/m^3.

1.21e Consider an arbitrarily shaped, charged finite conductor imbedded in an homogeneous dielectric region of infinite extent that also contains a volume-charge density distribution. Starting from Eq. 1.21(8) show that (6) results. Make use of the identity in Prob. 1.11a and the fact that the surface bounding the dielectric includes both the surface of the conductor and that at infinity.

2

STATIONARY
MAGNETIC FIELDS

2.1 Introduction

Magnetic effects have many similarities to electric effects, but there are also important differences. Magnetic forces were first observed through the attraction of iron to naturally occurring magnetic materials such as lodestone. The compass, apparently developed in China, was introduced into Europe around 1190 A.D., and had a profound effect upon navigation thereafter. In 1600 William Gilbert, physician to Queen Elizabeth I published an important book, *De Magnete*, presenting a rational and thorough summary of the magnetic effects known to that date, with discussions of some of the similarities and differences with the electric effects then known. Had discoveries stopped at that point, we could immediately adapt the development of the preceding chapter to magnetic fields, the two kinds of magnetic "charges" being called north and south *poles*. The important difference is that magnetic charges have so far been found only in pairs, not isolated, so that we would be concerned with fields from dipoles, as in Example 1.8d.

Discoveries did not stop, however. In 1820, Hans Christian Oersted, during a class demonstration of an electric battery, observed that the electric current in a wire caused a nearby compass needle to be deflected, thus establishing clearly the first of several important relationships between electric and magnetic effects. André-Marie Ampère very quickly extended the experiments and developed a quantitative law for the phenomena. Others who contributed both to the understanding and the practical use of electromagnets within a very short period were Jean-Baptiste Biot, Felix Savart, Joseph Henry, and Michael Faraday. The force produced by magnetic fields (either from permanent magnets or electromagnets) on electric currents was also clearly established through these many experiments. These relationships between electric currents and magnetic fields will constitute the starting point for our development of magnetic fields in this chapter. The relationships are somewhat more complicated than those of the preceding chapter, primarily because both the current that acts as the source of field and the current element acting as a probe to measure it, are vectors whose directions must be introduced into the laws.

As with electric fields, the distributions studied in this chapter, although called "static," are applicable to many time-varying phenomena. These "quasistatic" problems are among the most important uses of the laws, and in some cases are valid for extremely rapid rates of change. Still we must remember that other phenomena enter—and are likely to be important—when the fields change with time. These will be studied in the following chapter.

Before beginning the detailed development, let us look briefly at a few examples of important static or quasistatic magnetic field problems. There was the prompt application of Oersted's observation to useful electromagnets. One of Henry's early magnets supported more than a ton of iron with the current driven only by a small battery. Electromagnets are now routinely used in loading or unloading scrap iron and many other applications. The development of practical superconductors in the 1960s has made possible magnets with high fields in large volumes with additional advantages of stability and light weight. Large currents can be made to flow in the magnet winding since there is no voltage drop and no heating. The need to refrigerate to a temperature of 4 K is compensated sufficiently for a number of special applications. They are used extensively in high-energy physics, where the need is for large volumes of strong field. Fusion research depends on massive superconductive magnets for containment of the ionized gases of a plasma. Motors and generators for special applications such as ship propulsion are being made lighter and smaller by using superconductors.

Moving charges constitute currents and magnetic fields produce forces on them as they travel through a vacuum or a semiconductor. Thus magnetic field coils are used for deflection and focusing of beams of electrons in television picture tubes and electron microscopes. The magnetic deflection of flowing charge carriers in a semiconductor is known as the *Hall* effect; it is used for measurement of the semiconductor properties or, with a known semiconductor, may be used as a probe for measurement of magnetic field.

Coils are used to provide inductance needed for high-frequency circuits and the magnetic fields can be found from the currents as in static calculations when the sizes involved are small compared with wavelength. (However, current distributions are complicated at high frequencies by distributed capacitance in the windings.) Just as we noted in Sec. 1.1 for electric fields, the distribution of magnetic field in the cross section of a transmission line is essentially the same as calculated using static field concepts even though the fields can actually be varying at billions of times per second.

STATIC MAGNETIC FIELD LAWS AND CONCEPTS

2.2   Concept of a Magnetic Field

As with the electrostatic fields of the preceding chapter, we use the measurable quantity, force, to define a magnetic field. We noted in Sec. 2.1 that magnetic forces may arise either from permanent magnets or from current flow. Since the approach from currents is more general—and on the whole more important—we start by consideration of the force between current elements. Permanent magnets may then be included, at least conceptually, by considering the effects of these as arising from atomic currents of the magnetic materials.

    The force arising from the interaction of two current elements depends on the magnitude of the currents, the medium, and the distance between currents analogously to the force between electric charges. However, current has direction so the force law between the two currents will be more complicated than that for charges. Consequently, it is convenient to proceed by first defining the quantity we will call the magnetic field, and then give in another section the law (Ampère's) that describes how currents contribute to that magnetic field. A vector field quantity **B**, usually known as the *magnetic flux density*, is defined in terms of the force **df** produced on a small current element of length **dl** carrying current I, such that

$$df = I \, dl \, B \sin \theta \tag{1}$$

where θ is the angle between **dl** and **B**. The direction relations of the vectors are so defined that the vector force **df** is along a perpendicular to the plane containing **dl** and **B**, and has the sense determined by the advance of a right-hand screw if **dl** is rotated into **B** through the smaller angle. (See Fig. 2.2.) It is convenient to express this information more compactly through the use of the *vector product*. The vector product (also called *cross product*) of two vectors (denoted by a cross) is defined as a vector having a magnitude equal to the product of the magnitudes of the two vectors and the sine of the angle between them, a direction perpendicular to the plane containing the two vectors, and a sense given by the advance of a right-hand screw if the first is rotated into the

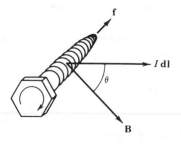

Fig. 2.2 Right-hand screw rule for force on a current element in a magnetic field.

second through the smaller angle. Relation (1) may then be written

$$d\mathbf{f} = I \, d\mathbf{l} \times \mathbf{B} \qquad (2)$$

The quantity known as the magnetic field vector or magnetic field intensity is denoted $\mathbf{H}$ and is related to the vector $\mathbf{B}$ defined by the force law (2) through a constant of the medium known as the *permeability*, μ:

$$\mathbf{B} = \mu \mathbf{H} \qquad (3)$$

Many technologically important materials such as iron and ferrite are nonlinear and/or anisotropic, in which case μ is not a scalar constant, but to keep this introductory treatment simple, the media will first be assumed to be homogeneous, isotropic, and linear. A somewhat more general form of (3) will be given in Sec. 2.3.

In SI units, force is in newtons (N). Current is in amperes (A), and magnetic flux density B is in tesla (T), which is a weber per meter squared or volt second per meter squared and is 10^4 times the common cgs unit, gauss. Magnetic field H is in amperes per meter and μ is in henrys per meter. Conversion factors to other cgs units are in Appendix 1. The value of μ for free space is

$$\mu_0 = 4\pi \times 10^{-7} \text{ H/m}$$

2.3　Ampère's Law

Ampère's law, deduced experimentally from a series of ingenious experiments,[1] describes how the magnetic field vector defined in Sec. 2.2 is calculated from a system of direct currents. Consider an unbounded, homogeneous, isotropic medium with a small line element of length dl' carrying a current I' located at a point in space defined by a vector $\mathbf{r}'$ from an arbitrary origin as in Fig. 2.3a. The magnitude of the magnetic field at some other point P in space defined by the vector $\mathbf{r}$ from the origin is

$$dH(\mathbf{r}) = \frac{I'(\mathbf{r}') \, dl' \sin \phi}{4\pi R^2}$$

where $R = |\mathbf{r} - \mathbf{r}'|$, the distance from the current element to the point of observation. The angle ϕ is that between the direction of the current defined by

---

[1] For a description, see J. C. Maxwell, *A Treatise on Electricity and Magnetism*, 3rd ed., Oxford, 1892, Part IV, Chapter 2. The law is now more frequently named after Biot and Savart, but the assignment remains somewhat arbitrary. Following Oersted's announcement of the effect of currents on permanent magnets in 1820, Ampère immediately announced similar forces of currents on each other. Biot and Savart presented the first quantitative statement for the special case of a straight wire; Ampère later followed with his formulation for more general current paths. The form given here is a derived form borrowing from all that work. For more of the history see E. T. Whittaker, *A History of the Theories of the Aether and Electricity*, Philosophical Library, New York, 1952, or P. F. Mottelay, *Bibliographical History of Electricity and Magnetism*, Charles Griffin & Co., London, 1922.

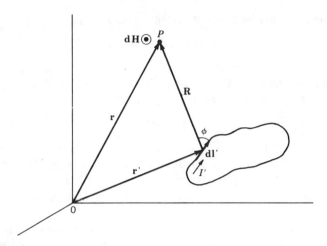

Fig. 2.3a Coordinates for calculation of magnetic field from current element.

dl′ and the vector **R** = **r** − **r′** from the current element to the point of observation. The direction of **dH(r)** is perpendicular to the plane containing **dl** and **R** and the sense is determined by the advance of a right-hand screw if **dl** is rotated through the smallest angle into the vector **R**. Thus, with the current direction shown in Fig. 2.3a, **dH** at P is outward from the page. We see then that the cross product can be used to write the vector form of Ampère's law:

$$\mathbf{dH(r)} = \frac{I'(\mathbf{r'})\,\mathbf{dl'} \times \mathbf{R}}{4\pi R^3} \tag{1}$$

To obtain the total magnetic field of the current elements along a current path, (1) is integrated over the path

$$\mathbf{H(r)} = \int \frac{I'(\mathbf{r'})\,\mathbf{dl'} \times \mathbf{R}}{4\pi R^3} \tag{2}$$

It is of interest to examine further the relation between **B** and **H**. We see that the field **H** is directly related to the currents, without regard for the nature of the medium as long as it fills all space homogeneously. The force on a current element was seen in Sec. 2.2 to depend upon magnetic flux density. The influence of the medium in relating **B** and **H** comes about in the following way. The electronic orbital and spin motions in the atoms can be thought of as circulating currents on which a force is exerted by **B** and which produce a field **M** (called *magnetization*) that adds to **H**. This is analogous to the response of a dielectric medium shown in Fig. 1.3c. Then **B** is related to **H** as though there were only free space but with the added field of the atomic currents

$$\mathbf{B} = \mu_0(\mathbf{H} + \mathbf{M}) \tag{3}$$

Magnetization **M** may have a permanent contribution (to be considered in Sec. 2.15), but here we neglect this and assume the material isotropic so that **M** is parallel to **H**. We can then write

$$\mathbf{B} = \mu_0(1 + \chi_m)\mathbf{H} = \mu\mathbf{H} = \mu_r\mu_0\mathbf{H}$$

where χ_m is called the *magnetic susceptibility*, μ is the permeability introduced in Sec. 2.2, and μ_r is the relative permeability. Many materials have nonlinear behavior so χ_m and μ are in general functions of the field strength. For diamagnetic materials $\chi_m < 0$ and for paramagnetic, ferromagnetic, and ferrimagnetic materials $\chi_m > 0$. Most materials commonly considered to be dielectrics or metals have either diamagnetic or paramagnetic behavior and typically $|\chi_m| < 10^{-5}$ so we treat them as free space, taking $\mu = \mu_0$. Ferromagnetic and ferrimagnetic materials usually have χ_m and μ/μ_0 much greater than unity.

Example 2.3a
Field on Axis of Circular Loop

As an example of the application of the law, the magnetic field is computed for a point on the axis of a circular loop of wire carrying dc current I (Fig. 2.3b). The element **dl'** has magnitude $a\,d\phi'$ and is always perpendicular to **R**. Hence the contribution dH from an element is

$$dH = \frac{Ia\,d\phi'}{4\pi(a^2 + z^2)} \tag{4}$$

As one integrates about the loop, the direction of **R** changes, and so the direction of **dH** changes, generating a conical surface as ϕ goes through 2π radians (rad).

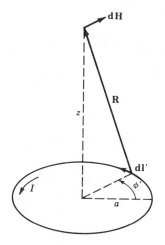

Fig. 2.3b Magnetic field from element of a circular current loop (Ex. 2.3a).

The radial components of the various contributions cancel, and the axial components

$$dH_z = dH \sin \theta = \frac{a\, dH}{(a^2 + z^2)^{1/2}}$$

add directly. Integrating in ϕ amounts to multiplying by 2π; thus

$$H_z = \frac{Ia^2}{2(a^2 + z^2)^{3/2}} \tag{5}$$

Note that for a point at the center of the loop, $z = 0$,

$$H_z\bigg|_{z=0} = \frac{I}{2a} \tag{6}$$

Example 2.3b

Field of a Finite Straight Line of Current

Let us find the magnetic field $\mathbf{H}$ at a point P a perpendicular distance r from the center of a finite length of current I, as shown in Fig. 2.3c. It is easy to see from the right-hand rule that there is only an H_ϕ component. Its magnitude is given by the integral of (1) over the length $2a$

$$H_\phi = \int_{-a}^{a} \frac{I \sin \phi \, dz}{4\pi R^2}$$

We can see from Fig. 2.3c that $\sin \phi = r/R$ and $R = (r^2 + z^2)^{1/2}$. Thus,

$$H_\phi = \frac{Ir}{4\pi} \int_{-a}^{a} \frac{dz}{(r^2 + z^2)^{3/2}} = \frac{I}{2\pi r} \frac{1}{[(r/a)^2 + 1]^{1/2}} \tag{7}$$

which becomes $I/2\pi r$ if $|a| \to \infty$. This same result is found in Ex. 2.4a by a different method.

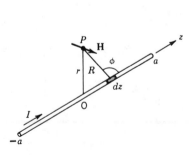

Fig. 2.3c Calculation of magnetic field of straight section of current (Ex. 2.3b).

_____ **Example 2.3c** _____
Field in an Infinite Solenoid

Let us here model the long, tightly wound solenoid shown in Fig. 2.3d by an equivalent current sheet to facilitate calculation of the magnetic field inside. We assume that though the wire makes a small helical angle with a cross-sectional plane, we can adequately model it with a circumferential current. The current flowing around the solenoid per meter is nI where n is the number of turns per meter and I is the current in each turn. Then, in a differential length of the sheet

Fig. 2.3d Tightly wound solenoid of n turns per meter and its representation by a current sheet of nI A/m (Ex. 2.3c).

model, there is a current $nI\,dz$. We will calculate, for simplicity, the field on the axis. But one can show, by means that will come later (see Ex. 2.4d) that the field is uniform throughout the inside of the solenoid. We can adapt (4) for the present calculation by taking I in (4) to be $nI\,dz$. Then the total field on the axis for the infinitely long solenoid is given by

$$H_z = \int_{-\infty}^{\infty} \frac{nIa^2\,dz}{2(a^2 + z^2)^{3/2}} \tag{8}$$

In evaluating the integral in (8), one first takes symmetrical finite limits as in (7) and then lets the limits go to infinity with the result

$$H_z = nI \tag{9}$$

For a solenoid of finite length, it is easy to modify (8) to obtain on-axis field (Prob. 2.3c) but difficult to perform the integrals for fields not on the axis.

_____

2.4   The Line Integral of Magnetic Field

Although Ampère's law describes how magnetic field may be computed from a given system of currents, other derived forms of the law may be more easily applied to certain types of problems. In this and the following sections, some of these forms will be presented, with examples of their application. The sketch of the derivations of these forms, because they are more complex than for the corresponding electrostatic forms, will be left to Appendix 3.

One of the most useful forms of the magnetic field laws derived from Ampère's law is that which states that a line integral of static magnetic field taken about any given closed path must equal the current enclosed by that path. In the vector notation,

$$\oint \mathbf{H} \cdot \mathbf{dl} = \int_S \mathbf{J} \cdot \mathbf{dS} = I \qquad . \qquad (1)$$

Equation (1) is often referred to as *Ampère's circuital law*. The sign convention for current on the right side of (1) is taken so that it is positive if it has the sense of advance of a right-hand screw rotated in the direction of circulation chosen for the line integration. This is simply a statement of the well-known right-hand rule relating directions of current and magnetic field.

Equation (1) is rather analogous to Gauss's law in electrostatics in the sense that it is an important general relation and is also useful for problem solving if there is sufficient symmetry in the problem. If the product $\mathbf{H} \cdot \mathbf{dl}$ is constant along some path, $\mathbf{H}$ can be found simply by dividing I by the path length.

Example 2.4a
Magnetic Field About a Line Current

An important example is that of a long, straight, round conductor carrying current I. If an integration is made about a circular path of radius r centered on the axis of the wire, the symmetry reveals that magnetic field is circumferential and does not vary with angle as one moves about the path. Hence the line integral is just the product of circumference and the value of H_ϕ. This must equal the current enclosed:

$$\oint \mathbf{H} \cdot \mathbf{dl} = 2\pi r H_\phi = I$$

or

$$H_\phi = \frac{I}{2\pi r} \quad \text{A/m} \qquad (2)$$

as was found by a different method in Ex. 2.3b. The sense relations are given in Fig. 2.4a.

Example 2.4b
Magnetic Field Between Coaxial Cylinders

A coaxial line (Fig. 2.4b) carrying current I on the inner conductor, and $-I$ on the outer (the return current), has the same type of symmetry as the isolated wire, and a circular path between the two conductors encloses current I, so that the

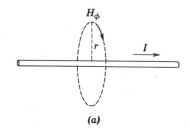

(a)

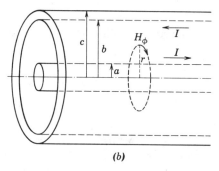

(b)

Fig. 2.4 (a) and (b) Magnetic field about line current and between coaxial cylinders (Exs. 2.4a and b).

result (1) applies directly for the region between conductors:

$$H_\phi = \frac{I}{2\pi r} \qquad a < r < b \tag{3}$$

Outside the outer conductor, a circular path encloses both the going and return currents, or a net current of zero. Hence the magnetic field outside is zero.

_____ **Example 2.4c** _____
Magnetic Field Inside a Uniform Current

Let us find the magnetic field inside the round inner conductor in Fig. 2.4b assuming a uniform distribution of current. We will apply (2) but with I replaced by $I(r)$, the current enclosed by a circle at radius r. The total current in the wire is $I(a) = I$ and the current density is $I/\pi a^2$. The current $I(r)$ is

$$I(r) = \left(\frac{r}{a}\right)^2 I \tag{4}$$

and using (2),

$$H_\phi(r) = \frac{I(r)}{2\pi r} = \frac{Ir}{2\pi a^2} \tag{5}$$

_____ Example 2.4d _____
Magnetic Field of a Solenoid

In Ex. 2.3c we showed that the magnetic field H_z on the axis of an infinitely long solenoid of n turns per meter carrying a current I A is nI. Now let us use the integral relation (1) to show that the field outside is zero and that inside is uniformly nI. Figure 2.4c shows the section through the solenoid in a plane containing the axis. Let us consider the integration paths shown by broken lines to be 1 m long in the z direction for simplicity of notation. Any radial component of **H** produced by a current element is canceled by that of a symmetrically placed element so **H · dl** is zero on the sides BD and AE of the integration path.

Taking the line integral around path $ABDEA$ and setting it equal to the enclosed current gives

$$\oint \mathbf{H} \cdot \mathbf{dl} = nI + \int_D^E \mathbf{H} \cdot \mathbf{dl} = nI \tag{6}$$

since H on the axis is nI. From (6) the integral from D to E is zero. Since the placement of the outside path DE is arbitrary, external H must be zero.

The line integral around path $ABCFA$ encloses no current so the integral along the arbitrarily positioned path CF must be equal in magnitude to, and of opposite sign from, that along AB. Thus, the internal field is everywhere z-directed and has the value

$$H_z = nI \tag{7}$$

Note that these symmetry arguments cannot be made for a solenoid of finite length, but the results given here are reasonably accurate for a solenoid having a length much greater than its diameter, except near the ends.

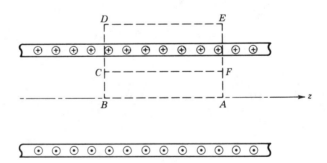

Fig. 2.4c Section through axis of infinite solenoid for Ex. 2.4d.

2.5   Inductance from Flux Linkages; External Inductance

The important circuit element which describes the effect of magnetic energy storage for an electric circuit is the inductor. It is of primary concern for dynamic, that is, time-varying, problems, but the inductance calculated from static concepts is often useful up to very high frequencies. This is the quasistatic use discussed in the introduction to this chapter. In a manner similar to the capacitance definition of Sec. 1.9 inductance can be defined in terms of flux linkage by

$$L = \frac{1}{I} \int_S \mathbf{B} \cdot d\mathbf{S} \tag{1}$$

where the surface S must be specified. Consider, for example, the loop of wire shown in Fig. 2.5a. The current I produces magnetic flux in the shaded area S bounded by the loop. Also, some of the flux produced by the current is inside the wire itself. It is convenient to separate the inductances related to these two components of flux and call them, respectively, *external inductance* and *internal inductance*. Examples of calculations of external inductance for simple structures are given below and an example of an internal inductance calculation is presented in Sec. 2.17.

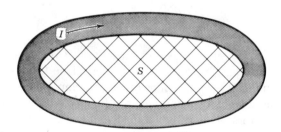

Fig. 2.5a Loop of wire; cross hatching shows surface used for calculation of external inductance.

_____ **Example 2.5a** _____
External Inductance of a Parallel-Plane Transmission Line

Here we find the external inductance for a unit length of a parallel-plane structure which is wide enough compared with the conductor spacing that the fields between the conductors are, to a reasonable degree of accuracy, those of infinite parallel planes, as suggested in Fig. 2.5b. Notice that the flux tubes (bounded by the field lines) spread out greatly outside the edges of the conductors. Thus, there is a strong reduction of flux density **B** and, therefore, also **H**.

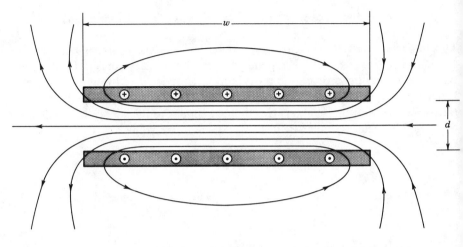

Fig. 2.5*b* Magnetic field of parallel-plane transmission line of finite width.

The line integral of **H** around one of the conductors has its predominant contribution from the field H_0 between the conductors:

$$I = \oint \mathbf{H} \cdot \mathbf{dl} \cong H_0 w \qquad (2)$$

where I is the total current in one conductor and w is the conductor width. This result applies to any path in the cross-sectional plane (Fig. 2.5*b*) between and parallel to the conductors, so H_0 can be considered approximately uniform.

The external inductance for a unit length is found by applying (1) to the surface between the conductors which is shown shaded in Fig. 2.5*c*. Since I is independent of z and H_0 is nearly constant through the space between the conductors and is

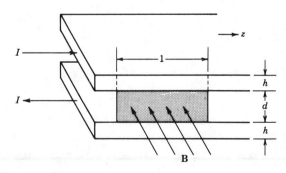

Fig. 2.5*c* Surface for calculation of external inductance of a parallel-plane transmission line.

perpendicular to the shaded surface, (1) becomes

$$L = \frac{1}{I}\mu_0\left(\frac{I}{w}\right)d = \mu_0\frac{d}{w} \quad \text{H/m} \tag{3}$$

This relation is based on the neglect of fringing fields and is most accurate for small d/w.

_____ Example 2.5b _____
External Inductance of a Coaxial Transmission Line

For a coaxial line as pictured in Fig. 2.5d with axial current I flowing in the inner conductor and returning in the outer, the magnetic field is circumferential and for $a < r < b$ is (Ex. 2.4b):

$$H_\phi = \frac{I}{2\pi r} \tag{4}$$

For a unit length the magnetic flux between radii a and b is, by integration over the shaded area in Fig. 2.5d,

$$\int_S \mathbf{B} \cdot d\mathbf{S} = \int_a^b \mu\left(\frac{I}{2\pi r}\right) dr = \frac{\mu I}{2\pi}\ln\frac{b}{a} \tag{5}$$

So, from (1), the inductance per unit length is

$$L = \frac{\mu}{2\pi}\ln\frac{b}{a} \quad \text{H/m} \tag{6}$$

For high frequencies, there is not much penetration of fields into conductors as will be seen in Chapter 3, so this is then the main contribution to inductance. The internal inductance for low frequencies will be considered in Sec. 2.17.

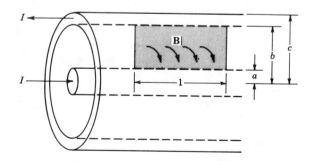

Fig. 2.5d Surface for calculation of external inductance of a coaxial transmission line.

DIFFERENTIAL FORMS FOR MAGNETOSTATISTICS
AND THE USE OF POTENTIAL

2.6   The Curl of a Vector Field

To write differential equation forms for laws having to do with line integrals, it will be necessary to make use of the vector operation called *curl*. This is defined in terms of a line integral taken around an infinitesimal path, divided by the area enclosed by that path. It is seen to have some similarities to the operation of divergence of Sec. 1.11, which was defined as the surface integral taken about an infinitesimal surface divided by the volume enclosed by that surface. Unlike the divergence, however, the curl operation results in a vector because the orientation of the surface element about which the integral is taken must be defined. This additional complication seems to be enough to make curl a more difficult concept for a beginning student. The student should attempt to obtain as much physical significance as possible from the definitions to be given, but at the same time should recognize that full appreciation of the operation will come only with practice in its use.

The curl of a vector field is defined as a vector function whose component at a point in a particular direction is found by orienting a small area normal to the desired direction at that point, and finding the limit of the line integral divided by the area:

$$[\text{curl } \mathbf{F}]_i \triangleq \lim_{\Delta S_i \to 0} \frac{\oint \mathbf{F} \cdot \mathbf{dl}}{\Delta S_i} \tag{1}$$

where i denotes a particular direction, ΔS_i is normal to that direction, and the line integral is taken in the right-hand sense with respect to the positive i direction. In rectangular coordinates, for example, to compute the z component of the curl, the small area $\Delta S = \Delta x \, \Delta y$ is selected in the $x-y$ plane in order to be normal to the z direction (Fig. 2.6a). The right-hand sense of integration about the path with respect to the positive z direction is as shown by the arrows of the figure. The line integral is then

$$\oint \mathbf{F} \cdot \mathbf{dl} = \Delta y F_y \bigg|_{x+\Delta x} - \Delta x F_x \bigg|_{y+\Delta y} - \Delta y F_y \bigg|_x + \Delta x F_x \bigg|_y$$

We find F_y at $x + \Delta x$ and F_x at $y + \Delta y$ by truncated Taylor series expansions

$$F_x \bigg|_{y+\Delta y} \cong F_x \bigg|_y + \Delta y \frac{\partial F_x}{\partial y} \bigg|_y \, ; \qquad F_y \bigg|_{x+\Delta x} \cong F_y \bigg|_x + \Delta x \frac{\partial F_y}{\partial x} \bigg|_x \tag{2}$$

So

$$\oint \mathbf{F} \cdot \mathbf{dl} \cong \left(\frac{\partial F_y}{\partial x} - \frac{\partial F_x}{\partial y} \right) \Delta x \, \Delta y$$

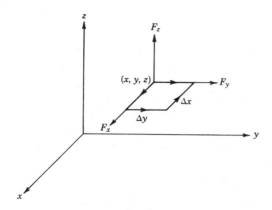

Fig. 2.6a Path for line integral in definition of curl.

Then using the definition (1), we get

$$[\operatorname{curl} \mathbf{F}]_z = \frac{\partial F_y}{\partial x} - \frac{\partial F_x}{\partial y} \tag{3}$$

because the expansions (2) become exact in the limit. Similarly, by taking the elements of area in the y–z plane and x–z plane, respectively, we find

$$[\operatorname{curl} \mathbf{F}]_x = \frac{\partial F_z}{\partial y} - \frac{\partial F_y}{\partial z} \tag{4}$$

$$[\operatorname{curl} \mathbf{F}]_y = \frac{\partial F_x}{\partial z} - \frac{\partial F_z}{\partial x} \tag{5}$$

These components may be multiplied by the corresponding unit vectors and added to form the vector representing the curl:

$$\operatorname{curl} \mathbf{F} = \hat{\mathbf{x}}\left[\frac{\partial F_z}{\partial y} - \frac{\partial F_y}{\partial z}\right] + \hat{\mathbf{y}}\left[\frac{\partial F_x}{\partial z} - \frac{\partial F_z}{\partial x}\right] + \hat{\mathbf{z}}\left[\frac{\partial F_y}{\partial x} - \frac{\partial F_x}{\partial y}\right] \tag{6}$$

If this form is compared with the form of the cross product and the definition of the vector operator $\mathbf{V}$, Eq. 1.10(7), the above can logically be written as

$$\operatorname{curl} \mathbf{F} \equiv \mathbf{V} \times \mathbf{F} = \begin{vmatrix} \hat{\mathbf{x}} & \hat{\mathbf{y}} & \hat{\mathbf{z}} \\ \dfrac{\partial}{\partial x} & \dfrac{\partial}{\partial y} & \dfrac{\partial}{\partial z} \\ F_x & F_y & F_z \end{vmatrix} \tag{7}$$

In deriving curl for other coordinate systems, the variation of line elements with coordinates must be considered, just as the variation of surface elements with coordinates in spherical coordinates was considered in Sec. 1.11. (See Appendix 2.) Results for circular cylindrical and spherical coordinates are given on the inside back cover.

The name *curl* (or *rotation* as it is sometimes called) has some physical significance in the sense that a finite value for the line integral taken in the vicinity of a point is obtained if the curl is finite. The name should not be associated with the curvature of the field lines, however, for a field consisting of closed circles may have zero curl nearly everywhere, and a straight-line field varying in certain ways may have a finite curl. The following examples illustrate these points.

Example 2.6a

Curl-Free Field with Circular Field Lines

The magnetic field in the region surrounding a current in a long straight round wire was seen in Eq. 2.4(2) to be $H_\phi = I/2\pi r$. If we write this in rectangular coordinates using $\sin \phi = y/r$, $\cos \phi = x/r$, and $r^2 = x^2 + y^2$, we get

$$H_x = -H_\phi \sin \phi = -\frac{I}{2\pi} \frac{y}{x^2 + y^2} \tag{8}$$

$$H_y = H_\phi \cos \phi = \frac{I}{2\pi} \frac{x}{x^2 + y^2} \tag{9}$$

$$H_z = 0 \tag{10}$$

as can be seen from Fig. 2.6b. Since there is no z component and no dependence on z, (6) shows immediately that the x and y components of the curl are zero

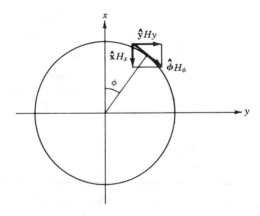

Fig. 2.6b Resolution of H_ϕ of a line current into rectangular components (Ex. 2.6a).

Substituting (8) and (9) into (6) with $\mathbf{F} = \mathbf{H}$ we obtain

$$\text{curl } \mathbf{H} = \hat{z}\left(\frac{\partial H_y}{\partial x} - \frac{\partial H_x}{\partial y}\right) = 0 \tag{11}$$

This result is found more naturally and directly for this problem using the expression for the curl in cylindrical coordinates found inside the back cover:

$$\nabla \times \mathbf{H} = \hat{r}\left[\frac{1}{r}\frac{\partial H_z}{\partial \phi} - \frac{\partial H_\phi}{\partial z}\right] + \hat{\phi}\left[\frac{\partial H_r}{\partial z} - \frac{\partial H_z}{\partial r}\right] + \hat{z}\left[\frac{1}{r}\frac{\partial(rH_\phi)}{\partial r} - \frac{1}{r}\frac{\partial H_r}{\partial \phi}\right] \tag{12}$$

Since there is only an H_ϕ and no z dependence, the first two components vanish. The r and ϕ components are the transverse ones corresponding to x and y components. Since there is no H_r and rH_ϕ does not depend upon r, we see that $\nabla \times \mathbf{H} = 0$ as seen above.

Example 2.6b
Field with Nonvanishing Curl

The magnetic field inside a uniform current with circular symmetry was seen in Ex. 2.4c to be $H_\phi(r) = Ir/2\pi a^2$. As in the preceding example, we see that the symmetries indicate the presence of only a z component of the curl in (12). Also, the second term in the z component is zero. Thus

$$\nabla \times \mathbf{H} = \hat{z}\frac{1}{r}\frac{\partial(rH_\phi)}{\partial r} = \hat{z}\frac{I}{\pi a^2} \tag{13}$$

Example 2.6c
Nonvanishing Curl in Field of Straight Parallel Vectors

A theoretically stable electron flow in a type of microwave electron tube called a planar magnetron has an electron velocity distribution described by $\mathbf{v} = \hat{z}y$ and is shown in Fig. 2.6c. We see there a vector function with all vectors straight and parallel. It is immediately evident by substitution of $\mathbf{v}$ in (6) that

$$\text{curl } \mathbf{v} = \hat{x}[\text{curl } \mathbf{v}]_x = \hat{x} \neq 0 \tag{14}$$

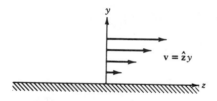

Fig. 2.6c Electron flow in planar magnetron (Ex. 2.6c).

It is instructive to see, by using the line-integral definition of the curl (1) why this result obtains. All vectors and their spatial variations are in the y–z plane, and (6) shows there can be only an x component of the curl. Then we can write for a small area $\Delta S = \Delta y\,\Delta z$

$$[\text{curl } \mathbf{v}]_x = \lim_{\Delta S \to 0} \frac{[v_z(y + \Delta y) - v_z(y)]\,\Delta z}{\Delta y\,\Delta z} \tag{15}$$

We see that the curl is nonzero because the velocity is larger on one side of the loop than on the other.

_____ **Example 2.6d** _____
Curl of the Gradient of a Scalar

Here we show the useful fact that the curl of the gradient of a scalar is zero. If we write

$$\mathbf{F} = \nabla \xi = \hat{\mathbf{x}} \frac{\partial \xi}{\partial x} + \hat{\mathbf{y}} \frac{\partial \xi}{\partial y} + \hat{\mathbf{z}} \frac{\partial \xi}{\partial z}$$

and substitute it in (6), we get

$$\nabla \times \mathbf{F} = \hat{\mathbf{x}} \left(\frac{\partial^2 \xi}{\partial y\,\partial z} - \frac{\partial^2 \xi}{\partial z\,\partial y} \right) + \hat{\mathbf{y}} \left(\frac{\partial^2 \xi}{\partial z\,\partial x} - \frac{\partial^2 \xi}{\partial x\,\partial z} \right) + \hat{\mathbf{z}} \left(\frac{\partial^2 \xi}{\partial x\,\partial y} - \frac{\partial^2 \xi}{\partial y\,\partial x} \right) \tag{16}$$

Since the order of the partial derivative operations is arbitrary $\nabla \times \mathbf{F} = 0$. A particularly important example is the electrostatic field. The fact that $\nabla \times \mathbf{E} = 0$ follows immediately from either $\mathbf{E} = -\nabla \Phi$ or $\oint \mathbf{E} \cdot \mathbf{dl} = 0$. We shall see in Chapter 3 that these properties of $\mathbf{E}$ do not apply for time-varying fields.

_____

2.7   Curl of Magnetic Field

Now let us use the formulations of the last two sections to derive a new relation for magnetic field. The line integral of $\mathbf{H}$ around an area ΔS_i is substituted in the definition of the curl, Eq. 2.6(1), to get

$$[\text{curl } \mathbf{H}]_i = \lim_{\Delta S_i \to 0} \frac{\oint \mathbf{H} \cdot \mathbf{dl}}{\Delta S_i}$$

But $\oint \mathbf{H} \cdot \mathbf{dl}$ is the current through the area ΔS_i by Eq. 2.4(1) so

$$[\text{curl } \mathbf{H}]_i = \lim_{\Delta S_i \to 0} \frac{\int_{\Delta S_i} \mathbf{J} \cdot \mathbf{dS}}{\Delta S_i} = J_i \tag{1}$$

This relation holds for all three orthogonal components. If these are multiplied by the corresponding unit vectors and added, we get the vector relation

$$\text{curl } \mathbf{H} \triangleq \nabla \times \mathbf{H} = \mathbf{J} \tag{2}$$

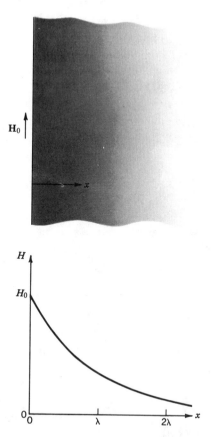

Fig. 2.7 Penetration of magnetic field into a thick sheet of superconducting material.

This can be thought of as the equivalent of Eq. 2.4(1) for a differential path taken around a point. Note that the curl **H** found in Eq. 2.6(13) is the current density, as required by (2).

_____ **Example 2.7** _____
Current Density at Superconductor Surface

If a sheet of superconductive material[2] is in a magnetic field $\mathbf{H} = \hat{z}H_z$ parallel to its surface, there is a penetration of **H** only a very short distance into the superconductor as shown in Fig. 2.7. The decay of H_z with distance is given by

$$H_z = H_0 e^{-x/\lambda} \tag{3}$$

[2] Superconductors include lead, tin, niobium, and numerous other elements, alloys, and compounds. They have zero dc resistance and other special properties below their critical temperatures. See for example A. C. Rose-Innes and E. H. Rhoderick, _Introduction to Superconductivity_, Pergamon Press, Oxford, 1978.

where H_0 is the value at the surface and λ, called the *penetration depth*, is a property of the material. We can find the corresponding current density using (2) and the expansion in Eq. 2.6(6):

$$J_y = [\operatorname{curl} \mathbf{H}]_y = -\frac{\partial H_z}{\partial x} = \frac{H_0}{\lambda} e^{-x/\lambda}$$

Thus, the current also is found only near the surface.

2.8   Relation Between Differential and Integral Forms of the Field Equations

The differential form relating magnetic field to current density was derived from the integral form through the definition of curl. One can proceed in reverse by using Stokes's theorem, which states that for a vector function $\mathbf{F}$,

$$\oint \mathbf{F} \cdot d\mathbf{l} = \int_S (\operatorname{curl} \mathbf{F}) \cdot d\mathbf{S} \equiv \int_S (\nabla \times \mathbf{F}) \cdot d\mathbf{S} \tag{1}$$

This theorem is made plausible by looking at a general surface as in Fig. 2.8a, breaking it into elemental areas. For each differential area, the contribution $(\nabla \times \mathbf{F}) \cdot d\mathbf{S}$ gives the line integral about that area by the definition of curl. If contributions from infinitesimal areas are summed over the surface, the line integral must disappear for all internal areas, since a boundary is first traversed in one direction and then later in the opposite direction in determining the contribution from an adjacent area. The only places where these contributions do not

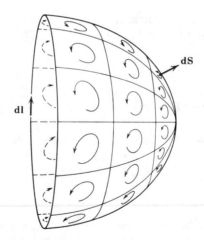

Fig. 2.8a Subdivision of arbitrary surface for proof of Stokes's theorem.

disappear are along the outer boundary, so that the result of the summation is then the line integral of the vector around the boundary as stated in (1). It is recognized that the process is similar to the transformation from the differential to the integral form of Gauss's law through the divergence theorem in Sec. 1.11. Then writing Stokes's theorem for magnetic field, we have

$$\oint \mathbf{H} \cdot \mathbf{dl} = \int_S (\nabla \times \mathbf{H}) \cdot \mathbf{dS} \tag{2}$$

But, by Eq. 2.7(2), the curl may be replaced by the current density:

$$\oint \mathbf{H} \cdot \mathbf{dl} = \int_S \mathbf{J} \cdot \mathbf{dS} \tag{3}$$

The right side represents the current flow through the surface of which the path for the line integration on the left is a boundary. Hence (3) is exactly equivalent to Eq. 2.4(1).

_____ **Example 2.8a** _____
Demonstration of Stokes's Theorem

Let us demonstrate Stokes's theorem for a magnetic field that is part of an electromagnetic wave in a certain kind of transmission structure. The field at a particular instant of time is described by

$$\mathbf{H} = \hat{\mathbf{y}} A \cos \frac{\pi x}{a} \tag{4}$$

We will apply (2) to the area shown in Fig. 2.8b where the field distribution (4) is illustrated. The line integral of (4) along the broken path is

$$\oint \mathbf{H} \cdot \mathbf{dl} = \int_0^a H_x \, dx + \int_0^1 H_y \, dy + \int_a^0 H_x \, dx + \int_1^0 H_y \, dy$$

$$= 0 + A \cos \pi + 0 - A \cos 0 = -2A \tag{5}$$

where the facts that $H_x = 0$ and $H_y \neq f(y)$ are used.

The curl of **H** in rectangular coordinates is

$$\nabla \times \mathbf{H} = \hat{\mathbf{z}} \frac{\partial H_y}{\partial x} = -\hat{\mathbf{z}} A \frac{\pi}{a} \sin \frac{\pi x}{a} \tag{6}$$

The integral of (6) over the surface bounded by the broken line in Fig. 2.8b is

$$\int_S (\nabla \times \mathbf{H}) \cdot \mathbf{dS} = \int_0^a - A \frac{\pi}{a} \sin \frac{\pi x}{a} \, dx = A \cos \frac{\pi x}{a} \Big|_0^a$$

$$= -2A \tag{7}$$

Since (5) and (7) give the same results, Stokes's theorem (1) is illustrated.

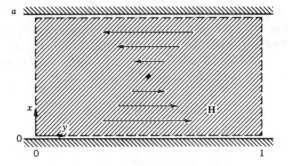

Fig. 2.8b Area for integration of field **H** to demonstrate the validity of Stokes's theorem (Ex. 2.8a).

_____ **Example 2.8b** _____
Proof That $\nabla \cdot \nabla \times \mathbf{F} = 0$

The proof that $\nabla \cdot \nabla \times \mathbf{F} = 0$ can be done by using the expressions in rectangular coordinates as was done for $\nabla \times \nabla\psi$ in Ex. 2.6d. Here we take a different approach that uses Stokes's theorem. Since Stokes's theorem applies to any surface we may treat the surface shown in Fig. 2.8c and let the bounding line shrink to zero so the surface becomes a closed one. Then the line integral on the left side of (2) vanishes and we have

$$\oint_S (\nabla \times \mathbf{F}) \cdot d\mathbf{S} = 0 \tag{8}$$

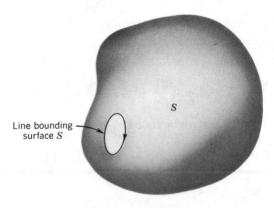

Fig. 2.8c Surface used in Ex. 2.8b.

Then we may apply the divergence theorem (Sec. 1.11) to the vector $\mathbf{V} \times \mathbf{F}$:

$$\oint_S (\mathbf{V} \times \mathbf{F}) \cdot d\mathbf{S} = \int_V \mathbf{V} \cdot \mathbf{V} \times \mathbf{F} \, dV \tag{9}$$

Since we saw in (8) that the left side is zero with an arbitrary choice of surface, the integrand on the right side must vanish:

$$\mathbf{V} \cdot \mathbf{V} \times \mathbf{F} = 0 \tag{10}$$

which was to be shown. This is a useful relation in the study of electromagnetic fields.

2.9   Vector Magnetic Potential

We introduce here another potential, which is often used as a conveniently calculated quantity from which the magnetic field can be found. An integral expression for the flux density can be obtained from Eq. 2.3(2) by multiplying by μ for homogeneous media:

$$\mathbf{B(r)} = \int \frac{\mu I'(\mathbf{r'}) \, d\mathbf{l'} \times \mathbf{R}}{4\pi R^3} \tag{1}$$

It is shown in Appendix 3 that this can be broken into two steps by making use of certain vector equivalences. The result gives

$$\mathbf{B(r)} = \mathbf{V} \times \mathbf{A(r)} \tag{2}$$

where

$$\mathbf{A(r)} = \int \frac{\mu I'(\mathbf{r'}) \, d\mathbf{l}}{4\pi R} \tag{3}$$

The current may be given as a vector density $\mathbf{J}$ in current per unit area spread over a volume V'. Then since $I = J \, dS$ where dS is the differential area element perpendicular to $\mathbf{J}$, and $d\mathbf{l}$ is in the direction of $\mathbf{J}$, $dS \, dl$ forms a volume element dV and the equivalent to (3) is

$$\mathbf{A(r)} = \int_V \frac{\mu \mathbf{J(r')} \, dV'}{4\pi R} \tag{4}$$

In both (3) and (4), R is the distance from a current element of the integration to the point at which $\mathbf{A}$ is to be computed. The function $\mathbf{A}$, introduced as an intermediate step, is computed as an integral over the given currents from (3) or (4) and then differentiated in the manner defined by (2) to yield the magnetic field. Function $\mathbf{A}$ is called the *magnetic vector potential*. Note that each element of $\mathbf{A}$ has the direction of the current element producing it. It is analogous to the potential function of electrostatics, which is found in terms of an integral over charges and then differentiated in a certain way to yield the electric field. It is different, however, because it is a vector, and does not have the simple physical significance of work done in moving through the field that electrostatic potential has. Some physical pictures can be formed but the student should not worry about these until more familiarity with the function has been developed through certain examples.

_____ **Example 2.9a** _____
Vector Potential and Magnetic Field of a Current Element

Here we show that the magnetic flux density of a current element found using (3) and (2), in that order, is the same as the integrand of (1), which expresses Ampère's law. The magnetic vector potential **A** exists throughout the region surrounding the given current element, as shown in Fig. 2.9a. From (3) we find

$$\mathbf{A} = \hat{\mathbf{z}}A_z = \hat{\mathbf{z}}\frac{\mu I\,dz}{4\pi r} \tag{5}$$

since the origin of coordinates is positioned at the current element. As noted earlier **dA** is parallel to the current element producing it. It is most convenient to use spherical coordinates in this example. From the figure we see that $A_r = A_z \cos\theta$ and $A_\theta = -A_z \sin\theta$. The curl in spherical coordinates (from the inside of the back cover) reduces to

$$\mathbf{B} = \nabla \times \mathbf{A} = \frac{\hat{\boldsymbol{\phi}}}{r}\left[\frac{\partial}{\partial r}(rA_\theta) - \frac{\partial A_r}{\partial\theta}\right] \tag{6}$$

since $A_\phi = 0$ and $\partial/\partial\phi = 0$, by symmetry. Substituting A_r and A_ϕ using (5), we find

$$\mathbf{B} = \nabla \times \mathbf{A} = \hat{\boldsymbol{\phi}}\left(\frac{\mu I\,dz}{4\pi}\right)\frac{\sin\theta}{r^2} \tag{7}$$

Notice that $\mathbf{dl'} \times \mathbf{R}$ is $dz\,r\sin\theta$ so (7) is equivalent to the integrand in (1).

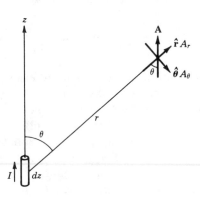

Fig. 2.9a Vector potential in region surrounding a current element.

_____ Example 2.9b _____
Vector Potential and Field of a
Parallel-Wire Transmission Line

Let us consider a parallel-wire transmission line of infinite length carrying current I in one conductor and its return in the other distance $2a$ away. The coordinate system is set up as in Fig. 2.9b. Since the field quantities do not vary with z, it is convenient to calculate them in the plane $z = 0$. The conductors will first be taken as extending from $z = -L$ to $z = L$ to avoid indeterminacies in the integrals. Since current is only in the z direction, $\mathbf{A}$ by (3) will be in the z direction also. The contribution to A_z from both wires is

$$A_z = \int_{-L}^{L} \frac{\mu I\, dz'}{4\pi\sqrt{(x - a)^2 + y^2 + z'^2}} - \int_{-L}^{L} \frac{\mu I\, dz''}{4\pi\sqrt{(x + a)^2 + y^2 + z''^2}}$$

$$= \frac{2\mu}{4\pi}\left[\int_0^L \frac{I\, dz'}{\sqrt{(x - a)^2 + y^2 + z'^2}} - \int_0^L \frac{I\, dz''}{\sqrt{(x + a)^2 + y^2 + z''^2}}\right]$$

The integrals may be evaluated [3]

$$A_z = \frac{I\mu}{2\pi}\{\ln[z' + \sqrt{(x - a)^2 + y^2 + z'^2}]$$

$$- \ln[z'' + \sqrt{(x + a)^2 + y^2 + z''^2}]\}_0^L$$

Now, as L is allowed to approach infinity, the upper limits of the two terms cancel. Hence

$$A_z = \frac{I\mu}{4\pi}\ln\left[\frac{(x + a)^2 + y^2}{(x - a)^2 + y^2}\right] \tag{8}$$

If (2) is then applied, using the expression for curl in rectangular coordinates, we find

$$H_x = \frac{1}{\mu}\frac{\partial A_z}{\partial y} = \frac{I}{2\pi}\left[\frac{y}{(x + a)^2 + y^2} - \frac{y}{(x - a)^2 + y^2}\right] \tag{9}$$

$$H_y = -\frac{1}{\mu}\frac{\partial A_z}{\partial x} = \frac{I}{2\pi}\left[\frac{(x - a)}{(x - a)^2 + y^2} - \frac{(x + a)}{(x + a)^2 + y^2}\right] \tag{10}$$

---

[3] Most integrals of this text can be found in standard handbooks such as the *CRC Handbook of Chemistry and Physics* (any recent edition), M. R. Spiegel, *Mathematical Handbook*, Schaum's Outline Series, McGraw-Hill, 1968; or *Handbook of Mathematical Functions*, M. Abramowitz and I. A. Stegun, eds., National Bureau of Standards Applied Mathematics, U.S. Government Printing Office, Washington, D.C., 1964. One of the most complete listings is I. S. Gradshteyn and I. M. Ryzhik, *Table of Integrals, Series, and Products*, Translated by Alan Jeffrey, Academic Press, New York, 1980.

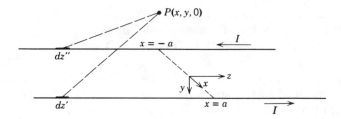

Fig. 2.9b Parallel-wire transmission line.

2.10   Distant Field of Current Loop: Magnetic Dipole

The magnetic field on the axis of a loop of current was derived in Ex. 2.3a. Here we will find the magnetic vector potential and field at locations not restricted to the axis but distant from the loop. The arrangement to be analyzed is shown in Fig. 2.10. For any point (r, θ, ϕ) at which **A** is to be found, some current elements $I\,\mathbf{dl'}$ are oriented such that they produce components of **A** in directions other than the ϕ direction. However, by the symmetry of the loop, equal and opposite amounts of such components exist. As a result **A** is ϕ directed and is independent of the value of ϕ at which it is to be found. For convenience, we choose to

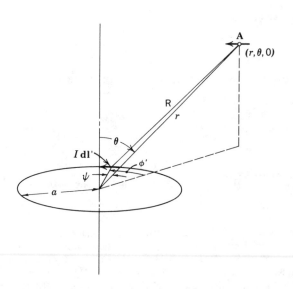

Fig. 2.10 Coordinates for calculation of magnetic-dipole fields

calculate **A** at the point $(r, \theta, 0)$. The ϕ-directed contribution of a differential element of current is

$$dA_\phi = \frac{\mu I \, dl' \cos \phi'}{4\pi R} \tag{1}$$

where R is the distance from the element dl' to $(r, \theta, 0)$. The total is found as the integral around the loop

$$A_\phi = \frac{\mu I}{4\pi} \oint \frac{dl' \cos \phi'}{R} = \frac{\mu I a}{4\pi} \int_0^{2\pi} \frac{\cos \phi' \, d\phi'}{R} \tag{2}$$

where a is the radius of the loop. The distance R can be expressed in terms of the radius from the origin to $(r, \theta, 0)$ as

$$R^2 = r^2 + a^2 - 2ra \cos \psi \tag{3}$$

To get $ra \cos \psi$ we note that $r \cos \psi$ is the projection of r onto the extension of the radius line to dl'. Therefore

$$ra \cos \psi = ra \sin \theta \cos \phi' \tag{4}$$

Substituting (4) into (3) and assuming $r \gg a$, we find

$$R \approx r\left(1 - 2\frac{a}{r} \sin \theta \cos \phi'\right)^{1/2}$$

or

$$R^{-1} \approx r^{-1}\left(1 + \frac{a}{r} \sin \theta \cos \phi'\right) \tag{5}$$

Utilizing this expression in (2), we find

$$A_\phi = \frac{\mu I a}{4\pi r} \int_0^{2\pi} \left(\cos \phi' + \frac{a}{r} \sin \theta \cos^2 \phi'\right) d\phi'$$

$$= \frac{\mu I a}{4\pi r} \frac{a\pi \sin \theta}{r} = \frac{\mu(I\pi a^2) \sin \theta}{4\pi r^2} \tag{6}$$

As was noted at the outset the result applies to any value of ϕ. The components of **B**, found by substituting (6) in Eq. 2.9(2), are

$$B_r = \frac{\mu I \pi a^2}{2\pi r^3} \cos \theta \tag{7}$$

$$B_\theta = \frac{\mu I \pi a^2}{4\pi r^3} \sin \theta \tag{8}$$

$$B_\phi = 0 \tag{9}$$

The group of terms $I\pi a^2$ can be given a special significance by comparison of (7) to (9) with the fields of an electric dipole, Eq. 1.10(10). The identity of the functional form of the fields has led to defining the magnitude of the *magnetic dipole moment* as

$$m = I\pi a^2 \tag{10}$$

The dipole direction is along the $\theta = 0$ axis in Fig. 2.10 for the direction of I shown. The vector potential can be written in terms of the magnetic dipole moment **m** as

$$\mathbf{A} = \frac{-\mu_0}{4\pi} \mathbf{m} \times \nabla\left(\frac{1}{r}\right) \tag{11}$$

where the partial derivatives in the gradient operation are with respect to the point of observation of **A**.

2.11   Divergence of Magnetic Flux Density

As given by Eq. 2.9(2) (derived in Appendix 3), the magnetic flux density **B** can be expressed as the curl of another vector **A** when the sources of **B** are currents. We have shown in Ex. 2.8b that the divergence of the curl of any vector is zero. Thus,

$$\nabla \cdot \mathbf{B} = 0 \tag{1}$$

A major difference between electric and magnetic fields is now apparent. The magnetic field must have zero divergence everywhere. That is, when the magnetic field is due to currents, there are no sources of magnetic flux which correspond to the electric charges as sources of electric flux. Fields with zero divergence such as these are consequently often called *source-free* fields.

Magnetic field concepts are often developed from an exact parallel with electric fields by considering the concept of isolated magnetic poles as sources of magnetic flux, corresponding to the charges of electrostatics. The result of zero divergence then follows because such poles have so far been found in nature only as equal and opposite pairs. Physicists continue to search for isolated magnetic poles; if they are found, a magnetic charge density ρ_m will simply be added to the equations giving a finite $\nabla \cdot \mathbf{B}$.

2.12   Differential Equation for Vector Magnetic Potential

The differential equation for magnetic field in terms of current density was developed in Sec. 2.7:

$$\nabla \times \mathbf{H} = \mathbf{J}$$

If the relation for **B** as the curl of vector potential **A** is substituted,

$$\nabla \times \nabla \times \mathbf{A} = \mu \mathbf{J} \tag{1}$$

This may be considered a differential equation relating **A** to current density. It is more common to write it in a different form utilizing the Laplacian of a vector function defined in rectangular coordinates as the vector sum of the Laplacians of the three scalar components:

$$\nabla^2 \mathbf{A} = \hat{x}\nabla^2 A_x + \hat{y}\nabla^2 A_y + \hat{z}\nabla^2 A_z \tag{2}$$

It may then be verified that, for rectangular coordinates

$$\nabla \times \nabla \times \mathbf{A} = -\nabla^2 \mathbf{A} + \nabla(\nabla \cdot \mathbf{A}) \tag{3}$$

For other than rectangular coordinate systems, separation in the form (2) cannot be done so simply and (3) may be taken as the definition of ∇^2 of a vector. With $\nabla \cdot \mathbf{A} = 0$, (3) and (1) give

$$\nabla^2 \mathbf{A} = -\mu \mathbf{J} \tag{4}$$

This is a vector equivalent of the Poisson equation first met in Sec. 1.12. It includes three component scalar equations which are exactly of the Poisson form.

Example 2.12
Vector Potential and Field of Uniform Circularly Symmetric Current Density Flowing Axially

Let us show that the appropriate form for the vector potential in a uniform circularly symmetric flow of z-directed current is

$$A_z = -\frac{\mu J_0}{4}(x^2 + y^2) \tag{5}$$

From (4) and (2),

$$J_z = -\frac{1}{\mu}\nabla^2 A_z = -\frac{1}{\mu}\left(\frac{\partial^2 A_z}{\partial x^2} + \frac{\partial^2 A_z}{\partial y^2}\right) = J_0 \tag{6}$$

Or, going the other way, we can see that (5) is the appropriate form for vector potential in a cylindrical conductor carrying a current of constant density J_0. The magnetic field found from (5) is

$$\mathbf{H} = \frac{1}{\mu}(\nabla \times \mathbf{A}) = \frac{-J_0}{2}(\hat{x}y - \hat{y}x) \tag{7}$$

In cylindrical coordinates, this is

$$\mathbf{H} = \hat{\phi}\frac{J_0 r}{2} \tag{8}$$

which is the value of Eq. 2.4(5).

2.13   Scalar Magnetic Potential for Current-Free Regions

In many problems concerned with the finding of magnetic fields, at least a part of the region is current-free. The curl of the magnetic field vector $\mathbf{H}$ is then zero for such current-free regions [Eq. 2.7(2)]. Any vector with zero curl may be represented as the gradient of a scalar (see Ex. 2.6d). Thus the magnetic field can be expressed for such points as

$$\mathbf{H} = -\nabla\Phi_m \tag{1}$$

where the minus sign is conventionally taken only to complete the analogy with electrostatic fields. The vector potential applies to both current-carrying and current-free regions; but it is usually more convenient for the latter to use this scalar potential.

Since the divergence of the magnetic flux density $\mathbf{B}$ is everywhere zero,

$$\nabla \cdot \mu\nabla\Phi_m = 0 \tag{2}$$

Thus, for a homogeneous medium, Φ_m satisfies Laplace's equation

$$\nabla^2\Phi_m = 0 \tag{3}$$

It will be observed from (1) that

$$\Phi_{m2} - \Phi_{m1} = -\int_1^2 \mathbf{H} \cdot \mathbf{dl} \tag{4}$$

Thus, if the path of integration encircles a current, Φ_m does not have a unique value. For if 1 and 2 are the same point in space and the path of integration encloses a current I, two values of Φ_m, differing by I, will be assigned to the point.

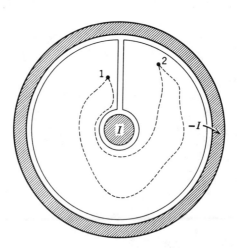

Fig. 2.13 Simply connected region between coaxial cylinders.

To make the scalar magnetic potential unique, we must restrict attention to regions which do not entirely encircle currents. Suitable regions are called "simply connected" because any two paths connecting a pair of points in the region form a loop which does not enclose any exterior points. An example of a simply connected region between coaxial conductors is shown in Fig. 2.13. The restriction to a simply connected region is not a serious limitation once it is understood.

The importance of the scalar potential is that it satisfies Laplace's equation for which exist numerous methods of solution. The graphical methods given in Chapter 1 for electrostatic fields are directly applicable, as are the more powerful numerical methods, conformal transformations, and method of separation of variables to be studied in Chapter 7.

2.14   Boundary Conditions for Static Magnetic Fields

The boundary conditions at an interface between two regions with different permeabilities can be found in the same way as was done for static electric fields in Sec. 1.14. Consider a volume in the shape of a pillbox enclosing the boundary between the two media as shown in Fig. 2.14. The surfaces ΔS of the volume are considered to be arbitrarily small so that the normal flux density B_n does not vary across the surface. Also the thickness of the pillbox is vanishingly small so that

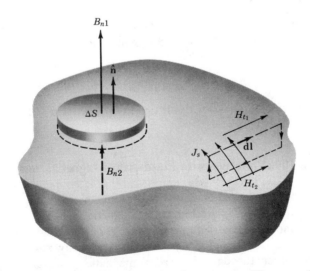

Fig. 2.14 Magnetic fields at boundary between two different media.

there is negligible flux flowing through the side wall. The net outward flux from the box is

$$B_{n1} \, \Delta S = B_{n2} \, \Delta S$$

or

$$B_{n1} = B_{n2} \tag{1}$$

where the sense of B_n is as shown in the figure.

The relation between transverse magnetic fields may be found by integrating the magnetic field **H** along a line enclosing the interface plane as shown in Fig. 2.14:

$$\oint \mathbf{H} \cdot \mathbf{dl} = H_{t1} \, \Delta l - H_{t2} \, \Delta l = J_s \, \Delta l \tag{2}$$

where J_s is a surface current in amperes per meter width flowing in the direction shown. The lengths Δl of the sides are arbitrarily small so H_t may be considered uniform. The other legs of the integration path are effectively reduced to zero length. From (2)

$$H_{t1} - H_{t2} = J_s \tag{3}$$

There is a discontinuity of the tangential field at the boundary between two regions equal to any surface current which may exist on the boundary. With direction information included, where $\hat{n}$ is the unit vector normal to the surface,

$$\hat{n} \times (\mathbf{H}_1 - \mathbf{H}_2) = \mathbf{J}_s \tag{4}$$

In problems involving the scalar magnetic potential, continuity of H_t where $J_s = 0$ is ensured by taking Φ_m to be continuous across the boundary. Where surface currents exists, (4) leads to

$$\hat{n} \times (\nabla\Phi_{m2} - \nabla\Phi_{m1}) = \mathbf{J}_s \tag{5}$$

as may be seen by combining (3) with the definition of Φ_m, Eq. 2.13(1).

2.15   Materials with Permanent Magnetization

Permanent magnets have a remnant value of magnetization [defined in Eq. 2.3(3)] when all applied fields are removed. Magnetic materials are discussed in more detail in Chapter 13, but here we consider some examples with permanent magnetization M_0 and no true current flow. There are two ways of analyzing such problems, either through the scalar magnetic potential or the vector potential.

Use of Scalar Magnetic Potential Since current density **J** is zero, we may derive **H** from a scalar potential as in Sec. 2.13,

$$\mathbf{H} = -\nabla\Phi_m \tag{1}$$

Now using the definition of magnetization from Eq. 2.3(3),

$$\mathbf{H} = \frac{\mathbf{B}}{\mu_0} - \mathbf{M} \tag{2}$$

If the divergence of (2) is taken, with $\nabla \cdot \mathbf{B} = 0$ utilized, we can write

$$\nabla^2\Phi_m = -\frac{\rho_m}{\mu_0} \tag{3}$$

where

$$\rho_m = -\mu_0 \nabla \cdot \mathbf{M} \tag{4}$$

In this formulation we see that we have a Poisson equation for potential Φ_m, with an equivalent magnetic charge density in the region proportional to the divergence of magnetization. For a uniform magnetization, the divergence is zero and Φ_m satisfies Laplace's equation. At the boundaries of the magnet, however, integration of (3) would show that there is an equivalent magnetic surface charge density ρ_{sm} given by

$$\rho_{sm} = \mu_0 \hat{\mathbf{n}} \cdot \mathbf{M} \tag{5}$$

The arguments for this are similar to those for surface charge density ρ_s when there is a discontinuity in **D**, as explained in Sec. 1.14. We will illustrate this through an example after giving a formulation using the vector potential.

Use of Vector Magnetic Potential If we write **B** as curl of vector potential **A** as in Eq. 2.9(2) and use the definition of magnetization,

$$\mathbf{B} = \mu_0(\mathbf{H} + \mathbf{M}) = \nabla \times \mathbf{A} \tag{6}$$

we can take the curl of this equation, using $\nabla \times \mathbf{H} = 0$ since $\mathbf{J} = 0$, to write

$$\nabla \times \nabla \times \mathbf{A} = \mu_0 \mathbf{J}_{eq} \tag{7}$$

where

$$\mathbf{J}_{eq} = \nabla \times \mathbf{M} \tag{8}$$

So by comparison with Eq. 2.12(1), the problem is equivalent to one with internal currents in free space proportional to the curl of magnetization. Inside a region of uniform magnetization, the curl is zero and there are no internal currents. At the boundary of the magnetic material, a surface integral of (8) over the area enclosed by the path used to get Eq. 2.14(2) and application of Stokes's theorem gives

$$\oint \mathbf{M} \cdot d\mathbf{l} = \int_S \mathbf{J}_{eq} \cdot d\mathbf{S}$$

In the same way as for Eq. 2.14(4), this gives an equivalent surface current

$$(\mathbf{J}_{eq})_S = \mathbf{M} \times \hat{\mathbf{n}} \tag{9}$$

since $\mathbf{M} = 0$ outside the magnetic material. So in this formulation the magnet is replaced by a system of volume and surface currents from which magnetic field may be found through use of the vector potential, or directly by using Ampère's law. The second example to follow illustrates this procedure.

Example 2.15a
Uniformly Magnetized Sphere

Consider first a sphere of magnetic material with uniform magnetization M_0 in the z direction as in Fig. 2.15a using the method with scalar magnetic potential. Since M is uniform, there is no volume charge by (4), but if space surrounds the sphere, there is a surface magnetic charge density at $r = a$ given by

$$\rho_{sm} = -\mu_0 M_0 \cos\theta \tag{10}$$

Solutions of (3), in spherical coordinates with a variation corresponding to (10) and $\rho_m = 0$, are

$$\Phi_{m1} = \frac{Cr}{a} \cos\theta \qquad r < a$$

$$\tag{11}$$

$$\Phi_{m2} = \frac{Ca^2}{r^2} \cos\theta \qquad r > a$$

as can be verified by substitution in the expression for $\nabla^2\Phi = 0$ in spherical coordinates on the inside back cover. The surface magnetic charge given by (10) gives a discontinuity in derivative,

$$\mu_0 \left| \frac{\partial\Phi_{m2}}{\partial r} - \frac{\partial\Phi_{m1}}{\partial r} \right|_{r=a} = -\mu_0 M_0 \cos\theta \tag{12}$$

from which

$$C = \frac{M_0 a}{3} \tag{13}$$

Thus for $r < a$, using (1) in spherical coordinates

$$H = -\frac{M_0}{3}[\hat{\mathbf{r}} \cos\theta - \hat{\boldsymbol{\theta}} \sin\theta] = -\hat{\mathbf{z}}\frac{M_0}{3} \tag{14}$$

which is a uniform field within the sphere. For $r > a$,

$$H = \frac{M_0 a^3}{3r^3}[2\hat{\mathbf{r}} \cos\theta + \hat{\boldsymbol{\theta}} \sin\theta] \tag{15}$$

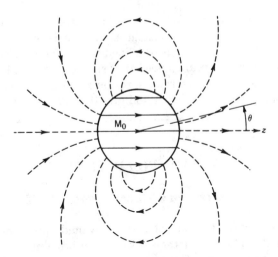

Fig. 2.15a Sphere of radius a with uniform magnetization $\hat{z}M_0$. Field lines (**H** or **B**) outside the sphere shown dashed.

which are curves (shown dashed in Fig. 2.15a) similar to those outside a magnetic dipole (Sec. 2.10).

_____ **Example 2.15b** _____
Round Rod with Uniform Magnetization

A circular cylindrical bar magnet of length l having uniform magnetization in the axial direction is shown in Fig. 2.15b. Using the second formulation given above, we see from (8) that there are no equivalent volume currents since $\nabla \times \mathbf{M} = 0$, but there is a surface current at the discontinuous boundary $r = a$,

$$\mathbf{J}_s = \hat{\phi}M_0 \tag{16}$$

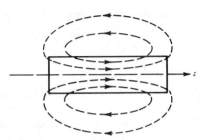

Fig. 2.15b Cylinder of radius a, length l with magnetization $\hat{z}M_0$. Flux density lines **B** shown dashed. (**H** lines are of the same form outside the magnet.)

We see that this problem is then identical to that of the solenoid of length l with current per unit length given by (16) insofar as the calculation of $\mathbf{A}$ (and hence $\mathbf{B}$) is concerned. As noted in Ex. 2.3c, it is difficult to calculate field lines for an off-axis point, but $\mathbf{B}$ lines will appear somewhat as shown dashed in Fig. 2.15b. Lines of magnetic field $\mathbf{H}$ will be of the same form outside the magnet, but will be of different form inside through the vector addition $\mathbf{H} = \mathbf{B}/\mu_0 - \mathbf{M}$.

MAGNETIC FIELD ENERGY

2.16 Energy of a Static Magnetic Field

In considering the energy of a magnetic field, it would appear by analogy with Sec. 1.21 that we should consider the work done in bringing a group of current elements together from infinity. This point of view is correct in principle, but not only is it more difficult to carry out than for charges because of the vector nature of currents, but it also requires consideration of time-varying effects as shown in references deriving the relation from this point of view.[4] We will consequently set down the result at this point, waiting for further discussion until we derive a most important general energy relationship in Chapter 3. The general relation for nonlinear materials is

$$dU_H = \int_V \mathbf{H} \cdot d\mathbf{B}\, dV \tag{1}$$

where dU_H is the energy added to the system when $\mathbf{B}$ is changed by a differential amount (possibly different amounts for different positions within the volume). For linear materials, $\mathbf{H}$ is proportional to $\mathbf{B}$ so (1) may be integrated over $\mathbf{B}$ to give

$$U_H = \frac{1}{2} \int_V \mathbf{B} \cdot \mathbf{H}\, dV = \int_V \frac{\mu}{2} H^2\, dV \tag{2}$$

The analogy to Eq. 1.21(6) is apparent, and here also we interpret the energy of a system of sources as actually stored in the fields produced by those sources. The result is consistent with the inductive circuit energy term, $\frac{1}{2}LI^2$, when circuit concepts hold and will be utilized in the following article.

Example 2.16a
Energy Storage in Superconducting Solenoid

It has been proposed to use energy stored in large superconducting coils to meet peaks in electric power demand. Superconducting coils are chosen because their

---

[4] J. A. Stratton, *Electromagnetic Theory*, McGraw-Hill, New York, 1941, pp. 118–124.

zero dc resistance allows carrying very large currents with zero power loss (though of course, refrigeration power must be supplied). To be useful such a storage system must be capable of providing about 50 MW for 6 hours—that is, storing an energy of about 10^6 MJ. Let us assume the coil is a solenoid and that the field is uniform, and find the required coil properties and current. The field from Eq. 2.4(6) is $H_z = nI$ so $B_z = \mu nI$. The energy from (2), for volume V, is

$$U_H = \tfrac{1}{2}\mu(nI)^2 V$$

For a realistic current of 1000 A and flux density of 15 T, a coil of 27-m diameter and 20-m length with 1.2×10^4 turns/m would give the required energy. The most promising coil shape is actually a toroid but it would have dimensions and currents of the same magnitude as calculated in this example.

Example 2.16b
Energy Dissipation in Hysteretic Materials

We will see here how energy loss in hysteretic materials can be explained in terms of their nonlinear B–H relations. A typical hysteretic relation is shown in Fig. 2.16. We will assume an isotropic material so that $\mathbf{B} \cdot \mathbf{H} = BH$. The energy required for one traversal of the loop by varying H from a large negative value to a large positive value and back again can be found from (1). The differential energy is shown as a shaded bar on the hysteresis loop in Fig. 2.16. When the field is decreased, a portion of the energy indicated by the part of the bar outside the loop is returned to the field. The result of integrating around the loop is that total expended energy per unit volume is equal to the area of the loop.

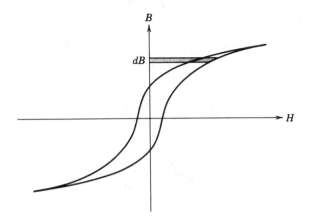

Fig. 2.16 Hysteretic, nonlinear B–H relation.

2.17 Inductance from Energy Storage; Internal Inductance

It was shown in Sec. 2.16 that the magnetic energy may be found by integrating an energy density of $\frac{1}{2}\mu H^2$ throughout the volume of significant fields. From a circuit point of view, this is known to be $\frac{1}{2}LI^2$, where I is the instantaneous current flow through the inductance. Equating these two forms gives

$$\tfrac{1}{2}LI^2 = \int_V \frac{\mu}{2} H^2 \, dV \tag{1}$$

The form of (1) is useful as an alternate to the flux linkage method of calculating information given in Sec. 2.5. It is especially convenient for problems that would require consideration of partial linkages if done by the method of flux linkages. Problems of calculating internal inductance, defined in Sec. 2.5, are of this nature.

_____ **Example 2.17** _____
Internal Inductance of Conductors with Uniform Current Distribution in a Coaxial Transmission Line

As an example of the use of the energy method of inductance calculation, we will find the internal inductances for the two conductors of a coaxial transmission line under the assumption that the current is distributed uniformly in the conductors. The result for the inner conductor applies more generally to any straight, round wire with a uniform current distribution. The magnetic field in the inner conductor (Ex. 2.4c) is

$$H_\phi = \frac{Ir}{2\pi a^2} \qquad r < a \tag{2}$$

For a unit length, utilizing (1),

$$\tfrac{1}{2}LI^2 = \int_0^a \frac{\mu}{2}\left(\frac{Ir}{2\pi a^2}\right)^2 2\pi r \, dr = \frac{\mu I^2}{4\pi a^4} \cdot \frac{a^4}{4} \tag{3}$$

or

$$L = \frac{\mu_0}{8\pi} \quad \text{H/m} \tag{4}$$

The magnetic field in the outer conductor (Prob. 2.4a) is

$$H(r) = \frac{I}{2\pi(c^2 - b^2)}\left(\frac{c^2}{r} - r\right) \tag{5}$$

Substituting (5) in (1) we find

$$L = \frac{\mu}{2\pi}\left[\frac{c^4 \ln c/b}{(c^2 - b^2)^2} + \frac{b^2 - 3c^2}{4(c^2 - b^2)}\right] \quad \text{H/m} \tag{6}$$

For frequencies low enough to assume uniform current distribution in the conductors, the total inductance per unit length for the coaxial line is the sum of (4), (6), and Eq. 2.5(6).

PROBLEMS

2.2a Assuming that each electron constituting the current in a differential length of conductor is acted on by a force $-e\mathbf{v} \times \mathbf{B}$, show that the total force is equal to that given by Eq. 2.2(1).

2.2b Write the torque about an axis in terms of vector notation (suitably defining a vector to represent torque) when a force $\mathbf{F}$ acts at vector distance $\mathbf{r}$ from the axis.

2.2c Show the following:

$$\mathbf{A} \times (\mathbf{B} + \mathbf{C}) = \mathbf{A} \times \mathbf{B} + \mathbf{A} \times \mathbf{C}$$

$$\mathbf{A} \times (\mathbf{B} \times \mathbf{C}) = \mathbf{B}(\mathbf{A} \cdot \mathbf{C}) - \mathbf{C}(\mathbf{A} \cdot \mathbf{B})$$

$$\mathbf{A} \cdot (\mathbf{B} \times \mathbf{C}) = \mathbf{B} \cdot (\mathbf{C} \times \mathbf{A}) = \mathbf{C} \cdot (\mathbf{A} \times \mathbf{B})$$

2.2d Utilizing the stated conversion factors of Appendix 1, and the fact that B/H is $\mu_0 = 4\pi \times 10^{-7}$ H/m in the rationalized mks system, show that B/H is unity for vacuum in the emu system of units.

2.3a Find from Ampère's law the magnetic field at a point distance r from an infinite straight wire carrying dc current I_0.

2.3b dc current I_0 flows in a square loop of wire having sides of length $2a$. Find the magnetic field on the axis at a point z from the plane of the loop.

2.3c Represent a solenoid of finite length L and radius a having n turns per meter by a continuous sheet of circumferential current. Find the axial magnetic field at the center of the solenoid and determine the length for which the field is one-half that of the infinite solenoid.

2.3d Show that the magnetic field on axis of a long solenoid at the ends is half the value for an infinite solenoid.

2.3e* An arrangement that can provide a region of relatively uniform fields consists of a pair of parallel, coaxial loops; the uniform-field region is on the axis midway between the loops. Show, using power-series expansions, that the largest region of uniform field along the axis can be achieved by making the loop radii a equal to the spacing d of the loops, the so-called *Helmholtz coil* configuration.

2.4a For the coaxial line of Fig. 2.4b, find the magnetic field for $b < r < c$, assuming that current is distributed uniformly over the cross section.

2.4b A certain kind of electron beam of circular cross section contains a current density $J_z = J_0[1 - (r/a)^4]$. Find $H_\phi(r)$ inside the beam.

---

* The asterisk on problems denotes ones longer or harder than the average; a double asterisk will denote unusually difficult or lengthy problems.

2.4c Express the magnetic field about a long line current in rectangular coordinate components, taking the wire axis as the z axis, and evaluate $\oint \mathbf{H} \cdot \mathbf{dl}$ about a square path in the x–y plane from $(-1, -1)$ to $(1, -1)$ to $(1, 1)$ to $(-1, 1)$ back to $(-1, -1)$. Also evaluate the integral about the path from $(-1, 1)$ to $(1, 1)$ to $(1, 2)$ to $(-1, 2)$ back to $(-1, 1)$. Comment on the two results.

2.4d A long thin wire carries a current I_1 in the positive z direction along the axis of a cylindrical coordinate system as shown in Fig. P2.4d. A thin rectangular loop of wire lies in a plane containing the axis. The loop contains the region $0 \le z \le b$, $R - a/2 \le r \le R + a/2$ and carries a current I_2 which has the direction of I_1 on the side nearest the axis. Find the vector force on each side of the loop and the resulting force on the entire loop.

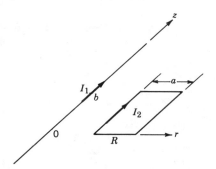

Fig. P2.4d Wire arrangement for Prob. 2.4d.

2.4e* Consider a round straight wire carrying a uniform current density J throughout, except for a round cylindrical void parallel with the wire axis so that the cross section is constant. Call the radius of the wire c, the radius of the hole b, and the distance of the center of the hole from the center of the wire a. Take $b < a < c$ and $b < c - a$. Use superposition to find the field H as a function of position along a radial line through the center of the hole for all values of radius from the center of the wire.

2.4f A demonstration can be given that a thin metal tube can be crushed by magnetic forces by passing current through it. Take the radius of the tube to be 2 cm and the magnetic field at which failure of the metal occurs as 9 Wb/m². (i) What is the maximum current that could flow axially along the tube before it would be crushed by the magnetic forces arising from this current? (ii) What is the force per unit area on the surface of the tubing under this condition?

2.5 A coaxial transmission line with inner conductor of radius a and outer conductor having inner radius b has a coaxial cylindrical ferrite of permeability μ_1 extending from $r = a$ to $r = d$ (with $d < b$), and air from radius d to b. Find the inductance per unit length.

2.6a Find the curl of a vector field $\mathbf{F} = \hat{x}x^2z^2 + \hat{y}y^2z^2 + \hat{z}x^2y^2$.

2.6b By using the rectangular coordinate forms show that

$$\nabla \times (\psi \mathbf{F}) = \psi \nabla \times \mathbf{F} - \mathbf{F} \times \nabla \psi$$

where $\mathbf{F}$ is any vector function and ψ any scalar function.

2.6c Derive the expression for curl in the spherical coordinate system.

2.7 For the coaxial line of Fig. 2.4b, express the magnetic field found in Ex. 2.4b and Prob. 2.4a in rectangular coordinates and find the curl in the four regions, $r < a$, $a < r < b$, $b < r < c$, $r > c$. Comment on the results.

2.8 Show that $\nabla \times \nabla \psi \equiv 0$ by integrating over an arbitrary surface and applying Stokes's theorem.

2.9a Check the results Eqs. 2.9(9) and (10) by adding vectorially the magnetic field from the individual wires, using the result of Ex. 2.4a.

2.9b** A square loop of thin wire lies in the x–y plane extending from $(-a, -a)$ to $(a, -a)$ to (a, a) to $(-a, a)$ back to $(-a, -a)$ and carries current I in that sense of circulation. Find **A** and **H** for any point (x, y, z).

2.9c* A circular loop of thin wire carries current I. Find **A** for a point distance z from the plane of the loop, and radius r from the axis, for $r/z \ll 1$. Use this to find the expression for magnetic field on the axis.

2.9d Show that the same value of **B** is given by either $\mathbf{B} = \nabla \times \mathbf{A}$ or $\mathbf{B} = \nabla \times \mathbf{A}'$ where $\mathbf{A}' = \mathbf{A} + \nabla \psi$ and ψ is any scalar function.

2.9e* A very long thin conducting sheet having a width b carries a direct current I in the direction of its length. Show that if the sheet is assumed to lie in the x–z plane with the z axis along its centerline, the magnetic field about the strip will be given by

$$H_x = -\frac{I}{2\pi b}\left(\tan^{-1}\frac{b/2 + x}{y} + \tan^{-1}\frac{b/2 - x}{y}\right)$$

$$H_y = \frac{I}{4\pi b}\ln\left[\frac{(b/2 + x)^2 + y^2}{(b/2 - x)^2 + y^2}\right]$$

2.9f For an infinite single-wire line of current, show that A_z as calculated in Ex. 2.9b is infinite. Then show that if vector potential is calculated for a finite length $-L < z < L$ and **B** calculated from this before letting L approach infinity, the correct value of **B** is obtained.

2.10 Show that the torque on a small loop of current can be expressed as $\tau = \mathbf{m} \times \mathbf{B}$.

2.12a Deduce the appropriate form for vector potential outside a long straight round wire carrying current I, and show that $\nabla^2 \mathbf{A}$ is equal to zero there.

2.12b Use the rectangular coordinate forms to prove Eq. 2.12(3).

2.12c Show that the magnetic field given by Eq. 2.12(7) is circumferential and give its value.

2.12d A certain current density is said to produce within itself a vector potential having the form

$$\mathbf{A} = \frac{C}{x^2 + y^2}\hat{\mathbf{z}}$$

where C is a constant. Find the divergence of **A**, the current density and magnetic field, assuming the medium to be free space.

2.12e We saw in Ex. 2.7 that magnetic field in a superconductor decays from the surface as

$$H_z = H_0 e^{-x/\lambda}$$

where z is parallel to the plane of the surface and x is perpendicular to the surface. Find the corresponding vector potential $\mathbf{A}$, and from it the current density comparing with the result of Ex. 2.7.

2.13a Show whether either of the following vector fields can be obtained from a scalar potential, and give the potential function if applicable:

$$\mathbf{F} = \hat{x}3y^2z + \hat{y}6xyz + \hat{z}3y^2x$$

$$\mathbf{F} = \hat{x}3y + \hat{y}2x + \hat{z}4$$

2.13b Find the form of scalar magnetic potential for the region between conductors as shown in Fig. 2.13, defined for the region $0 \leq \phi < 2\pi$; similarly for the region outside the outer conductor. Current I flows in the inner conductor and the return current in the outer one.

2.14* Consider the boundary between free space and a plane superconductor with nearby parallel line current I at $x = d$. It is the nature of a superconductor that when placed in a weak magnetic field, currents flow in such a way as to eliminate flux inside the superconductor so that B_n at the surface is zero, as is the tangential H inside the superconductor. Show that fields in the free-space region $x > 0$ can be found by replacing the superconductor with an image current at $x = -d$ carrying current $-I$. Find the magnetic field at $x = 0+$ and from this the surface current density $\mathbf{J}_s$.

2.15a* Work the problem of Fig. 2.15a using equivalent current sources and the vector potential $\mathbf{A}$.

2.15b For the problem of Fig. 2.15b, what magnetic charge distribution would be obtained for the formulation in terms of equivalent magnetic charges? How would this be modified if magnetization is inhomogeneous as defined below?

$$\mathbf{M} = \hat{z}M_0(1 + kz)$$

2.16a Find the energy stored per unit length between the conductors of a coaxial line carrying current I in the inner conductor and equal return current in the outer.

2.16b Show that Eq. 2.16(1) leads to (2) for linear, isotropic materials.

2.16c Assume that the material having the B–H relation shown in Fig. 2.16 saturates at $B = 1000$ G and estimate graphically the energy per unit volume for one complete traversal of the hysteresis loop.

2.17a Find the inductance per unit length for the arrangement of Prob. 2.5 from energy considerations.

2.17b Find the internal inductance per unit length for the parallel-plane transmission line of Fig. 2.5c if current is assumed of uniform density in each of the conductors.

3

MAXWELL'S EQUATIONS

3.1 Introduction

The laws of static electric and magnetic fields have been studied in Chapters 1 and 2. It has been noted that these are useful in predicting effects for many time-varying problems, but there are important dynamic effects not described by the static relations, so other time-varying problems require a more complete formulation. One important dynamic effect is the generation of electric fields by time-varying magnetic fields, as expressed through Faraday's law. A second is the complementary effect whereby time-varying electric fields produce magnetic fields. This latter effect is expressed through the concept of *displacement current*, introduced by Maxwell.

Faraday's law is well known to us through its importance in transformers, motors, generators, induction heaters, and similar devices. The effect can be simply demonstrated by moving a coil of wire through the field of a strong permanent magnet and noting the trace on an oscilloscope connected to the coil (Fig. 3.1a). With readily available magnets, practical numbers of turns in the coil, movement by hand will generate millivolts, and such voltages are readily observed on the oscilloscope. An alternative demonstration utilizes an electromagnet with its flux threading a fixed coil. A switch to turn on and off the current in the electromagnet causes buildup and decay of the magnetic field and generates the voltage to be observed.

The above-described demonstrations and useful devices utilize induced effects in conductors. An interesting example to show that changing magnetic fields induce electric fields in space is that of the betatron. This useful particle accelerator, illustrated in Fig. 3.1b, accelerates electrons or other charged

111

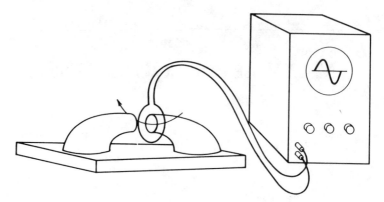

Fig. 3.1a Experimental arrangement to demonstrate the induced voltage predicted by Faraday's law. Coil can be moved with sufficient speed by hand to display the induced voltage on a simple oscilloscope.

particles by means of a circumferential electric field induced by a changing magnetic field between poles N and S of an electromagnet. The charges are in an evacuated chamber, clearly illustrating that Faraday's law applies in space as well as along conductors.

The second dynamic effect, referred to above as a displacement current effect, is probably best known to us through the concept of a capacitance current. We

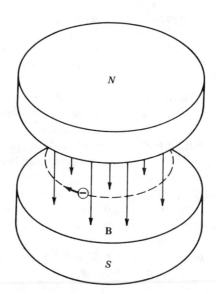

Fig. 3.1b Schematic illustration of a betatron, which is used to accelerate electrons by means of an electric field induced by a changing magnetic field.

may, however, think of this only as current in the conductors to capacitor terminals, supplying the time rate of change of charge on capacitor plates. We shall see that the changing electric flux in the dielectric between plates contributes to magnetic fields just as does conduction current, and acts to complete the current path. Displacement currents also exist in the vicinity of moving charges, and so are important in vacuum tubes or solid-state electron devices. For example, time-varying effects in the Schottky barrier of Ex. 1.4a or the *p–n* junction of Sec. 1.16 produce displacement currents in the respective depletion regions. Effects of these displacement currents must be understood in the analysis of devices using such junctions.

There is a far-reaching consequence of the fact that changing magnetic flux density produces a change in electric field and vice versa; it leads to propagation of electromagnetic waves. In general, wave phenomena result when there are two forms of energy, and the presence of a time rate of change of one leads to a change of the other. In a sound wave, for example, an initial pressure variation in air (potential energy) in one location causes a motion of the air molecules (kinetic energy) that varies both in time and space. This builds up excess pressure at another position, and the effect continues. Similarly, if we change the magnetic field (or flux density) at one position, this generates a change of electric field in both time and space, by Faraday's law. The subsequent change of electric field produces a change of magnetic field through the displacement current, and so on. In energy terms, the energy interchanges between electric and magnetic types as the wave progresses.

Electromagnetic waves exist in nature in the radiation that takes place when atoms or molecules change from one energy state to a lower one, with frequencies from the microwave through visible into x-ray regions of the spectrum. (Still lower frequencies are generated by lightning and other natural fluctuations.) These natural radiations are utilized in astronomy and radio astronomy. Telecommunications, navigational guidance, radar, and power transmission depend upon our ability to generate, guide, store, radiate, receive, and detect electromagnetic waves. This involves many kinds of structures whose properties the designer must be able to predict. The complete set of laws for time-varying electromagnetic phenomena is known as Maxwell's equations and are central to such predictions.

LARGE-SCALE AND DIFFERENTIAL FORMS OF MAXWELL'S EQUATIONS

3.2 Voltages Induced by Changing Magnetic Fields

Faraday discovered experimentally that a voltage is induced in a conducting circuit when the magnetic field linking that circuit is altered. The voltage is proportional to the time rate of change of magnetic flux linking the circuit. For a circuit of n turns, the induced voltage V can be written

$$V = n\frac{d\psi_m}{dt} \tag{1}$$

where ψ_m is the magnetic flux linking each turn of the coil. This equation may be used directly to find the voltage induced in the secondary coil of a transformer, for example, or to find the voltage induced in a single circuit because of a time-varying current interacting with the self-inductance of that circuit. For an electric machine, such as a generator or a motor, the change in flux linkages to be used in (1) may be found from the movement of the coil of the machine through a spatially variable magnetic field. Faraday's experiments included both stationary and moving systems. The question of moving systems may be approached in several ways and will be discussed more in the following section.

One very important generalization of (1) is to a path in space or other nonconducting medium. Such an extension is plausible since the resistance of the path does not appear in the equation. Nevertheless the extension should have experimental verification and it does. Much of the experimental support comes from the wave behavior to be studied in the remainder of the book. As described in Sec. 3.1, the betatron[1] accelerates charged particles in vacuum by means of an electric field induced by a changing magnetic field, as predicted by Faraday's law. (See Prob. 3.2c.)

Before defining Faraday's law more precisely, we should be clear about several definitions. By *voltage* between two points along a specified path, we mean the negative line integral of electric field between the points along that path. For static fields, we saw that the line integral is independent of the path and equal to the *potential difference* between the two points, but this is not true when there are contributions from Faraday's law. When there is a contribution from changing magnetic flux, the voltage about a closed path is frequently called the *electromotive force (emf)* of that path.

$$\text{emf} \equiv \text{voltage about closed path} \equiv -\oint \mathbf{E} \cdot \mathbf{dl} \tag{2}$$

It is equal, by Faraday's law, to the time rate of change of magnetic flux through the path. For a circuit which is not moving,

$$\oint \mathbf{E} \cdot \mathbf{dl} = -\frac{\partial \psi_m}{\partial t} = -\frac{\partial}{\partial t}\int_s \mathbf{B} \cdot \mathbf{dS} \tag{3}$$

where the flux ψ_m is found by evaluating the normal component of flux density $\mathbf{B}$ over any surface which has the desired path as its boundary, as indicated by the last term in (3). The negative sign is introduced in the law to agree with the sense

[1] D. W. Kerst and R. Serber, *Phys. Rev.* **60**, 53 (1941).

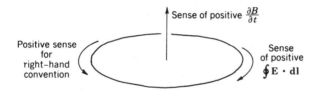

Fig. 3.2a Sense relations for Faraday's law.

relations revealed by experiment, using the usual right-hand convention in the integrals of (3). Thus, as in Fig. 3.2a, if the *rate of change* of flux is positive in the directions shown by the vertical arrow, the line integral will be positive in the direction shown—opposite to the conventional right-hand positive direction. If there are several turns, the line integral of (3) is taken about all of them, and if flux through each is the same, we have the form first stated in (1).

To transform (3) to differential equation form, we can apply Stokes's theorem, Sec. 2.8, to the left side of (3) and move the time differentiation inside the integral:

$$\int_S (\nabla \times \mathbf{E}) \cdot d\mathbf{S} = -\int_S \frac{\partial \mathbf{B}}{\partial t} \cdot d\mathbf{S} \tag{4}$$

For this equation to be valid for an arbitrary surface, the integrands must be equal so that

$$\nabla \times \mathbf{E} = -\frac{\partial \mathbf{B}}{\partial t} \tag{5}$$

Faraday's law (4) of course reduces to the static case when time derivatives are zero and, as we saw in Ex. 2.6d, the line integral of electric field about a closed path is then zero. For the time-varying field it is not in general zero, showing that work can be done in taking a charge about a closed path in such a field. This work comes from the changing stored energy of the magnetic field.

_____ **Example 3.2** _____
Air Breakdown from Induced emf

Consider the possibility of an ionizing breakdown in air because of electric fields generated by changing magnetic fields. An axially symmetric electromagnet (Fig. 3.2b) has radius 0.20 m and has essentially uniform field up to this radius and negligible field beyond it. Suppose that it is desired to raise magnetic field from zero to 10 T (tesla) linearly with time in as short a time τ as possible without such breakdown. Because of the axial symmetry we can write Faraday's law for a loop of radius r as

$$2\pi r |E_\phi| = \pi r^2 \frac{\partial B_z}{\partial t} = \pi r^2 \frac{B_{max}}{\tau} \tag{6}$$

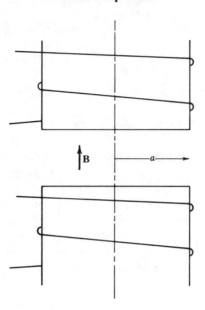

Fig. 3.2b Electromagnet with time-varying current producing a time-varying magnetic field.

Electric field is thus maximum at the outer radius of 0.20 m. Take breakdown strength of air as 3×10^6 V/m. Then

$$\tau = \frac{rB_{max}}{2|E_\phi|} = \frac{0.2 \times 10}{2 \times 3 \times 10^6} = \frac{1}{3} \ \mu s \tag{7}$$

3.3   Faraday's Law for a Moving System

For the use of Eq. 3.2(1) for a moving system, one must find the change of magnetic flux threading a circuit as it moves through the field. A simple and classical example is that of an elemental ac generator as pictured in Fig. 3.3a. This indicates a single rectangular loop rotating at constant angular frequency Ω in the uniform field B_0 between the two pole pieces. When the plane of the loop is at angle ϕ with respect to the x axis, the magnetic flux passing through it is

$$\psi_m = 2B_0 al \sin \phi \tag{1}$$

But angle ϕ changes with time and may be written Ωt. Thus

$$\psi_m = 2B_0 al \sin \Omega t \tag{2}$$

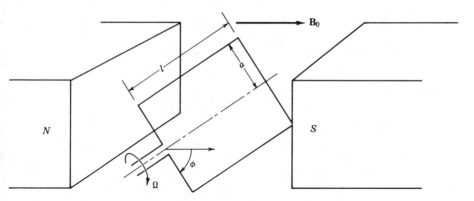

Fig. 3.3a Elemental generator with rotating loop between permanent magnets.

And if voltage is the rate of change of this flux (neglecting signs),

$$V = \frac{d\psi_m}{dt} = 2\Omega B_0 al \cos \Omega t \tag{3}$$

Thus we see the sinusoidal ac voltage produced by this basic generator. Now let us introduce a slightly different point of view to get the same answer. A point of view used effectively by Faraday is that the electric field of the moving conductor is generated by its motion through and hence "cutting" the lines of force. Faraday gave much physical significance to the flux tubes and lines of force. This point of view can be developed rigorously by writing the time derivative on the right side of Eq. 3.2(3) as a total derivative instead of a partial derivative,

$$\oint \mathbf{E} \cdot d\mathbf{l} = -\frac{d}{dt} \int_S \mathbf{B} \cdot d\mathbf{S} \tag{4}$$

For a closed path moving in space with velocity $\mathbf{v}$, this may be transformed by a vector transformation developed by Helmholtz,[2] which, with $\nabla \cdot \mathbf{B} = 0$, is

$$\oint \mathbf{E} \cdot d\mathbf{l} = -\int_S \left[\frac{\partial \mathbf{B}}{\partial t} - \nabla \times (\mathbf{v} \times \mathbf{B}) \right] \cdot d\mathbf{S} \tag{5}$$

The first term on the right is the one we have seen before. The second term gives an added contribution to emf, and by use of Stokes's theorem, may be written as the line integral of $\mathbf{v} \times \mathbf{B}$ about the closed path. The result may be interpreted as a *motional electric field* given at each point of the circuit path by

$$\mathbf{E}_m = \mathbf{v} \times \mathbf{B} \tag{6}$$

In the example of Fig. 3.3a, the motional field in the upper conductor, by (6), is

$$E_{mz} = (a\Omega)B_0 \cos \phi \tag{7}$$

[2] See, for example, C. T. Tai, *Proc. IEEE* **60**, 936 (1972).

That in the lower conductor is the negative of this. There is no motional field along the side elements since $v \times B$ gives a contribution normal to the wires for these sides. Thus the line integral about the loop yields

$$\oint E \cdot dl = la\Omega B_0 \cos \Omega t - (-la\Omega B_0 \cos \Omega t) = 2la\Omega B_0 \cos \Omega t \qquad (8)$$

which is identical to (3).

The differential form of Faraday's law, Eq. 3.2(5), may be transformed to a set of moving coordinates with the same result. This may be done by a Galilean transformation for low velocities, and by a Lorentz transformation at relativistic velocities.[2] Although relativity is beyond the scope of this text, it is important to know that Maxwell's equations are consistent with the theory of relativity, although Einstein developed that theory later. Their invariance to Lorentz transformations in fact had much to do with the development of the theory of special relativity.

_____ **Example 3.3** _____

Rectangular Loop Moving Through Inhomogeneous Field

If a loop of wire is moved through a region of static magnetic field which is a function of position, the flux threading the loop changes as the loop moves and an emf is generated. Consider the rectangular loop of wire (Fig. 3.3b) translated in the x direction with velocity v through a z-directed static magnetic field which varies linearly with x, $B_z = Cx$. If the left-hand edge is at $x = 0$ at $t = 0$, it is at $x = vt$ at time t and the magnetic flux threading the loop is

$$\int_S B \cdot dS = b \int_{vt}^{vt+a} Cx\, dx = bC \frac{x^2}{2}\Big|_{vt}^{vt+a} = \frac{baC}{2}(2vt + a) \qquad (9)$$

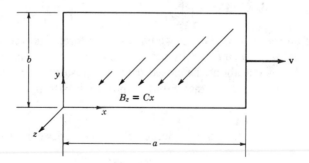

Fig. 3.3b Rectangular loop of wire moving through magnetic field which varies with distance.

The induced emf is then the time rate of change of this flux,

$$-\oint \mathbf{E} \cdot \mathbf{dl} = \frac{d}{dt} \int_S \mathbf{B} \cdot \mathbf{dS} = baCv \tag{10}$$

This result can be checked by finding the motional field in the four sides using (6). The field $\mathbf{v} \times \mathbf{B}$ is normal to the wires along the top and bottom. On the left it is $-vCx$ and on the right side $-vC(x + a)$. Thus the integral is

$$-\oint \mathbf{E} \cdot \mathbf{dl} = vCb(x + a) - vCbx = baCv \tag{11}$$

agreeing with the result (10).

3.4   Continuity of Charge and the Concept of Displacement Current

Faraday's law is but one of the fundamental laws for changing fields. Let us assume for the moment that certain of the laws derived for static fields in Chapters 1 and 2 can be extended simply to time-varying fields. We will write the divergence of electric and magnetic fields in exactly the same form as in statics with the understanding that all field and source quantities are functions of time as well as of space. For the curl of electric field we take the result of Faraday's law, Eq. 3.2(5). For the curl of magnetic field, we take for the time being the form from statics, Eq. 2.7(2).

$$\mathbf{V} \cdot \mathbf{D} = \rho \tag{1}$$

$$\mathbf{V} \cdot \mathbf{B} = 0 \tag{2}$$

$$\mathbf{V} \times \mathbf{E} = -\frac{\partial \mathbf{B}}{\partial t} \tag{3}$$

$$\mathbf{V} \times \mathbf{H} = \mathbf{J} \tag{4}$$

An elimination among these equations can be made to give an equation relating charge and current. We would expect this to show that, however ρ varies with space or time, total charge is conserved. If current flows out of any volume, the amount of charge inside must decrease, and, if current flows in, charge inside increases. Considering a smaller and smaller volume, in the limit the outward flow of current per unit time and per unit volume (which is recognized as the divergence of current density) must give the negative of the time rate of change of charge per unit volume at that point:

$$\mathbf{V} \cdot \mathbf{J} = -\frac{\partial \rho}{\partial t} \tag{5}$$

However, if we take the divergence of $\mathbf{J}$ from (4),

$$\nabla \cdot \mathbf{J} = \nabla \cdot (\nabla \times \mathbf{H}) \equiv 0$$

which does not agree with the continuity argument and equation (5). Maxwell, by reasoning similar to this, recognized that (4), borrowed from statics, is not complete for time-varying fields. He postulated an added term $\partial \mathbf{D}/\partial t$:

$$\nabla \times \mathbf{H} = \mathbf{J} + \frac{\partial \mathbf{D}}{\partial t} \qquad (6)$$

Continuity is now satisfied, as may be shown by taking the divergence of (6) and substituting from (1)

$$\nabla \cdot \mathbf{J} = -\frac{\partial}{\partial t}(\nabla \cdot \mathbf{D}) = -\frac{\partial \rho}{\partial t}$$

The term added to form (6) contributes to the curl of magnetic field in the same way as an actual conduction current density (motion of charges in conductors), or convection current density (motion of charges in space). Because it arises from the displacement vector $\mathbf{D}$, it has been named the *displacement current* term. There is an actual time-varying displacement of bound charges in a material dielectric, but note that displacement current can be nonzero even in vacuum. Thus (6) could be written

$$\nabla \times \mathbf{H} = \mathbf{J}_c + \mathbf{J}_d \qquad (7)$$

where $\mathbf{J}_c$ = conduction or convection current density in amperes per square meter and $\mathbf{J}_d$ = displacement current density = $\partial \mathbf{D}/\partial t$ amperes per square meter.

The displacement current term is important within the dielectric of a capacitor whenever the capacitive voltage changes with time. It also always plays a role when moving charges induce currents in nearby electrodes. Both of these phenomena will be explored in the following article. Displacement current is negligible for many other low-frequency problems. For example, it is negligible in comparison with conduction currents in good conductors up to optical frequencies. (This point will be explored more in Sec. 3.16.) But displacement current becomes important in more and more situations as the frequency of time-varying phenomena is increased. It is essential, along with the Faraday's law terms for electric field, to the understanding of all electromagnetic wave phenomena.

3.5   Physical Pictures of Displacement Current

The displacement current term enables one to explain certain things that would have proved inconsistent had only conduction or convection current been included in the magnetic field laws. Consider, for example, the circuit including the ac generator and the capacitor of Fig. 3.5a. Suppose that it is required to evaluate the line integral of magnetic field around the loop a–b–c–d–a. The law from statics states that the result obtained should be the current enclosed—that is, the current through any surface of which the loop is a boundary. If we take as the arbitrary surface through which current is to be evaluated one which cuts the wire A, as does S_1, a finite value is clearly obtained for the line integral. But suppose that the surface selected is one which does not cut the wire, but instead passes between the plates of the capacitor, as does S_2. If conduction current alone were included, the computation would have indicated no current passing through this surface and the result would be zero. The path around which the integral is evaluated is the same in each case, and it would be quite annoying to possess two different results. It is the displacement current term which appears at this point to preserve the continuity of current between the plates of the capacitor, giving the same answer in either case.

To show how this continuity is preserved, consider an ideal parallel-plate capacitor of capacitance C, spacing d, area of each plate A, and applied voltage $V_0 \sin \omega t$. From circuit theory the charging current is

$$I_c = C\frac{dV}{dt} = \omega C V_0 \cos \omega t \tag{1}$$

The field inside the capacitor has a magnitude $E = V/d$, so the displacement current density is

$$J_d = \varepsilon \frac{\partial E}{\partial t} = \omega \varepsilon \frac{V_0}{d} \cos \omega t \tag{2}$$

Total displacement current flowing between the plates is the area of the plate multiplied by the density of displacement current:

$$I_d = A J_d = \omega\left(\frac{\varepsilon A}{d}\right) V_0 \cos \omega t \tag{3}$$

The factor in parentheses is recognized as the electrostatic capacitance for the ideal parallel-plate capacitor, so (1) and (3) are equal. This value for total displacement current flowing between the capacitor plates is then exactly the same as the value of charging current flowing in the leads, calculated by the usual circuit methods above, so the displacement current does act to complete the circuit, and the same result would be obtained by the use of either S_1 or S_2 of Fig. 3.5a, as required.

Inclusion of displacement current is necessary for a valid discussion of another example in which a charge region q (Fig. 3.5b) moves with velocity **v**. If the line integral of magnetic field is to be evaluated about some loop A at a given instant, it should be possible to set it equal to the current flow for that instant through any surface of which A is a boundary. If the displacement current term were ignored, we could use any one of the infinite number of possible surfaces, as S_1, having no charge passing through, and obtain the result zero. If one of the

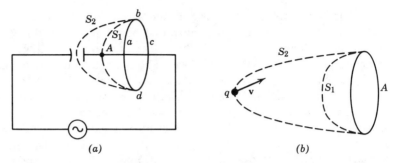

(a)   *(b)*

Fig. 3.5 Illustrations of how displacement current completes the circuit: *(a)* in a circuit with capacitor; *(b)* near a moving charge.

surfaces is selected, as S_2, through which charge is passing at that instant, however, there is a contribution from convection current, and a nonzero result. The apparent inconsistency is resolved when one notes that the electric field arising from the moving charge must vary with time, and thus will actually give rise to a displacement current term through both of the surfaces S_1 and S_2. The sum of displacement and convection currents for the two surfaces is the same at the given instant.

_____ **Example 3.5** _____

Induced Currents by a Slab of Charge Moving in a Planar Diode

In a planar vacuum diode, as sketched in Fig. 3.5c, the cathode has been pulsed to produce a slab of charge moving from cathode to anode. The density of charge is taken as a uniform ρ_0. Width is w and at time t the left-hand edge is at $x = x'$

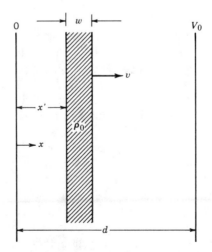

Fig. 3.5c Slab of charge moving between parallel plates.

moving with velocity v. We note that the electric field is a constant E_{x1} for $x < x'$, varies linearly for $x' < x < x' + w$, and is again constant for $x > x' + w$. Its integral is

$$-V_0 = E_{x1}d + \frac{\rho_0}{\varepsilon} w(d - x') - \frac{\rho_0 w^2}{2\varepsilon}$$

Then the electric field for the three regions $0 < x < x'$, $x' < x < x' + w$, and $x' + w < x < d$ are, respectively (Prob. 3.5d),

$$E_{x1} = -\frac{V_0}{d} + \frac{\rho_0}{\varepsilon}\left[\frac{x'w}{d} + w\left(\frac{w}{2d} - 1\right)\right] \tag{4}$$

$$E_{x2} = -\frac{V_0}{d} + \frac{\rho_0}{\varepsilon}\left[x'\left(\frac{w}{d} - 1\right) + w\left(\frac{w}{2d} - 1\right) + x\right] \tag{5}$$

$$E_{x3} = -\frac{V_0}{d} + \frac{\rho_0 w}{\varepsilon d}\left(x' + \frac{w}{2}\right) \tag{6}$$

But x' is a function of time and differentiation with respect to time gives velocity v. Thus displacement current density for the three regions is

$$\varepsilon \frac{\partial E_{x1}}{\partial t} = \rho_0 \frac{w}{d}\frac{\partial x'}{\partial t} = \frac{\rho_0 w v}{d} \tag{7}$$

$$\varepsilon \frac{\partial E_{x2}}{\partial t} = \rho_0 v\left(\frac{w}{d} - 1\right) \tag{8}$$

$$\varepsilon \frac{\partial E_{x3}}{\partial t} = \rho_0 \frac{\partial x'}{\partial t} = \frac{\rho_0 w v}{d} \tag{9}$$

To the displacement current density of the region within the charge, given by (8), we add the convection current density $\rho_0 v$ so that the sum of convection and conduction currents is the same for each region, and this will also be the current per unit area induced in the plane electrodes.

3.6   Maxwell's Equations in Differential Equation Form

Rewriting the group of equations of Sec. 3.4 with the displacement current term added, we have

$$\mathbf{\nabla \cdot D} = \rho \tag{1}$$

$$\mathbf{\nabla \cdot B} = 0 \tag{2}$$

$$\mathbf{\nabla \times E} = -\frac{\partial \mathbf{B}}{\partial t} \tag{3}$$

$$\mathbf{\nabla \times H} = \mathbf{J} + \frac{\partial \mathbf{D}}{\partial t} \tag{4}$$

This set of equations, together with certain auxiliary relations and definitions, is the basic set of equations of classical electricity and magnetism, governing all electromagnetic phenomena in the range of frequencies from zero through the highest frequency radio waves (and many phenomena at light frequencies), and in the range of sizes above atomic size. The equations were first written (not in the above notation) by Maxwell in 1863, and are known as Maxwell's equations. The material in the sections preceding this should not be considered a derivation of the laws, for they cannot in any real sense be derived from less fundamental laws. Their ultimate justification comes, as with all experimental laws, in that they have predicted correctly, and continue to predict, all electromagnetic phenomena over a wide range of physical experience.

The foregoing set of equations is a set of differential equations, relating the time and space rates of change of the various field quantities at a point in space and time. The use of these will be demonstrated in the following chapters. Equivalent large-scale equations will be given in the following section.

The major definitions and auxiliary relations that must be added to complete the information are as follows:

1. *The Force Law*   This is, from one point of view, merely the definition of the electric and magnetic fields. For a charge q moving with velocity $\mathbf{v}$ through an electric field $\mathbf{E}$ and a magnetic field of flux density $\mathbf{B}$, the force is

$$\mathbf{f} = q[\mathbf{E} + \mathbf{v} \times \mathbf{B}] \quad \text{N} \tag{5}$$

2. *The Definition of Conduction Current (Ohm's Law)*   For a conductor,

$$\mathbf{J} = \sigma\mathbf{E} \quad \text{A/m}^2 \tag{6}$$

where σ is conductivity in siemens per meter.

3. *The Definition of Convection Current*   For a charge density ρ moving with velocity $\mathbf{v}_\rho$, the current density is

$$\mathbf{J} = \rho\mathbf{v}_\rho \quad \text{A/m}^2 \tag{7}$$

4. *Definition of Permittivity (Dielectric Constant)*   The electric flux density $\mathbf{D}$ is related to the electric field intensity $\mathbf{E}$ by the relation.

$$\mathbf{D} = \varepsilon\mathbf{E} = \varepsilon_r\varepsilon_0\mathbf{E} \tag{8}$$

where ε_0 is the permittivity of space $\approx 8.854 \times 10^{-12}$ F/m and ε_r characterizes the effect of the atomic and molecular dipoles in the material.

As with static fields (Sec. 1.3) ε, or ε_r, is in the most general case, anisotropic and a function of space, of time, and of the strength of the applied field. But for many materials it is a scalar constant and unless specifically noted otherwise, the text will be concerned with homogeneous, isotropic, linear, and time-invariant materials for which ε is a scalar constant.

5. *Definition of Permeability*   The magnetic flux density **B** is related to the magnetic intensity **H** by

$$\mathbf{B} = \mu\mathbf{H} = \mu_r\mu_0\mathbf{H} \tag{9}$$

where μ_0 is the permeability of space $= 4\pi \times 10^{-7}$ H/m and μ_r measures the effect of the magnetic dipole moments of the atoms comprising the medium (Sec. 2.3). In general μ and μ_r are anisotropic and functions of space, time, and magnetic field strength, but unless otherwise noted they will be considered scalar constants, representing homogeneous, isotropic, linear, and time-invariant materials.

_____ **Example 3.6** _____
Nonarbitrariness of Forms Which Satisfy Maxwell's Equations

Much of the work for the remainder of the text will be in finding forms that are solutions of Maxwell's equations. The interrelationship among electric and magnetic field components defined by Maxwell's equations means that we cannot select arbitrary functions for any one component. To illustrate the point, let us take magnetic field in a region of free space as

$$\mathbf{B} = \hat{z}C_0 x \sin \omega t \tag{10}$$

The corresponding electric field from (3) is

$$\mathbf{E} = \hat{y}\left[-\frac{\omega C_0 x^2}{2} \cos \omega t + F(y, z, t) \right] \tag{11}$$

Let us see if the other curl relationship (4) is satisfied. For space, $\mathbf{J} = 0$ and $\mu = \mu_0$, $\varepsilon = \varepsilon_0$. The rectangular-coordinate component of (4) relating H_z and E_y is

$$\frac{\partial H_z}{\partial x} = -\frac{\partial D_y}{\partial t} = -\varepsilon_0 \frac{\partial E_y}{\partial t} \tag{12}$$

Substitution of (10) and (11) in (12) gives

$$\frac{C_0}{\mu_0} \sin \omega t = -\varepsilon_0\left[\frac{\omega^2 C_0 x^2}{2} \sin \omega t + \frac{\partial F(y, z, t)}{\partial t} \right] \tag{13}$$

Neither the left side nor the last term on the right is a function of x so that there is no way in which the term varying with x^2 can be canceled and (12) is not satisfied.

_____

3.7   Maxwell's Equations in Large-Scale Form

It is also convenient to have the information of Maxwell's equations in large-scale or integral form applicable to overall regions of space and paths of finite size. This is of course the type of relation that we started with in the discussion of Faraday's law, Sec. 3.2, when we derived the differential expression from it. The large-scale equivalents for Eqs. 3.6(1)–3.6(4) are

$$\oint_S \mathbf{D} \cdot \mathbf{dS} = \int_V \rho \, dV \tag{1}$$

$$\oint_S \mathbf{B} \cdot \mathbf{dS} = 0 \tag{2}$$

$$\oint \mathbf{E} \cdot \mathbf{dl} = -\frac{\partial}{\partial t} \int_S \mathbf{B} \cdot \mathbf{dS} \tag{3}$$

$$\oint \mathbf{H} \cdot \mathbf{dl} = \int_S \mathbf{J} \cdot \mathbf{dS} + \frac{\partial}{\partial t} \int_S \mathbf{D} \cdot \mathbf{dS} \tag{4}$$

Equations (1) and (2) are obtained by integrating respectively Eqs. 3.6(1) and 3.6(2) over a volume and applying the divergence theorem. Equations (3) and (4) are obtained by integrating, respectively, Eqs. 3.6(3) and 3.6(4) over a surface and applying Stokes's theorem. For example, integrating Eq. 3.6(1),

$$\int_V \mathbf{\nabla} \cdot \mathbf{D} \, dV = \int_V \rho \, dV$$

and applying the divergence theorem,

$$\oint_S \mathbf{D} \cdot \mathbf{dS} = \int_V \rho \, dV$$

Equation (1) is seen to be the familiar form of Gauss's law utilized so much in Chapter 1. Now that we are concerned with fields that are a function of time, the interpretation is that the electric flux flowing out of any closed surface at a given instant is equal to the charge enclosed by the surface at that instant.

Equation (2) states that the surface integral of magnetic field or total magnetic flux flowing out of a closed surface is zero for all values of time, expressing the fact that magnetic charges have not been found in nature. Of course the law does not prove that such charges will never be found; if they are, a term on the right similar to the electric charge term in (1) will simply be added, and a corresponding magnetic current term will be added to (3).

Equation (3) is Faraday's law of induction, stating that the line integral of electric field about a closed path (electromotive force) is the negative of the time rate of change of magnetic flux flowing through the path. The law was discussed in some detail in Sec. 3.2.

Equation (4) is the generalized Ampère's law including Maxwell's displacement current term, and it states that the line integral of magnetic field about a closed path (magnetomotive force) is equal to the total current (conduction, convection, and displacement) flowing through the path. The physical significance of this complete law has been discussed in Secs. 3.4–3.5.

3.8 Maxwell's Equations for the Time-Periodic Case

By far the most important time-varying case is that involving steady-state ac fields varying sinusoidally in time. Many engineering applications use sinusoidal fields. Other functions of time, such as the pulses utilized in a digitally coded system, may be considered a superposition of steady-state sinusoids of different frequencies. Fourier analysis (Fourier series for periodic functions and the Fourier integral for aperiodic functions) provides the mathematical basis for this superposition. Rather than using real sinusoidal functions directly, it is found convenient to introduce the complex exponential $e^{j\omega t}$. Electrical engineers are familiar with the advantages of this approach in the analysis of ac circuits and physicists use the complex exponential in a variety of physical problems with sinusoidal behavior. The advantage, which comes from the fact that derivatives and integrals of $e^{j\omega t}$ are proportional to $e^{j\omega t}$ so that the function can be canceled from all equations, is even more important for the vector field problems than for scalar problems such as the circuit example. It is assumed that the reader has used this technique before in circuit analysis or other physical problems, but if review is needed, the use in analysis of a simple electrical circuit may be referred to in Appendix 4. Formally, the set of equations 3.6(1)–3.6(4) are easily changed to the complex form by replacing $\partial/\partial t$ by $j\omega$:

$$\mathbf{V} \cdot \mathbf{D} = \rho \tag{1}$$

$$\mathbf{V} \cdot \mathbf{B} = 0 \tag{2}$$

$$\mathbf{V} \times \mathbf{E} = -j\omega\mathbf{B} \tag{3}$$

$$\mathbf{V} \times \mathbf{H} = \mathbf{J} + j\omega\mathbf{D} \tag{4}$$

And the auxiliary relations, Eqs. 3.6(6)–3.6(9), remain

$$\mathbf{J} = \sigma\mathbf{E} \text{ for conductors} \tag{5}$$

$$\mathbf{D} = \varepsilon\mathbf{E} = \varepsilon_r\varepsilon_0\mathbf{E} \tag{6}$$

$$\mathbf{B} = \mu\mathbf{H} = \mu_r\mu_0\mathbf{H} \tag{7}$$

Equations 3.6(5) and 3.6(7) should be used with instantaneous values because of the nonlinear terms in the equations. The constitutive parameters, μ and ε, are in general functions of frequency. Materials for which frequency dependence is important are called *dispersive*.

It must be recognized that the symbols in the equations of this article have a different meaning from the same symbols used in Sec. 3.6. There they represented the instantaneous values of the indicated vector and scalar quantities. Here they represent the complex multipliers of $e^{j\omega t}$, giving the in-phase and out-of-phase parts with respect to the chosen reference. The complex scalar quantities are commonly referred to as phasors, and by analogy the complex vector multipliers of $e^{j\omega t}$ may be called vector phasors. It would seem less confusing to use a different notation for the two kinds of quantities, but one quickly runs out of symbols. The difference is normally clear from the context, and when there is danger of confusion, we will use functional notation to denote the time-varying quantities.

If we wish to obtain the instantaneous values of a given quantity from the complex value, we insert the $e^{j\omega t}$ and take the real part. For example, for the scalar ρ suppose that the complex value of ρ is

$$\rho = \rho_r + j\rho_i \tag{8}$$

where ρ_r and ρ_i are real scalars. The instantaneous value of ρ is then

$$\rho(t) = \text{Re}[(\rho_r + j\rho_i)e^{j\omega t}] = \rho_r \cos \omega t - \rho_i \sin \omega t \tag{9}$$

Or, alternatively, if ρ is given in magnitude and phase,

$$\rho = |\rho|e^{j\theta_\rho} \tag{10}$$

where

$$|\rho| = \sqrt{\rho_r^2 + \rho_i^2}$$

$$\theta_\rho = \tan^{-1}\frac{\rho_i}{\rho_r}$$

The true time-varying form is

$$\rho(t) = \text{Re}[|\rho|e^{j(\omega t + \theta_\rho)}] = |\rho| \cos(\omega t + \theta_\rho) \tag{11}$$

For a vector quantity, such as **E**, the complex value may be written

$$\mathbf{E} = \mathbf{E}_r + j\mathbf{E}_i \tag{12}$$

where $\mathbf{E}_r$ and $\mathbf{E}_i$ are real vectors. Then

$$\mathbf{E}(t) = \text{Re}[(\mathbf{E}_r + j\mathbf{E}_i)e^{j\omega t}] = \mathbf{E}_r \cos \omega t - \mathbf{E}_i \sin \omega t \tag{13}$$

Note that $\mathbf{E}_r$ and $\mathbf{E}_i$ have the same directions in space only for certain special cases. When they are in the same direction, the vector phasor (12) can be expressed as a vector "magnitude" and a scalar phase angle, but in the general case, when they are in different directions, the six scalar quantities defining the two vectors must be specified (Prob. 3.8a).

_____ **Example 3.8** _____
A Phasor Solution of Maxwell's Equations

As an example of phasor solutions, let us consider the following fields, which we will later find to be important as standing waves:

$$B_z = -jD_0 \sin(\omega\sqrt{\mu_0\varepsilon_0}\,x) \tag{14}$$

$$E_y = \frac{D_0}{\sqrt{\mu_0\varepsilon_0}} \cos(\omega\sqrt{\mu_0\varepsilon_0}\,x) \tag{15}$$

Let us show that these do satisfy the phasor forms of Maxwell's equations. The needed rectangular coordinate components of (3) and (4), with $J = 0$, are

$$\frac{\partial E_y}{\partial x} = -j\omega B_z \tag{16}$$

$$\frac{\partial H_z}{\partial x} = \frac{1}{\mu_0}\frac{\partial B_z}{\partial x} = -j\omega\varepsilon_0 E_y \tag{17}$$

Substitution of (14) and (15) gives

$$-\frac{\omega\sqrt{\mu_0\varepsilon_0}}{\sqrt{\mu_0\varepsilon_0}} D_0 \sin(\omega\sqrt{\mu_0\varepsilon_0}\,x) = (-j)^2\omega D_0 \sin(\omega\sqrt{\mu_0\varepsilon_0}\,x) \tag{18}$$

$$-\frac{jD_0\,\omega\sqrt{\mu_0\varepsilon_0}}{\mu_0} \cos(\omega\sqrt{\mu_0\varepsilon_0}\,x) = -\frac{j\omega\varepsilon_0 D_0}{\sqrt{\mu_0\varepsilon_0}} \cos(\omega\sqrt{\mu_0\varepsilon_0}\,x) \tag{19}$$

The divergence of (14) and (15) are also found to be zero,

$$\frac{\partial B_z}{\partial z} \equiv 0 \quad \text{and} \quad \frac{\partial E_y}{\partial y} \equiv 0 \tag{20}$$

So the forms (14) and (15) are solutions of the phasor Maxwell equations for this source-free region. If we wish the time-varying forms, we insert $e^{j\omega t}$ and take the real part,

$$B_z(x, t) = \text{Re}[B_z e^{j\omega t}] = D_0 \sin(\omega\sqrt{\mu_0\varepsilon_0}\,x) \sin\omega t \tag{21}$$

$$E_y(x, t) = \text{Re}[E_y e^{j\omega t}] = \frac{D_0}{\sqrt{\mu_0\varepsilon_0}} \cos(\omega\sqrt{\mu_0\varepsilon_0}\,x) \cos\omega t \tag{22}$$

EXAMPLES OF USE OF MAXWELL'S EQUATIONS

3.9   Maxwell's Equations and Plane Waves

In order to make the information of Maxwell's equations still more concrete, let us show how the equations predict the propagation of uniform plane electromagnetic waves. Such waves illustrate the interplay of electric and magnetic effects and are also of great fundamental and practical importance. Let us begin from the time-varying forms of Sec. 3.6. We postulate a simple medium with constant, scalar permittivity and permeability, and one with no free charges and currents ($\rho = 0$, $\mathbf{J} = 0$). Maxwell's equation are then

$$\mathbf{V} \cdot \mathbf{D} = 0 \tag{1}$$

$$\mathbf{V} \cdot \mathbf{B} = 0 \tag{2}$$

$$\mathbf{V} \times \mathbf{E} = -\frac{\partial \mathbf{B}}{\partial t} = -\mu \frac{\partial \mathbf{H}}{\partial t} \tag{3}$$

$$\mathbf{V} \times \mathbf{H} = \frac{\partial \mathbf{D}}{\partial t} = \varepsilon \frac{\partial \mathbf{E}}{\partial t} \tag{4}$$

For uniform plane waves, we assume variation in only one direction. Take this as the z direction of a rectangular coordinate system. Then $\partial/\partial x = 0$ and $\partial/\partial y = 0$. Let us start with the two curl equations (3) and (4) in rectangular coordinates. With the specialization defined above,

$$\mathbf{V} \times \mathbf{E} = -\mu \frac{\partial \mathbf{H}}{\partial t} \quad \text{leads to} \quad \begin{cases} -\dfrac{\partial E_y}{\partial z} = -\mu \dfrac{\partial H_x}{\partial t} & (5) \\[2mm] \dfrac{\partial E_x}{\partial z} = -\mu \dfrac{\partial H_y}{\partial t} & (6) \\[2mm] 0 = -\mu \dfrac{\partial H_z}{\partial t} & (7) \end{cases}$$

$$\mathbf{V} \times \mathbf{H} = \varepsilon \frac{\partial \mathbf{E}}{\partial t} \quad \text{leads to} \quad \begin{cases} -\dfrac{\partial H_y}{\partial z} = \varepsilon \dfrac{\partial E_x}{\partial t} & (8) \\[2mm] \dfrac{\partial H_x}{\partial z} = \varepsilon \dfrac{\partial E_y}{\partial t} & (9) \\[2mm] 0 = \varepsilon \dfrac{\partial E_z}{\partial t} & (10) \end{cases}$$

Equations (7) and (10) show that the time-varying parts of H_z and E_z are zero. Thus the fields of the wave are entirely transverse to the direction of propagation. The remaining equations break into two independent sets, with (5) and (9) relating E_y and H_x, and (6) and (8) relating E_x and H_y.

The propagation behavior is illustrated by either set. Choose the set with E_x and H_y, differentiating (6) partially with respect to z and (8) with respect to t:

$$\frac{\partial^2 E_x}{\partial z^2} = -\mu \frac{\partial^2 H_y}{\partial z\, \partial t}, \qquad -\frac{\partial^2 H_y}{\partial t\, \partial z} = \varepsilon \frac{\partial^2 E_x}{\partial t^2}$$

Substitution of the second equation into the first yields

$$\frac{\partial^2 E_x}{\partial z^2} = \mu\varepsilon \frac{\partial^2 E_x}{\partial t^2} \tag{11}$$

The important partial differential equation (11) is a classical form known as the *one-dimensional wave equation*, having solutions that demonstrate propagation of a function (a "wave") in the z direction with velocity

$$v = \frac{1}{\sqrt{\mu\varepsilon}} \tag{12}$$

To show this, test a solution of the form

$$E_x(z, t) = f_1\!\left(t - \frac{z}{v}\right) + f_2\!\left(t + \frac{z}{v}\right) \tag{13}$$

Differentiating

$$\frac{\partial E_x}{\partial t} = f_1' + f_2' \qquad \frac{\partial E_x}{\partial z} = -\frac{1}{v} f_1' + \frac{1}{v} f_2'$$

$$\frac{\partial^2 E_x}{\partial t^2} = f_1'' + f_2'' \qquad \frac{\partial^2 E_x}{\partial z^2} = \frac{1}{v^2} f_1'' + \frac{1}{v^2} f_2''$$

where the prime denotes differentiation of the function with respect to the entire argument, and double prime the corresponding second derivative. Comparison of the two second derivatives show that (11) is satisfied by such a solution with v given by (12). The first term of the solution in (13) represents a function f_1 moving in the z direction with velocity v. To show this consider the function $f_1(t)$ at $z = 0$ as illustrated in Fig. 3.9. In order to keep on a constant reference of this wave, we must maintain the argument $t - z/v$ equal to a constant. This implies a velocity $dz/dt = v$. Similarly to keep on a constant reference of the second term in (12) we must keep $t + z/v$ a constant, implying a velocity $dz/dt = -v$. Thus the second

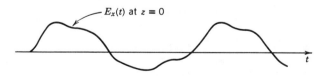

$E_x(t)$ at $z = 0$

Fig. 3.9 A general wave of electric field versus time at reference plane $z = 0$. This will produce a wave moving in the z direction according to $E_x(t - z/v)$ so the argument decreases with increasing positive z.

term represents the function f_2 traveling in the negative z direction, with velocity v. These moving functions may be thought of as "waves," so that the name "wave equation" is explained.

The velocity v defined by (12) is found to be the velocity of light for the medium. In particular, for free space

$$v = c = \frac{1}{\sqrt{\mu_0 \varepsilon_0}} = (4\pi \times 10^{-7} \times 8.85419 \times 10^{-12})^{-1/2} = 2.9979 \times 10^8 \text{ m/s}$$

$$(14)$$

(Note that to three significant figures this is the conveniently remembered value 3×10^8 m/s, corresponding to ε_0 taken as $1/36\pi \times 10^{-9}$ F/m.) This equivalence between the velocity of light and the predicted velocity of electromagnetic waves helped Maxwell to establish light as an electromagnetic phenomenon.

For a medium with relative permittivity ε_r and relative permeability μ_r the velocity of the plane wave is then

$$v = \frac{c}{\sqrt{\mu_r \varepsilon_r}} \tag{15}$$

Example 3.9
Sinusoidal Wave

The most common and useful wave solution is one varying sinusoidally in space and time. Consider the function

$$E_x(z, t) = A \sin \omega \left(t - \frac{z}{v} \right) \tag{16}$$

which is a special case of (13). To show that it satisfies (11), perform the differentiations:

$$\frac{\partial^2 E_x}{\partial z^2} = -\frac{\omega^2}{v^2} A \sin \omega \left(t - \frac{z}{v} \right) \tag{17}$$

$$\mu \varepsilon \frac{\partial^2 E_x}{\partial t^2} = -\mu \varepsilon \omega^2 A \sin \omega \left(t - \frac{z}{v} \right) \tag{18}$$

But since $v^2 = (\mu\varepsilon)^{-1}$, this is a solution. To show that it is a "wave," we see that we can stay on a maximum or crest of the function if we set

$$\omega \left(t - \frac{z}{v} \right) = (4n + 1) \frac{\pi}{2}, \qquad n = 0, 1, 2, \dots$$

or

$$z = vt - \frac{(4n + 1)\pi v}{2\omega} \tag{19}$$

so that the crest does move in the z direction with velocity v as time progresses.

3.10   Uniform Plane Waves with Steady-State Sinusoids

In order to show the usefulness of the complex phasor approach for steady-state sinusoids, let us continue with this important special case. Replacement of Eqs. 3.9(6) and 3.9(8) with the complex phasor equivalents, obtained by replacing time derivatives with $j\omega$, yields

$$\frac{dE_x}{dz} = -j\omega\mu H_y \tag{1}$$

$$-\frac{dH_y}{dz} = j\omega\varepsilon E_x \tag{2}$$

Here we have utilized the total derivative with respect to z since that is now the only variable. Differentiation of (1) with respect to z and substitution of (2) yields

$$\frac{d^2 E_x}{dz^2} = -\omega^2 \mu\varepsilon E_x \tag{3}$$

This is the equivalent of the wave equation, Eq. 3.9(11), but now written in phasor form. It is called a *one-dimensional Helmholtz equation*. It could also be obtained by replacing $\partial^2/\partial t^2$ with $-\omega^2$ in Eq. 3.9(11). Solution is in terms of exponentials, as can be verified by substituting in (3):

$$E_x = c_1 e^{-jkz} + c_2 e^{jkz} \tag{4}$$

where

$$k = \omega\sqrt{\mu\varepsilon} \tag{5}$$

The constant k will be met frequently in wave problems. It is a constant of the medium for a particular angular frequency ω and is frequently called the *wave number*. It may also be written in terms of the velocity v defined by Eq. 3.9(12):

$$k = \frac{\omega}{v} \tag{6}$$

The first term of (4) is one that changes its phase linearly with z, becoming increasingly negative or lagging as one moves in the positive z direction. This behavior is consistent with the interpretation that the sinusoid is traveling in the positive z direction with velocity v, resulting in a *phase constant* k rad/m. The second term of (4) is delayed (becomes more negative) in phase as one moves in the *negative* z direction and so represents a negatively traveling wave with the same phase constant.

To show the exact correspondence of this approach with that of Sec. 3.9, let us convert the phasor form to a time-varying form by the rules given in Sec. 3.8. We multiply the phasor by the exponential $e^{j\omega t}$ and take the real part of the product:

$$E_x(z,t) = \text{Re}[E_x e^{j\omega t}] = \text{Re}[c_1 e^{-jkz} e^{j\omega t} + c_2 e^{jkz} e^{j\omega t}] \tag{7}$$

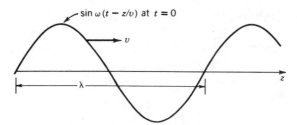

Fig. 3.10a Sinusoidal function plotted versus distance for one instant of time. For a positively traveling wave, the function progresses in the positive z direction with velocity v.

For simplicity take c_1 and c_2 to be real. Then

$$E_x(z, t) = c_1 \cos(\omega t - kz) + c_2 \cos(\omega t + kz) \qquad (8a)$$

$$= c_1 \cos \omega\left(t - \frac{z}{v}\right) + c_2 \cos \omega\left(t + \frac{z}{v}\right) \qquad (8b)$$

Following the interpretation of Sec. 3.9, we then see two real sinusoids, the first traveling in the positive z direction with velocity v and the second traveling in the negative z direction with the same velocity. The result is then exactly as in Sec. 3.9.

Figure 3.10a shows the sinusoidal variation of E_x with z at a particular instant (say $t = 0$). This pattern moves to the right with velocity v if it is a positively traveling wave, and to the left if it is negatively traveling. The distance between two planes with the same magnitude and direction of E_x is called *wavelength*, λ, and is found by the distance for which phase changes by 2π:

$$\lambda = \frac{2\pi}{k} = \frac{2\pi v}{\omega} = \frac{v}{f} \qquad (9)$$

where f is frequency. Figure 3.10b shows electric field vectors of a sinusoidal wave.

Let us also look at the magnetic fields. Returning to the complex forms, we use the solution (4) in the differential equation (1):

$$H_y = -\frac{1}{j\omega\mu}\frac{dE_x}{dz} = \frac{k}{\omega\mu}[c_1 e^{-jkz} - c_2 e^{jkz}] \qquad (10)$$

Using the definition of k from (5),

$$H_y = \sqrt{\frac{\varepsilon}{\mu}}[c_1 e^{-jkz} - c_2 e^{jkz}] \qquad (11)$$

The instantaneous equivalent of this is

$$H_y(z, t) = \text{Re}[H_y(z)e^{j\omega t}] = \sqrt{\frac{\varepsilon}{\mu}}[c_1 \cos(\omega t - kz) - c_2 \cos(\omega t + kz)]$$

$$(12)$$

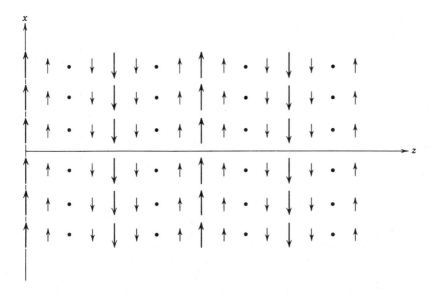

Fig. 3.10b Vectors showing magnitude and direction of electric field in a sinusoidal, uniform, plane wave filling the half-space $0 \leq z$.

So E_x/H_y is $\sqrt{\mu/\varepsilon}$ for the positively traveling wave and is $-\sqrt{\mu/\varepsilon}$ for the negatively traveling wave. The consequences of these relationships for problems of wave transmission and reflection will be seen in Chapter 6.

3.11   The Wave Equation in Three Dimensions

The one-dimensional example studied in the preceding two sections is important because it illustrates wave behavior simply, and also because it is a useful model for many important practical problems. Nevertheless we have to be concerned with wave behavior in two or three dimensions also. In order to derive the equation governing such phenomena, let us still specialize to simple media in which ε and μ are scalar constants, and assume no free charges or convection currents within the region of concern. We may then return to the special form of Maxwell's equations given as Eqs. 3.9(1) to 3.9(4). Take the curl of Eq. 3.9(3), interchanging time and space partial derivatives,

$$\nabla \times \nabla \times \mathbf{E} = -\mu \frac{\partial}{\partial t}(\nabla \times \mathbf{H}) \tag{1}$$

The left side is expanded by a vector identity. (See inside back cover.) The curl of magnetic field on the right side utilizes Eq. 3.9(4).

$$-\nabla^2 \mathbf{E} + \nabla(\nabla \cdot \mathbf{E}) = -\mu \frac{\partial}{\partial t}\left(\varepsilon \frac{\partial \mathbf{E}}{\partial t}\right) = -\mu\varepsilon \frac{\partial^2 \mathbf{E}}{\partial t^2} \tag{2}$$

For a source-free dielectric, $\mathbf{V} \cdot \mathbf{D} = 0$ and if ε is not a function of space coordinates, $\mathbf{V} \cdot \mathbf{E} = 0$ also. Then

$$\nabla^2 \mathbf{E} = \mu\varepsilon \frac{\partial^2 \mathbf{E}}{\partial t^2} \tag{3}$$

This is the three-dimensional wave equation to be derived. It will be found useful in a variety of problems to be considered later, as in the analysis of propagating modes of a waveguide, the resonant modes of a cavity resonator, or the radiating waves from an antenna. Note that the vector equation breaks into three scalar equations, and for rectangular coordinates it separates into three scalar wave equations of the same form:

$$\nabla^2 E_x = \mu\varepsilon \frac{\partial^2 E_x}{\partial t^2} \tag{4}$$

and similarly for E_y and E_z. Note that if $\partial/\partial x = 0$ and $\partial/\partial y = 0$, ∇^2 is just $\partial^2/\partial z^2$ and we have the one-dimensional wave equation studied in Sec. 3.9,

$$\frac{\partial^2 E_x}{\partial z^2} = \mu\varepsilon \frac{\partial^2 E_x}{\partial t^2} \tag{5}$$

The wave equation applies also to magnetic field for the simple medium considered here, as can be shown by taking the curl of Eq. 3.9(4) and substituting Eq. 3.9(3), to obtain

$$\nabla^2 \mathbf{H} = \mu\varepsilon \frac{\partial^2 \mathbf{H}}{\partial t^2} \tag{6}$$

In complex or phasor notation these reduce to three-dimensional Helmholtz equations, obtained by replacing $\partial^2/\partial t^2$ with $-\omega^2$ in (3) and (6).

$$\text{phasor forms} \begin{cases} \nabla^2 \mathbf{E} = -k^2 \mathbf{E} & (7) \\ \nabla^2 \mathbf{H} = -k^2 \mathbf{H} & (8) \\ k^2 = \omega^2 \mu\varepsilon & (9) \end{cases}$$

_____ **Example 3.11** _____
Resonant Wave Solution for a Rectangular Box

Let us see under what conditions the following phasor function is a solution of the three-dimensional Helmholtz equation:

$$E_x = C \cos k_x x \sin k_y y \sin k_z z \tag{10}$$

We use the x component of (7), expressed in rectangular coordinates,

$$\frac{\partial^2 E_x}{\partial x^2} + \frac{\partial^2 E_x}{\partial y^2} + \frac{\partial^2 E_x}{\partial z^2} = -k^2 E_x \tag{11}$$

Carrying out the differentiations,

$$-k_x^2 E_x - k_y^2 E_x - k_z^2 E_x = -k^2 E_x$$

or

$$k_x^2 + k_y^2 + k_z^2 = k^2 \qquad (12)$$

So the phase constants in the three directions must be related by the condition (12). We shall see in Chapter 10 that this relation, combined with boundary conditions at the conducting walls, gives the conditions for resonance of waves in a rectangular cavity resonator.

3.12 Power Flow in Electromagnetic Fields— Poynting's Theorem

The preceding sections have shown how electromagnetic waves may propagate through space or a dielectric. We know from experience that such waves can carry energy. The sun's rays, which are now known to be electromagnetic waves, warm us. The radio waves from a distant antenna bring power—admittedly small—to drive the first amplifier stage of a receiver. For lumped electrical circuits we express power through voltage and current. For electromagnetic fields, we can find a similar but more general relationship giving power and energy relationships in terms of the fields. The resulting theorem, Poynting's theorem, is one of the most fundamental and useful relationships of electromagnetic theory.

We start with the time-varying forms (Sec. 3.6) and write the two curl equations of Maxwell,

$$\mathbf{\nabla} \times \mathbf{E} = -\frac{\partial \mathbf{B}}{\partial t} \qquad (1)$$

$$\mathbf{\nabla} \times \mathbf{H} = \mathbf{J} + \frac{\partial \mathbf{D}}{\partial t} \qquad (2)$$

An equivalence of vector operations (inside cover) shows that

$$\mathbf{H} \cdot (\mathbf{\nabla} \times \mathbf{E}) - \mathbf{E} \cdot (\mathbf{\nabla} \times \mathbf{H}) = \mathbf{\nabla} \cdot (\mathbf{E} \times \mathbf{H}) \qquad (3)$$

If products involving (1) and (2) are taken as indicated, (3) becomes

$$-\mathbf{H} \cdot \frac{\partial \mathbf{B}}{\partial t} - \mathbf{E} \cdot \frac{\partial \mathbf{D}}{\partial t} - \mathbf{E} \cdot \mathbf{J} = \mathbf{\nabla} \cdot (\mathbf{E} \times \mathbf{H}) \qquad (4)$$

This may now be integrated over the volume of concern,

$$\int_V \left(\mathbf{H} \cdot \frac{\partial \mathbf{B}}{\partial t} + \mathbf{E} \cdot \frac{\partial \mathbf{D}}{\partial t} + \mathbf{E} \cdot \mathbf{J} \right) dV = -\int_V \mathbf{\nabla} \cdot (\mathbf{E} \times \mathbf{H}) \, dV$$

From the divergence theorem, Sec. 1.11, the volume integral of div(**E** × **H**) equals the surface integral of **E** × **H** over the boundary.

$$\int_V \left(\mathbf{H} \cdot \frac{\partial \mathbf{B}}{\partial t} + \mathbf{E} \cdot \frac{\partial \mathbf{D}}{\partial t} + \mathbf{E} \cdot \mathbf{J} \right) dV = -\oint_S (\mathbf{E} \times \mathbf{H}) \cdot \mathbf{dS} \tag{5}$$

This is the important *Poynting's theorem* and in this form is valid for general media since we have so far made no specializations with respect to the medium. In order to interpret it further, let us specialize to a linear, isotropic, time-invariant medium in which μ and ε are scalars and not functions of time or of field strength, although they may be functions of position. Then

$$\frac{1}{2} \frac{\partial (\mathbf{D} \cdot \mathbf{E})}{\partial t} = \frac{1}{2} \frac{\partial (\varepsilon E^2)}{\partial t} = \mathbf{E} \cdot \frac{\partial \mathbf{D}}{\partial t}$$

$$\frac{1}{2} \frac{\partial (\mathbf{B} \cdot \mathbf{H})}{\partial t} = \mathbf{H} \cdot \frac{\partial \mathbf{B}}{\partial t}$$

Equation (5) then becomes

$$\int_V \left[\frac{\partial}{\partial t} \left(\frac{\mathbf{B} \cdot \mathbf{H}}{2} \right) + \frac{\partial}{\partial t} \left(\frac{\mathbf{D} \cdot \mathbf{E}}{2} \right) + \mathbf{E} \cdot \mathbf{J} \right] dV = -\oint_S (\mathbf{E} \times \mathbf{H}) \cdot \mathbf{dS} \tag{6}$$

The terms $\varepsilon E^2/2$ was shown (Sec. 1.21) to represent the energy storage per unit volume for an electrostatic field. If this interpretation is extended by definition to any electric field,[3] the second term of (6) represents the time rate of increase of the stored energy in the electric fields of the region. Similarly, if $\mu H^2/2$ is defined as the density of energy storage for a magnetic field, the first term represents the time rate of increase of the stored energy in the magnetic fields of the region. The third term represents either the ohmic power loss if **J** is a conduction current density, or the power required to accelerate charges if **J** is a convection current arising from moving charges. Both of these cases will be illustrated in the examples at the end of this section. Also, if there is an energy source, $\mathbf{E} \cdot \mathbf{J}$ is negative for that source and represents energy flow out of the region. All the net energy change must be supplied externally. Thus the term on the right represents the energy flow into the volume per unit time. Changing sign, the rate of energy flow out through the enclosing surface is

$$W = \oint_S \mathbf{P} \cdot \mathbf{dS} \tag{7}$$

where

$$\mathbf{P} = \mathbf{E} \times \mathbf{H} \tag{8}$$

and is called the Poynting vector.

---

[3] For an excellent discussion of the arbitrariness of these definitions, refer to J. A. Stratton, *Electromagnetic Theory*, McGraw-Hill, New York, 1941, p. 133.

Although it is known from the proof only that total energy flow out of a region per unit time is given by the total surface integral (6), it is often convenient to think of the vector **P** defined by (8) as the vector giving direction and magnitude of energy flow density at any point in space. Though this step does not follow strictly, it is a most useful interpretation and one which is justified for the majority of applications. (But see Prob. 3.12a.)

_____ **Example 3.12a** _____
Ohmic Loss

To demonstrate the interpretation of the theorem, let us take the simple example of a round wire carrying direct current I_z (Fig. 3.12). If R is the resistance per unit length, the electric field in the wire is known from Ohm's law to be

$$E_z = I_z R$$

The magnetic field at the surface, or at any radius r outside the wire, is

$$H_\phi = \frac{I_z}{2\pi r} \tag{9}$$

The Poynting vector $\mathbf{P} = \mathbf{E} \times \mathbf{H}$ is everywhere radial, directed toward the axis:

$$P_r = -E_z H_\phi = -\frac{RI_z^2}{2\pi r} \tag{10}$$

We then make an integration over a cylindrical surface of unit length and radius equal to that of the wire (there is no flow through the ends of the cylinder since **P** has no component normal to the ends). All the flow is through the cylindrical surface, giving a power flow inward of amount

$$W = 2\pi r(-P_r) = I_z^2 R \tag{11}$$

We know that this result does represent the correct power flow into the conductor, being dissipated in heat. If we accept the Poynting vector as giving the correct _density_ of power flow at each point, we must then picture the battery or other source of energy as setting up the electric and magnetic fields, so that the

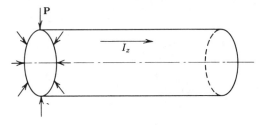

Fig. 3.12 Round wire with Poynting vector directed radially inward in order to supply power for ohmic losses.

energy flows through the field and into the wire through its surface. The Poynting theorem cannot be considered a proof of the correctness of this interpretation, for it says only that the *total* power balance for a given region will be computed correctly in this manner, but the interpretation is nevertheless a useful one.

_____ **Example 3.12b** _____
Cases of Zero Poynting Flow

It is also instructive to consider the cases for which there will be no power flow through the electromagnetic field. Accepting the foregoing interpretation of the Poynting vector, we see that it will be zero when either **E** or **H** is zero, or when the two vectors are mutually parallel. Thus , for example, there is no power flow in the vicinity of a system of static charges which has electric field but no magnetic field. Another very important case is that of a perfect conductor, which by definition must have a zero tangential component of electric field at its surface. Then **P** can have no component normal to the conductor and there can be no power flow into the perfect conductor.

_____ **Example 3.12c** _____
Moving Charges

Let us next consider the example in which **J** is a convection current. For simplicity take a region containing particles of charge value q, mass m, and velocity $\mathbf{v}_\rho$. The convection current density is

$$\mathbf{J} = \rho\mathbf{v}_\rho = nq\mathbf{v}_\rho \tag{12}$$

where n is the density of particles. From the force law the acceleration of charges is

$$\mathbf{F} = q\mathbf{E} = m\frac{d\mathbf{v}_\rho}{dt} \tag{13}$$

and the third term in the Poynting theorem (6) is

$$\int_V \mathbf{E} \cdot \mathbf{J}\, dV = \int_V \frac{m}{q}\frac{d\mathbf{v}_\rho}{dt} \cdot (nq\mathbf{v}_\rho)\, dV = \int_V \frac{mn}{2}\frac{d(v_\rho^2)}{dt}\, dV \tag{14}$$

which we recognize to be the rate of change of the total kinetic energy of the charge group. In this example we will not try to work out the right side of (6), but the **E** in that term is related to the accelerating field, and the **H** is that from the convection current and the Poynting theorem will always be satisfied.[4]

_____

[4] If the charges move through the surface surrounding the region, the net kinetic energy transport by the charge stream through the surface is also included. This is actually contained in the third term on the left as shown by L. Tonks, *Phys. Rev.* **54**, 863 (1938).

_____ **Example 3.12d** _____
Poynting Flow in a Plane Wave

Finally we look at the Poynting theorem applied to the plane electromagnetic wave studied in preceding sections. The form of a sinusoidally varying wave with E_x and H_y propagating in the positive z direction was shown to be

$$E_x = E_0 \cos(\omega t - kz) \tag{15}$$

$$H_y = \sqrt{\frac{\varepsilon}{\mu}} E_0 \cos(\omega t - kz) \tag{16}$$

The Poynting vector is then in the z direction, which is consistent with our interpretation that power is flowing in that direction:

$$P_z = E_x H_y = \sqrt{\frac{\varepsilon}{\mu}} E_0^2 \cos^2(\omega t - kz) \tag{17}$$

By the use of a trigonometric identity this is also

$$P_z = \sqrt{\frac{\varepsilon}{\mu}} E_0^2 \left[\frac{1}{2} + \frac{1}{2} \cos 2(\omega t - kz) \right] \tag{18}$$

Notice that there is a constant term showing that the wave carries an average power, as expected. There is also a time-varying portion representing the redistribution of stored energy in space as maxima and minima of fields pass through a given region.

_____

3.13 Poynting's Theorem for Phasors

Because of the importance of phasors for sinusoidal electromagnetic fields, we need the Poynting theorem in phasor form also. It might seem that we could simply substitute in the time-varying theorem, Eq. 3.12(5), replacing $\partial/\partial t$ by $j\omega$, but this does not work since the expression is nonlinear, involving products of the fields. We start with Maxwell's equations in complex form and derive the complex Poynting theorem by steps parallel to those used for the theorem in time-varying field quantities. The two curl equations in complex phasor form are (Sec. 3.8)

$$\mathbf{\nabla} \times \mathbf{E} = -j\omega \mathbf{B} \tag{1}$$

$$\mathbf{\nabla} \times \mathbf{H} = \mathbf{J} + j\omega \mathbf{D} \tag{2}$$

Consider the vector identity

$$\mathbf{\nabla} \cdot (\mathbf{E} \times \mathbf{H}^*) = \mathbf{H}^* \cdot (\mathbf{\nabla} \times \mathbf{E}) - \mathbf{E} \cdot (\mathbf{\nabla} \times \mathbf{H}^*) \tag{3}$$

where the asterisk denotes the complex conjugate. Equations (1) and (2) may now be substituted in this identity.

$$\mathbf{V} \cdot (\mathbf{E} \times \mathbf{H}^*) = \mathbf{H}^* \cdot (-j\omega\mathbf{B}) - \mathbf{E} \cdot (\mathbf{J}^* - j\omega\mathbf{D}^*) \tag{4}$$

The above expression is integrated through volume V and the divergence theorem utilized,

$$\int_V \mathbf{V} \cdot (\mathbf{E} \times \mathbf{H}^*) \, dV = \oint_S (\mathbf{E} \times \mathbf{H}^*) \cdot d\mathbf{S} = -\int_V [\mathbf{E} \cdot \mathbf{J}^* + j\omega(\mathbf{H}^* \cdot \mathbf{B} - \mathbf{E} \cdot \mathbf{D}^*)] \, dV \tag{5}$$

Equation (5) is the general Poynting theorem as it applies to complex phasors. To interpret, consider an isotropic medium in which all losses occur through conduction currents $\mathbf{J} = \sigma\mathbf{E}$ so that σ, μ, and ε are real scalars. Then (5) becomes

$$\oint_S (\mathbf{E} \times \mathbf{H}^*) \cdot d\mathbf{S} = -\int_V \sigma\mathbf{E} \cdot \mathbf{E}^* \, dV - j\omega \int_V [\mu\mathbf{H} \cdot \mathbf{H}^* - \varepsilon\mathbf{E} \cdot \mathbf{E}^*] \, dV \tag{6}$$

The first volume integral on the right side represents power loss in the conduction currents and is just twice the average power loss. (See Appendix 4.) Thus the real part of the complex Poynting flow on the left side can be related to this power loss. Or, interpreting the Poynting vector itself as a density of power flow as in Sec. 3.12,

$$\mathbf{P}_{av} = \tfrac{1}{2} \operatorname{Re}(\mathbf{E} \times \mathbf{H}^*) \quad \text{W/m}^2 \tag{7}$$

The second volume integral on the right of (6) is proportional to the difference between average stored magnetic energy in the volume and average stored electric energy. Taking into account a factor of $\tfrac{1}{2}$ in the energy expressions and another $\tfrac{1}{2}$ for averaging of squares of sinusoids, we can then interpret the imaginary part of (6) as

$$\operatorname{Im} \int_S (\mathbf{E} \times \mathbf{H}^*) \cdot d\mathbf{S} = 4\omega(U_{Eav} - U_{Hav}) \tag{8}$$

where U_{Eav} is average stored energy in electric fields and U_{Hav} that in magnetic fields. So the imaginary part of the Poynting flow through the surface can be thought of as *reactive power* flowing back and forth to supply the instantaneous changes in net stored energy in the volume.

Example 3.13
Average Power in Uniform Plane Waves

To illustrate the average Poynting vector for the plane-wave case, let us take the field expressions for plane waves derived in complex form in Sec. 3.10,

$$E_x = c_1 e^{-jkz} + c_2 e^{jkz} \tag{9}$$

$$H_y = \sqrt{\frac{\varepsilon}{\mu}} \, [c_1 e^{-jkz} - c_2 e^{jkz}] \tag{10}$$

The complex Poynting vector is then

$$\mathbf{E \times H^*} = \sqrt{\frac{\varepsilon}{\mu}}\,[c_1 e^{-jkz} + c_2 e^{jkz}][c_1^* e^{jkz} - c_2^* e^{-jkz}]\hat{\mathbf{z}} \tag{11}$$

and the average power density, by (7) is in the z direction and equal to

$$P_{\text{av}} = \frac{1}{2}\sqrt{\frac{\varepsilon}{\mu}}\,[c_1 c_1^* - c_2 c_2^*] \quad \text{W/m}^2 \tag{12}$$

This equation states that the average power is simply the average power of the positively traveling wave minus that of the negatively traveling wave. The cross-product terms of (11) contribute only to reactive power—that is, to the interchange of stored energy within the wave.

3.14   Continuity Conditions for ac Fields at a Boundary; Uniqueness of Solutions

In the study of static fields, certain boundary and continuity conditions were stated for such fields and were found essential in the solution of the field problems by the use of the differential equations. Similarly, for the use of Maxwell's equations in differential equation form, we need corresponding boundary and continuity conditions. Although the boundary conditions for a differential equation cannot be deduced from the differential equation itself, it is presumed that the physics embodied in Maxwell's equations applies to the region of the boundary and will be used to infer the appropriate conditions. Uniqueness arguments supply the final proof of the sufficiency of the conditions deduced.

Consider first Faraday's law in large-scale form, Eq. 3.2(3), applied to a path formed by moving distance Δl along one side of the boundary between any two materials, and returning on the other side, an infinitesimal distance into the second medium (Fig. 3.14a). The line integral of electric field is

$$\oint \mathbf{E \cdot dl} = (E_{t1} - E_{t2})\,\Delta l \tag{1}$$

Since the path is an infinitesimal distance on either side of the boundary, it encloses zero area, and therefore the contribution from changing magnetic flux is zero so long as rate of change of magnetic flux density is finite. Consequently,

$$(E_{t1} - E_{t2})\,\Delta l = 0 \quad \text{or} \quad E_{t1} = E_{t2} \tag{2}$$

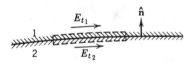

Fig. 3.14a Continuity of tangential electric field components at a dielectric boundary.

Similarly, the generalized Ampère law in large-scale form, Eq. 3.7(4), may be applied to a like path with its two sides on the two sides of the boundary. Again there is zero area enclosed by the path, and, so long as current density and rate of change of electric flux density are finite, the integral is zero. Thus, in like manner to the above.

$$H_{t1} = H_{t2} \tag{3}$$

Or in vector form, utilizing the unit vector $\hat{n}$ normal to the boundary as shown in Fig. 3.14a, (2) and (3) can be written

$$\hat{n} \times (\mathbf{E}_1 - \mathbf{E}_2) = 0 \tag{4}$$

$$\hat{n} \times (\mathbf{H}_1 - \mathbf{H}_2) = 0 \tag{5}$$

Thus tangential components of electric and magnetic field must be equal on the two sides of any boundary between physically real media. The condition (3) may be modified for an idealized case such as the perfect conductor where the current densities are allowed to become infinite. This case is discussed separately in Sec. 3.15.

The integral form of Gauss's law is Eq. 3.7(1). If two very small elements of area ΔS are considered (Fig. 3.14b), one on either side of the boundary between any two materials, with a surface charge density ρ_s existing on the boundary, the application of Gauss's law to this elemental volume gives

$$\Delta S(D_{n1} - D_{n2}) = \rho_s \Delta S$$

or

$$D_{n1} - D_{n2} = \rho_s \tag{6}$$

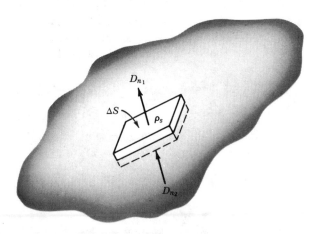

Fig. 3.14b Diagram showing how discontinuity in normal components of electric flux density at a boundary is related to surface charge density.

For a charge-free boundary,

$$D_{n1} = D_{n2} \quad \text{or} \quad \varepsilon_1 E_{n1} = \varepsilon_2 E_{n2} \tag{7}$$

That is, for a charge-free boundary, normal components of electric flux density are continuous; for a boundary with charges, they are discontinuous by the amount of the surface charge density.

Since there is no magnetic charge term on the right of Eq. 3.7(2), a development corresponding to the above shows that always the magnetic flux density is continuous:

$$B_{n1} = B_{n2} \quad \text{or} \quad \mu_1 H_{n1} = \mu_2 H_{n2} \tag{8}$$

For the time-varying case, which is of greatest importance to our study, the conditions on normal components are not independent of those given for the tangential components. The reason is that the former are derived from the divergence equations (or their equivalent in large-scale form), and these may be obtained from the two curl equations in the time-varying case (Probs. 3.6b and 3.6c). The conditions on tangential components were derived from the large-scale equivalents of the curl equations. Hence, for the ac solutions, it is necessary only to apply the continuity conditions on tangential components of electric and magnetic fields at a boundary between two media, and the conditions on normal components may be used as a check; if the normal components of $\mathbf{D}$ turn out to be discontinuous, (6) tells the amount of surface charge that is induced on the boundary.

Uniqueness   The procedure to prove the uniqueness of solutions of Maxwell's equations follows the philosophy in Sec. 1.17. One assumes two possible solutions with the same given tangential fields on the boundary of the region of interest. The difference field is formed, and found to satisfy a Poynting's theorem of the form of Eq. 3.12(5). Stratton[5] shows that for linear, isotropic (but possibly inhomogeneous), media, specification of tangential $\mathbf{E}$ and $\mathbf{H}$ on the boundary and of initial values of all fields at time zero, is sufficient to specify fields uniquely within the region at all later times. The argument can be extended to anisotropic materials and certain classes of nonlinear materials, but not to materials that have multivalued relations between $\mathbf{D}$ and $\mathbf{E}$ (or $\mathbf{B}$ and $\mathbf{H}$) or to "active" materials that produce oscillations. In steady-state problems we are not generally concerned with the specifications of initial conditions.

Although the discussion has been given for a region with closed boundaries, uniqueness arguments also apply to open regions extending to infinity, provided certain radiation conditions are satisfied by the fields. These require that the products $r\mathbf{E}$ and $r\mathbf{H}$ remain finite as r approaches infinity,[6] and are satisfied by

[5] J. A. Stratton, Ref. 3, pp. 486–488.

[6] S. Silver, *Microwave Antenna Theory and Design*, McGraw-Hill, New York, 1949, p. 85.

fields arising from real charge and current sources contained within a finite region. The extension to open regions is important to the potential formulation of the last part of this chapter.

3.15 Boundary Conditions at a Perfect Conductor for ac Fields

It is a good approximation in many practical problems to treat good conductors (such as copper and other metals) as though of infinite conductivity when finding the form of fields outside the conductor. We will study the effect of large but finite conductivity on fields *within* the conductor in Sec. 3.16. When we do, we will find that all fields and currents concentrate in a thin region or "skin" near the surface for time-varying fields, and this region approaches zero thickness as the conductivity approaches infinity. Thus, for the perfect conductor (infinite conductivity), we find that all fields are zero inside the conductor and any current flow must be only on the surface. The physical properties of perfect conductors are discussed in Sec. 13.4. If the electric field is zero within the perfect conductor, continuity of tangential electric field at a boundary requires that the surface tangential electric field be zero just outside the boundary also:

$$E_t = 0 \tag{1}$$

and Eq. 3.14(6) gives the normal electric flux density as

$$D_n = \rho_s \tag{2}$$

Furthermore, since magnetic fields also vanish inside the conductor, the statement of continuity of magnetic flux lines, Eq. 3.14(8) indicates that

$$B_n = 0 \tag{3}$$

at the conductor surface. As was pointed out in the last article, however, the continuity condition on normal **B** is not independent of the condition on tangential **E** in the time-varying case. Thus, in the ac solution, (3) follows from (1), but may sometimes be useful as a check, or as an alternative boundary condition.

The tangential component of magnetic field is likewise zero inside the perfect conductor but is not in general zero just outside. This discontinuity would appear to violate the condition of Eq. 3.14(3), but it will be recalled that a condition for that proof was that current density must remain finite. For the perfect conductor, the finite current **J** per unit width is assumed to flow on the surface as a current sheet of zero thickness, so that current density is infinite. The discontinuity in tangential magnetic field is found by a construction similar to that of Fig. 3.14a. The current enclosed by the path is the current per unit width **J** flowing on the

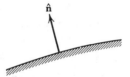

Fig. 3.15 Conducting boundary with the normal unit vector.

surface of the conductor perpendicular to the direction of the tangential magnetic field at the surface. Then

$$\oint \mathbf{H} \cdot \mathbf{dl} = H_t\, dl = J_s\, dl$$

or

$$J_s = H_t \quad \text{A/m} \tag{4}$$

where J_s is current per unit width, called a surface current density. The direction and sense relations for (4) are given most conveniently by the vector form of the law below.

To write the relations of (1)–(4) in vector notation, a unit vector $\hat{\mathbf{n}}$, normal to the conductor at any given point and pointing from the conductor into the region where fields exist, is defined (Fig. 3.15). Then conditions (1) to (4) become (5) to (8) below:

$$\hat{\mathbf{n}} \times \mathbf{E} = 0 \tag{5}$$

$$\hat{\mathbf{n}} \cdot \mathbf{B} = 0 \tag{6}$$

$$\rho_s = \hat{\mathbf{n}} \cdot \mathbf{D} \tag{7}$$

$$\mathbf{J}_s = \hat{\mathbf{n}} \times \mathbf{H} \tag{8}$$

For an ac problem, (5) represents the only required boundary condition at a perfect conductor. Equation (6) serves as a check or sometimes as an alternative to (5). Equations (7) and (8) are used to give the charge and current induced on the conductor by the presence of the electromagnetic fields.

3.16 Penetration of Electromagnetic Fields into a Good Conductor

Maxwell's equations have been illustrated by showing the wave behavior of electromagnetic fields in good dielectrics. A second extremely important class of materials used in many electromagnetic problems is that of "good conductors." Let us examine the basic behavior of electromagnetic fields in such conductors. The development in this and the following section will be for steady-state sinusoids using phasor notation, with the usual understanding that more general time variations may be broken up into a series or continuous distribution of such sinusoids. The conductors of concern are those satisfying Ohm's law,

$$\mathbf{J} = \sigma \mathbf{E} \tag{1}$$

The constant σ is the conductivity of the conductor. At optical frequencies metals are not well represented by a real constant σ, but the approximation is valid for microwaves and millimeter waves (Sec. 13.2). Substitution of (1) into the Maxwell equation 3.8(4) gives

$$\mathbf{\nabla} \times \mathbf{H} = (\sigma + j\omega\varepsilon)\mathbf{E} \tag{2}$$

It is easy to show that the assumption of Ohm's law implies the absence of free charges. Since the divergence of the curl of any vector is zero,

$$\mathbf{\nabla} \cdot \mathbf{\nabla} \times \mathbf{H} = (\sigma + j\omega\varepsilon)\mathbf{\nabla} \cdot \mathbf{E} = 0$$

where we have assumed homogeneity of σ and ε. Thus

$$\mathbf{\nabla} \cdot \mathbf{D} = \rho = 0 \tag{3}$$

The simple picture of the situation in a conductor is that mobile electrons drift through a lattice of positive ions encountering frequent collisions. On the average, over a volume large compared with the atomic dimensions but small compared with dimensions of interest in the system under study, the net charge is zero even though some of the charges are moving through the element and causing current flow. The net movement or "drift" in such cases is found proportional to the electric field.

For metals and other good conductors, it is found that displacement current is negligible in comparison with conduction current for radio frequencies, and in fact is not measurable until frequencies are well into the infrared. For the present we concentrate on the important cases for which $\omega\varepsilon$ in (2) is negligible in comparison with σ.

Thus, to summarize, the following specializations are appropriate to Maxwell's equations applied to good conductors, and may in fact be taken as a definition of a good conductor.

1. Conduction current is given by Ohm's law, $\mathbf{J} = \sigma\mathbf{E}$.
2. Displacement current is negligible in comparison with current, $\omega\varepsilon \ll \sigma$.
3. As a consequence of (1), the free-charge term is zero for homogeneous conductors.

To derive the differential equation which determines the penetration of the fields into the conductor we first take the curl of the Maxwell curl equation for electric field, Eq. 3.8(3) and make use of a vector identity (see inside back cover) and the definition of permeability to obtain

$$\mathbf{\nabla} \times \mathbf{\nabla} \times \mathbf{E} = \mathbf{\nabla}(\mathbf{\nabla} \cdot \mathbf{E}) - \nabla^2\mathbf{E} = -j\omega\mu\mathbf{\nabla} \times \mathbf{H} \tag{4}$$

Then using (3) and substituting (2) in (4) with displacement current neglected, we find

$$\nabla^2 \mathbf{E} = j\omega\mu\sigma\mathbf{E} \tag{5}$$

Equations with forms identical with (5) can be found in a similar way for magnetic field and current density.

$$\nabla^2 \mathbf{H} = j\omega\mu\sigma\mathbf{H} \tag{6}$$

$$\nabla^2 \mathbf{J} = j\omega\mu\sigma\mathbf{J} \tag{7}$$

We first consider the differential equations (5)–(7) for the simple but useful example of a plane conductor of infinite depth, with no field variations along the width or length dimension. This case is frequently taken as that of a conductor filling the half-space $x > 0$ in a rectangular coordinate system with the y–z plane coinciding with the conductor surface, and is then spoken of as a "semi-infinite solid." In spite of the infinite depth requirement, the analysis of this case is of importance to many conductors of finite extent, and with curved surfaces, because at high frequencies the depth over which significant fields are concentrated is very small. Radii of curvature and conductor depth may then be taken as infinite in comparison. Moreover, any field variations along the surface due to curvature, edge effects, or variations along a wavelength are ordinarily so small compared with the variations into the conductor that they may be neglected.

For the uniform field situation shown in Fig. 3.16*a* with the electric field vector in the z direction, we assume no variations with y or z and (5) becomes

$$\frac{d^2 E_z}{dx^2} = j\omega\mu\sigma E_z = \tau^2 E_z \tag{8}$$

where

$$\tau^2 \triangleq j\omega\mu\sigma \tag{9}$$

Since $\sqrt{j} = (1 + j)/\sqrt{2}$ (taking the root with the positive sign),

$$\tau = (1 + j)\sqrt{\pi f\mu\sigma} = \frac{1 + j}{\delta} \tag{10}$$

where

$$\delta = \frac{1}{\sqrt{\pi f\mu\sigma}} \quad \text{m} \tag{11}$$

A complete solution of (8) is in terms of exponentials:

$$E_z = C_1 e^{-\tau x} + C_2 e^{\tau x} \tag{12}$$

The field will increase to the impossible value of infinity at $x = \infty$ unless C_2 is zero. The coefficient C_1 may be written as the field at the surface if we let $E_z = E_0$ when $x = 0$. Then

$$E_z = E_0 e^{-\tau x} \tag{13}$$

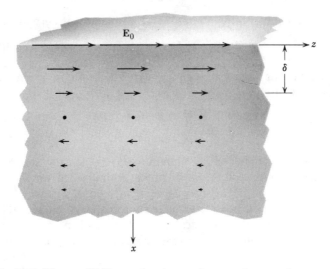

Fig. 3.16a Plane solid illustrating decay of current into conductor.

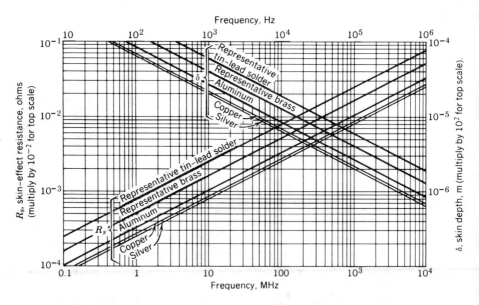

Fig. 3.16b Skin depth and surface resistivity versus frequency for several conductors.

Or, in terms of the quantity δ defined by (10) and (11)

$$E_z = E_0 e^{-x/\delta} e^{-jx/\delta} \tag{14}$$

Since the magnetic field and the current density are governed by the same differential equation as the electric field, forms identical with (14) apply; that is,

$$H_y = H_0 e^{-x/\delta} e^{-jx/\delta} \tag{15}$$

$$J_z = J_0 e^{-x/\delta} e^{-jx/\delta} \tag{16}$$

where H_0 and J_0 are the magnitudes of the magnetic field and current density at the surface.

It is evident from the forms of (14) to (16) that the magnitudes of the fields and current decrease exponentially with penetration into the conductor, and δ has the significance of the depth at which they have decreased to $1/e$ (about 36.9%) of their values at the surface. The quantity δ is accordingly called *depth of penetration* or *skin depth*. It is shown as a function of frequency for some common materials in Fig. 3.16*b*. The phases of the current and fields lag behind their surface values by x/δ rad at depth x into the conductor.

_____ **Example 3.16** _____
Skin Depth in Audio Transformer with Iron Core

An audiofrequency transformer has a core made of iron with $\sigma = 0.5 \times 10^7$ and $\mu = 1000\mu_0$. It is designed to work up to 15 kHz. Let us find the skin depth at this highest design frequency. From (11)

$$\delta = (\pi \times 15 \times 10^3 \times 10^3 \times 4\pi \times 10^{-7} \times 0.5 \times 10^7)^{-1/2} = (30\pi^2 \times 10^6)^{-1/2}$$

$$= 0.058 \times 10^{-3} \text{ m} = 0.058 \text{ mm} \tag{17}$$

Note that this is more than 30 times smaller than for a material of the same conductivity but a relative permeability of unity.

3.17 Internal Impedance of a Plane Conductor

The decay of fields into a good conductor may be looked at as the attenuation of a plane wave as it propagates into the conductor, or from the point of view that induced fields from the time-varying currents tend to counter the applied fields. The latter point of view is especially applicable to circuits, in which case we think of the field at the surface as the applied field. Currents (resulting from σE) concentrate near this surface and the ratio of surface electric field to current flow gives an *internal impedance* for use in circuit problems. By *internal*, we mean the contribution to impedance from the fields penetrating into the conductor. This gives in general a resistance term and an internal inductance, the latter to be added to any external inductance contribution arising from fields outside the conductor.

The total current flowing in the plane conductor is found by integrating the current density, Eq. 3.16(16) from the surface to the infinite depth. For a unit width,

$$J_{sz} = \int_0^\infty J_z \, dx = \int_0^\infty J_0 e^{-(1+j)(x/\delta)} \, dx = \frac{J_0 \delta}{(1+j)} \tag{1}$$

The electric field at the surface is given by the current density at the surface,

$$E_{z0} = \frac{J_0}{\sigma} \tag{2}$$

Internal impedance for a unit length and unit width is defined as

$$Z_s = \frac{E_{z0}}{J_{sz}} = \frac{1+j}{\sigma\delta} \tag{3}$$

Define

$$Z_s = R_s + j\omega L_i \tag{4}$$

Then

$$R_s = \frac{1}{\sigma\delta} = \sqrt{\frac{\pi f \mu}{\sigma}} \tag{5}$$

$$\omega L_i = \frac{1}{\sigma\delta} = R_s \tag{6}$$

The resistance and internal reactance of such a plane conductor are equal at any frequency. The internal impedance Z_s thus has a phase angle of 45 degrees. Equation (5) gives another interpretation of depth of penetration δ, for this equation shows that the skin-effect resistance of the semi-infinite plane conductor is exactly the same as the dc resistance of a plane conductor of depth δ. That is, resistance of this conductor with exponential decrease in current density is exactly the same as though current were uniformly distributed over a depth δ.

The resistance R_s of the plane conductor for a unit length and unit width, is called the *surface resistivity*. For a finite area of conductor, the resistance is obtained by multiplying R_s by length, and dividing by width since the width elements are essentially in parallel. Thus the dimension of R_s is ohms or, as it is sometimes called, ohms per square. Like the depth of penetration δ, R_s as defined by (5) may also be a useful parameter in the analyses of conductors of other than plane shape, and may be thought of as a constant of the material at frequency f. Values of R_s and δ for several materials are summarized in Table 3.17 and plotted versus frequency in Fig. 3.16b.

Table 3.17
Skin effect properties of typical metals

	Conductivity σ (S/m)	Permeability μ (H/m)	Depth of Penetration δ (m)	Surface Resistivity R_s (Ω)
Silver	6.17×10^7	$4\pi \times 10^{-7}$	$\dfrac{0.0642}{\sqrt{f}}$	$2.52 \times 10^{-7}\sqrt{f}$
Copper	5.80×10^7	$4\pi \times 10^{-7}$	$\dfrac{0.0660}{\sqrt{f}}$	$2.61 \times 10^{-7}\sqrt{f}$
Aluminum	3.72×10^7	$4\pi \times 10^{-7}$	$\dfrac{0.0826}{\sqrt{f}}$	$3.26 \times 10^{-7}\sqrt{f}$
Representative brass	1.57×10^7	$4\pi \times 10^{-7}$	$\dfrac{0.127}{\sqrt{f}}$	$5.01 \times 10^{-7}\sqrt{f}$ ·
Representative solder	0.706×10^7	$4\pi \times 10^{-7}$	$\dfrac{0.185}{\sqrt{f}}$	$7.73 \times 10^{-7}\sqrt{f}$

_____ **Example 3.17** _____
Approximate Internal Impedance of a Coaxial Line

The usefulness of this concept for practical problems may now be illustrated by
considering the coaxial transmission line of Fig. 3.17. We select as a circuit path
one which follows the outer surface of the inner conductor, AB, traversing
radially across to C and then following the inner surface of the outer conductor
CD, returning back radially to A. The difference between voltages V_{DA} and V_{CB}
will arise in part because of the inductance calculated from flux within the path
$ABCDA$; that is, that external to the conductors. We consequently call this the
external inductance and recognize it as that found for a coaxial line in Chapter 2.
But there is also a voltage contribution along the path AB due to the internal

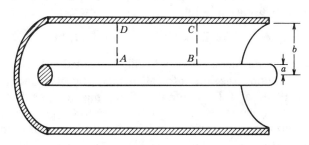

Fig. 3.17 Section of coaxial transmission line. Line integral of electric field about path
$ABCD$ relates to magnetic flux associated with external inductance.

impedance of the inner conductor, and one along CD arising from internal impedance of the outer conductor.

If radius of curvature a and b are large in comparison with skin depth δ, and if thickness of the outer tubular conductor is large compared with δ, both conductors may be treated to a good degree of approximation by the planar analysis of this and the preceding section. Current concentrates on the outer surface of the inner conductor and the inner surface of the outer conductor, adjacent to the region of the fields. The inner conductor, if curvature is negligible, then appears as a plane of width equal to its circumference, $2\pi a$. Internal impedance per unit length is then

$$Z_{i1} = \frac{Z_{s1}}{2\pi a} \quad \Omega/\text{m}$$

The outer conductor, with these approximations, appears as a plane of width equal to its inner circumference, $2\pi b$. Its thickness does not enter since it is presumed much larger than δ so that fields have died to a negligible value at the outer surface. Internal impedance per unit length from this part is then

$$Z_{i2} = \frac{Z_{s2}}{2\pi b} \quad \Omega/\text{m}$$

The sum of these two gives the total contribution to impedance from fields within the conductors and can be used in the transmission line analysis of Chapter 5.

3.18 Power Loss in a Plane Conductor

To find average power loss per unit area of the plane conductor, we may apply the Poynting theorem of Sec. 3.13. The field components E_z and H_y produce a power flow in the x direction, or into the conductor. Utilization of the field values at the surface gives the total power flowing from the field into the conductor. In complex phasor form, using Eq. 3.13(7):

$$W_L = \tfrac{1}{2}\,\text{Re}[\mathbf{E}_0 \times \mathbf{H}_0^*] = -\tfrac{1}{2}\,\text{Re}(E_{z0}H_{y0}^*) \tag{1}$$

The surface value of magnetic field can readily be related to the surface current, as can be seen by taking the line integral of magnetic field about some path $ABCD$ of Fig. 3.18 (C and D at infinity). Since magnetic field is in the $-y$ direction for this simple case, there is no contribution to $\mathbf{H} \cdot \mathbf{dl}$ along the sides BC and DA; there is no contribution along CD since field is zero at infinity. Hence, for a width w,

$$\oint_{ABCD} \mathbf{H} \cdot \mathbf{dl} = \int_A^B \mathbf{H} \cdot \mathbf{dl} = -wH_{y0} \tag{2}$$

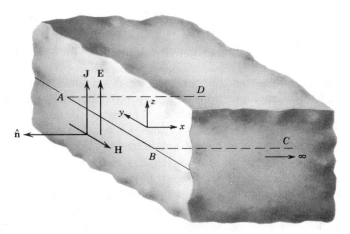

Fig. 3.18 Surface of plane conductor illustrating how magnetic field at surface relates to current flow per unit width.

This line integral of magnetic field must be equal to the conduction current enclosed, since displacement current has been shown to be negligible in a good conductor. The current is just the width w times the current per unit width, J_z. Then utilizing (2),

$$-wH_{y0} = wJ_{sz} \quad \text{or} \quad J_{sz} = -H_{y0} \tag{3}$$

This may be written in a vector form which includes the magnitude and sense information of (3), and the fact that **J** and **H** are mutually perpendicular,

$$\mathbf{J}_s = \hat{\mathbf{n}} \times \mathbf{H}, \tag{4}$$

where $\hat{\mathbf{n}}$ is a unit vector perpendicular to the conductor surface, pointing into the adjoining dielectric region and **H** is the magnetic field at the surface. Note that (4) is of the same form as for perfect conductors, Eq. 3.15(8). Then using (1), (3), and Eq. 3.17(3) we obtain

$$W_L = \tfrac{1}{2} \operatorname{Re}[Z_s J_s J_s^*] = \tfrac{1}{2} R_s |J_s|^2 \quad \text{W/m}^2 \tag{5}$$

This is a form that might have been expected in that it gives loss in terms of resistance multiplied by square of current magnitude. An alternate derivation (Prob. 3.18a) is by the integration of power loss at each point of the solid from the known conductivity and current density function.

Equation (5) will be found of the greatest usefulness throughout this text for the computation of power loss in the walls of waveguides, cavity resonators, and other electromagnetic structures. Although the walls of these structures are not plane solids of infinite depth, the results of this section may be applied for all practical purposes whenever the conductor thickness and radii of curvature are much greater than δ, depth of penetration. This includes most important cases at high frequencies. In these cases the quantities that are ordinarily known are the fields at the surface of the conductor.

POTENTIALS FOR TIME-VARYING FIELDS

3.19   A Possible Set of Potentials for Time-Varying Fields

As we have seen, time-varying electromagnetic fields are related to each other and to the charge and current sources through the set of differential equations known as Maxwell's equations. It is sometimes convenient to introduce some intermediate functions, known as *potential functions*, which are directly related to the sources, and from which the electric and magnetic fields may be derived. Such functions were found useful for static fields, and in the case of the electrostatic potential, the potential itself had useful physical significance. The physical interpretation was less clear in the case of the magnetic vector potential, but it does provide a useful simplification in the analysis of some problems. In this and following sections we look for similar potential functions for the time-varying fields. It turns out that there are many possible sets. We select a commonly used set known as *retarded potentials* which reduce to the potentials used for statics in the limit of no time variations.

We might try at first to use the forms found for statics, $\mathbf{E} = -\nabla\Phi$ and $\mathbf{B} = \nabla \times \mathbf{A}$ with all quantities functions of time. We are faced with this problem: the electric field for time-varying conditions cannot be derived only as the gradient of scalar potential since this would require that it have zero curl, and it may actually have a nonzero curl of value $-\partial\mathbf{B}/\partial t$; it cannot be derived alone as the curl of a vector potential since this would require that it have zero divergence, and it may have a finite divergence of value ρ/ε.

Since the divergence of magnetic field is zero in the general case as it was in the static, it seems that $\mathbf{B}$ may still be set equal to the curl of some magnetic vector potential, $\mathbf{A}$.

$$\mathbf{B} = \nabla \times \mathbf{A} \tag{1}$$

This relation may now be substituted in the Maxwell equation 3.6(3) and the result written

$$\nabla \times \left(\mathbf{E} + \frac{\partial\mathbf{A}}{\partial t} \right) = 0 \tag{2}$$

This equation states that the curl of a certain vector quantity is zero. But this is the condition that permits a vector to be derived as the gradient of a scalar, say Φ. That is,

$$\mathbf{E} + \frac{\partial\mathbf{A}}{\partial t} = -\nabla\Phi$$

or

$$\mathbf{E} = -\nabla\Phi - \frac{\partial\mathbf{A}}{\partial t} \tag{3}$$

Equations (1) and (3) are then valid relationships between fields and potential functions A and Φ. Note that no specializations on the medium have been made to this point. It is found that the potential functions are most useful, though, for linear, isotropic, homogeneous media, so in the remaining part of this discussion, we take μ and ε as scalar constants appropriate to such media. With this specialization we substitute (3) in Gauss's law, Eq. 3.6(1), to obtain

$$-\nabla^2\Phi - \frac{\partial}{\partial t}(\nabla \cdot A) = \frac{\rho}{\varepsilon} \tag{4}$$

Then substituting $B = \nabla \times A$ and (3) in Eq. 3.6(4), we find

$$\nabla \times \nabla \times A = \mu J + \mu\varepsilon\left[-\nabla\left(\frac{\partial\Phi}{\partial t}\right) - \frac{\partial^2 A}{\partial t^2}\right]$$

Using the vector identity

$$\nabla \times \nabla \times A \equiv \nabla(\nabla \cdot A) - \nabla^2 A$$

this becomes

$$\nabla(\nabla \cdot A) - \nabla^2 A = \mu J - \mu\varepsilon\nabla\left(\frac{\partial\Phi}{\partial t}\right) - \mu\varepsilon\frac{\partial^2 A}{\partial t^2} \tag{5}$$

Equations (4) and (5) can be simplified by further specification of A. That is, there are any number of vector functions whose curl is the same. One may specify also the divergence of A according to convenience.[7] If the divergence of A is chosen as[8]

$$\nabla \cdot A = -\mu\varepsilon\frac{\partial\Phi}{\partial t} \tag{6}$$

(4) and (5) then simplify to

$$\nabla^2\Phi - \mu\varepsilon\frac{\partial^2\Phi}{\partial t^2} = -\frac{\rho}{\varepsilon} \tag{7}$$

$$\nabla^2 A - \mu\varepsilon\frac{\partial^2 A}{\partial t^2} = -\mu J \tag{8}$$

Thus the potentials A and Φ, defined in terms of the sources J and ρ by the differential equations (7) and (8), may be used to derive the electric and magnetic fields by (1) and (3). It is easy to see that they do reduce to the corresponding

[7] Specification of divergence and curl of a vector, with appropriate boundary conditions, determines the vector uniquely through the Helmholtz theorem. See, for example, R. E. Collin, *Field Theory of Guided Waves*, McGraw-Hill, New York, 1960, Appendix A1.
[8] This choice is known as the Lorentz condition or Lorentz gauge and leads to the symmetry of (7) and (8). Other useful gauges are the *Coulomb* and *London* gauges. See for example, A. M. Portis, *Electromagnetic Fields: Sources and Media*, Wiley, New York, 1978.

expressions of statics, for, if time derivatives are allowed to go to zero, the set of
equations (1), (3), (7), and (8) becomes

$$\nabla^2 \Phi = -\frac{\rho}{\varepsilon} \qquad \mathbf{E} = -\nabla \Phi \tag{9}$$

$$\nabla^2 \mathbf{A} = -\mu \mathbf{J} \qquad \mathbf{B} = \nabla \times \mathbf{A} \tag{10}$$

which are recognized as the appropriate expressions from Chapters 1 and 2.

3.20 The Retarded Potentials as Integrals over Charges and Currents

The potential functions $\mathbf{A}$ and Φ for time-varying fields are defined in terms of the
currents and charges by the differential equations 3.19(7) and (8). General
solutions of the equation give the potentials as integrals over the charges and
currents, as in the static case. The following discussion applies to the very
important case of a region extending to infinity with a linear, isotropic, and
homogeneous medium.

From Chapters 1 and 2 the integrals for the static potentials, which may be
considered the solutions of Eqs. 3.19(9) and 3.19(10), are

$$\Phi = \int_V \frac{\rho \, dV}{4\pi\varepsilon r} \tag{1}$$

$$\mathbf{A} = \mu \int_V \frac{\mathbf{J} \, dV}{4\pi r} \tag{2}$$

A mathematical development to yield the corresponding integral solutions of the
inhomogeneous wave equations 3.19(7) and 3.19(8), is moderately difficult,[9] so
only a plausibility argument is given here. It can be shown that the solutions are

$$\Phi(x, y, z, t) = \int_V \frac{\rho(x', y', z', t - R/v) \, dV'}{4\pi\varepsilon R} \tag{3}$$

$$\mathbf{A}(x, y, z, t) = \mu \int_V \frac{\mathbf{J}(x', y', z', t - R/v) \, dV'}{4\pi R} \tag{4}$$

where

$$v = (\mu\varepsilon)^{-1/2} \tag{5}$$

(for free space, $v = c = 2.9987 \times 10^8$ m/s) and R is the distance between source
point (x', y', z') and field point (x, y, z),

$$R = [(x - x')^2 + (y - y')^2 + (z - z')^2]^{1/2} \tag{6}$$

---

[9] J. D. Jackson, *Classical Electrodynamics*, 2nd ed., Wiley, New York, 1975.

In the above, $t - R/v$ denotes that, for an evaluation of Φ at time t, the value of charge density ρ at time $t - R/v$ should be used. That is, for each element of charge $\rho \, dV$, the equation says that the contribution to potential is of the same form as in statics, (1), except that we must recognize a finite time of propagating the effect from the charge element to the point P at which potential is being computed, distance R away. The effect travels with velocity $v = 1/\sqrt{\mu\varepsilon}$, which, as we have seen, is just the velocity of a simple plane wave through the medium as predicted from the homogeneous wave equation. Thus, in computing the total contribution to potential Φ at a point P at a given instant t, we must use the values of charge density from points distance R away at an earlier time, $t - R/v$, since for a given element it is that effect which just reaches P at time t. A similar interpretation applies to the computation of $\mathbf{A}$ from currents in (4). Because of this "retardation" effect, the potentials Φ and $\mathbf{A}$ are called the *retarded potentials*. Once the phenomenon of wave propagation predicted from Maxwell's equations is known, this is about the simplest revision of the static formulas (1) and (2) that could be expected.

Example 3.20
Field from an ac Current Element

One of the simplest examples illustrating the meaning of this retardation, and one that will be met again in the study of radiating systems, is that of a very short wire carrying an ac current varying sinusoidally in time between two small spheres on which charges accumulate (Fig. 3.20). For a filamentary current in a small wire, the differences in distance from P to various points of a given cross section of the wire are unimportant, so that two parts of the volume integral in (1) may be done by integrating current density over the cross section to yield the total current in the wire. Thus, for any filamentary current,

$$\mathbf{A} = \mu \int \frac{I(t - r/v) \, \mathbf{dl}}{4\pi r} \tag{7}$$

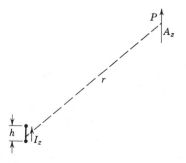

Fig. 3.20 Retarded potential from small current element.

For the particular case of Fig. 3.20, current is in the z direction only; so, by the above, A is also. If h is so small compared with r and wavelength, the remaining integration of (6) is performed by multiplying the current by h:

$$A_z = \frac{\mu h}{4\pi r} I_z\left(t - \frac{r}{v}\right) \tag{8}$$

Finally, if the current in the small element has the form

$$I_z = I_0 \cos \omega t \tag{9}$$

substitution in (8) gives A_z as

$$A_z = \frac{\mu h I_0}{4\pi r} \cos \omega\left(t - \frac{r}{v}\right) \tag{10}$$

From this value of $\mathbf{A}$, the magnetic and electric fields may be derived. This will be done when we return to radiation in Chapter 12.

3.21 The Retarded Potentials for the Time-Periodic Case

If all electromagnetic quantities of interest are varying sinusoidally in time, in the complex notation with $e^{j\omega t}$ understood, the set of Eqs. 3.19(1) and 3.19(3), 3.20(3) and 3.20(4), and 3.19(6) becomes

$$\mathbf{B} = \nabla \times \mathbf{A} \tag{1}$$

$$\mathbf{E} = -\nabla\Phi - j\omega\mathbf{A} \tag{2}$$

$$\Phi(x, y, z) = \int_V \frac{\rho(x', y', z')e^{-jkR}\,dV'}{4\pi\varepsilon R} \tag{3}$$

$$\mathbf{A}(x, y, z) = \mu \int_V \frac{\mathbf{J}(x', y', z')e^{-jkR}\,dV'}{4\pi R} \tag{4}$$

$$\nabla \cdot \mathbf{A} = -j\omega\mu\varepsilon\Phi \tag{5}$$

where $k = \omega/v = \omega\sqrt{\mu\varepsilon}$, and R is the distance between source and field points. Note that the retardation in this case is taken care of by the factor e^{-jkR} and amounts to a shift in phase of each contribution to potential according to the distance R from the contributing element to the point P at which potential is to be computed. (From here on we will frequently leave out the functional notation of coordinates, with the understanding that potentials are computed for the field point, and the integration is over all source points.)

For these steady-state sinusoids the relation between **A** and Φ [Eq. (5)] fixes Φ uniquely once **A** is determined. Thus, it is not necessary to compute the scalar potential Φ separately. Both **E** and **B** may be written in terms of **A** alone:

$$\mathbf{B} = \nabla \times \mathbf{A} \tag{6}$$

$$\mathbf{E} = -\frac{j\omega}{k^2}\nabla(\nabla \cdot \mathbf{A}) - j\omega\mathbf{A} \tag{7}$$

$$\mathbf{A} = \mu \int_V \frac{\mathbf{J}e^{-jkR}\,dV'}{4\pi R} \tag{8}$$

It is then necessary only to specify the current distribution over the system, to compute the vector potential **A** from it by (8), and then find the electric and magnetic fields by (6) and (7). It may appear that the effects of the charges of the system are being left out, but of course the continuity equation

$$\nabla \cdot \mathbf{J} = -j\omega\rho \tag{9}$$

relates the charges to the currents, and in fact, in this steady-state sinusoidal case, fixes ρ uniquely once the distribution of **J** is given. So an equivalent but lengthier procedure would be that of computing the charge distribution from the specified current distribution by means of the continuity equation (9), then using the complete set of equations (1) to (4).

PROBLEMS

3.2a A magnetic field of the approximate form $\mathbf{B} = \hat{z}C_0 x \sin \omega t$ passes through a rectangular loop in the x–y plane following the path $(0, 0)$ to $(a, 0)$ to (a, b) to $(0, b)$ to $(0, 0)$. Show that if Faraday's law in microscopic form is used to give **E**, the macroscopic form of Faraday's law is satisfied.

3.2b Find the emf around a circular loop in the $z = 0$ plane if the loop is threaded by an axial magnetic field varying with r, ϕ, and t approximately as

$$B_z = C_0 r \sin \phi \sin \omega t$$

Now find the emf for a circuit consisting of a half-circle of radius a and a straight wire from $r = a$, $\phi = 0$ to $r = a$, $\phi = \pi$.

3.2c* The betatron makes use of the electric field produced by a time-varying magnetic field in space to accelerate charged particles. Suppose that the magnetic field of a betatron has an axial component in circular cylindrical coordinates of the following form:

$$B_z(r, t) = Ctr^n \qquad (t \geq 0)$$

---

* The asterisk on problems denotes ones longer or harder than the average; a double asterisk will denote unusually difficult or lengthy problems.

Find the induced electric field in magnitude and direction at a particular radius a. From this find velocity v on a charge q after a time t, assuming that the charge stays in a path of constant radius a, and calculate the magnetic force on the charge. For what value(s) of n will this magnetic force just balance the centrifugal force (mv^2/a), where m is mass of the particle, so that it can remain in the path of constant radius as assumed?

3.2d Current I in a long, round wire varies sinusoidally with time. Assume no variation with z (coordinate along the wire) or ϕ (angular coordinate around the wire). Find magnetic field outside the wire, assuming it related to current flow at any instant as in the statics form. (This is called the "quasistatic" approximation.) From the differential equation form of Faraday's law find the electric field in the space outside the wire generated by the changing magnetic field of the form found. Use phasor forms and take the electric field at the surface of the wire ($r = a$) as the product of current and internal impedance Z_i per unit length.

3.3a To demonstrate Faraday's law in class, we often move a coil by hand through the poles of a permanent magnet and observe the generated voltage on an oscilloscope. One magnet used has a flux density of 0.1 T and pole pieces about 2 cm in diameter. Estimate the velocity you can conveniently obtain by hand motion and find how many turns you need to produce peak voltages around 10 mV. Sketch the waveform expected as the coil is moved through the region between poles.

3.3b In the generator of Fig. 3.3a, the poles are reshaped so that magnetic flux density is inhomogeneous. Take the magnetic field direction as the z direction and the vertical direction of the figure as the x direction. Assume the inhomogeneous field to have a quadratic variation with x,

$$B_z = B_m\left(1 - \frac{x^2}{a^2}\right)$$

Find emf generated in the rotating loop by considering rate of change of flux, and also by use of the motional electric field in the wires. Plot the waveform of this wave in time.

3.3c In Prob. 3.3b, it may seem surprising that motional field depends only on the value of B at the instantaneous position of the conductors, whereas flux enclosed depends upon integration of field throughout the region of inhomogeneous variation, yet both give identical answers. Explain why this is so for any arbitrary variation with x.

3.3d In the generator of Fig. 3.3a, the rectangular loop is replaced by a circular loop of radius a, rotated about a line in the plane of the loop and passing through the center. This axis is normal to $\mathbf{B}$ as in Fig. 3.3a. Take B_0 as uniform and find emf generated in this loop as it is rotated with angular velocity Ω about the defined axis.

3.3e The rectangular loop of Ex. 3.3 is moved with constant velocity v in the x direction through an inhomogeneous magnetic field which varies sinusoidally with x,

$$B_z = C\sin\left(\frac{\pi x}{L}\right)$$

Find the induced emf, both from rate of change of flux and by use of the motional electric field. Find values for the special cases $a/L = \frac{1}{2}, 1, 2$.

3.4 Conduction current density for copper with a field of 0.1 V/m (1 mV/cm) applied is 5.8×10^6 A/m^2. What number density of electrons is required to produce the same value of *convection* current density if the electrons have been accelerated in vacuum through a potential of 1 kV? What electric field magnitude would be required to produce the same magnitude of *displacement* current density in space for sinusoidally varying waves as follows: (1) a power wave of frequency 60 Hz; (2) a microwave beam of frequency 3 GHz; (3) a laser beam of wavelength 1.06 μm?

3.5a For a coaxial cylindrical capacitor of radii a and b and length l, evaluate the total displacement current flowing across any cylindrical surface of radius r ($a < r < b$), taking the voltage variation as sinusoidal in time, and the variation of electric field with radius the same as in statics. Show that the result is independent of radius and equal to the charging current for the capacitor.

3.5b Repeat Prob. 3.5a for the spherical surface of radius r lying between the concentric conductors of radii a and b of a spherical capacitor.

3.5c Starting from Eq. 3.4(7), prove that for a closed surface

$$\oint_S (\mathbf{J}_c + \mathbf{J}_d) \cdot d\mathbf{S} = 0$$

From this, show that the sum of convection and displacement currents is the same for both of the surfaces S_1 and S_2 in Fig. 3.5b.

3.5d Obtain the expressions for electric field Eqs. 3.5(4), 3.5(5), and 3.5(6) from the divergence equation.

3.5e* In a large class of electron devices, of which the klystron is a good example, current in the output circuit is induced because of the time-varying conduction current crossing the output gap. The ac current is superposed on the dc beam and moves across the gap approximately at the dc velocity v_0 of the electrons so that convection current density in phasor form may be written

$$J_c(x) = J_0 + J_1 e^{-j\omega x/v_0}$$

Suppose the output gap may be represented by parallel-plane electrodes as in Ex. 3.5. The results of that example may then be used for the induced current for each elemental slice of length dx, and total induced current for the gap may be obtained by integrating contributions over the total length d of the gap. Carry out the integration to find the induced current in the output gap and notice how it depends upon transit angle, $\omega d/v_0$.

3.6a Check the dimensional consistency of equations (1) through (9) of Sec. 3.6.

3.6b Show that, in a charge-free, current-free dielectric, the two divergence equations, 3.6(1) and 3.6(2), may be derived from the two curl equations, (3) and (4), so far as time-varying parts of the field are concerned.

3.6c Show that, if the equation for continuity of charge is assumed, the two divergence equations, 3.6(1) and 3.6(2), may be derived from the curl equations, (3) and (4), so far as ac components of the field are concerned, for regions with finite ρ and $\mathbf{J}$. This fact has made it quite common to refer to the two curl equations alone as Maxwell's equations.

3.6d Check to see which if any of following could be a field consistent with Maxwell's equations.

i) $\mathbf{B} = \hat{z}xt$ (rectangular coordinates)

ii) $\mathbf{E} = \hat{r}C/r$ (circular cylindrical coordinates)

iii) $\mathbf{E} = \hat{r}(C/r)\cos(\omega t - \omega\sqrt{\mu\varepsilon}z)$ (circular cylindrical coordinates)

3.7a A conducting spherical balloon is charged with a constant charge Q, and its radius made to vary sinusoidally in time in some manner from a minimum value, r_{min}, to a maximum value, r_{max}. It might be supposed that this would produce a spherically symmetric, radially outward propagating electromagnetic wave. Show that this does not happen by finding the electric field at some radius $r > r_{max}$.

3.7b A capacitor formed by two circular parallel plates has an essentially uniform axial electric field produced by a voltage $V_0 \sin \omega t$ across the plates. Utilize the symmetry to find the magnetic field at radius r between the plates. Show that the axial electric field could not be exactly uniform under this time-varying condition.

3.7c Suppose that there were free magnetic charges of density ρ_m, and that a continuity relation similar to Eq. 3.4(5) applied to such charges. Find the magnetic current term that would have to be added to Maxwell's equations in such a case. Give the units of ρ_m and the magnetic current density.

3.8a Under what conditions can a complex vector quantity $\mathbf{E}$ be represented by a vector magnitude and phase angle,

$$\mathbf{E} = \mathbf{E}_0 e^{j\theta_E}$$

where $\mathbf{E}_0$ is a real vector and θ_E a real scalar?

3.8b** Consider a case in which the complex field vectors can be represented by single values of magnitude and phase,

$$\mathbf{E} = \mathbf{E}_0(x, y, z)e^{j\theta_1(x,y,z)}$$

$$\mathbf{H} = \mathbf{H}_0(x, y, z)e^{j\theta_2(x,y,z)}$$

$$\mathbf{J} = \mathbf{J}_0(x, y, z)e^{j\theta_3(x,y,z)}$$

$$\rho = \rho_0(x, y, z)e^{j\theta_4(x,y,z)}$$

Substitute in Maxwell's equations in the complex form, and separate real and imaginary parts to obtain the set of differential equations relating $\mathbf{E}_0, \mathbf{H}_0, \ldots, \theta_4$. Check the result by using the corresponding instantaneous expressions,

$$\mathbf{E}_{inst.} = \text{Re}[\mathbf{E}_0 e^{j\theta_1}e^{j\omega t}] = \mathbf{E}_0(x, y, z) \cos[\omega t + \theta_1(x, y, z)], \text{ etc.}$$

substituting in Maxwell's equations for general time variations, eliminating the time variations, and again getting the set of equations relating $\mathbf{E}_0, \ldots, \theta_4$.

3.8c* Check to see which, if any, of the following could be phasor representations of fields consistent with Maxwell's equations:

(i) $\mathbf{E} = \hat{x}Ce^{-j\omega\sqrt{\mu\varepsilon}z}$ (rectangular coordinates)

(ii) $\mathbf{H} = \hat{\phi}(C/r)e^{-j\omega\sqrt{\mu\varepsilon}z}$ (circular cylindrical coordinates)

(iii) $\mathbf{E} = \hat{\theta}(C/r)e^{-j\omega\sqrt{\mu\varepsilon}r\cos\theta}$ (spherical coordinates)

3.9a Plot the sinusoidal solution 3.9(16) versus $\omega z/v$ for various times, $\omega t = 0, \pi/4, \pi/2, 3\pi/4, \pi$, and 2π and interpret as a traveling wave.

3.9b A uniform plane wave is excited by a waveshape E_x rectangular in time. That is, $E_x = C$ for $mT < t < (m + \frac{1}{2})T$, $m = 0, 1, 2, 3$, and zero otherwise. Plot E_x versus distance z for $t = T/4, 3T/4, 5T/4, 7T/4$.

3.9c* A uniform plane wave has electric field at $z = 0$ given as $E_x(0, t) = \cos \omega t + \frac{1}{2}\cos 3\omega t$. Plot E_x versus distance for a few periods in an ideal dielectric with no dispersion. Repeat for a dielectric in which wave velocity at frequency 3ω is $\frac{1}{3}$ that at frequency ω.

3.10a In Sec. 3.10 a wave with E_x and H_y is analyzed. The other set of fields (or the other *polarization* as it will be called in Chapter 6) relates E_y and H_x. With the same assumptions as to uniformity in the $x-y$ planes, find the wave solutions for this set in phasor form.

3.10b A radio wave that may be considered a uniform plane wave propagates through the ionosphere and interacts with some charged particles which are moving with a velocity a tenth the velocity of light in the direction normal to the magnetic field of the wave. Find the ratio of the magnetic force of the wave on the charges to the electric force.

3.10c Show that Faraday's law is satisfied in the plane wave with sinusoidal variations by taking a line integral of electric field from $z = 0$, $x = 0$ to $z = 0$, $x = a$ to $z = d$, $x = a$ to $z = d$, $x = 0$ back to $(0, 0)$ and relating it to the magnetic flux through that path.

3.10d Show that the generalized Ampère's law is satisfied for the plane wave with sinusoidal variations by taking a line integral of magnetic field from $z = 0$, $y = 0$ to $z = 0$, $y = b$ to $z = d$, $y = b$ to $z = d$, $y = 0$ back to $(0, 0)$ and relating it to displacement current through that path.

3.11a What are the relations among the constants required for each of the following to be a solution of the three-dimensional Helmholtz equation?

i) $E_x = C \sin k_x x \sin k_y y \sin k_z z$

ii) $E_x = C \sinh K_x x \sin k_y y \sin k_z z$

iii) $E_x = C \sinh K_x x \sinh K_y y \sinh K_z z$

3.11b Show that the wave equation may be written directly in terms of any of the components of H or E in rectangular coordinates, or for the axial components of H or E in any coordinate system, but not for other components, such as radial and tangential components in cylindrical coordinates, or any component in spherical coordinates. That is,

$$\nabla^2 E_x = \mu\varepsilon \frac{\partial^2 E_x}{\partial t^2} \qquad \nabla^2 H_z = \mu\varepsilon \frac{\partial^2 H_z}{\partial t^2}, \text{ etc}$$

but

$$\nabla^2 E_r \neq \mu\varepsilon \frac{\partial^2 E_r}{\partial t^2} \qquad \nabla^2 H_\phi \neq \mu\varepsilon \frac{\partial^2 H_\phi}{\partial t^2}, \text{ etc.}$$

3.11c Check to see under what conditions the following is a solution of the Helmholtz equation in circular cylindrical coordinates:

$$E = \hat{r}\left(\frac{C}{r}\right)e^{-jk_z z}$$

Note that the vector form of ∇^2 must be used; see Prob. 3.11b.)

3.12a Describe the Poynting vector and discuss its interpretation for the case of a static point charge Q located at the center of a small loop of wire carrying direct current I.

3.12b Assuming current density constant over the conductor cross section in the Ex. 3.12a, find the Poynting vector within the wire and interpret this in terms of the distribution of dissipation.

3.12c Interpret the Poynting vector about a parallel-plate capacitor charged from zero to some final charge Q. Repeat for an inductor in which current builds up from zero to some final value. Repeat for each of these cases as charge and current is made to decay from a given value to zero.

3.12d Show that the power flow in the uniform plane wave of Ex. 3.12d equals the product of the average energy density and the velocity v of the wave.

3.12e In each of the following, use a plane wave model to estimate the quantities for various laser systems:

(i)   A small helium–neon laser ($\lambda = 633$ nm) typically produces 1 mW in a beam 1 mm in diameter. Estimate strengths of electric and magnetic fields in the laser beam.

(ii) It is fairly easy to focus the power for a medium-power CO_2 laser ($\lambda = 10.6\,\mu m$) so that there is breakdown in air. Taking breakdown strength as at lower frequencies about 3×10^6 V/m, estimate the power density in such a laser beam.

(iii) The highest power lasers are now built for laser fusion experiments. One of the very high power ones (Nd–glass with $\lambda = 1.06\,\mu m$) has produced 10.2 kJ in 0.9 ns and is designed for focusing on targets around 0.5 mm in diameter. Estimate electric field strength in the beam at the target.

3.13a Find the imaginary part of Eq. 3.13(11) and simplify by letting $C_1 = A_1 e^{j\phi_1}$ and $C_2 = A_2 e^{j\phi_2}$ where A_1, A_2, ϕ_1, and ϕ_2 are real. Explain the variation with z in view of Eq. 3.13(8).

3.13b The field a large distance from a dipole radiator has the form, in spherical coordinates,

$$E_\sigma = \sqrt{\frac{\mu}{\varepsilon}}\, H_\phi = \left(\frac{A}{r}\right) e^{-jkr} \sin\theta$$

Find the average power radiated through a large sphere of radius r.

3.14 Space is filled by two dielectrics, ε_1 filling the half-space $x > 0$ and ε_2 filling the half-space $x < 0$. Determine whether or not there can exist a uniform plane wave with E_x and H_y only and no variations with x or y, propagating in the z direction in this composite dielectric. The propagation factor may be e^{-jkz} with any value of k. Note that the wave, if it exists, must satisfy the wave equation in each region and the continuity conditions at the plane between the two regions.

3.16 Find the variation of Poynting vector for a plane wave within a good conductor and interpret.

3.17a Show that R_s as defined by $R_s = \sqrt{\pi f \mu / \sigma}$ does have the dimensions of ohms.

3.17b Iron and tin have the same order of conductivity σ, around 10^7 S/m. For slab conductors of each of these used at 60 Hz, at 1 kHz, and at 1 MHz, find the surface resistance of the two materials if relative permeability of the iron is 500.

3.17c Find the magnetic field **H** for any point x in the plane conductor in terms of J_0 by first finding the electric field, and then utilizing the appropriate one of Maxwell's equations to give **H**. Show that J_{sz} of Eq. 3.17(1) is equal to $-H_y$ at the surface.

3.18a The average power loss per unit volume at any point in the conductor is $|J_z|^2/2\sigma$. Show that Eq. 3.18(5) may be obtained by integrating over the conductor depth to obtain the total power loss per unit area.

3.18b A uniform plane wave of frequency 1 GHz has a power density of 1 MW/m^2 and falls upon an aluminum sheet. It can be shown that upon reflection from a good conductor, magnetic field at the surface of the conductor is essentially double that in the incident wave. Estimate the power absorbed in the aluminum per unit area and note it as a fraction of the incident power.

3.19a Show that **E** and **H** satisfy the following differential equations in a homogeneous medium containing charges and currents:

$$\nabla^2 \mathbf{E} - \mu\varepsilon \frac{\partial^2 \mathbf{E}}{\partial t^2} = \frac{1}{\varepsilon}\nabla\rho + \mu\frac{\partial \mathbf{J}}{\partial t}$$

$$\nabla^2 \mathbf{H} - \mu\varepsilon \frac{\partial^2 \mathbf{H}}{\partial t^2} = -\nabla \times \mathbf{J}$$

3.19b A potential function commonly used in electromagnetic theory is the Hertz vector potential Π, so defined that electric and magnetic fields are derived from it as follows, for a homogeneous medium:

$$\mathbf{H} = \varepsilon\frac{\partial}{\partial t}\,\nabla \times (\Pi)$$

$$\mathbf{E} = \nabla(\nabla \cdot \Pi) - \mu\varepsilon\frac{\partial^2 \Pi}{\partial t^2}$$

where

$$\nabla^2 \Pi - \mu\varepsilon\frac{\partial^2 \Pi}{\partial t^2} = -\frac{\mathbf{P}}{\varepsilon}$$

and **P**, the *polarization vector* associated with sources, is so defined that

$$\mathbf{J} = \frac{\partial \mathbf{P}}{\partial t}, \qquad \rho = -\nabla \cdot \mathbf{P}$$

Show that **E** and **H** derived in this manner are consistent with Maxwell's equations.

3.19c An alternative to the *Lorentz gauge*, which defines $\nabla \cdot \mathbf{A}$ by Eq. 3.19(6), is the *Coulomb gauge* which selects it to be $\nabla \cdot \mathbf{A} = 0$. Give the differential equations relating Φ and **A** to sources ρ and **J** in this case.

3.19d As noted in the text, Eqs. 3.19(1) and 3.19(3) apply to general media. Find the differential equations corresponding to 3.19(7) and 3.19(8) (using Lorentz gauge) if the medium is inhomogeneous, with μ and ε functions of position.

3.20a By analogy with the integral solutions for **A** and Φ, write the integral for the Hertz vector Π in terms of the polarization **P**. (See Prob. 3.19b.)

3.20b* From continuity of charge, find the values of the charges that must exist at the ends of the small current element of Ex. 3.20. Find scalar potential Φ from these charges, using Eq. 3.20(3). Show that Φ and the **A** of Eq. 3.20(8) are related by the Lorentz condition 3.19(6).

3.20c* Utilizing the result of Prob. 3.20b, find electric and magnetic fields in spherical coordinates for the small current element of Ex. 3.20, with sinusoidal current variation given by Eq. 3.20(9).

3.21 Find the relation between the Hertz potential Π of Prob. 3.19b and the vector potential **A** when time variations are taken of the form $e^{j\omega t}$.

4

THE ELECTROMAGNETICS
OF CIRCUITS

4.1 Introduction

Much of the engineering design and analysis of electromagnetic interactions is done through the mechanism of lumped-element circuits. In these, the energy-storage elements (inductors and capacitors) and the dissipative elements (resistors) are connected to each other and to sources or active elements within the circuit by conducting paths of negligible impedance. There may be mutual couplings, either electrical or magnetic, but in the ideal circuit these couplings are planned and optimized. The advantage of this approach is that functions are well separated and cause and effect relationships readily understandable. Powerful methods of synthesis, analysis, and computer optimization of such circuits have consequently been developed.

Most of the individual elements in an electrical circuit are small compared with wavelength so that fields of the elements are *quasistatic*; that is, although varying with time, the electric or magnetic fields have the spatial forms of static field distributions. There are important distributed effects in many real circuits, but often they can be represented by a few properly chosen lumped coupling elements. But in some circuits, of which the transmission lines are primary examples, the distributed effects are the major ones and must be considered from the beginning. In some cases in which the lumped idealizations described above do not strictly apply, lumped-element *models* can nevertheless be deduced and are useful for analysis because of the powerful circuit methods that have been developed.

We have introduced the lumped-circuit concepts, inductance and capacitance, in our studies of static fields. We have also seen how the skin effect phenomenon in conductors changes both resistance and inductance at high frequencies. We now wish to examine circuits and circuit elements more carefully from the point of view of electromagnetics. It is easy to see the idealizations required to derive Kirchhoff's laws from Maxwell's equations. It is also possible to make certain extensions of the concepts when the simplest idealizations do not apply. In particular, introduction of the retardation concepts show that circuits may radiate energy when comparable in size with wavelength. The amount of radiated power may be estimated from these extended circuit ideas for some configurations. But for certain classes of circuits it becomes impossible to make the extensions without a true field analysis. We shall look at both types of circuits in this chapter.

THE IDEALIZATIONS IN CLASSICAL CIRCUIT THEORY

4.2 Kirchhoff's Voltage Law

Kirchhoff's two laws provide the basis for classical circuit theory. We begin with the voltage law as a way of reviewing the basic element values of lumped-circuit theory. The law states that for any closed loop of a circuit, the algebraic sum of the voltages for the individual branches of the loop is zero:

$$\sum_i V_i = 0 \tag{1}$$

The basis for this law is Faraday's law for a closed path, written as

$$-\oint \mathbf{E} \cdot \mathbf{dl} = \frac{\partial}{\partial t} \int_s \mathbf{B} \cdot \mathbf{dS} \tag{2}$$

and the definition of voltages between two reference points of the loop,

$$V_{ba} = -\int_a^b \mathbf{E} \cdot \mathbf{dl} \tag{3}$$

To illustrate the relation between the circuit expression (1) and the field expressions (2) and (3), consider first a single loop with applied voltage $V_0(t)$ and passive resistance, inductance, and capacitance elements in series, Fig. 4.2a. A convention for positive voltage at the source is selected as shown by the $+$ $-$ signs on the voltage generator, which means by (3) that field of the source is directed from b to a when V_0 is positive. A convention for positive current is also chosen, as shown by the arrow on $I(t)$. The interpretation of (1) by circuit theory for this basic circuit is then known to be

$$V_0(t) - RI(t) - L\frac{dI(t)}{dt} - \frac{1}{C}\int I(t)\,dt = 0 \tag{4}$$

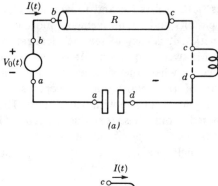

Fig. 4.2 (a) Series circuit with resistor, inductor and capacitor. (b) Detail of inductor.

In order to compare, we break the closed line integral of (2) into its contributions over the several elements,

$$-\int_a^b \mathbf{E} \cdot \mathbf{dl} - \int_b^c \mathbf{E} \cdot \mathbf{dl} - \int_c^d \mathbf{E} \cdot \mathbf{dl} - \int_d^a \mathbf{E} \cdot \mathbf{dl} = \frac{\partial}{\partial t} \int_S \mathbf{B} \cdot \mathbf{dS} \qquad (5)$$

or

$$V_0(t) + V_{cb} + V_{dc} + V_{ad} = \frac{\partial}{\partial t} \int_S \mathbf{B} \cdot \mathbf{dS} \qquad (6)$$

The right side of (6) is not zero as is the right side of (4), but we recognize it as the contribution to emf generated by any rate of change of magnetic flux within the path defined as the circuit. If not entirely negligible, it can be considered as arising from an inductance of the loop which can be added to the lumped element L, or a mutually induced coupling if the flux is from an external source. Thus we will from here on consider it as negligible, or included in L so that the right side of (6) is zero. (Mutual effects are added later.) We now examine separately the three voltage terms related to the passive components R, L, and C.

Resistance Element   The field expression to be applied to the resistive material is the differential form of Ohm's law,

$$\mathbf{J} = \sigma \mathbf{E} \qquad (7)$$

so that the voltage V_{cb} is

$$V_{cb} = -\int_b^c \mathbf{E} \cdot \mathbf{dl} = -\int_b^c \frac{\mathbf{J}}{\sigma} \cdot \mathbf{dl} \qquad (8)$$

where the path is taken along some current flow path of the conductor. Conductivity σ may vary along this path. At dc or low frequencies, current I is uniformly distributed over the cross section A of the conductor, which can also vary with position. Thus

$$V_{cb} = -\int_b^c \frac{I \, dl}{\sigma A} = -IR \tag{9}$$

where

$$R = \int_b^c \frac{dl}{\sigma A} \tag{10}$$

This last is the usual dc or low-frequency resistance. The situation is more complicated at higher frequencies because of the effect of the changing magnetic fields on currents within the conductor. Current distribution over the cross section is then nonuniform, and the particular path along the conductor must be specified. In the plane skin effect analysis of Chapter 3, current was related to electric field at the surface in order to define a surface impedance. We shall return to this concept later in the chapter for conductors of circular cross section.

Inductance Element   The voltage across the terminals of the inductive element comes from the time rate of change of magnetic flux within the inductor, shown in the figure as a coil. Assuming first that resistance of the conductor of the coil is negligible, let us take a closed line integral of electric field along the conductor of the coil, returning by the path across the terminals, Fig. 4.2b. Since the contribution along the part of the path which follows the conductor is zero, all the voltage appears across the terminals:

$$-\oint \mathbf{E} \cdot \mathbf{dl} = -\int_{c(\text{cond.})}^d \mathbf{E} \cdot \mathbf{dl} - \int_{d(\text{term.})}^c \mathbf{E} \cdot \mathbf{dl} = -\int_{d(\text{term.})}^c \mathbf{E} \cdot \mathbf{dl} \tag{11}$$

By Faraday's law, this is the time rate of change of magnetic flux enclosed

$$-\int_{d(\text{term.})}^c \mathbf{E} \cdot \mathbf{dl} = -V_{dc} = \frac{\partial}{\partial t} \int_S \mathbf{B} \cdot \mathbf{dS} \tag{12}$$

Inductance L is defined as the magnetic flux linkage per unit of current, Sec. 2.5

$$L = \left[\int \mathbf{B} \cdot \mathbf{dS} \right] \bigg/ I \tag{13}$$

so the voltage contributed by this term, assuming L independent of time, is

$$V_{cd} = \frac{\partial}{\partial t}(LI) = L \frac{dI}{dt} \tag{14}$$

Note that in computing flux enclosed by the path, we add a contribution each time we follow another turn around the flux. Thus for N turns, the contribution to induced voltage is just N times that of one turn, provided the same flux links each turn. This enters into the calculation of L and will be seen specifically when we find inductances of a coil.

If there is finite resistance in the turns of the coil, the second term of (11) is not zero but is the resistance of the coil, R_L, multiplied by current; therefore (11) becomes

$$-\oint \mathbf{E} \cdot \mathbf{dl} = -R_L I - V_{dc} = \frac{\partial}{\partial t} \int_S \mathbf{B} \cdot \mathbf{dS}$$

or

$$V_{cd} = R_L I + L \frac{dI}{dt} \tag{15}$$

Thus, as expected, we simply add another series resistance to take care of finite conductivity in the conductors of the coil.

Capacitive Element   The ideal capacitor is one in which we store only electric energy; magnetic fields are negligible so there is no contribution to voltage from changing magnetic fields but only from the charges on plates of the capacitor. The problem is then quasistatic and voltage is synonymous with potential difference between capacitor plates. So, in contrast to the inductor, we can take any path between the terminals of the capacitor for evaluation of voltage V_{da}, provided it does not stray into regions influenced by magnetic fields from other elements. We also take the definition of capacitance from electrostatics (Sec. 1.9) as the charge on one plate divided by the potential difference,

$$C = \frac{Q}{V} \tag{16}$$

Thus, from continuity,

$$I = \frac{dQ}{dt} = \frac{d}{dt}(CV_{da}) = C\frac{dV_{da}}{dt} \tag{17}$$

The last term in the above implies a capacitance which is not changing with time. Integration of (17) with time leads to

$$V_{da} = \frac{1}{C}\int I\, dt \tag{18}$$

If the dielectric of the capacitor is lossy, there are conduction currents to add to (17), which are represented in the circuit as a conductance $G_C = 1/R_C$ in parallel with C; the value of R_C may be calculated from (10) by using conductivity of the dielectric and area of the capacitor plates.

Induced Voltages from Other Parts of the Circuit   In addition to voltages induced by charges and currents of the circuit path being considered, there may

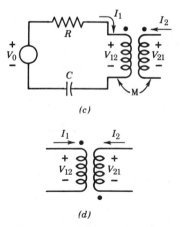

(c)

(d)

Fig. 4.2 (c) Circuit with a mutual inductor. (d) Designation of mutual coupling with negative M.

be induced voltages from other portions of the circuit. In particular, if the magnetic field from one part of the circuit links another part, an induced voltage is produced through Faraday's law when this magnetic field changes with time. This coupling is represented in the circuit by means of a mutual inductor M, as shown in Fig. 4.2c. The value of M is defined as the magnetic flux ψ_{12} linking path 1, divided by the current I_2

$$M = M_{12} = \frac{\psi_{12}}{I_2} \tag{19}$$

The voltage induced in the first path is then

$$V_{12} = \frac{d\psi_{12}}{dt} = M\frac{dI_2}{dt} \tag{20}$$

and the circuit equation (4) is modified to be

$$V_0 - RI_1 - L\frac{dI_1}{dt} - M\frac{dI_2}{dt} - \frac{1}{C}\int I_1\, dt = 0 \tag{21}$$

The mutual inductance M may be either positive or negative depending upon the sense of flux with respect to the defined positive reference for I_2. The sign of M is designated on a circuit diagram by the placing of dots, and with sign conventions for currents and voltages as shown; those on Fig. 4.2c denote positive M; negative M would be designated as in Fig. 4.2d.

Except for certain materials (to be considered in Chapter 13) there is a reciprocal relation showing that the same M gives the voltage induced in circuit 2 by time-varying current in circuit 1,

$$V_{21} = \frac{d\psi_{21}}{dt} = M\frac{dI_1}{dt} \tag{22}$$

All mutual effects to be considered in this chapter have this reciprocal relationship.

In summary, we find that if losses in inductor and capacitor are ignored, the field approach, with understandable approximations, leads to the definitions for the three induced voltage terms for the passive elements used in the circuit approach, Eq. (4). Moreover, the definitions (10), (13), and (16) are the usual quasistatic definitions for these elements. If losses are present, a series resistance is added to L and a shunt conductance to C, again as is commonly done in the circuit approach. Coupling between circuit paths by magnetic flux adds mutual inductance elements. We next examine the Kirchhoff current law and the extension through this to multimesh circuits.

4.3   Kirchhoff's Current Law and Multimesh Circuits

The current law of Kirchhoff states that the algebraic sum of currents flowing out of a junction is zero. Thus, referring to Fig. 4.3a,

$$\sum_{n=1}^{N} I_n(t) = 0 \tag{1}$$

It is evident that the idea behind this law is that of continuity of current, so we refer to the continuity equation implicit in Maxwell's equations, Eq. 3.4(5), or its large-scale equivalent,

$$\oint_S \mathbf{J} \cdot \mathbf{dS} = -\frac{\partial}{\partial t} \int_V \rho \, dV \tag{2}$$

If we apply this to a surface S surrounding the junction, the only conduction current flowing out of the surface is that in the wires, so the left side of (2) becomes just the algebraic sum of the currents flowing out of the wires, as in (1).

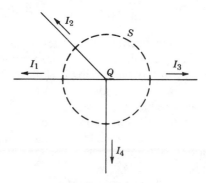

Fig. 4.3a Current flow from a junction.

The right side is the negative time rate of change of charge Q, if any, accumulating at the junction. So (2) may be written

$$\sum_{n=1}^{N} I_n(t) = -\frac{dQ(t)}{dt} \tag{3}$$

A comparison of (1) and (3) shows an apparent difference, but it is only one of interpretation. If Q is nonzero, we know that we take care of this in a circuit problem by adding one or more capacitive branches to yield the capacitive current dQ/dt at the junction. That is, in interpreting (3), the current terms on the left are taken only as convection or conduction currents, whereas in (1) displacement or capacitance currents are included. With this understanding, (1) and (3) are equivalent.

With the two laws, the circuit analysis illustrated in the preceding section can be extended to circuits with several meshes. As a simple example, consider the low-pass filter of Fig. 4.3b or 4.3c. Although currents and voltages are taken as time varying, we drop the functional notation for simplicity. Figure 4.3b illustrates the standard method utilizing mesh currents I_1 and I_2. Note that the net current through C is $(I_1 - I_2)$ which automatically satisfies the current law at node b. The voltage law is then written about each loop as follows:

$$V_0 - R_s I_1 - L_1 \frac{dI_1}{dt} - \frac{1}{C}\int (I_1 - I_2)\, dt = 0 \tag{4}$$

$$-\frac{1}{C}\int (I_2 - I_1)\, dt - L_2 \frac{dI_2}{dt} - R_L I_2 = 0 \tag{5}$$

The two equations are then solved by appropriate means to give I_1 and I_2 for a given V_0.

A second standard method of circuit analysis uses node voltages V_a, V_b, and V_c as shown in Fig. 4.3c. These are defined with respect to some reference, here taken as the lower terminal of the voltage generator, denoted 0. Then Kirchhoff's voltage law is automatically satisfied, for if we add voltages around the first loop we have

$$V_0 + (V_a - V_0) + (V_b - V_a) + (0 - V_b) \equiv 0 \tag{6}$$

Kirchhoff's current law is then applied at each of the three nodes as follows:

Node a: $\dfrac{V_a - V_0}{R_s} + \dfrac{1}{L_1}\displaystyle\int (V_a - V_b)\, dt = 0$ (7)

Node b: $\dfrac{1}{L_1}\displaystyle\int (V_b - V_a)\, dt + \dfrac{1}{L_2}\displaystyle\int (V_b - V_c)\, dt + C\dfrac{dV_b}{dt} = 0$ (8)

Node c: $\dfrac{1}{L_2}\displaystyle\int (V_c - V_b)\, dt + \dfrac{V_c}{R_L} = 0$ (9)

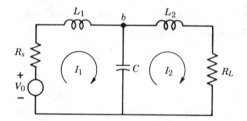

Fig. 4.3b Low-pass filter; loop current analysis.

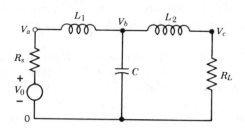

Fig. 4.3c Low-pass filter; node voltage analysis.

Solution of these by appropriate means yields the three node voltages in terms of the given voltage V_0. Note that no equation for the reference node need be written as it is contained in the above.

In the above we seem to be treating voltage as a potential difference when we take voltage of a node with respect to the chosen reference, but note that this is only after the circuit is defined and we are only breaking up $\int \mathbf{E} \cdot \mathbf{dl}$ into its contributions over the various branches. As illustrated in the preceding section, we do have to define the path carefully whenever there are inductances or other elements with contributions to voltage from Faraday's law.

Finally, a word about sources. The voltage generator most often met in lumped-element circuit theory is a highly localized one. For example, the electrons and holes of a semiconductor diode or transistor may induce electric fields between the conducting electrodes fabricated on the device. The entire device is typically small compared with wavelength so that the electric field, although time varying, may be written as the gradient of a time-varying scalar potential. The integral of electric field at any instant thus yields an instantaneous potential difference V_s between the electrodes, which is the source voltage (or $V_s - IZ_s$ if current flows). The induced effects from a modulated electron stream passing across a klystron gap are similar, as are those from many other practical devices. There are interesting field problems in the analysis of induced effects from such devices, but from the point of view of the circuit designer, they are simply point sources representable by the V_s used in the circuits.

A quite different limiting case is that in which the fields driving the circuit are not localized but are distributed. An important example is that of a receiving

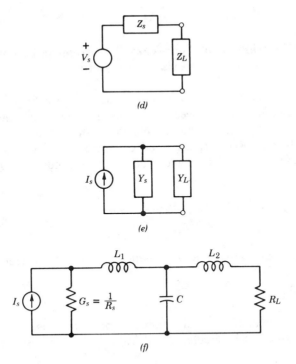

Fig. 4.3 (*d*) Thévenin circuit configuration. (*e*) Norton circuit form. (*f*) Equivalent of circuit in (*c*) using Norton source.

antenna with the fields set down by a distant transmitting antenna. If voltage is taken as the line integral of electric field along the antenna, applied voltage clearly depends upon the circuit configuration and orientation with respect to the applied field. Although quite different from the case with a localized source, it is found that circuit theory is useful here also. A formulation in terms of the retarded potentials will be applied to this case in Sec. 4.11.

Current generators are natural to use as sources in place of voltage generators if emphasis is on the current induced between electrodes of the point source or small-gap device. Similarly for the distributed source, if applied magnetic field at the circuit conductor is given, induced current can be calculated and a current representation is natural. However one has a choice in any case since the Thévenin and Norton theorems[1] show that the two representations of Figs. 4.3*d* and 4.3*e* are equivalent, with the relations

$$Y_s = Z_s^{-1}, \quad I_s = V_s Y_s \tag{10}$$

Thus an equivalent to Fig. 4.3*c* is that of Fig. 4.3*f*, utilizing a current generator.

[1] C. A. Desoer and E. S. Kuh, *Basic Circuit Theory*, McGraw-Hill, New York, 1969.

SKIN EFFECT IN PRACTICAL CONDUCTORS

4.4   Distribution of Time-Varying Currents in Conductors of Circular Cross Section

In order to study the resistive term at frequencies high enough so that current distribution is not uniform, we need to first find the current distribution. This was done in Sec. 3.16 for plane conductors. We now wish to do this for the useful case of round conductors. Recall that a good conductor is defined as one for which displacement current is negligible in comparison with conduction current so that

$$\mathbf{V} \times \mathbf{H} = \mathbf{J} = \sigma \mathbf{E} \tag{1}$$

The Faraday's law equation is (in phasor form)

$$\mathbf{V} \times \mathbf{E} = -j\omega\mu\mathbf{H} \tag{2}$$

From these two we derived the differential equation for current density, Eq. 3.16(7):

$$\nabla^2 \mathbf{J} = j\omega\mu\sigma\mathbf{J} \tag{3}$$

We now take current in the z direction and no variations with z or angle ϕ. Equation (3), expressed in circular cylindrical coordinates (inside back cover), is then

$$\frac{d^2 J_z}{dr^2} + \frac{1}{r}\frac{dJ_z}{dr} + T^2 J_z = 0 \tag{4}$$

where

$$T^2 = -j\omega\mu\sigma$$

or

$$T = j^{-1/2}\sqrt{\omega\mu\sigma} = j^{-1/2}\frac{\sqrt{2}}{\delta} \tag{5}$$

where δ is the useful parameter called "depth of penetration" or "skin depth." The differential equation (4) is a Bessel equation. Equations of this type will be studied in detail in Chapter 7, but for the present we write the two independent solutions as

$$J_z = AJ_0(Tr) + BH_0^{(1)}(Tr) \tag{6}$$

For a solid wire, $r = 0$ is included in the solution, and then it is necessary that $B = 0$ since a study of $H_0^{(1)}(Tr)$ shows that this is infinite at $r = 0$. Therefore,

$$J_z = AJ_0(Tr) \tag{7}$$

The arbitrary constant A may be evaluated in terms of current density at the surface, which is σE_0, with E_0 the surface electric field.

$$J_z = \sigma E_0 \quad \text{at} \quad r_0$$

Then (7) becomes

$$J_z = \frac{\sigma E_0}{J_0(Tr_0)} J_0(Tr) \tag{8}$$

A study of the series definitions of the Bessel functions with complex argument shows that J_0 is complex. It is convenient to break the complex Bessel function into real and imaginary parts, using the definitions

$$\text{Ber}(v) \equiv \text{real part of } J_0(j^{-1/2}v)$$

$$\text{Bei}(v) \equiv \text{imaginary part of } J_0(j^{-1/2}v)$$

That is,

$$J_0(j^{-1/2}v) \equiv \text{Ber}(v) + j\,\text{Bei}(v) \tag{9}$$

$\text{Ber}(v)$ and $\text{Bei}(v)$ are tabulated in many references.[2] Using these definitions and (5), (8) may be written

$$J_z = \sigma E_0 \frac{\text{Ber}(\sqrt{2}r/\delta) + j\,\text{Bei}(\sqrt{2}r/\delta)}{\text{Ber}(\sqrt{2}r_0/\delta) + j\,\text{Bei}(\sqrt{2}r_0/\delta)} \tag{10}$$

In Fig. 4.4a the magnitude of the ratio of current density to that at the outside of the wire is plotted as a function of the ratio of radius to outer radius of wire, for different values of the parameter (r_0/δ). Also, for purposes of the physical picture, these are interpreted in terms of current distribution for a 1-mm diameter copper wire at different frequencies by the figures in parentheses.

As an example of the applicability of the plane analysis for curved conductors at high frequencies where δ is small compared with radii, we can take the present case of the round wire. If we are to neglect the curvature and apply the plane analysis, the coordinate x, distance below the surface, is $(r_0 - r)$ for a round wire. Then Eq. 3.16(16) gives

$$\left|\frac{J_z}{\sigma E_0}\right| \approx e^{-(r_0-r)/\delta} \tag{11}$$

In Fig. 4.4b are plotted curves of $|J_z/\sigma E_0|$ by using this formula, and comparisons are made with curves obtained from the exact formula (10). This is done for two cases, $r_0/\delta = 2.39$ and $r_0/\delta = 7.55$. In the latter, the approximate distribution

[2] H. B. Dwight, *Tables of Integrals*, 3rd ed., MacMillan, New York, 1961; N. W. McLachlan, *Bessel Functions for Engineers*, 2nd ed., Oxford Clarendon Press, New York, 1955; M. R. Spiegel, *Mathematical Handbook of Formulas and Tables*, Schaum's Outline Series, McGraw-Hill, New York, 1968.

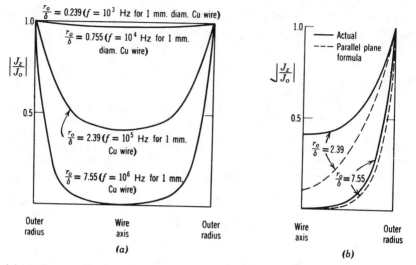

Fig. 4.4 (*a*) Current distribution in cylindrical wire for several frequencies. (*b*) Actual and approximate (parallel-plane formula) distribution in cylindrical wire. $J_o = \sigma E_o$.

agrees well with the exact; in the former it does not. Thus, if ratio of wire radius to δ is large, it seems that there should be little error in analyzing the wire from the results developed for plane solids. This point will be pursued in impedance calculations to follow.

4.5   Impedance of Round Wires

The internal impedance (resistance and contribution to reactance from magnetic flux inside the wire) of the round wire is found from total current in the wire and the electric intensity at the surface, according to the ideas of Sec. 4.2. Total current may be obtained from an integration of current density, as for the plane conductor in Sec. 3.17; however, it may also be found from the magnetic field at the surface, since the line integral of magnetic field around the outside of the wire must be equal to the total current in the wire:

$$\oint \mathbf{H} \cdot \mathbf{dl} = I$$

or

$$2\pi r_0 H_\phi|_{r=r_0} = I \tag{1}$$

Magnetic field is obtained from the electric field by Maxwell's equations:

$$\nabla \times \mathbf{E} = -j\omega\mu\mathbf{H} \tag{2}$$

For the round wire with no variations in z or ϕ, the fields E_z and H_ϕ alone are present, and only r derivatives remain, so (2) is simply

$$H_\phi = \frac{1}{j\omega\mu}\frac{dE_z}{dr} \tag{3}$$

An expression for current density has already been obtained in Eq. 4.4(8). Electric field is related to this through the conductivity σ:

$$E_z = \frac{J_z}{\sigma} = E_0\frac{J_0(Tr)}{J_0(Tr_0)} \tag{4}$$

By substituting in (3) and recalling that $T^2 = -j\omega\mu\sigma$,

$$H_\phi = \frac{E_0 T}{j\omega\mu}\frac{J_0'(Tr)}{J_0(Tr_0)} = -\frac{\sigma E_0}{T}\frac{J_0'(Tr)}{J_0(Tr_0)}$$

where $J_0'(Tr)$ denotes $[d/d(Tr)]J_0(Tr)$. From (1),

$$I = -\frac{2\pi r_0\sigma E_0}{T}\frac{J_0'(Tr_0)}{J_0(Tr_0)} \tag{5}$$

The internal impedance per unit length is

$$Z_i = \frac{E_z|_{r=r_0}}{I} = -\frac{TJ_0(Tr_0)}{2\pi r_0\sigma J_0'(Tr_0)} \tag{6}$$

Low-Frequency Expressions   For low frequencies, Tr_0 is small and series expansions of the Bessel functions show that (6) may be expanded as

$$Z_i \approx \frac{1}{\pi r_0^2\sigma}\left[1 + \frac{1}{48}\left(\frac{r_0}{\delta}\right)^2\right] + j\frac{\omega\mu}{8\pi} \tag{7}$$

The real or resistive part is

$$R_{lf} \approx \frac{1}{\pi r_0^2\sigma}\left[1 + \frac{1}{48}\left(\frac{r_0}{\delta}\right)^2\right] \tag{8}$$

The first term of this expression is the dc resistance, and the second a correction useful for r_0/δ as large as unity—that is, for radius equal to skin depth δ. The imaginary term of (7) may be interpreted as arising from an internal inductance term at low frequencies, so that this internal inductance is

$$(L_i)_{lf} \approx \frac{\mu}{8\pi} \quad \text{H/m} \tag{9}$$

The low-frequency internal inductance is the same as that found by energy methods in Sec. 2.17.

High-Frequency Expressions   For high frequencies, the complex argument Tr_0 is large. It may be shown that $J_0(Tr_0)/J_0'(Tr_0)$ approaches $-j$ and the high-frequency approximation to (6) is

$$(Z_i)_{hf} = \frac{j(j)^{-1/2}}{\sqrt{2\pi r_0 \sigma \delta}} = \frac{(1+j)R_s}{2\pi r_0} \tag{10}$$

or

$$(R)_{hf} = (\omega L_i)_{hf} = \frac{R_s}{2\pi r_0} \quad \Omega/m \tag{11}$$

So resistance and internal reactance are equal at high frequencies, and both are equal to the values for a plane solid of width $2\pi r_0$ just as assumed on physical grounds in Sec. 3.17 where $R_s = (\sigma \delta)^{-1}$.

Expression for Arbitrary Frequency   To interpret (6) for arbitrary frequencies, it is useful to break into real and imaginary parts using the Ber and Bei functions, defined in Eq. 4.4(9), and their derivatives. That is,

$$\text{Ber } v + j \text{ Bei } v = J_0(j^{-1/2}v)$$

Also let

$$\text{Ber}' v + j \text{ Bei}' v = \frac{d}{dv}(\text{Ber } v + j \text{ Bei } v)$$

$$= j^{-1/2} J_0'(j^{-1/2}v)$$

Then (6) may be written

$$Z_i = R + j\omega L_i = \frac{jR_s}{\sqrt{2}\pi r_0}\left[\frac{\text{Ber } q + j \text{ Bei } q}{\text{Ber}' q + j \text{ Bei}' q}\right]$$

where

$$R_s = \frac{1}{\sigma \delta} = \sqrt{\frac{\pi f \mu}{\sigma}} \qquad q = \frac{\sqrt{2}r_0}{\delta}$$

or

$$R = \frac{R_s}{\sqrt{2}\pi r_0}\left[\frac{\text{Ber } q \text{ Bei}' q - \text{Bei } q \text{ Ber}' q}{(\text{Ber}' q)^2 + (\text{Bei}' q)^2}\right] \quad \Omega/m$$

$$\omega L_i = \frac{R_s}{\sqrt{2}\pi r_0}\left[\frac{\text{Ber } q \text{ Ber}' q + \text{Bei } q \text{ Bei}' q}{(\text{Ber}' q)^2 + (\text{Bei}' q)^2}\right] \quad \Omega/m \tag{12}$$

These are the expressions for resistance and internal reactance of a round wire at any frequency in terms of the parameter q, which is $\sqrt{2}$ times the ratio of wire radius to depth of penetration. Curves giving the ratios of these quantities to the dc and to the high-frequency values as functions of r_0/δ are plotted in Figs. 4.5a and 4.5b. A careful study of these will reveal the ranges of r_0/δ over which it is permissible to use the approximate formulas for resistance and reactance.

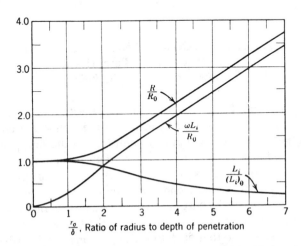

Fig. 4.5a Solid-wire skin effect quantities compared with dc values.

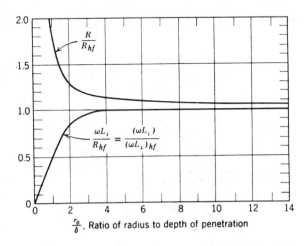

Fig. 4.5b Solid-wire skin effect quantities compared with values from high-frequency formulas.

CALCULATION OF CIRCUIT ELEMENTS

4.6 Self-Inductance Calculations

Self-inductance, as defined in Chapter 2, was related to field concepts in the first part of this chapter. We have shown examples of inductance calculations for simple configurations by the method of flux linkages (Sec. 2.5) and from an energy point of view (Sec. 2.17). We now give additional examples of each method.

Example 4.6a
External Inductance of Parallel-Wire
Transmission Line (Approximate)

Figure 4.6 shows two parallel conductors of radius R with their axes separated by distance $2d$. Current I flows in the z direction in the right-hand conductor and returns in the other. Magnetic field at any point (x, y) is the superposition of that from the two conductors. If conductors are far enough apart, the current distribution in either conductor is not much affected by the presence of the other so that magnetic field from each conductor may be taken as circumferential about its axis, and equal to the current divided by 2π times radius from the axis. For the $y = 0$ plane passing through the axes of the two wires, the contribution from both wires is vertical so that field, to the approximation described above, is

$$H_y(x, 0) \approx \frac{I}{2\pi(d + x)} + \frac{I}{2\pi(d - x)} \tag{1}$$

The magnetic flux between the two conductors (used in finding external inductance) is then found by integrating over this central plane. For a unit length in the z direction,

$$\psi_m \approx \frac{\mu I}{2\pi} \int_{-(d-R)}^{(d-R)} \left[\frac{1}{d + x} + \frac{1}{d - x} \right] dx = \frac{\mu I}{2\pi} [\ln(d + x) - \ln(d - x)]_{-(d-R)}^{(d-R)} \tag{2}$$

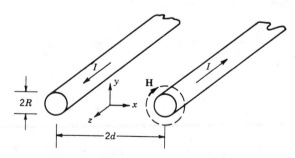

Fig. 4.6 Parallel-wire transmission line.

Inductance per unit length is then

$$L = \frac{\psi_m}{I} \approx \frac{\mu}{2\pi}\left[\ln\left(\frac{2d - R}{R}\right) - \ln\left(\frac{R}{2d - R}\right)\right] = \frac{\mu}{\pi}\ln\left(\frac{2d}{R} - 1\right) \qquad (3)$$

_____ **Example 4.6b** _____
External Inductance of Parallel-Wire
Transmission Line (Exact)

When spacing between conductors is comparable with wire radii, current distribution in the wires is affected and the result obtained above is modified. It can be shown either by a method of images, or by conformal transformations to be described in Chapter 7 that the exact magnetic flux function ψ_m [analogous to electric flux function in Eq. 1.6(1)] and scalar magnetic potential Φ_m for this problem are

$$\psi_m = -\frac{\mu I}{4\pi}\ln\left[\frac{(x - a)^2 + y^2}{(x + a)^2 + y^2}\right] \qquad (4)$$

$$\Phi_m = -\frac{I}{2\pi}\left[\tan^{-1}\frac{y}{(x - a)} - \tan^{-1}\frac{y}{(x + a)}\right] \qquad (5)$$

where

$$a = \sqrt{d^2 - R^2}$$

Taking the flux difference at $x = d - R$ and $x = -d + R$ (both at $y = 0$),

$$\Delta\psi_m = \psi_m(d - R, 0) - \psi_m(-d + R, 0) \qquad (6)$$

$$= -\frac{\mu I}{4\pi}\left\{\ln\left[\frac{d - R - a}{d - R + a}\right]^2 - \ln\left[\frac{-d + R - a}{-d + R + a}\right]^2\right\}$$

$$= -\frac{\mu I}{\pi}\ln\left|\frac{d - R - a}{d - R + a}\right| = -\frac{\mu I}{\pi}\ln\left|\frac{d - R - \sqrt{d^2 - R^2}}{d - R + \sqrt{d^2 - R^2}}\right| \qquad (7)$$

By multiplying numerator and denominator by $[(d - R) - \sqrt{d^2 - R^2}]$, this reduces to

$$\Delta\psi_m = -\frac{\mu I}{\pi}\ln\left\{\frac{d}{R} - \sqrt{\left(\frac{d}{R}\right)^2 - 1}\right\} = \frac{\mu I}{\pi}\cosh^{-1}\left(\frac{d}{R}\right) \qquad (8)$$

so

$$L = \frac{\Delta\psi_m}{I} = \frac{\mu}{\pi}\cosh^{-1}\left(\frac{d}{R}\right) \qquad (9)$$

_____ Example 4.6c _____
Internal Inductance of Plane Conductor with Skin Effect

As a third example, we utilize the energy method and calculate internal inductance. This example differs from that of Ex. 2.17, for which uniform current distribution was assumed. Moreover, we utilize the phasor forms of the skin effect formulation. The basis, as in Sec. 2.17, is the equation of the circuit form of energy storage to the field form,

$$\frac{1}{2}LI^2 = \int_V \frac{\mu}{2} H^2 \, dV \tag{10}$$

The magnetic field distribution for a semi-infinite conductor with sinusoidally varying currents was found to be [Eq. 3.16(15)]:

$$H_y = -\frac{\sigma \delta E_0}{(1+j)} e^{-(1+j)x/\delta} \tag{11}$$

where the coordinate system of Fig. 3.16a is used. The current per unit width, J_{sz}, is just the value of H_y at the surface

$$J_{sz} = -H_y(0) = \frac{\sigma \delta E_0}{(1+j)} \tag{12}$$

We may now apply (10) to the calculation of L. But first we recognize (10), as written, is for instantaneous I and H. To use with phasors, we must either convert to instantaneous forms or write the equivalent of (10) for time-average stored energies. The latter procedure is simpler and we find

$$\frac{1}{4}L|I|^2 = \int_V \frac{\mu}{4} |H|^2 \, dV \tag{13}$$

where the extra factor of $\frac{1}{2}$ on each side comes from the time average of squares of sinusoids. Taking a width w, so that current is wJ_{sz}, and a length l, and substituting (11) and (12) in (13)

$$\frac{L}{4} \frac{\sigma^2 \delta^2 E_0^2}{2} w^2 = wl \int_0^\infty \frac{\mu}{4} \frac{\sigma^2 \delta^2 E_0^2}{2} e^{-2x/\delta} \, dx$$

or

$$L = \frac{\mu l}{w} \int_0^\infty e^{-2x/\delta} \, dx = \frac{\mu l \delta}{2w} [-e^{-2x/\delta}]_0^\infty = \frac{\mu l \delta}{2w}$$

so

$$\omega L = \frac{l}{w} \cdot \frac{\omega \mu \sigma}{2\sigma} \cdot \sqrt{\frac{2}{\omega \mu \sigma}} = \frac{l}{w\sigma\delta} = \frac{R_s l}{w} \tag{14}$$

where relations for skin depth and surface resistivity have been substituted from Secs. 3.16 and 3.17. As found there, the internal reactance per square is equal to surface resistivity, R_s. This is multiplied by length l and divided by width w to give the internal reactance of the overall unit.

4.7 Mutual Inductance

The mutual inductance was defined in Sec. 4.2 as that arising from the induced voltage in one circuit due to current flowing in another circuit. We now discuss several approaches to its calculation, some of which may also be applied to calculation of self-inductance.

Flux Linkages The most direct approach is that from Faraday's law, finding the magnetic flux linking one circuit related to current in the other circuit, as in Eq. 4.2(19). Thus for two circuits 1 and 2 we write

$$M_{12} = \frac{\int_{S1} \mathbf{B}_2 \cdot d\mathbf{S}_1}{I_2} \tag{1}$$

where B_2 is the magnetic flux arising from current I_2 and integration is over the surface of circuit 1. By reciprocity $M_{21} = M_{12}$ (for isotropic magnetic materials), so the calculation may be made with the inducing current in either circuit. Consider for example the two parallel, coaxial conducting loops pictured in Fig. 4.7a. The magnetic field from a current in one loop has been found for a point on the axis in Ex. 2.3a

$$B_z(0, d) = \frac{\mu I_2 b^2}{2(b^2 + d^2)^{3/2}} \tag{2}$$

If loop 2 is small enough compared with spacing d, this will be relatively constant over the second loop and the relation (1) gives

$$M = \frac{\pi a^2 B_z(0, d)}{I_2} = \frac{\mu \pi a^2 b^2}{2(b^2 + d^2)^{3/2}} \tag{3}$$

The exact formula is found by integrating the field over the cross section, but we will approach the exact calculation by another method.

Use of Magnetic Vector Potential Since $\mathbf{B} = \nabla \times \mathbf{A}$, application of Stokes's theorem to (1) yields an equivalent expression in terms of the magnetic vector potential,

$$M = \frac{\int_{S1} (\nabla \times \mathbf{A}_2) \cdot d\mathbf{S}_1}{I_2} = \frac{\oint \mathbf{A}_2 \cdot d\mathbf{l}_1}{I_2} \tag{4}$$

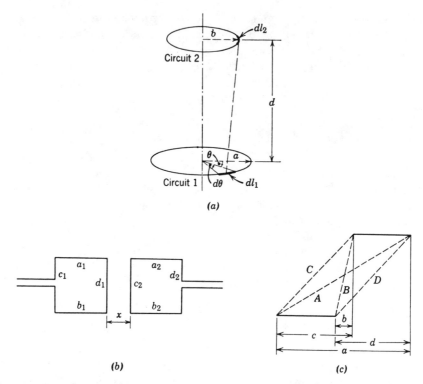

Fig. 4.7 (a) Two circular loops. (b) Two rectangular coupling loops. (c) Parallel current elements displaced from one another.

This form is useful in any problem for which the vector potential is more easily found than the magnetic field directly. (See Sec. 2.9 for a discussion of such problems.) It is especially useful for problems in which the circuit has straight-line segments, or can be approximated by such segments, as in the problem of the coupling of square loops pictured in Fig. 4.7b. Vector potential $\mathbf{A}$ is in the direction of the current element contributing to it by Eq. 2.9(5), so the contribution to $\mathbf{A}$ from horizontal sides a_1 and b_1 is only horizontal. These sides thus contribute to mutual inductance only through integration by (4) over the horizontal parts of circuit 2, a_2 and b_2. Similarly vertical currents in c_1 and d_1 contribute to mutual inductance only by integration over the two parallel (vertical) sides c_2 and d_2. The basic coupling element in such a configuration is then that of two parallel but displaced current elements as pictured in Fig. 4.7c. The contribution to mutual inductance from such elements (Prob. 4.7d) can be shown to be

$$M = \frac{\mu}{4\pi}\left[\ln\left\{\frac{(A + a)^a(B + b)^b}{(C + c)^c(D + d)^d}\right\} + (C + D) - (A + B)\right] \tag{5}$$

This point of view is quite useful for qualitative thinking about couplings in a circuit as well as for quantitative analysis.

Neumann's Form   Another standard form for calculation of mutual coupling of two filamentary circuits follows directly from the above. Following the point of view of the vector potential approach above, we write the vector potential **A** arising from current in circuit 2, assuming that current to be in line filaments and neglecting retardation.

$$\mathbf{A}_2 = \oint \frac{\mu I_2 \, \mathbf{dl}}{4\pi R} \tag{6}$$

where R is the distance between current element $\mathbf{dl}_2$ and the field point. Substitution in (4) yields

$$M = \frac{1}{I_2} \oint \oint \frac{\mu I_2 \, \mathbf{dl}_2 \cdot \mathbf{dl}_1}{4\pi R} = \frac{\mu}{4\pi} \oint \oint \frac{\mathbf{dl}_1 \cdot \mathbf{dl}_2}{R} \tag{7}$$

This standard form is due to Neumann. Note in particular its illustration of the reciprocity relation $M_{12} = M_{21}$ since integrations about circuits 1 and 2 may be taken in either order.

_____ Example 4.7a _____
Mutual Inductance of Coaxial Loops by Neumann's Form

For the coaxial loops of Fig. 4.7a, let $\mathbf{dl}_1$ be any element of circuit 1 and $\mathbf{dl}_2$ be any element of circuit 2. Then

$$\mathbf{dl}_1 \cdot \mathbf{dl}_2 = dl_2 a \, d\theta \cos \theta \tag{8}$$

$$R = \sqrt{d^2 + (a \sin \theta)^2 + (a \cos \theta - b)^2} \tag{9}$$

By substituting $\theta = \pi - 2\phi$ and

$$k^2 = \frac{4ab}{d^2 + (a + b)^2} \tag{10}$$

the integral will then be found to become

$$M = \mu \sqrt{abk} \int_0^{\pi/2} \frac{(2 \sin^2 \phi - 1) \, d\phi}{\sqrt{1 - k^2 \sin^2 \phi}} \tag{11}$$

which can be written as

$$M = \mu \sqrt{ab} \left[\left(\frac{2}{k} - k \right) K(k) - \frac{2}{k} E(k) \right] \tag{12}$$

where

$$E(k) = \int_0^{\pi/2} \sqrt{1 - k^2 \sin^2 \phi} \, d\phi \tag{13}$$

$$K(k) = \int_0^{\pi/2} \frac{d\phi}{\sqrt{1 - k^2 \sin^2 \phi}} \tag{14}$$

The definite integrals (13) and (14) are given in tables[3] as functions of k and are called *complete elliptic integrals* of the first and second kinds, respectively.

_____ Example 4.7b _____

Self-Inductance of Circular Loop Through Mutual Inductance Concepts

Neumann's form does not appear useful for the calculation of self-inductances of filamentary current paths, since radius R in (6) becomes zero at some point in the integration for such filaments. For a conductor of finite area, however, as in the round loop of wire pictured in Fig. 4.7d, one obtains the external contribution to self-inductance by calculating induced field at the surface of the conductor, say through the vector potential A as in (6). If wire radius a is small compared with loop radius r, this field is nearly the same as though current were concentrated along the center of the wire. Thus we conclude that the external inductance of the loop is well approximated by the mutual inductance between the two filaments of Fig. 4.7e. Utilizing Eq. 4.7(12) for the mutual inductance between two concentric circles of radii r and $(r - a)$ we then have

$$L_0 = \mu(2r - a)\left[\left(1 - \frac{k^2}{2}\right)K(k) - E(k)\right] \tag{15}$$

$$k^2 = \frac{4r(r - a)}{(2r - a)^2}$$

where $E(k)$ and $K(k)$ are as defined by Eqs. 4.7(13) and 4.7(14). If a/r is very small, k is nearly unity, and K and E may be approximated by

$$K(k) \cong \ln\left(\frac{4}{\sqrt{1 - k^2}}\right)$$

$$E(k) \cong 1$$

so

$$L_0 \cong r\mu\left[\ln\left(\frac{8r}{a}\right) - 2\right] \quad \text{H} \tag{16}$$

To find total L, values of internal inductance, as found in Sec. 4.5, must be added.

_____

[3] For example, Dwight or Spiegel, Ref. 2.

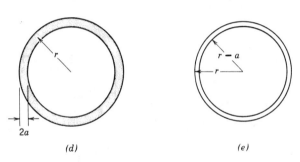

Fig. 4.7 (*d*) Conducting loop for which external self-inductance is to be found. (*e*) Filamentary loops, one through center of wire and other along inside edge, for which mutual inductance may be calculated.

4.8 Inductance of Practical Coils

A study of the inductance of coils at low frequencies involves no new concepts but only new troubles because of the complications in geometry. Certain special cases are simple enough for calculation by a straightforward application of previously outlined methods. For example, for a circular coil of N turns formed into a circular cross section (Fig. 4.8*a*) we may modify the formula for a circular loop of one turn, Eq. 4.7(16), provided the cross section is small compared with the coil radius. Magnetic field must be computed on the basis of a current NI; in addition, to compute the total induced voltage about the coil, N integrations must be made about the loop. Equation 4.7(16) is thus modified by a factor N^2. The external inductance for this coil is then

$$L_0 = N^2 R\mu\left[\ln\left(\frac{8R}{a}\right) - 2\right] \quad \text{H} \tag{1}$$

For the other extreme, the inductance of a very long solenoid (Fig. 4.8*b*) may be computed. If the solenoid is long enough, the magnetic field on the inside is essentially constant, as for the infinite solenoid:

$$H_z = \frac{NI}{l} \tag{2}$$

where N is the total number of turns and l the length. The flux linkages for N turns is then $N\pi R^2 \mu H_z$, and the inductance is

$$L_0 = \frac{\pi\mu R^2 N^2}{l} \quad \text{H} \tag{3}$$

For coils of intermediate length to radius ratio, empirical or semiempirical formulas frequently have to be used.[4] The famous Nagaoka formula applies a

[4] *Reference Data for Radio Engineers*, 5th ed., International Telephone and Telegraph Corporation, New York, 1968.

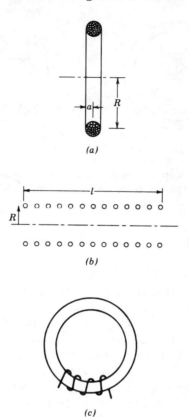

Fig. 4.8 (a) Coil of large radius-to-length ratio. (b) Solenoidal coil. (c) Solenoidal coil on high-permeability core.

correction factor K to the formula for the long solenoid, (3). The factor K or its equivalent is tabulated in the references cited and in many standard handbooks.[5] Another simple approximate form useful for $l > 0.8R$ is[6]

$$L_0 = \frac{\pi \mu R^2 N^2}{l + 0.9R} \quad \text{H} \tag{4}$$

If a coil is wound on a toroidal core of high permeability as shown in Fig. 4.8c, the flux essentially is restricted to the core region, independent of the length of the winding. The magnetic field intensity is again given by (2) and the inductance by (3) with $l = 2\pi r_0$ where r_0 is the mean radius of the toroid.

At higher frequencies the problem becomes more complicated. When turns are relatively close together, the assumption made previously in calculating internal

[5] For example, F. E. Terman, *Radio Engineers' Handbook*, McGraw-Hill, New York, 1943.
[6] H. A. Wheeler, *Proc. I.R.E.* **16**, 1398 (1928).

impedance (other portions of the circuit so far away that circular symmetry of current in the wire is not disturbed) certainly does not apply. Current elements in neighboring turns will be near enough to produce nearly as much effect upon current distribution in a given turn as the current in that turn itself. Values of skin effect resistance and internal inductance are then not as previously calculated. External inductance may also be different since changes in external fields result when current loses its symmetrical distribution with respect to the wire axis. In fact, the strict separation of internal and external inductance may not be possible for these coils, for a given field line may be sometimes inside and sometimes outside of the conductor. Finally, distributed capacitances may be important and further complicate matters (see following section).

Coils utilizing superconductors, which are materials giving zero resistance below some critical temperature near absolute zero, have recently become important because one can obtain with proper design very high values of uniform magnetic fields with them, without the use of iron. They may also be very efficient devices for storage of large energies. The electromagnetic principles of design are the same as given for other coils—and in fact the approximations may be better satisfied by the thin wires typically used in superconducting magnets. The mechanical forces of the large currents must be considered in the design, and the transient behavior of a superconductor is very different from that of an ordinary conductor. Isawa et al.[7] give examples of various coil configurations, with reference to the background literature.

4.9 Self and Mutual Capacitance

The concept of electrostatic capacitance between two conductors was introduced in Chapter 1 as the charge on one of the conductors divided by the potential difference between conductors. In the circuit analysis of Sec. 4.2 this definition was carried over as a quasistatic concept to give the usual capacitance term utilized in the analysis of circuits with time-varying excitation. Little more need be said about the simple two-conductor capacitor, but it is useful to collect expressions we have developed for some of the common capacitive elements.

Parallel planes with negligible fringing, A = area, d = spacing:

$$C = \frac{\varepsilon A}{d} \quad \text{F} \tag{1}$$

Concentric spheres of radii a and b $(b > a)$:

$$C = \frac{4\pi\varepsilon ab}{(b - a)} \quad \text{F} \tag{2}$$

[7] Y. Isawa and D. B. Montgomery, in *Applied Superconductivity*, V. L. Newhouse, ed., Academic Press, New York, 1975.

Coaxial cylinders of radii a and b $(b > a)$:

$$C = \frac{2\pi\varepsilon}{\ln(b/a)} \quad \text{F/m} \tag{3}$$

Parallel cylinders with wires of radius a, with axes separated by d (to be derived in Chapter 7):

$$C = \frac{\pi\varepsilon}{\cosh^{-1}(d/2a)} \quad \text{F/m} \tag{4}$$

If there are several conductors, the electric flux from one conductor may end on several of the others and induce charge on each of those. Consider for example the multiconductor problem diagrammed in Fig. 4.9a. Suppose conductor 1 is raised to a positive potential with the other three bodies, 0, 2, and 3, grounded. The electric flux from 1 will divide among the other three bodies and induce negative charges on each of these. The amounts of the separate charges may be used to define capacitances C_{10}, C_{12}, and C_{13} in the circuit representation of Fig. 4.9b. Similarly, raising conductor 2 to a nonzero potential and finding induced charges on grounded conductors 0, 1, and 3 determines C_{20} and C_{23}, and provides a check on C_{12}. Repetition of the process with conductor 3 at a nonzero potential gives the remaining element C_{30}, and provides a check on C_{13} and C_{23}. However it is usually not possible to measure the individual charges on the conductors which are tied together. Usually a capacitance current, dQ/dt, is measured by applying a time-varying voltage and the Q is the sum of the charges on electrodes connected together. Thus the three measurements described would yield $(C_{10} + C_{12} + C_{13})$, $(C_{20} + C_{12} + C_{23})$, and $(C_{30} + C_{13} + C_{23})$. Three additional measurements with linearly independent combinations of V_1, V_2, and V_3 are required to determine the six elements of the circuit.

A common problem is that of decreasing the capacitive coupling between two bodies—that is, of electrostatically shielding them from one another. Consider for example the conductors 1 and 3 of Fig. 4.9c. If a new conductor 2 is introduced and made to surround either body 1 or 3 completely, as in Fig. 4.9c, it is evident that a change in potential of 3 can in no way influence the charge on 1 so that mutual capacitance $C_{13} = 0$. If the new conductor surrounds 1, as shown, flux from 1 cannot end on ground either so that $C_{10} = 0$ also.

More often the added conductor may not completely enclose any body, so that the capacitance coupling may not be made zero, but may only be reduced from its original value. It can be shown that any finite conductor, as 2, introduced into the field acts to decrease the mutual capacitance C_{13} from its value prior to the introduction of 2, and hence provides some decrease in the capacitive coupling between 1 and 3. The reason is that fewer of the flux lines of the charge on 1 will terminate on 3 with a grounded conductor as shown by comparing Figs. 4.9d and 4.9e. However, if 2 is not connected to the ground (the infinite supply of charge), the effect of the added electrode will be to shorten the flux lines as seen in Fig. 4.9f. In terms of the equivalent circuit, the effective capacitance between 1 and 3

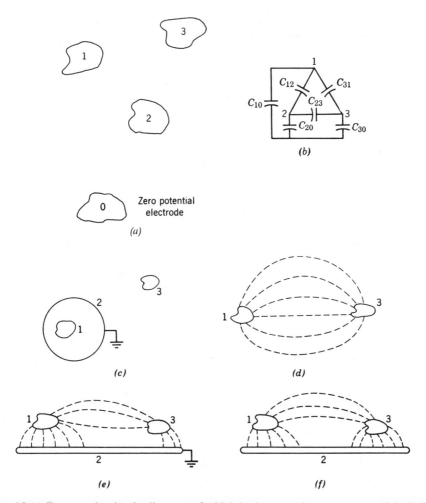

Fig. 4.9 (a) Four conducting bodies, one of which is chosen to have zero potential. (b) The equivalent circuit for (a). (c) Electrostatic shielding by a grounded sphere. (d) Flux lines between a pair of conductors without shielding. (e) Partial shielding by a grounded conducting plane. (f) Ungrounded nearby conductor increases coupling.

is seen from the equivalent circuit of Fig. 4.9b to be given by C_{13} in parallel with C_{12} and C_{23} in series:

$$(C_{13})_{\text{eff}} = C_{13} + \frac{C_{12}C_{23}}{C_{12} + C_{23}}$$

This value is generally greater than the value of C_{13} prior to the introduction of 2 (though it need not be if 2 lies along an equipotential surface of the original field); so, if insulated from ground, the additional conductor may act to increase the effective capacitive coupling between 1 and 3. It often happens that electrodes, although grounded for direct current, may be effectively insulated or floating at high frequencies because of impedance in the grounding leads. In such cases the new electrodes do not accomplish their shielding purposes but may in fact increase capacitive coupling.

CIRCUITS WHICH ARE NOT SMALL COMPARED WITH WAVELENGTH

4.10 Distributed Effects and Retardation

We now consider the generalizations to circuit theory when effects are distributed rather than lumped, and also when circuits become comparable in size with wavelength so that retardation from one part of the circuit to another must be considered. Considering first the distributed effects, we recognize that the fields contributing to circuit elements are always distributed in space and the representation by a lumped element is valid only when the region is small in comparison with wavelength and when only one type of energy storage (electric or magnetic) is important for that region. If the electric energy storage in parts of a primarily inductive element, or magnetic energy in a primarily capacitive element becomes important, the approach through classic circuit theory is to divide into subelements that can be treated as one or the other. For example, suppose there is electric field (capacitive) coupling between the turns of the inductor of Fig. 4.10a. A first approximation is that of adding a capacitive element across the terminals of L to represent all the electric energy storage of the element as shown in Fig. 4.10b. A still better approximation is that of adding a capacitive element between each pair of adjacent turns, as in Fig. 4.10c. But there may be coupling between nonadjacent turns and still other capacitances can be added as in Fig. 4.10d. The effect of these at high frequencies is to bypass some of the turns so that not all turns have the same current. This last effect would not be at all included in the simpler representation of Fig. 4.10b. Finally one might go to the limit and consider differential elements of the coil, attempting to find couplings to all other differential elements in order to write and solve a differential equation for current distribution. This process could be carried out only for simple configurations, and even the approach through a finite number of lumped elements as in c or d becomes complicated if there are many turns.

Consider next the retardation effect arising from the finite time of propagation of electromagnetic effects across the circuit. To simplify this discussion, we consider only sinusoidal excitation so that we can define a wavelength and discuss phase relationships. More general excitations can of course be broken into a series of sinusoids through Fourier analysis. Consider for example the

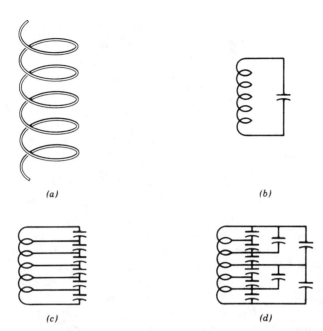

(a)   *(b)*

(c)   *(d)*

Fig. 4.10 (*a*) Coil. (*b*) Circuit with single capacitance representing electric-field coupling among turns. (*c*) Circuit representation with capacitance coupling shown between each adjacent turn. (*d*) Representation with capacitances added between nonadjacent turns.

simple single-loop antenna of Fig. 4.10*e*. At low frequencies, with diameter *d* small in comparison with wavelength, the time of propagation of effects from one part of the loop to another is negligible. Thus magnetic field produced by a current element at a point such as *A* travels to another point such as *B* in a negligible part of a cycle and so has negligible phase delay. The induced field from the time rate of change of the field is then 90 degrees out of phase with current in *B* and contributes to the inductive effect we expect for the loop at low

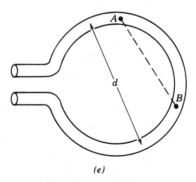

(e)

Fig. 4.10 (*e*) Loop antenna showing phase retardation between sources at *A* and induced fields at *B* when *d* is comparable with wavelength.

frequencies. At higher frequencies, with d comparable with wavelength, the finite time of propagation about the circuit must be considered. Current at B may then not be in phase with current at A, and the magnetic field at B arising from the element at A may not be in phase with either. The time rate of change of magnetic field induces an electric field which may then be not exactly 90 degrees out of phase with I_B. If there is an in-phase component, it represents energy transfer, which turns out to be a contribution to the energy radiated by this antenna. If current distribution is known, fields throughout the circuit can be calculated and the contribution to radiated power represented in the circuit by a so-called *radiation resistance*. But to find the actual current distribution, one really needs to solve the boundary-value problem represented by the conducting loop. For some antennas or other circuits comparable in size with wavelength, it is possible to make reasonable assumptions about current distribution and extend circuit theory in this way, but the extension must be done carefully. Additional discussion of this point will be given in the next section utilizing a retarded potential formulation for circuit theory.

One important circuit having both distributed and propagation effects is the uniform transmission line. It turns out that circuit theory can be extended to this case. Agreement with field solutions is exact for perfectly conducting transmission lines, and very good for lines with losses, as will be seen in Chapter 8. The circuit theory of transmission lines, to be developed in Chapter 5, thus is of very special importance.

4.11　Circuit Formulation Through the Retarded Potentials

The cause-and-effect relationships embodied in the retarded potentials of Sec. 3.19 can provide additional insights into the circuit formulation for electromagnetic problems, especially for circuits large in comparison with wavelength. This approach was first used by J. R. Carson.[8] A typical circuit follows a conductor for all or part of its path, so we start with Ohm's law in field form for a point along this path,

$$\mathbf{E} = \frac{\mathbf{J}}{\sigma} \tag{1}$$

where σ is the conductivity for the point under consideration and may vary as one moves about the circuit path. We next break up the field into an applied portion, $\mathbf{E}_0$, and an induced portion, $\mathbf{E}'$, the latter arising from the charges and the currents of the circuit itself. We also write $\mathbf{E}'$ in terms of the retarded potentials of Sec. 3.19

$$\mathbf{E}_0 + \mathbf{E}' = \mathbf{E}_0 - \nabla\Phi - \frac{\partial \mathbf{A}}{\partial t} = \frac{\mathbf{J}}{\sigma} \tag{2}$$

---

[8] J. R. Carson, *Bell System Tech. J.* **6**, 1 (1927).

where **A** and Φ are given as integrals over the charges and currents of the circuit, as defined in Eqs. 3.20(3) and 3.20(4).

The term $\mathbf{J}/\sigma$ in (2) is indeterminate over nonconducting portions of the path since both **J** and σ are zero for insulating portions and σ is generally undefined within any localized source. We consequently integrate (2) over conducting portions of the path, obtaining a cause-and-effect relationship which can be considered the general circuit equation:

$$\int \mathbf{E}_0 \cdot \mathbf{dl} - \int \frac{\mathbf{J}}{\sigma} \cdot \mathbf{dl} - \int \frac{\partial \mathbf{A}}{\partial t} \cdot \mathbf{dl} - \int \nabla \Phi \cdot \mathbf{dl} = 0 \qquad (3)$$

In a conventional circuit, the first term is applied voltage, the second a resistive term, the third an inductive term, and the fourth a capacitive term. The terms are discussed separately.

Applied Voltage   The first term of (3) can be identified as the applied voltage of circuit theory and is just the integral of applied electric field over the circuit path. In a circuit such as a receiving antenna, Fig. 4.11a, the applied field is clearly

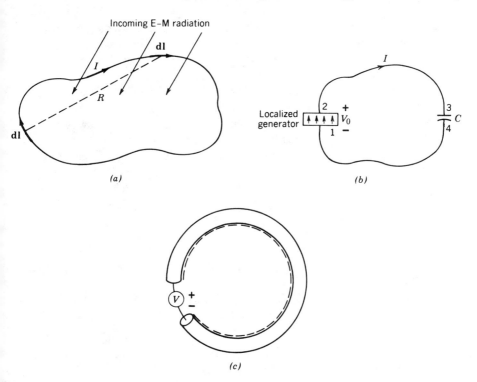

Fig. 4.11 (a) Closed filamentary loop excited by incoming electromagnetic wave. (b) Filamentary circuit with capacitor excited by a localized generator (point source). (c) Circular loop of round wire with circuit path along inner boundary.

distributed over the circuit through the mechanism of the incoming electromagnetic wave and the integration of $\mathbf{E}_0$ is about the complete path,

$$V_0 = \oint \mathbf{E}_0 \cdot d\mathbf{l} \tag{4}$$

For the localized sources, discussed in Sec. 4.3, for which electric field can be considered the gradient of a scalar potential, the integration of $\mathbf{E}_0$ from 2 to 1 about the circuit of Fig. 4.11b is the negative of that from 1 to 2 of the source since the closed line integral of the gradient is zero. The gap in the capacitor can be ignored in this step since the localized source produces negligible field there. Thus the source voltage is

$$V_0 = \int_{2(\text{circuit})}^{1} \mathbf{E}_0 \cdot d\mathbf{l} = -\int_{1(\text{source})}^{2} \mathbf{E}_0 \cdot d\mathbf{l} \tag{5}$$

In this class of problem, V_0 is independent of the circuit path, whereas in the receiving antenna class of problem discussed above, V_0 depends very much upon the circuit configuration and orientation with respect to V_0.

Internal Impedance Term   The second term in (3) is of exactly the same form as the ohmic term for the resistor in the circuit example of Sec. 4.2. There we showed that in the limit of dc this corresponds to the expected resistance term. Here it is understood that σ may vary over different parts of the circuit path and the integration brings in the total resistance of the circuit path. For ac circuits it turns out that this term may also include a contribution from the inductance internal to the conductor along which the circuit path is taken, as was seen for the round wire in Sec. 4.5. Thus for the important sinusoidal case with phasor representations for currents and voltages, this term gives a complex contribution resulting from internal reactance in addition to the resistance. That is, if internal impedance per unit length is defined as the ratio of surface electric field to the total current in the conductor,

$$Z_i' = \frac{E_s}{I} \tag{6}$$

the term under consideration becomes the total internal impedance Z_i, multiplied by current I:

$$\int \frac{\mathbf{J}}{\sigma} \cdot d\mathbf{l} = \int \mathbf{E}_s \cdot d\mathbf{l} = I \int Z_i' \, dl = IZ_i \tag{7}$$

where the integrals are taken over the conducting portions of the circuit from 2 to 3 and 4 to 1 in Fig. 4.11b.

External Inductance Term   The third term in (3) is the inductance term and, if the circuit path is properly selected, represents only the contribution from magnetic flux external to the conductor. Consider, for example, the loop of wire in Fig. 4.11c, and take the circuit path along the inner surface of the conductor. We will assume that the integral in the third term of (3) taken over the conducting portions of the circuit differs negligibly from an integral which would include the small gaps at the source and in any capacitors included in the circuit. This allows evaluation of that term with closed integrals. We take the path as stationary so that

$$\oint \frac{\partial \mathbf{A}}{\partial t} \cdot d\mathbf{l} = \frac{d}{dt} \oint \mathbf{A} \cdot d\mathbf{l} \tag{8}$$

From Stokes's theorem,

$$\oint \mathbf{A} \cdot d\mathbf{l} = \int_S (\nabla \times \mathbf{A}) \cdot d\mathbf{S} \tag{9}$$

But

$$\nabla \times \mathbf{A} = \mathbf{B} \tag{10}$$

so

$$\oint \frac{\partial \mathbf{A}}{\partial t} \cdot d\mathbf{l} = \frac{d}{dt} \int_S \mathbf{B} \cdot d\mathbf{S} \tag{11}$$

The surface integral of (11) is the magnetic flux linking the chosen circuit, exactly as in the approach through Faraday's law in Sec. 3.2. Thus the term may be defined as an inductance term, as before, recognizing that it is the contribution from flux threading the chosen circuit path (i.e., the external inductance):

$$\oint \frac{\partial \mathbf{A}}{\partial t} \cdot d\mathbf{l} = L \frac{dI}{dt} \tag{12}$$

Thus this provides an alternate way of calculating inductance,

$$L = \frac{1}{I} \oint \mathbf{A} \cdot d\mathbf{l} \tag{13}$$

The above assumes the circuit small compared with wavelength so that retardation is neglected. The extensions when this assumption is not valid are discussed shortly.

Capacitive Term   As with the other terms in (3) we must integrate the $\nabla \Phi$ over the conducting portions of the circuit, that is, from 2 to 3 and 4 to 1 in Fig. 4.11b. Here we assume that the fields arising from charges on the capacitor are negligible at the source, so we may use, as the range of integration, 4 to 3 through the source. Then, since the integral of the gradient of a scalar completely around a closed path (here including the capacitor gap) is zero, we may write

$$\int_{4(circuit)}^{3} \nabla\Phi \cdot d\mathbf{l} = -\int_{3(gap)}^{4} \nabla\Phi \cdot d\mathbf{l} = \Phi_3 - \Phi_4 \qquad (14)$$

In a lumped capacitor this potential difference is related to charge Q through the capacitance C,

$$\Phi_3 - \Phi_4 = \frac{Q}{C} \qquad (15)$$

so that this term is the capacitance term of circuit theory,

$$\int_{4(circuit)}^{3} \nabla\Phi \cdot d\mathbf{l} = \frac{Q}{C} = \frac{1}{C}\int I \, dt \qquad (16)$$

Circuits Comparable in Size with Wavelength   The formulation in terms of retarded potentials has been shown to reduce to the usual low-frequency circuit concepts as obtained in the earlier formulation using only fields, under the same assumptions. The present formulation is attractive in that it appears more readily extendible to large-dimension circuits, such as an antenna, when retardation effects are important. To illustrate, consider the circuit of Fig. 4.11a, for which current is assumed concentrated in a thin wire. Let us assume that ohmic resistance is negligible and that there is no capacitor so that there is only an applied voltage and a term from the potential **A**. We also take steady-state sinusoids and phasor notation for this discussion. Thus (3) becomes

$$\oint \mathbf{E}_0 \cdot d\mathbf{l} - j\omega \oint \mathbf{A} \cdot d\mathbf{l} = 0 \qquad (17)$$

and **A**, for the filamentary current, is

$$\mathbf{A} = \oint \frac{\mu I e^{-jkR}}{4\pi R} d\mathbf{l}' \qquad (18)$$

Substituting (18) in (17) and breaking up the exponential into its sinusoidal components,

$$\oint \mathbf{E}_0 \cdot d\mathbf{l} - j\omega \oint \oint \frac{\mu I(\cos kR - j\sin kR)}{4\pi R} d\mathbf{l} \cdot d\mathbf{l}' = 0 \qquad (19)$$

We see that even if current I were assumed entirely in phase about the circuit, finite values of kR would lead to both real and imaginary parts of the contribution from this term. The imaginary part is the inductive reactance, as found before for this term, except now modified by the integration of the retardation term. But there is a new term in phase with I which corresponds to the energy radiated from the circuit, and can be expressed as current times a radiation resistance.

Although the general modifications for large-dimension circuits are shown by this approach, it is difficult to carry much further since we really do not know the distribution of I about the circuit, and cannot find it without a field solution of the problem. In some antennas it is possible to make reasonable guesses about the current and proceed, but it is clear that these guesses must eventually be checked either through experiment or a field analysis. Also, as we have seen, the integration is to be only over conducting surfaces to avoid the indeterminacy of the second term of (3). For many antennas, the "gaps" are larger than the conductors, and fields definitely not quasistatic in the open regions, so this further limits the applicability of this approach. A specific example will be carried further in the following section.

4.12   Circuits with Radiation

To conclude this dicussion of the relationship between field theory and circuit theory, let us look at two specific circuits with radiation. As we saw in Sec. 4.11, a circuit that is not small in comparison with wavelength has retardation of induced fields from one part of the circuit to the other. The resulting phase changes produce components of induced field which are in phase with the currents, and an average power flow results. This power can be shown to be the radiation from the circuit. The phase shifts also produce some changes in the reactive impedance of the circuit, but this is usually a higher order effect. The term we are concerned with is the integration of induced effects, the second term of Eq. 4.11(19):

$$V_{\text{induced}} = j\omega \oint \oint \frac{\mu I(\cos kR - j \sin kR)}{4\pi R} \, \mathbf{dl} \cdot \mathbf{dl'} \tag{1}$$

we illustrate this with two examples.

_____ **Example 4.12a** _____
Small Circular Loop Antenna or Circuit

The first example is that of a circular loop, as in Fig. 4.12a, small enough so that current may be considered constant about the loop. If I is independent of position, it may be taken outside the integral (1) so that the induced term may be written

$$V_{\text{induced}} = (R_r + j\omega L)I \tag{2}$$

where

$$R_r = \oint \oint \frac{\omega\mu \sin kR}{4\pi R} \, \mathbf{dl} \cdot \mathbf{dl'}$$

$$\tag{3}$$

$$L = \oint \oint \frac{\mu \cos kR}{4\pi R} \, \mathbf{dl} \cdot \mathbf{dl'}$$

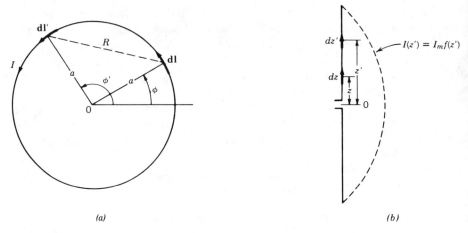

Fig. 4.12 (a) Circular loop with constant current I with coordinates for calculation of retardation effects. (b) Straight antenna of finite length with current distribution.

The value of $\mathbf{dl}$ is $\hat{\boldsymbol{\phi}}a\,d\phi$ and of $\mathbf{dl'}$ is $\hat{\boldsymbol{\phi}}'a\,d\phi'$. The angle between $\mathbf{dl}$ and $\mathbf{dl'}$ is $(\phi - \phi')$ and the distance R is $2a\sin[(\phi - \phi')/2]$. If the circuit is small in comparison with wavelength, $kR \ll 1$ and the sine term in (1) may be replaced by the first two terms of its Taylor series

$$R_r \approx \int_0^{2\pi}\int_0^{2\pi} \frac{\omega\mu}{4\pi R}\left\{kR - \left[\frac{k^3 R^3}{3!}\right]\right\} a^2 \cos(\phi - \phi')\,d\phi\,d\phi' \tag{4}$$

The first term of the above integrates to zero, so we see why it is necessary to retain at least two terms of the series. The second terms gives

$$R_r \approx \int_0^{2\pi}\int_0^{2\pi} \frac{-4\omega\mu k^3 a^4}{24\pi} \sin^2\left(\frac{\phi - \phi'}{2}\right)\cos(\phi - \phi')\,d\phi\,d\phi' \tag{5}$$

The integrals are readily evaluated to give

$$R_r = \frac{-\omega\mu k^3 a^4}{6\pi}(-\pi^2) = \frac{\eta\pi}{6}(ka)^4 \tag{6}$$

Thus radiation resistance increases as the fourth power of the ratio of radius to wavelength (but with the understanding that this ratio is always small.) For $a = 0.05\lambda$, the value is

$$R_r = \frac{120\pi^2}{6}(2\pi \times 0.05)^4 = 1.537\ \Omega \tag{7}$$

If $\cos kR$ in the expression for L is likewise expanded as a series,

$$L \approx \int_0^{2\pi}\int_0^{2\pi} \frac{\mu}{4\pi R}\left[1 - \frac{k^2 R^2}{2!} + \frac{k^4 R^4}{4!} + \cdots\right] a^2 \cos(\phi - \phi')\,d\phi\,d\phi' \tag{8}$$

The first term is recognized as the Neumann form for inductance of this loop (Sec. 4.7) and the remaining terms represent corrections to the inductance because of retardation. It is seldom necessary to calculate these last-mentioned corrections for circuits properly considered as lumped-element circuits.

_____ **Example 4.12b** _____

Radiation Resistance of a Straight Antenna by Circuit Methods

As a second example, consider a straight dipole antenna as shown in Fig. 4.12b with current distribution

$$I(z) = I_m f(z) \tag{9}$$

where $f(z)$ is real. But here the conductor does not form a closed circuit, and as explained earlier, the Carson formulation (Sec. 4.11) only applies unambiguously over the surface of conductors. Thus we first find retarded potential **A**, which has only a z component,

$$A_z = \int_{-l}^{l} \frac{\mu I_z e^{-jkR}}{4\pi R} dz' = \mu I_m \int_{-l}^{l} \frac{f(z')e^{-jk|z-z'|}}{4\pi|z-z'|} dz' \tag{10}$$

Electric field is given in terms of **A** by

$$\mathbf{E} = -j\omega \left[\mathbf{A} + \frac{1}{k^2} \nabla(\nabla \cdot \mathbf{A}) \right] = -j\omega \hat{z} \left[A_z + \frac{1}{k^2} \frac{\partial^2 A_z}{\partial z^2} \right] \tag{11}$$

The portion of **E** in phase with current causes the radiated power in this picture, and that clearly comes from the imaginary part of A_z. The integration of $I(z)E_{\text{in-phase}}$ over the antenna gives the total power transferred, or radiated, and this may be expressed in terms of a _radiation resistance_

$$W = \int_{-l}^{l} I(z)(E_z)_{\text{in-phase}} \, dz = \frac{I_m^2 R_r}{2} \tag{12}$$

Thus substituting (10) and (11)

$$R_r = \frac{2\omega\mu}{4\pi} \int_{-l}^{l} dz \, f(z) \int_{-l}^{l} f(z') \left\{ \left[\frac{\sin k|z-z'|}{|z-z'|} \right] + \frac{\partial^2}{\partial z^2} \left[\frac{\sin k|z-z'|}{|z-z'|} \right] \right\} dz' \tag{13}$$

In evaluating these integrals, series expansions of the $\sin k|z - z'|$ terms are often made. When carried out for the half-wave dipole [$l = \lambda/4$ and $f(z) = \cos kz$], R_r is found to be about 73.1 Ω in agreement with the value found by a Poynting integration to be utilized later. This method of finding radiation resistances of antennas is called the _induced emf method_. It is seldom easier than the Poynting integration but does show the relationship to circuit theory.

Note that for both examples, we had to assume a form for current distribution in order to proceed. This is a clear limitation as it can only be done with reasonable confidence in specific cases. When that is not possible, field theory must be invoked for the whole problem.

PROBLEMS

4.2a Many circuits contain nonlinear elements, that is, ones for which μ, ε, or σ, or some combination, are functions of the fields for at least a part of the circuit. Review the formulation of Sec. 4.2 to show this behavior explicitly. Is the general form Eq. 4.2(1) changed in such cases?

4.2b Some circuits contain time-varying elements, for which μ, ε, or σ, or a combination, are functions of time for at least a part of the circuit. Discuss these cases as in Prob. 4.2a.

4.2c The sign of mutual inductance coupling is designated on a circuit diagram by the placing of black dots. With the sign convention for positive voltage and current shown in Figs. 4.2c and 4.2d, the dot location in the former denotes positive M and in the latter negative M. Show that either can be represented by a "T-network" as in Fig. P4.2c, where the upper signs denote Fig. 4.2c and the lower Fig. 4.2d.

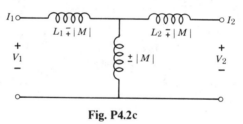

Fig. P4.2c

4.3a Show that the term on the right of Eq. 4.3(3) is just the displacement current flow into the surface S.

4.3b It has been pointed out that the mesh analysis utilizes Kirchhoff's voltage law explicitly but the current law only implicitly. Show that the current law is satisfied for each node of the circuit with mesh currents defined by Fig. 4.3b. Similarly show that the voltage law is satisfied by each mesh of Fig. 4.3c, with node voltages as shown.

4.3c The generator in the example of Figs. 4.3b and c is taken as a voltage generator in series with a source resistance. It can alternatively be taken as a current generator I_0 in parallel with a source conductance G_s. Make this substitution and write the new loop and node equations for the filter.

4.3d With a complex load impedance (admittance) connected to the source terminals as in Figs. 4.3d and e, show that the two source representations are equivalent in producing current in and voltage across this impedance, when the conditions of Eq. 4.3(10) are satisfied.

4.3e Show that for fixed V_s and Z_s, the maximum possible power is delivered to the load when it is a "conjugate match" to Z_s, that is, $Z_L = Z_s^*$ (or $Y_L = Y_s^*$).

4.4a Make a power series expansion of Eq. 4.4(8), retaining up to quadratic terms, to show the variation of magnitude and phase with r to this approximation. Up to about what r_0/δ will this be a reasonable approximation? (See Sec. 7.14.)

4.4b Utilize the asymptotic expansions of Bessel functions to derive the approximate expression Eq. 4.4(11). What phase variation is found in this approximation? (See Sec. 7.15.)

4.4c Obtain tables of the Ber and Bei functions and plot phase of current density versus r/r_0 for $r_0/\delta = 2.39$.

4.5a Show that the ratio of very high-frequency resistance to dc resistance of a round conductor of radius r_0 and material with depth of penetration δ can be written

$$\frac{R_{hf}}{R_0} = \frac{r_0}{2\delta}$$

4.5b Using the approximate formula 4.5(8), find the value of r_0/δ below which R differs from dc resistance R_0 by less than 2%. To what size wire does this correspond for copper at 10 kHz? For copper at 1 MHz? For brass at 1 MHz?

4.5c* For two z-invariant systems having the same shape of cross section and of good conductors of the same material, show that current distributions will be similar, and current densities equal in magnitude at similar points, if the applied voltage to the small system is $1/K$ in magnitude and K^2 in frequency that of the large system. Also show that the characteristic impedance of the small system will be K times that of the large system under these conditions. Check these conclusions for the case of two round wires of different radii. K is the ratio of linear dimensions ($K > 1$).

4.6a For the symmetrical parallel-wire line, plot normalized external inductance, $\pi L/\mu$ versus R/d from both the approximate and exact expressions and note the range over which the approximate formula gives good results.

4.6b Derive the formula for external inductance of the coaxial line in Fig. 2.4b by the energy method assuming the usual situation of a material with permeability μ_0 between the electrodes.

4.6c For a parallel-plane transmission line, find the dielectric thickness, in terms of the conductor thickness, for which the low-frequency internal inductance equals the external inductance. Take both conductor thicknesses to be the same.

4.6d A coaxial transmission line has a solid copper inner conductor of radius 0.20 cm and a tubular copper outer conductor of inner radius 1 cm, wall thickness 0.1 cm. Find the total impedance per unit length of line for a frequency of 3 GHz, including the internal impedance of both conductors.

4.6e Equivalent circuit for a differential length of coaxial transmission line. Take the lines C–B and D–A in Fig. 3.17 to be separated by a differential distance dz with z positive to the right. Write Faraday's law for the loop $ABCDA$ and use the capacitance expression given in Eq. 1.9(4) to show that the equivalent circuit shown in Fig. P4.6e is correct. (L_i is internal inductance per square and L_e is external inductance per unit length.)

4.7a A coaxial line has the radius of the inner conductor as a and inside radius of outer conductor as b, and is closed by a conducting plane at $z = 0$. A square loop is introduced for coupling, lying in a longitudinal plane and extending from $z = 0$ to $z = d$ in length and from $r = r_1$ to $r = r_2$ in radius ($a < r_1 < r_2 < b$). Find the mutual inductance between loop and line assuming $d \ll \lambda$ so that field is essentially independent of z.

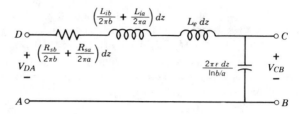

Fig. P4.6e

4.7b From tables of the complete elliptic integrals given in the references, plot the form of mutual inductance in Eq. 4.7(12) against d/a for $b/a = 1$.

4.7c Investigate the properties of the complete elliptic integrals for $k \ll 1$ and for $k \approx 1$, and obtain approximate expressions for mutual inductance for these two cases.

4.7d By integration of Eq. 4.7(4), show that the contribution to mutual inductance from two parallel line segments displaced as shown in Fig. 4.7c is as given by (5).

4.7e Apply Eq. 4.7(5) to the calculation of mutual inductance between two square loops used for coupling between open-wire transmission lines as shown in Fig. 4.7b. The length of each side is 0.03 m; the separation x is 0.01 m. Assume that the gaps at which the lines enter are small enough to be ignored.

4.7f Plot $L_0/a\mu$ for the circular loop of round wire versus a/r from approximate and "exact" expressions and note the range of usefulness of the former. Comment on the validity of the selected mutual approach for a/r approaching unity. (Note that tables of elliptic functions are required for this comparison.)

4.7g Suppose 1-mm-diameter copper wire is formed into a single circular loop having a radius of 10 cm. A voltage generator of 1-V rms and 10 MHz is connected to an infinitesimal gap in the loop. Find the current flowing in the loop, taking into account internal impedance as well as external inductance. Justify all approximations used.

4.8a Plot $L_0/\mu R$ versus R/l from the expression for a long solenoid, and the empirical expression 4.8(4) and compare. (If you have access to tables for the Nagaoka formula, add this curve also.)

4.8b You are given 20-m-length of small wire. A radius of at least 1 cm is desired for your coil. Design a coil with about as large inductance as you can within these constraints.

4.8c For an ideal infinite solenoid, magnetic flux is uniform everywhere inside the solenoid. For a finite coil, as pictured in Fig. 4.8b, there may be more flux through the central turns of the coil than those near the ends. Explain how you make a circuit model of the coil in view of these "partial flux linages."

4.9a Discuss qualitatively the case of Fig. 4.9e in which two bodies, 1 and 3, which are relatively far apart have a grounded conducting plane brought in their vicinity. Give an energy argument to show that C_{13} is decreased when the plane is added.

4.9b* Suppose that the bodies 1 and 3 of Figs. 4.9e and f are spheres of radii a separated by a distance d with $a/d \ll 1$. If the added plane is parallel to the line joining their centers and distance b from it ($a/b \ll 1$), find C_{13} before and after introduction of the plane when grounded, and the effective C_{13} with the plane present and insulated from ground.

4.9c Two cylindrical conductors of radius 1 cm have their axes 4 cm apart and each axis is 4 cm above a parallel ground plane. Make rough graphical field maps and estimate the capacitances (per unit length) for this three-conductor problem. (Think about how many plots you need and the best choice of potentials for each.)

4.10a Make an order of magnitude estimate of the capacitance between adjacent turns of a coil as pictured in Fig. 4.10a if the diameter of the coil is 2 cm, the diameter of the wire 1 mm, and the spacing between turns 1 mm. Clearly state your model for the calculation.

4.10b For a coil as in Fig. 4.10a, in the form of a fairly open helix, it is found that the phase delay of current along the coil is well estimated by assuming propagation *along the wire* at the velocity of light in the surrounding dielectric material. For a helix of 100 turns in air, each turn 1 cm in diameter and spaced 1 mm apart on centers (wire diameter being appreciably smaller than this),

(a) Find the phase difference between current at the end and that at the beginning of the helix for such a traveling wave at $f = 150 \, \text{MHz}$.

(b) Compare this with the phase difference in the retardation term, calculated along a direct path between the two ends.

4.12a What radius do you need to give a radiation resistance of 50 Ω from the expression derived from the small-loop circuit? Do you think the approximations reasonably satisfied for this size?

4.12b Using the method of Sec. 4.12 derive radiation resistance for a small square loop with sides d and uniform current assumed about the loop.

4.12c* As indicated in Ex. 4.12b, make a series expansion for the sine terms within the integral (retain three terms), assume $f(z') = \cos kz'$, and estimate R_r for the half-wave dipole, $l = \lambda/4$.

5

TRANSMISSION LINES

5.1 Introduction

In Chapter 3, we saw that the interchange of electric and magnetic energy gives rise to the propagation of electromagnetic waves in space. More specifically, the magnetic fields that change with time induce electric fields as explained by Faraday's law, and the time-varying electric fields induce magnetic fields, as explained by the generalized Ampère's law. This interrelationship also occurs along conducting or dielectric boundaries, and can give rise to waves that are guided by such boundaries. These waves are of paramount importance in guiding electromagnetic energy from a source to a device or system in which it is to be used. Dielectric guides, hollow-pipe waveguides, and surface guides are all important for such purposes, but one of the simplest systems to understand—and one very important in its own right—is the two-conductor transmission line. This system may be considered a distributed circuit and so is useful in establishing a relation between circuit theory and the more general electromagnetic theory expressed in Maxwell's equations. The concepts of energy propagation, reflections at discontinuities, standing versus traveling waves and the resonance properties of standing waves, phase and group velocity, and the effects of losses upon wave properties are easily extended from these transmission-line results to the more general classes of guiding structures.

A parallel two-wire system is a typical and important example of the transmission lines to be studied in this chapter. In any transverse plane, electric field lines pass from one conductor to the other, defining a voltage between conductors for that plane. Magnetic field lines surround the conductors, corresponding to current flow in one conductor and an equal but oppositely directed current flow in the other. Both voltage and current (and of course the fields from which they are derived) are functions of distance along the line. In the two following sections we set down the transmission-line equations from distributed-circuit theory, but then discuss its relation to field theory.

Transmission-line effects are not always desirable ones. A cable interconnecting two high-speed computers may be intended as a direct connection, but will at the very least introduce a time delay (around 5 ns/m in typical dielectric-filled cable). Moreover, if the interconnections are not impedance matched at the two ends there will be reflections of the waves (as we shall see later in the chapter). These "echoes" of pulses representing the digits could introduce serious errors. A still further complication is *dispersion*. In a real transmission line the velocity of propagation varies to some degree with frequency, so the frequency components which represent the pulse (by Fourier analysis) travel at different velocities and the pulse distorts as it travels. If dispersion is excessive, the pulses may be blurred enough so that individual digits cannot be clearly distinguished. All of these effects occur also in the interconnections of elements in printed circuits and even in semiconductor integrated circuits, but the close spacings in the last case limit performance only for extremely short pulses.

Transmission-line analysis is useful, by analogy, in studying a variety of wave phenomena, such as the propagation of acoustic waves and their reflection from materials with different acoustic properties. An especially interesting analog is that of the propagation of signals along a nerve of the human body.

TIME AND SPACE DEPENDENCE OF SIGNALS ON IDEAL TRANSMISSION LINES

5.2   Voltage and Current Variations Along an Ideal Transmission Line

We begin by considering the transmission line as a distributed circuit. In Sec. 2.5 we identified an inductance per unit length associated with the flux produced by the oppositely directed currents in a pair of parallel conductors. When the currents vary with time, there is a voltage change along the line. Likewise, the distributed capacitance produces displacement current between the conductors when the voltage is time varying and leads to change of the current flowing along the conductors. The interrelationship leads to the wave equation for voltage and current along an ideal lossless transmission line.

Figure 5.2 shows a representative two-conductor line and the circuit model for a differential length. It should be kept in mind that the external inductance per unit length of a parallel-conductor line is not associated with one conductor or the other. Also, the circuit model is simply a representation of a differential length of line; there is not a one-for-one identification of the two sides of the circuit with the two conductors of the modeled line.

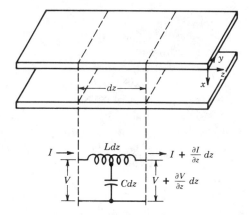

Fig. 5.2 Section of a representative transmission line and the equivalent circuit for a differential length.

Consider a differential length of line dz, having the distributed inductance, L per unit length, and the distributed capacitance, C per unit length. The length dz then has inductance $L\,dz$ and capacitance $C\,dz$. The change in voltage across this length is then equal to the product of this inductance and the time rate of change of current. For such a differential length, the voltage change along it at any instant may be written as the length multiplied by the rate of change of voltage with respect to length. Then

$$\text{voltage change} = \frac{\partial V}{\partial z}\,dz = -(L\,dz)\frac{\partial I}{\partial t} \tag{1}$$

Note that time and space derivatives are written as partial derivatives, since the reference point may be changed in space or time in independent fashion.

Similarly, the change in current along the line at any instant is merely the current that is shunted across the distributed capacitance. The rate of decrease of current with distance is given by the capacitance multiplied by time rate of change of voltage. Partial derivatives are again called for:

$$\text{current change} = \frac{\partial I}{\partial z}\,dz = -(C\,dz)\frac{\partial V}{\partial t} \tag{2}$$

The length dz may be canceled in (1) and (2):

$$\frac{\partial V}{\partial z} = -L\frac{\partial I}{\partial t} \tag{3}$$

$$\frac{\partial I}{\partial z} = -C\frac{\partial V}{\partial t} \tag{4}$$

Equations (3) and (4) are the fundamental differential equations for the analysis of the ideal transmission line. Note that they are identical in form with the pairs,

Eqs. 3.9(5) and 3.9(9) or Eqs. 3.9(6) and 3.9(8) found from Maxwell's equations for plane electromagnetic waves. As was done there, (3) and (4) can be combined to form a wave equation for either of the variables. For example, one can differentiate (3) partially with respect to distance and (4) with respect to time:

$$\frac{\partial^2 V}{\partial z^2} = -L \frac{\partial^2 I}{\partial z \, \partial t} \tag{5}$$

$$\frac{\partial^2 I}{\partial t \, \partial z} = -C \frac{\partial^2 V}{\partial t^2} \tag{6}$$

Partial derivatives are the same taken in either order (assuming continuous functions) so (6) may be substituted directly in (5):

$$\frac{\partial^2 V}{\partial z^2} = LC \frac{\partial^2 V}{\partial t^2} = \frac{1}{v^2} \frac{\partial^2 V}{\partial t^2} \tag{7}$$

where

$$v = (LC)^{-1/2} \tag{8}$$

All real signals are continuous functions, as required for (7) to apply. The discontinuous step waves used later as examples are to be understood as approximations to real signals. Equations (3) and (4) are known as the *telegraphist's equations*, and the differential equation (7) is the one-dimensional wave equation. A similar equation may be obtained in terms of current by differentiating (4) with respect to z, (3) with respect to t, and combining the results:

$$\frac{\partial^2 I}{\partial z^2} = \frac{1}{v^2} \frac{\partial^2 I}{\partial t^2} \tag{9}$$

We saw in Sec. 3.9 that an equation of the form (7) has a solution

$$V(z, t) = F_1\left(t - \frac{z}{v}\right) + F_2\left(t + \frac{z}{v}\right) \tag{10}$$

where F_1 and F_2 are arbitrary functions. A constant value of $F_1(t - z/v)$ would be seen by an observer moving in the $+z$ direction with a velocity v so $F_1(t - z/v)$ represents a wave traveling in the $+z$ direction with velocity v. Similarly, $F_2(t + z/v)$ represents a wave moving in the $-z$ direction with velocity v.

To find the current on the line in terms of the functions F_1 and F_2, substitute the expression for voltage given by (10) in the transmission line equation (3)

$$-L \frac{\partial I}{\partial t} = -\frac{1}{v} F_1'\left(t - \frac{z}{v}\right) + \frac{1}{v} F_2'\left(t + \frac{z}{v}\right) \tag{11}$$

This expression may be integrated partially with respect to t:

$$I = \frac{1}{Lv}\left[F_1\left(t - \frac{z}{v}\right) - F_2\left(t + \frac{z}{v}\right) \right] + f(z) \tag{12}$$

If this result were substituted in the other transmission line equation (4), it would be found that the function of integration, $f(z)$, could only be a constant. This is a possible superposed dc solution not of interest in studying the wave solution, so the constant will be ignored. Equation (12) may then be written

$$I = \frac{1}{Z_0}\left[F_1\left(t - \frac{z}{v}\right) - F_2\left(t + \frac{z}{v}\right)\right] \tag{13}$$

where

$$Z_0 = Lv = \sqrt{\frac{L}{C}} \quad \Omega \tag{14}$$

The constant Z_0 as defined by (14) is called the *characteristic impedance* of the line, and is seen from (10) and (12) to be the ratio of voltage to current for a single one of the traveling waves at any given point and given instant. The negative sign for the negatively traveling wave is expected since the wave propagates to the left, and by our convention current is positive if flowing to the right.[1]

_____ **Example 5.2** _____

Characteristic Impedance and Wave Velocity for
a Coaxial Line

Let us find expressions for the characteristic impedance and wave velocity for an ideal coaxial line and examine some typical values. We will assume the conductor spacing is large enough to neglect internal inductance. Using C from Eq. 1.9(4) and L from Eq. 2.5(6) in (14) we find

$$Z_0 = \frac{\ln b/a}{2\pi} \sqrt{\frac{\mu}{\varepsilon}} \tag{15}$$

where a and b are the radii of the inner and outer conductors at the dielectric surfaces, respectively. A common commercial coaxial cable is designated RG58/U. It has a dielectric of relative permittivity 2.26 and the radii are: $a = 0.406$ mm and $b = 1.48$ mm. Substituting these values in (15) and taking $\mu \cong \mu_0$ (Sec. 2.3) one finds that $Z_0 = 51.6\,\Omega$. This is slightly below the published normal value $Z_0 = 53.5\,\Omega$ due in part to the neglect of the frequency-dependent internal inductance of the conductors. There is not much variation of the relative permittivity among the various materials used as the dielectrics in coaxial lines

_____

[1] Since Z_0 as defined here is real, it is more logical to call it a "characteristic resistance," especially since the concept of impedance implies use with the phasor forms appropriate to steady-state sinusoidal excitation. That is an important special case to be considered later, but even for transmission lines used with pulses or other general signals, it is common to refer to the defined Z_0 as *characteristic impedance*.

and since the radius ratio comes in only in a logarithm, one finds that most commercial coaxial lines have characteristic impedance in a limited range, usually $50\,\Omega \lesssim Z_0 \lesssim 80\,\Omega$. The wave velocity (8) becomes

$$v = \frac{1}{\sqrt{\mu\varepsilon}} \tag{16}$$

which is the same as the velocity of plane waves in the same dielectric, Sec. 3.9. This result obtains for all two-conductor transmission lines when the internal inductance can be neglected.[2] The wave velocity is usually between about 0.5 and 0.7 of the velocity of light in vacuum, 3×10^8 m/s, for lines with plastic dielectrics.

5.3   Relation of Field and Circuit Analysis of Transmission Lines

Although we largely utilize the distributed-circuit model for transmission line analysis in this chapter, let us relate the equations obtained in Sec. 5.2 to field concepts of Chapter 3. First, let us take the special case of a parallel-plane transmission line, as indicated in Fig. 5.2, with the conducting planes assumed wide enough in the y direction so that fringing at the edges is not important. If the planes are also assumed perfectly conducting, it is clear that a portion of a uniform plane wave with E_x and H_y, as studied in Chapter 3, can be placed in the dielectric region between the planes and will satisfy the boundary condition that electric field enter normally to the perfectly conducting planes. The Maxwell equations for such a wave [Eqs. 3.9(6) and 3.9(8)] are

$$\frac{\partial E_x(z, t)}{\partial z} = -\mu \frac{\partial H_y(z, t)}{\partial t} \tag{1}$$

$$\frac{\partial H_y(z, t)}{\partial z} = -\varepsilon \frac{\partial E_x(z, t)}{\partial t} \tag{2}$$

If we define voltage as the line integral of $-\mathbf{E}$ between planes at a given z,

$$V(z, t) = -\int_1^2 \mathbf{E} \cdot d\mathbf{l} = -\int_0^a E_x\, dx = -aE_x(z, t) \tag{3}$$

Current for a width b, with positive sense defined for the upper plane, is related to the tangential magnetic field by

$$I(z, t) = -bH_y(z, t) \tag{4}$$

---

[2] It follows that knowledge of either L or C for such ideal lines determines the other.

With these substituted in Eqs. 5.2(3) and 5.2(4), we find the result identical to (1) and (2) if

$$C = \frac{\varepsilon b}{a} \quad \text{F/m}, \qquad L = \frac{\mu a}{b} \quad \text{H/m} \tag{5}$$

These are, respectively, the capacitance per unit length and inductance per unit length for such a system of parallel-plane conductors, calculated from static concepts. The field and circuit concepts are thus identical in this simple case.

We would also find field and distributed circuit approaches identical if we applied them to the coaxial transmission line with ideal conductors, or to two-conductor systems of other shape so long as conductors are perfect. This is because such systems can be shown to propagate *transverse electromagnetic* (TEM) waves, for which both electric and magnetic fields have only transverse components. The absence of an axial magnetic field means that there are no induced transverse electric fields and no corresponding contributions to the line integral $\int \mathbf{E} \cdot \mathbf{dl}$ taken between the two conductors, so long as the integration paths remain in the transverse plane; thus the voltage between conductors can be taken as *uniquely defined for that plane*. Similarly the absence of an axial electric field means that there is no displacement current contribution to $\oint \mathbf{H} \cdot \mathbf{dl}$ for paths in a given transverse plane, and if such a closed path surrounds one conductor, the integral will be just the conduction current flow in that conductor for that plane at that instant of time. Moreover, the transverse $\mathbf{E}$ and $\mathbf{H}$ fields can be shown to satisfy Laplace's equation *in the transverse plane* (Prob. 5.3), thus explaining the appropriateness of using Laplace solutions for the calculation of the L and C of the transmission line.

When the finite resistances of conductors are taken into account, the identity of circuit and field analysis is no longer an exact one, but has been shown to be a good approximation, for practical transmission lines. This field basis for TEM waves will be developed more in Chapter 8.

5.4 Reflection and Transmission at a Resistive Discontinuity

Most transmission-line problems are concerned with junctions between a given line and another of different characteristic impedance, a load resistance, or some other element that introduces a discontinuity. By Kirchhoff's laws, total voltage and current must be continuous across the discontinuity. The total voltage in the line may be regarded as the sum of voltage in a positively traveling wave, equal to V_+ at the point of discontinuity, and a voltage in a reflected or negatively traveling wave, equal to V_- at the discontinuity. The sum of V_+ and V_- must be V_L, the voltage appearing across the junction:

$$V_+ + V_- = V_L \tag{1}$$

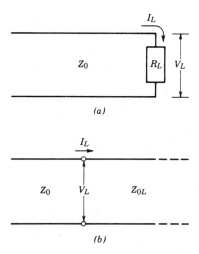

Fig. 5.4 (a) Ideal transmission line with a resistive load. (b) Ideal transmission line of characteristic impedance Z_0 with a second ideal line, of infinite length and characteristic impedance Z_{0L} as a load.

Similarly, the sum of current in the positively and negatively traveling waves of the line, at the point of discontinuity, must be equal to the current flowing into the junction:

$$I_+ + I_- = I_L \tag{2}$$

The simplest form of discontinuity is one in which a load resistance R_L is connected to the transmission line at the junction, as shown in Fig. 5.4a. Another case that is equivalent is that of Fig. 5.4b in which the first ideal line is connected to a second ideal line of infinite length and characteristic impedance Z_{0L}; here $R_L = Z_{0L}$. Still other forms of load circuits can produce an effective resistance R_L at the junction. In all these cases, $V_L = R_L I_L$. By utilizing the relations between voltage and current for the two traveling waves as found in section 5.2, equation (2) becomes

$$\frac{V_+}{Z_0} - \frac{V_-}{Z_0} = \frac{V_L}{R_L} \tag{3}$$

By eliminating between (1) and (3), the ratio of voltage in the reflected wave to that in the incident wave (*reflection coefficient*) and the ratio of the voltage on the load to that in the incident wave (*transmission coefficient*) may be found:

$$\rho \triangleq \frac{V_-}{V_+} = \frac{R_L - Z_0}{R_L + Z_0} \tag{4}$$

$$\tau \triangleq \frac{V_L}{V_+} = \frac{2R_L}{R_L + Z_0} \tag{5}$$

The most interesting, and perhaps the most obvious, conclusion from the foregoing relations is this: there is no reflected wave if the terminating resistance is exactly equal to the characteristic impedance of the line. All energy of the incident wave is then transferred to the load and τ of (5) is unity.

In Sec. 5.5 the definitions of reflection and transmission coefficients will be given for the case of sinusoidal signals and will include other than purely resistive loads.

The instantaneous incident power at the load is $W_T^+ = I_+ V_+ = V_+^2/Z_0$. The fractional power reflected is, therefore, the constant

$$\frac{W_T^-}{W_T^+} = \rho^2 \tag{6}$$

The remainder of the power goes into the load resistor or the second line so

$$\frac{W_{TL}}{W_T^+} = 1 - \rho^2 \tag{7}$$

Example 5.4a
Pulse on Short-Circuited Line

Let us consider a signal in the form of a pulse having a constant value V_0 between $t = 0$ and $t = t_1/5$ and zero otherwise. The pulse is fed into an ideal transmission line of length l such that $l = vt_1$. We will analyze what happens when the pulse reaches the end of the line, which will be taken as short circuited.

Drawing (i) in Fig. 5.4c shows the pulse moving along the line at $t \approx 0.3t_1$. The arrows connecting the charges on the conductors are electric field vectors. The integral of the electric field is the voltage between conductors. The current flows in the conductors only where there is voltage, and current continuity is accounted for by displacement currents $\varepsilon\, \partial E/\partial t$ at the leading and trailing edges of the pulse.

At time t_1 the leading edge of the pulse reaches the end of the line. Drawing (ii) in Fig. 5.4c shows the pulse shortly after $t = t_1$. The short circuit requires that the voltage be zero. To maintain the voltage at zero during the time that the incident pulse is at the termination $t_1 < t < t_2$, a negative-z-traveling (reflected) wave having opposite voltage polarity and equal amplitude is generated as shown in drawing (iii). Note that this result is predicted by (4) for $R_L = 0$. Also, the zero voltage on the load agrees in (5) with $R_L = 0$. Notice that the polarity of the current in the reflected wave is the same as in the incident wave, as could be argued from Eqs. 5.2(10) and 5.2(12) with the fact that $V_- = -V_+$. The total voltage on the line at the time used for drawings (ii) and (iii) is their superposition; this is shown in drawing (iv) where it is seen that the voltages are almost

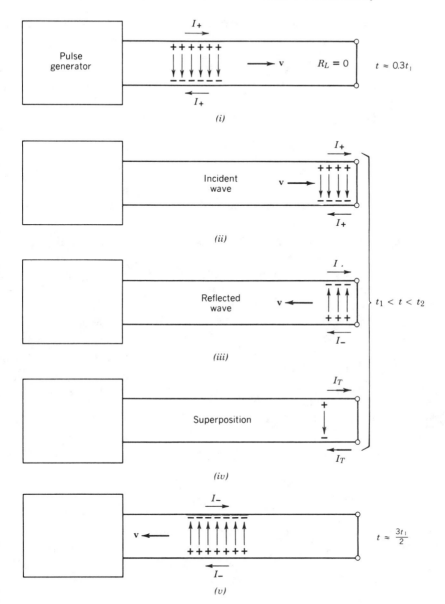

Fig. 5.4c Reflection of a pluse at the end of a shorted line. The times t_1 and t_2 are those for which the leading edge and trailing edges, respectively, reach the end of the line. Drawings (ii), (iii), and (iv) are for various instants during the period in which the reflected wave is generated to maintain the voltage at zero across the short circuit.

completely canceled. At a still later time, the reflected pulse is seen on its way to
the generator [drawing (v)]. At $t = 2t_1$, the pulse will reach the pulse generator.
What happens there depends on the impedance seen looking into the generator.
Typically, the characteristic impedance of the transmission line equals the output
impedance of the generator (it is *matched*). In that case, the reflected pulse is
absorbed in the generator. Otherwise, another reflection takes place.

Example 5.4b
Pulse Reflections on a Transmission Line
Interconnecting Two Computers

The aim of this example is to show the importance of transmission-line matching
in controlling reflections for a transmission line used to interconnect two
computers. Consider the two computers shown in Fig. 5.4d interconnected by a
coaxial cable 100 m in length, and with a velocity of propagation 2×10^8 m/s so
that there is a time delay of 500 ns for a pulse to propagate from input of the
cable to its output. Consider a portion of the digitally coded signal, made up of
10-ns pulses with basic spacing 20 ns as sketched in Fig. 5.4e. This is sketched
versus distance in Fig. 5.4f at 5 ns before the first pulse edge reaches computer
no. 2. If input impedance of the second computer matches the characteristic
impedance of the line, there is no reflection and the entire signal is accepted. But
suppose its input impedance is 100 Ω and the characteristic impedance of the
cable is 50 Ω. By (4), reflection coefficient is

$$\rho = \frac{R_L - Z_0}{R_L + Z_0} = \frac{100 - 50}{100 + 50} = \frac{1}{3}$$

So the train of pulses is reflected, at $\frac{1}{3}$ amplitude, toward computer no. 1 as
sketched versus distance 110 ns after pictured in Fig. 5.4f. If there is impedance
mismatch at the terminals of computer no. 1 additional reflection will take place
when the signal returns there and a spurious signal will be superposed on
whatever desired code is being sent between the computers at that time. Since the

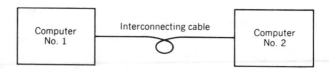

Fig. 5.4d Transmission line cable for transmitting digital signals between two computers.

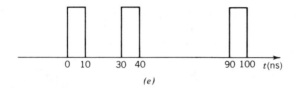

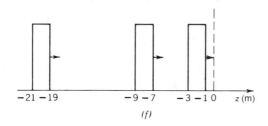

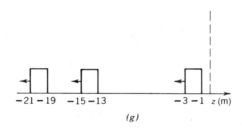

Fig. 5.4 (e) A portion of a pulse-coded signal versus time for the computer interconnection of Fig. 5.4d. (f) Voltage versus distance for a time 5 ns before leading edge of first pulse reaches input of computer no. 2 (defined as $z = 0$). (g) Voltage versus distance for reflected signal 110 ns after instant of sketch (f).

reflected "echo" is of lower amplitude than the original signal, differentiation is possible on the basis of amplitude level, but there is an obvious advantage in matching impedances well enough that reflected signals are small.

_____ **Example 5.4c** _____
Square Wave in z Applied to Infinite Line

As a third example, consider an infinite line suddenly charged at $t = 0$ with a square wave in distance from $z = -1$ m to $z = +1$ m. Voltage distribution at $t = 0$ is then as in Fig. 5.4h. This may be considered a superposition of two such square waves, each of amplitude $V_0/2$. One of these moves to the right and the other to the left, each with velocity v. Thus at $t = 1.667$ ns (taking $v = 3 \times 10^8$ m/s) the two partial waves and their sum are shown in Fig. 5.4i. Figure 5.4j shows the corresponding current distribution, taking into account the different sign relations between current and voltage for positively and negatively traveling waves. Figure 5.4k shows voltage and current distributions at $t = 5$ ns.

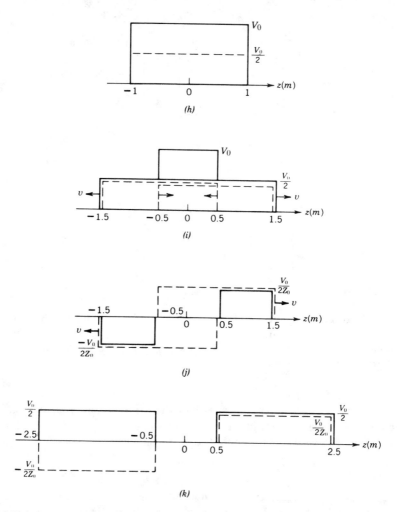

Fig. 5.4 Voltage and current distributions for Ex. 5.4c. (h) Voltage distribution at $t = 0$. (i) Traveling waves (dashed) and total voltage versus z (solid) at $t = 1.667$ ns. (j) Traveling waves of current (dashed) and total current (solid) versus z at $t = 1.667$ ns. (k) Voltage (solid) and current (dashed) versus z at $t = 5$ ns.

SINUSOIDAL WAVES ON IDEAL TRANSMISSION LINES WITH DISCONTINUITIES

5.5 Reflection and Transmission Coefficients and Impedance and Admittance Transformations for Sinusoidal Voltages

The preceding discussion has involved little restriction on the type of variation with time of the voltages applied to the transmission lines. Many practical problems are concerned with sinusoidal time variations. If a sinusoidal voltage is supplied to a line, it can be represented at $z = 0$ by

$$V(0, t) = V \cos \omega t \tag{1}$$

The corresponding wave traveling in the positive z direction is

$$V_+(z, t) = |V_+| \cos \omega \left(t - \frac{z}{v_p} \right)$$

and that traveling in the negative z direction is

$$V_-(z, t) = |V_-| \cos \left[\omega \left(t + \frac{z}{v_p} \right) + \theta_\rho \right]$$

The total voltage is the sum of the two traveling waves:

$$V(z, t) = |V_+| \cos \omega \left(t - \frac{z}{v_p} \right) + |V_-| \cos \left[\omega \left(t + \frac{z}{v_p} \right) + \theta_\rho \right] \tag{2}$$

The corresponding current, from Eq. 5.2(13), is

$$I(z, t) = \frac{|V_+|}{Z_0} \cos \omega \left(t - \frac{z}{v_p} \right) - \frac{|V_-|}{Z_0} \cos \left[\omega \left(t + \frac{z}{v_p} \right) + \theta_\rho \right] \tag{3}$$

In Sec. 5.2 we saw that a constant point on a wave described by $F(t - z/v)$ is seen by an observer moving with a velocity v in the positive z direction. The argument of a sinusoid is called its *phase* so the velocity for which phase is constant is called the *phase velocity* v_p.

For sinusoidal time variations, it is useful to rewrite (2) and (3) in phasor form:

$$V = V_+ e^{-j\beta z} + V_- e^{j\beta z} \tag{4}$$

$$I = \frac{1}{Z_0} [V_+ e^{-j\beta z} - V_- e^{j\beta z}] \tag{5}$$

where

$$\beta = \frac{\omega}{v_p} \tag{6}$$

We may take V_+ as the reference for zero phase so that it is real. Then V_- is in general complex and equal to $|V_-|e^{j\theta_\rho}$ with θ_ρ being the phase angle between reflected and incident waves at $z = 0$ as in the instantaneous form (2).

The quantity β is called the *phase constant* of the line since βz measures the instantaneous phase at a point z with respect to $z = 0$. Moreover, voltage (or current) is observed to be the same at any two points along the line that are separated in z such that βz differs by multiples of 2π. The shortest distance between points of like current or voltage is called a *wavelength* λ. By the foregoing reasoning.

$$\beta\lambda = 2\pi$$

or

$$\beta = \frac{2\pi}{\lambda} \tag{7}$$

The expressions for reflection and transmission coefficients in Eqs. 5.4(4) and 5.4(5) can now be written in a special form for sinusoidal waves. It is convenient to choose the origin of the z coordinate at the discontinuity to be analyzed, as shown in Fig. 5.5 for three representative discontinuities. It is assumed that a

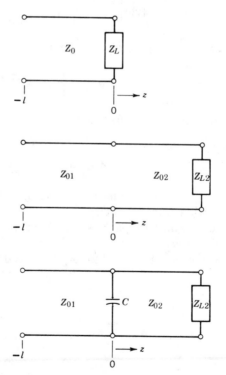

Fig. 5.5 Representative situations where a line of length l with a discontinuity at $z = 0$ is the subject of the analysis.

previous analysis gave the equivalent value of impedance looking in the $+z$ direction at $z = 0$ in the two lower lines. We will see below how this is done. The ratio of total phasor voltage to total phasor current at any point is the definition of impedance. We set the impedance at $z = 0$, the load impedance Z_L, equal to the ratio (4) to (5). Solving for the ratio V_-/V_+, the reflection coefficient for sinusoidal waves is obtained:

$$\rho = \frac{V_-}{V_+} = \frac{Z_L - Z_0}{Z_L + Z_0} \tag{8}$$

Also, (4) and (5) can be combined to give the transmission coefficient:

$$\tau = \frac{V_L}{V_+} = \frac{2Z_L}{Z_L + Z_0} \tag{9}$$

The load voltage V_L is the total voltage at $z = 0$.

The expressions for power reflected and transmitted at a discontinuity, given for real functions of time in Eqs. 5.4(6)–(7), can be adapted to sinusoidal signals of the complex exponential type. Since power in this case is $VI^*/2$ and for a single wave in a loss-free line this is $VV^*/2Z_0 = |V|^2/2Z_0$, the fractional reflected power is

$$\frac{W_T^-}{W_T^+} = \frac{|V_-|^2}{|V_+|^2} = |\rho|^2 \tag{10}$$

and the remainder goes into the load:

$$\frac{W_{TL}}{W_T^+} = 1 - |\rho|^2 \tag{11}$$

Now let us find expressions for the input impedance and admittance at $-l$. We find impedance by dividing (4) by (5) for $z = -l$:

$$Z_i = Z_0 \left[\frac{e^{j\beta l} + \rho e^{-j\beta l}}{e^{j\beta l} - \rho e^{-j\beta l}} \right] \tag{12}$$

Or, substituting (8),

$$Z_i = Z_0 \left[\frac{Z_L \cos \beta l + j Z_0 \sin \beta l}{Z_0 \cos \beta l + j Z_L \sin \beta l} \right] \tag{13}$$

By defining admittances $Y_i = 1/Z_i$, $Y_L = 1/Z_L$, and $Y_0 = 1/Z_0$, we can find an expression of the same form for Y_i:

$$Y_i = Y_0 \left[\frac{Y_L \cos \beta l + j Y_0 \sin \beta l}{Y_0 \cos \beta l + j Y_L \sin \beta l} \right] \tag{14}$$

_____ **Example 5.5** _____
Cascaded Thin-Film Lines

Suppose the second diagram of Fig. 5.5 represents two thin-film transmission lines in a microwave integrated circuit. The load Z_{L2} represents a device having a real impedance of 20 Ω at the signal frequency, 18 GHz. Line 2, of characteristic impedance $Z_{02} = 30$ Ω, has a length $l_2 = 2$ mm. Line 1, of characteristic impedance $Z_{01} = 20$ Ω, has a length $l_1 = 1.5$ mm. The phase velocities for both lines are the same, 2×10^8 m/s. Let us find the impedance at the input to line 1.

First we must solve for the way the load impedance Z_{L2} transforms along the line attached to it. To do this we apply formula (13) to that section of line. For both lines (6) gives

$$\beta = \frac{\omega}{v_p} = \frac{(2\pi)(18 \times 10^9) \text{ rad/s}}{2 \times 10^8 \text{ m/s}} = 566 \text{ rad/m}$$

For variety we will take angles in degrees here as both degrees and radians are commonly used in transmission-line calculations

$$Z_{i2} = 30 \left[\frac{20 \cos 64.9° + j30 \sin 64.9°}{30 \cos 64.9° + j20 \sin 64.9°} \right]$$

$$= 36.7 + j11.8 \text{ Ω}$$

Now we can find Z_{i1} at the input to line 1 using Z_{i2} as the load Z_{L1}:

$$Z_{i1} = 20 \left[\frac{(36.7 + j11.8) \cos 48.6° + j20 \sin 48.6°}{20 \cos 48.6° + j(36.7 + j11.8) \sin 48.6°} \right]$$

$$= 18.9 - j14.6 \text{ Ω}$$

Systems with several lines of different characteristic impedances in cascade can be analyzed as we have in this example. In each case the analysis starts with the point farthest from the signal source, transforming the impedance back successively to the next discontinuity until the input is reached. In general, Z_0, β, and l will be different for each section.

_____

5.6 Standing-Wave Ratio

Let us examine the phases of the voltages in Eq. 5.5(4). One coefficient, say V_+, can be chosen to be real by choice of origin of time. The reflection coefficient, Eq. 5.5(8), which is a complex number, can be written in the form $|\rho|e^{j\theta_\rho}$ so V_- in Eq. 5.5(4) can be replaced with $V_+|\rho|e^{j\theta_\rho}$ giving

$$V = V_+ e^{-j\beta z} + V_+|\rho|e^{j(\theta_\rho + \beta z)} \tag{1}$$

Let us write this as a real function of time with $-z$ replaced by l, the distance from the end of the line:

$$V(t, -l) = V_+ \cos(\omega t + \beta l) + V_+ |\rho| \cos(\omega t - \beta l + \theta_\rho)$$

Considering any particular instant, say $t = 0$, we can readily see that the argument of the cosine in the incident wave (first term), which is its phase, increases with distance from $z = 0$ and that the phase of the reflected wave (second term) decreases. These phases are shown in the top drawings of Fig. 5.6 where it is clear that at some distance z_0 the phases are the same. At z_π they differ by π rad, one having decreased by $\pi/2$ and the other having increased by $\pi/2$. At $z_{2\pi}$ they differ by 2π rad, and so on. So there are a series of locations where the two sinusoids are in phase and another series of locations where they are π rad out of phase. Where they are in phase, they add directly at each instant and where π rad out of phase, they subtract. At the former locations the total voltage has its maximum amplitude and at the latter, its minimum. Analysis of the sum of the incident and reflected waves given in (1) (see Prob. 5.6f) shows that the total

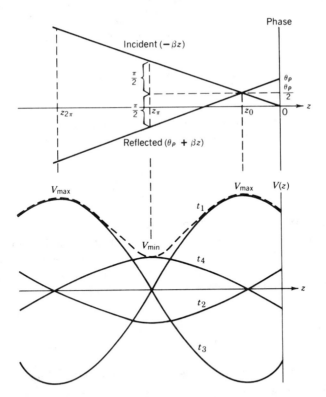

Fig. 5.6 The upper graph shows the phases of the incident and reflected waves at $t = 0$ on a line with a reflection coefficient $\rho = |\rho|e^{j\theta_\rho}$. The total voltage $V(z)$ is shown for selected times in the lower graph. The broken line gives the voltage amplitude along the line.

voltage can also be represented as the sum of a standing wave and a traveling wave. The total voltage is shown in the lower drawing of Fig. 5.6 for several particular times in a cycle selected to show the voltage when it has its maximum and minimum peak values. The broken line shows the voltage *amplitude* along the transmission line.

The maximum voltage is

$$V_{max} = |V_+| + |V_-| \tag{2}$$

and the minimum, found a quarter-wavelength from the maximum, is

$$V_{min} = |V_+| - |V_-| \tag{3}$$

The standing-wave ratio is then defined as the ratio of the maximum voltage amplitude to the minimum voltage amplitude,

$$S = \frac{V_{max}}{V_{min}} \tag{4}$$

By substituting (2) and (3) and the definition of reflection coefficient, Eq. 5.5(8), we find

$$S = \frac{|V_+| + |V_-|}{|V_+| - |V_-|} = \frac{1 + |\rho|}{1 - |\rho|} \tag{5}$$

It is seen that standing-wave ratio is directly related to the magnitude of reflection coefficient ρ, giving the same information as this quantity. The inverse relation is

$$|\rho| = \frac{S - 1}{S + 1} \tag{6}$$

Figure 5.6 is plotted for $S = 3$, corresponding to $|\rho| = \frac{1}{2}$.

Because of the negative sign appearing in the current equation, Eq. 5.5(5), it is evident that, at the position where the two traveling-wave terms add in the voltage relations, they subtract in the current relation, and vice versa. The maximum voltage position is then a minimum current position. The value of the minimum current is

$$I_{min} = \frac{|V_+| - |V_-|}{Z_0} \tag{7}$$

At this position impedance is purely resistive and has the maximum value it will have at any point along the line:

$$Z_{max} = Z_0 \left[\frac{|V_+| + |V_-|}{|V_+| - |V_-|} \right] = Z_0 S \tag{8}$$

At the position of the voltage minimum, current is a maximum, and impedance is a minimum and real:

$$I_{max} = \frac{|V_+| + |V_-|}{Z_0} \tag{9}$$

$$Z_{min} = Z_0 \left[\frac{|V_+| - |V_-|}{|V_+| + |V_-|} \right] = \frac{Z_0}{S} \tag{10}$$

_____ Example 5.6 _____
Slotted-Line Impedance Measurement

A *slotted-line* is an instrument that can be used to measure impedances. It is a transmission line containing a movable probe with which the standing-wave ratio, and the position of a voltage minimum or maximum can be found. The unknown impedance is connected to the end of the slotted line.

Suppose a measurement on a slotted-line of characteristic impedance $Z_0 = 50\ \Omega$ reveals a standing-wave ratio $S = 3$ and the closest voltage minimum is 0.33λ from the unknown load impedance. Let us see how Z_L is deduced. In Fig. 5.6, we see that at z_π where the minimum is found, the phase of the incident wave, $-\beta z$ attains the value $(\theta_p + \pi)/2$ so

$$l_{min} = -z_{min} = \frac{\theta_p + \pi}{2\beta}$$

and

$$\theta_p = 2\beta(0.33\lambda) - \pi = 2\left(\frac{2\pi}{\lambda}\right)(0.33\lambda) - \pi = 1.0 \text{ rad}$$

where we have used Eq. 5.5(7). Using the given S and (6) we find $|\rho|$ to be 0.5. Then

$$\rho = 0.5e^{j1.0}$$

From Eq. 5.5(8) we can get Z_L in terms of ρ and Z_0. Thus

$$Z_L = Z_0 \left[\frac{1 + \rho}{1 - \rho} \right] = 50 \left[\frac{1 + 0.5e^{j1.0}}{1 - 0.5e^{j1.0}} \right] = 52.8 + j59.3\ \Omega$$

5.7   The Smith Transmission-Line Chart

Many graphical aids for transmission line computations have been devised. Of these, the most generally useful has been one presented by P. H. Smith,[3] which

___

[3] P. H. Smith, *Electronic Applications of the Smith Chart*, Krieger, Melbourne, FL, 1983.

consists of loci of constant resistance and reactance plotted on a polar diagram in which radius corresponds to magnitude of reflection coefficient. The chart enables one to find simply how impedances are transformed along the line, or to relate impedance to reflection coefficient or to standing-wave ratio and position of a voltage minimum. By combinations of operations, it enables one to understand the behavior of complex impedance-matching techniques and to devise new ones. It is much used in displaying the locus of impedance of many useful devices as frequency is varied. Although computer programs are available for transmission-line calculations, its role in displaying and understanding matching mechanisms remains useful.

The discussion of the chart will begin with Eq. 5.5(12) which gives impedance in terms of reflection coefficient. If we define a normalized impedance

$$\zeta(l) = (r + jx) \triangleq \frac{Z_i}{Z_0} \tag{1}$$

and a complex variable w equal to the reflection coefficient at the end of the line, shifted in phase to correspond to the input position l,

$$w = u + jv \triangleq \rho e^{-2j\beta l} \tag{2}$$

Equation 5.5(12) may then be written

$$\zeta(l) = \frac{1 + w}{1 - w} \tag{3}$$

or

$$r + jx = \frac{1 + (u + jv)}{1 - (u + jv)} \tag{4}$$

This equation may be separated into real and imaginary parts as follows:

$$r = \frac{1 - (u^2 + v^2)}{(1 - u)^2 + v^2} \tag{5}$$

$$x = \frac{2v}{(1 - u)^2 + v^2} \tag{6}$$

or

$$\left(u - \frac{r}{1 + r}\right)^2 + v^2 = \frac{1}{(1 + r)^2} \tag{7}$$

$$(u - 1)^2 + \left(v - \frac{1}{x}\right)^2 = \frac{1}{x^2} \tag{8}$$

If we then wish to plot the loci of constant resistance r on the w plane (u and v serving as rectangular coordinates), (7) shows that they are circles with centers on the u axis at $[r/(1 + r), 0]$ and with radii $1/(1 + r)$. The curves for $r = 0, \frac{1}{2}, 1, 2, \infty$

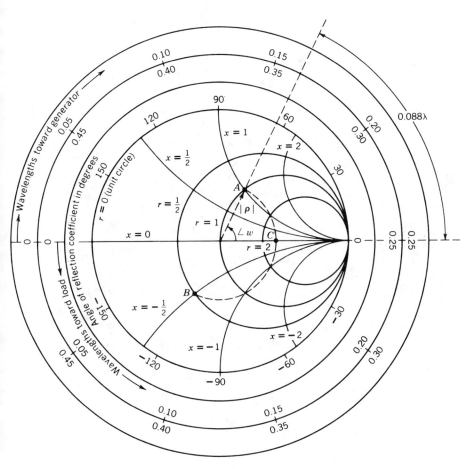

Fig. 5.7a Polar transmission-line chart for impedances in plane of the complex variable $w = u + jv = |\rho|e^{j(\theta_\rho - 2\beta l)}$. If the point at which w is measured is the load, $\angle w = \angle \rho$. Points A, B, and C and associated broken lines relate to Ex. 5.7a, 5.7b, and 5.7c.

are sketched in Fig. 5.7a. From (8), the curves of constant x plotted on the w plane are also circles, with centers at $(1, 1/x)$ and with radii $1/|x|$. Circles for $x = 0$, $\pm\frac{1}{2}$, ± 1, ± 2, ∞ are sketched in Fig. 5.7a. Any point on a given transmission line will have some impedance with positive resistance part, and so will correspond to a particular point on the inside of the unit circle of the w plane. Several uses of the chart will follow. Many extensions and combinations of the ones to be cited will be obvious to the reader. A chart with more divisions is given in Fig. 5.7b.

To Find Reflection Coefficient Given Load Impedance, and Conversely The point within the unit circle of the Smith chart corresponding to a particular position on a transmission line may of course be located at once if the normalized

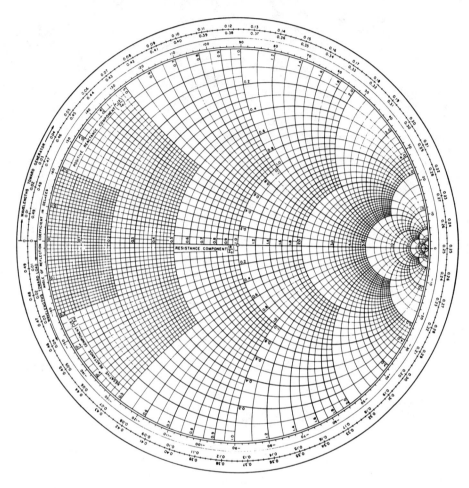

Fig. 5.7*b*

impedance corresponding to that position is known. This is done with a reasonable degree of accuracy by utilizing the orthogonal families of circles giving resistance and reactance as described above. Thus the point A of Fig. 5.7*a* is the intersection of the circles $r = 1$ and $x = 1$, and corresponds to a position with normalized impedance $1 + j1$. It is clear from (2) that $|w| = |\rho|$ and from (7) that $|w| = (u^2 + v^2)^{1/2} = 1$ on the $r = 0$ circle, which is the outer edge of the graph. A measure of the radius to some point on the graph as a fraction of the radius to the $r = 0$ circle thus gives $|\rho|$ directly. If the point on the Smith chart is the normalized load impedance, $l = 0$, and $\angle w = \angle \rho$ so the phase angle of the reflection coefficient can be read directly. One can, of course, reverse the process to find Z_L if ρ is given.

_____ **Example 5.7a** _____
Reflection Coefficient from Load Impedance

Suppose a transmission line of characteristic impedance $Z_0 = 70\ \Omega$ is terminated with a load $Z_L = 70 + j70\ \Omega$. The normalized load impedance is $\zeta(0) = 1 + j1$, shown as the point A in Fig. 5.7a. The magnitude of ρ is 0.45 and $\angle\rho = \angle w = 1.11$ rad so $\rho = 0.45e^{j1.11}$.

To Transform Impedances Along the Line　　As position l along a loss-free line, measured relative to the load, is changed, only the phase angle of w changes, as can be seen from (2) wherein ρ is a complex number, the reflection coefficient at the load. Thus, change of position along an ideal line is represented on the chart by movement along circles centered at the origin of the w plane. The angle through which w changes is proportional to the length of the line, and by (2) is just twice the electrical length of line βl. (Most charts have a scale around the outside calibrated in fractions of a wavelength, so that the angle need not be computed explicitly. See Fig. 5.7b.) Finally, the direction in which one moves is also defined by (2). If one moves toward the generator (increasing l), the angle of w becomes increasingly negative, which corresponds to clockwise motion about the chart. Motion toward the load corresponds to decreasing l in (2) and thus corresponds to counterclockwise motion about the chart.

_____ **Example 5.7b** _____
Impedance Transformation

Consider the line and load of Ex. 5.7a for which the normalized load impedance is $1 + j1$ and is shown at point A in Fig. 5.7a. If the line is a quarter-wave long (90 electrical degrees), we move through an angle of 180 degrees at constant radius on the chart toward the generator (clockwise) to point B. The normalized input impedance is then read as $0.5 - j0.5$ for point B. If input impedance is given and load impedance desired, the reverse of this procedure can obviously be used.

To Find Standing-Wave Ratio and Position of Voltage Maximum from a Given Impedance, and Conversely　　If we wish the standing-wave ratio of an ideal transmission line terminated in a known load impedance, we make use of the information found in the preceding section. Equation 5.6(8) shows that the location of maximum impedance is also the location of maximum voltage and the impedance there is real. We see that

$$S = \frac{Z_{max}}{Z_0} = \zeta_{max} \qquad (9)$$

and can see from Fig. 5.7*a* that the point where impedance is real and maximum along any ideal transmission line (represented by a $|\rho| = $ constant circle) lies on the right side along the horizontal axis (*u* axis). Thus, in following about the circle on the chart determined by the given load impedance, we note its crossing of the right-hand *u* axis of the *w* plane. The value of the normalized resistance of this point is then the standing-wave ratio; the angle moved through to this position from the load impedance fixes the position of the voltage maximum.

The reversal of this procedure to determine the load impedance, if standing-wave ratio and position of a voltage maximum are given, is straightforward, as is the extension to finding position of voltage minimum, or finding input impedance in place of load impedance.

Example 5.7c
Determination of Standing-Wave Ratio and Location of Voltage Maximum

Let us continue our analysis of the ideal transmission line and load discussed in Exs. 5.7a and 5.7b. The normalized load impedance is $1 + j1$ plotted at point *A*

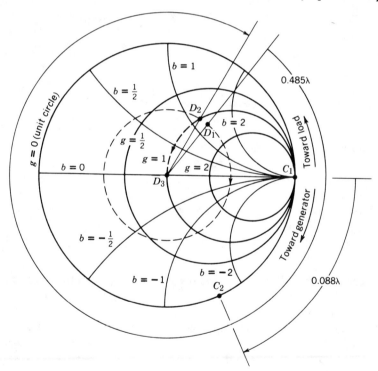

Fig. 5.7c Polar transmission line chart for admittances. The constructions involving points C_1, C_2 and D_1–D_3 relate to Ex. 5.7d.

in Fig. 5.7a. Moving along the line away from the load (clockwise), one arrives at the pure-resistance point C by going 0.088 wavelength. The value of maximum normalized resistance, which equals the standing wave ratio S by (9) is read as 2.6.

Use as an Admittance Diagram   Since admittance transforms along the ideal line in exactly the same manner as impedance, Eq. 5.5(14), it is evident that exactly the same chart may be used for transformation of admittances with the same procedure as for impedances described in the above. Admittance is read for impedance, conductance for resistance, and susceptance for reactance as seen in Fig. 5.7c. The differences to remember are: the right-hand u axis now represents an admittance maximum, and therefore a current maximum instead of a voltage maximum; the phase of the reflection coefficient read as described above and corresponding to a given normalized load admittance is that for current in the reflected wave compared with current in the incident wave and is therefore different by π from that based on voltages. (See Prob. 5.5d.)

_____ **Example 5.7d** _____
Admittance Analysis of Variable Shorted-Stub Tuner

In this example we will use the Smith chart for admittances to analyze a mismatched line that employs a variable shorted-stub tuner to produce a unity standing-wave ratio in the line leading up to the stub. Figure 5.7d shows the arrangement. Suppose $Z_0 = 50\,\Omega$ in the main line and $Z_{0s} = 70\,\Omega$ in the stub.

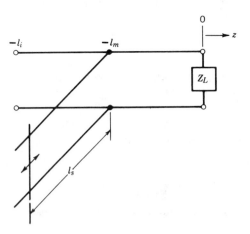

Fig. 5.7d Variable shorted-stub tuner connected in shunt to provide matching at $-l_m$ and therefore $S = 1$ for $z < -l_m$.

Assume $Z_L = 20 - j20\,\Omega$. We will find the appropriate stub location $z = -l_m$ and the stub length l_s. The admittance chart rather than the impedance form is used because of the convenience in handling shunt circuits in the admittance formulation.

The load admittance is $Y_L = 1/Z_L = 0.025 + j0.025$ S and the characteristic admittance is $Y_0 = 1/Z_0 = 0.020$ S. The normalized load admittance is $1.25 + j1.25$; this is at point D_1 in Fig. 5.7c.

The input admittance of the shorted stub is seen from Eq. 5.5(14) to be purely imaginary since $Y_i = -jY_0 \cot \beta l$. This suggests that it should be placed at a point along the main line $-z = l_m$ where the admittance has a normalized real part $g = 1$ and an imaginary part b that can be canceled by Y_{is}. Following a path of constant $|\rho| = |w|$ toward the generator, we come to the $g = 1$ curve where $b = 1.13$ (point D_2) corresponding to an unnormalized susceptance of $B = (0.020)(1.13) = 0.0226$ S. That is seen to be 0.485λ from the load. A stub with an input susceptance of -0.0226 S is connected in shunt to cancel out the imaginary part of the admittance. This moves us on the admittance chart from D_2 to D_3 where $g = 1$ and the line is perfectly matched for waves approaching the stub location from the generator.

Now let us find the length l_s of the stub. Since $Z_{0s} = 70\,\Omega$, $Y_{0s} = 0.014$ S. The normalized input susceptance of the stub must be $b_{is} = -0.0226/0.014 = -1.61$. The shorted end of the stub has an infinite admittance. That is at C_1 on the Smith admittance chart in Fig. 5.7c at the right end of the real axis. To transform this admittance to the normalized susceptance -1.61 we move clockwise as shown to point C_2. The length of the stub must be 0.088λ.

The line to the right of the stub appears as a conductance in parallel with a capacitance; addition of the shunt stub provides an inductance which makes the combination appear as a parallel tuned circuit. It should be clear that this matching technique leaves a standing wave in the line between the load and the stub as well as in the stub.

To Transform Impedance Along Cascaded Lines   It is often useful to find the input impedance of cascaded lines of differing characteristics as in Ex. 5.5a. The Smith chart involves impedances normalized to the characteristic impedance so a study of impedance transformation in a cascade of lines requires renormalization for each line. One starts at the load and transforms, line by line, back toward the generator.

_____ **Example 5.7e** _____
Impedance Transformation Along Cascaded Lines

For the cascaded ideal lines in Fig. 5.7e, let us find the fraction of the power incident in line 1 in a 10-GHz signal that is absorbed in the load. This is given by

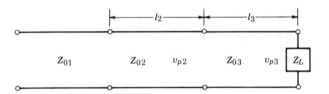

Fig. 5.7e Cascaded transmission lines with parameters for Ex. 5.7e. $Z_{01} = 50\,\Omega$, $Z_{02} = 70\,\Omega$, $Z_{03} = 50\,\Omega$, $Z_L = 100\,\Omega$, $l_2 = 3\,\text{mm}$, $l_3 = 2\,\text{mm}$, $v_{p2} = 1.5 \times 10^8\,\text{m/s}$, $v_{p3} = 2 \times 10^8\,\text{m/s}$.

a knowledge of $|\rho|$ in line 1 using Eq. 5.5(11) and recognizing that the power passing the junction at the end of line 1 must be absorbed in the load since we are assuming ideal lossless lines.

Using the parameters for line 3 given in Fig. 5.7e we find the normalized load impedance $\zeta_{L3} = Z_L/Z_{03} = 2$. The wavelength $\lambda_3 = v_3/f = 2\,\text{cm}$ so $l_3 = 0.1\lambda_3$. The load impedance ζ_{L3} is marked as point E_1 on the Smith chart in Fig. 5.7f. We move along the constant $|w|$ circle toward the generator by 0.1λ. The point E_2 is at the normalized input impedance $\zeta_{i3} = 0.98 - j0.70$.

To transform the impedance along line 2, we must first denormalize ζ_{i3} and then normalize it to line 2 to get the load impedance ζ_{L2}; thus, $\zeta_{L2} = Z_{03}\zeta_{i3}/Z_{02} = 0.70 - j0.50$. This point is marked E_3. The wavelength in line 2 is $\lambda_2 = v_2/f = 1.5\,\text{cm}$ so $l_2 = 0.2\lambda_2$. We move along a constant $|w|$ circle clockwise from E_3 by 0.2λ to the input of line 2 (marked E_4) where $\zeta_{i2} = 0.65 + j0.46$.

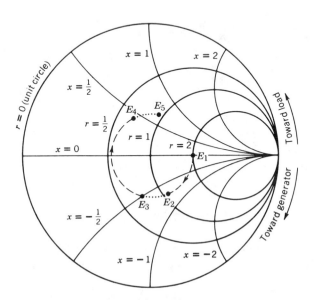

Fig. 5.7f Determination of $|\rho|$ for Ex. 5.7e.

To find the reflection coefficient at the load point for line 1, we renormalize ζ_{i2} so $\zeta_{L1} = Z_{02}\zeta_{i2}/Z_{01} = 0.91 + j0.64$. This is the point E_5. Measuring the distance from E_5 to the origin and dividing by the radius of the $r = 0$ line, we obtain $|\rho_1| = 0.32$. Then

$$\frac{W_{TL}}{W_{T1}^+} = 1 - |\rho_1|^2 = 0.90$$

is the fraction of the incident power in line 1 that is dissipated in the load.

5.8 Purely Standing Wave on an Ideal Line

An important special case of standing waves on a transmission line, introduced in general in Sec. 5.6, is one in which all the incident energy is reflected. The reflected wave has the same amplitude as the incident wave so $S = \infty$. It is clear from Eq. 5.5(8) that $|\rho| = 1$ so that $|V_-| = |V_+|$ if any of the following conditions exist: (1) short-circuit load, $Z_L = 0$; (2) open-circuit load, $Z_L = \infty$; (3) purely reactive load. The last is less obvious than the others but is easily shown (Prob. 5.8b). In each case $|V_-| = |V_+|$ because the load cannot dissipate power and it must, therefore, be fully reflected.

Suppose that a transmission line, shorted at one end, is excited by a sinusoidal voltage at the other. Let us select the position of the short as the reference, $z = 0$. The short imposes the condition that, at $z = 0$, voltage must always be zero. From Eq. 5.5(4):

$$V(0) = V_+ + V_- = 0$$

If $V_- = -V_+$ is substituted in Eqs. 5.5(4) and 5.5(5),

$$V = V_+[e^{-j\beta z} - e^{j\beta z}] = -2jV_+ \sin \beta z \tag{1}$$

$$I = \frac{V_+}{Z_0}[e^{-j\beta z} + e^{j\beta z}] = 2\frac{V_+}{Z_0} \cos \beta z \tag{2}$$

These results, typical for standing waves, show the following.

1. Voltage is always zero not only at the short, but also at multiples of $\lambda/2$ to the left. That is,

$$V = 0 \quad \text{at} \quad -\beta z = n\pi \quad \text{or} \quad z = -n\frac{\lambda}{2}$$

2. Voltage is a maximum at all points for which βz is an odd multiple of $\pi/2$. These are at distances odd multiples of a quarter wavelength from the short circuit. Figure 5.8 shows this and also the time evolution of the voltage found in the usual way from (1) and (2), choosing the zero of time to make V_+ real.

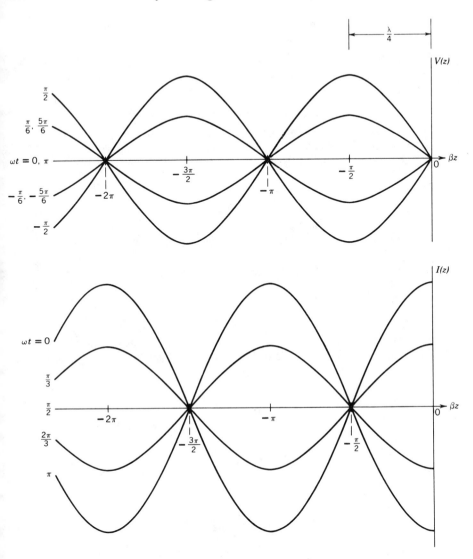

Fig. 5.8 Time evolution of voltage and current on a shorted transmission line. The zeros and extrema remain at the same locations.

3. Current is a maximum at the short circuit and at all points where voltage is zero; it is zero at all points where voltage is a maximum. Figure 5.8 shows the time variation of current along the line.

4. Current and voltage are not only displaced in their space patterns, but also are 90 degrees out of time phase, as indicated by the j appearing in (1) and as seen in Fig. 5.8.

5. The ratio between the maximum current on the line and the maximum voltage is Z_0, the characteristic impedance of the line.
6. The total energy in any length of line a multiple of a quarter-wavelength long is constant, merely interchanging between energy in the electric field of the voltages and energy in the magnetic field of the currents.

To check the energy relation just stated, let us calculate the magnetic energy of the currents at a time when the current pattern is a maximum and voltage is zero everywhere along the line. Current is given by (2). The energy is calculated for a quarter-wavelength of the line, assuming V_+ to be real.

$$U_M = \frac{L}{2} \int_{-\lambda/4}^{0} |I|^2 \, dz = \frac{L}{2} \int_{-\lambda/4}^{0} \frac{4V_+^2}{Z_0^2} \cos^2 \beta z \, dz$$

$$= \frac{2V_+^2 L}{Z_0^2} \left[\frac{z}{2} + \frac{1}{4\beta} \sin 2\beta z \right]_{-\lambda/4}^{0}$$

Since $\beta = 2\pi/\lambda$ by Eq. 5.5(7), the foregoing is

$$U_M = \frac{V_+^2 L \lambda}{4Z_0^2} \tag{3}$$

The maximum energy stored in the distributed capacitance effect of the line is calculated for the quarter-wavelength when the voltage pattern is a maximum and current is everywhere zero. Voltage is given by (1).

$$U_E = \frac{C}{2} \int_{-\lambda/4}^{0} |V|^2 \, dz = \frac{C}{2} \int_{-\lambda/4}^{0} 4V_+^2 \sin^2 \beta z \, dz$$

$$= 2CV_+^2 \left[\frac{z}{2} - \frac{1}{4\beta} \sin 2\beta z \right]_{-\lambda/4}^{0} = \frac{CV_+^2 \lambda}{4} \tag{4}$$

By the definition of Z_0, (3) may also be written

$$U_M = \frac{V_+^2 L \lambda}{4L/C} = \frac{V_+^2 C \lambda}{4} = U_E \tag{5}$$

Thus the maximum energy stored in magnetic fields is exactly equal to that stored in electric fields 90 degrees later in phase. It can also be shown that the sum of electric and magnetic energy at any other part of the cycle is equal to this same value.

Expressions (1) and (2) are also valid for a transmission line with short circuits both at $z = 0$ and another point where $z = n(-\pi/\beta)$, for any integer n. With some way to couple energy into a section of line short circuited at both ends, at a frequency such that the above criterion on z is satisfied (recall that $\beta = \omega/v$), there will be voltages and currents satisfying (1) and (2). At each such frequency, the line is said to have a *resonance*. This idea will be developed further in Sec. 5.10.

NONIDEAL TRANSMISSION LINES

5.9   Power Loss and Attenuation in Low-Loss Lines

In this section we use physical arguments to find expressions for approximate power loss and attenuation of a traveling wave on a low-loss transmission line. Let us take the formula for a traveling wave and assume it is known that the major effect of small but finite losses in the line is to produce an attenuating exponential multiplier:

$$V = V_+ e^{-\alpha z} e^{-j\beta z} \tag{1}$$

$$I = I_+ e^{-\alpha z} e^{-j\beta z} \tag{2}$$

The average power transfer at any position is then given by $W_T = \frac{1}{2}\,\mathrm{Re}(VI^*)$:

$$W_T = \tfrac{1}{2} V_+ I_+ e^{-2\alpha z} \tag{3}$$

It is assumed here that Z_0 is real so I_+ is in phase with V_+. With losses, Z_0 actually becomes complex, as we will see in Sec. 5.11. But for small attenuation, there is no first-order effect on the attenuation constant found here. However, both external and internal inductance should be included in calculating Z_0.

The rate of decrease of the average power (3) with distance along the line must equal the average power loss w_L in the line, per unit length:

$$\frac{\partial W_T}{\partial z} = -w_L = -2\alpha\left(\frac{1}{2}V_+ I_+ e^{-2\alpha z}\right) = -2\alpha W_T$$

or

$$\alpha = \frac{w_L}{2W_T} \tag{4}$$

This is an important formula which, by the nature of the development, applies to the attenuation of a traveling wave along any uniform system.

To apply (4) to a transmission line with series resistance R and shunt conductance G (see equivalent circuit in Fig. 5.9), we first calculate the average power loss per unit length, part of which comes from the current flow through the

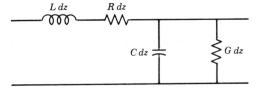

Fig. 5.9 Circuit model of a differential length of a transmission line with losses in the conductor (represented by $R\,dz$) and in the dielectric (represented by $G\,dz$).

resistance and another part from voltage appearing across the shunt conductance. For convenience we calculate w_L and W_T at $z = 0$:

$$w_L = \frac{I_+^2 R}{2} + \frac{V_+^2 G}{2} = \frac{V_+^2}{2}\left[G + \frac{R}{Z_0^2}\right] \tag{5}$$

The average power transferred by the wave at $z = 0$ is

$$W_T = \frac{1}{2}V_+ I_+ = \frac{1}{2}\frac{V_+^2}{Z_0} \tag{6}$$

So (4) gives attenuation as

$$\alpha = \frac{1}{2}\left[GZ_0 + \frac{R}{Z_0}\right] \quad \text{Np/m} \tag{7}$$

The *neper* (Np) is a unit-free name for attenuation that measures the decay of voltage amplitude. One neper per meter indicates that the *amplitude* has decayed to $1/e$ of its incoming value in 1 m. The *decibel* (dB) is an alternative measure describing the rate of *power* decrease by the formula $10\log_{10}W_{T2}/W_{T1}$.[4] Attenuation in decibels per meter is 8.686 times the attenuation in nepers per meter.

--- **Example 5.9** ---
Attenuation in a Thin-Film Transmission Line

Let us find the attenuation in an aluminum thin-film parallel-plane transmission line for a signal of 18 GHz. The structure has the form in Fig. 2.5c and fringing fields will be neglected. The metal thickness h is 2.0 μm with a dielectric thickness d also of 2.0 μm. The width of the conductors is typical of photolithographically defined lines, $w = 10\ \mu$m. The relative permittivity of the dielectric is 3.8 and it is assumed to be lossless. From Eq. 1.9(3) the capacitance per unit length is $C = \varepsilon w/d = 1.67 \times 10^{-10}$ F/m. From Eq. 2.5(3) the external inductance per unit length is $L = \mu_0 d/w = 2.51 \times 10^{-7}$ H/m. To see how to treat the internal inductance and resistance of the conductors, the depth of penetration must be compared with the film thickness. From Table 3.17 the depth of penetration for aluminum is $\delta = 0.0826/\sqrt{f}$ so for 18 GHz, $\delta = 0.616\ \mu$m. Thus the aluminum films are 3.2 times the depth of penetration so they are well approximated by arbitrarily thick layers. Then the internal inductance and resistance is given by

---

[4] Decibels are often used for ratios of voltage amplitudes using the formula $20\log_{10}V_2/V_1$. This is only correct if the voltages are across identical impedances as in the case of voltages at points along a transmission line.

the surface impedance. From Table 3.17 the surface resistance is $R_s = 3.26 \times 10^{-7}\sqrt{f}$ so from Eqs. 3.17(4) and 3.17(6) the surface impedance is

$$Z_s = 3.26 \times 10^{-7}\sqrt{f}(1 + j) \qquad (8)$$

and the internal impedance per unit length for both electrodes is

$$Z_i = \frac{2Z_s}{w} = 8.75 \times 10^3(1 + j) \quad \Omega \qquad (9)$$

The characteristic impedance is found from $Z_0 = \sqrt{L/C}$ where L includes both external and internal inductances. The latter is found from (9) by dividing by ω so $L_i = 7.74 \times 10^{-8}$ H. Adding this to the external inductance and substituting the sum and the capacitance into Z_0 we find $Z_0 = 44.3\,\Omega$. Note that if we had neglected internal inductance, Z_0 would have been calculated as $38.4\,\Omega$. The resistance per unit length R is the real part of (9). Substituting Z_0 and R in (7) we get the attenuation constant:

$$\alpha = \frac{R}{2Z_0} = 0.988 \text{ Np/cm} \qquad (10)$$

From this we see that the wave attenuates by about a factor of e in a distance of 1 cm.

5.10   Input Impedance and Quality Factor for Resonant Transmission Lines

Resonant systems play a very significant role in communication systems for impedance matching and filtering and we have already seen some aspects of this in the problem of Ex. 5.7d where resonant sections of lines were used for matching impedances. In Sec. 5.8 we analyzed standing waves on short-circuited ideal lines. In the present section we consider a low-loss line shorted at either one or both ends so that standing waves similar to those discussed in Sec. 5.8 occur. The line is supplied by a voltage source connected at a voltage maximum in either of the ways shown in Fig. 5.10. We will find approximate expressions for the resistance seen by the source and for the quality factor Q of the line, considered as a resonant circuit.

For an ideal line, there are points of voltage maximum and zero current at odd multiples of a quarter-wavelength from the short-circuited ends so the impedance is infinite there. When losses are present, however, the impedance at these positions is high but finite, representing the energy dissipated in the losses of the line. Let us find these losses approximately for a line of n quarter-wavelengths using the expressions for voltage and current derived for an ideal line, Eqs. 5.8(1)

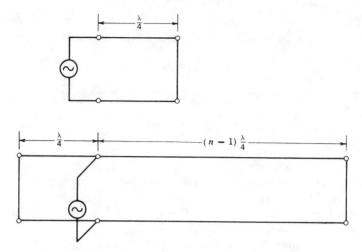

Fig. 5.10 Resonant transmission lines driven by voltage sources at positions of maximum voltage.

and 5.8(2), assuming that they are not greatly changed by the small losses. The average power dissipated in the shunt conductance is then

$$W_G = \int_0^{n\lambda/4} (2V_+ \sin \beta z)^2 \frac{G}{2}\, dz = \left(\frac{4V_+^2 G}{4}\right)\left(\frac{n\lambda}{4}\right) \tag{1}$$

and the average power dissipated in the series resistance is

$$W_R = \int_0^{n\lambda/4} \left(\frac{2V_+ \cos \beta z}{Z_0}\right)^2 \frac{R}{2}\, dz = \left(\frac{4V_+^2 R}{4Z_0^2}\right)\left(\frac{n\lambda}{4}\right) \tag{2}$$

The input resistance (at a voltage maximum) must be such that the voltage appearing across this resistance will produce losses equal to the sum of (1) and (2). The magnitude of voltage there is $2V_+$. Thus

$$\frac{1}{2}\frac{(2V_+)^2}{R_i} = \frac{nV_+^2 \lambda}{4}\left(G + \frac{R}{Z_0^2}\right)$$

or

$$R_i = \frac{8Z_0}{n\lambda[GZ_0 + (R/Z_0)]} \tag{3}$$

A general expression for the quality factor Q of any resonant system is

$$Q \triangleq \frac{\omega_0\,(\text{energy stored})}{\text{average power loss}} = \frac{\omega_0 U}{W_L} \tag{4}$$

For a resonant transmission line of n quarter-wavelengths, the stored energy for each quarter-wavelength is taken as that for the ideal line, Eq. 5.8(5), and the power loss is given by the sum of (1) and (2). The result is

$$Q = \frac{\omega_0 U}{W_L} = \frac{4\omega_0 C V_+^2 n\lambda}{4V_+^2 n\lambda[G + (R/Z_0^2)]} = \frac{\omega_0 C Z_0}{G Z_0 + (R/Z_0)} \tag{5}$$

We see that Q is independent of n; this results from the fact that both the stored energy and power dissipated are proportional to the length. Thus Q is a property of the line, independent of the number of resonant quarter-wavelengths.

The input resistance for a shorted quarter-wavelength section or at the maximum-voltage point of a line $n\lambda/4$ (n even) long shorted at both ends can be re-written using (3) and (5) with Eqs. 5.2(8) and 5.2(14) and 5.5(6) and 5.5(7):

$$R_i = \frac{8Q}{n\lambda\omega_0 C} = \frac{4QZ_0}{n\pi} \tag{6}$$

The input resistance measures the power supplied to maintain a given voltage level; Q increases as the losses decrease leading to a higher input resistance.

If the frequency and, therefore λ is changed so the distance from the input point to the short circuit differs from $\lambda/4$, the input impedance acquires a reactive component of first-order importance. With the same amount of frequency change, the voltage and current patterns do not change much so the resistive part (6) does not change much.

It will be convenient to complete the analysis in terms of admittance. The susceptance that arises is in parallel with the conductance equivalent of (6). Let us calculate it for the lower circuit in Fig. 5.10. It consists of two susceptances in parallel, that for the $\lambda/4$ section on the left and that of the remaining $(n-1)\lambda/4$ portion on the right. For each section, the load admittance is infinite so Eq. 5.5(14) gives, in the lossless approximation:

$$jB_i = -jY_0(\cot \beta l_L + \cot \beta l_R) \tag{7}$$

where $Y_0 = 1/Z_0$ and l_L and l_R are the lengths of the left and right sides of the line. Letting $\beta = \omega/v_p = \omega_0(1 + \delta)/v_p = \beta_0 + \beta_0\delta$ and taking $\beta_0 l_L = \pi/2$ and $\beta_0 l_R = (n-1)\pi/2$, the cotangents in (7) can be approximated for small δ to give

$$B_i = Y_0\left[\frac{\pi}{2}\delta + \frac{(n-1)\pi}{2}\delta\right] = \frac{n\pi}{2}\delta Y_0 \tag{8}$$

Then using (6) and (8), we have the admittance at the feed point in the $n\lambda/4$ line in the lower circuit of Fig. 5.10:

$$Y_i = \frac{n\pi}{2} Y_0\left(\frac{1}{2Q} + j\delta\right) \tag{9}$$

For the upper circuit in Fig. 5.10, the $\cot \beta l_R$ terms in (7) and (8) are missing so (9) describes that circuit with $n = 1$. From this we see that the fractional frequency shift for which the susceptance becomes equal to the conductance, a common measure of circuit sharpness, is

$$\delta_1 = \frac{1}{2Q} \tag{10}$$

or

$$Q = \frac{\omega_0}{2\,\Delta\omega_1} = \frac{f_0}{2\,\Delta f_1} \tag{11}$$

where $2\,\Delta f_1$ is the frequency width between points where the admittance magnitude reaches $\sqrt{2}$ times its value at resonance ($\omega = \omega_0$). Thus Q, as defined by (4), is useful as a measure of sharpness of frequency response, as it is for lumped-element circuits. Resonant low-loss transmission lines can have Q's of thousands in the UHF range of frequencies.

Example 5.10
Open-Ended Parallel-Plane Transmission Line

Consider standing waves in an ideal open-ended transmission line as in Fig. 5.2, but with conductors spaced by a thin dielectric. The losses include dissipation in the conductors and dielectric (Sec. 5.9) as well as radiation out the sides and ends. For the purposes of this example we will concentrate on the radiation out the ends (in the direction of wave propagation).

The power leakage at the ends requires, according to Eqs. 5.5(8) and 5.5(11), a finite Z_L. The more closely spaced the plates, the lower the Z_0 and the greater the discontinuity at the open ends. For a large Z_L/Z_0 ratio, considering first one end,

$$\rho = \frac{1 - Z_0/Z_L}{1 + Z_0/Z_L} \approx \left(1 - \frac{Z_0}{Z_L}\right)^2 \approx 1 - 2\frac{Z_0}{Z_L} = 1 - \xi \tag{12}$$

where $\xi = 2Z_0/Z_L$. The actual value for the complex Z_L can be found approximately from advanced electromagnetic theory.[5]

If this form for ρ is put into Eqs. 5.5(4) and 5.5(5), one finds

$$V = 2V_+ \cos \beta z - \xi V_+ e^{j\beta z} \tag{13}$$

$$I = -j\frac{2V_+}{Z_0} \sin \beta z + \xi \frac{V_+}{Z_0} e^{j\beta z} \tag{14}$$

We see from (13) that without the perturbation of the leakage term in (13) the voltage maximum of the standing-wave pattern is found at $z = 0$. With the frequency chosen so that $\beta l = m\pi$, where l is the length of the line and m is an integer (l is a multiple of $\lambda/2$), another maximum is found at the other end.

---

[5] N. Marcuvitz, *Waveguide Handbook*, McGraw-Hill, New York, 1951, pp. 179–183.

Likewise the minima of the standing wave of current are at the ends. Small traveling-wave components are present because of the power loss.

The quality factor Q is defined by (4). Again we calculate the stored energy using the fields with zero power leakage as an approximation, Eqs. 5.8(4)–(5). The average power lost out the two ends, by symmetry, is $W_L = 2(\frac{1}{2}) \text{Re}(VI^*)$ evaluated at $z = 0$. Use of (5) then gives

$$Q_e = \frac{m\pi}{4Z_0 \, \text{Re}(1/Z_L)} \tag{15}$$

It is easily seen that the Q found in (5) for dielectric and conductor losses applies to the open-circuited lines as well. Radiation out the sides is a more difficult problem but can also be represented by a Q. If all Q's are large they may be treated independently, as we have done, and combined as $Q = [Q_1^{-1} + Q_2^{-1} + Q_3^{-1}]^{-1}$. Thin-film open-ended transmission lines are made with Q values as high as about 1000.

5.11 Transmission Lines with General Forms of Distributed Impedances; ω–β Diagram

In the preceding section, we considered the possibility of adding series resistance and shunt conductance to the ideal transmission line of Fig. 5.2. For an exact analysis of this, and generalization to other forms of distributed impedances, let us take the line shown in Fig. 5.11a with a distributed series impedance Z per

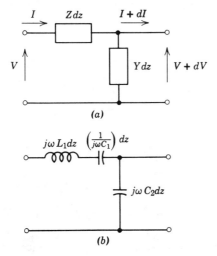

(a)

(b)

Fig. 5.11 (a) Differential length of general transmission line. (b) High-pass transmission line.

unit length, and a distributed shunt admittance Y per unit length. For steady-state sinusoids, using complex notation, the differential equations for voltage and current variations with distance are then

$$\frac{dV}{dz} = -ZI \tag{1}$$

$$\frac{dI}{dz} = -YV \tag{2}$$

Differentiation of (1) and substitution in (2) then yields

$$\frac{d^2V}{dz^2} = \gamma^2 V \tag{3}$$

where

$$\gamma = \sqrt{ZY} \tag{4}$$

The solution to (3) may be written in terms of exponentials, as can be verified by substitution of the expression

$$V = V_+ e^{-\gamma z} + V_- e^{\gamma z} \tag{5}$$

From (1), the corresponding solution for current is

$$I = \frac{1}{Z_0}[V_+ e^{-\gamma z} - V_- e^{\gamma z}] \tag{6}$$

where

$$Z_0 = \frac{Z}{\gamma} = \sqrt{\frac{Z}{Y}} \tag{7}$$

The characteristic impedance, Z_0, is in general complex, indicating that the voltage and current for a single traveling wave are not in phase. The quantity γ is called the *propagation constant* and is also generally complex,

$$\gamma = \alpha + j\beta = \sqrt{ZY} \tag{8}$$

so that if (5) is written in terms of α and β,

$$V = V_+ e^{-\alpha z} e^{-j\beta z} + V_- e^{\alpha z} e^{j\beta z} \tag{9}$$

Thus α tells the rate of exponential attenuation of each wave and is correspondingly called the attenuation constant. The constant β tells the amount of phase shift per unit length for each wave and is called the phase constant as in the loss-free case.

The formula for reflection coefficient derived in Eq. 5.5(8) applies to this case also, remembering that Z_0 is complex. To find the input impedance at $z = -l$ in terms of a given reflection coefficient $\rho = V_+/V_-$ at $z = 0$, division of (5) by (6) yields

$$Z_i = Z_0\left[\frac{V_+e^{\gamma l} + V_-e^{-\gamma l}}{V_+e^{\gamma l} - V_-e^{-\gamma l}}\right] = Z_0\left[\frac{1 + \rho e^{-2\gamma l}}{1 - \rho e^{-2\gamma l}}\right] \tag{10}$$

This may be put in terms of load impedance by substituting Eq. 5.5(8):

$$Z_i = Z_0\left[\frac{Z_L\cosh\gamma l + Z_0\sinh\gamma l}{Z_0\cosh\gamma l + Z_L\sinh\gamma l}\right] \tag{11}$$

Transmission Line with Series and Shunt Losses A very important case in practice is one in which losses must be considered in the transmission line. In general there may be distributed series resistance in the conductors of the line, and distributed shunt conductance because of leakage through the dielectric of the line. Distributed impedance and admittance are then

$$Z = R + j\omega L, \qquad Y = G + j\omega C \tag{12}$$

where L includes both external and internal inductance. These may be used as the values of Z and Y in (4) and (7) to determine propagation constant and characteristic impedance. The formula (10) applies to impedance transformations, and the Smith transmission line chart may be utilized with a modification which recognizes that γ is complex. The procedure is as in Sec. 5.7 except that, in moving along the line toward the generator, one moves not along a circle but along a spiral of radius decreasing according to the exponential $e^{-2\alpha l}$.

For many important problems, losses are finite but relatively small. If $R/\omega L \ll 1$ and $G/\omega C \ll 1$, the following approximations are obtained by retaining up to second-order terms in the binominal expansions of (4) and (7), with (12) substituted.

$$\alpha \approx \frac{R}{2\sqrt{L/C}} + \frac{G\sqrt{L/C}}{2} \tag{13}$$

$$\beta \approx \omega\sqrt{LC}\left[1 - \frac{RG}{4\omega^2 LC} + \frac{G^2}{8\omega^2 C^2} + \frac{R^2}{8\omega^2 L^2}\right] \tag{14}$$

$$Z_0 \approx \sqrt{\frac{L}{C}}\left[\left(1 + \frac{R^2}{8\omega^2 L^2} - \frac{3G^2}{8\omega^2 C^2} + \frac{RG}{4\omega^2 LC}\right) + j\left(\frac{G}{2\omega C} - \frac{R}{2\omega L}\right)\right] \tag{15}$$

In using the foregoing approximate formulas, it is often sufficient to retain only first-order correction terms, in which case β reduces to its ideal value of $2\pi/\lambda$, α is computed from (13), and Z_0 includes a first-order reactive part, given by the last part of (15). Note that the first-order effects of the two loss factors tend to cancel in Z_0 whereas they add in α.

Several of the important formulas for loss-free, low-loss, and general lines are summarized in Table 5.11a. Formulas for properties of lines of several different cross-sectional forms are listed in Table 5.11b.

Table 5.11a
General formulas for transmission lines

Quantity	General Line	Ideal Line	Approximate Results for Low-Loss Lines (See α and β below)
Propagation constant $\gamma = \alpha + j\beta$	$\sqrt{(R+j\omega L)(G+j\omega C)}$	$j\omega\sqrt{LC}$	(See α and β below)
Phase constant β	$\mathrm{Im}(\gamma)$	$\omega\sqrt{LC} = \dfrac{\omega}{v} = \dfrac{2\pi}{\lambda}$	$\omega\sqrt{LC}\left[1 - \dfrac{RG}{4\omega^2 LC} + \dfrac{G^2}{8\omega^2 C^2} + \dfrac{R^2}{8\omega^2 L^2}\right]$
Attenuation constant α	$\mathrm{Re}(\gamma)$	0	$\dfrac{R}{2Z_0} + \dfrac{GZ_0}{2}$
Characteristic impedance Z_0	$\sqrt{\dfrac{R+j\omega L}{G+j\omega C}}$	$\sqrt{\dfrac{L}{C}}$	$\sqrt{\dfrac{L}{C}}\left[1 + j\left(\dfrac{G}{2\omega C} - \dfrac{R}{2\omega L}\right)\right]$
Input impedance Z_i	$Z_0\left[\dfrac{Z_L\cosh\gamma l + Z_0\sinh\gamma l}{Z_0\cosh\gamma l + Z_L\sinh\gamma l}\right]$	$Z_0\left[\dfrac{Z_L\cos\beta l + jZ_0\sin\beta l}{Z_0\cos\beta l + jZ_L\sin\beta l}\right]$	$Z_0\left[\dfrac{\alpha l\cos\beta l + j\sin\beta l}{\cos\beta l + j\alpha l\sin\beta l}\right]$
Impedance of shorted line	$Z_0\tanh\gamma l$	$jZ_0\tan\beta l$	$Z_0\left[\dfrac{\cos\beta l + j\alpha l\sin\beta l}{\alpha l\cos\beta l + j\sin\beta l}\right]$
Impedance of open line	$Z_0\coth\gamma l$	$-jZ_0\cot\beta l$	

Impedance of quarter-wave line	$Z_0\left[\dfrac{Z_L\sinh\alpha l + Z_0\cosh\alpha l}{Z_0\sinh\alpha l + Z_L\cosh\alpha l}\right]$	$\dfrac{Z_0^2}{Z_L}$	$Z_0\left[\dfrac{Z_0 + Z_L\alpha l}{Z_L + Z_0\alpha l}\right]$
Impedance of half-wave line	$Z_0\left[\dfrac{Z_L\cosh\alpha l + Z_0\sinh\alpha l}{Z_0\cosh\alpha l + Z_L\sinh\alpha l}\right]$	Z_L	$Z_0\left[\dfrac{Z_L + Z_0\alpha l}{Z_0 + Z_L\alpha l}\right]$
Voltage along line $V(z)$	$V_i\cosh\gamma z - I_i Z_0\sinh\gamma z$	$V_i\cos\beta z - jI_i Z_0\sin\beta z$	
Current along line $I(z)$	$I_i\cosh\gamma z - \dfrac{V_i}{Z_0}\sinh\gamma z$	$I_i\cos\beta z - j\dfrac{V_i}{Z_0}\sin\beta z$	
Reflection coefficient ρ	$\dfrac{Z_L - Z_0}{Z_L + Z_0}$	$\dfrac{Z_L - Z_0}{Z_L + Z_0}$	
Standing-wave ratio	$\dfrac{1+|\rho|}{1-|\rho|}$	$\dfrac{1+|\rho|}{1-|\rho|}$	

R, L, C Distributed resistance, inductance, conductance, capacitance per unit length.

l Length of line.

Subscript i Denotes input end quantities.

Subscript L Denotes load end quantities.

z Distance along line from input end.

λ Wavelength measured along line.

v Phase velocity of line equals velocity of light in dielectric of line for an ideal line.

Table 5.11b

Formulas for specific transmission line configurations

	coaxial	two-wire	shielded pair $p=\dfrac{s}{d}$ $q=\dfrac{s}{D}$	Formulas for $a\ll b$
Capacitance C, farads/meter	$\dfrac{2\pi\epsilon}{\ln\left(\dfrac{r_0}{r_i}\right)}$	$\dfrac{\pi\epsilon}{\cosh^{-1}\left(\dfrac{s}{d}\right)}$	— — — —	$\dfrac{\epsilon b}{a}$
External inductance L, henrys/meter	$\dfrac{\mu}{2\pi}\ln\left(\dfrac{r_0}{r_i}\right)$	$\dfrac{\mu}{\pi}\cosh^{-1}\left(\dfrac{s}{d}\right)$	— — — —	$\mu\dfrac{a}{b}$
Conductance G, siemens/meter	$\dfrac{2\pi\sigma}{\ln\left(\dfrac{r_0}{r_i}\right)}=\dfrac{2\pi\omega\epsilon''}{\ln\left(\dfrac{r_0}{r_i}\right)}$	$\dfrac{\pi\sigma}{\cosh^{-1}\left(\dfrac{s}{d}\right)}=\dfrac{\pi\omega\epsilon''}{\cosh^{-1}\left(\dfrac{s}{d}\right)}$	— — — —	$\dfrac{\sigma b}{a}=\dfrac{\omega\epsilon'' b}{a}$
Resistance R, ohms/meter	$\dfrac{R_s}{2\pi}\left(\dfrac{1}{r_0}+\dfrac{1}{r_i}\right)$	$\dfrac{2R_s}{\pi d}\left[\dfrac{s/d}{\sqrt{(s/d)^2-1}}\right]$	$\dfrac{2R_{s2}}{\pi d}\left[1+\dfrac{1+2p^2}{4p^4}(1-4q^2)\right]$ $+\dfrac{8R_{s3}}{\pi D}q^2\left[1+q^2-\dfrac{1+4p^2}{8p^4}\right]$	$\dfrac{2R_s}{b}$
Internal inductance L_i, henrys/meter (for high frequency)	$\longleftarrow$		$\dfrac{R}{\omega}$	$\longrightarrow$
Characteristic impedance at high frequency Z_0, ohms	$\dfrac{\eta}{2\pi}\ln\left(\dfrac{r_0}{r_i}\right)$	$\dfrac{\eta}{\pi}\cosh^{-1}\left(\dfrac{s}{d}\right)$	$\dfrac{\eta}{\pi}\left\{\ln\left[2p\left(\dfrac{1-q^2}{1+q^2}\right)\right]-\dfrac{1+4p^2}{16p^4}(1-4q^2)\right\}$	$\eta\dfrac{a}{b}$
Z_0 for air dielectric	$60\ln\left(\dfrac{r_0}{r_i}\right)$	$120\cosh^{-1}\left(\dfrac{s}{d}\right)\cong 120\ln\left(\dfrac{2s}{d}\right)$ if $s/d\gg1$	$120\left\{\ln\left[2p\dfrac{(1-q^2)}{(1+q^2)}\right]-\dfrac{1+4p^2}{16p^4}(1-4q^2)\right\}$	$120\pi\dfrac{a}{b}$
Attenuation due to conductor α_c	$\longleftarrow$		$\dfrac{R}{2Z_0}$	$\longrightarrow$
Attenuation due to dielectric α_d	$\longleftarrow$		$\dfrac{GZ_0}{2}=\dfrac{\sigma\eta}{2}=\dfrac{\pi}{\lambda}\left(\dfrac{\epsilon''}{\epsilon'}\right)$	$\longrightarrow$
Total attenuation dB/meter	$\longleftarrow$		$8.686(\alpha_c+\alpha_d)$	$\longrightarrow$
Phase constant for low-loss lines β	$\longleftarrow$		$\omega\sqrt{\mu\epsilon'}=\dfrac{2\pi}{\lambda}$	$\longrightarrow$

All units above are mks.

$\left.\begin{array}{l}\epsilon=\epsilon'-j\epsilon''=\text{permittivity, farads/meter}\\ \mu=\text{permeability, henrys/meter}\\ \eta=\sqrt{\mu/\epsilon}\text{ ohms}\end{array}\right\}$ for the dielectric

$\epsilon''=$ loss factor of dielectric $=\sigma\cdot/\omega$
$R_s=$ skin effect surface resistivity of conductor, ohms
$\lambda=$ wavelength in dielectric

Formulas for shielded pair obtained from Green, Leibe, and Curtis, *Bell System Tech. Journ.*, **15**, pp. 248–284 (April 1936).

Filter-Type Distributed Circuits   Suppose the distributed series impedance of the general transmission line is formed by inductance and capacitance in series, as shown in Fig. 5.11b. The propagation constant γ is then

$$\gamma = \sqrt{j\omega C_2\left(j\omega L_1 + \frac{1}{j\omega C_1}\right)} = j\omega\sqrt{L_1 C_2\left(1-\frac{\omega_c^2}{\omega^2}\right)} \tag{16}$$

where

$$\omega_c = (L_1 C_1)^{-1/2} \tag{17}$$

The interesting characteristic of this system is that for the lower range of frequencies, $\omega < \omega_c$, γ is purely real, representing an attenuation without losses in the system,

$$\gamma = \alpha = \omega \sqrt{L_1 C_2 \left(\frac{\omega_c^2}{\omega^2} - 1\right)}, \qquad \omega < \omega_c \tag{18}$$

The attenuation in this circuit occurs below the cutoff frequency defined by (17), so that the system is a distributed high-pass filter. The reactive attenuation which occurs arises essentially because of continuous reflections in the system, and is of the same nature as the attenuation in a loss-free, lumped-element filter in the attenuating band.

For frequencies above ω_c, the propagation constant γ is purely imaginary so $\gamma = j\beta$ and is given by (16). It is found useful to plot relations between β and ω with β on the abscissa and ω on the ordinate. These are called ω–β *diagrams*. Figure 5.11c shows the ω–β relation (16) for the line discussed here. Notice that β goes to zero for $\omega = \omega_c$ and does not exist for $\omega < \omega_c$. We saw in (18) that, for this line, there is only attenuation for $\omega < \omega_c$. An important reason for choosing the coordinates of the ω–β diagram as done is that the phase velocity at any frequency which, from Eq. 5.5(6) is $v_p = \omega/\beta$, can be seen immediately as the slope of a line to the origin from the curve, as illustrated in Fig. 5.11c. We see that the phase velocity for the line under consideration is

$$v_p = \frac{1}{\sqrt{L_1 C_2}} \left[1 - \frac{\omega_c^2}{\omega^2}\right]^{-1/2} \tag{19}$$

which is seen to vary strongly for frequencies just above cutoff. Signals with several frequency components propagating in this range will thus have large *dispersion*, as will be discussed more in Sec. 5.12.

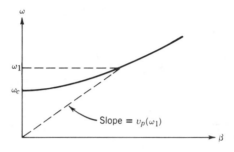

Fig. 5.11c The ω–β diagram for a transmission line having the differential circuit shown in Fig. 5.11b.

5.12   Group and Energy Velocities

A function of time with arbitrary wave shape may be expressed as a sum of sinusoidal waves by Fourier analysis. If it happens that v_p is the same for each frequency component and there is no attenuation, the component waves will add in proper phase at each point along the line to reproduce the original wave shape exactly, but delayed by the time of propagation z/v_p. The velocity v_p in this case describes the rate at which the wave moves down the line and could be said to be the velocity of propagation. This case occurs, for example, in the ideal loss-free transmission line already studied for which v_p is a constant equal to $(LC)^{-1/2}$. If v_p changes with frequency, there is said to be *dispersion* and a signal may change shape as it travels. This causes distortion of analog signals and limits data rates because of the spreading of pulses in digital signals.

In the nondispersive situation described above, a wave of a given shape propagates along a line without distortion. The velocity of the group of frequency components is the same as the phase velocity of any one. In many transmission systems the velocity of the envelope of the wave such as that in Fig. 5.12a can be different from the phase velocities of the frequency components, and it is useful to introduce a so-called *group velocity* to describe the motion of the group, or envelope of the wave. This is the typical case when a high-frequency carrier is modulated by a digital or analog signal.

Let us consider the simplest possible group, a wave having two equal-amplitude sinusoidal components of slightly different frequency. The voltage at $z = 0$ with unity-amplitude components is

$$V(t) = \sin(\omega_0 - d\omega)t + \sin(\omega_0 + d\omega)t \tag{1}$$

Then for a lossless line, the voltage at any point is

$$V(t, z) = \sin[(\omega_0 - d\omega)t - (\beta_0 - d\beta)z] + \sin[(\omega_0 + d\omega)t - (\beta_0 + d\beta)z] \tag{2}$$

in which β is to be regarded as a function of ω; $d\beta$ corresponds to $d\omega$. Expression (2) can be put in the form

$$V(t, z) = 2\cos[(d\omega)t - (d\beta)z]\sin(\omega_0 t - \beta_0 z) \tag{3}$$

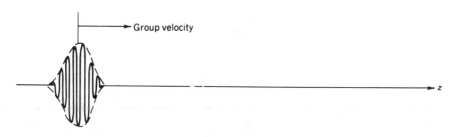

Fig. 5.12a Envelope or group velocity.

From (3) we see that the voltage in this wave group has the form shown in Fig. 5.12b for one instant of time. The sinusoid of center frequency moves at the phase velocity $v_p = \omega_0/\beta_0$ whereas the envelope, described by $\cos[(d\omega)t - (d\beta)z]$ has the form of a wave but it moves at a different velocity. This is found by keeping the argument of the cosine term a constant:

$$v_g = \frac{d\omega}{d\beta} \tag{4}$$

Velocity v_g is called the group velocity and is shown in Fig. 5.12b. Notice that v_g is the slope of ω–β curve at the center frequency of the group. Forming the derivative $dv_p/d\omega$ using $v_p = \omega/\beta$, another useful form for group velocity can be derived:

$$v_g = \frac{v_p}{1 - (\omega/v_p)(dv_p/d\omega)} \tag{5}$$

For more complicated groups than the two-frequency one considered above, we will show later by Fourier analysis that the envelope of a modulated wave retains its shape so long as v_g is constant (i.e., the ω–β curve is linear) over the range of frequencies required to represent the wave. In that case, a pulse such as

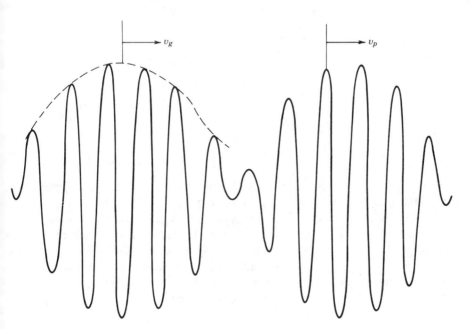

Fig. 5.12b Phase and group velocities for a group of two sinusoids of slightly different frequencies.

that illustrated in Fig. 5.12*a* would retain its *envelope* shape and propagate with delay time τ_d over distance *l*.

$$\tau_d = \frac{l}{v_g} = l\frac{d\beta}{d\omega} \tag{6}$$

If $d\beta/d\omega$ is not constant over the frequency band of the signal, there is a broadening or distortion of the envelope. This is known as *group dispersion* and will be seen in several later examples.

Group velocity is often referred to as the 'velocity of energy travel." This concept has validity for many important cases, but is not universally true.[6,7] To illustrate the basis for the concept, let us define here a separate velocity v_E based on energy flow so that power transfer is stored energy multiplied by this velocity. That is,

$$v_E = \frac{W_T}{u_{av}} \tag{7}$$

where W_T is average power flow in a single wave and u_{av} is average energy storage per unit length. If this definition is applied to the ideal transmission line of Sec. 5.5, we find that $v_E = v_g = v_p$. More interesting is the case of the filter-type circuit of Fig. 5.11*b*, which is a case of *normal dispersion* $(dv_p/d\omega < 0)$. Here

$$W_T = \frac{1}{2}Z_0II^* = \frac{II^*}{2}\left[\frac{L_1}{C_2}\left(1 - \frac{\omega_c^2}{\omega^2}\right)\right]^{1/2}$$

$$u_{av} = \frac{1}{2}\left(\frac{C_2VV^*}{2} + \frac{L_1II^*}{2} + \frac{C_1}{2}\frac{II^*}{\omega^2C_1^2}\right)$$

$$= \frac{L_1}{4}\left(\frac{C_2Z_0^2}{L_1} + 1 + \frac{\omega_c^2}{\omega^2}\right)II^* = \frac{L_1II^*}{2}$$

so

$$v_E = (L_1C_2)^{-1/2}\left[1 - \frac{\omega_c^2}{\omega^2}\right]^{1/2} \tag{8}$$

This is equal to group velocity $d\omega/d\beta$, as can be found by differentiating Eq. 5.11(16), and is different from phase velocity.

The identity of group velocity and energy velocity can also be shown to be true for simple waveguides, and it also applies to many other cases of normal dispersion. It does not usually apply to systems with *anomalous dispersion* $(dv_p/d\omega > 0)$, including the simple transmission line with losses. In any event the concept of an energy velocity is useful only when there is limited dispersion so that the input signal can be recognized at the output.

[6] J. A. Stratton, *Electromagnetic Theory*, McGraw-Hill, New York, 1941, pp. 330–340.
[7] L. Brillouin, *Wave Propagation in Periodic Structures*, McGraw-Hill, New York, 1946; Dover, New York, 1953.

5.13   Backward Waves

A wave in which phase velocity and group velocity have opposite signs is known as a *backward wave*. Conditions for these may seem unexpected or rare, but they are not. Consider for instance the distributed system of Fig. 5.13a in which there are series capacitances and shunt inductances—the dual of the simple transmission line of Sec. 5.2. From Sec. 5.11,

$$\gamma = j\beta = \sqrt{ZY} = \sqrt{\left(\frac{1}{j\omega C}\right)\left(\frac{1}{j\omega L}\right)} = -\frac{j}{\omega\sqrt{LC}} \tag{1}$$

This ω–β relation is shown in Fig. 5.13b. The phase and group velocities are

$$v_p = \frac{\omega}{\beta} = -\omega^2\sqrt{LC} \tag{2}$$

and

$$v_g = \frac{d\omega}{d\beta} = \omega^2\sqrt{LC} \tag{3}$$

So it is seen that this very simple transmission system satisfies the conditions for backward waves. If energy is made to flow in the positive z direction, group

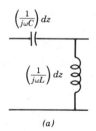

(a)

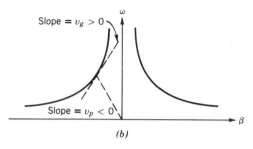

(b)

Fig. 5.13 (a) Equivalent circuit for a transmission line which propagates backward waves. (b) ω–β relation for line of (a) showing phase and group velocity directions discussed in the text.

velocity will be in this direction, as this is a case where v_g does represent energy flow (see Prob. 5.13c). However, the phase becomes increasingly negative or "lagging" in the direction of propagation because of the C–L configuration. Thus there is a negative phase velocity.

There are many other filter-type circuits having other combinations of series and shunt inductances and capacitances on which can exist waves with increasingly lagging phase in the direction of energy propagation. Also, all periodic circuits (Sec. 9.10) have equal numbers of forward and backward "space harmonics."

5.14 Transmission Lines Used with Pulses

Transmission lines are increasingly used for digital or pulse-coded information. We considered some simple examples in Sec. 5.4, and will now return to consider somewhat more complicated ones.

────────────────────── **Example 5.14a** ──────────────────────
Pulse Forming Line

One way of forming pulses of a desired length is by charging a transmission line of length l to a dc voltage V_0 and then connecting to a resistor as shown in Fig. 5.14a. (In practice, the line is often approximated by lumped elements, so the lumped circuit approximation is shown in Fig. 5.14b.) If resistance is matched to characteristic impedance, a pulse of height $V_0/2$ is formed for a time $2t_1$, where t_1 is one-way propagation time down the line and the line completely discharged. It may then be recharged and the process repeated. It may at first seem puzzling that voltage across the resistor is not just V_0 when the switch is closed, but this is because a traveling wave to the right is excited by the connection. Voltage across the resistor just after closing the switch is then the sum of dc voltage and the voltage of the positive wave, V_+

$$V_R = V_0 + V_+ \tag{1}$$

The current flowing in the resistor is just the negative of current in the positively traveling wave since there is no current associated with the dc voltage:

$$I_R = -I_+ = -\frac{V_+}{Z_0} \tag{2}$$

Since $V_R = RI_R$, combination of (1) and (2) gives

$$V_R = \left(\frac{R}{R + Z_0}\right)V_0 = -\frac{R}{Z_0}V_+ \tag{3}$$

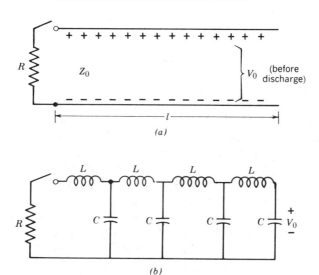

(a)

(b)

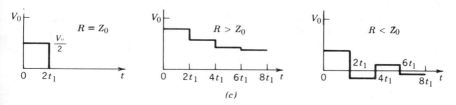

(c)

Fig. 5.14 (a) Transmission line of length l (time delay one way $= t_1$) charged to potential V_0 and connected to resistance R at $t = 0$. (b) Lumped element approximation to (a). (c) Wave shapes of resistor voltage for different relations of R to Z_0.

Thus if $R = Z_0$, $V_R = V_0/2$ and $V_+ = -V_0/2$. Current in the wave started to the right, $I_+ = -V_0/2Z_0$ by (2). When this current reaches the open end, $z = l$, there must then be a reflected wave with current $I_- = V_0/2Z_0$ so that net current is zero at the open end as required. This requires a voltage in the reflected wave $V_- = -Z_0 I_- = -V_0/2$, which travels to the left canceling the remaining voltage on the line and bringing zero voltage and current to the resistance at $t = 2t_1$. From then on all is still. Thus in the case of $R = Z_0$, the wave to the right discharges half the voltage initially on the line, and the wave to the left the other half, yielding a rectangular pulse as shown in Fig. 5.14c.

If $R \neq Z_0$, the wave started to the right upon closing of the switch is other than $V_0/2$, so cancellation of the voltage on the charged line in one round trip does not occur, and there are further reflections when the wave returns to the input. Figure 5.14c sketches the form of resistor current for $R > Z_0$ and for $R < Z_0$. (See also Prob. 5.14b.)

Example 5.4b considered reflections for pulses much shorter than the travel time down the transmission line. For many interconnections, especially those within integrated circuits, the delay time is shorter than pulse width. Reflections because of mismatch may still cause a problem, causing structure within the pulse. To illustrate the point, consider the first part of the pulse as a step function of voltage V_0 applied to an ideal transmission line at $z = -l$, with terminating impedances at end $z = 0$ so high that it may be considered an open circuit. The front of the wave travels along the line in the positive z direction and reaches the end at $t = t_1$. Since current must be zero at the open-circuited end $z = 0$, a reflected step wave must be generated with current opposite to that of the positively traveling wave, starting at the instant of arrival of the latter at $z = 0$. From Eq. 5.2(13), we know that current in the negatively traveling wave is $-V/Z_0$, so it follows that $V_- = V_+$.

As time goes on, the conditions to be met are that the total voltage at the input of the line ($z = -l$) must be the value V_0 of the pulse applied there and the current at the open end must be zero. These conditions can be satisfied by the sum of two square waves, one traveling in the positive z direction and one in the negative z direction. This is the direct analog of the sinusoidal case already discussed extensively in this chapter. Figure 5.14d shows the voltages of the two

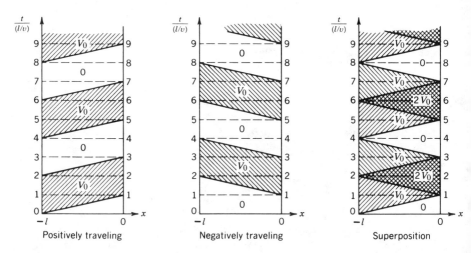

Fig. 5.14d Positively and negatively traveling wave components and their superposition to match boundary conditions on a transmission line with a pulse of voltage V_0 applied to the line at $z = -l$ and the line open-circuited at $z = 0$. The applied pulse length is much longer than the travel time along the line.

individual waves and the superposed, or total, voltage. Notice that the total voltage is always equal to V_0 at $z = -l$, as required by the boundary condition. As discussed in the above paragraph, the condition where the negatively traveling wave has the same polarity of voltage as the positively traveling wave gives a zero total current. Thus, for either the situation with both waves having zero value or both with V_0, the boundary condition at the open end is satisfied. It is seen in Fig. 5.14d, that the sum of the two waves meets that requirement. Since the two waves satisfy the transmission line equations and their sum satisfies the boundary conditions, they constitute the unique solution.

Notice that a square wave of voltage is produced at $z = 0$ with voltage zero for intervals of $2L/v$, interspersed with intervals of the same value with $V = 2V_0$.

5.15 Nonuniform Transmission Lines

For a transmission line with varying spacing or size of conductors, as illustrated in Fig. 5.15, a natural extension of the transmission line analysis would lead one to consider impedance and admittance as varying with distance in the transmission line equations. Actually, fields may be distorted so that the formulation is not this simple, but it is a good approximation in a number of important cases, and the methods discussed apply to some wave problems with spatial variations of the medium (i.e., inhomogeneous materials). The remainder of this section will consider cases where such nonuniform transmission line theory yields a good approximation.

If impedance and admittance per unit length vary with distance, the transmission line equations corresponding to Eqs. 5.11(1)–(2) are

$$\frac{dV(z)}{dz} = -Z(z)I(z) \tag{1}$$

$$\frac{dI(z)}{dz} = -Y(z)V(z) \tag{2}$$

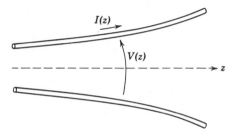

Fig. 5.15 Nonuniform transmission line.

Differentiate (1) with respect to z, denoting z differentiation by primes.

$$V'' = -[ZI' + Z'I] \tag{3}$$

To obtain a differential equation in voltage alone, I may be substituted from (1) and I' from (2). The result is

$$V'' - \left(\frac{Z'}{Z}\right)V' - (ZY)V = 0 \tag{4}$$

Similar procedure, starting with differentiation of (2), yields a second-order differential equation in I:

$$I'' - \left(\frac{Y'}{Y}\right)I'' - (ZY)I = 0 \tag{5}$$

If Z' and Y' are zero, (4) and (5) reduce, as they should, to the equations for a uniform line (Sec. 5.11). When these derivatives are nonzero, representing the nonuniform line discussed, the equations may be solved numerically or with analog equipment for arbitrary variations of Z and Y with distance. A few forms of the variation permit analytic solutions, including the "radial transmission line," where either Z or Y are proportional to z, and their product is constant. Another important case is the "exponential line," which is taken as the example for this article.

_____ **Example 5.15** _____
Line with Exponentially Varying Properties

Let us consider a loss-free exponential line with Z and Y varying as follows:

$$Z = j\omega L_0 e^{qz}; \qquad Y = j\omega C_0 e^{-qz} \tag{6}$$

These variations yield constant values of ZY, Z'/Z, and Y'/Y so that (4) and (5) become equations with constant coefficients,

$$V'' - qV' + \omega^2 L_0 C_0 V = 0 \tag{7}$$

$$I'' + qI' + \omega^2 L_0 C_0 I = 0 \tag{8}$$

These have solutions of the exponentially propagating form,

$$V = V_0 e^{-\gamma_1 z}; \qquad I = I_0 e^{-\gamma_2 z} \tag{9}$$

where

$$\gamma_1 = -\frac{q}{2} \pm \sqrt{\left(\frac{q}{2}\right)^2 - \omega^2 L_0 C_0} \tag{10}$$

$$\gamma_2 = +\frac{q}{2} \pm \sqrt{\left(\frac{q}{2}\right)^2 - \omega^2 L_0 C_0} \tag{11}$$

We see the interesting property of "cutoff" again, for γ_1 and γ_2 are purely real for low frequencies $\omega < \omega_c$ where

$$\omega_c^2 L_0 C_0 = \left(\frac{q}{2}\right)^2 \tag{12}$$

The attenuation represented by these real values, like that for the loss-free filter-type lines, is reactive. This represents no power dissipation but only a continuous reflection of the wave. For $\omega > \omega_c$, however, the values of γ have both real and imaginary parts, which is a different behavior from that of the loss-free filters. Again the real parts represent no power dissipation (see Prob. 5.15b). The values of γ approach purely imaginary values representing phase change only for $\omega \gg \omega_c$.

The greatest use of this type of line is in matching between lines of different characteristic impedance. Unlike the resonant matching sections (Prob. 5.5c), this type of matching is insensitive to frequency. Note the variation of characteristic impedance

$$Z(z) = \frac{V(z)}{I(z)} = \frac{V_0 e^{-\gamma_1 z}}{I_0 e^{-\gamma_2 z}} = \frac{V_0}{I_0} e^{-(\gamma_1 - \gamma_2)z} = Z_0(0)e^{qz} \tag{13}$$

Thus Z_0 can be changed by an appreciable factor if qz is large enough. The transmission line approximation will become poor, however, if there is too large a change of Z and Y in a wavelength, or in a distance comparable to conductor spacing.

The design of nonuniform matching sections is explored in detail in Ref. 8.

PROBLEMS

5.2a Sketch the function

$$V(z, t) = \cos \omega\left(t + \frac{z}{v}\right) + \frac{1}{2}\cos 2\omega\left(t + \frac{z}{v}\right)$$

versus $\omega z/v$ for values of $\omega t = 0$, $\pi/2$, π, $3\pi/2$ and explain how this shows traveling-wave behavior.

5.2b (i) Derive an expression for the characteristic impedance of the parallel-plate line in Fig. 5.2 having a width w and spacing a neglecting the internal inductance of the conductors. Thin-film transmission lines in some computer circuits can be modeled approximately by the parallel-plane line. The line width is usually about $5\,\mu m$ and the spacing is by means of dielectric of $1\,\mu m$ thickness and relative permittivity of 2.5 (as is usually true for dielectrics, the relative permeability can be taken as ≈ 1.0).

   (ii) Calculate the characteristic impedance Z_0 and wave velocity v.

   (iii) Suppose the dielectric thickness is halved and find the new values of Z_0 and v. A better model for such lines is given in Chapter 8.

[8] R. E. Collin, *Foundations for Microwave Engineering*, McGraw-Hill, New York, 1966, Chapter 5.

5.2c The capacitance per unit length of a parallel-wire line having radii R with distance $2d$ between axes is $C = \pi\varepsilon/\cosh^{-1}(d/R)$. Find characteristic impedance of a line with air dielectric and spacing between axes 1 cm if (i) wire radius is 2 mm; (ii) wire radius is 0.5 mm.

5.2d Calculate propagation time along the following transmission lines interconnecting computer elements:
(i)   A thin-film line on GaAs ($\varepsilon_r = 11$) between circuit elements 100 μm apart.
(ii) Transmission line interconnecting two devices on a silicon computer chip 1 mm apart, $\varepsilon_r = 12$.
(iii) Coaxial cable 100 m long with $\varepsilon_r = 2.4$, used to interconnect computer terminal and central processor.

5.2e A second type of solution to the wave equation, to be studied later in the chapter, is the *standing wave* solution. Find under what conditions the following such solution satisfies Eq. 5.2(7):

$$V(z, t) = V_m \cos \omega t \sin \beta z$$

Find the current $I(z, t)$ corresponding to this voltage distribution.

5.2f Expand the cosine and sine in the expression of Prob. 5.2e in terms of complex exponentials, $\cos x = \frac{1}{2}(e^{jx} + e^{-jx})$, and so on, and after multiplying out, show that the products can be interpreted as traveling waves of the form of Eq. 5.2(10).

5.2g Examine the expression for characteristic impedance of a coaxial transmission line, Eq. 5.2(15), and explain why it is difficult to obtain high characteristic impedances for such a line without having unreasonable dimensions or very high losses. Plot dc resistance per unit length for such a line versus Z_0 if the outer conductor is a tubular copper conductor of inner radius 1 cm and wall thickness 1 mm, and the inner conductor a solid copper cylinder.

5.2h Repeat Prob. 5.2g but plot ac resistance at 100 MHz using the approximation of Ex. 3.17.

5.2i Use Eqs. 5.2(3) and (4) to show that the spatial rate of change of power flow on an ideal line is equal to the negative of the time rate of change of the stored energy per unit length.

5.3 To show some simple properties of transverse electromagnetic (TEM) waves, utilize Maxwell's equations in rectangular coordinates (though the boundaries need not be rectangular). Take the dielectric as source-free and without losses. Show that if $H_z = 0$ and $E_z = 0$,
(a) Propagation must be at the velocity of light in the dielectric.
(b) Both **E** and **H** (which are transverse) satisfy the Laplace equation in x and y.

5.4a Calculate the reflection coefficient, transmission coefficient, and the power quantities W_T^-/W_T^+ and W_{TL}/W_T^+ for $R_L = 0, \frac{1}{2}Z_0, Z_0, 2Z_0$, and ∞.

5.4b Derive an expression for W_{TL}/W_T^+ in terms of τ. Substitute the value of τ in terms of R_L and Z_0 and show that the result is the same as given by Eq. 5.4(7).

5.4c Analyze, as in Ex. 5.4a and with drawings like those in Fig. 5.4c, the case of a pulse of length $t_1/5$ reaching a termination at $l = vt_1$ with $R_L = 2Z_0$. Find an expression for the energy dissipated in the load in terms of the voltage of the incident pulse.

5.4d A transmission line of characteristic impedance $Z_{01} = 50\,\Omega$ and length $l = 200$ m is connected to a second line of characteristic impedance $Z_{02} = 100\,\Omega$ and infinite length.

Velocity of propagation in both lines is 2×10^8 m/s. Voltage $V_0 = 100$ V is suddenly applied at $t = 0$. Sketch current versus distance z at $t = 1.3 \, \mu s$. Calculate power in the incident wave, the reflected wave, and in the wave transmitted into line 2, showing that there is a power balance.

5.4e Plot the reflected wave from the terminal of computer no. 2, as in Fig. 5.4g, if $Z_0 = 50 \, \Omega$ and $R = 10 \, \Omega$.

5.4f At $t = 0$ a charge distribution is suddenly placed in the central portion of an infinite line as in Ex. 5.4c except that the voltage distribution in z is triangular, with maximum voltage V_0 at $z = 0$, falling to zero at $z = \pm 1$ m. Find voltage and current distributions at $t = 1.667$ ns and at $t = 5$ ns, as in Ex. 5.4c.

5.5a The alternative approach to derivation of the phasor forms for voltage and current along a transmission line is to replace $\partial/\partial t$ by $j\omega$ in Eqs. 5.2(3) and 5.2(4). Write such equations and show that Eqs. 5.5(4) and 5.5(5) satisfy them.

5.5b Find the special cases of Eq. 5.5(13) for a shorted line; an open line; a half-wave line with load impedance Z_L; a quarter-wave line with load impedance Z_L.

5.5c When two transmission lines are to be connected in cascade, a reflection of the wave to be transmitted from one to the other will occur if they do not have the same characteristic impedances. Show that a quarter-wave length line inserted between the cascaded lines will cause the first line to see its characteristic impedance Z_{01} as a termination and thus eliminate reflection in transfer if $\beta_2 l_2 = \pi/2$ and $Z_{02} = \sqrt{Z_{01} Z_{03}}$ where Z_{02} and Z_{03} are the characteristic impedances of the quarter-wave section and the final line, respectively.

5.5d Derive an expression for a reflection coefficient for current $\rho_I = I_-/I_+$ and show that it differs in phase from the voltage reflection coefficient by π rad.

5.5e A television receiving line of negligible loss is one-third of a wavelength long and has characteristic impedance of $100 \, \Omega$. The detuned receiver acts as a load of $100 + j100 \, \Omega$. Find the input impedance. Sketch phasor diagram showing the values of V_+, V_-, I_+, and I_- at both the load and input, and check the calculated results for impedance from this diagram.

5.5f A strip transmission line of characteristic impedance $20 \, \Omega$ is used at a frequency of 10 GHz with a load that is a microwave diode with conductance 0.05 S in parallel with a 1-pF capacitor. The line is one-eighth wavelength long at the design frequency. Find reflection coefficient at the load and the input admittance.

5.5g If $Z_L \ll Z_0$ and the line not near a multiple of quarter-wavelength in length, show that the first-order terms of a binomial expansion for Eq. 5.5(13) give the following approximate expression for input impedance:

$$Z_i \approx jZ_0 \tan \beta l + Z_L \sec^2 \beta l$$

5.5h If $Z_L \gg Z_0$ and the line not near a multiple of quarter-wavelength in length, show that the first-order terms of a binomial expansion for Eq. 5.5(13) give the following approximate expression for input impedance:

$$Z_i \approx -jZ_0 \cot \beta l + \frac{Z_0^2}{Z_L} \csc^2 \beta l$$

5.6a An impedance of $100 + j100\,\Omega$ is placed as a load on a transmission line of characteristic impedance $50\,\Omega$. Find the reflection coefficient in magnitude and phase and the standing-wave ratio of the line.

5.6b Suppose that reflection coefficient is given in magnitude and phase as $|\rho|e^{j\phi}$ at the load at $z = 0$. Find the value of (negative) z for which voltage is a maximum. Show that current is in phase with voltage at this position, so that impedance there is real, as stated. Calculate the position of maximum voltage for the numerical values of Prob. 5.6a.

5.6c A slotted line measurement shows a standing-wave ratio of 1.5 with voltage minimum 0.1λ in front of the load. Find magnitude and phase of reflection coefficient at the load, and the input impedance for a length 0.2λ of the line.

5.6d An ideal transmission line is terminated by a resistance with value half the characteristic impedance, $R = Z_0/2$. What resistance can you put in parallel with the line $\lambda/4$ in front of the load to eliminate reflections on the generator side of that resistance? Can you find a value for such a parallel resistance if the load resistance is $2Z_0$?

5.6e Give two designs for a power splitter consisting of one 50-Ω input line T-connected to two 50-Ω lines with matched terminations, using quarter-wave transformers (see Prob. 5.5c) as necessary to ensure unity standing-wave ratio at the input at the design frequency and an equal power split. Plot power reflected in the input line as a function of frequency.

5.6f Show that Eq. 5.6(1) can be written in phasor notation in the form of a standing wave plus a traveling wave. Rewrite as a real function of time and, for the example in Fig. 5.6, calculate $V(z)$ at a value of ωt shifted in phase by $\pi/4$ rad from ωt_1.

5.7a A 50-Ω line is terminated in a load impedance of $75 - j69\,\Omega$. The line is 3.5 m long and is excited by a source of energy at 50 MHz. Velocity of propagation along the line is 3×10^8 m/s. Find the input impedance, the reflection coefficient in magnitude and phase, the value of standing-wave ratio, and the position of a voltage minimum, using the Smith chart.

5.7b The standing-wave ratio on an ideal 70-Ω line is measured as 3.2, and a voltage minimum is observed 0.23 wavelength in front of the load. Find the load impedance using the Smith chart.

5.7c Show that the Smith chart for admittance has the form in Fig. 5.7c, developing equations corresponding to Eqs. 5.7(1)–(8).

5.7d Repeat Prob. 5.7b using the chart to determine load admittance. Check to see if the result is consistent with the impedance found in Prob. 5.7b.

5.7e A 70-Ω line is terminated in an impedance of $50 + j10\,\Omega$. Find the position and value of a reactance that might be added in series with the line at some point to eliminate reflections of waves incident from the source end.

5.7f Repeat Prob. 5.7e to determine the position and value of shunt susceptance to be placed on the line for matching.

5.7g* A 50-Ω transmission line is terminated with a load of $Z_L = 20 + j30$. A *double-stub tuner* consisting of a pair of shorted 50-Ω transmission lines connected in shunt to the main line at points spaced by 0.25λ is located with one stub at 0.2λ from the load. Find the lengths of the stubs to give unity standing-wave ratio at 0.45λ from the load.

5.7h* Two antennas have impedances of $100 + j100\,\Omega$ for a particular frequency and are fed by transmission lines of characteristic impedance $300\,\Omega$. By inspection of the Smith chart, show that it is possible to choose different lengths for the two feed lines so that when combined in series at the input, the series combination perfectly matches the 300-Ω line to which they are connected. Give the lengths of the two lines in fractions of a wavelength. By study of the procedure you have used, state whether or not this compensation approach will work if the two antennas have arbitrary but equal impedances.

5.7i* The problem is as in Prob. 5.7h except that the two transmission lines are connected in parallel.

5.7j* A certain coaxial line has an alternating dielectric of vacuum and a material with $\varepsilon = 4\varepsilon_0$ and $\mu = \mu_0$ and is terminated at the end of a vacuum section by an impedance equal to the characteristic impedance of the vacuum regions. At frequency f_0 the dielectric and vacuum regions are each $\lambda/2$ long (λ appropriate to each region). Show on a Smith chart the path of impedance variation along the line for operating frequencies f_0 and $2f_0$. Also plot the standing-wave ratio as a function of distance from the load.

5.8a Write the instantaneous expressions for voltage and current represented by the complex values in Eqs. 5.8(1) and 5.8(2). Make an integration of total energy, electric plus magnetic, for a quarter-wavelength of the line and show that it is independent of time.

5.8b Show that $|\rho| = 1$ for a purely reactive load on an ideal transmission line. Find and sketch as in Fig. 5.8 suitably normalized $V(z, t)$ and $I(z, t)$ for a line terminated in a pure inductive reactance equal in magnitude to the characteristic impedance.

5.8c Calculate the instantaneous power flow for a short-circuited line $W_T(t, z) = V(t, z)I(t, z)$ and plot the results in a diagram like Fig. 5.8. Discuss this $W_T(t, z)$ in connection with conclusion 6 reached from (1) and (2) in Sec. 5.8.

5.8d One of the limitations of energy-storage systems for large energies is the breakdown of air, unless the system can be evacuated. For air with breakdown strength 3×10^6 V/m, estimate the maximum energy storage in a resonant air-filled half-wave parallel-plate line. Spacing between plates is 2 cm and characteristic impedance is 50 Ω. Is there any advantage with respect to breakdown over the use of a parallel-plate capacitor? Discuss an inductor as an energy-storage system from the same point of view.

5.9a Calculate for frequencies 1 MHz and 1 GHz the attenuation in decibels per meter for an air-filled coaxial transmission line with copper conductors using the skin-effect approximations for high-frequency resistance of Ex. 3.17. The inner conductor is a solid cylinder of radius 2 mm and the outer conductor is a thick tubular cylinder of inner radius 1 cm.

5.9b Repeat Prob. 5.9a if the transmission line is now filled with a polystyrene dielectric. The equivalent conductances of polystyrene at 1 Mz and 1 GHz are, respectively, $\sigma_d = 10^{-8}$ S/m and 2.8×10^{-5} S/m.

5.10a Plot Eq. 5.10(7) normalized to Y_0 as a function of frequency near where $n = 1$ and see over what range (8) is a reasonable approximation.

5.10b A half-wavelength coaxial transmission line with air dielectric has copper conductors with the dimensions of Prob. 5.9a, and is designed for resonance at 6 GHz. Find the bandwidth $2 \Delta f_1$ in Eq. 5.10(11).

5.11a For the filter-type circuit studied in Sec. 5.11, find expressions for characteristic impedance Z_0. Show that this is real in the propagating region and imaginary in the attenuating region. What does this signify with respect to power flow in a single traveling wave?

5.11b Use the formula for input impedance of a transmission line with losses to check Eq. 5.10(3), making approximations consistent with $R/\omega L \ll 1$ and $G/\omega C \ll 1$.

5.11c A certain continuous transmission line has an equivalent circuit consisting of series inductance L_1 H/m and a shunt element consisting of capacitance C F/m and inductance L_2 H/m in parallel. Let $\omega_1^2 = 1/L_2 C$ and

(i) Obtain expressions for γ, α, and β in terms of L_1, L_2, ω_1, and ω.

(ii) Plot γ^2 versus ω.

(iii) Replot with ω versus α, where α is real, and versus β, where β is real.

5.11d Defining α_c from conductor losses and α_d from dielectric losses by identifying the appropriate parts of Eq. 5.11(13), write the approximate expressions, Eqs. 5.11(14) and (15) for β and Z_0 in terms of α_c and α_d.

5.11e If the transmission line of Ex. 5.9 is made with thinner films, the fields in the films can be almost uniform. With that approximation for films 0.2 μm thick, calculate the characteristic impedance and attenuation. Comment on the effects of using thinner films.

5.11f* Show that the variation of complex power along a lossy transmission line carrying sinusoidal waves can be written as $(d/dz)(\frac{1}{2}VI^*) + \frac{1}{2}(RII^* + GVV^*) + (j/2)(XII^* - BVV^*) = 0$ where X and B are the imaginary parts of Z and Y, respectively. Also, show that a direct term-by-term identification with the complex Poynting theorem can be made by multiplying the above expression by dz and considering the volume for Eq. 3.13(6) to be of dz thickness and infinite width.

5.12a Is phase or group velocity the larger for normal dispersion $(dv_p/d\omega < 0)$? For anomalous dispersion $(dv_p/d\omega > 0)$?

5.12b Find the phase and group velocities for a transmission line with small losses.

5.12c Consider a transmission line with very high leakage conductance G per unit length so that series resistance R and shunt capacitance C are negligible. Find phase and group velocities.

5.12d For Probs. 5.12b and 5.12c, show that an energy velocity as defined by Eq. 5.12(7) is not equal to group velocity.

5.12e Certain water waves of large amplitude have phase velocities given by $v_p = g/\omega$ where g is acceleration due to gravity and ω is angular frequency. Determine the ratio of group velocity to phase velocity for such a wave.

5.12f For Prob. 5.11e, plot v_p and v_g versus ω and show that $v_p v_g = 1/L_1 C$ for all frequencies.

5.13a Plot the ω–β diagram for a distributed transmission system with series inductance L_1 per unit length, and a shunt admittance made up of L_2 in parallel with C_2, for a unit length of the system. Is this a backward or forward wave system? Show cutoff frequency and illustrate phase and group velocity on the plot.

5.13b Repeat Prob. 5.13a for a system with the shunt admittance made up of L_2 and C_2 in series.

5.13c Calculate the velocity of energy propagation, as defined in Sec. 5.12 for the backward-wave line of Fig. 5.13a. Show that it does correspond to group velocity for this line and discuss the concepts of normal and anomalous dispersion for backward waves.

5.14a An ideal open ended line of length l is charged to dc voltage V and shorted at its input at time $t = 0$. Sketch the current wave shape through the short as a function of time.

5.14b A charged cable is connected suddenly to a load resistor equal to its 50-Ω characteristic impedance. If its length is 3 m and its phase velocity is one-half the velocity

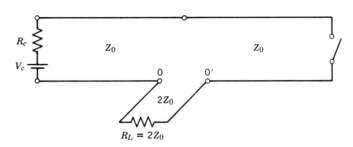

Fig. P5.14c Blumlein pulse generator.

of light, how long is the pulse in the load? Sketch the waveform at the midpoint of the line assuming the cable is initially charged to 100 V and the load is 25 Ω instead of 50.

5.14c* The circuit shown in Fig. P5.14c is a so-called Blumlein pulse generator (A. T. Starr, *Radio and Radar Technique*, Pitman & Sons, London, 1953) and has the property that it produces a voltage pulse equal to the voltage to which the lines are initially charged by the source V_c. The resistor R_c can be considered essentially infinite. At a time τ after the switch is closed, a voltage V_c appears across the output line terminals 0–0' and that voltage remains across the terminals 0–0' for a time 2τ. The initially charged lines are of equal length. Analyze the behavior of the circuit to show the above described behavior, treating the output line as a lumped resistor $R_L = 2Z_0$.

5.15a Show that Eqs. 5.15(9) with definitions (10) and (11) do give the solutions of (7) and (8) with the variations (6). Describe ways in which the exponential variation of L and C might be achieved, at least approximately.

5.15b Show that average power transfer is independent of z in the loss-free exponential line considered here for frequencies above cutoff, $\omega > \omega_c$.

5.15c There are two values of Eqs. 5.15(10) and 5.15(11) representing positively and negatively traveling waves, as expected. Write the complete solutions for $V(z)$ and $I(z)$, showing both waves, using as constants the voltage amplitudes in positively and negatively traveling waves at $z = 0$. Note the interchange of behavior of positive and negative waves if the sign of q is changed, and explain physically.

5.15d Modify the analysis for the exponential line to include losses, retaining constancy of ZY, Z'/Z, and Y'/Y, and interpret the effect on attenuation constant.

6

PLANE WAVE PROPAGATION AND REFLECTION

6.1   Introduction

The first example of the application of Maxwell's equations in Chapter 3 was that of electromagnetic wave propagation in a simple dielectric medium. We now return to the plane wave example and extend it in this chapter, before considering the more general guided, resonant, and radiating waves.

Plane waves are good approximations to real waves in many practical situations. Radio waves at large distances from the transmitter, or from diffracting objects, have negligible curvature and are well represented by plane waves. Much of optics utilizes the plane wave approximation. More complicated electromagnetic wave patterns can be considered as a superposition of plane waves, so in this sense the plane waves are basic building blocks for all wave problems. Even when that approach is not followed, the basic ideas of propagation, reflection, and refraction, which are met simply here, help the understanding of other wave problems. The methods developed in the preceding chapter on transmission lines will be very valuable for such problems. A large part of this chapter is concerned with the reflection and refraction phenomena when waves pass from one medium to another, with examples for both radio waves and light.

PLANE WAVE PROPAGATION

6.2   Uniform Plane Waves in a Perfect Dielectric

The uniform plane wave was given in Chapter 3 as the first example of the use of Maxwell's equations. We now discuss its properties in more detail, restricting attention to media for which μ and ε are constants. For a uniform plane wave, variations in two directions, say x and y, are assumed to be zero with the remaining (z) direction taken as the direction of propagation. As in Sec. 3.9, Maxwell's equations in rectangular coordinates then reduce to

$$\nabla \times \mathbf{E} = -\mu \frac{\partial \mathbf{H}}{\partial t} \qquad \nabla \times \mathbf{H} = \varepsilon \frac{\partial \mathbf{E}}{\partial t}$$

In component form these are:

$$\frac{\partial E_y}{\partial z} = \mu \frac{\partial H_x}{\partial t} \qquad (1) \qquad\qquad \frac{\partial H_y}{\partial z} = -\varepsilon \frac{\partial E_x}{\partial t} \qquad (4)$$

$$\frac{\partial E_x}{\partial z} = -\mu \frac{\partial H_y}{\partial t} \qquad (2) \qquad\qquad \frac{\partial H_x}{\partial z} = \varepsilon \frac{\partial E_y}{\partial t} \qquad (5)$$

$$0 = \mu \frac{\partial H_z}{\partial t} \qquad (3) \qquad\qquad 0 = \varepsilon \frac{\partial E_z}{\partial t} \qquad (6)$$

As noted in Sec. 3.9, the above equations (3) and (6) show that both E_z and H_z are zero, except possibly for constant (static) parts which are not of interest in the wave solution. That is, electric and magnetic fields of this simple wave are transverse to the direction of propagation.

In Sec. 3.9 we showed that combination of the above equations (2) and (4) leads to the one-dimensional wave equation in E_x,

$$\frac{\partial^2 E_x}{\partial z^2} = \mu\varepsilon \frac{\partial^2 E_x}{\partial t^2} \qquad (7)$$

which has a general solution

$$E_x = f_1\left(t - \frac{z}{v}\right) + f_2\left(t + \frac{z}{v}\right) \qquad (8)$$

where

$$v = \frac{1}{\sqrt{\mu\varepsilon}} \qquad (9)$$

which is the velocity of light for the medium. The first term of (8) can be interpreted as a wave propagating with velocity v in the positive z direction, the second representing a wave propagating with the same velocity in the negative z direction. That is,

$$E_{x+} = f_1\left(t - \frac{z}{v}\right), \qquad E_{x-} = f_2\left(t + \frac{z}{v}\right) \tag{10}$$

Using either (2) or (4), magnetic field H_y was found to be

$$H_y = H_{y+} + H_{y-} = \frac{E_{x+}}{\eta} - \frac{E_{x-}}{\eta} \tag{11}$$

where

$$\eta = \sqrt{\frac{\mu}{\varepsilon}} \tag{12}$$

The quantity η is thus seen to be the ratio of E_x to H_y in a single traveling wave of this simple type, and as defined by (12) it may also be considered a constant of the medium, and will be a useful parameter in the analysis of more complicated waves. It has dimensions of ohms and is known as the *intrinsic impedance* of the medium. For free space

$$\eta_0 = \sqrt{\frac{\mu_0}{\varepsilon_0}} = 376.73 \approx 120\pi \ \Omega \tag{13}$$

Now looking at the remaining two components, E_y and H_x, combination of (1) and (5) leads to the wave equation in E_y

$$\frac{\partial^2 E_y}{\partial z^2} = \mu\varepsilon \frac{\partial^2 E_y}{\partial t^2} \tag{14}$$

which also has solutions in the form of positively and negatively traveling waves as in (8). We write this

$$E_y = f_3\left(t - \frac{z}{v}\right) + f_4\left(t + \frac{z}{v}\right) = E_{y+} + E_{y-} \tag{15}$$

Either (1) or (5) then show that magnetic field is

$$H_x = -\frac{E_{y+}}{\eta} + \frac{E_{y-}}{\eta} \tag{16}$$

To stress the relationship of electric and magnetic fields for the waves we write the results of (11) and (16) as

$$\frac{E_{x+}}{H_{y+}} = -\frac{E_{y+}}{H_{x+}} = \eta, \qquad \frac{E_{x-}}{H_{y-}} = -\frac{E_{y-}}{H_{x-}} = -\eta \tag{17}$$

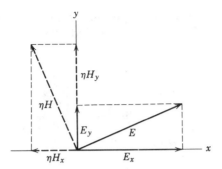

Fig. 6.2a Relations between **E** and **H** for a wave propagating in positive z direction (out of page).

These results show a number of things. First, relations (17) are sufficient to require that **E** and **H** shall be perpendicular to one another in each of the traveling waves. They also require that the value of E at any instant must be η times the value of H at that instant, for each wave. Finally we note that, if $\mathbf{E} \times \mathbf{H}$ is formed, it points in the positive z direction for the positively traveling parts of (17) and in the negative z direction for the negatively traveling part, as expected. These relations are indicated for a positively traveling wave in Fig. 6.2a.

The energy relations are also of interest. The stored energy in electric fields per unit volume is

$$u_E = \frac{\varepsilon E^2}{2} = \frac{\varepsilon}{2}(E_x^2 + E_y^2) \tag{18}$$

and that in magnetic fields is

$$u_H = \frac{\mu H^2}{2} = \frac{\mu}{2}(H_x^2 + H_y^2) \tag{19}$$

By (17), u_E and u_H are equal, so the energy density at each point at each instant is equally divided between electric and magnetic energy. The Poynting vector for the positive traveling wave is

$$P_{z+} = E_{x+}H_{y+} - E_{y+}H_{x+} = \frac{1}{\eta}(E_{x+}^2 + E_{y+}^2) \tag{20}$$

and is always in the positive z direction except at particular planes where it may be zero for a given instant. Similarly, the Poynting vector for the negatively traveling wave is always in the negative z direction except where it is zero. The time-average value of the Poynting vector must be the same for all planes along the wave since no energy can be dissipated in the perfect dielectric, but the instantaneous values may be different at two different planes, depending on whether there is a net instantaneous rate of increase or decrease of stored energy between those planes.

In Sec. 3.10 we also studied the important phasor forms for a plane wave with E_x and H_y. Extending that analysis to include the remaining components, we have

$$E_x(z) = E_1 e^{-jkz} + E_2 e^{jkz} \tag{21}$$

$$\eta H_y(z) = E_1 e^{-jkz} - E_2 e^{jkz} \tag{22}$$

$$E_y(z) = E_3 e^{-jkz} + E_4 e^{jkz} \tag{23}$$

$$\eta H_x(z) = -E_3 e^{-jkz} + E_4 e^{jkz} \tag{24}$$

where

$$k = \frac{\omega}{v} = \omega\sqrt{\mu\varepsilon} \quad \text{m}^{-1} \tag{25}$$

This constant is the phase constant for the uniform plane wave, since it gives the change in phase per unit length for each wave component. It may also be considered a constant of the medium at a particular frequency defined by (25), known as the *wave number*, and will be found useful in the analysis of all waves, as will be seen.

The *wavelength* is defined as the distance the wave propagates in one period. It is then the value of z which causes the phase factor to change by 2π:

$$k\lambda = 2\pi \quad \text{or} \quad k = \frac{2\pi}{\lambda} \tag{26}$$

or

$$\lambda = \frac{2\pi}{\omega\sqrt{\mu\varepsilon}} = \frac{v}{f} \tag{27}$$

This is the common relation between wavelength, phase velocity, and frequency. The *free-space wavelength* is obtained by using the velocity of light in free space in (27) and is frequently used at the higher frequencies as an alternative to giving the frequency. It is also common in the optical range of frequencies to utilize *a refractive index n* given by

$$n = \frac{c}{v} = \sqrt{\frac{\mu}{\mu_0}\frac{\varepsilon}{\varepsilon_0}} \tag{28}$$

For most materials in the optical range $\mu = \mu_0$ so that n is just the square root of the relative permittivity for that frequency.

To summarize the properties for a single wave of this simple type, which may be described as a *uniform plane wave*:

1. Velocity of propagation, $v = 1/\sqrt{\mu\varepsilon}$.
2. No electric or magnetic field in direction of propagation.
3. Electric field normal to magnetic field.

4. Value of electric field η times that of magnetic field at each instant.
5. Direction of propagation given by direction of **E** × **H**.
6. Energy stored in electric field per unit volume at any instant and any point is equal to energy stored in magnetic field.
7. Instantaneous value of Poynting vector given by $E^2/\eta = \eta H^2$, where E and H are the instantaneous values of total electric and magnetic field strengths.

_____ **Example 6.2** _____
Propagation of a Modulated Wave in a Nondispersive Medium

If a radio wave of angular frequency ω_0 is amplitude modulated by a sine wave of angular frequency ω_m, the resulting function may be written

$$E(t) = A[1 + m \cos \omega_m t] \cos \omega_0 t \tag{29}$$

Suppose this function at $z = 0$ excites a uniform plane wave propagating in the positive z direction. To obtain the form of the propagating wave it is straightforward to replace t by $t - z/v$ to obtain

$$E(z, t) = A\left[1 + m \cos \omega_m\left(t - \frac{z}{v}\right)\right]\cos \omega_0\left(t - \frac{z}{v}\right) \tag{30}$$

The interpretation of this expression is that the entire function propagates in the z direction with velocity v as illustrated in Fig. 6.2b. All this is correct provided

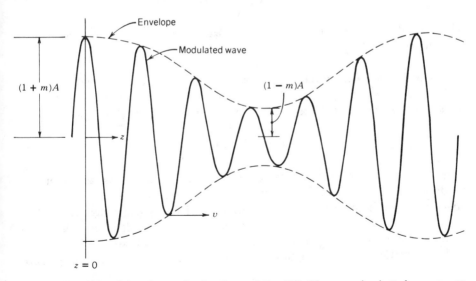

Fig. 6.2b Simple modulated wave having form of Eq. (30). The wave is plotted versus z at time $t = 0$. In a nondispersive medium, envelope and modulated portion move with velocity v in z direction.

that the medium is nondispersive (i.e., that v is independent of frequency). If there is dispersion, an expansion of (29) shows that different frequency components are present and that each component then propagates at its appropriate v. The result then is that the envelope moves with a different velocity than the modulated wave as shown in the discussion of group velocity, Sec. 5.12.

6.3   Polarization of Plane Waves

If several plane waves have the same direction of propagation, it is straight-forward to superpose these for a linear medium. The orientations of the field vectors in the individual waves and the resultant, are described by the *polarization*[1] of the waves. In this discussion we are primarily concerned with sinusoidal waves of the same frequency.

Let us take a positively traveling wave only, use phasor representation, and assume there are both x and y components of electric field. The general expression for such a wave is then

$$\mathbf{E} = (\hat{\mathbf{x}}E_1 + \hat{\mathbf{y}}E_2 e^{j\psi})e^{-jkz} \tag{1}$$

where E_1 and E_2 are taken as real and ψ is the phase angle between x and y components. The corresponding magnetic field is

$$\mathbf{H} = \frac{1}{\eta}(-\hat{\mathbf{x}}E_2 e^{j\psi} + \hat{\mathbf{y}}E_1)e^{-jkz} \tag{2}$$

The several classes of polarization then depend upon the phase and relative amplitudes E_1 and E_2.

Linear or Plane Polarization   If the two components are in phase, $\psi = 0$, they add at every plane z to give an electric vector in some fixed direction defined by angle α with respect to the x axis, as pictured in Fig. 6.3a.

$$\alpha = \tan^{-1}\frac{E_y}{E_x} = \tan^{-1}\frac{E_2}{E_1} \tag{3}$$

This angle is real and hence the same for all values of z and t. Since $\mathbf{E}$ maintains its direction in space, this polarization is called *linear*. It is also called *plane polarization* since the electric vector defines a plane as it propagates in the z direction. In communications engineering it is common to describe polarization by the plane of the electric vector so that the term "vertical polarization" implies that $\mathbf{E}$ is vertical. In older optics texts the magnetic field defined the plane of

---

[1] The term polarization is used in electromagnetics both for this purpose and also for the unrelated concept of the contribution of atoms and molecules to dielectric properties as described in Secs. 1.3 and 13.2. Usually the intended usage is clear from the context.

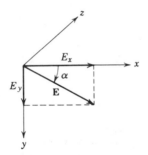

Fig. 6.3a Components of a linearly (plane) polarized wave.

polarization but it is now also common in optics to use electric field.[2] To avoid ambiguity in either case it is best to specify explicitly, "polarized with electric field in the vertical plane."

Circular Polarization   A second important special case arises when amplitudes E_1 and E_2 are equal and phase angle is $\pm \pi/2$. Equation (1) then becomes

$$\mathbf{E} = (\hat{\mathbf{x}} \pm j\hat{\mathbf{y}})E_1 e^{-jkz} \tag{4}$$

The magnitude of $\mathbf{E}$ is seen to be $\sqrt{2}E_1$ from the above, and it may be inferred that it rotates in circular manner, but to see this clearly let us go to the instantaneous forms,

$$\mathbf{E}(z, t) = \text{Re}[(\hat{\mathbf{x}} \pm j\hat{\mathbf{y}})E_1 e^{j\omega t} e^{-jkz}] \tag{5}$$

$$= E_1[\hat{\mathbf{x}} \cos(\omega t - kz) \mp \hat{\mathbf{y}} \sin(\omega t - kz)]$$

The sum of the squares of instantaneous E_x and E_y,

$$E_x^2(z, t) + E_y^2(z, t) = E_1^2[\cos^2(\omega t - kz) + \sin^2(\omega t - kz)] = E_1^2 \tag{6}$$

does define the equation of a circle. The instantaneous angle α with respect to the x axis is

$$\alpha = \tan^{-1} \frac{E_y(z, t)}{E_x(z, t)} = \tan^{-1}\left(\mp \frac{\sin(\omega t - kz)}{\cos(\omega t - kz)}\right) = \mp(\omega t - kz) \tag{7}$$

In a given z plane, the vector thus rotates with constant angular velocity with $\alpha = \mp \omega t$. For a fixed time, the vector traces out a spiral in z as pictured in Fig. 6.3b. The propagation may be pictured as the movement of this "corkscrew" in the z direction with velocity v.

Note that $\psi = +\pi/2$ leads to $\alpha = -\omega t$ (for $z = 0$) and $\psi = -\pi/2$ leads to rotation in the opposite direction. The first case is called the left-hand or counterclockwise sense of circular polarization (looking in the direction of

[2] M. Born and E. Wolf, *Principles of Optics*, 6th ed., Macmillan, New York, 1980, p. 28.

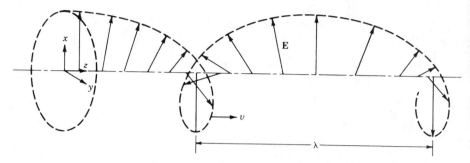

Fig. 6.3b Circularly polarized wave. Terminus of the electric field vectors forms a spiral of period equal to the wavelength at any instant of time. This spiral moves in the z direction with velocity v, so that the vector in a given z plane traces out a circle as time progresses.

propagation) and the latter right-hand or clockwise. The magnetic field for the circularly polarized wave, using $E_1 = E_2$ and $\psi = \pm \pi/2$, is

$$\mathbf{H} = \frac{E_1}{\eta}(\mp j\hat{\mathbf{x}} + \hat{\mathbf{y}})e^{-jkz} \tag{8}$$

Elliptic Polarization   For the general case, with $E_1 \neq E_2$, or $E_1 = E_2$ but ψ other than 0 or $\pm \pi/2$, the terminus of the electric field traces out an ellipse in a given z plane so that the condition of polarization is called *elliptical*. To see this, again let us take instantaneous forms of (1),

$$\mathbf{E}(z, t) = \mathbf{Re}[(\hat{\mathbf{x}}E_1 + \hat{\mathbf{y}}E_2 e^{j\psi})e^{j\omega t}e^{-jkz}] \tag{9}$$

$$= \hat{\mathbf{x}}E_1 \cos(\omega t - kz) + \hat{\mathbf{y}}E_2 \cos(\omega t - kz + \psi)$$

or in a given z plane, say $z = 0$,

$$E_x(z, t) = E_1 \cos \omega t \tag{10}$$

$$E_y(z, t) = E_2 \cos(\omega t + \psi)$$

These are the parametric equations of an ellipse. If $\psi = \pm \pi/2$ the major and minor axis of the ellipse are aligned with the x and y axes, but for general ψ the ellipse is tilted as illustrated in Fig. 6.3c.

Unpolarized Wave   We sometimes also speak of an unpolarized wave in which there is a component in any arbitrary direction for each instant of time. Note that this concept applies only to the superposition of waves of different frequency, or of random phases, since superposition of any number of components of the same frequency and defined phase reduces to one of the three cases described above. We may meet unpolarized waves when we have a frequency spectrum (as in sunlight), or a random variation of phase between components, as in the propagation of a radio wave through the ionosphere.

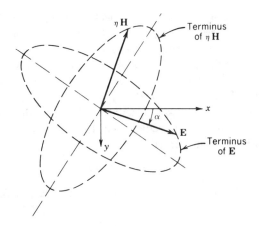

Fig. 6.3c Elliptically polarized wave. The locus of the terminus of electric and magnetic field vectors is in each case an ellipse for a given z plane as time progresses.

Example 6.3
Linearly Polarized Wave as Superposition of
Two Circularly Polarized Waves

Just as the circularly polarized wave may be looked at as the superposition of two linearly polarized waves, so may a linearly polarized wave be considered a superposition of two oppositely circulating circular polarized waves. To show this, let us add right-hand and left-hand circularly polarized waves of the same amplitude. Using (4),

$$\mathbf{E} = (\hat{\mathbf{x}} + j\hat{\mathbf{y}})E_1 e^{-jkz} + (\hat{\mathbf{x}} - j\hat{\mathbf{y}})E_1 e^{-jkz} = 2\hat{\mathbf{x}}E_1 e^{-jkz}$$

and magnetic field, using (8) is

$$\mathbf{H} = \frac{E_1}{\eta}(-j\hat{\mathbf{x}} + \hat{\mathbf{y}})e^{-jkz} + \frac{E_1}{\eta}(j\hat{\mathbf{x}} + \hat{\mathbf{y}})e^{-jkz} = \frac{2E_1}{\eta}\hat{\mathbf{y}}e^{-jkz}$$

The results are the expressions for $\mathbf{E}$ and $\mathbf{H}$ in a wave polarized with electric field in the x direction. To obtain $\mathbf{E}$ in the y direction, we have only to subtract the two circularly polarized parts.

6.4   Waves in Imperfect Dielectrics and Conductors

Materials properties and their effect on wave propagation will be dealt with extensively in Chapter 13. Here we consider only isotropic, linear materials and bring in the effect of losses on the response of applied fields. The effect of losses can enter through a response to either the electric or magnetic field, or both. An example of a material where the response to the magnetic field leads to losses is the so-called *ferrite* (or *ferrimagnetic* materials) and in this case the permeability must be a complex quantity $\mu = \mu' - j\mu''$. For most materials of interest in wave studies, magnetic response is very weak and the permeability is a real constant that differs very little from the permeability of free space; this will be assumed throughout this text unless otherwise specified. A study of the response of bound electrons in atoms and ions in molecules (Sec. 13.2) shows that the total current density resulting from their motion is

$$\mathbf{J} = j\omega\varepsilon\mathbf{E} = j\omega(\varepsilon' - j\varepsilon_b'')\mathbf{E}$$

If, in addition, the material contains free electrons or holes, there is a conduction current density

$$\mathbf{J} = \sigma\mathbf{E}$$

At sufficiently high frequencies, σ can be complex (Sec. 13.3) but we will assume that frequency is low enough to consider it real (satisfactory through the microwave range). Then the total current density is

$$\mathbf{J} = j\omega\left(\varepsilon' - j\varepsilon_b'' - j\frac{\sigma}{\omega}\right)\mathbf{E} \tag{1}$$

In materials called *dielectrics*, there are usually few free electrons and any free-electron current component is included in ε'' and σ is taken as zero. On the other hand, in materials considered conductors such as normal metals and semiconductors, the conduction current dominates and the effect of the bound electrons ε_b'' is subsumed in the conductivity.

Concentrating on dielectric materials with real permeability and complex permittivity, we may write

$$\nabla \times \mathbf{H} = j\omega\varepsilon\mathbf{E} = j\omega(\varepsilon' - j\varepsilon'')\mathbf{E} \tag{2}$$

where conduction currents are included in the loss factor ε''.

The wave number, Eq. 6.2(25), is complex in this case and is

$$k = \omega\sqrt{\mu(\varepsilon' - j\varepsilon'')} \tag{3}$$

The wave number k may be separated into real and imaginary parts:

$$jk = \alpha + j\beta = j\omega\sqrt{\mu\varepsilon'\left[1 - j\left(\frac{\varepsilon''}{\varepsilon'}\right)\right]} \tag{4}$$

where

$$\alpha = \omega\sqrt{\left(\frac{\mu\varepsilon'}{2}\right)\left[\sqrt{1 + \left(\frac{\varepsilon''}{\varepsilon'}\right)^2} - 1\right]} \tag{5}$$

and

$$\beta = \omega \sqrt{\left(\frac{\mu\varepsilon'}{2}\right)\left[\sqrt{1 + \left(\frac{\varepsilon''}{\varepsilon'}\right)^2} + 1\right]} \tag{6}$$

According to Sec. 6.2, the exponential propagation factor for phasor waves is e^{-jkz} which becomes, when k is complex,

$$e^{-jkz} = e^{-\alpha z}e^{-j\beta z}$$

Thus the wave attenuates as it propagates through the material and the attenuation depends upon the dielectric losses and the conduction losses, as would be expected.

The intrinsic impedance, or ratio of electric to magnetic field for a uniform plane wave, becomes

$$\eta = \sqrt{\frac{\mu}{\varepsilon}} = \sqrt{\frac{\mu}{\varepsilon'[1 - j(\varepsilon''/\varepsilon')]}} \tag{7}$$

An important parameter appearing in (4) through (7) is the ratio $\varepsilon''/\varepsilon'$. For low-loss materials such as the examples given in Table 6.4a, this ratio is much less than unity. We may refer to such a material as an *imperfect dielectric*. Under these conditions, the attenuation constant (5) and the phase constant (6) may be approximated by expanding both as binomial series,

$$\alpha \approx \frac{k\varepsilon''}{2\varepsilon'} \tag{8}$$

$$\beta \approx k\left[1 + \frac{1}{8}\left(\frac{\varepsilon''}{\varepsilon'}\right)^2\right] \tag{9}$$

Table 6.4a
Properties of common dielectrics

Material[a]	$\varepsilon'/\varepsilon_0$			Loss tangent, $10^4\,\varepsilon''/\varepsilon'$		
	$f = 10^6$	$f = 10^8$	$f = 10^{10}$	$f = 10^6$	$f = 10^8$	$f = 10^{10}$
Glass, Corning 707	4.00	4.00	4.00	8	12	21
Fused quartz	3.78	3.78	3.78	2	1	1
Ruby mica	5.4	5.4	—	3	2	—
Ceramic Alsimag 393	4.95	4.95	4.95	10	10	9.7
TiO$_2$	100	100	100	3	2.5	5
Polystyrene	2.56	2.55	2.54	0.7	1	4.3
Neoprene	5.7	3.4	—	950	1600	—

[a] Low-loss plastics with a wide range of permittivities exist: from Eccofoam PS [$(\varepsilon'/\varepsilon_0) = 1.02$, $(\varepsilon''/\varepsilon') < 10^{-4}$] to Eccoceram Hi K [$(\varepsilon'/\varepsilon_0) = 30$, $(\varepsilon''/\varepsilon') = 8 \times 10^{-4}$]. A useful microwave absorbing material, Eccosorb, is obtainable with $1.5 < (\varepsilon'/\varepsilon_0) < 50$ and $0.08 < (\varepsilon''/\varepsilon') \lesssim 1$. The values for these plastics are averages for 1–30 GHz.

where $k = \omega\sqrt{\mu\varepsilon'}$. It is seen that there is a small increase of the phase constant and, therefore, decrease of the phase velocity when losses are present in a dielectric. The intrinsic impedance in this case is, from (7),

$$\eta \approx \eta'\left\{\left[1 - \frac{3}{8}\left(\frac{\varepsilon''}{\varepsilon'}\right)^2\right] + j\frac{\varepsilon''}{2\varepsilon'}\right\} \tag{10}$$

where $\eta' = \sqrt{\mu/\varepsilon'}$.

In a material such as a semiconductor where the losses are predominantly conductive,

$$\nabla \times \mathbf{H} = (j\omega\varepsilon' + \sigma)\mathbf{E} = j\omega\varepsilon'\left[1 - j\frac{\sigma}{\omega\varepsilon'}\right]\mathbf{E} \tag{11}$$

For radian frequencies much larger than σ/ε' the loss term is small and the approximations (8) to (10) may be used with $\varepsilon''/\varepsilon'$ replaced by $\sigma/\omega\varepsilon'$. For frequencies much below σ/ε' the material may be considered a *good conductor*. If we make use of the fact that

$$\frac{\sigma}{\omega\varepsilon'} \gg 1$$

in (4), we find

$$jk = j\omega\sqrt{\frac{\mu\sigma}{j\omega}} = (1 + j)\sqrt{\pi f\mu\sigma} = \frac{1 + j}{\delta} \tag{12}$$

where δ is the depth of penetration defined by Eq. 3.16(11) and used extensively in Chapter 3. The propagation factor for the wave shows that the wave decreases in magnitude exponentially, and has decreased to $1/e$ of its original value after propagating a distance equal to depth of penetration of the material. The phase factor corresponds to a very small phase velocity

$$v_p = \frac{\omega}{\beta} = \omega\delta = c\frac{2\pi\delta}{\lambda_0} \tag{13}$$

where c is the velocity of light in free space; λ_0 is free-space wavelength. Since δ/λ_0 is usually very small, this phase velocity is usually much less than the velocity of light.

Equation (7) gives, for a good conductor,

$$\eta = \sqrt{\frac{j\omega\mu}{\sigma}} = (1 + j)\sqrt{\frac{\pi f\mu}{\sigma}} = (1 + j)R_s \tag{14}$$

where R_s is the surface resistivity or high-frequency skin effect resistance per square of a plane conductor of great depth. Equation (14) shows that electric and magnetic fields are 45 degrees out of time phase for the wave propagating in a good conductor. Also, since R_s is very small ($0.014\,\Omega$ for copper at 3 GHz), the ratio of electric field to magnetic field in the wave is small.

Table 6.4b
Material parameters for microwave frequencies and below

Material	Conductivity, S/m	$\dfrac{\varepsilon'}{\varepsilon_0}$	Frequency at which $\sigma = \omega\varepsilon'$, Hz
Copper	5.80×10^7	—	(Optical)
Platinum	0.94×10^7	—	(Optical)
Germanium (lightly doped)	10^2	16	1.1×10^{11}
Sea water	4	81	8.9×10^8
Fresh water	10^{-3}	81	2.2×10^5
Silicon (lightly doped)	10	12	1.5×10^{10}
Representative wet earth	10^{-2}	30	6.0×10^6
Representative dry earth	10^{-3}	7	2.6×10^6

The above mentioned results for the good conductor agree with those found in Chapter 3 by using what appeared to be a different analysis. In Sec. 3.16 the assumption that the conduction current greatly exceeds the displacement current is made at the outset with the result that the differential equation is not the wave equation but the so-called *diffusion equation*. The assumptions in both approaches are identical, however, so it is to be expected that the results would be the same.

Table 6.4b lists several materials for which the dominant losses are those resulting from finite conductivity. The values listed for conductivity and real part of permittivity are those applicable up through the microwave frequency range. The frequencies at which displacement currents equal conduction currents are listed and serve to indicate the range of frequencies in which the two above-mentioned approximations are valid.

PLANE WAVES NORMALLY INCIDENT ON DISCONTINUITIES

6.5 Reflection of Normally Incident Plane Waves from Perfect Conductors

If a uniform plane wave is normally incident on a plane perfect conductor located at $z = 0$, we know that there must be some reflected wave in addition to the incident wave. The boundary conditions cannot be satisfied by a single one of the traveling wave solutions, but will require just enough of the two so that the resultant electric field at the conductor surface is zero for all time. From another point of view, we know from the Poynting theorem that energy cannot pass the perfect conductor, so all energy brought by the incident wave must be returned in a reflected wave. In this simple case the incident and reflected waves are of equal amplitudes, and together form a standing wave pattern whose properties will now be studied.

Let us consider a single-frequency, uniform plane wave and select the orientation of axes so that total electric field lies in the x direction. The phasor electric field, including waves traveling in both the positive and negative z directions (Fig. 6.5) is

$$E_x = E_+ e^{-jkz} + E_- e^{jkz}$$

If $E_x = 0$ at $z = 0$ as required by the perfect conductor, $E_- = -E_+$.

$$E_x = E_+(e^{-jkz} - e^{jkz}) = -2jE_+ \sin kz \tag{1}$$

The relation of the magnetic field to the electric field for the incident and reflected waves is given by Eq. 6.2(11). Hence

$$H_y = \left(\frac{E_+}{\eta} e^{-jkz} - \frac{E_-}{\eta} e^{jkz} \right)$$

$$= \frac{E_+}{\eta}(e^{-jkz} + e^{jkz}) = \frac{2E_+}{\eta} \cos kz \tag{2}$$

Equations (1) and (2) state that, although total electric and magnetic fields for the combination of incident and reflected waves are still mutually perpendicular in space and related in magnitude by η, they are now in time quadrature. The pattern is a standing wave since a zero of electric field is always at the conductor surface, and also always at $kz = -n\pi$ or $z = -n\lambda/2$. Magnetic field has a maximum at the conductor surface, and there are other maxima each time there are zeros of electric field. Similarly, zeros of magnetic field and maxima of electric field are at $kz = -(2n + 1)\pi/2$, or $z = -(2n + 1)\lambda/4$. This situation is sketched in Fig. 6.5, which shows a typical standing wave pattern such as was found for the

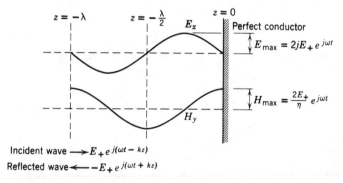

Fig. 6.5 Standing wave patterns of electric and magnetic fields when a plane wave is reflected from a perfect conductor each shown at instants of time differing by one-quarter of a cycle.

shorted transmission line in Sec. 5.8. At an instant in time, occurring twice each cycle, all the energy of the line is in the magnetic field; 90 degrees later the energy is stored entirely in the electric field. The *average* value of the Poynting vector is zero at every cross-sectional plane; this emphasizes the fact that on the average as much energy is carried away by the reflected wave as is brought by the incident wave.

These points are stressed by finding the instantaneous forms for fields,

$$E_x(z, t) = \text{Re}[-2jE_+ \sin kz e^{j\omega t}] = 2E_+ \sin kz \sin \omega t \tag{3}$$

$$H_y(z, t) = \text{Re}\left[\frac{2E_+}{\eta} \cos kz e^{j\omega t}\right] = \frac{2E_+}{\eta} \cos kz \cos \omega t \tag{4}$$

6.6 Transmission-Line Analogy of Wave Propagation: The Impedance Concept

In the problem of wave reflections from a perfect conductor, we found all the properties previously studied for standing waves on an ideal transmission line. The analogy between the plane wave solutions and the waves along an ideal line is in fact an exact and complete one. It is desirable to make use of this whether we start with a study of classical transmission-line theory and then undertake the solution of wave problems, or proceed in the reverse order. In either case the algebraic steps worked out for the solution of one system need not be repeated in analyzing the other; any aids (such as the Smith chart or computer programs) developed for one may be used for the other; any experimental techniques applicable to one system will in general have their counterparts in the other system. We now show the basis for this analogy.

Let us write side by side the equations for the field components in positively and negatively traveling uniform plane waves and the corresponding expressions found in Chapter 5 for an ideal transmission line. For simplicity we orient the axes so that the wave has E_x and H_y components only:

$$E_x(z) = E_+ e^{-jkz} + E_- e^{jkz} \tag{1} \qquad V(z) = V_+ e^{-j\beta z} + V_- e^{j\beta z} \tag{5}$$

$$H_y(z) = \frac{1}{\eta}[E_+ e^{-jkz} - E_- e^{jkz}] \tag{2} \qquad I(z) = \frac{1}{Z_0}[V_+ e^{-j\beta z} - V_- e^{j\beta z}] \tag{6}$$

$$k = \omega\sqrt{\mu\varepsilon} \tag{3} \qquad \beta = \omega\sqrt{LC} \tag{7}$$

$$\eta = \sqrt{\frac{\mu}{\varepsilon}} \tag{4} \qquad Z_0 = \sqrt{\frac{L}{C}} \tag{8}$$

We see that, if in the field equations we replace E_x by voltage V, H_y by current I, permeability μ by inductance per unit length L, and dielectric constant ε by capacitance per unit length C, we get exactly the transmission-line equations (5) to (8). To complete the analogy, we must consider the continuity conditions at a discontinuity between two regions. For the boundary between two dielectrics, we know that total tangential electric and magnetic field components must be continuous across this boundary. For the case of normal incidence (other cases will be considered separately later), E_x and H_y are the tangential components, so these continuity conditions are in direct correspondence to those of transmission lines which require that total voltage and current be continuous at the junction between two transmission lines.

To exploit this analogy fully, it is desirable to consider the ratio of electric to magnetic fields in the wave analysis, analogous to the ratio of voltage to current which is called impedance and used so extensively in the transmission-line analysis. It is a good idea to use ratios such as this in the analysis, quite apart from the transmission-line analogy or the name given these ratios, but in this case it will be especially useful to make the identification with impedance because of the large body of technique existing under the heading of "impedance matching" in transmission lines, most of which may be applied to problems in plane wave reflections. Credit for properly evaluating the importance of the wave impedance concept to engineers and making its use clear belongs to S. A. Schelkunoff.[3]

At any plane z, we shall define the field or wave impedance as the ratio of total electric field to total magnetic field at that plane

$$Z(z) = \frac{E_x(z)}{H_y(z)} \tag{9}$$

For a single positively traveling wave this ratio is η at all planes, so that η, which has been called the intrinsic impedance of the medium, might also be thought of as a *characteristic wave impedance* for uniform plane waves. For a single negatively traveling wave the ratio (9) is $-\eta$ for all z. For combinations of positively and negatively traveling waves, it varies with z. The input value Z_i distance l in front of a plane at which the "load" value of this ratio is given as Z_L may be found from the corresponding transmission line formula, Eq. 5.5(13), taking advantage of the exact analogy. The intervening dielectric has intrinsic impedance η:

$$Z_i = \eta \left[\frac{Z_L \cos kl + j\eta \sin kl}{\eta \cos kl + jZ_L \sin kl} \right] \tag{10}$$

It may be argued that in wave problems the primary concern is with reflections and not with impedances directly. This is true, but as in the transmission-line case there is a one-to-one correspondence between reflection coefficient and impedance mismatch ratio. The analogy may again be invoked to adapt Eqs. 5.5(8) and 5.5(9) to give the reflection and transmission coefficients for a dielectric medium

[3] S. A. Schelkunoff, *Bell Sys. Tech. J.* **17**, 17 (1938).

of intrinsic impedance η when it is terminated with some known load value of field impedance Z_L:

$$\rho = \frac{E_-}{E_+} = \frac{Z_L - \eta}{Z_L + \eta} \tag{11}$$

$$\tau = \frac{E_2}{E_{1+}} = \frac{2Z_L}{Z_L + \eta} \tag{12}$$

We see from this that there is no reflection when $Z_L = \eta$ (i.e., when impedances are matched). There is complete reflection $|\rho| = 1$, when Z_L is zero, infinity, or is purely imaginary (reactive). Other important uses of formulas (10) to (12) will follow in succeeding sections.

_____ **Example 6.6a** _____
Field Impedance in Front of Conducting Plane

If we calculate $Z(z)$ for the plane wave normally incident on a plane conductor, as studied in Sec. 6.5, we recognize that the plane acts as a short circuit since it constrains E_x to be zero there, corresponding to zero voltage in the transmission-line analogy. If we take $Z_L = 0$ in (10), we obtain

$$Z_i = j\eta \tan kl \tag{13}$$

The same result is obtained by taking the ratio of phasor electric and magnetic field at $z = -l$ from Eqs. 6.5(1) and 6.5(2). Like the impedance of a shorted transmission line, this is always imaginary (reactive). It is zero at $kl = n\pi$ and infinite at $kl = (2n + 1)\pi/2$.

_____ **Example 6.6b** _____
Elimination of Wave Reflections from Conducting Surfaces

To eliminate wave reflections from a plane perfectly conducting surface, it is clear that coating the conductor with a thin film having conductivity σ does not help since the perfect conductor merely shorts this out. One can move a quarter-wavelength in front of the conductor surface and place a sheet with resistance η Ω/square. As seen from (13), this is at a point of infinite impedance due to wave reflections so that there is no shunting effect. The wave impinging on the front (if the sheet is thin compared with its depth of penetration) then sees wave impedance η and is perfectly matched. The needed conductivity of the sheet is then

$$\frac{1}{\sigma d} = \eta, \qquad d \ll \delta \tag{14}$$

The arrangement just described is not very convenient to make, and is sensitive to frequency of operation and to angle of incidence. For this reason coatings of "microwave dark rooms" more often use a nonuniform-line approach to matching, with a porous, lossy material, with pyramidal taperings on the surface facing the incident wave.

6.7   Normal Incidence on a Dielectric

If a uniform plane wave is normally incident on a single dielectric boundary from a medium with $\sqrt{\mu_1/\varepsilon_1} = \eta_1$ to one with $\sqrt{\mu_2/\varepsilon_2} = \eta_2$, the wave reflection and transmission may be found from the concepts and equations of Sec. 6.6. Select the direction of the electric field as the x direction, and the direction of propagation of the incident wave as the positive z direction, with the boundary at $z = 0$ (Fig. 6.7a). The medium to the right is assumed to be effectively infinite in

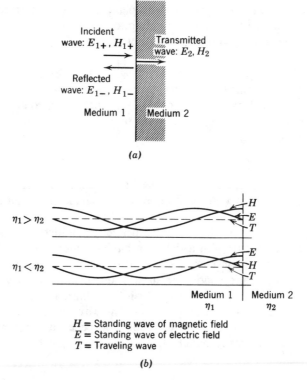

(a)

(b)

H = Standing wave of magnetic field
E = Standing wave of electric field
T = Traveling wave

Fig. 6.7 Reflection and transmission of a plane wave from a plane boundary between two dielectric media.

extent, so that there is no reflected wave in that region. The field impedance there is then just the intrinsic impedance η_2 for all planes, and in particular this becomes the known load impedance at the plane $z = 0$. Applying Eq. 6.6(11) to give the reflection coefficient for medium 1 referred to $z = 0$,

$$\rho = \frac{E_{1-}}{E_{1+}} = \frac{\eta_2 - \eta_1}{\eta_2 + \eta_1} \tag{1}$$

The transmission coefficient giving the amplitude of electric field transmitted into the second dielectric, from Eq. 6.6(12), is

$$\tau = \frac{E_2}{E_{1+}} = \frac{2\eta_2}{\eta_2 + \eta_1} \tag{2}$$

The fraction of incident power density reflected is

$$\frac{P_{1-}}{P_{1+}} = \left(\frac{E_{1-}^2}{2\eta_1}\right)\left(\frac{E_{1+}^2}{2\eta_1}\right)^{-1} = |\rho|^2 \tag{3}$$

And the fraction of the incident power density transmitted into the second medium is

$$\frac{P_2}{P_{1+}} = 1 - |\rho|^2 \tag{4}$$

From (1), we see that there is no reflection if there is a match of impedances, $\eta_1 = \eta_2$. This would of course occur for the trivial case of identical dielectrics, but also for the case of different dielectrics if they could be made with the same *ratio* of μ to ε. This latter case is not usually of importance since we do not commonly find high-frequency dielectric materials with permeability different from that of free space, but it is interesting since we might not intuitively expect a reflection-less transmission in going from free space to a dielectric with both permittivity and permeability increased by, say, 10 times.

In the general case there will be a finite value of reflection in the first region, and from (1) we can show that the magnitude of ρ is always less than unity. (It approaches unity as η_2/η_1 approaches zero or infinity.) The reflected wave can then be combined with a part of the incident wave of equal amplitude to form a standing wave pattern as in the case of complete reflection studied in Sec. 6.5. The remaining part of the incident wave can be thought of as a traveling wave carrying the energy that passes on into the second medium. The combination of the traveling and standing wave parts then produces a space pattern with maxima and minima, but with the minima not zero in general. As for corresponding transmission lines, it is convenient to express the ratio of ac amplitude at the electric field maximum to the minimum ac amplitude (occurring a quarter-wavelength away) as a standing wave ratio S:

$$S = \frac{|E_x(z)|_{\max}}{|E_x(z)|_{\min}} = \frac{1 + |\rho|}{1 - |\rho|} \tag{5}$$

By utilizing (1), it may be shown for real η that

$$S = \begin{cases} \eta_2/\eta_1 & \text{if } \eta_2 > \eta_1 \\ \eta_1/\eta_2 & \text{if } \eta_1 > \eta_2 \end{cases} \tag{6}$$

Since η_1 and η_2 are both real for perfect dielectrics, ρ is real and the plane $z = 0$ must be a position of a maximum or minimum. It is a maximum of electric field if ρ is positive, since reflected and incident waves then add, so it is a maximum of electric field and minimum of magnetic field if $\eta_2 > \eta_1$. The plane $z = 0$ is a minimum of electric field and a maximum of magnetic field if $\eta_1 > \eta_2$. These two cases are sketched in Fig. 6.7b.

_____ **Example 6.7a** _____
Reflection from Quartz and Germanium
at Infrared Wavelengths

Let us find reflection of plane waves normally incident from air onto quartz, and also onto germanium, at a wavelength of $2\,\mu\text{m}$, neglecting losses. Equation (1) may be written in terms of refractive indices [Eq. 6.2(28)]. Since $\mu_1 = \mu_2$ and $n_1 = \sqrt{\varepsilon_1/\varepsilon_0}$, $n_2 = \sqrt{\varepsilon_2/\varepsilon_0}$, (1) becomes

$$\rho = \frac{n_1 - n_2}{n_1 + n_2} \tag{7}$$

Using data in Fig. 13.2b, n_2 for quartz is about 1.5 at $2\,\mu\text{m}$ and n_1 may be taken as unity. Reflection coefficient ρ is then -0.2 and fraction of incident power reflected, ρ^2, is 0.04 or 4%. For germanium n_2 is 4 and ρ is found to be -0.6 so that ρ^2 is 0.36 or 36%.

_____ **Example 6.7b** _____
Reflection from a Good Conductor

From Eq. 6.4(14) the characteristic wave impedance of a conductor is seen to be $(1 + j)R_s$ with R_s the surface resistivity defined in Sec. 3.17. Thus for a plane wave normally incident from a dielectric of intrinsic impedance η onto such a conductor, (1) yields

$$\rho = \frac{(1 + j)R_s - \eta}{(1 + j)R_s + \eta} = -\frac{1 - (1 + j)R_s/\eta}{1 + (1 + j)R_s/\eta} \tag{8}$$

Examination of values for typical conductors from Table 3.17 shows that R_s/η is normally very small. Thus a binomial expansion of (8) with first-order terms only retained gives

$$\rho \approx -1 + \frac{2(1 + j)R_s}{\eta} \tag{9}$$

Then to first order,

$$|\rho|^2 \approx 1 - \frac{4R_s}{\eta} \tag{10}$$

so the fraction of power transmitted into the conductor is approximately $4R_s/\eta$. For air into copper at 100 MHz, this is $4 \times (2.61 \times 10^{-3})/377 = 2.76 \times 10^{-5}$ or only 0.0028%.

6.8   Reflection Problems with Several Dielectrics

We shall next be interested in considering the case of several parallel dielectric discontinuities with a uniform wave incident in some material to the left, as pictured in the case for three dielectric materials in Fig. 6.8a. We might at first be tempted to treat the problem by considering a series of wave reflections, the incident wave breaking into one part reflected and one part transmitted at the first plane; of the part transmitted into region 2 some is transmitted at the second plane and some is reflected back toward the first plane; of the latter part some is transmitted and some reflected, and so on through an infinite series of wave reflections. This lengthy procedure can be avoided by considering total quantities at each stage of the discussion, and again the impedance formulation is useful in the solution.

If the region to the right has only a single outwardly propagating wave, the wave or field impedance at any plane in this medium is η_3, which then becomes the load impedance to place at $z = l$. The input impedance for region 2 is then given at once by Eq. 6.6(10), and since this is the impedance at $z = 0$, it may also be considered the load impedance for region 1:

$$Z_{L1} = Z_{i2} = \eta_2 \left(\frac{\eta_3 \cos k_2 l + j\eta_2 \sin k_2 l}{\eta_2 \cos k_2 l + j\eta_3 \sin k_2 l} \right) \tag{1}$$

The reflection coefficient in region 1, referred to $z = 0$, is given by Eq. 6.6(11):

$$\rho = \frac{Z_{L1} - \eta_1}{Z_{L1} + \eta_1} \tag{2}$$

The fraction of power reflected and transmitted is given by Eqs. 6.7(3) and 6.7(4), respectively.

If there are more than the two parallel dielectric boundaries, the process is simply repeated, the input impedance for one region becoming the load value for the next, until one arrives at the region in which reflection is to be computed. It is of course desirable in many cases to utilize the Smith chart described in Sec. 5.7 in place of (1) to transform load to input impedances, and to compute reflection

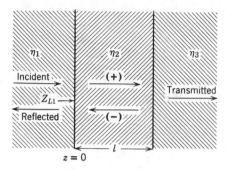

Fig. 6.8a Wave reflections from a system with a dielectric medium interfaced between two other media.

coefficient or standing wave ratio once the impedance mismatch ratio is known, just as the chart is used in transmission-line calculations.

We now wish to consider several special cases which are of importance.

_____ **Example 6.8a** _____
Half-Wave Dielectric Window

If the input and output dielectrics are the same in Fig. 6.8a ($\eta_1 = \eta_3$), and the intervening dielectric window is some multiple of a half-wavelength referred to medium 2 so that $k_2 l = m\pi$, then (1) gives

$$Z_{L1} = \eta_3 = \eta_1 \tag{3}$$

so that by (2) reflection at the input face is zero.

A window such as the above will of course give reflections for frequencies other than that for which its thickness is a multiple of a half-wavelength. For example, a Corning 707 glass window with $\varepsilon_r = 4$ and thickness 0.025 m corresponds to $k_2 l = \pi$ at a frequency of 3 GHz. Let us calculate the reflection at 4 GHz using the Smith chart. Normalized load impedance for region 2 is 377/188.5 = 2, so we enter the chart at point A of Fig. 6.8b. We then move on a circle of constant radius $4/3 \times \lambda/2$ or 0.667λ toward the generator. This is one complete circuit of the chart plus 0.167λ, ending at point B where we read normalized input impedance $Z_{i2}/Z_{02} = 0.62 - j0.38$. Renormalizing to the characteristic impedance of the input region, we multiply by 1.88.5/377 or $\frac{1}{2}$ and find $0.31 - j0.19$. This is entered at point C, for which we read from the radial scale a reflection coefficient $|\rho| = 0.55$, so that $|\rho|^2$ is 0.30 or 30%. This value can be checked by numerical calculation using (1) and (2).

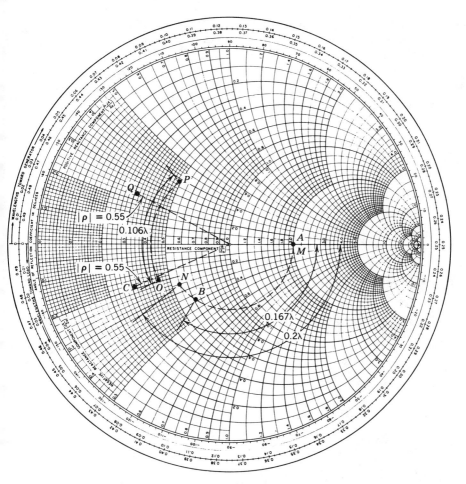

Fig. 6.8b Smith chart constructions for Ex. 6.8a and 6.8d.

_____ **Example 6.8b** _____
Electrically Thin Window

If $\eta_1 = \eta_3$ and $k_2 l$ is so small compared with unity that $\tan k_2 l \approx k_2 l$, (1) becomes

$$Z_{L1} \approx \eta_2 \left(\frac{\eta_1 + j\eta_2 k_2 l}{\eta_2 + j\eta_1 k_2 l} \right) \approx \eta_1 \left[1 + jk_2 l \left(\frac{\eta_2}{\eta_1} - \frac{\eta_1}{\eta_2} \right) \right] \tag{4}$$

Substituting in (2), we see that

$$\rho \approx j \frac{k_2 l}{2} \left(\frac{\eta_2}{\eta_1} - \frac{\eta_1}{\eta_2} \right) \tag{5}$$

The magnitude of reflection coefficient is thus proportional to the electrical length of the dielectric window for small values of $k_2 l$; and the fraction of incident power reflected is proportional to the square of this length.

For example, a polystyrene window ($\varepsilon_r \approx 2.54$) 3 mm in thickness, with a normally incident plane wave at 3 GHz from air, would give $\rho = -j0.145$ so that 2% of incident power is reflected.

_____ **Example 6.8c** _____
Quarter-Wave Coating for Eliminating Reflections

Another important case is that of a quarter-wave coating placed between two different dielectrics. If its intrinsic impedance is the geometric mean of those on the two sides, it will eliminate all wave reflections for energy passing from the first medium into the third. To show this, let

$$k_2 l = \frac{\pi}{2} \qquad \eta_2 = \sqrt{\eta_1 \eta_3} \tag{6}$$

From (1),

$$Z_{L1} = \frac{\eta_2^2}{\eta_3} = \frac{\eta_1 \eta_3}{\eta_3} = \eta_1$$

and from (2)

$$\rho = 0 \tag{7}$$

This technique is used, for example, in coating optical lenses to decrease the amount of reflected light, and is exactly analogous to the technique of matching transmission lines of different characteristic impedances by introducing a quarter-wave section having characteristic impedance the geometric mean of those on the two sides. In all cases the matching is perfect only at specific frequencies for which the length is an odd multiple of a quarter-wavelength, but is approximately correct for bands of frequencies about these values. Multiple coatings are used to increase the frequency band obtainable with a specified permissible reflection.[4]

As a numerical example of this technique, consider the coating needed to eliminate reflections from a 488-nm wavelength argon laser beam (taken as a plane wave) in going from air to fused silica with refractive index 1.46. It follows from (6) that refractive index of the coating should be the geometric mean of that of the air and window, or about 1.21. Note from Fig. 13.2b that there are few

[4] J. M. Stone, *Radiation and Optics: an Introduction to the Classical Theory*, McGraw-Hill, New York, 1963.

materials with this low index, but if one is found its thickness should be a quarter-wavelength measured in the coating, which is about 0.1 μm.

_____ Example 6.8d _____
Reflection from a Two-Ply Dielectric

Now take the two-ply dielectric of Fig. 6.8c in which $d_2 = 2$ mm, $\varepsilon_2/\varepsilon_0 = 2.54$ (polystyrene) and $d_3 = 3.0$ mm, $\varepsilon_3/\varepsilon_0 = 4$ (Corning 707 glass). The normally incident wave is at frequency 10 GHz ($\lambda_0 = 3$ cm). We will do this just on the Smith chart and enter points on Fig. 6.8b. The steps are clear extensions of the above.

Starting at the 3–4 interface:

$$\frac{Z_{L3}}{Z_{03}} = \frac{377}{188.5} = 2.00 \qquad \text{(point } M\text{)}$$

Moving toward generator

$$\frac{d_3}{\lambda_3} = \frac{0.30}{3.0/\sqrt{4}} = 0.2 \text{ wavelength} \qquad \text{(point } N\text{)}$$

We read $Z_{i3}/Z_{03} = 0.55 - j0.235$ and renormalize to get load on region 2:

$$\frac{Z_{L2}}{Z_{02}} = (0.55 - j0.235)\sqrt{\frac{2.54}{4}} = 0.438 - j0.187 \qquad \text{(point } O\text{)}$$

Moving toward generator

$$\frac{d_2}{\lambda_2} = \frac{0.20}{3.0/\sqrt{2.54}} = 0.106 \text{ wavelength} \qquad \text{(point } P\text{)}$$

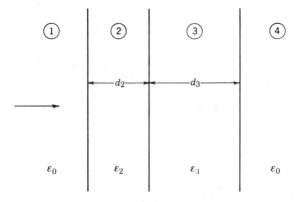

Fig. 6.8c A composite window with two slab dielectrics and free space on either side.

Finally we read $Z_{i2}/Z_{02} = 0.49 + j0.38$ and renormalize to air to get load on region 1:

$$\frac{Z_{L1}}{Z_{01}} = (0.49 + j0.38)\sqrt{\frac{1}{2.54}} = 0.31 + j0.24 \qquad \text{(point } Q\text{)}$$

Radius to Q (as fraction of radius of chart) gives $|\rho| = 0.55$ so that over 30% of incident power is reflected.

PLANE WAVES OBLIQUELY INCIDENT ON DISCONTINUITIES

6.9   Incidence at Any Angle on Perfect Conductors

We next remove the restriction to normal incidence which has been assumed in all the preceding examples. It is possible and desirable to extend the impedance concept to apply to this case also, but before doing this we shall consider the reflection of uniform plane waves incident at an arbitrary angle on a perfect conductor in order to develop certain ideas of the behavior at oblique incidence. It is also convenient to separate the discussion into two cases, polarization with electric field in the plane of incidence, and normal to the plane of incidence. Other cases may be considered a superposition of these two. The plane of incidence is defined by a normal to the surface on which the wave impinges, and a ray following the direction of propagation of the incident wave. That is, it is the plane of the paper as we have drawn sketches in this chapter. We consider here loss-free media.[5] Polarization with **E** in the plane of incidence may also be referred to as *transverse magnetic* (*TM*) since magnetic field is then transverse. When **E** is normal to the plane of incidence it may be called *transverse electric* (*TE*).[6]

Polarization with Electric Field in the Plane of Incidence (*TM*)   In Fig. 6.9a the ray drawn normal to the incident wave front makes an angle θ with the normal to the conductor. We know that, since energy cannot pass into the perfect conductor, there must be a reflected wave, and we draw its direction of propagation at some unknown angle θ'. The electric and magnetic fields of both incident and reflected wave must lie perpendicular to their respective directions of propagation by the properties of uniform plane waves (Sec. 6.2), so the electric fields may be

---

[5] For a treatment of waves in free space obliquely incident on the plane surface of a lossy medium, see R. B. Adler, L. J. Chu, and R. M. Fano, *Electromagnetic Energy Transmission and Radiation*, Wiley, New York, 1960, p. 442.

[6] An alternate designation, common in the scientific literature, employs P (for German "parallel") and S (for German "senkrecht," i.e., perpendicular), respectively, for the orientations of **E** in the two cases.

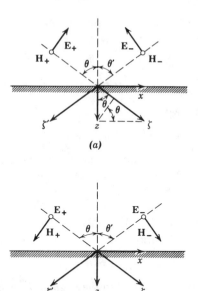

Fig. 6.9 Wave incident at an angle on a dielectric interface. (*a*) Polarization with electric field in plane of incidence. (*b*) Polarization with electric field perpendicular to plane of incidence.

drawn as shown by E_+ and E_-. The corresponding magnetic fields H_+ and H_- are then both normally out of the paper, so that $E \times H$ gives the direction of propagation for each wave. Moreover, with the senses as shown,

$$\frac{E_+}{H_+} = \frac{E_-}{H_-} = \eta \tag{1}$$

If we draw a ζ direction in the actual direction of propagation for the incident wave as shown, and a ζ' direction so that the reflected wave is traveling in the negative ζ' direction, we know that the phase factors for the two waves may be written as $e^{-jk\zeta}$ and $e^{jk\zeta'}$, respectively. The sum of incident and reflected waves at any point $x, z\ (z < 0)$ can be written

$$E(x, z) = E_+ e^{-jk\zeta} + E_- e^{jk\zeta'} \tag{2}$$

where E_+ and E_- are reference values at the origin. We now express all coordinates in terms of the rectangular system aligned with the conductor surface. The conversion of ζ and ζ' from the diagram is

$$\zeta = x \sin \theta + z \cos \theta \tag{3}$$

$$\zeta' = -x \sin \theta' + z \cos \theta'$$

so that, if these are substituted in the phase factors of (2), and the two waves broken into their x and z components, we have

$$E_x(x, z) = E_+ \cos \theta e^{-jk(x\sin\theta + z\cos\theta)} - E_- \cos \theta' e^{jk(-x\sin\theta' + z\cos\theta')} \tag{5}$$

$$E_z(x, z) = -E_+ \sin \theta e^{-jk(x\sin\theta + z\cos\theta)} - E_- \sin \theta' e^{jk(-x\sin\theta' + z\cos\theta')} \tag{6}$$

The magnetic field in the two waves is

$$H_y(x, z) = H_+ e^{-jk(x\sin\theta + z\cos\theta)} + H_- e^{jk(-x\sin\theta' + z\cos\theta')} \tag{7}$$

The next step is the application of the boundary condition of the perfect conductor, which is that, at $z = 0$, E_x must be zero for all x. From (5),

$$E_x(x, 0) = E_+ \cos \theta e^{-jkx\sin\theta} - E_- \cos \theta' e^{-jkx\sin\theta'} = 0 \tag{8}$$

This equation can be satisfied for all x only if the phase factors in the two terms are equal, and this in turn requires that

$$\theta = \theta' \tag{9}$$

That is, *the angle of reflection is equal to the angle of incidence.* With this result in (8), it follows that the two amplitudes must be equal:

$$E_+ = E_- \tag{10}$$

If the results (9) and (10) are substituted in (5), (6), and (7), we have the final expressions for field components at any point $z < 0$,

$$E_x(x, z) = -2jE_+ \cos \theta \sin(kz \cos \theta)e^{-jkx\sin\theta} \tag{11}$$

$$E_z(x, z) = -2E_+ \sin \theta \cos(kz \cos \theta)e^{-jkx\sin\theta} \tag{12}$$

$$\eta H_y(x, z) = 2E_+ \cos(kz \cos \theta)e^{-jkx\sin\theta} \tag{13}$$

The foregoing field has the character of a traveling wave with respect to the x direction, but that of a standing wave with respect to the z direction. That is, E_x is zero for all time at the conducting plane, and also in parallel planes distance nd in front of the conductor, where

$$d = \frac{\lambda}{2 \cos \theta} = \frac{1}{2f\sqrt{\mu\varepsilon} \cos \theta} \tag{14}$$

The ac amplitude of E_x is a maximum in planes an odd multiple of $d/2$ in front of the conductor. H_y and E_z are maximum where E_x is zero, are zero where E_x is maximum, and are everywhere 90 degrees out of time phase with respect to E_x. Perhaps the most interesting result from this analysis is that the distance between successive maxima and minima, measured normal to the plane, becomes *greater* as the incidence becomes more oblique. A superficial survey of the situation might lead one to believe that they would be at projections of the wavelength in this direction, which would become smaller with increasing θ. This point will be pursued more in the following section.

Polarization with Electric Field Normal to the Plane of Incidence (TE)   In this polarization (Fig. 6.9b), E_+ and E_- are normal to the plane of the paper, and H_+ and H_- are then as shown. Proceeding exactly as before, we can write the components of the two waves in the x, z system of coordinates as

$$E_y(x, z) = E_+ e^{-jk(x\sin\theta + z\cos\theta)} + E_- e^{jk(-x\sin\theta' + z\cos\theta')} \tag{15}$$

$$\eta H_x(x, z) = -E_+ \cos\theta e^{-jk(x\sin\theta + z\cos\theta)} + E_- \cos\theta' e^{jk(-x\sin\theta' + z\cos\theta')} \tag{16}$$

$$\eta H_z(x, z) = E_+ \sin\theta e^{-jk(x\sin\theta + z\cos\theta)} + E_- \sin\theta' e^{jk(-x\sin\theta' + z\cos\theta')} \tag{17}$$

The boundary condition at the perfectly conducting plane is that E_y is zero at $z = 0$ for all x, which by the same reasoning as before leads to the conclusion that $\theta = \theta'$ and $E_+ = -E_-$. The field components, (15) to (17), then become

$$E_y = -2jE_+ \sin(kz \cos\theta)e^{-jkx\sin\theta} \tag{18}$$

$$\eta H_x = -2E_+ \cos\theta \cos(kz \cos\theta)e^{-jkx\sin\theta} \tag{19}$$

$$\eta H_z = -2jE_+ \sin\theta \sin(kz \cos\theta)e^{-jkx\sin\theta} \tag{20}$$

This set again shows the behavior of a traveling wave in the x direction and a standing-wave pattern in the z direction with zeros of E_y and H_z and maxima of H_x at the conducting plane and at parallel planes distance nd away, with d given by (14).

6.10 Phase Velocity and Impedance for Waves at Oblique Incidence

Phase Velocity   Let us consider an incident wave, such as that of Sec. 6.9, traveling with velocity $v = 1/\sqrt{\mu\varepsilon}$ in a positive direction, which makes angle θ with a desired z direction aligned normally to some reflecting surface. We saw that it is possible to express the phase factor in terms of the x and z coordinates:

$$\mathbf{E}(x, z) = \mathbf{E}_+ e^{-jk\zeta} = \mathbf{E}_+ e^{-jk(x\sin\theta + z\cos\theta)} \tag{1}$$

For many purposes it is desirable to concentrate on the change in phase as one moves in the x direction, or in the z direction. We may then define the two phase constants for these directions:

$$\beta_x = k \sin\theta \tag{2}$$

$$\beta_z = k \cos\theta \tag{3}$$

Wave (1) in instantaneous form is then

$$\mathbf{E}(x, z, t) = \text{Re}[\mathbf{E}_+ e^{j(\omega t - \beta_x x - \beta_z z)}] \tag{4}$$

If we wish to keep the instantaneous phase constant as we move in the x direction, we keep $\omega t - \beta_x x$ constant (the last term does not change if we move only in the x direction), and the velocity required for this is defined as the phase velocity referred to the x direction:

$$v_{px} = \frac{\partial x}{\partial t}\Big|_{(\omega t - \beta_x x) = \text{const}} = \frac{\omega}{\beta_x}$$

or

$$v_{px} = \frac{\omega}{k \sin \theta} = \frac{1}{\sqrt{\mu\varepsilon} \sin \theta} = \frac{v}{\sin \theta} \tag{5}$$

Similarly for the z direction

$$v_{pz} = \frac{\omega}{\beta_z} = \frac{v}{\cos \theta} \tag{6}$$

where v is the velocity normal to its wave front, $1/\sqrt{\mu\varepsilon}$.

We see that in both cases the phase velocity is *greater* than the velocity measured normal to the wave front and will in fact be so for any oblique direction. There is no violation of relativistic principles by this result, since no material object moves at this velocity. It is the velocity of a fictitious point of intersection of the wave front and a line drawn in the selected direction. Thus in Fig. 6.10a if a plane of constant phase aa moves to $a'a'$ in a given interval of time, the distance moved normal to the wave front is XX', but the distance moved by this constant phase reference along the z direction is the greater distance YY'. Since

$$YY' = XX' \sec \theta$$

this picture would again lead to the result (6) for phase velocity in the z direction.

Thus it is the phase constant β in a particular direction that is reduced by cosine or sine of the angle between normal to the boundary and normal to the wave front whereas phase velocity is increased by the same factor. The concept of a phase velocity, and the understanding of why it may be greater than the

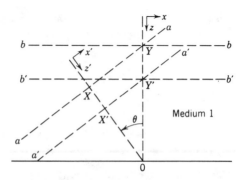

Fig. 6.10a Uniform plane wave moving at angle θ toward a plane.

velocity of light, is essential to the discussion of guided waves in later chapters, as well as to the remainder of this chapter.

Wave Impedance   In the problems of oblique incidence on a plane boundary between different media, it is also useful to define the wave or field impedance as the ratio of electric to magnetic field components in planes parallel to the boundary. The reason for this is the continuity of the *tangential* components of electric and magnetic fields at a boundary, and the consequent equality of the above defined ratio on the two sides of the boundary. That is, if the value of this ratio is computed as an input impedance for a region to the right in some manner, it is also the value of load impedance at that plane for the region to the left, just as in the examples of normal incidence.

Thus, for incident and reflected waves making angle θ with the normal as in Sec. 6.9, we may define a characteristic wave impedance referred to the z direction in terms of the components in planes transverse to that direction. From Eqs. 6.9(5) and 6.9(7) *for waves polarized with electric field in the plane of incidence,*

$$(Z_z)_{TM} = \frac{E_{x+}}{H_{y+}} = -\frac{E_{x-}}{H_{y-}} = \eta \cos \theta \tag{7}$$

$+$ and $-$ refer, respectively, to incident and reflected wave; the sign of the ratio is chosen for each wave to yield a positive impedance. From Eqs. 6.9(15) and 6.9(16) *for waves polarized with electric field normal to the plane of incidence,*

$$(Z_z)_{TE} = -\frac{E_{y+}}{H_{x+}} = \frac{E_{y-}}{H_{x-}} = \eta \sec \theta \tag{8}$$

we see that, for the first type of polarization, the characteristic wave impedance is always less than η, as we would expect since only a component of total electric field lies in the transverse x–y plane, whereas the total magnetic field lies in that plane. In the latter polarization, the reverse is true and Z_z is always greater than η.

The interpretation of the example of the last section from the foregoing point of view is then that the perfect conductor amounts to a zero impedance or short to the transverse field component E_x. We would then expect a standing wave pattern in the z direction with other zeros at multiples of a half-wavelength away, this wavelength being computed from phase velocity in the z direction. This is consistent with the interpretation of Eq. 6.9(14).

_____ **Example 6.10** _____
Diffraction Orders from a Bragg Grating

It is known that a periodic grating of wires, slots, or similar perturbations, reradiates, or diffracts, a plane wave incident upon it into various directions

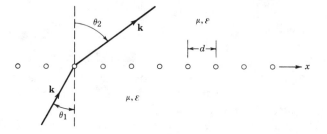

Fig. 6.10b Diffraction grating with plane wave incident at an angle.

described as the *diffraction orders* for the grating. The directions are defined as those for which phase constants along the grating match on the two sides, except that multiples of 2π difference may exist between grating elements and the contributions still add constructively. Consider a grating as in Fig. 6.10b, with a plane wave incident from the bottom at angle θ_1 from the normal. Phase constant in the x direction is $k \sin \theta_1$ so that the phase difference between induced effects in adjacent grating elements is $kd \sin \theta_1$. For the reradiated wave in the top, at angle θ_2 from the normal, there will then be constructive interference if

$$kd \sin \theta_1 = kd \sin \theta_2 \pm 2m\pi, \qquad m = 0, 1, 2, \ldots$$

or

$$\sin \theta_2 = \sin \theta_1 \pm \frac{m\lambda}{d}$$

Angle $\theta_2 = \theta_1$ for $m = 0$ (the *principal order*) but there can be other diffraction orders or *lobes* provided that $|\sin \theta_1 \pm m\lambda/d| \leq 1$, which requires $d > \lambda/2$.

6.11   Incidence at Any Angle on Dielectrics

Law of Reflection   For a uniform plane wave incident at angle θ_1 from the normal to the plane boundary between two dielectrics ε_1 and ε_2, Fig. 6.11, there is a reflected wave at some angle θ_1' with the normal, and a transmitted (*refracted*) wave into the second medium which is drawn at some angle θ_2 with the normal. For either type of polarization, the continuity condition on tangential components of electric and magnetic field at the boundary $z = 0$ must be satisfied for all values of x. As in the argument applied to the problem of reflection from the perfect conductor, this is possible for all x only if incident, reflected, and refracted waves all have the same phase factor with respect to the x direction.

$$k_1 \sin \theta_1 = k_1 \sin \theta_1' = k_2 \sin \theta_2 \tag{1}$$

The first pair in (1) gives the result

$$\theta'_1 = \theta_1 \tag{2}$$

or the angle of reflection is equal to the angle of incidence.

Snell's Law of Refraction　From the last pair of (1) we find a relation between the angle of refraction θ_2 and the angle of incidence θ_1:

$$\frac{\sin \theta_2}{\sin \theta_1} = \frac{k_1}{k_2} = \frac{v_2}{v_1} = \frac{n_1}{n_2} \tag{3}$$

This relation is a familiar one in optics and is known as Snell's law. The *refractive index n* is defined to be unity for free space so its value for any other dielectric is a measure of the phase velocity of electromagnetic waves in the medium, relative to free space. It is common to use the refractive index to characterize properties of dielectrics in the infrared and optical frequency ranges as explained in Sec. 6.2. At microwave and lower frequencies it is more common to express the velocities in (3) in terms of permittivity and permeability. For most dielectrics n_1/n_2 may be replaced by $(\varepsilon_1/\varepsilon_2)^{1/2}$ since $\mu_1 \approx \mu_2 \approx \mu_0$.

Reflection and Transmission for Polarization with E in Plane of Incidence　To compute the amount of the wave reflected and the amount transmitted, we may use the impedance concept as extended for oblique incidence in the last section. To show the validity of this procedure, we write the continuity conditions for total E_x and H_y, including both incident and reflected components in region 1:

$$E_{x+} + E_{x-} = E_{x2} \tag{4}$$

$$H_{y+} + H_{y-} = H_{y2} \tag{5}$$

Following Sec. 6.10, if we define wave impedances in terms of the tangential components for this *TM* polarization,

$$Z_{z1} = \frac{E_{x+}}{H_{y+}} = -\frac{E_{x-}}{H_{y-}} \tag{6}$$

$$Z_L = \frac{E_{x2}}{H_{y2}} \tag{7}$$

Equation (5) may be written

$$\frac{E_{x+}}{Z_{z1}} - \frac{E_{x-}}{Z_{z1}} = \frac{E_{x2}}{Z_L} \tag{8}$$

An elimination between (4) and (8) results in equations for reflection and transmission coefficients:

$$\rho = \frac{E_{x-}}{E_{x+}} = \frac{Z_L - Z_{z1}}{Z_L + Z_{z1}} \tag{9}$$

$$\tau = \frac{E_{x2}}{E_{x+}} = \frac{2Z_L}{Z_L + Z_{z1}} \tag{10}$$

For the present case, in which we assume there is no returning wave in medium 2, the load impedance Z_L is just the characteristic wave impedance for the refracted wave referred to the z direction, obtainable from (3) and Eq. 6.10(7),

$$Z_L = \eta_2 \cos \theta_2 = \eta_2 \sqrt{1 - \left(\frac{v_2}{v_1}\right)^2 \sin^2 \theta_1} \tag{11}$$

And the characteristic wave impedance for medium 1 referred to the z direction is

$$Z_{z1} = \eta_1 \cos \theta_1 \tag{12}$$

The second form of (11) is applicable even when the result for Z_L is complex. Note that, for dielectrics with $\mu_1 = \mu_2$,

$$\frac{\eta_2}{\eta_1} = \sqrt{\frac{\varepsilon_1}{\varepsilon_2}} = \frac{n_1}{n_2} \tag{13}$$

The total fields in region 1 may then be written as the sum of incident and reflected waves, utilizing (9) and the basic properties of uniform plane waves. We shall use H_{y+} (denoted H_+) of the incident wave as the reference component since it is parallel to the boundary.

$$E_x = \eta_1 H_+ \cos \theta_1 e^{-j\beta_x x}[e^{-j\beta_z z} + \rho e^{j\beta_z z}] \tag{14}$$

$$H_y = H_+ e^{-j\beta_x x}[e^{-j\beta_z z} - \rho e^{j\beta_z z}] \tag{15}$$

$$E_z = \eta_1 H_+ \sin \theta_1 e^{-j\beta_x x}[-e^{-j\beta_z z} + \rho e^{j\beta_z z}] \tag{16}$$

$$\beta_x = k_1 \sin \theta_1 \qquad \beta_z = k_1 \cos \theta_1 \tag{17}$$

This field again has the character of a traveling-wave field in the x direction and a standing-wave field in the z direction, but here the minima in the z direction do not in general reach zero. The ratio of maxima to minima could be expressed as a standing wave ratio and would be related to the magnitude of reflection coefficient by the usual expression, Eq. 6.7(5).

Reflection and Transmission for Polarization with E Normal to Plane of Incidence
For this (*TE*) polarization, the basic relations (9) and (10) between impedances and reflection or transmission may also be shown to apply. Note that they were first introduced in connection with transmission line waves in Chapter 5 but have now found usefulness for many wave problems through the impedance concept applied to wave phenomena:

$$\rho = \frac{E_{y-}}{E_{y+}} = \frac{Z_L - Z_{z1}}{Z_L + Z_{z1}} \tag{18}$$

$$\tau = \frac{E_{y2}}{E_{y+}} = \frac{2Z_L}{Z_L + Z_{z1}} \tag{19}$$

For this polarization, the proper wave impedances are obtained from Eq. 6.10(8):

$$Z_L = \eta_2 \sec \theta_2 = \eta_2 \left[1 - \left(\frac{v_2}{v_1} \right)^2 \sin^2 \theta_1 \right]^{-1/2} \tag{20}$$

$$Z_{z1} = \eta_1 \sec \theta_1 \tag{21}$$

The total fields in region 1 are (E_+ denotes the value of E_{y+} in the incident wave)

$$E_y = E_+ e^{-j\beta_x x}[e^{-j\beta_z z} + \rho e^{j\beta_z z}] \tag{22}$$

$$\eta_1 H_x = -E_+ \cos \theta_1 e^{-j\beta_x x}[e^{-j\beta_z z} - \rho e^{j\beta_z z}] \tag{23}$$

$$\eta_1 H_z = E_+ \sin \theta_1 e^{-j\beta_x x}[e^{-j\beta_z z} + \rho e^{j\beta_z z}] \tag{24}$$

$$\beta_x = k_1 \sin \theta_1 \qquad \beta_z = k_1 \cos \theta_1 \tag{25}$$

_____ **Example 6.11** _____
Reflection and Transmission of a Circularly Polarized Wave at Oblique Incidence

A circularly polarized wave, incident at an angle on a dielectric discontinuity as in Fig. 6.11, can be resolved into the two linearly polarized parts (Sec. 6.3) and relations of this section used. To be specific, let us assume such a wave incident at angle $\theta_1 = 60$ degrees from air onto fused quartz with $\varepsilon_r = 3.78$. By Snell's law,

$$\theta_2 = \sin^{-1} \left[\sqrt{\frac{\varepsilon_1}{\varepsilon_2}} \sin \theta_1 \right] = \sin^{-1} \frac{0.866}{\sqrt{3.78}} = \sin^{-1}(0.445) = 26.4°$$

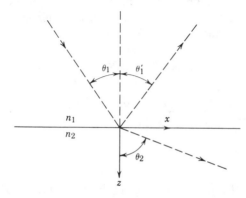

Fig. 6.11 Oblique incidence on boundary between two isotropic dielectrics.

For the TM component, wave impedances are

$$Z_{z1} = \eta_1 \cos \theta_1 = 377 \cos 60° = 188.5 \, \Omega$$

$$Z_{z2} = \eta_2 \cos \theta_2 = \frac{377}{\sqrt{3.78}} \cos 26.4° = 174 \, \Omega$$

from which we calculate ρ and τ from (9) and (10) as

$$\rho = \frac{174 - 188.5}{174 + 188.5} = -0.04$$

$$\tau = \frac{2 \times 174}{174 + 188.5} = 0.96$$

For the TE component,

$$Z_{z1} = \eta_1 \sec \theta_1 = 756 \, \Omega$$

$$Z_{z2} = \eta_2 \sec \theta_2 = 217 \, \Omega$$

$$\rho = 0.554$$

$$\tau = 0.448$$

So the second component is highly reflected, the first part weakly reflected, and both reflected and transmitted waves will be elliptically polarized.

6.12 Total Reflection

A study of the general results from Sec. 6.11 shows that there are several particular conditions of incidence of special interest. The first is one that leads to a condition of total reflection. From the basic formula for reflection coefficient, Eq. 6.11(9) or 6.11(18), we know that there is complete reflection ($|\rho| = 1$) if the load impedance Z_L is zero, infinity, or purely imaginary. To show the last condition, let $Z_L = jX_L$ and note that Z_{z1} is real:

$$|\rho| = \left| \frac{jX_L - Z_{z1}}{jX_L + Z_{z1}} \right| = \frac{\sqrt{X_L^2 + Z_{z1}^2}}{\sqrt{X_L^2 + Z_{z1}^2}} = 1 \tag{1}$$

The value of Z_L for TM polarization, given by Eq. 6.11(11), is seen to become zero for some critical angle $\theta = \theta_c$ such that

$$\sin \theta_c = \frac{v_1}{v_2} \tag{2}$$

The value of Z_L for TE polarization, given by Eq. 6.11(20), becomes infinite for this same condition. For both polarizations, Z_L is imaginary for angles of incidence greater than θ_c, so there is total reflection for such angles of incidence. For loss-free dielectrics having $\mu_1 = \mu_2$, (2) reduces to

$$\sin \theta_c = \sqrt{\frac{\varepsilon_2}{\varepsilon_1}} \tag{3}$$

It is seen that there are real solutions for the critical angle in this case only when $\varepsilon_1 > \varepsilon_2$, or when the wave passes from an optically dense to an optically rarer medium. From Snell's law, Eq. 6.11(3), we find that the angle of refraction is $\pi/2$ for $\theta = \theta_c$ and is imaginary for greater angles of incidence. So from this point of view also we expect no transfer of energy into the second medium. Although there is no energy transfer, there are finite values of field in the second region as required by the continuity conditions at the boundary. Fields die off exponentially with distance from the boundary as the phase constant β_z, becomes imaginary.

Although the reflected wave has the same amplitude as the incident wave for angles of incidence greater than the critical, it does not in general have the same phase. The phase relation between E_{x-} and E_{x+} for the first type of polarization is also different from that between E_{y-} and E_{y+} for the second type of polarization incident at the same angle. Thus, if the incident wave has both types of polarization components, the reflected wave under these conditions is elliptically polarized (Sec. 6.3).

The phenomenon of total reflection is very important at optical frequencies, as it provides reflection with less loss than from conducting mirrors. The use in total reflecting prisms is a well-known example, and its importance to dielectric waveguides will be shown in later chapters.

_____ **Example 6.12** _____
Evanescent Decay in Second Medium

It was noted above that fields are nonzero in the second medium under conditions of total reflection, but die off exponentially from the boundary (i.e., are evanescent). Let us show this more specifically using expressions already developed. We have so far considered the propagation factor in the z direction as

$$e^{-j\beta_z z} \tag{4}$$

where, for medium 2,

$$\beta_z = k_2 \cos \theta_2 \tag{5}$$

and using Snell's law,

$$\beta_z = k_2 \sqrt{1 - \left(\frac{v_2}{v_1}\right)^2 \sin^2 \theta_1} \tag{6}$$

This becomes imaginary, which by (4) represents an attenuation factor $e^{-\alpha z}$, for $(v_2/v_1)\sin\theta_1 > 1$, with

$$\alpha = k_2 \sqrt{\left(\frac{v_2}{v_1}\right)^2 \sin^2\theta_1 - 1} \quad \text{Np/m} \tag{7}$$

Since $k_2 = 2\pi\sqrt{\varepsilon_{2r}}/\lambda$, α will be of order of magnitude 2π Np (or 54.5 dB) per wavelength.

6.13 Polarizing or Brewster Angle

Let us next ask under what conditions there might be no reflected wave when the uniform plane wave is incident at angle θ on the dielectric boundary. We know that this occurs for a matching of impedances between the two media, $Z_L = Z_{z1}$. For the wave with *TM* polarization and for a medium with $\mu_1 = \mu_2$, Eqs. 6.11(11) and 6.11(12) become

$$Z_L = \sqrt{\frac{\mu_1}{\varepsilon_2}} \sqrt{1 - \frac{\varepsilon_1}{\varepsilon_2}\sin^2\theta_1} \tag{1}$$

$$Z_{z1} = \sqrt{\frac{\mu_1}{\varepsilon_1}}\cos\theta_1 \tag{2}$$

These two quantities may be made equal for a particular angle $\theta_1 = \theta_p$ such that

$$\cos\theta_p = \sqrt{\frac{\varepsilon_1}{\varepsilon_2}} \sqrt{1 - \frac{\varepsilon_1}{\varepsilon_2}\sin^2\theta_p} \tag{3}$$

This equation has a solution

$$\theta_p = \sin^{-1}\sqrt{\frac{\varepsilon_2}{\varepsilon_1 + \varepsilon_2}} = \tan^{-1}\sqrt{\frac{\varepsilon_2}{\varepsilon_1}} = \tan^{-1}\left(\frac{n_2}{n_1}\right) \tag{4}$$

Note that (4) yields real values of θ_p for either $\varepsilon_1 > \varepsilon_2$ or $\varepsilon_2 > \varepsilon_1$, and so, for *TM* polarization there is always some angle for which there is no reflection; all energy incident at this angle passes into the second medium.

For *TE* polarization a study of Eqs. 6.11(20) and 6.11(21) shows that there is no angle yielding an equality of impedances for materials with different dielectric constants but like permeabilities. Hence, a wave incident at angle θ_p with both polarization components present has some of the second polarization component but none of the first reflected. The reflected wave at this angle is thus plane polarized with electric field normal to the plane of incidence, and the angle θ_p is correspondingly known as the *polarizing angle*. It is also alternatively known as the *Brewster angle*. Gas lasers sometimes use windows placed at the Brewster angle to allow oscillation for only one of the two possible polarizations, since for this polarization there is low reflection from the ends of the tube, and the external optical resonator governs the behavior.

_____ **Example 6.13** _____
Brewster Angle Windows on He–Ne Laser

Let us calculate the angle for Brewster angle windows on a helium–neon laser ($\lambda = 0.633$ μm) using fused quartz with refractive index $n_2 = 1.46$. From (4),

$$\theta_p = \tan^{-1} \frac{1.46}{1} = 55.6°$$

Note that this is the angle between the beam axis and the normal to the window (Fig. 6.13). If we find the angle in the glass by Snell's law

$$\theta_2 = \sin^{-1}\left[\frac{1}{1.46} \sin 55.6°\right] = 34.4°$$

Note that this is the proper Brewster angle for going from glass to air,

$$\theta_{p2} = \tan^{-1} \frac{1}{1.46} = 34.4°$$

So the exit from the window is without reflection also.

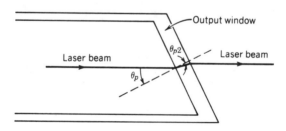

Fig. 6.13 Output window of a laser set at the Brewster (polarizing) angle to eliminate reflections for one polarization.

6.14   Multiple Dielectric Boundaries with Oblique Incidence

If there are several dielectric regions with parallel boundaries, the problem may be solved by successively transforming impedances through the several regions, using the standard transmission-line formula, Eq. 6.6(10), or a graphical aid such as the Smith chart. For each region the phase constant and characteristic wave impedance must include the function of angle from the normal as well as the properties of the dielectric material. Thus for the ith region, from the concepts of Sec. 6.11 the phase constant is

$$\beta_{zi} = k_i \cos \theta_i \tag{1}$$

and the characteristic wave impedance is

$$Z_{zi} = \eta_i \cos \theta_i \qquad \text{for } TM \text{ polarization} \tag{2}$$

$$Z_{zi} = \eta_i \sec \theta_i \qquad \text{for } TE \text{ polarization} \tag{3}$$

When the impedance is finally transformed to the surface at which it is desired to find reflection, the reflection coefficient is calculated from the basic reflection formula, Eq. 6.11(9), and the fraction of the incident power reflected is just the square of its magnitude. The angles in the several regions are found by successively applying Snell's law, starting from the first given angle of incidence.

_____ **Example 6.14** _____
Oblique Incidence on a Two-Ply Dielectric Window

We return to the two-ply dielectric of Ex. 6.8d and consider incidence of a plane wave at angle $\theta_1 = 40$ degrees from the normal (Fig. 6.14). From Snell's law, angles in the dielectric regions are

$$\theta_2 = \sin^{-1} \frac{\sin 40^\circ}{\sqrt{2.54}} = 23.76^\circ$$

$$\theta_3 = \sin^{-1} \sqrt{\frac{2.54}{4}} \sin 23.76^\circ = 18.73^\circ$$

$$\theta_4 = \sin^{-1}(\sqrt{4} \sin 18.73^\circ) = 40^\circ$$

This checks that the exit angle is equal to entrance angle as required (Prob. 6.14d). Let us take TE polarization. Wave impedance for the several regions are

$$Z_{z1} = Z_{z4} = \frac{377}{\cos 40^\circ} = 492 \, \Omega$$

$$Z_{z2} = \frac{377}{\sqrt{2.54} \cos 23.76^\circ} = 258.5 \, \Omega$$

$$Z_{z3} = \frac{377}{\sqrt{4} \cos 18.73^\circ} = 199.0 \, \Omega$$

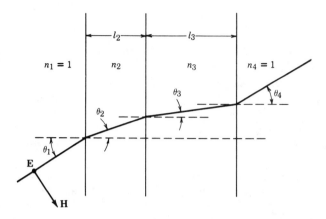

Fig. 6.14 Composite window with wave incident at an angle.

The length in wavelengths for each region is

$$\frac{l_2}{\lambda_2} = \frac{0.2 \times \sqrt{2.54} \cos 23.76°}{3} = 0.097$$

$$\frac{l_3}{\lambda_3} = \frac{0.3 \times \sqrt{4} \cos 18.73°}{3} = 0.189$$

So now, paralleling Ex. 6.8d,

$$\frac{Z_{L3}}{Z_{03}} = \frac{492}{199.0} = 2.47$$

Moving 0.189 wavelength toward generator we read

$$\frac{Z_{i3}}{Z_{03}} = 0.48 - j0.325$$

Renormalizing, $Z_{L2}/Z_{02} = (199/258.5)(0.48 - j0.325) = 0.37 - j0.25$. Moving 0.097 wavelength toward generator we now find

$$\frac{Z_{i2}}{Z_{02}} = 0.38 + j0.30$$

Renormalizing, $Z_{L1}/Z_{01} = (258.5/492.2)(0.38 + j0.30) = 0.20 + j0.158$. For this point we read $|\rho| = 0.66$ from which $|\rho|^2 = 0.435$.

PROBLEMS

6.2a Draw a sketch similar to Fig. 6.2*a* demonstrating the relations in a negatively traveling wave.

6.2b Consider an inhomogeneous dielectric in which ε is a function of z. (Permeability is μ_0 everywhere.) For a wave with E_x and H_y and with no variations in the x or y directions, find the differential equation which replaces the wave equation 6.2(7).

6.2c A step-function uniform plane wave is generated by suddenly impressing a constant electric field $E_x = C$ at $z = 0$ at time $t = 0$ and maintaining it thereafter. A perfectly conducting plane is placed normal to the z direction at $z = 600$ m. Sketch total E_x and ηH_y versus z at $t = 1$ μs and at $t = 3$ μs.

6.2d Write the instantaneous forms corresponding to phasor fields given by Eqs. 6.02(21)–(24). Find the instantaneous Poynting vector and show that the average part is equal to the average Poynting vector of the positively traveling wave minus that for the negatively traveling wave.

6.2e Review the concept of group velocity from Sec. 5.12. Describe the behavior of the modulated wave, Ex. 6.2, for a dispersive medium.

6.3a Check to show that Eqs. 6.3(1) and 6.3(2) satisfy Maxwell's equations under the specializations appropriate to plane waves.

6.3b Sketch the locus of **E** for the following special cases of Eq. 6.3(1), identifying the type of polarization for each: (i) $E_1 = 1$, $E_2 = 2$, $\psi = 0$; (ii) $E_1 = 1$, $E_2 = 2$, $\psi = \pi$; (iii) $E_1 = 1$, $E_2 = 1$, $\psi = \pi/2$; (iv) $E_1 = 1$, $E_2 = 2$, $\psi = \pi/2$; (v) $E_1 = 1$, $E_2 = 1$, $\psi = \pi/4$.

6.3c A circularly polarized wave of strength E_1 travels in the positive z direction with the reflected circularly polarized wave of strength E_1' returning,

$$\mathbf{E} = (\hat{\mathbf{x}} + j\hat{\mathbf{y}})E_1 e^{-jkz} + (\hat{\mathbf{x}} + j\hat{\mathbf{y}})E_1' e^{jkz}$$

Find the average Poynting vector.

6.3d Show that any arbitrary elliptically polarized wave may be broken up into two oppositely rotating circularly polarized components instead of the two plane-polarized components.

6.4a For a uniform plane wave of frequency 10 GHz propagating in polystyrene, calculate the attenuation constant, phase velocity, and intrinsic impedance.

6.4b Plot a curve showing attenuation constant in sea water from 10^4 to 10^9 Hz, assuming that the constants given do not vary over this range. Comment on the implications of the results to the problem of communicating by radio waves through sea water.

6.4c A horn antenna excited at 1 GHz is buried in dry earth. How thick could the earth covering be if half the power is to reach the surface? (Reflections at the surface are not at issue in this question.)

6.4d Plot attenuation in nepers per meter versus wavelength for nickel and silver over the wavelength range of Fig. 13.3*b*, using data of that figure.

6.4e Derive the expression for group velocity of a uniform plane wave propagating in a good conductor.

6.4f Determine the group velocity for uniform plane waves in a lossy dielectric, assuming σ and ε' independent of frequency; assuming $\varepsilon''/\varepsilon'$ independent of frequency.

6.5a Find the instantaneous Poynting vector for plane z for the standing wave of Sec. 6.5. Note the planes for which it is zero for all values of t and comment on the significance of these planes. Show that the average Poynting vector is zero as stated in Sec. 6.5.

6.5b Evaluate instantaneous values of stored energy in electric field and in magnetic field between the conductor and plane z in the standing wave of Sec. 6.5. Note planes for which the two forms of energy have the same maximum values (occurring at different times) and verify the statements concerning stored energy made in Sec. 6.5.

6.6a Write the formulas for a uniform plane wave with E_y and H_x only, and give the correspondence to voltage and current in the transmission-line equations.

6.6b Consider a lossy ferrite with both μ and ε complex, $\mu' - j\mu''$ and $\varepsilon' - j\varepsilon''$, respectively. Show that the transmission-line analogy for a plane wave in this material has both series resistance and shunt conductance. Determine expressions for these elements.

6.6c For transmission of a plane wave across a plane between two different dielectrics, Eq. 6.6(12) gives transmission coefficient in terms of electric field. Derive also:

$$\tau_H = \frac{H_2}{H_{1+}}$$

and

$$\tau_{PWR} = \frac{P_{z2}}{P_{z1+}}$$

6.6d A conducting film of impedance $377\,\Omega$/square is placed a quarter-wave in air from a plane conductor to eliminate wave reflections at 9 GHz. Assume negligible displacement currents in the film. Plot a curve showing the fraction of incident power reflected versus frequency for frequencies from 6 to 18 GHz.

6.7a A 10-GHz radar produces a substantially plane wave which is normally incident upon a still ocean. Find the magnitude and phase of reflection coefficient and percent of incident energy reflected.

6.7b For Prob. 6.7a, find magnitude and phase of τ and percent of incident energy transmitted into the body of the sea.

6.7c For a certain dielectric material of effectively infinite depth, reflections of an incident plane wave from free space are observed to produce a standing-wave ratio of 2.7 in the free space. The face is an electric field minimum. Find the dielectric constant.

6.7d Check the expression for power transmitted into a good conductor, given at the end of Ex. 6.7b, by assuming that magnetic field at the surface is the same as for reflection from a perfect conductor, and computing the conductor losses due to the currents associated with this magnetic field.

6.7e Write the total electric field and total magnetic field for $z < 0$ and for $z > 0$ for the case of reflection from a single dielectric boundary as studied in Sec. 6.7. Write both phasor and instantaneous forms.

6.8a Calculate the reflection coefficient and percent of incident energy reflected when a uniform plane wave is normally incident on a plexiglas radome (dielectric window) of thickness $\frac{3}{8}$ in., relative permittivity $\varepsilon_r = 2.8$, with free space on both sides. Frequency corresponds to free-space wavelength of 20 cm. Repeat for $\lambda_0 = 10$ cm; for 3 cm.

6.8b* For a sandwich-type radome consisting of two identical thin sheets (thickness 1.5 mm, relative permittivity $\varepsilon_r = 4$) on either side of a thicker foam-type dielectric (thickness 1.81 cm, relative permittivity $\varepsilon_r = 1.1$), calculate the reflection coefficient for waves striking at normal incidence. Take frequency 3×10^9 Hz; repeat for 6×10^9 Hz. *Suggestion:* Use the Smith chart.

6.8c What refractive index and what thickness do you need to make a quarter-wave antireflection coating between air and silicon at 10 GHz? At $\lambda_0 = 10 \, \mu m$?

6.8d For the 10-μm design of Prob. 6.8c, plot fraction of incident energy reflected versus wavelength from $\lambda_0 = 15 \, \mu m$ to $\lambda_0 = 5 \, \mu m$.

6.8e* Imagine two quarter-wave layers of intrinsic impedance η_2 and η_3 between dielectrics of intrinsic impedances η_1 and η_4. Show that perfect matching occurs if $\eta_2/\eta_3 = (\eta_1/\eta_4)^{1/2}$. For $\eta_4 = 4$, $\eta_3 = 3$, $\eta_2 = 1.5$, $\eta_1 = 1$, calculate reflection coefficient at a frequency 10% below that for perfect matching. Compare with the result for a single quarter-wave matching coating with $\eta = 2$.

6.8f** A dielectric window of polystyrene (see Table 6.4a) is made a half-wavelength thick (referred to the dielectric) at 10^8 Hz, so that there would be no reflections for normally incident uniform plane waves from space, neglecting losses in the dielectric. Considering the finite losses, compute the reflection coefficient and fraction of incident energy reflected from the front face. Also determine the fraction of the incident energy lost in the dielectric window.

6.8g A slab of dielectric of length l, constants ε' and ε'', is backed by a conducting plane at $z = l$ which may be considered perfect. Determine the expression for field impedance at the front face, $z = 0$. Find value for $\epsilon'/\epsilon_o = 4$, $\epsilon''/\epsilon' = 0.01$, $f = 3 \times 10^9$ Hz, $l = 1.25$ cm.

6.9a Write instantaneous forms for the field components of Eqs. 6.9(5)–(7). Find the average and instantaneous components of Poynting vector for both the x and z directions.

6.9b Repeat Prob. 6.9a for the other polarization, defined by Eqs. 6.9(18)–(20).

6.9c For the polarization with E_x, E_z, and H_y, find stored energy in electric fields and in magnetic fields, for unit area, between planes $z = 0$ and $z = \pi/(k \cos \theta)$ and interpret.

6.10a Show that a generalization of Eq. 6.10(1) for a wave oblique to all three axes can be written in the convenient form

$$E(x, y, z) = E_+ e^{-jk \cdot r}$$

where

$$k = \hat{x}k_x + \hat{y}k_y + \hat{z}k_z$$

$$r = \hat{x}x + \hat{y}y + \hat{z}z$$

6.10b A diffraction grating as in Ex. 6.10 has grating spacing $2 \, \mu m$. Find the number and angle of the diffracted orders for an argon laser beam with $\lambda_0 = 0.488 \, \mu m$ at: (i) normal incidence, $\theta_1 = 0$; (ii) incidence with $\theta_1 = 60$ degrees.

6.10c Find the modified relation for diffracted angle in Ex. 6.10 if there is dielectric ε_1 on the incident side of the grating and a different ε_2 on the exit side. ($\mu = \mu_0$ for both.)

6.11a Plot versus incident angle the phase and magnitude of ρ for a wave incident from air onto a ceramic with $\varepsilon_r = 4.95$.

6.11b Repeat Prob. 6.11a with the wave passing from the ceramic into air. (Note that reflection is total beyond some angle, as will be explained in Sec. 6.12.)

6.11c Obtain the special forms of ρ and τ for grazing incidence ($\theta = \pi/2 - \delta$ with $\delta \ll 1$) for both polarizations and $\varepsilon_2 > \varepsilon_1$.

6.11d For both polarizations, give the conditions for which the standing-wave pattern in z shows a minimum of tangential electric field at the boundary surface; repeat for a maximum of tangential E at the surface.

6.11e Write expressions for E_x, H_y, and E_z in region 2 as functions of x and z for polarization with $\mathbf{E}$ in the plane of incidence. Obtain the Poynting vector for each region and demonstrate the power balance. (Take $\varepsilon_2 > \varepsilon_1$ to avoid the special case to be treated in Sec. 6.12.)

6.12a A microwave transmitter is placed below the surface of a freshwater lake. Neglecting absorption, find the cone over which you could expect radiation to pass into the air.

6.12b Using data of Fig. 13.2b, find the critical angle for a wave of wavelength $3\ \mu m$ passing from fused silica into air and also for silicon into air. For both cases plot phase of reflection coefficient versus incident angle over the range $\theta_c \leq \theta \leq \pi/2$ for a wave polarized with $\mathbf{E}$ in the plane of incidence.

6.12c In a GaAs laser, the generated radiation ($\lambda_0 = 0.85\ \mu m$) is "trapped" or guided along the thin junction region by total reflection from the adjacent layers. (The mechanism of dielectric guiding will be explored more in Chapters 8 and 14). If the junction layer has refractive index $n = 3.60$, the upper layer has $n = 3.45$ and the lower layer $n = 3.50$, find critical angle at upper and lower surfaces.

6.12d* Defining ψ as the phase E_{x-}/E_{x+} and ψ' as the phase of E_{y-}/E_{y+}, find expressions for ψ and ψ' under conditions of total reflection. Show that the phase difference between these two polarization components, $\delta = \psi - \psi'$, is given by

$$\tan\left(\frac{\delta}{2}\right) = \frac{(\eta_2/\eta_1)[(v_2/v_1)^2 - 1]\sin^2\theta}{[(\eta_2/\eta_1)^2 - 1]\cos\theta\sqrt{(v_2/v_1)^2\sin^2\theta - 1}}$$

6.12e Optical fibers, now of importance in optical communications, guide light by the phenomenon of total reflection (as will be discussed more in Chapter 14). The evanescent fields outside of the guiding core could cause coupling or "crosstalk" between adjacent fibers. To estimate the size of this coupling, take a plane model with silica core having refractive index $n = 1.535$ and external cladding of a borosilicate glass having $n = 1.525$. The optical signal has free-space wavelength of $0.85\ \mu m$. How far away can the next core be placed if field is to be 10^{-3} the surface value at that plane, if (i) incident angle of waves within the core is 85 degrees; (ii) if incident angle is 89 degrees.

6.12f For several applications of total reflection (note Probs. 6.12c and 6.12e), refractive indices on the two sides of the boundary are not very different. If $n_2 = n_1(1 - \delta)$ where $\delta \ll 1$, show that $\theta_c \approx \pi/2 - \Delta$ where $\Delta = \sqrt{2\delta}$.

6.13a In Ex. 6.13, it was found that polarizing angle for entrance to the window is also correct for exit from the window. Prove this for general ratios of n_2/n_1.

6.13b Examine Fig. 13.2b to determine materials which might be suitable as Brewster windows for a CO_2 laser operating at $10.6\ \mu m$. Calculate the appropriate window angle for each such material.

6.13c A green ion laser beam, operating at $\lambda_0 = 0.545\ \mu m$, is generated in vacuum, and then passes through a glass window of refractive index 1.5 into water with $n = 1.34$. Design a window to give zero reflection at the two surfaces for a wave polarized with $\mathbf{E}$ in the plane of incidence.

6.13d Imagine a material with $\varepsilon_1 = \varepsilon_2$ but $\mu_1 \neq \mu_2$. Which polarization component would then yield a solution for incident angle giving no reflections? Give the angle.

6.14a An incident wave in medium 1 of permittivity ε_1 makes angle θ_1 with the normal. Find the proper length and permittivity of a medium 2 to form a "quarter-wave matching section" to a medium of permittivity ε_3.

6.14b* A uniform plane wave of free-space wavelength 3 cm is incident from space on a window of permittivity 3, and thickness equal to a half-wavelength referred to the dielectric material so that it gives no reflections for normal incidence. For general angles of incidence, plot the fraction of incident energy reflected versus θ for polarization with **E** in the plane of incidence, and also for polarization normal to the plane of incidence.

6.14c By use of the transmission-line analogies, determine the spacing between a thin film and a parallel perfect conductor, and the conductivity properties of that film if reflections are to be perfectly eliminated for a wave incident at an angle θ from the normal for the two types of polarization. (See Ex. 6.6b.)

6.14d For a series of parallel boundaries between different dielectrics, as in Fig. 6.14, show that the relation between exit angle and angle of incidence upon the first boundary is given by Snell's law utilizing indices of refraction of entrance and exit materials only, provided there is no intermediate surface at which total reflection occurs.

6.14e An optical instrument called an *ellipsometer* employs a beam of monochromatic light with elliptical polarization incident at an oblique angle on a surface whose properties are to be determined. Measurements on the incident and reflected beams yield a value of the complex ratio of reflection coefficients for components in the plane of incidence and normal to it. Suppose a measurement is to be made of the index of refraction and thickness of an unknown but lossless film on a silicon substrate of effectively infinite thickness. Assume the beam incident angle is 70 degrees and, for simplicity, take the silicon to be lossless and with refraction index $n_{Si} = 3.89$ at the frequency of the light. Find the ratio of reflection coefficients that will be measured by the ellipsometer in terms of the two unknown thin-film properties, thickness and refractive index.

7

TWO- AND THREE-DIMENSIONAL BOUNDARY VALUE PROBLEMS

7.1 Introduction

In the preceding chapters, numerous special techniques have been presented for solving static and dynamic field problems. Before continuing with the important problems of wave guiding, resonance, and interaction of fields with materials and radiation, it is necessary to develop some more general and somewhat more powerful techniques of problem solution. The methods developed in this chapter will usually be illustrated first through static examples before extending to dynamic problems, and in some cases are most useful for static or quasistatic problems. Even then such solutions are of use in certain time-varying problems, as we have seen in the case of circuits and transmission lines in the preceding chapters.

The approach in this chapter is through the solution of differential equations subject to boundary conditions. In certain cases the field distributions themselves are desired, but in other cases (as we saw in the calculation of circuit elements) these distributions are only steps along the way to other useful parameters.

The most general analytical method to be considered in this chapter is that of separation of variables, leading to orthogonal functions which may be superposed to represent very general field distributions. In developing this method we will spend some time on the special functions needed for circular cylindrical coordinates (Bessel functions) and for spherical coordinates (Legendre functions). A second powerful analytical method is that of conformal transformation. Although restricted to two-dimensional problems, and primarily (but not exclusively) useful for solutions of Laplace's equation, it is the most convenient way of solving many problems of importance in circuits and transmission lines.

Numerical solution of field problems becomes increasingly important with the continuing advances in computing power. This is a special field in itself, and a rapidly changing one, but we will give some idea of its basis and some elementary approaches to its use.

THE BASIC DIFFERENTIAL EQUATIONS
AND NUMERICAL METHODS

7.2   Roles of Helmholtz, Laplace, and Poisson Equations

We have seen how specific differential equations—the wave equation, the Helmholtz equation, or the diffusion equation—result from Maxwell's equations with certain specializations. We shall generally be concerned with such special cases, but let us look first at somewhat more general forms. We use the phasor forms, and limit ourselves to homogeneous, isotropic, and linear media. Starting with the Maxwell equation for curl $\mathbf{E}$, [Eq. 3.8(3)],

$$\nabla \times \mathbf{E} = -j\omega\mu\mathbf{H} \tag{1}$$

The curl of this is taken and expanded (inside back cover)

$$\nabla \times \nabla \times \mathbf{E} = -\nabla^2\mathbf{E} + \nabla(\nabla \cdot \mathbf{E}) = -j\omega\mu\nabla \times \mathbf{H} \tag{2}$$

The divergence of $\mathbf{E}$ and curl of $\mathbf{H}$ are substituted from the Maxwell equations Eqs. 3.8(1) and 3.6(4):

$$-\nabla^2\mathbf{E} + \nabla\left(\frac{\rho}{\varepsilon}\right) = -j\omega\mu[\mathbf{J} + j\omega\varepsilon\mathbf{E}]$$

or

$$\nabla^2\mathbf{E} + k^2\mathbf{E} = j\omega\mu\mathbf{J} + \frac{1}{\varepsilon}\nabla\rho \tag{3}$$

where $k^2 = \omega^2\mu\varepsilon$. By similar operations on the curl $\mathbf{H}$ equation, we obtain

$$\nabla^2\mathbf{H} + k^2\mathbf{H} = -\nabla \times \mathbf{J} \tag{4}$$

Equations (3) and (4) may be considered inhomogeneous Helmholtz equations. General solutions of these are difficult, but usually start from solutions of the corresponding homogeneous equations,[1]

$$\nabla^2\mathbf{E} + k^2\mathbf{E} = 0 \tag{5}$$

$$\nabla^2\mathbf{H} + k^2\mathbf{H} = 0 \tag{6}$$

Many of the problems we are concerned with have no sources except on the boundaries, so the Helmholtz equations considered in this chapter are the homogeneous ones, (5) and (6).

---

[1] J. D. Jackson, *Classical Electrodynamics*, 2nd. ed., Wiley, New York, 1975.

Note that the vector equations separate simply in rectangular coordinates,

$$\nabla^2 E_x + k^2 E_x = 0 \tag{7}$$

and similarly for E_y, E_z, H_x, H_y, and H_z. They do not separate so simply for curvilinear coordinates, as one can see by examining the expansion for ∇^2 of a vector in cylindrical and spherical coordinates, inside back cover. But for any cylindrical coordinate system, the axial component of (5) or (6) satisfies a simple Helmholtz equation,

$$\nabla^2 E_z + k^2 E_z = 0 \tag{8}$$

and similarly for H_z.

For quasistatic problems, the term in k^2 is negligible so that (5) and (6) reduce to Laplace equations,

$$\nabla^2 \mathbf{E} = 0 \tag{9}$$

$$\nabla^2 \mathbf{H} = 0 \tag{10}$$

These separate into coordinate components as discussed above. However, for quasistatic or purely static problems it is often more convenient to use the scalar potential functions defined by

$$\mathbf{E} = -\nabla\Phi, \qquad \mathbf{H} = -\nabla\Phi_m \tag{11}$$

with Φ and Φ_m satisfying Laplace equations,

$$\nabla^2\Phi = 0, \qquad \nabla^2\Phi_m = 0 \tag{12}$$

In certain cases we are concerned with static or quasistatic solutions for regions containing charges, in which case the Poisson equation applies to Φ (Sec. 1.12):

$$\nabla^2\Phi = -\frac{\rho}{\varepsilon} \tag{13}$$

Thus the Laplace, Helmholtz, and Poisson equations govern a large number of important problems and will be the ones used for illustration of solution methods in this chapter.

Boundary Conditions   As noted in Sec. 1.17, unique solutions of the Laplace or Poisson equations result if the function is specified on a boundary surrounding the region of interest. Specification of the normal derivative on such a boundary determines the solution within a constant. Section 3.14 pointed out that unique solutions of the Helmholtz equations (5) or (6) are obtained by specifying the tangential component of $\mathbf{E}$ or $\mathbf{H}$ on the closed boundary.

Superposition   Since ∇^2 is a linear operator (as are the other operators in Maxwell's equations), any two solutions are superposable and the sum is a solution provided that the medium itself is linear. We have made use of this fact previously, as in the superposition of linearly polarized waves to form a circularly polarized one. We will use the principle in many future examples. Here we give an example of its use in reasoning to a simple result.

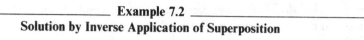

_____ **Example 7.2** _____
Solution by Inverse Application of Superposition

An interesting example of the use of superposition is the solution for the potential at the center of a symmetrical structure. For example, consider a homogeneous dielectric surrounded by the circular cylinder shown in Fig. 7.2, with a potential V_0 applied over a portion of the boundary subtending the angle α and zero potential on the remainder. Suppose that $\alpha = 2\pi/n$. If the potential at the center were found for n different sets of boundary conditions as shown in Fig. 7.2, where the only difference between these is that the section of the boundary to be at potential V_0 is rotated by the angle $k\alpha$, with k an integer, the sum of the n solutions would be the potential at the center of a cylinder with V_0 over the entire boundary; this is just V_0. Since every problem is identical except for a rotation by α, which would not affect the potential at the center, the potential at the center for the original problem must be V_0/n. This same technique could be applied to find the potential at the center point of a square, cube, equilateral polygon, sphere, and so on, with one portion at a given potential.

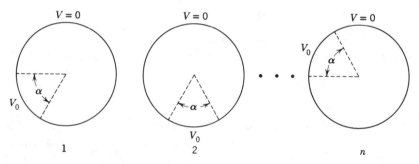

Fig. 7.2 Series of circular cylinders with sectors of angle α at the potential V_0 oriented at multiples of α with respect to each other.

7.3   Numerical Solution of the Laplace, Poisson, and Helmholtz Equations

Before turning to analytical methods, let us consider the numerical methods that are becoming increasingly attractive as digital computer speed and memory capacity continue to increase. Among the powerful methods are those using finite differences,[2] finite elements,[2] or Fourier transformations.[3] Still others will undoubtedly be developed as computing capabilities continue to increase. Here we illustrate the idea through the elemental difference equation approach and some of its extensions.

We consider first the Poisson equation with potential specified on the boundary. For simplicity we take a two-dimensional problem (no variations in z). The internal region is divided by a grid of mutually orthogonal lines with potential eventually to be determined at each of the grid points. Rectangular coordinates are used and potential at a point (x, y) expanded in a Taylor series,

$$\Phi(x + h, y) \approx \Phi(x, y) + h \frac{\partial \Phi(x, y)}{\partial x} + \frac{h^2}{2} \frac{\partial^2 \Phi(x, y)}{\partial x^2} \tag{1}$$

and

$$\Phi(x - h, y) \approx \Phi(x, y) - h \frac{\partial \Phi(x, y)}{\partial x} + \frac{h^2}{2} \frac{\partial^2 \Phi(x, y)}{\partial x^2} \tag{2}$$

By adding (1) and (2) and rearranging, we have the approximation:

$$\frac{\partial^2 \Phi(x, y)}{\partial x^2} = \frac{\Phi(x + h, y) - 2\Phi(x, y) + \Phi(x - h, y)}{h^2} \tag{3}$$

The second partial derivative with respect to y can be obtained in the same way. Then Poisson's equation in two dimensions

$$\frac{\partial^2 \Phi}{\partial x^2} + \frac{\partial^2 \Phi}{\partial y^2} = -\frac{\rho}{\varepsilon}$$

can be expressed in the approximate form

$$\Phi(x + h, y) + \Phi(x - h, y) + \Phi(x, y + h)$$
$$+ \Phi(x, y - h) - 4\Phi(x, y) = -\frac{\rho h^2}{\varepsilon} \tag{4}$$

where the distance increment h is taken, for simplicity, to be equal in the two directions. It is of interest to note that, if space charge is zero, the potential is the average of the potentials at the surrounding points.

Note that potential is known on boundary points so that a straightforward approach is to solve a set of equations such as (4) for the unknown potentials at the grid points in terms of the known values on boundary points. This is

[2] L. J. Segerlind, *Applied Finite Element Analysis*, Wiley, New York, 1976. V. Vemuri and W. J. Karplus, *Digital Computer Treatment of Partial Differential Equations*, Prentice-Hall, Englewood Cliffs, NJ, 1981.
[3] R. W. Hockney and J. W. Eastwood, *Computer Simulation Using Particles*, McGraw-Hill, New York, 1981.

sometimes done by a matrix inversion technique, but if memory capacity is a problem, it may be better to use a method for direct iterative adjustment of grid potentials. This starts from an initial guess and corrects by bringing in the given values on the boundary through successive passes through the grid. We illustrate this first by a simple averaging technique.

_____ **Example 7.3** _____
Numerical Solution of Laplace Equation by Simple Averaging

As an illustration of iteration with simple averaging, let us find the potentials for the grid of points in the structure in Fig. 7.3a. This is an infinite cylinder of square cross section with the potentials specified on the entire boundary. The space charge will be assumed to be zero so we will be solving Laplace's equation.

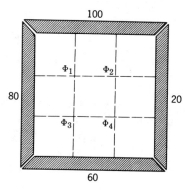

Fig. 7.3a Cylinders of square cross section and grid for difference equation solution.

The broken lines represent the grid to be used to approximate the region for the finite-difference solution. The coarse grid was chosen to simplify the example; a finer grid would be used in most practical problems. The four unknown potentials designated Φ_1 to Φ_4 are assumed initially to be the average of the boundary potentials, 65 V. The first calculation is to find Φ_1 as the average of the four surrounding potentials (80, 100, $\Phi_2 = 65$, $\Phi_3 = 65$). Therefore, in the column labeled Step 1 in Table 7.3, Φ_1 is given the value 77.50. Then Φ_2 is found as the average of 100, 20, 77.50, and 65, and this value is put in the table in Step 1. The procedure is repeated for Φ_3 and Φ_4. Then Step 2 proceeds in the same way. It is seen that after several steps the potentials converge to definite values. Since (4) is approximate, the potentials have converged to approximate answers. They would differ less from the correct solution if the grid were made finer. The correct potentials for the four points are also listed in the table.

Table 7.3
Iterative calculation for Example 7.3

| | Step | | | | | Correct |
	1	2	3	4	5	Potentials
Φ_1	77.50	79.07	77.89	77.60	77.51	75.2
Φ_2	65.63	63.28	62.70	62.55	62.51	60.5
Φ_3	70.63	68.28	67.70	67.52	67.52	65.4
Φ_4	54.06	52.89	52.60	52.51	52.51	50.7

Mesh Relaxation   The above method, in which successive averaging of the potentials leads to a final result not far from the correct potentials is convenient for small problems and, in many cases, has a satisfactory rate of convergence. A more generally useful method of calculation is based on a defined *residual* for each grid point that measures the amount by which the potential there differs from the value dictated by the potentials on the neighboring grid points. The residual for the kth pass through the grid is defined by

$$R^{(k)} = \Phi^{(i)}(x, y + h) + \Phi^{(i)}(x, y - h) + \Phi^{(i)}(x + h, y)$$

$$+ \Phi^{(i)}(x - h, y) - 4\Phi^{(k-1)}(x, y) + \frac{h^2 \rho}{\varepsilon} \tag{5}$$

where $i = k$ or $k - 1$ since, at any grid point, the potentials at some of the neighboring points will generally have already been adjusted on that pass through the grid, as was seen in the above illustration. On each pass (corresponding to Steps 1–5 in Table 7.4) and at each grid intersection, one calculates the residual $R^{(k)}$ and then the new potential according to

$$\Phi^{(k)} = \Phi^{(k-1)} + \Omega \frac{R^{(k)}}{4} \tag{6}$$

where Ω is called the *relaxation factor* because it determines the rate at which the potentials relax toward the correct solution. It is taken in the range $1 \le \Omega \le 2$. Selection of $\Omega = 1$, called *simple relaxation*, corresponds to the method of averaging illustrated above. When $\Omega > 1$, the procedure is called the method of *successive over-relaxation* (SOR). If Ω is fixed, it is usually taken near 2 for problems with many mesh points, but this can cause an initial increase in error, so it is often better to start with $\Omega = 1$ and increase it gradually with each iteration step. One procedure that is found useful for large grids is the *cyclic Chebyshev method* in which the following program of varying Ω is used:

$$\Omega^{(1)} = 1 \tag{7}$$

$$\Omega^{(2)} = \frac{1}{1 - \frac{1}{2}\eta_{\text{it}}} \tag{8}$$

where

$$\eta_{it} = \frac{1}{4}\left(\cos\frac{\pi}{n} + \cos\frac{\pi}{m}\right)^2 \tag{9}$$

for a grid with n mesh points in one direction and m points in the other.

$$\Omega^{(k+1)} = \frac{1}{1 - \frac{1}{4}\eta_{it}\Omega^{(k)}} \tag{10}$$

$$\Omega^{(\infty)} = \Omega_{opt} = \frac{2}{1 + \sqrt{1 - \eta_{it}}} \tag{11}$$

When this method is used, the mesh is swept like a checkerboard, with all the red squares being treated on the first pass with $\Omega^{(1)}$, all the black squares being treated on the second pass using $\Omega^{(2)}$, the reds on the third with $\Omega^{(3)}$, and so on. This method requires an additional memory cell in the computer for each mesh point but the speeded convergence usually makes it worthwhile. The convergence can be improved appreciably by having a good initial guess for the potentials, say by using results from a similar problem.

Boundary Conditions   In setting up the boundary conditions on a grid, the easiest situation is where potentials are specified on grid points. A more difficult problem is where the normal derivatives are specified. The derivatives are usually zero, as would be the case at an insulating surface in a conduction problem (Ex. 1.14) or along a line of symmetry used as an artificial boundary to reduce the required grid size. Examples of such boundaries are shown in Figs. 7.3b and 7.3c. In one case the structure has obvious symmetry in the x–y plane so that a solution need be found for only one-fourth of the structure. In the other example, the boundary may be taken along the axis of a cylindrically symmetric system. In the latter case, the difference equations can include the symmetry (Prob. 7.3c). In the former case, to make the normal derivative of potential zero at a boundary, an imaginary grid point is set up outside the boundary and its potential is kept the same as at the point symmetrically located just inside the boundary. Sometimes the boundary points do not lie on mesh points; in such cases, linear interpolation is used to set the potentials at mesh points nearest to the boundary.

Numerical Solutions of the Scalar Helmholtz Equation[4]   The general form of the Helmholtz equation (Sec. 7.2) is

$$\nabla^2\psi + u^2\psi = 0 \tag{12}$$

---

[4] P. Silvester, *Modern Electromagnetic Fields*, Prentice-Hall, Englewood Cliffs, N.J., 1968.

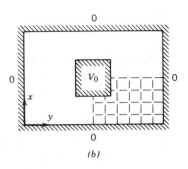

(b)

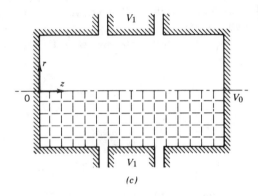

(c)

Fig. 7.3b,c Examples in rectangular and cylindrical coordinates where symmetry reduces the required size of the grid for finite-difference solutions.

In general, u^2 is an unknown constant. Assuming a grid in two-dimensional rectangular coordinates and following the procedures that led to (5), we have the residual at the kth step:

$$R^k(x, y) = \psi^{(i)}(x, y + h) + \psi^{(i)}(x, y - h) + \psi^{(i)}(x + h, y)$$
$$+ \psi^{(i)}(x - h, y) - (4 - u^2 h^2)\psi^{(k-1)}(x, y) \tag{13}$$

The change of the variable from one iteration step to the next in the successive overrelaxation method is governed by the relation

$$\psi^{(k)} = \psi^{(k-1)} + \Omega \frac{R^{(k)}}{4 - u^2 h^2} \tag{14}$$

The problem in applying (14) is that u is not known at this point but is itself determined by the boundary conditions (it is called an *eigenvalue* of the problem).

METHOD OF CONFORMAL TRANSFORMATION

7.4   Method of Conformal Transformation and Introduction to Complex-Function Theory

A very general mathematical attack for the two-dimensional field distribution problem utilizes the theory of functions of a complex variable. The method is in principle the most general for two-dimensional problems, and the work can be carried out to yield actual solutions for a wide variety of practical problems. For these reasons, the general method with some examples will be presented in this and the following sections.

In the theory of complex variables, we use the complex variable $Z = x + jy$, where both x and y are real variables. It is convenient to associate any given value of Z with a point in the x–y plane (Fig. 7.4a), and to call this plane the complex Z plane. Of course the coordinates may also be expressed in polar form in terms of r and θ:

$$r = \sqrt{x^2 + y^2}; \qquad \theta = \tan^{-1}\left(\frac{y}{x}\right)$$

Then

$$Z = x + jy = r(\cos\theta + j\sin\theta) = re^{j\theta} \tag{1}$$

Suppose that there is now a different complex variable W, where

$$W = u + jv = \rho e^{j\phi}$$

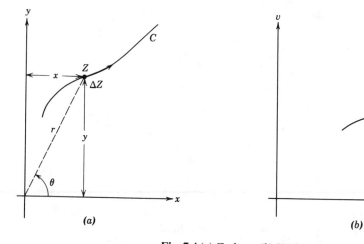

Fig. 7.4 (a) Z plane. (b) W plane.

such that W is some function of Z. This means that, for each assigned value of Z, there is a rule specifying a corresponding value of W. The functional relationship is written

$$W = f(Z) \tag{2}$$

If Z is made to vary continuously, the corresponding point in the complex Z plane moves about, tracing out some curve C. The values of W vary correspondingly, tracing out a curve C'. To avoid confusion, the values of W are usually shown on a separate graph, called the complex W plane (Fig. 7.4b).

Next consider a small change ΔZ in Z and the corresponding change ΔW in W. The derivative of the function will be defined as the usual limit of the ratio $\Delta W / \Delta Z$ as the element ΔZ becomes infinitesimal:

$$\frac{dW}{dZ} = \lim_{\Delta Z \to 0} \frac{\Delta W}{\Delta Z} = \lim_{\Delta Z \to 0} \frac{f(Z + \Delta Z) - f(Z)}{\Delta Z} \tag{3}$$

A complex function is said to be *analytic* or *regular* whenever the derivative defined above exists and is unique. The derivative may fail to exist at certain isolated (*singular*) points where it may be infinite or undetermined, somewhat as in real function theory. But it would appear that there is another ambiguity in respect to complex variables, since ΔZ may be taken in any arbitrary direction in the Z plane from the original point. For the derivative to be unique, the ratio $\Delta W / \Delta Z$ should turn out to be independent of this direction.

If this independence of direction is to result, a necessary condition is that we obtain the same result if Z is changed in the x direction alone, or in the y direction alone. For $\Delta Z = \Delta x$,

$$\frac{dW}{dZ} = \frac{\partial W}{\partial x} = \frac{\partial}{\partial x}(u + jv) = \frac{\partial u}{\partial x} + j\frac{\partial v}{\partial x} \tag{4}$$

For a change in the y direction, $\Delta Z = j\,\Delta y$,

$$\frac{dW}{dZ} = \frac{\partial W}{\partial(jy)} = \frac{1}{j}\frac{\partial}{\partial y}(u + jv) = \frac{\partial v}{\partial y} - j\frac{\partial u}{\partial y} \tag{5}$$

Two complex quantities are equal if and only if their real and imaginary parts are separately equal. Hence (4) and (5) yield the same result if

$$\frac{\partial u}{\partial x} = \frac{\partial v}{\partial y} \tag{6}$$

$$\frac{\partial v}{\partial x} = -\frac{\partial u}{\partial y} \tag{7}$$

These conditions, known as the *Cauchy–Riemann equations* are then necessary conditions for dW/dZ to be unique at a point, and the function $f(Z)$ analytic there. It can be shown that, if they are satisfied, the same result for dW/dZ is obtained for any arbitrary direction of the change ΔZ, so they are also sufficient conditions.

Example 7.4
Analyticity of Power Functions

$$W = Z^2 \tag{8}$$

$$u + jv = (x + jy)^2 = (x^2 - y^2) + j2xy$$

$$u = x^2 - y^2$$

$$v = 2xy$$

A check of the Cauchy–Riemann equations yields

$$\frac{\partial u}{\partial x} = \frac{\partial v}{\partial y} = 2x$$

$$\frac{\partial u}{\partial y} = -\frac{\partial v}{\partial x} = -2y$$

So they are satisfied everywhere in the finite Z plane, and the function is analytic everywhere there.

Actually, it is not necessary to apply the check when the functional relation is expressed explicitly between Z and W in terms of functions which possess a power-series expansion about the origin, as e^z, $\sin Z$, and so on. The reason is that each term in the series, $C_n Z^n$, can be shown to satisfy the Cauchy–Riemann conditions, and consequently a series of such terms also satisfies them.

7.5   Properties of Analytic Functions of Complex Variables

If Eq. 7.4(6) is differentiated with respect to x, Eq. 7.4(7) differentiated with respect to y, and the resulting equations added, there results

$$\frac{\partial^2 u}{\partial x^2} + \frac{\partial^2 u}{\partial y^2} = 0 \tag{1}$$

Similarly, if the order of differentiation is reversed, there results

$$\frac{\partial^2 v}{\partial x^2} + \frac{\partial^2 v}{\partial y^2} = 0 \tag{2}$$

These are recognized as Laplace equations in two dimensions. Thus both the real and the imaginary parts of an analytic function of a complex variable satisfy Laplace's equation, and would be suitable for use as the potential functions for two-dimensional electrostatic problems. The manner in which these are used in specific problems, and the limitations on this usefulness, are demonstrated by examples in this and the next section.

For a problem in which one of the two parts, u or v, is chosen as the potential function, the other becomes proportional to the flux function (Sec. 1.6). To show this, let us suppose that u is the potential function in volts for a particular problem. The electric field, obtained as the negative gradient of u, yields

$$E_x = -\frac{\partial u}{\partial x}; \qquad E_y = -\frac{\partial u}{\partial y} \tag{3}$$

By the equation for the total differential, the change in v corresponding to changes in the x and y coordinates of dx and dy is

$$dv = \frac{\partial v}{\partial x} dx + \frac{\partial v}{\partial y} dy$$

But, from Cauchy–Riemann conditions, Eqs. 7.4(6) and 7.4(7),

$$-dv = \frac{\partial u}{\partial y} dx - \frac{\partial u}{\partial x} dy = -E_y dx + E_x dy$$

or

$$-\varepsilon\, dv = -D_y dx + D_x dy \tag{4}$$

By inspection of Fig. 7.5a, this is recognized to be just the electric flux $d\psi$ between the curves v and $v + dv$, with the positive direction as shown by the arrow. Then

$$-d\psi = \varepsilon\, dv \tag{5}$$

And, except for a constant that can be set equal to zero by choosing the reference for flux at $v = 0$,

$$-\psi = \varepsilon v \quad \text{C/m} \tag{6}$$

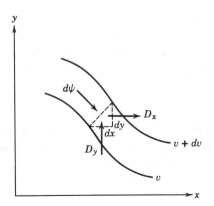

Fig. 7.5a Coordinates for the flux function.

Similarly, if v is chosen as the potential function in volts for some problem, εu is the flux function in coulombs per meter, with proper choice of the direction for positive flux.

We have seen that either u or v may be used as a potential function, and then the other may be used as the flux function, since both satisfy Laplace's equation. The utility of the concept, however, hinges on being able to find the analytic function $W = f(Z)$ such that u and v also satisfy the boundary conditions for the problem being considered.

Example 7.5

Electrodes in Parallel-Plane Diode

As an example, suppose we desire the distribution of potentials in the Z plane where the given boundary condition is

$$V = x^{4/3}, \qquad y = 0 \tag{7}$$

If we let

$$W = Z^{4/3} \tag{8}$$

it is clear that for $y = 0$, the real part of W is $u = x^{4/3}$. Furthermore, we see that dW/dZ exists and is unique except at $Z = 0$ (Prob. 7.4e). Thus u is a suitable potential function for this problem; the real part of (8) gives the potential distribution. It is most convenient for this particular function to express Z in polar coordinates

$$W = u + jv = r^{4/3}e^{j4\theta/3} \tag{9}$$

Thus

$$u = r^{4/3} \cos \tfrac{4}{3}\theta \tag{10}$$

$$v = r^{4/3} \sin \tfrac{4}{3}\theta$$

Equipotentials, found by setting u equal to a constant, are shown in Fig. 7.5b for $u = 0$ and 1. It is of interest to notice that the boundary function (7) has the same form as the potential in a plane diode with the cathode at $x = 0$ and with the anode potential unity at $x = 1.0$:

$$\Phi = x^{4/3}$$

Using these ideas, a plane diode can be truncated and the correct potentials produced on the free edge by placing electrodes along the equipotential lines as shown in Fig. 7.5b. This procedure is most important in designing electron guns with regular flow and the result is known as the Pierce gun.[5]

----

[5] P. T. Kirstein, G. S. Kino, and W. E. Waters, *Space-Charge Flow*, McGraw-Hill, New York, 1967.

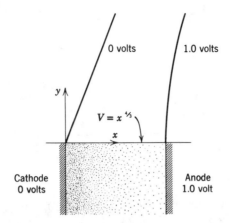

Fig. 7.5b Focusing for electron flow in a plane diode. The upper portion of the figure shows the electrodes outside the electron-flow region.

7.6   Conformal Mapping for Laplace's Equation

A somewhat different point of view toward the method in Sec. 7.5 follows if we refer to the Z and W planes introduced in Sec. 7.4. Since the functional relationship fixes a value of W corresponding to a given value of Z for a given function

$$W = f(Z)$$

any point (x, y) in the Z plane yields some point (u, v) in the W plane. As this point moves along some curve $x = F(y)$ in the Z plane, the corresponding point in the W plane traces out a curve $u = F_1(v)$. If it should move throughout a region in the Z plane, the corresponding point would move throughout some region in the W plane. Thus, in general, a point in the Z plane transforms to a point in the W plane, a curve transforms to a curve, and a region to a region, and the function that accomplishes this is frequently spoken of as a particular *transformation* between the Z and W planes.

When the function $f(Z)$ is analytic, as we have seen, the derivative dW/dZ at a point is independent of the direction of the change dZ from the point. The derivative may be written in terms of magnitude and phase:

$$\frac{dW}{dZ} = Me^{j\alpha} \tag{1}$$

or

$$dW = Me^{j\alpha}\, dZ \tag{2}$$

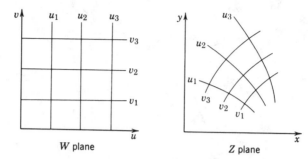

Fig. 7.6a A mapping of coordinate lines of the W plane in the Z plane.

By the rule for the product of complex quantities, the magnitude of dW is M times the magnitude of dZ, and the angle of dW is α plus the angle of dZ. So the entire infinitesimal region in the vicinity of the point W is similar to the infinitesimal region in the vicinity of the point Z. It is magnified by a scale factor M and rotated by an angle α. It is then evident that, if two curves intersect at a given angle in the Z plane, their transformed curves in the W plane intersect at the same angle, since both are rotated through the angle α. A transformation with these properties is called a *conformal transformation*.

In particular, the lines $u = $ constant and the lines $v = $ constant in the W plane intersect at right angles, so their transformed curves in the Z plane must also be orthogonal (Fig. 7.6a). We already know that this should be so, since the constant v lines have been shown to represent flux lines when the constant u lines are equipotentials, and vice versa. From this point of view, the conformal transformation may be thought of as one that takes a uniform field in the W plane (represented by the equispaced constant u and constant v lines) and transforms it so that it fits the given boundary conditions in the Z plane, always keeping the required properties of an electrostatic field.

Frequently the transformation is done in steps. That is, the uniform field is transformed first into some intermediate complex plane by $Z_1 = f(W)$, then perhaps into a second intermediate complex plane $Z_2 = g(Z_1)$, and then finally into a plane $Z_3 = h(Z_2)$ in which the boundary conditions are satisfied. In general, there can be any number of steps. Of course, these functions can be combined into a single transformation, the inverse of which can then be understood on the basis of finding a function with real or imaginary part satisfying the given boundary conditions as discussed in Sec. 7.5.

There are few circumstances in which knowledge of the required boundary conditions will lead directly to the transformation that gives the solution. For help in finding the required form there are tables of conformal transformations[6] which show how one field maps into another. The mapping functions given in the tables may be used individually or combined in a series of steps to transform the

---

[6] For example, see H. Kober, *Dictionary of Conformal Representations,* Dover, New York, 1952.

uniform field into a field that fits the given problem. Some examples of the simpler transformations will be given to illustrate the method.

_____ **Example 7.6a** _____
The Power Function: Field Near a Conducting Corner

As a first example, consider W expressed as Z raised to some power:

$$W = Z^p \tag{3}$$

It is convenient to use the polar form for Z [Eq. 7.4(1)]:

$$W = (re^{j\theta})^p = r^p e^{jp\theta}$$

or

$$u = r^p \cos p\theta \tag{4}$$

$$v = r^p \sin p\theta \tag{5}$$

From the conformal-mapping point of view, the field in the W plane is uniform. The parallel lines of equal potential (say, v equals constant) in the W plane can be mapped into the Z plane by setting v equals constant in (5). From the viewpoint of Sec. 7.5 one does not take explicit consideration of the existence of the W plane but simply recognizes that v is a solution of Laplace's equation and tries to adjust constants such that constant-v lines fit the equipotentials of the given problem. When only one step of transformation is required, the viewpoints are wholly equivalent.

If v is chosen as the potential function, the form of one curve of constant v (equipotential) is evident by inspection, for v is zero at $\theta = 0$, and also at $\theta = \pi/p$. Thus, if two semi-infinite conducting planes at zero potential intersect at angle α, where

$$p = \frac{\pi}{\alpha} \tag{6}$$

they coincide with this equipotential, and boundary conditions are satisfied. The form of the curves of constant u and of constant v within the angle then give the field configuration near a conducting corner. The field is assumed to result from the presence of an electrode with nonzero potential that either fits one of the constant-v lines or is far enough away that its shape causes no significant deviation of the u and v lines in the region of interest.

The equipotentials in the vicinity of the corner can be plotted by choosing given values of v, and plotting the polar equation of r versus θ from (5) with p given by (6). Similarly, the flux or field lines can be plotted by selecting several values of u and plotting the curves from (4). The form of the field, plotted in this manner, for corners with $\alpha = \pi/4, \pi/2,$ and $3\pi/2$ are shown in Figs. 7.6b, 7.6c, and

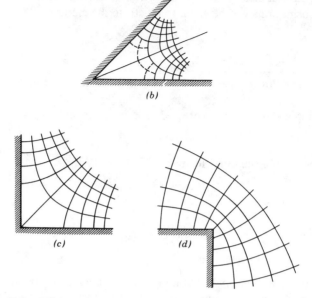

Fig. 7.6b–d Field near conducting corners of 45, 90, and 270 degrees.

7.6d, respectively. These plots are of considerable help in judging the correct form of the field in a graphical field map having one or more conducting boundaries.

_____ **Example 7.6b** _____
The Logarithmic Transformation:
Circular Conducting Boundaries

Consider next the logarithmic function,

$$W = C_1 \ln Z + C_2 \tag{7}$$

The logarithm of a complex number is readily found if the number is in the polar form:

$$\ln Z = \ln(re^{j\theta}) = \ln r + j\theta \tag{8}$$

so

$$W = C_1(\ln r + j\theta) + C_2$$

Take the constants C_1 and C_2 as real. Then

$$u = C_1 \ln r + C_2 \tag{9}$$

$$v = C_1 \theta \tag{10}$$

If u is to be chosen as the potential function, we recognize the logarithmic potential forms found previously for potential about a line charge, a charged cylinder, or between coaxial cylinders. The flux function, $\psi = -\varepsilon v$, is then proportional to angle θ, as it should be for a problem with radial electric field lines.

To evaluate the constants for a particular problem, take a coaxial line with an inner conductor of radius a at potential zero, and an outer conductor of radius b at potential V_0. Substituting in (9), we have

$$0 = C_1 \ln a + C_2$$

$$V_0 = C_1 \ln b + C_2$$

Solving, we have

$$C_1 = \frac{V_0}{\ln(b/a)} \qquad C_2 = -\frac{V_0 \ln a}{\ln(b/a)}$$

so (7) can be written

$$W = V_0 \left[\frac{\ln(Z/a)}{\ln(b/a)} \right] \tag{11}$$

or

$$\Phi = u = V_0 \left[\frac{\ln(r/a)}{\ln(b/a)} \right] \quad \text{V} \tag{12}$$

$$\psi = -\varepsilon v = \frac{-\varepsilon V_0 \theta}{\ln(b/a)} \quad \text{C/m} \tag{13}$$

In the foregoing, the reference for the flux function came out automatically at $\theta = 0$. If it is desired to use some other reference, the constant C_2 is taken as complex, and its imaginary part serves to fix the reference $\psi = 0$.

_____ **Example 7.6c** _____
The Inverse-Cosine Transformation: Hyperbolic and Elliptic Conducting Boundaries

Consider the function

$$W = \cos^{-1} Z \tag{14}$$

or

$$x + jy = \cos(u + jv) = \cos u \cosh v - j \sin u \sinh v$$

$$x = \cos u \cosh v$$

$$y = -\sin u \sinh v$$

It then follows that

$$\frac{x^2}{\cosh^2 v} + \frac{y^2}{\sinh^2 v} = 1 \tag{15}$$

$$\frac{x^2}{\cos^2 u} - \frac{y^2}{\sin^2 u} = 1 \tag{16}$$

Equation (15) for constant v represents a set of confocal ellipses with foci at ± 1, and (16) for constant u represents a set of confocal hyperbolas orthogonal to the ellipses. These are plotted in Fig. 7.6e. With a proper choice of the region and the function (either u or v) to serve as the potential function, the foregoing transformation could be made to give the solution to the following problems:

1. Field around a charged elliptic cylinder, including the limiting case of a flat strip.
2. Field between two confocal elliptic cylinders, or between an elliptic cylinder and a flat strip conductor extending between the foci.
3. Field between two confocal hyperbolic cylinders, or between a hyperbolic cylinder and a plane conductor extending from the focus to infinity.
4. Field between two semi-infinite conducting plates, coplanar and with a gap separating them. (This is a limiting case of 3.)
5. Field between an infinite conducting plane and a perpendicular semi-infinite plane separated from it by a gap.

To demonstrate how the result is obtained for a particular one of these, consider problem 5, illustrated by Fig. 7.6f. The infinite plane is taken at potential zero, and the perpendicular semi-infinite plane is taken at potential V_0. In using the results of the foregoing general transformation, we must now put in scale factors. To avoid confusion with the preceding, let us denote the variables for this specific problem by primes:

$$W' = C_1 \cos^{-1} kZ' + C_2 \tag{17}$$

The constant C_1 is put in to fix the proper scale of potential, the constant k to fix the scale of size, and the additive constant C_2 to fix the reference for the potential. By comparing with (14).

$$Z = kZ'$$

$$W' = C_1 W + C_2$$

The constants C_1 and C_2 may be taken as real for this problem. Then

$$u' = C_1 u + C_2 \tag{18}$$

By comparing Figs. 7.6e and 7.6f, we want Z' to be a when Z is unity, so $k = 1/a$. Also, when $u = 0$, we want $u' = V_0$; and, when $u = \pi/2$, $u' = 0$. Substitution of these values in (18) yields

$$C_1 = -\frac{2V_0}{\pi} \qquad C_2 = V_0$$

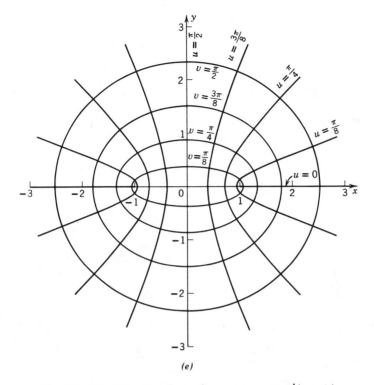

(e)

Fig. 7.6e Plot of the transformation $u + jv = \cos^{-1}(x + jy)$.

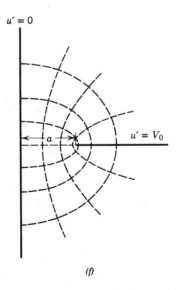

(f)

Fig. 7.6f Field between perpendicular planes with a finite gap.

So the transformation with proper scale factors for this problem is

$$W' = u' + jv' = V_0\left[1 - \frac{2}{\pi}\cos^{-1}\left(\frac{Z'}{a}\right)\right] \tag{19}$$

where u' is the potential function in volts, and $\varepsilon v'$ is the flux function in coulombs per meter. A few of the equipotential and flux lines with these scale factors applied are shown on Fig. 7.6f.

Example 7.6d
Parallel Conducting Cylinders

Consider next the function

$$W = K_1 \ln\left(\frac{Z - a}{Z + a}\right) \tag{20}$$

This may be written in the form

$$W = K_1[\ln(Z - a) - \ln(Z + a)]$$

By comparing with the logarithmic transformation of Ex. 7.6b which, among other things, could represent the field about a single line charge, it follows that this expression can represent the field about two line charges, one at $Z = a$, and the other of equal strength but opposite sign at $Z = -a$. However, it is more interesting to show that this form can also yield the field about parallel cylinders of any radius.

Taking K_1 as real,

$$u = \frac{K_1}{2} \ln\left[\frac{(x - a)^2 + y^2}{(x + a)^2 + y^2}\right] \tag{21}$$

$$v = K_1\left[\tan^{-1}\frac{y}{(x - a)} - \tan^{-1}\frac{y}{(x + a)}\right] \tag{22}$$

Thus lines of constant u can be obtained from (21) by setting the argument of the logarithm equal to a constant:

$$\frac{(x - a)^2 + y^2}{(x + a)^2 + y^2} = K_2$$

As this may be put in the form

$$\left[x - \frac{a(1 + K_2)}{1 - K_2}\right]^2 + y^2 = \frac{4a^2 K_2}{(1 - K_2)^2} \tag{23}$$

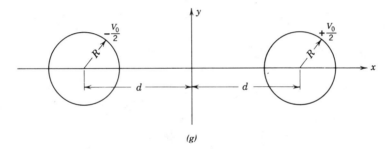

Fig. 7.6g Two parallel conducting cylinders.

the curves of constant u are circles with centers at

$$x = \frac{a(1 + K_2)}{1 - K_2}$$

and radii $(2a\sqrt{K_2})/(1 - K_2)$. If u is taken as the potential function, any one of the circles of constant u may be replaced by an equipotential conducting cylinder. Thus, if R is the radius of such a conductor with center at $x = d$ (Fig. 7.6g), the values of a and the particular value of K_2 (denoted K_0) may be obtained by setting

$$\frac{a(1 + K_0)}{1 - K_0} = d \qquad \frac{2a\sqrt{K_0}}{1 - K_0} = R$$

Solving

$$a = \pm\sqrt{d^2 - R^2} \tag{24}$$

$$\sqrt{K_0} = \frac{d}{R} + \sqrt{\frac{d^2}{R^2} - 1} \tag{25}$$

The constant K_1 in the transformation depends upon the potential of the conducting cylinder. Let this be $V_0/2$. Then, by the definition of K_2 ($=K_0$ on conducting cylinder) and (25),

$$\frac{V_0}{2} = K_1 \ln\sqrt{K_0} = K_1 \ln\left(\frac{d}{R} + \sqrt{\frac{d^2}{R^2} - 1}\right)$$

or

$$K_1 = \frac{V_0}{2 \ln[(d/R) + \sqrt{(d^2/R^2) - 1}]} = \frac{V_0}{2 \cosh^{-1}(d/R)} \tag{26}$$

Substituting in (21), the potential at any point (x, y) is

$$\Phi = u = \frac{V_0}{4 \cosh^{-1}(d/R)} \ln\left[\frac{(x - a)^2 + y^2}{(x + a)^2 + y^2}\right] \tag{27}$$

For $\Phi > 0$ with $x > 0$, $a < 0$ if K_1 is positive so negative sign must be chosen in (24). The flux function $\psi = -\varepsilon v$ is

$$\psi = -\varepsilon v = \frac{\varepsilon V_0}{2 \cosh^{-1}(d/R)} \left[\tan^{-1} \frac{y}{(x + a)} - \tan^{-1} \frac{y}{(x - a)} \right] \qquad (28)$$

Although we have not put in the left-hand conducting cylinder explicitly, the odd symmetry of the potential from (27) will cause this boundary condition to be satisfied also if the left-hand cylinder of radius R with center at $x = -d$ is at potential $-V_0/2$.

If we wish to use the result to obtain the capacitance per unit length of a parallel wire line, we obtain the charge on the right-hand conductor from Gauss's law by finding the total flux ending on it. In passing once around the conductor, the first term of (28) changes by 2π, and the second by zero. So

$$q = 2\pi \frac{\varepsilon V_0}{2 \cosh^{-1}(d/R)} \quad \text{C/m}$$

or

$$.C = \frac{q}{V_0} = \frac{\pi \varepsilon}{\cosh^{-1}(d/R)} \quad \text{F/m} \qquad (29)$$

A similar procedure can be used to find the external inductance of the parallel-wire line. In that case the roles of u and v are opposite from the above electric field problem, with v being proportional to the magnetic scalar potential. The result given in Eq. 4.6(9) for inductance is

$$L = \frac{\mu}{\pi} \cosh^{-1}\left(\frac{d}{R}\right) \qquad (30)$$

From (29) and (30) we see that $LC = \mu \varepsilon$ as was shown to be the case for other two-conductor lines in Chapter 5. That this is a general result is shown in Sec. 8.12.

7.7 The Schwarz Transformation for General Polygons

In the examples in Sec. 7.6 specific functions have been set down, and the electrostatic problems solvable by these deduced from a study of their properties. In a practical problem, the reverse procedure is usually required, for the specific equipotential conducting boundaries will be given and it will be desired to find the complex function useful in solving the problem. The greatest limitation on the method of conformal transformations is that, for general shaped boundaries,

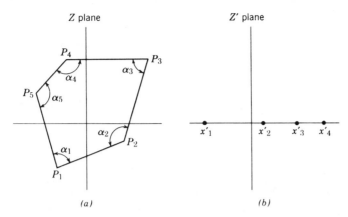

Fig. 7.7 (a) General polygon in Z plane. (b) Polygon of figure tranformed into straight line in Z' plane. Vertex x_5' is at infinity.

there is no straightforward procedure by which one can always arrive at the desired transformation if the two-dimensional physical problem is given. There is such a procedure, however, when the boundaries consist of straight-line sides with angle intersections.

The *Schwarz transformation* takes an arbitrary polygon in the Z plane into a series of segments along the real axis in a Z' plane as shown in Figs. 7.7a and 7.7b. The segments correspond to the sides of the polygon.[7] The transformation may be found by integrating the derivative,

$$\frac{dZ}{dZ'} = K(Z' - x_1')^{(\alpha_1/\pi) - 1}(Z' - x_2')^{(\alpha_2/\pi) - 1} \cdots (Z' - x_n')^{(\alpha_n/\pi) - 1} \qquad (1)$$

Each factor in (1) may be thought of as straightening out the boundary at one of the vertices as the transform of Ex. 7.6a did for the single corner. The setting down of (1) for a specific problem is usually easy, but the difficulties come in its integration.

Although we have spoken of the figure to be transformed as a polygon, in the practical application of the method, one or more of the vertices may be at infinity, and part of the boundary may be at a different potential from the remaining part. Then the real axis in the Z' plane consists of two parts at different potentials. This latter electrostatic problem may be solved by a transformation from the Z' to the W plane, and thus the transformation from the Z to the W plane is given with the Z' plane only as an intermediate step. Another sort of problem in which the method is useful is that in which a thin charged wire lies on the interior of a conducting polygon, parallel to the elements of the polygon. By the Schwarz transformation, the polygon boundary is transformed to the real axis and the

[7] For more details see R. V. Churchill, J. W. Brown, and R. F. Berkey, *Complex Variables and Applications*, 3rd Ed., McGraw-Hill, New York, 1974.

Table 7.7

$Z = x + jy$; $W = u + jv$, where u = flux function, v = potential

$$Z = \frac{b}{\pi}\left[\cosh^{-1}\left(\frac{\alpha^2 + 1 - 2\alpha^2 e^{kW}}{1 - \alpha^2}\right) - \alpha \cosh^{-1}\left(\frac{2e^{-kW} - (\alpha^2 + 1)}{1 - \alpha^2}\right)\right]$$

$$\alpha = \frac{a}{b} \qquad k = \frac{\pi}{V_0}$$

$$Z = \frac{b}{\pi}\left[\ln\left(\frac{1 + S}{1 - S}\right) - 2\alpha \tan^{-1}\left(\frac{S}{\alpha}\right)\right]$$

$$\alpha = \frac{a}{b} \qquad S = \sqrt{\frac{e^{kW} + \alpha}{e^{kW} - 1}}$$

wire corresponds to some point in the upper half of the Z' plane. This electrostatic problem can be solved by the method of images, and so the original problem can be solved in this case also.

Results for some important problems that have been solved by the Schwarz technique are given in Table 7.7.

Example 7.7
Fringing Field in Parallel-Plate Capacitor

To illustrate how the concept of a polygon can be applied, consider the parallel-plate capacitor structure in Fig. 7.7c with $\Phi = V_0$ on the infinite bottom plane and $\Phi = 0$ on the plane D–C. For the purposes of the Schwarz transformation, the structure may be considered a polygon with interior angles α_1, α_2, and α_3 and sides of infinite length. Application of the transformation puts all the boundaries along the real axis as in Fig. 7.7d where $\Phi = V_0$ for $x'_2 < 0$ and $\Phi = 0$ for $x'_2 > 0$. As was shown in Ex. 7.6b, a subsequent logarithmic transformation converts such

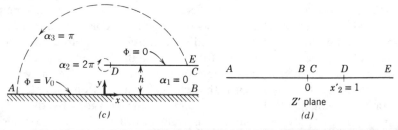

Fig. 7.7 (c) Edge of parallel-plate capacitor with one plane of infinite extent (equivalent to one-half of a symmetrical parallel-plate capacitor). (d) Transformation of the capacitor of (c) into a single plane.

a set of boundary potentials into a parallel-plane uniform field in the W plane. Combining the two transformations, one finds that the flux lines and potential lines, u and v, respectively, are implicitly defined in terms of position x, y in the Z plane by the relation

$$Z = \frac{h}{\pi}\left(e^{\pi W/V_0} - 1 - \frac{\pi W}{V_0} + j\pi\right) \tag{2}$$

One can select values of u and v and calculate the corresponding points x and y, thus plotting out the field shown in Fig. 1.9a.

7.8   Conformal Mapping for Wave Problems

We have seen that in conformal transformations for statics complicated boundaries are transformed to simple ones. Conformal transformations of wave problems can similarly simplify complicated boundaries.[8] The transformations can be made only in two dimensions so the fields must be independent of the third dimension. The simplification of the boundaries is also normally accompanied by increased complexity of the dielectric so this trade-off only occasionally helps.

Let us assume that there exists an analytic function $W = u + jv = f(Z) = f(x + jy)$ which transforms the given boundary shapes in the Z plane to lines of constant u and v in the W plane. We will first determine the relation between $\nabla^2_{xy}\psi$ and $\nabla^2_{uv}\psi$ in order to transform the scalar Helmholtz equation, Eq. 7.2(8),

$$\frac{\partial^2\psi}{\partial x^2} + \frac{\partial^2\psi}{\partial y^2} + k^2\psi = 0 \tag{1}$$

from the Z plane to the W plane. To transform the derivatives, we apply the chain rule. First,

$$\frac{\partial\psi}{\partial x} = \frac{\partial\psi}{\partial u}\frac{\partial u}{\partial x} + \frac{\partial\psi}{\partial v}\frac{\partial v}{\partial x} \tag{2}$$

Applying the chain rule a second time leads to

$$\frac{\partial^2\psi}{\partial x^2} = \frac{\partial^2\psi}{\partial u^2}\left(\frac{\partial u}{\partial x}\right)^2 + \frac{\partial^2\psi}{\partial v^2}\left(\frac{\partial v}{\partial x}\right)^2 + 2\frac{\partial^2\psi}{\partial u\,\partial v}\frac{\partial u}{\partial x}\frac{\partial v}{\partial x} \tag{3}$$

and similarly for $\partial^2\psi/\partial y^2$:

$$\frac{\partial^2\psi}{\partial y^2} = \frac{\partial^2\psi}{\partial u^2}\left(\frac{\partial u}{\partial y}\right)^2 + \frac{\partial^2\psi}{\partial v^2}\left(\frac{\partial v}{\partial y}\right)^2 + 2\frac{\partial^2\psi}{\partial u\,\partial v}\frac{\partial u}{\partial y}\frac{\partial v}{\partial y} \tag{4}$$

---

[8] F. E. Borgnis and C. H. Papas, *Handbuch der Physik* (S. Flügge, Ed.) **16**, 358 (1958) (in English).

Making use of the Cauchy–Riemann conditions in Eqs. 7.4(6) and 7.4(7) in the second terms of the right sides of (3) and (4) and in the last term of (4), and adding (3) and (4),

$$\frac{\partial^2\psi}{\partial x^2} + \frac{\partial^2\psi}{\partial y^2} = \left(\frac{\partial^2\psi}{\partial u^2} + \frac{\partial^2\psi}{\partial v^2}\right)\left[\left(\frac{\partial u}{\partial x}\right)^2 + \left(\frac{\partial u}{\partial y}\right)^2\right] \tag{5}$$

Note from Eq. 7.4(4) and 7.4(7) that

$$\left|\frac{dW}{dZ}\right|^2 = \left(\frac{\partial u}{\partial x}\right)^2 + \left(\frac{\partial v}{\partial x}\right)^2 = \left(\frac{\partial u}{\partial x}\right)^2 + \left(\frac{\partial u}{\partial y}\right)^2 \tag{6}$$

Thus

$$\nabla^2_{xy}\psi = \left|\frac{dW}{dZ}\right|^2 \nabla^2_{uv}\psi \tag{7}$$

The quantity $|dZ/dW|$ is a scale factor which relates a differential length $|dW|$ in the W plane to the corresponding length $|dZ|$ in the Z plane, as we discussed in Sec. 7.6. The Helmholtz equation (1) is thus transformed to the W plane giving

$$\nabla^2_{uv}\psi + \left|\frac{dZ}{dW}\right|^2 \psi = 0 \tag{8}$$

Note that, in general, $|dZ/dW|$ is a function of the coordinates so that (8) is equivalent to the Helmholtz equation in an inhomogeneous medium.

Boundary conditions in the Z plane consisting of zero values of ψ or its normal derivatives carry over unchanged to the corresponding boundaries in the W plane since the orthogonality of coordinates is conserved. If a nonzero normal derivative is specified on a boundary, the scale factor $|dW/dZ|$ enters the conversion of the boundary condition through the relation between gradients in the two planes:

$$\nabla_{uv}\psi = \left|\frac{dZ}{dW}\right|\nabla_{xy}\psi \tag{9}$$

_____ **Example 7.8** _____
Curved Dielectric Waveguide

A layer of a dielectric material imbedded in materials of lower permittivity can serve to guide electromagnetic waves, as will be studied in more detail in Chapter 9. The phenomenon of total internal reflection analyzed in Sec. 6.12 supplies a qualitative understanding of dielectric waveguides. Here we see how wave propagation in a curved layer, as in Fig. 7.8a, can be treated using conformal mapping.

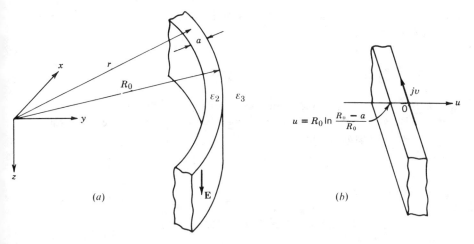

Fig. 7.8 (*a*) Curved dielectric waveguide in Z plane. (*b*) Curved dielectric guide transformed into W plane.

The wave is assumed to be polarized with its electric field in the z direction and E_z is independent of z. Then identifying E_z with ψ, we can write from Eq. 7.2(8), with $\partial/\partial z = 0$:

$$\nabla^2_{xy} E_z(x, y) + k^2 E_z(x, y) = 0 \tag{10}$$

Using the transformation of Ex. 7.6b in slightly different form,

$$W = R_0 \ln \frac{Z}{R_0} \tag{11}$$

for which

$$\left| \frac{dZ}{dW} \right| = e^{u/R_0} \tag{12}$$

(8) becomes

$$\nabla^2_{uv} E_z(u, v) + k^2 e^{2u/R_0} E_z(u, v) = 0 \tag{13}$$

From (11) it is easily seen that

$$u = R_0 \ln \frac{r}{R_0} \tag{14}$$

and

$$v = R_0 \theta \tag{15}$$

Therefore, the edge of the guiding layer at $r = R_0$ is the $u = 0$ line in the W plane and the other edge of the guide at $r = R_0 - a$ is at $u = R_0 \ln(R_0 - a)/R_0$. The result is that the curved layer in the Z plane becomes the planar region shown in Fig. 7.8b. Note that this layer in the W plane is inhomogeneous. Equation (13) has the usual form of the Helmholtz equation only if we identify a new wave number k' by

$$k' = \omega \sqrt{\mu\varepsilon \exp \frac{2u}{R_0}} \tag{16}$$

Since k' is a function of u, one cannot substitute it directly for k in the usual wave solution. However, the problem is solvable and has been used to study the leakage of energy from bends in dielectric waveguides.[9]

PRODUCT-SOLUTION METHOD

7.9 Laplace's Equation in Rectangular Coordinates

One of the most powerful techniques for solution of linear partial differential equations is that of separation of variables. This leads to solutions which are products of three functions (for three-dimensional problems), each function depending upon one coordinate variable only. Such solutions might not seem very general, but they may be added to form a series which can represent very general functions. Moreover, single-product solutions of the wave equation represent modes which can propagate individually. These are of great practical importance in waveguides and resonant systems and are studied extensively in following chapters.

As the simplest example of the method of separation of variables, let us first consider two-dimensional problems in the rectangular coordinates x and y, as we have in the transformation method of the past section. Laplace's equation in these coordinates is

$$\frac{\partial^2 \Phi}{\partial x^2} + \frac{\partial^2 \Phi}{\partial y^2} = 0 \tag{1}$$

We wish to study product solutions of the form

$$\Phi(x, y) = X(x)Y(y) \tag{2}$$

where we see that we have a function of x alone times a function of y alone. From this point on $X(x)$ will be replaced by X and $Y(y)$ by Y. Substituting in (1), we have

$$X''Y + XY'' = 0 \tag{3}$$

---

[9] M. Heiblum and J. H. Harris, *IEEE J. of Quantum Electronics* **QE-11**, 75 (1975).

The double prime denotes the second derivative with respect to the independent variable in the function. Now to separate into the sum of functions of one variable only, divide (3) by (2)

$$\frac{X''}{X} + \frac{Y''}{Y} = 0 \tag{4}$$

Next follows the key argument for this method. Equation (4) is to hold for all values of the variables x and y. Since the second term does not contain x, and so cannot vary with x, the first term cannot vary with x either. A function of x alone which does not vary with x is a constant. Similarly, the second term must be a constant. Let us denote the first as k_x^2 and the second as k_y^2. Then

$$k_x^2 + k_y^2 = 0 \tag{5}$$

and

$$X'' - k_x^2 X = 0$$
$$Y'' - k_y^2 Y = 0 \tag{6}$$

We recognize that these are in the standard form having real exponentials or hyperbolic functions as solutions. Let us write them in hyperbolic form and substitute in (2):

$$\Phi(x, y) = (A \cosh k_x x + B \sinh k_x x)(C \cosh k_y y + D \sinh k_y y) \tag{7}$$

It is clear from (5) that either k_x^2 or k_y^2 must be negative and therefore either k_x or k_y must be imaginary while the other is real. Furthermore, their magnitudes must be the same. Thus (7) can have either of two forms,

$$\Phi(x, y) = (A \cosh kx + B \sinh kx)(C \cos ky + D' \sin ky) \tag{8}$$

or

$$\Phi(x, y) = (A \cos kx + B' \sin kx)(C \cosh ky + D \sinh ky) \tag{9}$$

where, since $|k_x| = |k_y|$, we have used the single symbol k. The primes are used to indicate that the constants have changed. The choice between (8) and (9) is dictated by the nature of the boundary conditions. If the potential is required to have repeated zeros as a function of y, then (8) is used; if repeated zeros are specified for the x variation, (9) is chosen. If the boundaries extend to infinity in one direction, real exponentials are used in place of hperbolic functions. It may be noted from (6) that for $k_x = jk_y = 0$ the general solution has the form

$$\Phi(x, y) = (A_1 x + B_1)(C_1 y + D_1) \tag{10}$$

It is typical for product solutions that when the separation constants go to zero the functional forms of the solutions change. We will see in subsequent sections how the constants are evaluated using the boundary conditions.

For the three-dimensional case in rectangular coordinates, the procedure is simply extended. Laplace's equation is

$$\frac{\partial^2 \Phi}{\partial x^2} + \frac{\partial^2 \Phi}{\partial y^2} + \frac{\partial^2 \Phi}{\partial z^2} = 0 \tag{11}$$

Consider solutions of the form

$$\Phi(x, y, z) = X(x)Y(y)Z(z) \tag{12}$$

where each term on the right side is a function of just one of the independent space variables. Substituting (12) in (11), we have

$$X''YZ + XY''Z + XYZ'' = 0$$

and dividing by Φ, we see that

$$\frac{X''}{X} + \frac{Y''}{Y} + \frac{Z''}{Z} = 0 \tag{13}$$

We use the same argument as was used in the two-dimensional case. If the second two terms do not vary with x, neither can the first. Since it is a function of x alone and does not vary with x, it must be a constant. Similar arguments apply for the second and third terms. If we let the first term be k_x^2, the second k_y^2, and the third k_z^2, (13) becomes

$$k_x^2 + k_y^2 + k_z^2 = 0 \tag{14}$$

and differential equations of the form (6) apply for X, Y, and Z. So the general solution, written as the product of X, Y, and Z, and sometimes called a *rectangular harmonic*, is

$$\Phi(x, y, z) = [A \cosh k_x x + B \sinh k_x x][C \cosh k_y y + D \sinh k_y y]$$
$$\times [E \cosh k_z z + F \sinh k_z z] \tag{15}$$

It is clear that at least one of k_x^2, k_y^2, or k_z^2 must be negative for (14) to hold, so at least one of k_x, k_y, or k_z must be imaginary. If repeated potential zeros are required in the x and y directions, the functions of x and y must be trigonometric functions so k_x and k_y are imaginary. There are various other combinations which may be useful. In some cases it is advantageous to replace the hyperbolic functions by real exponentials as mentioned earlier for the two-dimensional solutions.

In (15) there appear to be nine constants, to be evaluated using the six possible boundary conditions, two for each of the three coordinate directions. If, however one divides the first bracket by B, the second by D, and the third by F and multiplies the entire by BDF, it becomes clear that there are just four independent multiplicative constants. From (14) we see that there are only two independent separation constants so the total number of unknowns equals the number of boundary conditions.

7.10   Static Field Described by a Single Rectangular Harmonic

Let us see what boundaries would be required in order to have some one of the forms of Sec. 7.9 as a solution. Take the special case of Eq. 7.9(9) with $A = 0$, $C = 0$. The product of remaining constants, $B'D$, may be denoted as a single constant C_1:

$$\Phi = C_1 \sin kx \sinh ky \tag{1}$$

It is evident from (1) that potential is zero at $y = 0$ for all x. Hence one boundary can be a zero-potential conducting plane at $y = 0$. Similarly, potential is zero along the plane $x = 0$, and also at other parallel planes defined by $kx = n\pi$. Let us confine attention to the region $0 < kx < \pi$ and $0 < y < \infty$. The intersecting zero-potential planes of interest then form a rectangular conducting trough. Let its depth in the x direction be a. Then $ka = \pi$ or

$$k = \frac{\pi}{a} \tag{2}$$

If there is to be a finite field in the region, there must be some electrode at a potential other than zero. Without knowing its shape for the moment, let us take the value of y at which it crosses the midplane $x = a/2$ as $y = b$, and the potential of the electrode as V_0. Then, from (1),

$$V_0 = C_1 \sin \frac{\pi}{2} \sinh \frac{\pi b}{a} = C_1 \sinh \frac{\pi b}{a}$$

or, substituting in (1), we have

$$\Phi = \frac{V_0 \sinh(\pi y/a)}{\sinh(\pi b/a)} \sin \frac{\pi x}{a} \tag{3}$$

The potential at any point x, y may be computed from (3). In particular, the form that the electrode at potential V_0 must take, can be found from (3) by setting $\Phi = V_0$ yielding

$$\sinh \frac{\pi y}{a} = \frac{\sinh(\pi b/a)}{\sin(\pi x/a)} \tag{4}$$

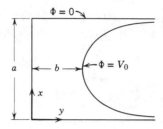

Fig. 7.10a Electrodes and potentials for which a single harmonic is the complete solution.

Equation (4) can be plotted to show the form of the electrode. This is done for a value $b/a = \frac{1}{2}$ in Fig. 7.10a. Actually, the electrodes should extend to infinity, but if they are extended a large but finite distance, the solution studied here will represent the potential very well everywhere except near edges.

Alternatively, a straight boundary at $y = b$ could be supplied with a sinusoidal distribution of potential and a single harmonic would describe the potential at all points in the box. Thus, for example, if $ka = 3\pi$ and the boundary potential were

$$\Phi(x, b) = V_0 \sin \frac{3\pi}{a} x \tag{5}$$

then the harmonic

$$\Phi = \frac{V_0 \sinh(3\pi/a)y}{\sinh(3\pi b/a)} \sin \frac{3\pi}{a} x \tag{6}$$

shown in Fig. 7.10b satisfies the boundary conditions and describes the potential at all points.

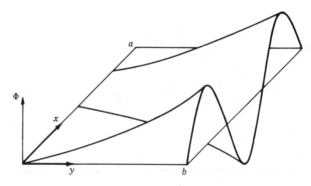

Fig. 7.10b Potential in a two-dimensional box with a sinusoidal distribution of potential on one side.

7.11   Fourier Series and Integral

In the preceding section, we saw that a single-product solution could only satisfy very special forms of boundary conditions. For more general boundaries a sum of such solutions must be used. This is one example of situations where Fourier series or integrals are useful in forming solutions for field problems. We provide here a review of the Fourier tools with the assumption that the reader has already a measure of familiarity with them.

Fourier Series   Fourier series are used to represent periodic functions. For the independent variable x, the required periodicity is expressed by

$$f(x) = f(x + L) \tag{1}$$

where L is the *period* of the function. We assume that the function can be represented by a constant plus the sum of infinite series of sine and cosine functions of harmonics of a fundamental spatial frequency k:

$$f(x) = a_0 + a_1 \cos kx + a_2 \cos 2kx + a_3 \cos 3kx + \cdots$$
$$+ b_1 \sin kx + b_2 \sin 2kx + b_3 \sin 3kx + \cdots \tag{2}$$

where the phase factor k is related to the period L in the usual way:

$$kL = 2\pi \quad \text{rad} \tag{3}$$

To evaluate the unknown constants in (2) for a given function $f(x)$, we make use of so-called *orthogonality* properties of sinusoids. These are

$$\int_{-L/2}^{L/2} \cos nkx \cos mkx \, dx = 0 \qquad m \neq n \tag{4}$$

$$\int_{-L/2}^{L/2} \sin nkx \sin mkx \, dx = 0 \qquad m \neq n \tag{5}$$

and

$$\int_{-L/2}^{L/2} \sin mkx \cos nkx \, dx = 0 \qquad \begin{cases} m \neq n \\ m = n \end{cases} \tag{6}$$

However,

$$\int_{-L/2}^{L/2} \cos^2 mkx \, dx = \int_{-L/2}^{L/2} \sin^2 mkx \, dx = \frac{L}{2} \tag{7}$$

To make use of these properties, we multiply each term in (2) by $\cos nkx$ and integrate over one period. Every term on the right vanishes because of the properties in (4)–(6) except the one containing $\cos nkx$; that term gives $a_n L/2$ according to (7). Thus

$$a_n = \frac{2}{L} \int_{-L/2}^{L/2} f(x) \cos nkx \, dx \tag{8}$$

Similarly, multiplication of (2) by $\sin nkx$ and integrating from $-L/2$ to $L/2$ leaves only the term involving $\sin nkx$ on the right-hand side and its coefficient, by (7), is

$$b_n = \frac{2}{L} \int_{-L/2}^{L/2} f(x) \sin nkx \, dx \tag{9}$$

Finally, to obtain the constant term a_0, every term is integrated directly over a period and all the terms on the right side disappear except that containing a_0 so that

$$a_0 = \frac{1}{L} \int_{-L/2}^{L/2} f(x) \, dx \tag{10}$$

This merely states that a_0 is the average of the function $f(x)$.

For a general function, an infinite number of terms is required in the Fourier series representation. But often a sufficient degree of approximation to the desired wave shape is obtained when only a finite number of terms is used. For functions with sharp discontinuities, however, many terms may be required near the sharp corners, and the theory of Fourier series shows that the series does not converge to the function in the neighborhood of the discontinuity (Gibbs phenomenon). The derivative of the series also does not converge to the derivative of the function, but the integral of the series always converges to that of the function.

―――――――――――― **Example 7.11a** ――――――――――――
**Fourier Series Representation of a Function over
a Finite Interval**

In static field problems, one commonly has the boundary potential specified over a finite interval, such as along a straight boundary at a constant value of one coordinate. For the purposes of matching the given boundary potential, it is desirable to express it in a Fourier series. This can be done even though the function is not periodic, having been specified only over a finite interval. The point of view is that the interval of length a may be considered a period or an integral fraction of a period, and a periodic function defined to agree with the given function over the given interval, repeating itself outside that interval. A Fourier series may then be written for this periodic function which will give desired values in the interval, and, although it also gives values outside the interval, that is of no consequence since the original function is not defined there.

The interval is commonly selected as a half-period since the function extended outside the interval may then be made either even or odd, and the corresponding

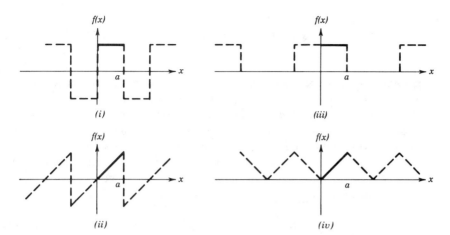

Fig. 7.11a Examples of functions specified over a finite interval (solid line) and odd and even continuations (broken lines).

Fourier series will then have respectively either cosine terms alone or sine terms alone. Figure 7.11a shows by solid lines some possible examples of functions specified over the interval $0 < x < a$. Their extensions outside that interval as either odd or even functions are shown by the broken lines. Note that in one case the interval is $L/4$. The choice of whether to consider the function continued as an odd or an even function depends upon the form used to represent the potential in the problem. Thus, for example, in Eq. 7.10(3) the potential is expressed in terms of $\sin \pi x/a$ and the appropriate series for the representation of the boundary potential will be in sines.

$$f(x) = \sum_{n=1}^{\infty} b_n \sin \frac{n\pi x}{a} \tag{11}$$

where we have made use of the fact that $a = L/2$, one-half period. The coefficients are found from (9) noting that the contribution to the integral from the negative and positive intervals are equal. Thus, with $a = L/2$,

$$b_n = \frac{2}{a} \int_0^a f(x) \sin \frac{n\pi x}{a} \, dx \tag{12}$$

Suppose, for example, that the specified function is $f(x) = C$ over the interval $0 < x < a$ and it is desired to expand it in a sine series. Then (12) yields

$$b_n = \frac{2}{a} \int_0^a C \sin \frac{n\pi x}{a} \, dx = \frac{2C}{n\pi} \left[-\cos \frac{n\pi x}{a} \right]_0^a \tag{13}$$

and the series is

$$f(x) = \frac{2C}{\pi} \left[2 \sin \frac{\pi x}{a} + \frac{2}{3} \sin \frac{3\pi x}{a} + \frac{2}{5} \sin \frac{5\pi x}{a} + \cdots \right] \tag{14}$$

The series (14) has the required value $f(x) = C$ over the interval $0 < x < a$ but also represents the dashed portion of waveform (i) in Fig. 7.11a outside that interval.

Fourier Integral   In some problems the function of interest is defined over the entire range and is aperiodic. An example is a square function that is constant in some range $-a \leq x \leq a$ and zero elsewhere, as shown in Fig. 7.11b. This could be considered the limiting case of a periodic series of square pulses where the period L goes to infinity. The spacing of the components $(n + 1)k - nk = 2\pi/L$ from (3) becomes vanishingly small as the period L approaches infinity and in the limit the spectrum of component sinusoidal waves becomes a continuum.[10]
    In the limiting, aperiodic case the series (2) is replaced by an integral

$$f(x) = \frac{1}{2\pi} \int_{-\infty}^{\infty} g(k)e^{jkx}\, dk \tag{15}$$

and the function $g(k)$ which takes the place of a_n and b_n of (8)–(10) is given by [11]

$$g(k) = \int_{-\infty}^{\infty} f(x)e^{-jkx}\, dx \tag{16}$$

The theory of Fourier integrals shows that for (15) to give the same $f(x)$ that appears in (16) the function must be continuous or have only a finite number of finite discontinuities in any finite interval and must be absolutely integrable, that is

$$\int_{-\infty}^{\infty} |f(x)|\, dx < \infty \tag{17}$$

These conditions do not place strong limitations on the utility of the transform pair (15) and (16).

Example 7.11b

Fourier Transform of a Rectangular Pulse

Using (16) we find the spectrum of spatial frequency components for the rectangular function in Fig. 7.11b:

$$g(k) = \int_{-a}^{a} Ce^{-jkx}\, dx = C\left[\frac{e^{-jkx}}{-jk}\right]_{-a}^{a} = 2Ca\left(\frac{\sin ka}{ka}\right) \tag{18}$$

---

[10] See, for example, D. Sakrison, *Communication Theory: Transmission of Waveforms and Digital Information*, Wiley, New York, 1968, p. 30, or for a detailed treatise, R. Bracewell, *The Fourier Transform and its Applications*, McGraw-Hill, New York, 1965.
[11] The placement of 2π in the pair (23) and (24) is arbitrary and is done in various ways in the literature.

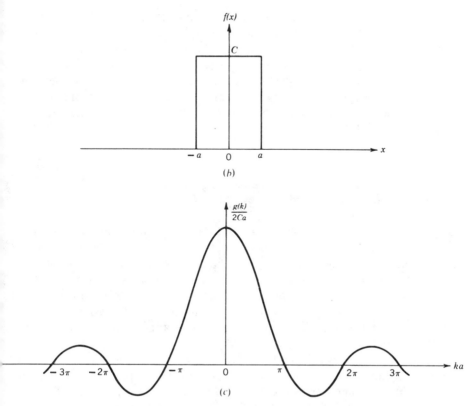

Fig. 7.11 (*b*) Inverse Fourier transform of function in Fig. 7.11*c*. (*c*) Fourier transform of rectangular function in Fig. 7.11*b*. Value of $g(0)/2Ca$ is unity.

This very important function occurs frequently in practice and is shown in Fig. 7.11*c*. The function $f(x)$ in Fig. 7.11*b* is called the *inverse* Fourier transform of that in Fig. 7.11*c*. The two are called a *transform pair*.

7.12 Series of Rectangular Harmonics for Two- and Three-Dimensional Static Fields

We saw in Sec. 7.10 that product solutions (harmonics) satisfying zero boundary conditions on three of the four boundaries in a two-dimensional rectangular structure can be found. However, the use of a single harmonic as the expression for the potential requires either a fourth boundary of complicated shape or a simple flat one with a sinusoidal variation of potential along it. To solve problems with an arbitrary variation of potential along a flat boundary on a coordinate line, one may use a sum of harmonics, each of which satisfies the zero conditions on three boundaries and has a weighting in the sum such that it equals the given potential at the fourth boundary. Then the given potential is expanded in a Fourier series of either sines or cosines, chosen to match the functions in the sum of harmonics. The harmonic series is evaluated at the fourth boundary and compared, term by term, with the Fourier series to evaluate the weighting coefficients in the former. These procedures are sometimes slightly modified by use of symmetries and superposition, as seen in the following examples and problems.

--------------------------------- Example 7.12a ---------------------------------
Two-Dimensional Problem with Specified Boundary Potentials

As an example of a problem which cannot be solved by using a single one of the solutions of Sec. 7.9, but can be by means of a series of these solutions, consider the two-dimensional region of Fig. 7.12a bounded by a zero-potential plane at $y = 0$, a zero-potential plane at $x = 0$, a parallel zero-potential plane at $x = a$, and a plane conducting lid of potential V_0 at $y = b$. In the ideal problem, the lid is separated from the remainder of the rectangular box by infinitesimal gaps. In a practical problem, it would only be expected that these gaps should be small compared with the rest of the box.

In selecting the proper forms from Sec. 7.9, we will choose the form having sinusoidal solutions in x since potential is zero at $x = 0$ and also at $x = a$, and sinusoids have repeated zeros. So the form of Eq. 7.9(9) is suitable. Moreover, $\Phi = 0$ at $y = 0$ for all x of interest, so the function of y must go to zero at $y = 0$, showing that $C = 0$. Similarly, since $\Phi = 0$ at $x = 0$ for all y of interest, $A = 0$. Then Φ is again zero at $x = a$, so $ka = m\pi$, or

$$k = \frac{m\pi}{a}$$

Denoting the product of the remaining constants $B'D$ as C_m, we have

$$\Phi = C_m \sin \frac{m\pi x}{a} \sinh \frac{m\pi y}{a}$$

This form satisfies the Laplace equation and the boundary conditions at $y = 0$, at $x = 0$, and at $x = a$, but a single term of this form cannot satisfy the boundary condition along the plane lid at $y = b$, as the study in Sec. 7.10 has shown. A series of such solutions also satisfies Laplace's equation and the boundary conditions at $y = 0$, at $x = 0$, and at $x = a$:

$$\Phi = \sum_{m=1}^{\infty} C_m \sin \frac{m\pi x}{a} \sinh \frac{m\pi y}{a} \tag{1}$$

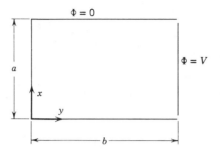

Fig. 7.12a Two dimensional box for Ex. 7.12a.

For the sum (1) to give the required constant potential V_0 along the plane $y = b$ over the interval $0 < x < a$, we require

$$V_0 = \sum_{m=1}^{\infty} C_m \sin \frac{m\pi x}{a} \sinh \frac{m\pi b}{a}, \qquad 0 < x < a \tag{2}$$

But this is recognized as a Fourier expansion in sines of the constant function V_0 over the interval $0 < x < a$. This expansion was carried out in Ex. 7.11a to yield

$$f(x) = V_0 = \sum_{m=1}^{\infty} a_m \sin \frac{m\pi x}{a} \qquad 0 < x < a \tag{3}$$

with

$$a_m = \begin{cases} \dfrac{4V_0}{m\pi}, & m \text{ odd} \\[2mm] 0, & m \text{ even} \end{cases} \tag{4}$$

Comparison of (3) with (2) shows that

$$C_m \sinh \frac{m\pi b}{a} = a_m \tag{5}$$

Substitution of the results of (5) and (4) in (1) gives

$$\Phi = \sum_{m \text{ odd}} \frac{4V_0}{m\pi} \frac{\sinh(m\pi y/a)}{\sinh(m\pi b/a)} \sin \frac{m\pi x}{a} \tag{6}$$

This series is rapidly convergent except at corners of $x \to 0$, a, and $y \to b$, so it can be used for reasonably convenient calculation of potential elsewhere.

We note that the evaluation of the constants in the general solution depended upon the fact that the boundary potentials were specified on surfaces in the coordinate system. Furthermore, nonzero conditions, potential or normal derivative of potential, must exist on some part of the boundary to yield a nonzero solution. As will be clarified in the next example, superposition may be used to solve problems where the boundary conditions involve several sides.

—————————————— Example 7.12b ——————————————
Two-Dimensional Problem Requiring Superposition

In this example we see a situation in which the given boundary conditions can be satisfied only by including an average term in the series expansion of the boundary potential; this requires inclusion of a product solution with zero separation constant, which can be handled by superposition. Consider the problem of finding the potentials in *conducting* rectangular solid of infinite extent in the z direction shown in Fig. 7.12b. The surrounding region contains free space, the potential at $y = 0$ is zero, and that along the edge $y = b$ is given by $\Phi = V_0 x/a$.

This problem requires a solution with repetition in the x direction since the boundary conditions at the sides $x = 0, a$ are the same; the appropriate general form is that in Eq. 7.9(9). At $x = 0, a$ the x component of current density must be zero since no current can flow in the free space outside the conductor. Since $\mathbf{J} = \sigma\mathbf{E}$, then E_x must also be zero at $x = 0, a$ and therefore also $\partial\Phi/\partial x$, given by

$$\frac{\partial\Phi}{\partial x} = K(-A \sin Kx + B' \cos Kx)(C \cosh Ky + D \sinh Ky) \tag{7}$$

must be zero at $x = 0, a$ for all y. Since $\cos Kx = 1$ at $x = 0$, $B' = 0$. Also $\sin Ka = 0$ if $Ka = m\pi$ so that $K = m\pi/a$. To match the boundary condition $\Phi = 0$ at $y = 0$, requires $C = 0$. Thus the potential in the mth harmonic is

$$\Phi_m = C_m \cos\frac{m\pi}{a}x \sinh\frac{m\pi}{a}y \tag{8}$$

The potential on the boundary at $y = b$ should be expanded in a series of cosines so that term-by-term matching with a series of terms like (8) can be done. The appropriate periodic continuation of the given boundary potential is shown in Fig. 7.11a(iv). It is seen to have an average value of $V_0/2$ which will be present in the series expansion. Applying Eqs. 7.11(8)–(10),

$$\Phi(x, b) = \frac{V_0}{2} - \sum_{m\,\text{odd}}\frac{4V_0}{(m\pi)^2}\cos\frac{m\pi}{a}x \tag{9}$$

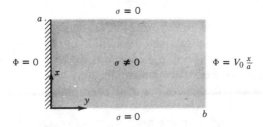

$\sigma = 0$

$\Phi = 0$ $\sigma \neq 0$ $\Phi = V_0\frac{x}{a}$

$\sigma = 0$

Fig. 7.12b Two-dimensional conductive solid imbedded in a nonconductive medium.

In addition to the hyperbolic functions, we need a linear term satisfying the condition on $\partial\Phi/\partial x$ at $x = 0, a$ and $\Phi = 0$ at $y = 0$. This is found from Eq. 7.9(10) to be

$$\Phi_1 = A_1 y \tag{10}$$

We divide the problem into two parts, one with a constant potential $V_0/2$ on the boundary at $y = b$ and one with the potential described by the cosine series in (9) on that boundary. In both problems the potential at $y = 0$ is zero and $\partial\Phi/\partial x = 0$ along the sides at $x = 0, a$. The solution of the first is (10) with $A_1 = V_0/2$ or

$$\Phi_1 = \frac{V_0 y}{2b} \tag{11}$$

The solution of the other is found by equating a series of terms like that in (8) evaluated at the boundary $y = b$ with the series in (9):

$$\sum_m C_m \cos \frac{m\pi x}{a} \sinh \frac{m\pi b}{a} = - \sum_{m\,\mathrm{odd}} \frac{4V_0}{(m\pi)^2} \cos \frac{m\pi x}{a} \tag{12}$$

The result for the second potential is

$$\Phi_2 = - \sum_{m\,\mathrm{odd}} \frac{4V_0}{(m\pi)^2} \frac{\sinh(m\pi y/a)}{\sinh(m\pi b/a)} \cos \frac{m\pi x}{a} \tag{13}$$

The complete solution is the superposition of (11) and (13), $\Phi = \Phi_1 + \Phi_2$.

_____ **Example 7.12c** _____
**Three-Dimensional Rectangular Box with Potential Specified
on One Face**

The method discussed above can be extended to three dimensions. As an example, we consider a box with zero potential on all faces except on the side $z = c$, where it is V_0. The box extends from the origin of coordinates to $x = a$, $y = b$, and $z = c$. The appropriate general form of the space harmonic is Eq. 7.9(15) with k_x and k_y imaginary and k_z real. To meet the zero-potential boundary conditions at $x = 0$, $y = 0$, and $z = 0$, the constants A, C, and E must be zero. Also, to satisfy the zero-potential condition at $x = a$ and $y = b$, the corresponding separation constants must be $m\pi/a$ and $n\pi/b$, respectively. Then from Eq. 7.9(14) we get $k_z = [(m\pi/a)^2 + (n\pi/b)^2]^{1/2}$. The general form of the potential must be a doubly infinite sum of the resulting functions:

$$\Phi = \sum_n \sum_m C_{mn} \sin \frac{m\pi}{a} x \sin \frac{n\pi}{b} y \sinh \sqrt{\left(\frac{m\pi}{a}\right)^2 + \left(\frac{n\pi}{b}\right)^2}\, z \tag{14}$$

If (14) is evaluated at the boundary $z = c$, the series becomes

$$V(x, y) = \sum_n \sum_m D_{mn} \sin \frac{m\pi}{a} x \sin \frac{n\pi}{b} y \tag{15}$$

where

$$D_{mn} = C_{mn} \sinh \sqrt{\left(\frac{m\pi}{a}\right)^2 + \left(\frac{n\pi}{b}\right)^2} \, c \tag{16}$$

The coefficients D_{mn} can be evaluated by multiplying (15) by $\sin(p\pi x/a) \sin(q\pi y/b)$ and integrating over x from 0 to a and over y from 0 to b. Application of the orthogonality conditions Eqs. 7.11(5) and 7.11(7) yields

$$D_{mn} = \frac{4}{ab} \int_0^b \int_0^a V(x, y) \sin \frac{m\pi}{a} x \sin \frac{n\pi}{b} y \, dx \, dy \tag{17}$$

For the special case of $V(x, y) = V_0$, (14), (16), and (17) give

$$\Phi = \sum_n \sum_m \frac{16 V_0}{nm\pi^2} \frac{\sin(m\pi/a)x \, \sin(n\pi/b)y \, \sinh\sqrt{(m\pi/a)^2 + (n\pi/b)^2} \, z}{\sinh\sqrt{(m\pi/a)^2 + (n\pi/b)^2} \, c} \tag{18}$$

where m and n are odd in the summation.

7.13   Cylindrical Harmonics for Static Fields

In a large class of problems of major interest, the field distribution is desired for regions with boundaries lying along the surfaces of a cylindrical coordinate system. Examples are the familiar electrostatic electron lenses found in many cathode-ray tubes or certain coaxial transmission line problems for which static solutions are useful. As has been pointed out in Sec. 7.12, the ability to evaluate the constants in product solutions depends upon having boundaries on coordinate surfaces. Therefore, the fields for this type of problem are found by separating variables in cylindrical coordinates.

A variety of types of solution are found, depending upon symmetries assumed. In general, Laplace's equation in cylindrical coordinates has the form

$$\frac{1}{r} \frac{\partial}{\partial r}\left(r \frac{\partial \Phi}{\partial r}\right) + \frac{1}{r^2} \frac{\partial^2 \Phi}{\partial \phi^2} + \frac{\partial^2 \Phi}{\partial z^2} = 0 \tag{1}$$

Axial Symmetry with Longitudinal Invariance   In Ex. 1.8a we saw that for this case

$$\Phi(r) = C_1 \ln r + C_2 \tag{2}$$

Longitudinal Invariance   It was shown in Prob. 7.9b that the solutions for this case, called circular harmonics, are given by

$$\Phi(r, \phi) = (C_1 r^n + C_2 r^{-n})(C_3 \cos n\phi + C_4 \sin n\phi) \tag{3}$$

Note that, for $n = 0$, axial symmetry exists but (3) breaks down and the solution is given by (2).

Axial Symmetry.   Since it is assumed that there are no variations with ϕ, Laplace's equation (1) becomes

$$\frac{\partial^2 \Phi}{\partial r^2} + \frac{1}{r}\frac{\partial \Phi}{\partial r} + \frac{\partial^2 \Phi}{\partial z^2} = 0 \tag{4}$$

To solve this equation, let us try to find solutions of the product form

$$\Phi(r, z) = R(r)Z(z) \tag{5}$$

Substituting in the differential equation (4), we have

$$R''Z + \frac{1}{r}R'Z + RZ'' = 0$$

where R'' denotes d^2R/dr^2, Z'' denotes d^2Z/dz^2, and so on. The variables are separated by dividing by (5)

$$\frac{Z''}{Z} = -\left[\frac{R''}{R} + \frac{1}{r}\frac{R'}{R}\right]$$

By the standard argument for the method of separation of variables, the left side, which is a function of z alone, and the right side, which is a function of r alone, must be equal to each other for all values of the variables r and z. Both sides must then be equal to a constant. Let this constant be T^2. Two ordinary differential equations then result as follows:

$$\frac{1}{R}\frac{d^2R}{dr^2} + \frac{1}{rR}\frac{dR}{dr} = -T^2 \tag{6}$$

$$\frac{1}{Z}\frac{d^2Z}{dz^2} = T^2 \tag{7}$$

Equation (6) can be written as

$$\frac{d^2R}{dr^2} + \frac{1}{r}\frac{dR}{dr} + T^2R = 0 \tag{8}$$

Equation (8) is the simplest form of the so-called *Bessel equation*. A sketch of the solution will be given. One method of finding a solution is to substitute a series and find the conditions on the terms of the series for it to be a valid solution of the differential equation. Thus to solve (8), the function R must be assumed to be a power series in r:

$$R = a_0 + a_1 r + a_2 r^2 + a_3 r^3 + \cdots$$

or

$$R = \sum_{p=0}^{\infty} a_p r^p \tag{9}$$

Substitution of this function in (8) shows that it is a solution if the constants are as follows:

$$a_p = a_{2m} = C_1 (-1)^m \frac{(T/2)^{2m}}{(m!)^2}$$

(C_1 is any arbitrary constant.) That is,

$$R = C_1 \sum_{m=0}^{\infty} \frac{(-1)^m (Tr/2)^{2m}}{(m!)^2} = C_1 \left[1 - \left(\frac{Tr}{2} \right)^2 + \frac{(Tr/2)^4}{(2!)^2} - \cdots \right] \tag{10}$$

is a solution to the differential equation (8). It is easy to show that (10) is convergent so that values may be calculated for any value of the argument (Tr). Such calculations have been made over a wide range of the values of the argument and the results are tabulated.

If T^2 is positive, the function defined by the series is denoted by $J_0(Tr)$ and called a Bessel function of the first kind and of zero order. This function is defined by

$$J_0(v) \triangleq 1 - \left(\frac{v}{2} \right)^2 + \frac{(v/2)^4}{(2!)^2} - \cdots = \sum_{m=0}^{\infty} \frac{(-1)^m (v/2)^{2m}}{(m!)^2} \tag{11}$$

The particular solution (10) may then be written simply as

$$R = C_1 J_0(Tr)$$

The differential equation (8) is of second order and so must have a second solution with a second arbitrary constant. (The sine and cosine constitute the two solutions for the simple harmonic motion equation.) This solution cannot be obtained by the power series method outlined above, since a general study of differential equations would show that at least one of the two independent solutions of (8) must have a singularity at $r = 0$. Once one solution is found there is a technique for obtaining a linearly independent solution for this class of equation[12] and several different forms are possible. One form for the second

---

[12] See for example, E. T. Whittaker and Watson, *A Course in Modern Analysis*, 4th ed., University Press, Cambridge, 1963, p. 369.

solution (any of which may be called Bessel functions of second kind, order zero) easily found in tables is

$$N_0(v) = \frac{2}{\pi} \ln\left(\frac{\gamma v}{2}\right) J_0(v)$$

$$-\frac{2}{\pi} \sum_{m=1}^{\infty} \frac{(-1)^m (v/2)^{2m}}{(m!)^2} \left(1 + \frac{1}{2} + \frac{1}{3} + \cdots + \frac{1}{m}\right) \tag{12}$$

The constant $\ln \gamma = 0.5772\ldots$ is Euler's constant. In general, then,

$$R = C_1 J_0(Tr) + C_2 N_0(Tr) \tag{13}$$

is the solution to (8), and the corresponding solution to (7) is

$$Z(z) = C_3 \sinh Tz + C_4 \cosh Tz \tag{14}$$

It should be noted from (12) that $N_0(Tr)$, the second solution to R, becomes infinite at $r = 0$, so it cannot be present in any problem for which $r = 0$ is included in the region over which the solution applies.

If T^2 is negative, we let $T^2 = -\tau^2$ or $T = j\tau$, where τ is real. The series (10) is still a solution and T in (10) may be replaced by $j\tau$. Since all powers of the series are even, imaginaries disappear, and a new series is obtained which is also real and convergent. That is,

$$J_0(jv) = 1 + \left(\frac{v}{2}\right)^2 + \frac{(v/2)^4}{(2!)^2} + \frac{(v/2)^6}{(3!)^2} + \cdots \tag{15}$$

Values of $J_0(jv)$ may be calculated for various values of v from such a series; these are also tabulated in the references, and are usually denoted $I_0(v)$. Thus one solution to (8) with $T = j\tau$ is

$$R = C'_1 J_0(j\tau r) \triangleq C'_1 I_0(\tau r) \tag{16}$$

There must also be a second solution which is commonly denoted $K_0(\tau r)$, so that the general solution to (8) with $T^2 = -\tau^2$ may be written

$$R = C'_1 I_0(\tau r) + C'_2 K_0(\tau r) \tag{17}$$

The second solution K_0 becomes infinite at $r = 0$ just as does N_0, and so is not required in problems which include the axis $r = 0$ in the range over which the solution is to apply. The solution to the z equation (7) when $T^2 = -\tau^2$ is

$$Z = C'_3 \sin \tau z + C'_4 \cos \tau z \tag{18}$$

Summarizing, either of the following forms satisfies Laplace's equation in the two cylindrical coordinates r and z:

$$\Phi(r, z) = [C_1 J_0(Tr) + C_2 N_0(Tr)][C_3 \sinh Tz + C_4 \cosh Tz] \tag{19}$$

$$\Phi(r, z) = [C'_1 I_0(\tau r) + C'_2 K_0(\tau r)][C'_3 \sin \tau z + C'_4 \cos \tau z] \tag{20}$$

As was the case with the rectangular harmonics, the two forms are not really different since (19) includes (20) if T is allowed to become imaginary, but the two separate ways of writing the solution are useful, as will be demonstrated in later examples. The case with no assumed symmetries is discussed in the following section.

7.14   Bessel Functions

In Sec. 7.13 an example of a Bessel function was shown as a solution of the differential equation 7.13(8) which describes the radial variations in Laplace's equation for axially symmetric fields where a product solution is assumed. This is just one of a whole family of functions which are solutions of the general Bessel differential equation.

Bessel Functions with Real Arguments   For certain problems, as, for example, the solution for field between the two halves of a longitudinally split cylinder, it may be necessary to retain the ϕ variations in the equation. The solution may be assumed in product form again, $RF_\phi Z$, where R is a function of r alone F_ϕ of ϕ alone and Z of z alone, Z has solutions in hyperbolic functions as before, and F_ϕ may also be satisfied by sinusoids:

$$Z = C \cosh Tz + D \sinh Tz \tag{1}$$

$$F_\phi = E \cos v\phi + F \sin v\phi \tag{2}$$

The differential equation for R is then slightly different from the zero-order Bessel equation obtained previously:

$$\frac{d^2 R}{dr^2} + \frac{1}{r}\frac{dR}{dr} + \left(T^2 - \frac{v^2}{r^2}\right)R = 0 \tag{3}$$

It is apparent at once that Eq. 7.13(8) is a special case of this more general equation, that is, $v = 0$. A series solution to the general equation carried through as in Sec. 7.13 shows that the function defined by the series

$$J_v(Tr) = \sum_{m=0}^{\infty} \frac{(-1)^m (Tr/2)^{v+2m}}{m!\Gamma(v+m+1)} \tag{4}$$

is a solution to the equation.

$\Gamma(v + m + 1)$ is the gamma function of $(v + m + 1)$ and, for v integral, is equivalent to the factorial of $(v + m)$. Also for v nonintegral, values of this gamma function are tabulated. If v is an integer n,

$$J_n(Tr) = \sum_{m=0}^{\infty} \frac{(-1)^m (Tr/2)^{n+2m}}{m!(n+m)!} \tag{5}$$

It can be shown that $J_{-n} = (-1)^n J_n$. A few of these functions are plotted in Fig. 7.14a. Similarly, a second independent solution[13] to the equation is

$$N_v(Tr) = \frac{\cos v\pi J_v(Tr) - J_{-v}(Tr)}{\sin v\pi} \tag{6}$$

and $N_{-n} = (-1)^n N_n$. As may be noted in Fig. 7.14b these are infinite at the origin. A complete solution to (3) may be written

$$R = AJ_v(Tr) + BN_v(Tr) \tag{7}$$

The constant v is known as the order of the equation. J_v is then called a Bessel function of first kind, order v; N_v is a Bessel function of second kind, order v. Of most interest for this chapter are cases in which $v = n$, an integer.

It is useful to keep in mind that, in the physical problem considered here, v is the number of radians of the sinusoidal variation of the potential per radian of angle about the axis. For different applications of Bessel functions, v has other significances.

The functions $J_v(v)$ and $N_v(v)$ are tabulated in the references.[14–17] Some care should be observed in using these references, for there is a wide variation in notation for the second solution, and not all the functions used are equivalent, since they differ in the values of arbitrary constants selected for the series. The $N_v(v)$ is chosen here because it is the form most common in current mathematical physics, and also the form most commonly tabulated. It is equivalent to the $Y_v(v)$ used by Watson and by McLachlan. Of course, it is quite proper to use any one of the second solutions throughout a given problem, since all the differences will be absorbed in the arbitrary constants of the problem, and the same final numerical result will be obtained; but it is necessary to be consistent in the use of only one of these throughout any given analysis.

It is of interest to observe the similarity between (3) and the simple harmonic equation, the solutions of which are sinusoids. The difference between these two differential equations lies in the term $(1/r)(dR/dr)$ which produces its major effect as $r \to 0$. Note that for regions far removed from the axis as, for example, near the outer edge of Fig. 1.19a, the region bounded by surfaces of a cylindrical coordinate system approximates a cube. For these reasons, it may be expected

---

[13] If v is nonintegral, J_{-v} is not linearly related to J_v, and it is then proper to use either J_{-v} or N_v as the second solution; for v integral, N_v must be used. Equation (6) is indeterminate for v integral but is subject to evaluation by usual methods.

[14] E. Jahnke, F. Emde, and F. Lösch, *Tables of Higher Functions*, 6th ed. revised by F. Lösch, McGraw-Hill, New York, 1960.

[15] G. N. Watson, *Theory of Bessel Functions*, 2nd ed., Cambridge University Press, Macmillan, New York, 1944.

[16] N. W. McLachlan, *Bessel Functions for Engineers*, 2nd ed., Oxford Clarendon Press, New York, 1955.

[17] M. Abramowitz and I. A. Stegun (eds.), *Handbook of Mathematical Functions*, National Bureau of Standards, U.S. Government Printing Office, Washington, D.C., 1964.

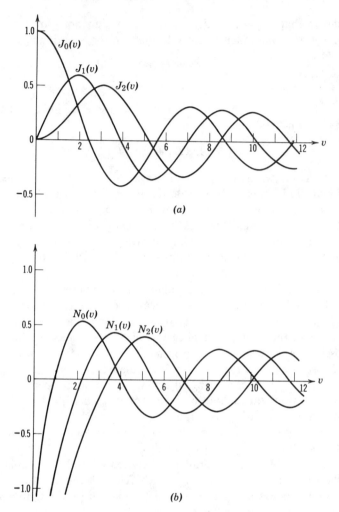

Fig. 7.14 (a) Bessel functions of the first kind. (b) Bessel functions of the second kind.

that, away from the origin, the Bessel functions are similar to sinusoids. That this is true may be seen in Figs. 7.14a and 7.14b. For large values of the arguments, the Bessel functions approach sinusoids with magnitude decreasing as the square root of radius, as will be seen in the asymptotic forms, Eqs. 7.15(1) and 7.15(2).

Hankel Functions   It is sometimes convenient to take solutions to the simple harmonic equation in the form of complex exponentials rather than sinusoids. That is, the solution of

$$\frac{d^2Z}{dz^2} + K^2Z = 0 \tag{8}$$

can be written as

$$Z = Ae^{+jKz} + Be^{-jKz} \tag{9}$$

where

$$e^{\pm jKz} = \cos Kz \pm j \sin Kz \tag{10}$$

Since the complex exponentials are linear combinations of cosine and sine functions, we may also write the general solution of (8) as

$$Z = A'e^{jKz} + B' \sin Kz$$

or other combinations.

Similarly, it is convenient to define new Bessel functions which are linear combinations of the $J_\nu(Tr)$ and $N_\nu(Tr)$ functions. By direct analogy with the definition (10) of the complex exponential, we write

$$H_\nu^{(1)}(Tr) = J_\nu(Tr) + jN_\nu(Tr) \tag{11}$$

$$H_\nu^{(2)}(Tr) = J_\nu(Tr) - jN_\nu(Tr) \tag{12}$$

These are called Hankel functions of the first and second kinds, respectively. Since they both contain the function $N_\nu(Tr)$, they are both singular at $r = 0$. Negative and positive orders are related by

$$H_{-\nu}^{(1)}(Tr) = e^{j\pi\nu}H_\nu^{(1)}(Tr)$$

$$H_{-\nu}^{(2)}(Tr) = e^{-j\pi\nu}H_\nu^{(2)}(Tr)$$

For large values of the argument, these can be approximated by complex exponentials with magnitude decreasing as square root of radius. For example,

$$H_\nu^{(1)}(Tr) \underset{Tr \to \infty}{=} \sqrt{\frac{2}{\pi Tr}}\, e^{j(Tr - \pi/4 - \nu\pi/2)}$$

This asymptotic form suggests that Hankel functions may be useful in wave propagation problems as the complex exponential is in plane wave propagation. It is also sometimes convenient to use Hankel functions as alternate independent solutions in static problems. Complete solutions of (3) may be written in a variety of ways using combinations of Bessel and Hankel functions.

Bessel and Hankel Functions of Imaginary Arguments   If T is imaginary, $T = j\tau$ (3) becomes

$$\frac{d^2R}{dr^2} + \frac{1}{r}\frac{dR}{dr} - \left(\tau^2 + \frac{\nu^2}{r^2}\right)R = 0 \tag{13}$$

The solution to (3) is valid here if T is replaced by $j\tau$ in the definitions of $J_\nu(Tr)$ and $N_\nu(Tr)$. In this case $N_\nu(j\tau r)$ is complex and so requires two numbers for each value of the argument whereas $j^{-\nu}J_\nu(j\tau r)$ is always a purely real number. It is convenient to replace $N_\nu(j\tau r)$ by a Hankel function. The quantity $j^{\nu-1}H_\nu^{(1)}(j\tau r)$ is also purely real and so requires tabulation of only one value for each value of the argument. If v is not an integer, $j^v J_{-v}(j\tau r)$ is independent of $j^{-v} J_v(j\tau r)$ and may be used as a second solution. Thus, for nonintegral v two possible complete solutions are

$$R = A_2 J_v(j\tau r) + B_2 J_{-v}(j\tau r) \tag{14}$$

and

$$R = A_3 J_v(j\tau r) + B_3 H_v^{(1)}(j\tau r) \tag{15}$$

where powers of j are included in the constants. For $v = n$, an integer, the two solutions in (14) are not independent but (15) is still a valid solution.

It is common practice to denote these solutions as

$$I_{\pm v}(v) = j^{\mp v} J_{\pm v}(jv) \tag{16}$$

$$K_v(v) = \frac{\pi}{2} j^{v+1} H_v^{(1)}(jv) \tag{17}$$

where $v = \tau r$.

As is noted in Sec. 7.15 some of the formulas relating Bessel functions and Hankel functions must be changed for these modified Bessel functions. Special cases of these functions were seen as $I_0(\tau r)$ and $K_0(\tau r)$ in Sec. 7.13 for the axially symmetric field. The forms of $I_v(\tau r)$ and $K_v(\tau r)$ for $v = 0, 1$ are shown in Fig. 7.14c. As is suggested by these curves, the asymptotic forms of the modified Bessel functions are related to growing and decaying real exponentials, as will be seen in Eqs. 7.15(5) and 7.15(6). It is also clear from the figure that $K_v(\tau r)$ is singular at the origin.

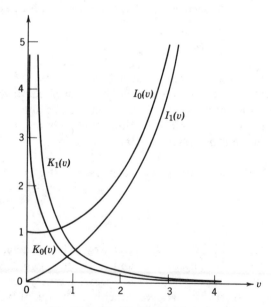

Fig. 7.14c Modified Bessel functions.

7.15　Bessel Function Zeros and Formulas[18]

The first several zeros of the low-order Bessel functions and of the derivatives of Bessel functions are given in Tables 7.15a and 7.15b, respectively.

Table 7.15a
Zeros of Bessel functions

J_0	J_1	J_2	N_0	N_1	N_2
2.405	3.832	5.136	0.894	2.197	3.384
5.520	7.016	8.417	3.958	5.430	6.794
8.654	10.173	11.620	7.086	8.596	10.023

Table 7.15b
Zeros of derivatives of Bessel functions

J_0'	J_1'	J_2'	N_0'	N_1'	N_2'
0.000	1.841	3.054	2.197	3.683	5.003
3.832	5.331	6.706	5.430	6.942	8.351
10.173	8.536	9.969	8.596	10.123	11.574

Asymptotic Forms

$$J_v(v) \underset{v\to\infty}{\to} \sqrt{\frac{2}{\pi v}} \cos\left(v - \frac{\pi}{4} - \frac{v\pi}{2}\right) \tag{1}$$

$$N_v(v) \underset{v\to\infty}{\to} \sqrt{\frac{2}{\pi v}} \sin\left(v - \frac{\pi}{4} - \frac{v\pi}{2}\right) \tag{2}$$

$$H_v^{(1)}(v) \underset{v\to\infty}{\to} \sqrt{\frac{2}{\pi v}} e^{j[v - (\pi/4) - (v\pi/2)]} \tag{3}$$

$$H_v^{(2)}(v) \underset{v\to\infty}{\to} \sqrt{\frac{2}{\pi v}} e^{-j[v - (\pi/4) - (v\pi/2)]} \tag{4}$$

$$j^{-v}J_v(jv) = I_v(v) \underset{v\to\infty}{\to} \sqrt{\frac{1}{2\pi v}} e^v \tag{5}$$

$$j^{v+1}H_v^{(1)}(jv) = \frac{2}{\pi} K_v(v) \underset{v\to\infty}{\to} \sqrt{\frac{2}{\pi v}} e^{-v} \tag{6}$$

[18] More extensive tabulations are found in Refs. 14–17.

Derivatives   The following formulas which may be found by differentiating the appropriate series, term by term, are valid for any of the functions $J_v(v)$, $N_v(v)$, $H_v^{(1)}(v)$, $H_v^{(2)}(v)$. Let $R_v(v)$ denote any one of these, and R_v' denote $(d/dv)[R_v(v)]$.

$$R_0' = -R_1(v) \tag{7}$$

$$R_1'(v) = R_0(v) - \frac{1}{v} R_1(v) \tag{8}$$

$$vR_v'(v) = vR_v(v) - vR_{v+1}(v) \tag{9}$$

$$vR_v'(v) = -vR_v(v) + vR_{v-1}(v) \tag{10}$$

$$\frac{d}{dv}[v^{-v}R_v(v)] = -v^{-v}R_{v+1}(v) \tag{11}$$

$$\frac{d}{dv}[v^v R_v(v)] = v^v R_{v-1}(v) \tag{12}$$

Note that

$$R_v'(Tr) = \frac{d}{d(Tr)}[R_v(Tr)] = \frac{1}{T}\frac{d}{dr}[R_v(Tr)] \tag{13}$$

For the I and K functions, different forms for the foregoing derivatives must be used. They may be obtained from these formulas by substituting Eqs. 7.14(16) and 7.14(17) in the preceding expressions. Some of these are

$$vI_v'(v) = vI_v(v) + vI_{v+1}(v)$$
$$vI_v'(v) = -vI_v(v) + vI_{v-1}(v) \tag{14}$$

$$vK_v'(v) = vK_v(v) - vK_{v+1}(v)$$
$$vK_v'(v) = -vK_v(v) - vK_{v-1}(v) \tag{15}$$

Recurrence formulas   By recurrence formulas, it is possible to obtain the values for Bessel functions of any order, when the values of functions for any two other orders, differing from the first by integers, are known. For example, subtract (10) from (9). The result may be written

$$\frac{2v}{v}R_v(v) = R_{v+1}(v) + R_{v-1}(v) \tag{16}$$

As before, R_v may denote J_v, N_v, $H_v^{(1)}$, $H_v^{(2)}$, but not I_v or K_v. For these, the recurrence formulas are

$$\frac{2v}{v}I_v(v) = I_{v-1}(v) - I_{v+1}(v) \tag{17}$$

$$\frac{2v}{v}K_v(v) = K_{v+1}(v) - K_{v-1}(v) \tag{18}$$

Integrals   Integrals that will be useful in solving later problems are given below. R_ν denotes J_ν, N_ν, $H_\nu^{(1)}$, $H_\nu^{(2)}$:

$$\int v^{-\nu} R_{\nu+1}(v)\, dv = -v^{-\nu} R_\nu(v) \tag{19}$$

$$\int v^\nu R_{\nu-1}(v)\, dv = v^\nu R_\nu(v) \tag{20}$$

$$\int v R_\nu(\alpha v) R_\nu(\beta v)\, dv = \frac{v}{\alpha^2 - \beta^2}$$
$$\times [\beta R_\nu(\alpha v) R_{\nu-1}(\beta v) - \alpha R_{\nu-1}(\alpha v) R_\nu(\beta v)], \qquad \alpha \neq \beta \tag{21}$$

$$\int v R_\nu^2(\alpha v)\, dv = \frac{v^2}{2} [R_\nu^2(\alpha v) - R_{\nu-1}(\alpha v) R_{\nu+1}(\alpha v)]$$
$$= \frac{v^2}{2} \left[R_\nu'^2(\alpha v) + \left(1 - \frac{v^2}{\alpha^2 v^2} \right) R_\nu^2(\alpha v) \right] \tag{22}$$

7.16   Expansion of a Function as a Series of Bessel Functions

In Sec. 7.11 a study was made of the method of Fourier series by which a function may be expressed over a given region as a series of sines or cosines. It is possible to evaluate the coefficients in such a case because of the orthogonality property of sinusoids. A study of the integrals, Eqs. 7.15(21) and 7.15(22), shows that there are similar orthogonality expressions for Bessel functions. For example, these integrals may be written for zero-order Bessel functions, and, if α and β are taken as p_m/a and p_q/a, where p_m and p_q are the mth and qth roots of $J_0(v) = 0$, that is, $J_0(p_m) = 0$ and $J_0(p_q) = 0$, $p_m \neq p_q$, then Eq. 7.15(21) gives

$$\int_0^a r J_0\left(\frac{p_m r}{a} \right) J_0\left(\frac{p_q r}{a} \right) dr = 0 \tag{1}$$

So, a function $f(r)$ may be expressed as an infinite sum of zero-order Bessel functions,

$$f(r) = b_1 J_0\left(p_1 \frac{r}{a} \right) + b_2 J_0\left(p_2 \frac{r}{a} \right) + b_3 J_0\left(p_3 \frac{r}{a} \right) + \cdots$$

or

$$f(r) = \sum_{m=1}^\infty b_m J_0\left(\frac{p_m r}{a} \right) \tag{2}$$

The coefficients b_m may be evaluated in a manner similar to that used for Fourier coefficients by multiplying each term of (2) by $rJ_0(p_m r/a)$ and integrating from 0 to a. Then by (1) all terms on the right disappear except the mth term:

$$\int_0^a rf(r)J_0\left(\frac{p_m r}{a}\right) dr = \int_0^a b_m r\left[J_0\left(\frac{p_m r}{a}\right)\right]^2 dr$$

From Eq. 7.15(22),

$$\int_0^a b_m rJ_0^2\left(\frac{p_m r}{a}\right) dr = \frac{a^2}{2} b_m J_1^2(p_m) \tag{3}$$

or

$$b_m = \frac{2}{a^2 J_1^2(p_m)} \int_0^a rf(r)J_0\left(\frac{p_m r}{a}\right) dr \tag{4}$$

In the above, as in the Fourier series, the orthogonality relations enabled us to obtain coefficients of the series under the assumption that the series is a proper representation of the function to be expanded, but two additional points are required to show that the representation is valid. The series must of course converge, and the set of orthogonal functions must be *complete*—that is, sufficient to represent an arbitrary function over the interval of concern. These points have been shown for the Bessel series of (2), and for other orthogonal sets of functions to be used in this text.[19]

Expansions similar to (2) can be made with Bessel functions of other orders and types (Prob. 7.16a).

---------- Example 7.16 ----------

Bessel Function Expansion for Constant in Range $0 < r < a$

If the function $f(r)$ in (4) is a constant V_0 in the range $0 < r < a$, we have

$$b_m = \frac{2V_0}{a^2 J_1^2(p_m)} \int_0^a rJ_0\left(\frac{p_m r}{a}\right) dr \tag{5}$$

Using Eq. 7.15(20) with $R = J$, $v = 1$, and $v = p_m r/a$, the integral in (5) becomes

$$\left(\frac{a}{p_m}\right)^2 \int_0^a \left(\frac{p_m r}{a}\right)J_0\left(\frac{p_m r}{a}\right)d\left(\frac{p_m r}{a}\right) = \left[\left(\frac{a}{p_m}\right)^2\left(\frac{p_m r}{a}\right)J_1\left(\frac{p_m r}{a}\right)\right]_0^a$$

$$= \frac{a^2}{p_m} J_1(p_m) \tag{6}$$

---

[19] See, for example, Ref. 12, pp. 374–378.

and the series expansion (2) for the constant V_0 is

$$f(r) = \sum_{m=1}^{\infty} \frac{2V_0}{p_m J_1(p_m)} J_0\left(\frac{p_m r}{a}\right) \tag{7}$$

or, using the values of the zeros of J_0 in Table 7.15a,

$$f(r) = \frac{0.832V_0}{J_1(2.405)} J_0\left(\frac{2.405r}{a}\right) + \frac{0.362V_0}{J_1(5.520)} J_0\left(\frac{5.520r}{a}\right)$$

$$+ \frac{0.231V_0}{J_1(8.654)} J_0\left(\frac{8.654r}{a}\right) + \cdots \tag{8}$$

Further evaluation of (8) requires reference to tables in Refs. 14–17 or numerical evaluation of Eq. 7.13(11).

7.17 Fields Described by Cylindrical Harmonics

We will consider here the two basic types of boundary value problems which exist in axially symmetric cylindrical systems. These can be understood by reference to Fig. 7.17a. In one type both Φ_1 and Φ_2, the potentials on the ends, are zero and a nonzero potential Φ_3 is applied to the cylindrical surface. In the second type $\Phi_3 = 0$ and either (or both) Φ_1 or Φ_2 are nonzero. The gaps between ends and side are considered negligibly small. For simplicity, the nonzero potentials will be taken to be independent of the coordinate along the surface. In the first type, a Fourier series of sinusoids is used to expand the boundary potentials as was done in the rectangular problems. In the second situation, a series of Bessel functions is used to expand the boundary potential along the radial coordinate.

Nonzero Potential on Cylindrical Surface   Since the boundary potentials are axially symmetric, zero-order Bessel functions should be used. The repeated zeros along the z coordinate dictate the use of sinusoidal functions of z. The potential in Eq. 7.13(20) is the appropriate form. Certain of the constants can be evaluated immediately. Since $K_0(\tau r)$ is singular on the axis, C_2 must be identically zero to give a finite potential there. The $\cos \tau z$ equals unity at $z = 0$ but the potential must be zero there so $C_4' = 0$. As in the problem discussed in Sec. 7.10 the repeated zeros at $z = l$ require that $\tau = m\pi/l$. Therefore the general harmonic which fits all boundary conditions except $\Phi = V_0$ at $r = a$ is

$$\Phi_m = A_m I_0\left(\frac{m\pi r}{l}\right) \sin\left(\frac{m\pi z}{l}\right) \tag{1}$$

Figure 7.17b shows a sketch of this harmonic for $m = 1$ and with the nonzero boundary potential on the cylinder. It is clear that we have here the problem of

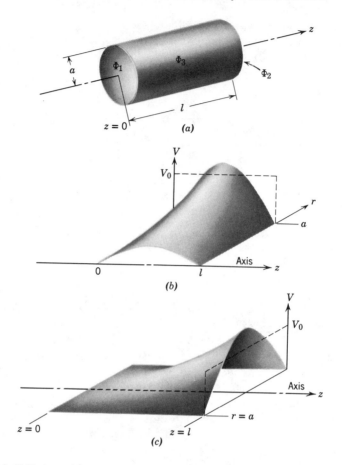

Fig. 7.17 (a) Cylinder with conducting boundaries. (b) One harmonic component for matching boundary conditions when nonzero potential is applied to cylindrical surface in (a). (c) One harmonic component for matching boundary conditions when nonzero potential is applied to end surface in (a).

expanding the boundary potential in sinusoids just as in the rectangular problem of Sec. 7.12. Following the procedure used there we obtain

$$\Phi(r, z) = \sum_{m\,\text{odd}} \frac{4V_0}{m\pi} \frac{I_0(m\pi r/l)}{I_0(m\pi a/l)} \sin \frac{m\pi z}{l} \tag{2}$$

Nonzero Potential on End In this situation if we refer to Fig. 7.17a, we see that $\Phi_1 = \Phi_3 = 0$ and $\Phi_0 = V_0$. In selecting the proper form for the solution from Sec. 7.13, the boundary condition that $\Phi = 0$ at $r = a$ for all values of z indicates that the R function must become zero at $r = a$. Thus we select the J_0 functions since the I_0's do not ever become zero. (The corresponding second solution, N_0, does not appear since potential must remain finite on the axis.) The value of T in Eq. 7.13(19) is determined from the condition that $\Phi = 0$ at $r = a$ for all values of z. Thus, if p_m is the mth root of $J_0(v) = 0$, T must be p_m/a. The corresponding solution for Z is in hyperbolic functions. The coefficient of the hyperbolic cosine term must be zero since Φ is zero at $z = 0$ for all values of r. Thus a sum of all cylindrical harmonics with arbitrary amplitudes which satisfy the symmetry of the problem and the boundary conditions so far imposed may be written

$$\Phi = \sum_{m=1}^{\infty} B_m J_0\left(\frac{p_m r}{a}\right) \sinh\left(\frac{p_m z}{a}\right) \tag{3}$$

One of the harmonics and the required boundary potentials are shown in Fig. 7.17c.

The remaining condition is that, at $z = l$, $\Phi = 0$ at $r = a$, and $\Phi = V_0$ for $r < a$. Here we can use the general technique of expanding the boundary potential in a series of the same form as that used for the potentials inside the region, as regards the dependence on the coordinate along the boundary. In Ex. 7.16 we expanded a constant over the range $0 < r < a$ in J_0 functions so that result can be used here to evaluate the constants in (3). Evaluating (3) at the boundary $z = l$, we have

$$\Phi\bigg|_{z=l} = \sum_{m=1}^{\infty} B_m \sinh\left(\frac{p_m l}{a}\right) J_0\left(\frac{p_m r}{a}\right) \tag{4}$$

Equations (4) and 7.16(7) must be equivalent for all values of r. Consequently, coefficients of corresponding terms of $J_0(p_m r/a)$ must be equal. The constant B_m is now completely determined, and the potential at any point inside the region is

$$\Phi = \sum_{m=1}^{\infty} \frac{2V_0}{p_m J_1(p_m) \sinh(p_m l/a)} \sinh\left(\frac{p_m z}{a}\right) J_0\left(\frac{p_m r}{a}\right) \tag{5}$$

7.18   Spherical Harmonics

Consider next Laplace's equation in spherical coordinates for regions with symmetry about the axis so that variations with azimuthal angle ϕ may be neglected. Laplace's equation in the two remaining spherical coordinates r and θ then becomes (obtainable from form of inside back cover)

$$\frac{\partial^2(r\Phi)}{\partial r^2} + \frac{1}{r \sin \theta} \frac{\partial}{\partial \theta}\left(\sin \theta \frac{\partial \Phi}{\partial \theta}\right) = 0 \tag{1}$$

or

$$r \frac{\partial^2 \Phi}{\partial r^2} + 2\frac{\partial \Phi}{\partial r} + \frac{1}{r}\frac{\partial^2 \Phi}{\partial \theta^2} + \frac{1}{r \tan \theta}\frac{\partial \Phi}{\partial \theta} = 0 \tag{2}$$

Assume a product solution,

$$\Phi = R\Theta$$

where R is a function of r alone, Θ of θ alone,

$$rR''\Theta + 2R'\Theta + \frac{1}{r}R\Theta'' + \frac{1}{r\tan\theta}R\Theta' = 0$$

and

$$\frac{r^2 R''}{R} + \frac{2rR'}{R} = -\frac{\Theta''}{\Theta} - \frac{\Theta'}{\Theta\tan\theta} \tag{3}$$

From the previous logic, if the two sides of the equations are to be equal to each other for all values of r and θ, both sides can be equal only to a constant. Since the constant may be expressed in any nonrestrictive way, let it be $m(m + 1)$. The two resulting ordinary differential equations are then

$$r^2 \frac{d^2 R}{dr^2} + 2r \frac{dR}{dr} - m(m + 1)R = 0 \tag{4}$$

$$\frac{d^2\Theta}{d\theta^2} + \frac{1}{\tan\theta}\frac{d\Theta}{d\theta} + m(m + 1)\Theta = 0 \tag{5}$$

Equation (4) has a solution which is easily verified to be

$$R = C_1 r^m + C_2 r^{-(m+1)} \tag{6}$$

A solution to (5) in terms of simple functions is not obvious, so, as with the Bessel equation, a series solution may be assumed. The coefficients of this series must be determined so that the differential equation (5) is satisfied and the resulting series made to define a new function. There is one departure here from an exact analog with the Bessel functions, for it turns out that a proper selection of the arbitrary constants will make the series for the new function terminate in a finite number of terms if m is an integer. Thus, for any integer m, the polynomial defined by

$$P_m(\cos\theta) = \frac{1}{2^m m!}\left[\frac{d}{d(\cos\theta)}\right]^m (\cos^2\theta - 1)^m \tag{7}$$

is a solution to the differential equation (5). The equation is known as Legendre's equation; the solutions are called Legendre polynomials of order m. Their forms for the first few values of m are tabulated below and are shown in Fig. 7.18a. Since they are polynomials and not infinite series, their values can be calculated

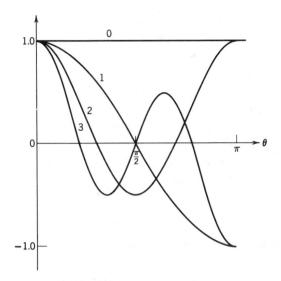

Fig. 7.18a Legendre polynomials.

easily if desired, but values of the polynomials are also tabulated in many references.

$$P_0(\cos \theta) = 1$$

$$P_1(\cos \theta) = \cos \theta$$

$$P_2(\cos \theta) = \tfrac{1}{2}(3 \cos^2 \theta - 1)$$

$$P_3(\cos \theta) = \tfrac{1}{2}(5 \cos^3 \theta - 3 \cos \theta) \tag{8}$$

$$P_4(\cos \theta) = \tfrac{1}{8}(35 \cos^4 \theta - 30 \cos^2 \theta + 3)$$

$$P_5(\cos \theta) = \tfrac{1}{8}(63 \cos^5 \theta - 70 \cos^3 \theta + 15 \cos \theta)$$

It is recognized that $\Theta = C_1 P_m(\cos \theta)$ is only one solution to the second-order differential equation (5). There must be a second independent solution, which may be obtained from the first in the same manner as for Bessel functions, but it turns out that this solution becomes infinite for $\theta = 0$. Consequently it is not needed when the axis of spherical coordinates is included in the region over which the solution applies. When the axis is excluded, it must be included and the references[20] should be consulted.

---

[20] W. R. Smythe, *Static and Dynamic Electricity*, 3rd ed., McGraw-Hill, New York, 1969, and E. W. Hobson, *Spherical and Ellipsoidal Harmonics*, Cambridge, 1931.

An orthogonality relation for Legendre polynomials is quite similar to those for sinusoids and Bessel functions which led to the Fourier series and expansion in Bessel functions, respectively.

$$\int_0^\pi P_m(\cos\theta)P_n(\cos\theta)\sin\theta\,d\theta = 0, \qquad m \neq n \tag{9}$$

$$\int_0^\pi [P_m(\cos\theta)]^2 \sin\theta\,d\theta = \frac{2}{2m+1} \tag{10}$$

It follows that, if a function $f(\theta)$ defined between the limits of 0 to π is written as a series of Legendre polynomials,

$$f(\theta) = \sum_{m=0}^{\infty} \alpha_m P_m(\cos\theta), \qquad 0 < \theta < \pi \tag{11}$$

the coefficients must be given by the formula

$$\alpha_m = \frac{2m+1}{2} \int_0^\pi f(\theta)P_m(\cos\theta)\sin\theta\,d\theta \tag{12}$$

Example 7.18a
High-Permeability Sphere in Uniform Field

We will examine the field distribution in and around a sphere of permeability $\mu \neq \mu_0$ when it is placed in an otherwise uniform magnetic field in free space. The uniform field is disturbed by the sphere as indicated in Fig. 7.18b. The reason for choosing this example is threefold. It shows, first of all, an application of spherical harmonics. Second, it is an example of a situation in which the

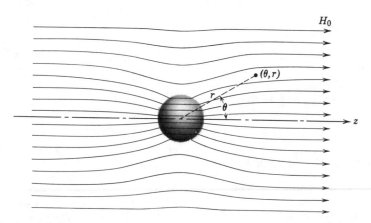

Fig. 7.18b Sphere of magnetic material in an otherwise uniform magnetic field.

constants in series solutions for two regions are evaluated by matching across a boundary. Finally, it is an example of a magnetic boundary-value problem.

Since there are no currents in the region to be studied, we may use the scalar magnetic potential introduced in Sec. 2.13. The magnetic intensity is given by

$$\mathbf{H} = -\nabla\Phi_m \tag{13}$$

As the problem is axially symmetric and the axis is included in the region of interest, the solutions $P_m(\cos\theta)$ are applicable. The series solutions with these restrictions are

$$\Phi_m(r, \theta) = \sum_m P_m(\cos\theta)[C_{1m}r^m + C_{2m}r^{-(m+1)}] \tag{14}$$

The procedure is to write general forms for the potential inside and outside the sphere and match these across the boundary. Since the potential must remain finite at $r = 0$, the coefficients of the negative powers of r must vanish for the interior. The series becomes, for the inside region,

$$\Phi_m = \sum_m A_m r^m P_m(\cos\theta) \tag{15}$$

Outside, the potential must be such that it gives a uniform magnetic field H_0 at infinity. The potential form which satisfies this condition is

$$\Phi_m = -H_0 r \cos\theta \tag{16}$$

That this gives a uniform field may be seen by noting that $dz = dr\cos\theta$ so

$$H_z = -\frac{\partial\Phi_m}{\partial z} = -\frac{1}{\cos\theta}\frac{\partial\Phi_m}{\partial r} = H_0 \tag{17}$$

Terms of the series (14) having negative powers of r may be added to (16), since they all vanish at infinity. Then the form of the solution outside the sphere is

$$\Phi_m = -H_0 r \cos\theta + \sum_m B_m P_m(\cos\theta)r^{-(m+1)} \tag{18}$$

It was pointed out in Sec. 2.14 that Φ_m is continuous across boundaries without surface currents. Therefore, the terms in (15) and (18) having the same form of θ dependence are equated, giving

$$
\begin{aligned}
A_0 &= B_0 a^{-1} & m &= 0 \\
A_1 a &= B_1 a^{-2} - H_0 a & m &= 1 \\
&\ \ \vdots \\
A_m a^m &= B_m a^{-(m+1)} & m &> 1
\end{aligned}
\tag{19}
$$

Furthermore, the normal flux density is continuous at the boundary so

$$\mu_0 \frac{\partial\Phi_m}{\partial r}\bigg|_{r=a+} = \mu\frac{\partial\Phi_m}{\partial r}\bigg|_{r=a-} \tag{20}$$

Substituting (15) and (18) in (20) and equating terms with the same θ dependence, we find

$$B_0 = 0 \qquad\qquad m = 0$$

$$\mu A_1 = -2\mu_0 B_1 a^{-3} - \mu_0 H_0 \qquad m = 1$$

$$\vdots \tag{21}$$

$$\mu m A_m a^{m-1} = -\mu_0(m+1)B_m a^{-(m+2)} \qquad m > 1$$

From (19) and (21) we see that $A_0 = B_0 = 0$, and that for $m > 1$, all coefficients must be zero to satisfy the two sets of conditions. The only remaining terms are those with $m = 1$. These two equations may be solved to give A_1 and B_1 in terms of H_0. Substituting the results in (18) gives for $r > a$

$$\Phi_m = \left[\left(\frac{\mu - \mu_0}{2\mu_0 + \mu}\right)\frac{a^3}{r^3} - 1\right]H_0 r \cos\theta \tag{22}$$

from which H can be found by using (13) for $r > a$. Substitution of A_1 into (15) gives for $r < a$

$$\Phi_m = -\left(\frac{3\mu_0}{2\mu_0 + \mu}\right)H_0 r \cos\theta \tag{23}$$

Applying (13), we find the field inside to be

$$\mathbf{H} = \hat{z}\left(\frac{3\mu_0}{2\mu_0 + \mu}\right)H_0 \tag{24}$$

It is of interest to observe that the field inside the homogeneous sphere is uniform. Finally, multiplication of (24) by μ gives the flux density

$$\mathbf{B} = \hat{z}\left(\frac{3\mu_0}{2(\mu_0/\mu) + 1}\right)H_0 \tag{25}$$

From (25) we see that for $\mu \gg \mu_0$ the maximum possible value of the flux density is

$$\mathbf{B} = \hat{z}3\mu_0 H_0 \tag{26}$$

_____ **Example 7.18b** _____

Expansion in Spherical Harmonics When Field is Given Along an Axis

It is often relatively simple to obtain the field or potential along an axis of symmetry by direct application of fundamental laws, yet difficult to obtain it at any point off this axis by the same technique. Once field is found along an axis of symmetry, expansions in spherical harmonics give its value at any other point.

Suppose potential, or any component of field which satisfies Laplace's equation, is given for every point along an axis in such a form that it may be expanded in a power series in z, the distance along this axis,

$$\Phi\Big|_{\text{axis}} = \sum_{m=0}^{\infty} b_m z^m, \qquad 0 < z < a \tag{27}$$

If this axis is taken as the axis of spherical coordinates, $\theta = 0$, the potential off the axis may be written

$$\Phi(r, \theta) = \sum_{m=0}^{\infty} b_m r^m P_m(\cos \theta) \tag{28}$$

This is true since it is a solution of Laplace's equation and does reduce to the given potential (27) for $\theta = 0$ where all $P_m(\cos \theta)$ are unity.

If potential is desired outside of this region, the potential along the axis must be expanded in a power series good for $a < z < \infty$.

$$\Phi\Big|_{\theta=0} = \sum_{m=1}^{\infty} c_m z^{-(m+1)}, \qquad z > a \tag{29}$$

Then Φ at any point outside is given by comparison with the second series of (14):

$$\Phi = \sum_{m=0}^{\infty} c_m P_m(\cos \theta) r^{-(m+1)}, \qquad r > a \tag{30}$$

For example, the magnetic field H_z was found along the axis of a circular loop of wire carrying current I in Sec. 2.3 as

$$H_z = \frac{a^2 I}{2(a^2 + z^2)^{3/2}} = \frac{I}{2a[1 + (z^2/a^2)]^{3/2}} \tag{31}$$

The binomial expansion

$$(1 + u)^{-3/2} = 1 - \frac{3}{2}u + \frac{15}{8}u^2 - \frac{105}{48}u^3 + \cdots$$

is good for $0 < |u| < 1$. Applied to (31), this gives

$$H_z\Big|_{\text{axis}} = \frac{I}{2a}\left[1 - \frac{3}{2}\left(\frac{z^2}{a^2}\right) + \frac{15}{8}\left(\frac{z^2}{a^2}\right)^2 - \frac{105}{48}\left(\frac{z^2}{a^2}\right)^3 + \cdots\right]$$

Since H_z, axial component of magnetic field, satisfies Laplace's equation (Sec. 7.2), H_z at any point r, θ with $r < a$ is given by

$$H_z(r, \theta) = \frac{I}{2a}\left[1 - \frac{3}{2}\left(\frac{r^2}{a^2}\right)P_2(\cos \theta) + \frac{15}{8}\left(\frac{r^4}{a^4}\right)P_4(\cos \theta) + \cdots\right] \tag{32}$$

7.19 Product Solutions for the Helmholtz Equation in Rectangular Coordinates

The technique used in the preceding sections for finding product solutions to Laplace's equation will be applied here to the scalar Helmholtz equation. Whereas the single-product solution for static problems was seen in Sec. 7.10 to be of little value, such solutions will be seen in the next chapter to be of great importance as waveguide propagation modes and will be analyzed extensively there.

Let us consider the scalar Helmholtz equation. Here we make the assumption that the dependent variable depends on z in the manner of a wave, as $e^{-\gamma z}$. The variable ψ remaining in the equation is, therefore, the coefficient of $e^{(j\omega t - \gamma z)}$. Written with the Laplacian explicitly in rectangular coordinates, we have

$$\frac{\partial^2 \psi}{\partial x^2} + \frac{\partial^2 \psi}{\partial y^2} = -k_c^2 \psi \tag{1}$$

where $k_c^2 = \gamma^2 + \omega^2 \mu \varepsilon$. Let us assume that the solution can be written as the product solution $\psi = X(x)Y(y)$. Substituting this form in (1),

$$X''Y + XY'' = -k_c^2 XY$$

or

$$\frac{X''}{X} + \frac{Y''}{Y} = -k_c^2 \tag{2}$$

The primes indicate derivatives. If this equation is to hold for all values of x and y, since x and y may be changed independently of each other, each of the ratios X''/X and Y''/Y can be only a constant. There are then several forms for the solutions, depending upon whether these ratios are both taken as negative constants, both positive, or one negative and one positive. If both are taken as negative,

$$\frac{X''}{X} = -k_x^2$$

$$\frac{Y''}{Y} = -k_y^2$$

The solutions to these ordinary differential equations are sinusoids, and by (2) the sum of k_x^2 and k_y^2 is k_c^2. Thus

$$\psi = XY \tag{3}$$

where

$$X = A \cos k_x x + B \sin k_x x$$

$$Y = C \cos k_y y + D \sin k_y y \tag{4}$$

$$k_x^2 + k_y^2 = k_c^2$$

Either or both of k_x and k_y may be imaginary in which case the corresponding sinusoid becomes a hyperbolic function. Values of the constants k_x and k_y are determined by conditions on ψ at the boundaries in the x–y plane. Examples of the application of these general forms will be seen extensively in the following chapter where the dependent variable ψ is identified as E_z or H_z.

7.20 Product Solutions for the Helmholtz Equation in Cylindrical Coordinates

In cylindrical structures, such as coaxial lines or waveguides of circular cross section, the wave components are most conveniently expressed in terms of cylindrical coordinates. Assuming that the z dependence is in the wave form $e^{-\gamma z}$, the scalar Helmholtz equation becomes

$$\frac{\partial^2 \psi}{\partial r^2} + \frac{1}{r}\frac{\partial \psi}{\partial r} + \frac{1}{r^2}\frac{\partial^2 \psi}{\partial \phi^2} = -k_c^2 \psi \tag{1}$$

where $k_c^2 = \gamma^2 + \omega^2 \mu \varepsilon$. For this partial differential equation, we shall again substitute an assumed product solution and separate variables in order to obtain two ordinary differential equations.

   Assume

$$\psi = RF_\phi$$

where R is a function of r alone and F_ϕ is a function of ϕ alone.

$$R''F_\phi + \frac{R'F_\phi}{r} + \frac{F_\phi'' R}{r^2} = -k_c^2 R F_\phi$$

Separating variables, we have

$$r^2 \frac{R''}{R} + \frac{rR'}{R} + k_c^2 r^2 = \frac{-F_\phi''}{F_\phi}$$

The left side of the equation is a function of r alone; the right of ϕ alone. If both sides are to be equal for all values of r and ϕ, both sides must equal a constant. Let this constant be v^2. There are then the two ordinary differential equations:

$$\frac{-F''}{F} = v^2 \tag{2}$$

and

$$r^2 \frac{R''}{R} + \frac{rR'}{R} + k_c^2 r^2 = v^2$$

or

$$R'' + \frac{1}{r}R' + \left(k_c^2 - \frac{v^2}{r^2}\right)R = 0 \tag{3}$$

The solution to (2) is in sinusoids. By comparing with Eq. 7.14(3) we see that solutions to (3) may be written in terms of Bessel functions of order v.

$$\psi = R F_\phi \tag{4}$$

where

$$R = A J_v(k_c r) + B N_v(k_c r) \tag{5}$$

$$F_\phi = C \cos v\phi + D \sin v\phi$$

Either or both of the Bessel functions may be replaced by Hankel functions [Eqs. 7.14(11), 7.14(12)] when one desires to look at waves as though propagation were in the radial direction. Thus, for example,

$$R = A_1 H_v^{(1)}(k_c r) + B_1 H_v^{(2)}(k_c r) \tag{6}$$

$$F_\phi = C \cos v\phi + D \sin v\phi$$

If k_c is imaginary, the ordinary Bessel functions can be replaced by the modified Bessel functions, Eqs. 7.14(16) and (17). In the examples in the following chapter, the variable ψ will be identified with E_z or H_z.

PROBLEMS

7.2a Find the form of differential equation satisfied by E_r in cylindrical coordinates for a charge-free, homogeneous dielectric region. Repeat for E_ϕ.

7.2b Show that none of the spherical components of electric field satisfy Laplace's equation for quasistatic problems in which $\nabla^2 \mathbf{E} = 0$.

7.2c* Show that the rectangular component E_z of electrostatic field satisfies Laplace's equation expressed in spherical coordinates.

7.2d Derive Laplace's equation for $\mathbf{H}$, $\mathbf{A}$, and Φ_m in a current-free region with static fields and for $\mathbf{J}$ and $\mathbf{E}$ in a homogeneous conductor with dc currents.

7.2e Use superposition to find the potential on the axis of an infinite cylinder with a potential specified as $\Phi(\phi) = V_0 \sin \phi/2$, for $0 \le \phi \le 2\pi$ on the boundary.

7.2f A spherical surface is at zero potential except for a sector in the region $0 < \phi < \pi/3$, $0 < \theta < \pi/2$. Find the potential at the center of the sphere.

7.3a Subdivide the region in Fig. 1.19c into a mesh of squares of sides $a/2$. Terminate the region on the right a distance a from the corner and on the left a distance $3a/2$ from the corner. Take $b = 2a$ and $V_0 = 100$ V. Consider the potentials at the left and right edges of the above-defined grid to be fixed at the values found in linear variation from top to bottom. Start with all interior grid points at 50 V. Find the potentials at the mesh points assuming zero space charge and applying the cyclic Chebyshev method.

7.3b Set up the difference equation for a three-dimensional potential distribution in rectangular coordinates. Consider a cubical box with the following potentials on the various sides: top, 80 V; right side, 60 V; bottom, 0 V; left side, 100 V; front, 40 V; back, 100 V. Define a grid of the same coarseness as in Fig. 7.3a and assume initial potentials for all interior grid points to be the average of the boundary potentials. Calculate the first set of corrected potentials by the three-dimensional equivalent of the simple scheme used for Table 7.3.

7.3c Derive the difference equation for potential in cylindrical coordinates with axial symmetry assumed ($\partial \Phi / \partial \phi = 0$).

7.3d* An electron beam accelerated from zero potential passes normally through a pair of parallel wire grids. Model the beam as infinitely broad and without transverse variation. Set up a one-dimensional difference equation for the potential between the grids. Divide the 5 mm space between grids into five segments. Take both grid potentials to be 1000 V and the beam current to be 10^4 A/m^2. Assume 1000 V as a first guess for all difference-equation grid points. Take three steps of potential adjustment with space charge based on the first guess. Recalculate space charge based on the new potentials and again iterate the potential three times. Repeat recalculations of space charge and potentials until the latter differ by no more than three percent between recalculations of space charge. Use the simple iterative form with $\Omega = 1$.

7.3e Solve for the potentials at the grid points in the problem in Fig. 7.3a by direct inversion of the set of difference equations expressing Laplace's equation for all grid points. Compare the results with those in Table 7.3 and discuss differences. Does direct inversion give the exact values of potentials at the grid points? Explain your answer.

7.4a Check by the Cauchy–Riemann equations the analyticity of the general power term $W = C_n Z^n$, and a series of such terms,

$$W = \sum_{n=1}^{\infty} C_n Z^n$$

7.4b Check the following functions by the Cauchy–Riemann equations to determine if they are analytic:

$$W = \sin Z$$
$$W = e^Z$$
$$W = Z^* = x - jy$$
$$W = ZZ^*$$

7.4c Check the analyticity of the following, noting isolated points where the derivatives may not remain finite:

$$W = \ln Z$$
$$W = \tan Z$$

7.4d Take the change ΔZ in any general direction $\Delta x + j \Delta y$. Show that, if the Cauchy–Riemann conditions are satisfied, Eq. 7.4(3) yields the same result for the derivative as when the change is in the x direction or the y direction alone.

7.4e If by following a path around some point in the Z plane, the variable W takes on different values when the same Z is reached, the point around which the path is taken is called a *branch point*. Evaluate $W = Z^{1/2}$ and $W = Z^{4/3}$ along a path of constant radius around the origin to show that $Z = 0$ is a branch point for these functions. Discuss the analyticity of these functions at the branch point.

7.5a Plot the shape of the $u = \pm 0.5$ equipotentials for the $V = x^{4/3}$, $y = 0$ boundary condition used in Ex. 7.5.

7.5b A thin cylindrical shell of radius a has a potential described by $\Phi(a, \theta) = V_0 \cos 2\theta$. Use the method in Ex. 7.5 to find the potentials everywhere outside and inside the shell.

7.5c Consider a plane boundary of a half-space (at $y = 0$) along which the potential is independent of z and is $-V_0$ for $x \leq -a$ and V_0 for $x \geq a$, with linear variation between. Assume that the potential along the boundary can be approximated by $V_0 \tan \sqrt{2} x/a$ (approximation valid to $\approx 11\%$ of full scale). Use the method of Ex. 7.5 to find an implicit expression $f(x, y) = 0$ along a line of constant potential in the half-space.

7.5d Show that if u is the potential function, the field intensity E_y is equal to the imaginary part of dW/dZ and E_x equals the negative of the real part.

7.5e Use the results of Prob. 7.5d to find an expression for the slope of equipotential lines in terms of dW/dZ. Show that all equipotential lines except $u = 0$ are normal to the beam edge in the electron flow in Fig. 7.5b. ($W = Z^{4/3}$ is not analytic at $Z = 0$, as was shown in Prob. 7.4e, and the $W = 0$ line at $y = 0$ is a special case.) *Hint:* Write an expression for du in terms of partial derivatives and set $du = 0$ to get relations existing along an equipotential.

7.6a Plot a few equipotentials and flux lines in the vicinity of conducting corners of angles $\alpha = \pi/3$ and $3\pi/4$.

7.6b Evaluate the constant C_1 and C_2 in the logarithmic transformation so that u represents the potential function in volts about a line charge of strength q_l C/m. Let potential be zero at $r = a$.

7.6c Show that, if v is taken as the potential function in the logarithmic transformation, it is applicable to the region between two semi-infinite conducting planes intersecting at an angle α, but separated by an infinitesimal gap at the origin so that the plane at $\theta = 0$ may be placed at potential zero, and the plane at $\theta = \alpha$ at potential V_0. Evaluate the constants C_1 and C_2, taking the reference for zero flux at $r = a$. Write the flux function in coulombs per meter.

7.6d In the example of Prob. 7.6c, take the gradient of potential v to give the electric field. From this, find the electric flux density vector. Integrate this from radius a to r to give the total flux function, and compare with the result of the above problem.

7.6e Find the form of the curves of constant u and constant v for the functions $\sin^{-1} Z$, $\cosh^{-1} Z$, and $\sinh^{-1} Z$. Do these permit one to solve problems in addition to those from the function $\cos^{-1} Z$?

7.6f Apply the results of the $\cos^{-1}$ transformation to item 4 in Ex. 7.6c. Take the right-hand semi-infinite plane extending from $x = a$ to $x = \infty$ at potential V_0. Take the left-hand semi-infinite plane extending from $x = -a$ to $x = -\infty$ at potential zero. Evaluate the scale factors and additive constant.

7.6g* Apply the results of the transformation to item 2 of Ex. 7.6c. Take the elliptic cylindrical conductor of semimajor axis a and semiminor axis b at potential V_0. The inner conductor is a strip conductor extending between the foci, $x = \pm c$, where

$$c = \sqrt{a^2 - b^2}$$

Evaluate all required scale factors and constants. Find the total charge per unit length induced upon the outer cylinder, and the electrostatic capacitance of this two-conductor system.

7.6h* Modify the derivation in Ex. 7.6d to apply to the problem of parallel cylinders of unequal radius. Take the left-hand cylinder of radius R_1 with center at $x = -d_1$, the right-hand cylinder of radius R_2 with center at $x = d_2$, and a total difference of potential V_0 between cylinders. Find the electrostatic capacitance per unit length in terms of R_1, R_2, and $(d_1 + d_2)$.

7.6i Show that the lines of constant v in the transformation of Ex. 7.6d do represent a family of circles.

7.6j The important bilinear transformation is of the form

$$Z = \frac{aZ' + b}{cZ' + d}$$

Take a, b, c, and d as real constants, and show that any circle is the Z' plane is transformed to a circle in the Z plane by this transformation. (Straight lines are considered circles of infinite radius.)

7.6k Consider the special case of Prob. 7.6j with $a = R$, $b = -R$, $c = 1$, and $d = 1$. Show that the imaginary axis of the Z' plane transforms to a circle of radius R, center at the origin, in the Z plane. Show that a line charge at $x' = d$ and its image at $x' = -d$ in the Z' plane transform to points in the Z plane at radii r_1 and r_2 with $r_1 r_2 = R^2$. Compare with the result for imaging line charges in a cylinder (Sec. 1.18).

7.7a Explain why a factor in the Schwarz transformation may be left out when it corresponds to a point transformed to infinity in the Z' plane.

7.7b In equation 7.7(2), separate Z into real and imaginary parts. Show that the boundary condition for potential is satisfied along the two conductors. Obtain the asymptotic equations for large positive u and for large negative u, and interpret the results in terms of the type of field approached in these limits.

7.7c** Work the example of Prob. 7.6f by the Schwarz technique and show that the same result is obtained. This is the problem of two coplanar semi-infinite plane conductors separated by a gap $2a$, with the left-hand conductor at potential zero and the right-hand conductor at potential V_0.

7.7d* For the first example of Table 7.7, find the electrostatic capacitance in excess of what would be obtained if a uniform field existed in both of the parallel-plane regions.

7.7e* Plot the $V_0/2$ equipotential for Ex. 7.7.

7.8 Suppose that the wave-guiding structure in Fig. 7.8a is bounded on the outside by a dielectric $\varepsilon_3(r)$ which has the value ε_2 at R_0 and then decreases to an appreciably lower value as r is increased. As was seen in Sec. 6.12, waves incident on a plane boundary between two dielectrics from the higher ε side can be totally reflected. Find the limiting rate of decrease of ε_3 at R_0 which can permit total reflection of rays approaching the boundary, by studying the variation of the equivalent dielectric constant in the W plane.

7.9a Find the basic forms [in the sense that Eq. 7.9(9) is different from 7.9(8)] of Eq. 7.9(15) obtained by allowing k_x, k_y, k_z and various combinations to become imaginary.

7.9b The so-called circular harmonics are the product solutions to Laplace's equation in the two circular cylindrical coordinates r and ϕ. Apply the basic separation of variables technique to Laplace's equation in these coordinates to yield two ordinary differential equations. Show that the r and ϕ equations are satisfied respectively by the functions R and F_ϕ where

$$R = C_1 r^n + C_2 r^{-n}$$

$$F_\phi = C_3 \cos n\phi + C_4 \sin n\phi$$

7.9c An infinite rod of a magnetic material of relative permeability μ_r lies with its axis perpendicular to the direction of a uniform magnetic field in which it is immersed. Take the rod to be of circular cross section with radius a and use the expressions in Prob. 7.9b to find the fields inside and outside the rod. Note the uniformity of the field inside.

7.10a Plot the form of equipotentials for $\Phi = V_0/4$, $V_0/2$, and $3V_0/4$ for Fig. 7.10a.

7.10b Describe the electrode structure for which the single rectangular harmonic $C_1 \cosh kx \sin ky$ is a solution for potential. Take electrodes at potential V_0 passing through $|x| = a$ when $y = a/2$.

7.10c Describe the electrode structure and exciting potentials for which the single circular harmonic (Prob. 7.9b) $Cr^2 \cos 2\phi$ is a solution.

7.11a Simplify the general expressions for Fourier coefficients found in Sec. 7.11 for:
(i) Even functions of x.
(ii) Odd functions of x.

7.11b Obtain Fourier series in sines and cosines for the following periodic functions:
(i) A triangular wave defined by $f(x) = V_0(1 - 2x/L)$ from 0 to $L/2$, and $V_0[(2x/L) - 1]$ from $L/2$ to L.
(ii) A sawtooth wave defined by $f(x) = V_0 x/L$ for $0 < x < L$.
(iii) A sinusoidal pulse given by $f(x) = (V_m \cos kx - V_0)$ for $-\alpha < kx < \alpha$, $f(x) = 0$ for $-\pi < kx < -\alpha$ and also for $\alpha < kx < \pi$.

7.11c Suppose that a function is given over the interval 0 to a as $f(x) = \sin \pi x/a$. What do the cosine and sine representations yield? Explain how this single sine term can be represented in terms of cosines.

7.11d Find sine and cosine representations for the function e^{kx} defined over the interval $0 < x < a$.

7.11e* Plot $f(x)$ given by Eq. 7.11(14) in the neighborhood of the discontinuities using (i) five sine terms and (ii) ten sine terms and discuss differences from the rectangular function being represented.

7.11f A complex form of the Fourier series for a function $f(x)$ defined over the interval $0 < x < a$ is

$$f(x) = \sum_{n=-\infty}^{\infty} c_n e^{j2\pi n x/a}$$

Show that if this is valid, c_n must be given by

$$c_n = \frac{1}{a} \int_0^a f(x) e^{-j2\pi n x/a} \, dx$$

7.11g Find representations for the constant C over the interval $0 < x < a$ in the complex form of Prob. 7.11f, and compare the result with Eq. 7.11(14).

7.11h Find the Fourier integral representation for a decaying exponential, $f(x) = 0$ for $x < 0$ and $f(x) = ce^{-\alpha x}$ for $x > 0$.

7.12a Obtain a series solution for the two-dimensional box problem in which sides at $y = 0$ and $y = b$ are at potential zero, and end planes at $x = a$ and $x = -a$ are at potential V_0. *Hint:* Utilize the symmetry of the problem in the evaluation of constants.

7.12b Find the potential distribution for the box of Prob. 7.12a with the same boundary conditions except that the potential on the side at $y = 0$ should be V_1 and that at $y = b$ should be $-V_1$.

7.12c In a two-dimensional problem, parallel planes at $y = 0$ and $y = b$ extend from $x = 0$ to $x = \infty$, and are at zero potential. The one end plane at $x = 0$ is at potential V_0. Obtain a series solution. *Hint:* Use the exponential form for the solution in x, and consider the condition at infinity.

7.12d The fringing that occurs at the open ends of a pair of parallel plates as seen in Fig. 1.9a leads to a modification of the fields between the plates from the ideal uniform distribution. Consider $x = 0$ to be the ends of the plates, which are at $y = 0, b$. The analysis of Ex. 7.7 can show that the potential between the ends of the plates may be expressed approximately as $\Phi(0, y) = V_0[(y/b) + 0.06 \sin 2\pi y/b]$. Find the distance x at which the potential distribution between the plates is linear to within 1%, using the analysis of Prob. 7.12c.

7.12e A two-dimensional conducting rectangular solid is bounded on three sides by perfect conductors: at $y = 0$, $\Phi = 0$; at $x = 0$, $\Phi = 0$; at $y = b$, $\Phi = V_0$. It is bounded at $x = a$ by a dielectric with zero conductivity. Find an expression for the potential distribution inside the conducting solid.

7.12f Two concentric cylinders are located at $r = a$ and $r = b$. The inner ($r = a$) cylinder is split along its length into two halves which are at different potentials. Potential is $-V_0$ for $-\pi < \phi < 0$ and V_0 for $0 < \phi < \pi$. The cylinder at $r = b$ is at zero potential. Find the potential between the two cylinders.

7.12g The potential along the plane boundary of a half-space is in strips of width a and alternates between $-V_0$ to V_0. Take the boundary to be at $y = 0$ and the strips to be invariant in the z direction. The origin of the x coordinate lies in the gap between strips so that the potential is $-V_0$ for $-a < x < 0$ and V_0 for $0 < x < a$. Find the potential distribution for $y \geq 0$ and determine the surface charge density along the $y = 0$ plane.

7.12h* Infinite parallel conducting plates are located at $y = 0$ and $y = a$. A conducting strip at $x = 0$, $a/2 \leq y \leq a$, $-\infty < z < \infty$ is connected to the plate at $y = a$, thus introducing additional capacitance between the plates. Assume a linear potential variation from $0 \leq y \leq a/2$ at $x = 0$, and use superposition of boundary conditions to find an expression for the capacitance per meter in the z direction added by the strip at $x = 0$.

7.12i Consider a rectangular prism of width a in the x direction and b in the y direction with all four sides at zero potential extending from $z = 0$ to $z = \infty$. At $z = 0$ the prism has a cap with the following potential distribution:

$$V(x, y, 0) = \begin{cases} 0 \text{ for } 0 < x < a/2 \text{ all } y \\ V_0 \text{ for } a/2 < x < a \text{ all } y \end{cases}$$

Find the potentials within the prism.

7.12j* For a box as in Ex. 7.12c, find the potential distribution if the box is filled with a homogeneous, isotropic dielectric with permittivity ε_1 in the bottom half of the box $0 \leq z \leq c/2$ and free space in the remainder.

7.13 Demonstrate that the series Eq. 7.13(10) does satisfy the differential equation 7.13(8).

7.16a Write a function $f(r)$ in terms of nth-order Bessel functions over the range 0 to a and determine the coefficients.

7.16b Determine coefficients for a function $f(r)$ expressed over the range 0 to a as a series of zero-order Bessel functions as follows:

$$f(r) = \sum_{m=1}^{\infty} c_m J_0\left(\frac{p'_m r}{a}\right)$$

where p'_m denotes the mth root of $J'_0(v) = 0$ [i.e., $J_1(v) = 0$].

7.17a A cylinder divided into a set of rings with appropriately applied voltages may be used to set up a nearly uniform electric field along the axis with advantageous focusing properties for electron beams. Suppose the field at the radius a of the cylinder is given approximately by $E_z(a, z) = E_0(1 + \cos \alpha z)$ where $\alpha = 2\pi/p$ and p is the period of the rings. Find the potential variation along the rings $(r = a)$ and for $r < a$. Determine the field on the axis and the period required to have the periodic part of the field one percent of E_0.

7.17b* Show that the function

$$\Phi(r, z) = A I_0(\tau r) \cos \tau z$$

satisfies the requirement of solutions of Laplace's equation that there should be no relative maxima or minima.

7.17c Find the series for potential inside the cylindrical region with end plates $z = 0$ and $z = l$ at potential zero, and the cylinder of radius a in two parts. From $z = 0$ to $z = l/2$, it is at potential V_0; from $z = l/2$ to $z = l$, it is at potential $-V_0$.

7.17d The problem is as in Prob. 7.17c except that the cylinder is divided in three parts with potential zero from $z = 0$ to $z = b$ and also from $z = l - b$ to $z = l$. Potential is V_0 from $z = b$ to $z = l - b$.

7.17e Write the general formula for obtaining potential inside a cylindrical region of radius a, with two zero-potential end plates at $z = 0$ and $z = l$, provided potential is given as $\Phi = f(z)$ at $r = a$.

7.17f Suppose that the end plate at $z = l$ of Fig. 7.17a is divided into insulated rings and connected to sources in such a way that the potential approximates a single J_0 function of radius. Specifically,

$$\Phi(r, l) = C J_0\left(\frac{p_1 r}{a}\right)$$

The end plate at $z = 0$ and the cylindrical wall are at zero potential. Write the solution for $\Phi(r, z)$ at any point inside the cylindrical region.

7.17g Write the general formula for obtaining potential inside a cylinder of radius a which, with its plane base at $z = 0$, is at potential zero, provided that the potential is given across the plane surface at $z = l$, as

$$\Phi(r, l) = f(r)$$

7.17h Find the potential distribution inside a cylinder with zero potential on the cylindrical surface at $r = a$, on the end plate at $z = 0$, and where $a/2 < r < a$ on the end plate at $z = l$. It also has $\Phi(r, l) = V_0$ for $0 \le r < a/2$.

7.18a Apply the separation of variables technique to Laplace's equation in the three spherical coordinates, r, θ, and ϕ, obtaining the three resulting ordinary differential equations. Write solutions to the r equation and the ϕ equation.

7.18b Assume a spherical surface split into two thin hemispherical shells with a small gap between them. Assume a potential V_0 on one hemisphere and zero on the other and find the potential distribution in the surrounding space.

7.18c Write the general formulas for obtaining potential for $r < a$ and for $r > a$, when potential is given as a general function $f(\theta)$ over a thin spherical shell at $r = a$.

7.18d For Ex. 7.18b, write the series for H_z at any point r, θ with $r > a$.

7.18e A Helmholtz coil is used to obtain very nearly uniform magnetic field over a region through the use of coils of large radius compared with coil cross sections. Consider two such coaxial coils, each of radius a, one lying in the plane $z = d$, and the other in the plane $z = -d$. Take the current for each coil (considered as a single turn) as I. Obtain the series for H_z applicable to a region containing the origin, writing specific forms for the first three coefficients. Show that, if $a = 2d$, the first nonzero coefficient (other than the constant term) is the coefficient of r^4.

7.19* In Eqs. 7.19(3) and 7.19(4), let ψ be the axial electric field component E_z, and simplify by taking A and C zero in (4). Discuss the forms of solutions, and the question of finding physical boundary conditions for (i) both k_x and k_y real; (ii) k_x real but k_y imaginary; (iii) both k_x and k_y imaginary. For (ii) and (iii) would physical applicability of solutions be changed if either or both of A and C were nonzero?

8

WAVEGUIDES WITH CYLINDRICAL CONDUCTING BOUNDARIES

8.1 Introduction

A waveguide is a structure, or part of a structure, that causes a wave to propagate in a chosen direction with some measure of confinement in the planes traverse to the direction of propagation. If the waveguide boundaries change direction, within reasonable limits, the wave is constrained to follow it. For example, in a transmission line used to transfer energy from a transmitter to an antenna, the energy follows the path of the line, at least for paths with only small discontinuities. The guiding of the waves in all such systems is accomplished by an intimate connection between the fields of the wave and the currents and charges on the boundaries, or by some condition of reflection at the boundary.

In this chapter we concentrate on cylindrical structures with conducting boundaries. Multiconductor lines can be used for frequencies from dc up to the millimeter-wave range. At the highest frequencies, they are often in the form of metallic films on insulating substrates. Hollow conducting cylinders of various cross-sectional shapes are used in the microwave and millimeter-wave frequency ranges (approximately 1–100 GHz).

Generally, in waveguide analyses we are interested in the distribution of the electromagnetic fields; but of greatest importance is the dependence of the propagation constant upon frequency. From the propagation constant one finds wave velocities, phase variation, and attenuation along the guide, and the pulse dispersion properties of the guide.

Several different types of guide are analyzed in this chapter, including the simple parallel-plate structure (and some more-practical, related forms) and hollow-tube guides of rectangular and circular cross section. An introduction to means for exciting waves in waveguides is presented. The chapter concludes with a study of the general properties of waves in cylindrical waveguides with conducting boundaries.

GENERAL FORMULATION FOR GUIDED WAVES

8.2 Basic Equations and Wave Types for Uniform Systems

We consider here cylindrical systems with axes taken along the z axis. We also consider time-harmonic waves with time and distance variations described by $e^{(j\omega t - \gamma z)}$, as in the study of transmission-line waves. The character of the propagation constant γ tells much about the properties of the wave, such as the degree of attenuation and the phase and group velocities. The fields in the wave must satisfy the wave equation and the boundary conditions. We will assume that there is no net charge density in the dielectric and that any conduction currents are included by allowing permittivity and therefore $k^2 = \omega^2\mu\varepsilon$ to be complex. The wave equations, which reduce to the Helmholtz equations for phasor fields, (Sec. 3.11) are

$$\nabla^2 \mathbf{E} = -k^2 \mathbf{E}; \qquad \nabla^2 \mathbf{H} = -k^2 \mathbf{H}$$

The three-dimensional ∇^2 may be broken into two parts:

$$\nabla^2 \mathbf{E} = \nabla_t^2 \mathbf{E} + \frac{\partial^2 \mathbf{E}}{\partial z^2}$$

The last term is the contribution to ∇^2 from derivatives in the axial direction. The first term is the two-dimensional Laplacian in the transverse plane, representing contributions to ∇^2 from derivatives in this plane. With the assumed propagation function $e^{-\gamma z}$ in the axial direction,

$$\frac{\partial^2 \mathbf{E}}{\partial z^2} = \gamma^2 \mathbf{E}$$

The foregoing wave equations may then be written

$$\nabla_t^2 \mathbf{E} = -(\gamma^2 + k^2)\mathbf{E} \qquad (1)$$

$$\nabla_t^2 \mathbf{H} = -(\gamma^2 + k^2)\mathbf{H} \qquad (2)$$

Equations (1) and (2) are the differential equations that must be satisfied in the dielectric regions of the transmission lines or guides. The boundary conditions imposed on fields follow from the configuration and the electrical properties of the boundaries.

The usual procedure is to find two components of the fields, usually the z components of $\mathbf{E}$ and $\mathbf{H}$, that satisfy the wave equations (1) and (2) and the boundary conditions and then the other field components can be found from these by using Maxwell's equations. To facilitate finding the other components, it usually is most convenient to have them explicitly in terms of the z components of $\mathbf{E}$ and $\mathbf{H}$.

The curl equations with the assumed functions $e^{(j\omega t - \gamma z)}$ are written below for fields in the dielectric system, assumed here to be linear, homogeneous, and isotropic.

$$\nabla \times \mathbf{E} = -j\omega\mu\mathbf{H} \qquad\qquad \nabla \times \mathbf{H} = j\omega\varepsilon\mathbf{E}$$

$$\frac{\partial E_z}{\partial y} + \gamma E_y = -j\omega\mu H_x \qquad (3) \qquad\qquad \frac{\partial H_z}{\partial y} + \gamma H_y = j\omega\varepsilon E_x \qquad (6)$$

$$-\gamma E_x - \frac{\partial E_z}{\partial x} = -j\omega\mu H_y \qquad (4) \qquad\qquad -\gamma H_x - \frac{\partial H_z}{\partial x} = j\omega\varepsilon E_y \qquad (7)$$

$$\frac{\partial E_y}{\partial x} - \frac{\partial E_x}{\partial y} = -j\omega\mu H_z \qquad (5) \qquad\qquad \frac{\partial H_y}{\partial x} - \frac{\partial H_x}{\partial y} = j\omega\varepsilon E_z \qquad (8)$$

It must be remembered in all analysis to follow that these coefficients, E_x, H_x, E_y, and so on, are functions of x and y only, by our agreement to take care of the z and time functions in the assumed $e^{(j\omega t - \gamma z)}$.

From the foregoing equations, it is possible to solve for E_x, E_y, H_x, or H_y in terms of E_z and H_z. For example, H_x is found by eliminating E_y from (3) and (7), and a similar procedure gives the other components.

$$E_x = -\frac{1}{\gamma^2 + k^2}\left(\gamma\frac{\partial E_z}{\partial x} + j\omega\mu\frac{\partial H_z}{\partial y}\right) \qquad (9)$$

$$E_y = \frac{1}{\gamma^2 + k^2}\left(-\gamma\frac{\partial E_z}{\partial y} + j\omega\mu\frac{\partial H_z}{\partial x}\right) \qquad (10)$$

$$H_x = \frac{1}{\gamma^2 + k^2}\left(j\omega\varepsilon\frac{\partial E_z}{\partial y} - \gamma\frac{\partial H_z}{\partial x}\right) \qquad (11)$$

$$H_y = -\frac{1}{\gamma^2 + k^2}\left(j\omega\varepsilon\frac{\partial E_z}{\partial x} + \gamma\frac{\partial H_z}{\partial y}\right) \qquad (12)$$

For propagating waves, it is convenient to use the substitution $\gamma = j\beta$ where β is real if there is no attentuation. Rewriting the above with this substitution,

$$E_x = -\frac{j}{k_c^2}\left(\beta\frac{\partial E_z}{\partial x} + \omega\mu\frac{\partial H_z}{\partial y}\right) \qquad (13)$$

$$E_y = \frac{j}{k_c^2}\left(-\beta\frac{\partial E_z}{\partial y} + \omega\mu\frac{\partial H_z}{\partial x}\right) \qquad (14)$$

$$H_x = \frac{j}{k_c^2}\left(\omega\varepsilon\,\frac{\partial E_z}{\partial y} - \beta\,\frac{\partial H_z}{\partial x}\right) \tag{15}$$

$$H_y = -\frac{j}{k_c^2}\left(\omega\varepsilon\,\frac{\partial E_z}{\partial x} + \beta\,\frac{\partial H_z}{\partial y}\right) \tag{16}$$

$$\nabla_t^2 E_z = -k_c^2 E_z \tag{17}$$

$$\nabla_t^2 H_z = -k_c^2 H_z \tag{18}$$

where

$$k_c^2 \triangleq \gamma^2 + k^2 = k^2 - \beta^2 \tag{19}$$

In studying guided waves along uniform systems, it is common to classify the wave solutions into the following types:

1. Waves that contain neither electric nor magnetic field in the direction of propagation. Since electric and magnetic field lines both lie entirely in the transverse plane, these may be called *transverse electromagnetic waves* (abbreviated *TEM*). They are the usual *transmission-line waves* along a multiconductor guide.
2. Waves that contain electric field but no magnetic field in the direction of propagation. Since the magnetic field lies entirely in transverse planes, they are known as *transverse magnetic (TM)* waves. They have also been referred to in the literature as E waves, or waves of electric type.
3. Waves that contain magnetic field but no electric field in the direction of propagation. These are known as *transverse electric (TE)* waves, and have also been referred to as H waves or waves of magnetic type.
4. *Hybrid waves* for which boundary conditions require all field components. These may often be considered as a coupling of *TE* and *TM* modes by the boundary.

The above is not the only way in which the possible wave solutions may be divided, but is a useful way in that any general field distribution excited in an ideal guide may be divided into a number (possibly an infinite number) of the above types with suitable amplitudes and phases. The propagation constants of these tell how the individual waves change phase and amplitude as they travel down the guide, so that they may be superposed at any later position and time to give the total resultant field there. Since it is disadvantageous to have a signal carried by several waves traveling at different velocities because of the resultant distortion, waveguides are normally designed so that only one wave can propagate even if many are excited at the entrance to the guide. (This condition often cannot be met in optical guides, Chapter 14.)

CYLINDRICAL WAVEGUIDES OF
VARIOUS CROSS SECTIONS

8.3   Waves Guided by Perfectly Conducting Parallel Plates

One of the simplest wave guiding systems for analysis is that formed by a slab of dielectric with parallel-plane conductors on top and bottom. The fields are assumed to be the same as if the plates were of infinite width, which means that any edge effects or other variations along one transverse coordinate are neglected for a first-order analysis of this model. We saw in Chapter 5 that such a system may be considered a two-conductor transmission line, with upper and lower plates acting as the two conductors of the line. But we shall see that the system also guides waves of other types. Analysis of this system helps in the understanding of all the wave types before going on to more complicated boundaries for the guiding system. We consider the three classes of waves defined in the preceding section.

TEM **Waves**   The transverse electromagnetic waves have neither E_z nor H_z. From Eqs. 8.2(9)–(12) we see that all transverse components must be zero also *unless* $\gamma^2 + k^2 = 0$. The propagation constant for a *TEM* wave must then be

$$\gamma_{TEM} = \pm jk \tag{1}$$

That is, propagation is with the velocity of light in the dielectric medium. Since this argument does not make use of the specific configuration of parallel planes, it applies to *TEM* waves in any shape of guide, as will be discussed more later. Moreover, if $\gamma^2 + k^2$ is zero, we see from Eqs. 8.2(1) and 8.2(2) that both electric and magnetic fields satisfy Laplace's equation so that both have the spatial distribution of two-dimensional *static* fields. It is known that the static electric field between parallel plane conductors is uniform and normal to the planes so that we may write

$$E_x = E_0 \tag{2}$$

and magnetic field, from Eq. 8.2(4) with $E_z = 0$ is

$$H_y = \frac{\gamma}{j\omega\mu} E_x = \pm \frac{j\omega\sqrt{\mu\varepsilon}}{j\omega\mu} E_x = \pm \sqrt{\frac{\varepsilon}{\mu}} E_x \tag{3}$$

where the upper sign is for positively traveling waves and the lower sign for negatively traveling waves. When interpreted in terms of voltage and current, we find the same results as those obtained from the transmission-line analysis.

TM **Waves**   Transverse magnetic waves have finite E_z but no H_z so we may use Eq. 8.2(17). The transverse Laplacian is taken as d^2/dx^2 because of the neglect of the y derivatives,

$$\frac{d^2 E_z}{dx^2} = -k_c^2 E_z \tag{4}$$

$$k_c^2 = \gamma^2 + k^2 \tag{5}$$

Solution of (4) is in sinusoids:

$$E_z = A \sin k_c x + B \cos k_c x \tag{6}$$

Boundary conditions are next applied at the conducting planes at $x = 0$ and a. Since these are taken as perfectly conducting, $E_z = 0$ there. Placing $E_z = 0$ at $x = 0$ in (6) requires $B = 0$. The condition $E_z = 0$ at $x = a$ then requires either $A = 0$, in which case we have nothing left, or $\sin k_c a = 0$, which is satisfied by

$$k_c a = m\pi, \qquad m = 1, 2, 3, \dots \tag{7}$$

So a solution for E_z which satisfies boundary conditions is

$$E_z = A \sin \frac{m\pi x}{a} \tag{8}$$

The transverse field components are now obtained from E_z by means of Eqs. 8.2(9) to 8.2(12), letting $H_z = 0$ and $\partial/\partial y = 0$.

$$E_x = -\frac{\gamma}{k_c^2} \frac{dE_z}{dx} = -\frac{\gamma a}{m\pi} A \cos \frac{m\pi x}{a} \tag{9}$$

$$H_y = -\frac{j\omega\varepsilon}{k_c^2} \frac{dE_z}{dx} = -\frac{j\omega\varepsilon a}{m\pi} A \cos \frac{m\pi x}{a} \tag{10}$$

$$H_x = 0, \qquad E_y = 0 \tag{11}$$

It is seen that there is an infinite number of solutions for the various integral values of m, each with a different field distribution. These solutions are called the *modes* of the guide (in this case, *TM* modes).

Let us now examine the properties of the propagation constant. Solving (5) for γ we have

$$\gamma = \sqrt{k_c^2 - k^2} = \sqrt{\left(\frac{m\pi}{a}\right)^2 - \omega^2 \mu\varepsilon} \tag{12}$$

At sufficiently high frequencies, the second term in the radicand is larger than the first and we can rewrite (12) as

$$\gamma = j\beta = j\omega\sqrt{\mu\varepsilon} \sqrt{1 - \frac{(m\pi/a)^2}{\omega^2 \mu\varepsilon}} \tag{13}$$

We see that the phase constant approaches that for a plane wave as frequency approaches infinity. Lowering the frequency reduces β and it goes to zero at the frequency

$$\omega_c = \frac{m\pi}{a\sqrt{\mu\varepsilon}} = \frac{m\pi v}{a} \tag{14}$$

where v is the velocity of light $(\mu\varepsilon)^{-1/2}$ in the given material. We call ω_c the cutoff frequency since propagation only takes place where β is real. The variation of β with ω is often plotted as in Fig. 8.3a for the reasons discussed in Secs. 5.11 and 5.12. It is conveniently expressed in terms of the cutoff frequency

$$\gamma = j\beta = j\omega\sqrt{\mu\varepsilon} \sqrt{1 - \left(\frac{\omega_c}{\omega}\right)^2}, \qquad \omega \geq \omega_c \tag{15}$$

The cutoff point may usefully be expressed in terms of the wavelength at the cutoff frequency

$$\lambda_c = \frac{2\pi v}{\omega_c} = \frac{2a}{m} \tag{16}$$

That is, cutoff occurs when the spacing between plates is m half-wavelengths, measured at the velocity of light for the dielectric material. We see the important feature that propagating TM waves only can exist above a cutoff frequency (or, equivalently, at wavelengths shorter than the cutoff wavelength); at lower frequencies γ is real and is given by

$$\gamma = \alpha = \frac{m\pi}{a} \sqrt{1 - \left(\frac{\omega}{\omega_c}\right)^2}, \qquad \omega \leq \omega_c \tag{17}$$

As seen in Fig. 8.3a there is attenuation without phase shift for frequencies below the cutoff frequency of a given mode, phase shift without attenuation for

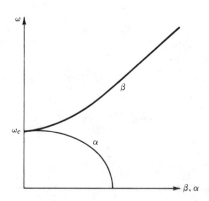

Fig. 8.3a Phase constant and attenuation as functions of frequency for a TM or TE waveguide mode.

frequencies above cutoff, and neither attenuation nor phase shift exactly at cutoff. Each of these modes then acts as a high-pass filter; the attenuation below cutoff, like that for a loss-free filter, is a reactive attenuation representing reflection but no dissipation for this nondissipative system.

For the propagating regime, $\omega > \omega_c$, we can define a phase velocity in the usual way,

$$v_p = \frac{\omega}{\beta} = \frac{v}{\sqrt{1 - (\omega_c/\omega)^2}} \tag{18}$$

Group velocity can be derived from this as in Sec. 5.12,

$$v_g = \frac{d\omega}{d\beta} = v\sqrt{1 - \left(\frac{\omega_c}{\omega}\right)^2} \tag{19}$$

Phase velocity is always greater than the velocity of light in the medium and group velocity is always less, both approaching each other and v at frequencies far above cutoff. A wavelength along the guide, λ_g, may also be defined as the distance for which phase shift increases by 2π,

$$\lambda_g = \frac{2\pi}{\beta} = \frac{\lambda}{\sqrt{1 - (\omega_c/\omega)^2}} \tag{20}$$

where λ is wavelength of a plane wave in the dielectric medium,

$$\lambda = \frac{2\pi v}{\omega} \tag{21}$$

The ratio of transverse electric field to transverse magnetic field of a single propagating wave may be defined as a characteristic wave impedance and is useful for certain types of reflection problems, much as the field impedance for plane waves was found to be in Chapter 6. For the TM wave, this impedance, from (9) and (10) with $\eta = (\mu/\varepsilon)^{1/2}$, is

$$Z_{TM} = \frac{E_x}{H_y} = \frac{\beta}{\omega\varepsilon} = \eta\sqrt{1 - \left(\frac{\omega_c}{\omega}\right)^2} \tag{22}$$

Note that this ratio is not a function of x and y. It is imaginary for frequencies below cutoff, so that a Poynting calculation will show no average power flow in that regime. It is real for frequencies above cutoff so that a Poynting calculation in that regime shows finite average power carried by the wave.

A plot of electric field lines in a TM_1 mode from the equations for E_x and E_z is shown in Fig. 8.3b (Prob. 8.3c). Note that induced charges on top and bottom plates are of the same sign in a given z plane for this wave, in contrast to the sign relations found for the TEM wave. Electric field lines starting on these charges turn and go axially down the guide, ending on charges of opposite sign a distance a half-guide wavelength down the guide. The entire pattern translates along the guide at the phase velocity.

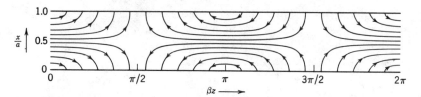

Fig. 8.3*b* Electric field lines of TM_1 wave between plane conductors.

TE **Waves** The transverse electric class of waves has nonzero H_z but no E_z. Equation 8.2(18) is then utilized,

$$\nabla_t^2 H_z = \frac{d^2 H_z}{dx^2} = -k_c^2 H_z \tag{23}$$

$$k_c^2 = \gamma^2 + k^2 = k^2 - \beta^2 \tag{24}$$

The solution will again be written in terms of sinusoids, but this time only the cosine term is retained since E_y, proportional to the derivative of H_z with x, must become zero at the perfectly conducting plane $x = 0$:

$$H_z = B \cos k_c x \tag{25}$$

From Eqs. 8.2(13) to 8.2(16), remembering that E_z is zero,

$$H_x = -\frac{j\beta}{k_c^2}\frac{dH_z}{dx} = \frac{j\beta}{k_c} B \sin k_c x \tag{26}$$

$$E_y = \frac{j\omega\mu}{k_c^2}\frac{dH_z}{dx} = -\frac{j\omega\mu}{k_c} B \sin k_c x \tag{27}$$

$$E_x = 0, \qquad H_y = 0 \tag{28}$$

Also, E_y must be zero at the conducting plane $x = a$, so k_c is determined from (27) as some multiple of π/a. As with the TM wave, this is identified from (24) as the value of k at cutoff

$$k_c = 2\pi f_c \sqrt{\mu\varepsilon} = \frac{m\pi}{a}, \qquad m = 1, 2, 3, \ldots \tag{29}$$

Propagation constant from (24) may then be written

$$\gamma = \alpha = \left(\frac{m\pi}{a}\right)\sqrt{1 - \left(\frac{\omega}{\omega_c}\right)^2}, \qquad \omega < \omega_c \tag{30}$$

$$\gamma = j\beta = jk\sqrt{1 - \left(\frac{\omega_c}{\omega}\right)^2}, \qquad \omega > \omega_c \tag{31}$$

The forms for attenuation constant in the cutoff range and phase constant in the propagation range are thus exactly the same as for the TM waves (Fig. 8.3a), and

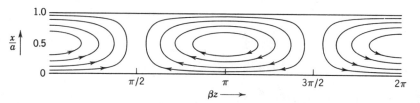

Fig. 8.3c Magnetic field lines of TE_1 wave between plane conductors.

by (14) and (29) conditions for cutoff are the same for TE modes as for TM modes of the same order. The expressions for phase velocity, group velocity, and guide wavelength in the propagation range follow from (31) and are exactly the same as (18) to (20). Wave or field impedance for the TE wave is

$$Z_{TE} = -\frac{E_y}{H_x} = \frac{j\omega\mu}{\gamma} = \frac{\eta}{\sqrt{1 - (\omega_c/\omega)^2}} \tag{32}$$

For frequencies below cutoff this wave impedance is imaginary, but for frequencies above cutoff it is real and always greater than η, as contrasted to the wave impedance for TM waves, which is always less than η.

The form of the field lines for the first-order TE mode is indicated in Fig. 8.3c. Here the magnetic field lines form closed curves surrounding the y-direction displacement current. There is no charge induced on the conducting plates and only a y component of current corresponding to the finite H_z tangential to the plates.

_____ **Example 8.3a** _____
Interpretation of Group Velocity as an Energy Velocity

Let us define a velocity of energy flow in terms of energy stored per unit length and average power flow,

$$v_E = \frac{W_T}{u} \tag{33}$$

Take the TM wave as example. Average power flow is found from the complex Poynting theorem for a width w:

$$W_T = w \int_0^a \frac{1}{2} Re(E_x H_y^*) \, dx$$

$$= \frac{w}{2} A^2 \beta \omega \varepsilon \left(\frac{a}{m\pi}\right)^2 \int_0^a \cos^2 \frac{m\pi x}{a} \, dx = \frac{waA^2}{4} \frac{a^2}{m^2\pi^2} \beta \omega \varepsilon \tag{34}$$

Time-average energy storage per unit length, including both electric and magnetic parts, is

$$u = w \int_0^a \left\{ \frac{\varepsilon}{4} [|E_x|^2 + |E_z|^2] + \frac{\mu}{4} |H_y|^2 \right\} dx \tag{35}$$

Using the fields from (8)–(10) this is

$$u = \frac{A^2 w}{4} \int_0^a \left\{ \varepsilon \left[\sin^2 \frac{m\pi x}{a} + \frac{\beta^2 a^2}{m^2 \pi^2} \cos^2 \frac{m\pi x}{a} \right] + \frac{\mu \omega^2 \varepsilon^2 a^2}{m^2 \pi^2} \cos^2 \frac{m\pi x}{a} \right\} dx$$

$$= \frac{A^2 wa\varepsilon}{8} \left[1 + \frac{\beta^2 a^2}{m^2 \pi^2} + \frac{k^2 a^2}{m^2 \pi^2} \right]$$

But from (13), $m^2 \pi^2 / a^2 + \beta^2$ is just k^2, so this reduces to

$$u = \frac{A^2 wa\varepsilon k^2 a^2}{4m^2 \pi^2} \tag{36}$$

so that energy velocity from (33) is

$$v_E = \frac{\beta\omega}{k^2} = \frac{1}{\sqrt{\mu\varepsilon}} \sqrt{1 - \left(\frac{\omega_c}{\omega}\right)^2} \tag{37}$$

which is exactly the same as expression (19) for group velocity. The same equivalence is found for *TE* waves. We will say more about the use of these velocities for signal propagation at the end of the chapter.

8.4 Guided Waves Between Parallel Planes as Superposition of Plane Waves

The modes of the parallel-plane guide, studied in the preceding section from the wave equation, can also be found to be superpositions of plane waves propagating at various angles. This picture is very helpful in developing a physical feeling for the different types of modes. The *TEM* wave is clearly just a portion of a uniform plane wave polarized with electric field in the *x* direction and propagating in the *z* direction,

$$E_x(x, z) = E_0 e^{-jkz} \tag{1}$$

Magnetic field for such a wave, from Sec. 6.2, is

$$H_y(x, z) = \frac{E_0}{\eta} e^{-jkz} \tag{2}$$

where $k = \omega \sqrt{\mu\varepsilon}$. These are of the same form as Eqs. 8.3(2) and 8.3(3).

For *TM* and *TE* waves, we superpose uniform plane waves propagating at an angle θ from the normal, as pictured in Fig. 8.4a. It was found in Sec. 6.9 that when plane waves strike a conducting plane at an angle, tangential electric field must be zero not only at that plane, but also at other planes which are multiples of a half-wavelength measured at phase velocity in the direction normal to the plane. That is, boundary conditions are satisfied if spacing a between plates is

$$a = \frac{m\lambda}{2 \cos \theta} \tag{3}$$

or

$$\cos \theta = \frac{m\lambda}{2a} = \frac{\lambda}{\lambda_c} = \frac{\omega_c}{\omega} \tag{4}$$

where

$$\lambda_c = \frac{2\pi v}{\omega_c} = 2a/m \tag{5}$$

The phase constant in the z direction is

$$\beta = k \sin \theta = k\sqrt{1 - \cos^2 \theta} \tag{6}$$

or, using (4),

$$\beta = k\sqrt{1 - \left(\frac{\omega_c}{\omega}\right)^2} \tag{7}$$

This is exactly the same as found from the detailed analysis in Sec. 8.3. Since β is the same, phase and group velocity will be also. We see from Fig. 8.4a that phase velocity is the velocity of the imaginary point P of intersection of the plane wave fronts with the z axis, and is greater than velocity v of the plane wave in the medium, as found in Sec. 8.3. Group velocity is the component of v in the z direction and is always less than v. At cutoff (Fig. 8.4b), the waves simply bounce laterally back and forth between planes without any forward progression so that group velocity is zero and phase velocity infinite. For frequencies much above cutoff, the angle is very flat (Fig. 8.4c) and the wave fronts are nearly normal to z, so that both v_p and v_g approach v.

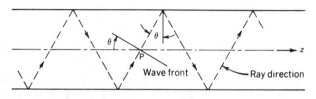

Fig. 8.4a Diagram showing ray directions and wave-front direction of a uniform plane wave component of *TM* or *TE* waves between parallel planes.

Fig. 8.4 (b) Special case of (a) when wave is at cutoff. (c) Special case for frequency far above cutoff so that $\theta \to \pi/2$ and wave front is nearly normal to guide axis.

Both *TM* and *TE* wave types have behavior as above. *TM* waves correspond to the plane waves polarized with electric field in the plane of incidence and *TE* to those polarized with magnetic field in the plane of incidence.

Example 8.4
Detailed Field Expressions for TM_m Wave from Plane Wave Components
---

To show that this point of view is exactly equivalent to that of Sec. 8.3, consider the field expressions given by Eqs. 6.9(11)–(13) for a wave polarized with electric field in the plane of incidence striking a conductor at an angle. We rewrite these with the coordinates x and z primed, since we take different coordinate systems from those in Sec. 6.9.

$$E_{x'}(x', z') = -2jE_+ \cos \theta \sin(kz' \cos \theta)e^{-jkx'\sin\theta} \tag{8}$$

$$E_{z'}(x', z') = -2E_+ \sin \theta \cos(kz' \cos \theta)e^{-jkx'\sin\theta} \tag{9}$$

$$\eta H_y(x', z') = 2E_+ \cos(kz' \cos \theta)e^{-jkx'\sin\theta} \tag{10}$$

To change to our present coordinate system, let $x' = z$ and $z' = -x$. Also let $k \sin \theta = \beta$ by (6), $k \cos \theta = m\pi/a$ from (3). The above then become

$$E_z(z, x) = +\frac{2jE_+ m\pi}{ka} \sin\left(\frac{m\pi x}{a}\right)e^{-j\beta z} \tag{11}$$

$$E_x(z, x) = \frac{2E_+ \beta}{k} \cos\left(\frac{m\pi x}{a}\right)e^{-j\beta z} \tag{12}$$

$$\eta H_y(z, x) = 2E_+ \cos\left(\frac{m\pi x}{a}\right)e^{-j\beta z} \tag{13}$$

If we next set the multiplier $(2jE_+ m\pi/ka)$ equal to a constant A and remember that the propagation factor $e^{-j\beta z}$ has been suppressed in Eqs. 8.3(8)–(10), we find that (11)–(13) are identical to Eqs. 8.3(8)–(10).

8.5   Parallel-Plane Guiding System with Losses

When the dielectric filling the region between conductors of the parallel-plane guide is imperfect, we may replace permittivity by its complex forms $\varepsilon' - j\varepsilon''$ (Sec. 6.4) in expressions of Sec. 8.3. We will be concerned here primarily with the effect on propagation constant. For the *TEM* wave we have

$$\gamma_{TEM} = j\omega\sqrt{\mu(\varepsilon' - j\varepsilon'')} \tag{1}$$

If $\varepsilon''/\varepsilon'$ is small in comparison with unity, a binomial expansion shows that the real part of this, which is the attenuation constant, is to first order

$$(\alpha_d)_{TEM} \cong \frac{\omega\sqrt{\mu\varepsilon'}}{2}\frac{\varepsilon''}{\varepsilon'} \tag{2}$$

For *TM* and *TE* waves from Eqs. 8.3(12) or 8.3(30)

$$\gamma_{TM,TE} = \left[\left(\frac{m\pi}{a}\right)^2 - \omega^2\mu(\varepsilon' - j\varepsilon'')\right]^{1/2}$$

$$\approx \left[\left(\frac{m\pi}{a}\right)^2 - \omega^2\mu\varepsilon'\right]^{1/2}\left\{1 + \frac{j\omega^2\mu\varepsilon''}{2}\left[\left(\frac{m\pi}{a}\right)^2 - \omega^2\mu\varepsilon'\right]^{-1}\right\} \tag{3}$$

The second expression is obtained by making a binomial expansion of the first. The attenuation constant (real part) for $\omega > \omega_c$ is then

$$(\alpha_d)_{TM,TE} = \frac{\omega\sqrt{\mu\varepsilon'}(\varepsilon''/\varepsilon')}{2\sqrt{1 - (\omega_c/\omega)^2}} \tag{4}$$

Retention of second-order terms gives a correction to phase constant, which may be important in considering the dispersive properties of the guided wave.

An exact solution is considerably more difficult to obtain when the finite conductivity of the conducting boundaries must be considered. The approach is to obtain field solutions in both dielectric and conductor regions, with proper continuity conditions applied at the boundary between them. This approach cannot even be carried out for some geometrical configurations, although it can for this simple system and shows that the approximations to be used in the following analysis are justified so long as the plane boundaries are made of much better conducting material than the intervening dielectric region.[1] The expression to be used is that giving attenuation in terms of power loss per unit length and average power transferred by the mode, Eq. 5.9(4). This is an exact expression, but the approximation comes in by calculating power transfer as that of the ideal

---

[1] More precisely, it is required that displacement current in the conductor be negligible in comparison with displacement current in the dielectric, and conduction current in the dielectric be small in comparison with conduction current in the conductor.

guide, and loss per unit length as that from currents of the ideal guide flowing in the real conductors. For the *TEM* wave, the power transfer for a width w is

$$(W_T)_{TEM} = \frac{awE_0^2}{2\eta} \tag{5}$$

The average power loss per unit area in the plates, if plates are thick compared with skin depth, is $(\frac{1}{2})R_s|J_{sz}|^2$, so for a unit length and width w, counting both plates

$$(w_L)_{TEM} = 2\frac{wR_s}{2}|H_y|^2 = wR_s\left(\frac{E_0}{\eta}\right)^2 \tag{6}$$

Attenuation constant is then calculated as

$$(\alpha_c)_{TEM} = \frac{w_L}{2W_T} = \frac{R_s}{\eta a} \tag{7}$$

This is the same as that which would be obtained from the transmission line formula, Eq. 5.9(7).

For the *TM* wave of order m, average power transfer from the Poynting theorem and results of Sec. 8.3 for $\omega > \omega_c$ is

$$
\begin{aligned}
(W_T)_{TM} &= w\int_0^a \frac{1}{2}(E_x H_y^*)\,dx \\
&= \frac{w}{2}\int_0^a \left(-\frac{j\beta aA}{m\pi}\cos\frac{m\pi x}{a}e^{-j\beta z}\right)\left(\frac{j\omega\varepsilon a}{m\pi}A\cos\frac{m\pi x}{a}e^{j\beta z}\right)dx \\
&= \frac{w}{2}\frac{\omega\varepsilon\beta a^2A^2}{m^2\pi^2}\int_0^a\cos^2\frac{m\pi x}{a}\,dx = \left(\frac{w\omega\varepsilon\beta a^2A^2}{2m^2\pi^2}\right)\frac{a}{2}
\end{aligned} \tag{8}
$$

The current flow in both upper and lower plates is

$$|J_{sz}| = |H_y|_{x=0} = |H_y|_{x=a} = \frac{\omega\varepsilon aA}{m\pi} \tag{9}$$

So average power loss per unit length for a width w is

$$(w_L)_{TM} = \frac{2wR_s}{2}|J_{sz}|^2 = wR_s\left(\frac{\omega\varepsilon aA}{m\pi}\right)^2 \tag{10}$$

The attenuation constant from conductor losses is then approximately

$$(\alpha_c)_{TM} = \frac{w_L}{2W_T} = \frac{2R_s\omega\varepsilon}{\beta a} = \frac{2R_s}{\eta a\sqrt{1-(\omega_c/\omega)^2}} \tag{11}$$

By a similar calculation we find attenuation for a *TE* mode to be

$$(\alpha_c)_{TE} = \frac{2R_s(\omega_c/\omega)^2}{\eta a\sqrt{1-(\omega_c/\omega)^2}} \tag{12}$$

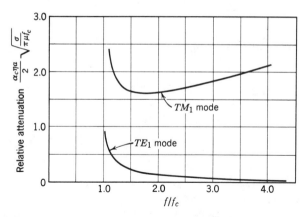

Fig. 8.5 Attenuation curves of waves between imperfectly conducting planes.

Curves of normalized attenuation versus frequency are shown in Fig. 8.5. There are several interesting features of these curves. Note first that expressions (11) and (12) approach infinity as $\omega \to \omega_c$, but the approximations break down at cutoff so that attenuation is actually finite although high there. The curve for the TM wave starts to decrease with increasing frequency above cutoff, but reaches a minimum (at $\omega = \sqrt{3}\omega_c$) and thereafter increases with increasing frequency because of the increase of R_s with frequency. The curve for the TE wave is always lower than that for the TM, and moreover continues to decrease with increasing frequency. This behavior arises because the currents in the conductors are the y-directed currents related to H_z, and this component of field approaches zero at high frequencies as the wave front becomes substantially normal to the axis, as explained in Sec. 8.4.

8.6   Stripline and Microstrip Transmission Systems

There are several different forms of wave guiding structures made from parallel metal strips on a dielectric substrate that have found use for microwave and millimeter-wave circuits as well as high-speed pulse circuits. In this section we will examine in some detail two types, called *stripline* and *microstrip*. These have forms that can be approximated by parallel-plane guides. Emphasis will be on the lowest-order mode, which is a TEM wave in the stripline and a quasi-TEM wave in the microstrip.

The stripline consists of a conducting strip lying between, and parallel to, two wide conducting planes, as shown in Fig. 8.6a. The region between the strip and the planes is filled with a uniform dielectric. Such a structure, with a uniform dielectric and more than one conductor, can support a TEM wave. If the strip width w is much greater than the spacing d and the two planes are at a common

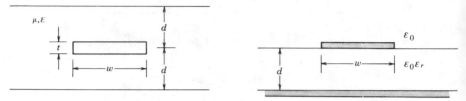

Fig. 8.6 (*a*) Symmetrical stripline. (*b*) Cross section of a microstrip.

potential, the structure is roughly approximated by two parallel-plane lines connected in parallel. More precise results are found from the capacitance per unit length. For a *TEM* wave, the phase velocity is $v_p = (\mu\varepsilon)^{-1/2}$ and from the transmission-line formalism it is also given by $v_p = (LC)^{-1/2}$. Since the capcitance per unit length can be expressed as $C = \varepsilon f$, where f is a shape factor, then $L = \mu(1/f)$ to satisfy the above forms of v_p. Therefore, the characteristic impedance is

$$Z_0 = \sqrt{\frac{L}{C}} = \frac{\eta}{f} = \frac{\sqrt{\mu\varepsilon}}{C} \tag{1}$$

Thus, for systems with uniform ε and μ the characteristic impedance can be found from the capacitance, which can be determined in a number of ways, as we saw in Chapters 1 and 7. An approximate expression for the characteristic impedance of the strip line, assuming a zero-thickness strip, has been found by conformal transformation[2] to be

$$Z_0 \approx \frac{\eta}{4} \frac{K(k)}{K(\sqrt{1 - k^2})} \tag{2}$$

where $\eta = \sqrt{\mu/\varepsilon}$ and k is given by

$$k = \left[\cosh\left(\frac{\pi w}{4d}\right)\right]^{-1} \tag{3}$$

and $K(k)$ is the complete elliptic integral of the first kind (see Ex. 4.7a). Collin has also estimated the effect of thickness t by variational methods. Velocity of propagation is just $1/\sqrt{\mu\varepsilon}$ as in all *TEM* waves, and the approximate attenuation if conductors have surface resistivity R_s is[2]

$$\alpha_c = \frac{R_s}{2Z_0 d}\left[\frac{\pi w/2d + \ln(8d/\pi t)}{\ln 2 + \pi w/4d}\right] \tag{4}$$

Attenuation from imperfect dielectrics is exactly as in Eq. 8.5(2).

Perhaps the most widely used thin-strip line is one formed with the strip lying on top of an insulator with a conductive backing. A different dielectric (usually

---

[2] R. E. Collin, *Field Theory of Guided Waves*, McGraw-Hill, New York, 1960, Sec. 4.3.

air) is above the insulator and strip (See Fig. 8.6*b*). In such an arrangement, there cannot be a true *TEM* wave. The reason is that such a wave, as shown in connection with Eq. 8.3(1), requires that $\gamma^2 + k^2 = 0$. The propagation constant γ is a single quantity for the wave, so $\gamma^2 + k_1^2$ and $\gamma^2 + k_2^2$ cannot both be zero if $k_1 \neq k_2$. Exact solution of this problem is complicated by the finite width of the strip and the two different dielectrics. As mentioned above for the stripline, the lowest-order approximation for the microstrip is a section of parallel-plane guide neglecting fringing. A much better, though still approximate, approach is to consider that the lowest-order wave is approximately a *TEM* wave so the distribution of fields in the transverse plane is the same as that for static fields.

In one formulation, the characteristic impedance is found first by using (1) for a given strip width *w* and spacing *d*, assuming the dielectric is everywhere free space. It is then corrected by an effective dielectric constant ε_r'. This effective dielectric constant is one which would give capacitance of the actual structure if it were to fill the entire region. Its value is found by an approximate conformal method; one of the several approximate formula is[3]

$$\varepsilon_r' = 1 + \frac{(\varepsilon_r - 1)}{2}\left[1 + \frac{1}{\sqrt{1 + 10d/w}}\right] \tag{5}$$

Using this value for ε_r', we have an equivalent uniform structure and the last term of (1) shows that the actual characteristic impedance is found from the one assuming a free-space dielectric by dividing by $\sqrt{\varepsilon_r'}$. The results for a variety of insulator dielectric constants is given in Fig. 8.6*c*. Also, since the above effective dielectric constant gives the capacitance for a uniformly filled structure, the phase velocity can be calculated in the usual way for a *TEM* wave using ε_r' given by (5).

At higher frequencies, the longitudinal components of the fields can no longer be neglected. Exact analyses can be made but are rather complicated. The dispersive properties have been expressed in a usefully accurate empirical formula.[4] The phase constant can be written as

$$\frac{\beta}{k_0} = \frac{\sqrt{\varepsilon_r} - \sqrt{\varepsilon_r'}}{1 + 4F^{-1.5}} + \sqrt{\varepsilon_r'} \tag{6}$$

where

$$F = \frac{4d\sqrt{\varepsilon_r - 1}}{\lambda_0}\left\{0.5 + \left[1 + 2\ln\left(1 + \frac{w}{d}\right)\right]^2\right\} \tag{7}$$

[3] M. V. Schneider, *Bell Sys. Tech. J.* **48**, 1421 (1969). H. A. Wheeler, *IEEE Trans. Microwave Theory and Techniques* **MTT-25**, 631 (1977).

[4] E. Yamashita, K. Atsuki, and T. Ueda, *IEEE Trans. Microwave Theory and Techniques* **MTT-27**, 1036 (1979).

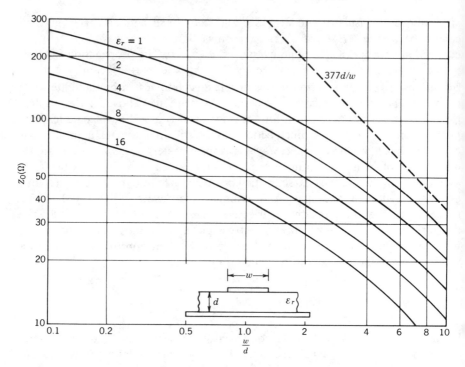

Fig. 8.6c Approximate characteristic impedance of stripline, adapted from Wheeler, Ref. 3. The dashed line is the parallel-plane approximation, for air dielectric. (© 1977 IEEE)

where k_0 and λ_0 are the free-space wave number and wavelength. The variation of the phase constant with frequency is accentuated where high permittivity dielectrics or small w/d ratios are used. Dispersion is discussed further in Sec. 8.16.

There are other types of waveguiding systems employing metal films on dielectrics in which all conductors are on the same side of the dielectric. These include the *slotline* (Fig. 8.6d), the *coplanar waveguide* (Fig. 8.6e), and *coplanar strips* (Fig. 8.6f). The slotline can sometimes be formed in the ground plane of a microstrip system, giving flexibility for forming some kinds of circuit elements. All of these coplanar types of lines afford the opportunity for convenient incorporation of lumped devices and short circuits, which is more difficult in the microstrip or stripline systems. For more details on coplanar, as well as microstrip lines, see Ref. 5.

---

[5] K. C. Gupta, R. Garg, and I. J. Bahl, *Microstrip Lines and Slotlines*, Artech House, Dedham, MA, 1979.

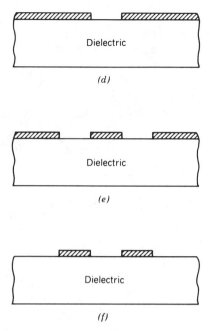

Fig. 8.6 Various forms of strip-type waveguiding systems. (*d*) Slotline. (*e*) Coplanar waveguide. (*f*) Coplanar strips.

8.7   Rectangular Waveguides

The most important of the hollow-pipe guides is that of rectangular cross section. As in Fig. 8.7*a*, a dielectric region of width *a* and height *b* extends indefinitely in the axial (*z*) direction and is closed by conducting boundaries on the four sides. In the ideal guide, both conductor and dielectric are loss-free. There can be no transverse electromagnetic wave (*TEM*) inside the hollow pipe since, as was

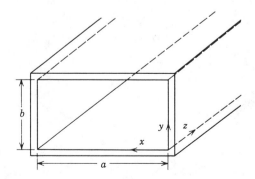

Fig. 8.7*a* Coordinate system for rectangular guide.

shown in Sec. 8.3, *TEM* waves have transverse variations like static fields, and no static fields can exist inside a region bounded by a single conductor. Transverse magnetic (*TM*) and transverse electric (*TE*) waves can exist and will be analyzed below.

TM **Waves**   Transverse magnetic waves have zero H_z but nonzero E_z. The differential equation governing E_z is Eq. 8.2(17), here expressed in rectangular coordinates:

$$\nabla_t^2 E_z = \frac{\partial^2 E_z}{\partial x^2} + \frac{\partial^2 E_z}{\partial y^2} = -k_c^2 E_z \tag{1}$$

This equation was solved in Sec. 7.19 by separation of variables procedures and found to have solutions of the form

$$E_z = (A' \sin k_x x + B' \cos k_x x)(C' \sin k_y y + D' \cos k_y y) \tag{2}$$

where

$$k_x^2 + k_y^2 = k_c^2 \tag{3}$$

The perfectly conducting boundary at $x = 0$ requires $B' = 0$ to produce $E_z = 0$ there. Similarly the ideal boundary at $y = 0$ requires $D' = 0$. We let $A'C'$ be a new constant A and have

$$E_z = A \sin k_x x \sin k_y y \tag{4}$$

Axial electric field E_z must also be zero at $x = a$ and $y = b$. This can only be so (except for the trivial solution $A = 0$) if $k_x a$ is an integral multiple of π:

$$k_x a = m\pi, \qquad m = 1, 2, 3, \ldots \tag{5}$$

Similarly, to make E_z zero at $y = b$, $k_y b$ must also be a multiple of π:

$$k_y b = n\pi, \qquad n = 1, 2, 3, \ldots \tag{6}$$

So the cutoff condition of the transverse magnetic wave with m variations in x and n in y (designated TM_{mn}) is found from (3):

$$\omega_{c_{m,n}} = \frac{k_{c_{m,n}}}{\sqrt{\mu\varepsilon}} = \frac{1}{\sqrt{\mu\varepsilon}}\left[\left(\frac{m\pi}{a}\right)^2 + \left(\frac{n\pi}{b}\right)^2\right]^{1/2} \tag{7}$$

Since k_c^2 is $k^2 - \beta^2$ as in Eq. 8.2(19), attenuation for frequencies below the cutoff frequency of a given mode, and phase constant for frequencies above the cutoff frequency have the same forms as for the parallel-plane guiding system:

$$\alpha = k_{c_{m,n}}\sqrt{1 - \left(\frac{\omega}{\omega_{c_{m,n}}}\right)^2}, \qquad \omega < \omega_{c_{m,n}} \tag{8}$$

$$\beta = k\sqrt{1 - \left(\frac{\omega_{c_{m,n}}}{\omega}\right)^2}, \qquad \omega > \omega_{c_{m,n}} \tag{9}$$

Phase and group velocities then also have the same forms as before [Eqs. 8.3(18) and 8.3(19)].

The remaining field components of the TM_{mn} wave are found from Eqs. 8.2(13)–(16) with $H_z = 0$ and E_z from (4):

$$E_x = -\frac{j\beta k_x}{k_{c_{m,n}}^2} A \cos k_x x \sin k_y y \tag{10}$$

$$E_y = -\frac{j\beta k_y}{k_{c_{m,n}}^2} A \sin k_x x \cos k_y y \tag{11}$$

$$H_x = \frac{j\omega\varepsilon k_y}{k_{c_{m,n}}^2} A \sin k_x x \cos k_y y \tag{12}$$

$$H_y = -\frac{j\omega\varepsilon k_x}{k_{c_{m,n}}^2} A \cos k_x x \sin k_y y \tag{13}$$

where k_x, k_y, $k_{c_{m,n}}$, and β are defined by (5), (6), (7), and (9), respectively, and all fields are multipled by the propagation term, $e^{-j\beta z}$. Plots of electric and magnetic field lines in the TM_{11} and TM_{21} modes are shown in Table 8.7. Note that electric field lines (shown solid) begin on charges on the guide walls at some fixed z plane, turn and go axially down the guide, and end on charges of opposite sign a half-guide wavelength down the guide. Magnetic field lines (shown dashed) surround the displacement currents represented by the changing electric fields as the pattern moves down the guide with velocity v_p. The pattern for the TM_{21} mode is that of two TM_{11} modes side by side and of opposite sense.

The attenuation resulting from losses in the conducting walls can be calculated for the TM_{mn} wave following the procedure used to find Eq. 8.5(11):

$$(\alpha_c)_{TM_{mn}} = \frac{2R_s}{b\eta\sqrt{1 - (f_c/f)^2}} \frac{[m^2(b/a)^3 + n^2]}{[m^2(b/a)^2 + n^2]} \tag{14}$$

We make two points before leaving this class of waves. Note from (4) and the definitions of k_x and k_y given by (5) and (6) that neither m nor n can be zero for the TM wave without its disappearing entirely. The second point is that we have required as boundary condition that E_z be zero along the perfectly conducting boundary, but should also be sure that other tangential components of electric

Table 8.7

Summary of wave types for rectangular guides[a]

[a] Electric field lines are shown solid and magnetic field lines are dashed.

field are zero there. From (10) and the definitions (5) and (6) we can see that E_x is zero as required at $y = 0$ and $y = b$; from (11) we see that E_y is zero at $x = 0$ and $x = a$. Thus all tangential components do satisfy the boundary conditions at the conductors. It can be shown (Prob. 8.7h) that imposition of the boundary condition on E_z necessarily causes the other tangential components of **E** to be zero on the boundaries because of the form of the relations 8.2(13)–(16).

TE Waves   Transverse electric waves have zero E_z and nonzero H_z so that the start is from Eq. 8.2(18), again expressed in rectangular coordinates,

$$\nabla_t^2 H_z = \frac{\partial^2 H_z}{\partial x^2} + \frac{\partial^2 H_z}{\partial y^2} = -k_c^2 H_z \tag{15}$$

Solution by the separation of variables techniques of Sec. 7.19 gives

$$H_z = (A'' \sin k_x x + B'' \cos k_x x)(C'' \sin k_y y + D'' \cos k_y y) \tag{16}$$

where

$$k_c^2 = k_x^2 + k_y^2 \tag{17}$$

Imposition of boundary conditions in this case is a little less direct, but from Eqs. 8.2(13) and 8.2(14) we find electric field components as

$$E_x = -\frac{j\omega\mu}{k_c^2} \frac{\partial H_z}{\partial y}$$

$$= -\frac{j\omega\mu k_y}{k_c^2}(A'' \sin k_x x + B'' \cos k_x x)(C'' \cos k_y y - D'' \sin k_y y) \tag{18}$$

$$E_y = \frac{j\omega\mu}{k_c^2} \frac{\partial H_z}{\partial x} \tag{19}$$

$$= \frac{j\omega\mu k_x}{k_c^2}(A'' \cos k_x x - B'' \sin k_x x)(C'' \sin k_y y + D'' \cos k_y y)$$

In order for E_x to be zero at $y = 0$ for all x, $C'' = 0$ and for $E_y = 0$ at $x = 0$ for all y, $A'' = 0$. Defining $B''D'' = B$, we have then

$$H_z = B \cos k_x x \cos k_y y \tag{20}$$

We also require E_x to be zero at $y = b$ so that $k_y b$ must be a multiple of π. E_y is zero at $x = a$ so that $k_x a$ is also a multiple of π:

$$k_x a = m\pi, \qquad k_y b = n\pi \tag{21}$$

In contrast to the TM waves, one but not both of m and n may be zero without the wave's vanishing. Although we found the boundary conditions by first calculating electric field, we can see from the way in which **E** is related to H_z that the derivative of H_z normal to the conducting boundary must be zero in order for the tangential electric field to be zero there, so boundary conditions can be imposed directly on the form (16) without requiring the explicit forms for E_x and E_y.

The forms of transverse electric field with the derived simplifications to (18) and (19) are

$$E_x = \frac{j\omega\mu k_y}{k_{c_{m,n}}^2} B \cos k_x x \sin k_y y \tag{22}$$

$$E_y = -\frac{j\omega\mu k_x}{k_{c_{m,n}}^2} B \sin k_x x \cos k_y y \tag{23}$$

Corresponding transverse magnetic field components from Eqs. 8.2(15) and 8.2(16) are

$$H_x = \frac{j\beta k_x}{k_{c_{m,n}}^2} B \sin k_x x \cos k_y y \tag{24}$$

$$H_y = \frac{j\beta k_y}{k_{c_{m,n}}^2} B \cos k_x x \sin k_y y \tag{25}$$

Since comparison of (21) with (5) and (6) shows that k_x and k_y have the same forms for TM and TE waves, cutoff frequency and propagation characteristics for a TE_{mn} mode found from (17) are exactly the same as for the same order TM_{mn} mode. That is, the expressions (7), (8), and (9) apply here without change. Modes that have different field distributions but the same cutoff frequencies are said to be *degenerate modes*.

Table 8.7 gives the field distribution for several different TE modes. Since electric field is confined to the transverse plane, we find that for each one of the TE modes shown, electric fields begin on charges for a portion of the boundary and end on charges of opposite sign on another portion, in the same x–y plane. Magnetic field lines surround the displacement currents represented by the changing transverse electric fields. For the TE_{10} mode having no variations in the vertical direction, electric fields go between top and bottom of the guide in straight lines, and magnetic fields lie entirely in planes parallel to top and bottom. The TE_{10} mode is so important that it will be discussed separately in the following section.

Figure 8.7b shows a line diagram indicating the cutoff frequencies of several of the lowest order modes referred to that of the so-called *dominant* TE_{10} mode for a guide with a side ratio $b/a = \frac{1}{2}$, which is close to the value used in most practical guides. Normally, such a guide is designed so that its cutoff frequency for the TE_{10} mode is somewhat (say, 30%) below the operating frequency. In this way only one mode can propagate so signal distortion caused by multimode propagation is avoided. Also, by not being too close to the cutoff frequency, dispersion caused by having different group velocities for different frequency components of the signal is minimized for the one propagating mode. Higher

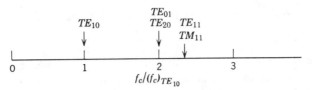

Fig. 8.7b Relative cutoff frequencies of waves in a rectangular guide ($b/a = \frac{1}{2}$).

order modes may be excited at the entrance to the guide but they are below their cutoff frequencies and die away in a short distance from the source.

The attenuation constant for TE_{mn} ($n \neq 0$) modes is found using the procedure leading to Eq. 8.5(12):

$$(\alpha_c)_{TE_{mn}} = \frac{2R_s}{b\eta\sqrt{1 - (f_c/f)^2}}\left\{\left(1 + \frac{b}{a}\right)\left(\frac{f_c}{f}\right)^2 + \left[1 - \left(\frac{f_c}{f}\right)^2\right]\left[\frac{(b/a)((b/a)m^2 + n^2)}{(b^2m^2/a^2) + n^2}\right]\right\}$$

(26)

And for TE_{m0} modes

$$(\alpha_c)_{TE_{m0}} = \frac{R_s}{b\eta\sqrt{1 - (f_c/f)^2}}\left[1 + \frac{2b}{a}\left(\frac{f_c}{f}\right)^2\right]$$

(27)

Figure 8.7c shows attenuation versus frequency for TM_{11} and TE_{10} modes in rectangular copper waveguides with various side ratios b/a found using (14) and (27), respectively.

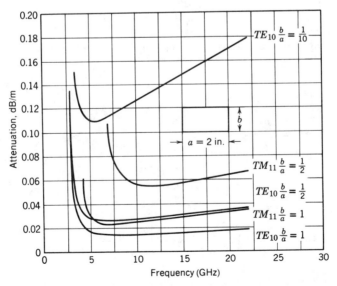

Fig. 8.7c Attenuation due to copper losses in rectangular waveguides of fixed width.

8.8 The TE_{10} Wave in a Rectangular Guide

One of the simplest of all the waves which may exist inside hollow-pipe waveguides is the dominant TE_{10} wave in the rectangular guide, which is one of the TE modes studied in the preceding section. This mode is of great engineering importance, partly for the following reasons:

1. Cutoff frequency is independent of one of the dimensions of the cross section. Consequently, for a given frequency this dimension may be made small enough so that the TE_{10} wave is the only wave which will propagate, and there is no difficulty with higher-order waves that end effects or discontinuities may cause to be excited.
2. The polarization of the field is definitely fixed, electric field passing from top to bottom of the guide. This fixed polarization may be required for certain applications.
3. For a given frequency the attenuation due to copper losses is not excessive compared with other wave types in guides of comparable size.

Let us now rewrite the expressions from the previous section for general TE waves in rectangular guides, Eqs. 8.7(20)–(24), setting $m = 1$, $n = 0$, in which case $k_y = 0$ and $k_c = k_x = \pi/a$.

$$H_z = B \cos k_x x \tag{1}$$

$$E_y = -\frac{j\omega\mu B}{k_x} \sin k_x x \tag{2}$$

$$H_x = \frac{j\beta B}{k_x} \sin k_x x \tag{3}$$

All other components are zero. This set may be rewritten in a useful alternate form,

$$E_y = -Z_{TE} H_x = E_0 \sin\left(\frac{\pi x}{a}\right) \tag{4}$$

$$H_z = \frac{jE_0}{\eta}\left(\frac{\lambda}{2a}\right)\cos\left(\frac{\pi x}{a}\right) \tag{5}$$

where

$$E_0 = -\frac{j\omega\mu B}{k_x} = -\frac{j2\eta a B}{\lambda} \tag{6}$$

$$Z_{TE} = \eta\left[1 - \left(\frac{\omega_c}{\omega}\right)^2\right]^{-1/2} = \eta\left[1 - \left(\frac{\lambda}{2a}\right)^2\right]^{-1/2} \tag{7}$$

$$\eta = \sqrt{\frac{\mu}{\varepsilon}}, \qquad \lambda = \frac{v}{f} = \frac{2\pi}{\omega\sqrt{\mu\varepsilon}} \tag{8}$$

Cutoff frequency, wavelength, and wavenumber are

$$f_c = \frac{1}{2a\sqrt{\mu\varepsilon}}, \qquad \lambda_c = 2a, \qquad k_c = \frac{\pi}{a} \tag{9}$$

Phase and group velocities, and wavelength measured along the guide, are

$$v_p = \frac{1}{\sqrt{\mu\varepsilon}\sqrt{1 - (\lambda/2a)^2}}, \qquad v_g = \frac{1}{\sqrt{\mu\varepsilon}}\sqrt{1 - \left(\frac{\lambda}{2a}\right)^2} \tag{10}$$

$$\lambda_g = \frac{v_p}{f} = \frac{2\pi}{\beta} = \frac{\lambda}{\sqrt{1 - (\lambda/2a)^2}} \tag{11}$$

The attenuation arising from an imperfect dielectric is obtained by replacing ε with $\varepsilon' - j\varepsilon''$ in the equation for γ. Since k_c in the TE_{m0} modes is of the same form as in the parallel-plane modes, Eq. 8.5(3) applies here and leads to a result equivalent to Eq. 8.5(4)

$$\alpha_d = \frac{k\varepsilon''/\varepsilon'}{2\sqrt{1 - (\lambda/2a)^2}} \tag{12}$$

To find attenuation if the conductor is imperfect, we first calculate the power transferred by the wave from the Poynting theorem,

$$W_T = \frac{1}{2}\text{Re}\int_0^a \int_0^b (-E_y H_x^*)\, dx\, dy \tag{13}$$

Utilizing the forms (4) we have

$$W_T = \frac{E_0^2 b}{2Z_{TE}}\int_0^a \sin^2 \frac{\pi x}{a}\, dx = \frac{E_0^2 ba}{4Z_{TE}} \tag{14}$$

Next we find approximate losses in the walls by using currents of the ideal mode in material of surface resistivity R_s. Current in a conductor is related to the tangential magnetic field H_z at the side walls $x = 0$ and $x = a$ so there is current per unit width $|J_{sy}| = |H_z|$ there. Both components H_x and H_z are tangential at top and bottom surfaces giving rise to surface current densities $|J_{sz}| = |H_x|$ and $|J_{sx}| = |H_z|$. Thus power loss per unit length is

$$(w_L)_{\text{SIDES}} = 2\left(\frac{bR_s}{2}|H_z|_{x=0}^2\right) = \frac{bR_s E_0^2 \lambda^2}{4\eta^2 a^2} \tag{15}$$

$$(w_L)_{\text{TOP AND BOTTOM}} = 2\frac{R_s}{2}\int_0^a (|H_x|^2 + |H_z|^2)\, dx$$

$$= R_s \int_0^a \left[\frac{E_0^2}{Z_{TE}^2}\sin^2 \frac{\pi x}{a} + \frac{E_0^2 \lambda^2}{4\eta^2 a^2}\cos^2 \frac{\pi x}{a}\right] dx$$

$$= \frac{a}{2}R_s\left(\frac{E_0^2}{Z_{TE}^2} + \frac{E_0^2 \lambda^2}{4\eta^2 a^2}\right) \tag{16}$$

Adding the two contributions (15) and (16) and substituting Z_{TE} from (7),

$$w_L = \frac{R_s E_0^2}{\eta^2}\left[\frac{b\lambda^2}{4a^2} + \frac{a}{2}\left(1 - \frac{\lambda^2}{4a^2} + \frac{\lambda^2}{4a^2}\right)\right] = \frac{R_s E_0^2}{2\eta^2}\left(a + \frac{b\lambda^2}{2a^2}\right)$$

The attenuation from conductor losses, Eq. 5.9(4), is then

$$\alpha_c = \frac{w_L}{2W_T} = \frac{R_s Z_{TE}}{\eta^2 ba}\left(a + \frac{b\lambda^2}{2a^2}\right) \tag{17}$$

or

$$\alpha_c = \frac{R_s}{b\eta\sqrt{1 - (\lambda/2a)^2}}\left[1 + \frac{2b}{a}\left(\frac{\lambda}{2a}\right)^2\right] \tag{18}$$

A study of the field distributions (1) to (3) or (4) and (5) shows the field patterns for this wave sketched in Table 8.7. First it is noted that no field components vary in the vertical or y direction. The only electric field component is the vertical one E_y passing between top and bottom of the guide. This is a maximum at the center and zero at the conducting walls, varying as a half-sine curve. The corresponding charges induced by the electric field lines ending on conductors are

1. Charges zero on side walls.
2. A charge distribution on top and bottom with $\rho_s = \varepsilon E_y$ on the bottom, $-\varepsilon E_y$ on the top.

The magnetic field forms closed paths surrounding the vertical electric displacement currents arising from E_y, so that there are components H_x and H_z. Component H_x is zero at the two side walls and a maximum in the center, following the distribution of E_y. Component H_z is a maximum at the side walls and zero at the center. Component H_x corresponds to a longitudinal current flow down the guide in the top, and opposite in the bottom; H_z corresponds to transverse currents in the top and bottom and vertical currents on the side walls. These current distributions are sketched in Fig. 8.8a.

This simple wave type is a convenient one to study in order to strengthen some of our physical pictures of wave propagation. Electric field is confined to the transverse plane and so passes between opposite charges of equal density on the top and bottom. The electric field E_y and the transverse magnetic field H_x are maximum at planes marked A. Halfway between those planes is the maximum rate of change of E_y and therefore the location of maximum displacement current. The currents concomitant to the tangential magnetic field feed into the displacement currents so there is continuity of current. The magnetic fields surround the electric displacement currents inside the guide and so must have an axial as well as a transverse component.

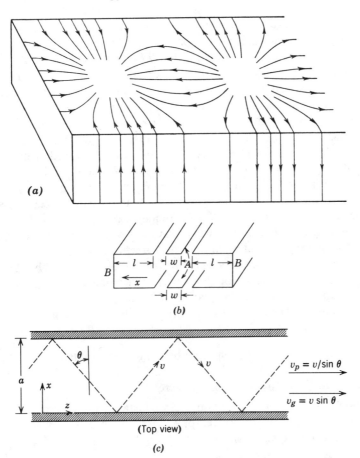

Fig. 8.8 (*a*) Current flow in walls of rectangular guide with TE_{10} mode. (*b*) Guide roughly divided into axial- and transverse-current regions. (*c*) Path of uniform plane wave component of TE_{10} wave in rectangular guide.

As a fairly crude way of looking at the problem, one might also think of this mode being formed by starting with a parallel plate transmission line *A* of width *w* to carry the longitudinal current in the center of the guide, and then adding shorted troughs *B* of depth *l* on the two sides to close the region, as pictured in Fig. 8.8*b*. Since one would expect the lengths *l* to be around a quarter-wavelength to provide a high impedance at the center, the overall width should be something over a half-wavelength, which we know to be true for propagation. The picture is only a rough one because the fields in the two regions are not separated, and propagation is not purely longitudinal in the center portion or transverse in the side portions.

A third viewpoint follows from that used in studying the higher order waves between parallel planes. There it was pointed out that one could visualize the

TM and *TE* waves in terms of plane waves bouncing between the two planes at such an angle that the interference pattern maintains a zero of electric field tangential to the two planes. Similarly, the TE_{10} wave in the rectangular guide may be thought of as arising from the interference between incident and reflected plane waves, polarized so that the electric vector is vertical, and bouncing between the two sides of the guide at such an angle with the sides that the zero electric field is maintained at the two sides. One such component uniform plane wave is indicated in Fig. 8.8c. As in the result of Sec. 8.4, when the width *a* is exactly $\lambda/2$, the waves travel exactly back and forth across the guide with no component of propagation in the axial direction. At slightly higher frequencies there is a small angle θ such that $a = \lambda/2\cos\theta$, and there is a small propagation in the axial direction, a very small group velocity in the axial direction $v\sin\theta$ and a very large phase velocity $v/\sin\theta$. At frequencies approaching infinity, θ approaches 90 degrees, so that the wave travels down the guide practically as a plane wave in space propagating in the axial direction.

All the foregoing points of view explain why the dimension *b* should not enter into the determination of cutoff frequency. Since the electric field is always normal to top and bottom, the placing of these planes plays no part in the boundary condition. However, this dimension *b* will be important from three other points of view.

1. It is desired to have a large spacing between the cutoff frequency of the TE_{10} mode and that of the next higher mode. If $b/a > \frac{1}{2}$ the TE_{01} mode has the next lowest f_c and that is raised by decreasing *b*.

2. The smaller *b* is (all other parameters constant), the greater is the electric field across the guide for a given power transfer, as is seen from (14), and so the danger of voltage breakdown is greater.

3. The smaller *b* is (all other parameters constant), the greater is the attenuation due to conductor losses. This may be seen by keeping fields constant as *b* approaches zero. Power transfer, by (14), and loss in the sides, by (15), approach zero, but loss in top and bottom remains constant at the value calculated in (16).

8.9   Circular Waveguides

Hollow-pipe waveguides of circular cross section are used in a number of instances, for example, when circular polarization is to be transmitted to certain classes of antennas. Also, as will be shown, the class of TE_{0n} modes (called *circular electric*) is interesting because of the low attenuation in this class at high frequencies. As before, we start with ideal dielectric and conducting boundary and make approximate modifications to these solutions when the materials have small losses. Before treating separately the *TM* and *TE* classes of waves, it is desirable to have the set of equations 8.2(13)–(16) transformed to circular cylindrical coordinates. A straightforward transformation gives

$$E_r = -\frac{j}{k_c^2}\left[\beta\frac{\partial E_z}{\partial r} + \frac{\omega\mu}{r}\frac{\partial H_z}{\partial \phi}\right] \tag{1}$$

$$E_\phi = \frac{j}{k_c^2}\left[-\frac{\beta}{r}\frac{\partial E_z}{\partial \phi} + \omega\mu\frac{\partial H_z}{\partial r}\right] \tag{2}$$

$$H_r = \frac{j}{k_c^2}\left[\frac{\omega\varepsilon}{r}\frac{\partial E_z}{\partial \phi} - \beta\frac{\partial H_z}{\partial r}\right] \tag{3}$$

$$H_\phi = -\frac{j}{k_c^2}\left[\omega\varepsilon\frac{\partial E_z}{\partial r} + \frac{\beta}{r}\frac{\partial H_z}{\partial \phi}\right] \tag{4}$$

where

$$k_c^2 = \gamma^2 + k^2 = k^2 - \beta^2 \tag{5}$$

TM Waves The transverse part of the Laplacian in Eq. 8.2(17) for E_z is expressed in circular cylindrical coordinates for this configuration, with the coordinate system as shown in Fig. 8.9a.

$$\nabla_t^2 E_z = \frac{1}{r}\frac{\partial}{\partial r}\left(r\frac{\partial E_z}{\partial r}\right) + \frac{1}{r^2}\frac{\partial^2 E_z}{\partial \phi^2} = -k_c^2 E_z \tag{6}$$

Separation of variables techniques in Sec. 7.20 led to the solution

$$E_z(r, \phi) = [A'J_n(k_c r) + B'N_n(k_c r)][C'\cos n\phi + D'\sin n\phi] \tag{7}$$

where J_n and N_n are nth order Bessel functions of first and second kind, respectively. The second kind, $N_n(k_c r)$, is infinite at $r = 0$ for any n and so cannot be included in the interior solution which includes the axis. Also, for simplicity, we choose the origin of ϕ so that we have just the $\cos n\phi$ variation. Letting $A'C' = A$,

$$E_z = AJ_n(k_c r)\cos n\phi \tag{8}$$

From (1) to (4) with $H_z = 0$, the remaining field components are

$$E_r = Z_{TM}H_\phi = -\frac{j\beta}{k_c}AJ_n'(k_c r)\cos n\phi \tag{9}$$

$$E_\phi = -Z_{TM}H_r = \frac{j\beta n}{k_c^2 r}AJ_n(k_c r)\sin n\phi \tag{10}$$

where

$$Z_{TM} = \frac{\beta}{\omega\varepsilon} \tag{11}$$

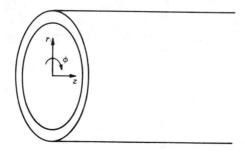

Fig. 8.9a Hollow-pipe circular cylindrical guide showing coordinate system.

The boundary condition imposed by the perfect conductor at $r = a$ requires E_z and E_ϕ to be zero there. We see from (8) that E_z is zero at the boundary if $k_c a$ is one of the zeros of the Bessel function,

$$k_c a = \omega_c \sqrt{\mu\varepsilon}\, a = \frac{2\pi a}{\lambda_c} = p_{nl} \qquad \text{where} \quad J_n(p_{nl}) = 0 \tag{12}$$

We see from (10) that this makes E_ϕ zero there also.

Equation (12) allows one to calculate cutoff frequency or wavelength for any mode order. For any n there is an infinite number of zeros of $J_n(k_c r)$ so there is a doubly infinite set of modes, denoted TM_{nl}. Note that the first subscript denotes angular variations and the second radial variations, differing from the usual cyclic order of coordinates. Field distributions of the TM_{01}, TM_{02}, and TM_{11} modes are given in Table 8.9 together with expressions for cutoff frequency and wavelength, and approximate attenuation from conductor losses if the cylindrical wall has surface resistivity R_s. Phase velocity, group velocity, guide wavelength, and attenuation for frequencies below cutoff are of the same forms in terms of cutoff frequency as we have seen for the parallel-plane and rectangular guide. Wave impedance, when β in terms of cutoff frequency is substituted in (11), also has the form found for TM modes in the guides studied earlier.

TE **Waves**   The differential equation for the nonzero H_z of *TE* waves, Eq. 8.2(18), expressed in cylindrical coordinates, is

$$\nabla_t^2 H_z = \frac{1}{r}\frac{\partial}{\partial r}\left(r\frac{\partial H_z}{\partial r}\right) + \frac{1}{r^2}\frac{\partial^2 H_z}{\partial \phi^2} = -k_c^2 H_z \tag{13}$$

The solution for this from Sec. 7.20 is

$$H_z(r, \phi) = B J_n(k_c r)\cos n\phi \tag{14}$$

Table 8.9
Summary of wave types for circular guides[a]

Wave Type	TM_{01}	TM_{02}	TM_{11}	TE_{01}	TE_{11}
Field distributions in cross-sectional plane, at plane of maximum transverse fields					
Field distributions along guide					
Field components present	E_z, E_r, H_ϕ	E_z, E_r, H_ϕ	$E_z, E_r, E_\phi, H_r, H_\phi$	H_z, H_r, E_ϕ	$H_z, H_r, H_\phi, E_r, E_\phi$
p_{nl} or p'_{nl}	2.405	5.52	3.83	3.83	1.84
$(k_c)_{nl}$	$\dfrac{2.405}{a}$	$\dfrac{5.52}{a}$	$\dfrac{3.83}{a}$	$\dfrac{3.83}{a}$	$\dfrac{1.84}{a}$
$(\lambda_c)_{nl}$	$2.61a$	$1.14a$	$1.64a$	$1.64a$	$3.41a$
$(f_c)_{nl}$	$\dfrac{0.383}{a\sqrt{\mu\varepsilon}}$	$\dfrac{0.877}{a\sqrt{\mu\varepsilon}}$	$\dfrac{0.609}{a\sqrt{\mu\varepsilon}}$	$\dfrac{0.609}{a\sqrt{\mu\varepsilon}}$	$\dfrac{0.293}{a\sqrt{\mu\varepsilon}}$
Attenuation due to imperfect conductors	$\dfrac{R_s}{a\eta}\dfrac{1}{\sqrt{1-(f_c/f)^2}}$	$\dfrac{R_s}{a\eta}\dfrac{1}{\sqrt{1-(f_c/f)^2}}$	$\dfrac{R_s}{a\eta}\dfrac{1}{\sqrt{1-(f_c/f)^2}}$	$\dfrac{R_s}{a\eta}\dfrac{(f_c/f)^2}{\sqrt{1-(f_c/f)^2}}$	$\dfrac{R_s}{a\eta}\dfrac{1}{\sqrt{1-(f_c/f)^2}}\left[\left(\dfrac{f_c}{f}\right)^2+0.420\right]$

[a] Electric field lines are shown solid and magnetic field lines are dashed.

Here we have left out the second solution and chosen the cosine variation with ϕ with the same justification as for the TM waves. The remaining field components, from (1) to (4) with $E_z = 0$, are

$$E_r = Z_{TE} H_\phi = \frac{j\omega\mu n}{k_c^2 r} B J_n(k_c r) \sin n\phi \qquad (15)$$

$$E_\phi = -Z_{TE} H_r = \frac{j\omega\mu}{k_c} B J_n'(k_c r) \cos n\phi \qquad (16)$$

where

$$Z_{TE} = \frac{\omega\mu}{\beta} \qquad (17)$$

In this case the boundary condition imposed by the perfect conductor at $r = a$ requires that $E_\phi = 0$ there, or

$$k_c a = \omega_c \sqrt{\mu\varepsilon} a = \frac{2\pi a}{\lambda_c} = p_{nl}' \qquad \text{where} \quad J_n'(p_{nl}) = 0 \qquad (18)$$

Again, expressions for phase and group velocity and wavelength along the guide are as before.

The field distributions of the TE_{01} and TE_{11} modes, and some data for these, are given in Table 8.9. Notice that the field distribution of the TE_{11} mode is quite a bit like that of the TE_{10} mode in the rectangular guide, with electric field going from top to bottom of the guide, so this is the one that would be primarily excited if a TE_{10}-mode rectangular guide were properly tapered and connected to the circular guide. It also has the lowest cutoff frequency of any mode in a given size circular pipe, as shown by Fig. 8.9b. In the TE_{01} mode, the electric field lines do not end on the guide walls, but form closed circles surrounding the axial time-varying magnetic field. This latter wave is especially interesting as a potential low-loss transmission system at high frequencies and will be considered more in the example.

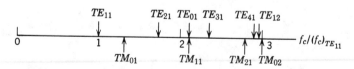

Fig. 8.9b Relative cutoff frequencies of waves in a circular guide.

_____ **Example 8.9** _____
Circular Electric TE_{01} Mode in Oversize Guide

The field expressions for the TE_{01} mode from the general forms (14)–(16) are

$$H_z = BJ_0(k_c r) \tag{19}$$

and

$$E_\phi = -Z_{TE} H_r = -\frac{j\omega\mu}{k_c} BJ_1(k_c r) \tag{20}$$

$$k_c a = p'_{01} = p_{11} = 3.83 \ldots \tag{21}$$

The average power transfer by the mode, from the Poynting theorem is

$$W_T = \int_0^a \frac{2\pi r}{2}(-E_\phi H_r^*)\, dr = \frac{\omega^2\mu^2 B^2\pi}{k_c^2 Z_{TE}} \int_0^a r J_1^2(k_c r)\, dr \tag{22}$$

The Bessel integral is evaluated by Eq. 7.15(22):

$$W_T = \frac{\omega^2\mu^2 B^2\pi}{k_c^2 Z_{TE}}\left[\frac{a^2}{2} J_0^2(p_{11})\right] \tag{23}$$

Conduction current in the guide walls is purely circumferential, related to the tangential H_z, so that wall losses per unit length are

$$w_L = 2\pi a \frac{R_s}{2}|H_z|^2_{r=a} = \pi a R_s B^2 J_0^2(p_{11}) \tag{24}$$

Attenuation per unit length, in terms of power transfer and loss, is then

$$\alpha_c = \frac{w_L}{2W_T} = \frac{k_c^2 Z_{TE} R_s}{\omega^2\mu^2 a} \tag{25}$$

When Z_{TE} is substituted from (17), this may be put in the form

$$\alpha_c = \frac{R_s(\omega_c/\omega)^2}{a\eta\sqrt{1-(\omega_c/\omega)^2}} \tag{26}$$

R_s is proportional to the square root of frequency, but the overall expression decreases with frequency. Thus we have the unusual result that attenuation in this mode, for a given size guide, decreases with increasing frequency, as shown in Fig. 8.9c which gives comparison with TE_{11} and TM_{01} modes for a 2-in.-diameter guide. The low attenuation is because the mode fields are very little

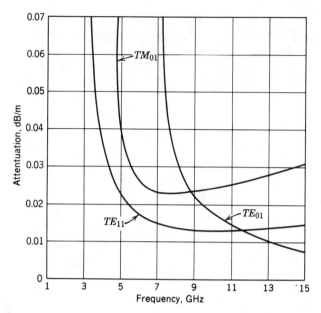

Fig. 8.9c Attenuation due to copper losses in circular waveguides; diameter $= 2$ in.

coupled to the guide walls at high frequencies. However, other modes may propagate, as shown by Fig. 8.9b, so that there are problems with mode conversion when such guides bend to go around corners. Practical ways of solving such problems have been developed[6] and this system has been used as a low-loss guiding system for millimeter waves.

8.10 Higher Order Modes on Coaxial Lines

The lowest-order mode on a coaxial line is a *TEM* wave; this was assumed implicitly in the transmission-line treatment of Ex. 5.2 where we used the capacitance and inductance found from static fields. As in the parallel-plane guide, higher order (*TM* and *TE*) modes can also exist. Normally, the line is designed in such a way that the cutoff frequencies of the higher order modes are well above the operating frequency. Even in that case, these modes can be of importance near discontinuities.

The general forms useful for the *TM* and *TE* modes in circular cylindrical coordinates are listed in Sec. 7.20. The boundary conditions require that E_z for *TM* waves be zero at r_0 and r_i (Fig. 8.10a).

---

[6] See, for example, S. E. Miller, *Bell System Tech. J.* **33**, 1209 (1954).

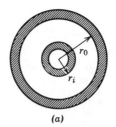

(a)

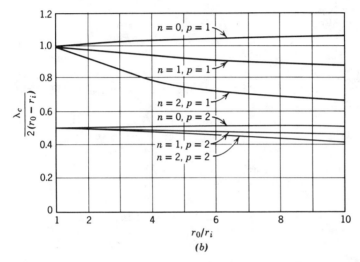

(b)

Fig. 8.10 (a) Cross section of a coaxial line. (b) Cutoff wavelength for some higher order *TM* waves in coaxial lines.

For *TM* waves,

$$A_n J_n(k_c r_i) + B_n N_n(k_c r_i) = 0$$

$$A_n J_n(k_c r_0) + B_n N_n(k_c r_0) = 0$$

or

$$\frac{N_n(k_c r_i)}{J_n(k_c r_i)} = \frac{N_n(k_c r_0)}{J_n(k_c r_0)} \tag{1}$$

For *TE* waves, the derivative of H_z normal to the two conductors must be zero at the inner and outer radii. [See discussion following Eq. 8.7(21).] Then, in place of (1),

$$\frac{N_n'(k_c r_i)}{J_n'(k_c r_i)} = \frac{N_n'(k_c r_0)}{J_n'(k_c r_0)} \tag{2}$$

Solutions to the transcendental equations (1) and (2) determine the values of k_c and hence cutoff frequency for any wave type and any particular values of r_i and r_0. Solution of the transcendental equations is accomplished by graphical or numerical methods. By analogy with the parallel-plane guide, we would expect to find certain modes with a cutoff such that the spacing between conductors is of the order of p half-wavelengths.

$$\lambda_c \approx \frac{2}{p}(r_0 - r_i), \qquad p = 1, 2, 3, \dots \tag{3}$$

This is verified by Fig. 8.10b for values of r_0/r_i near unity.

Probably more important is the lowest order TE wave with circumferential variations. This is analogous to the TE_{10} wave of a rectangular waveguide, and physical reasoning from the analogy leads one to expect cutoff for this wave type when the average circumference is about equal to wavelength. The field picture of the TE_{10} mode given in Sec. 8.8 should make this reasonable. Solution of (2) reveals this simple rule to be within about 4% accuracy for r_0/r_i up to 5. In general, for the nth order TE wave with circumferential variations,

$$\lambda_c \approx \frac{2\pi}{n}\left(\frac{r_0 + r_i}{2}\right), \qquad n = 1, 2, 3, \dots \tag{4}$$

There are, of course, other TE waves with further radial variations, and the lowest order of these has a cutoff about the same as the lowest order TM wave.

8.11 Excitation and Reception of Waves in Guides

The problems of exciting or receiving waves in a waveguide are not simple field problems. In this section we will give only a qualitative introduction to the manners of excitation of fields in various kinds of guides. Approaches to analysis and measurement of these junctions are given in Chapter 11. Reception of the energy of a wave uses the same kind of structure as excitation and is just the reverse process. To excite any particular desired wave, one should study the field pattern and use one of the following concepts.

1. Introduce the excitation in a probe or antenna oriented in the direction of electric field. The probe is most often placed near a maximum of the electric field of the mode pattern, but exact placing is a matter of impedance matching. Examples are shown in Figs. 8.11a and 8.11b.

2. Introduce the excitation through a loop oriented in a plane normal to magnetic field of the mode pattern (Fig. 8.11c).

3. Couple to the desired mode from another guiding system, by means of a hole or iris, the two guiding systems having some common field component over the extent of the hole. An example of coupling between waveguides using a large iris is shown in Fig. 8.11d. The coupling is sometimes done with a small hole as for coupling to resonant cavities (Sec. 10.10).

4. Introduce currents from one kind of transmission line into another, as in coupling from a coaxial line to microstrip shown in Fig. 8.11e.

5. For higher order waves combine as many of the exciting sources as are required, with proper phasings (Fig. 8.11f).

6. Gradually taper a transition between two types of guides, as for a TE_{10} wave in a rectangular guide to a TE_{11} in a circular guide.

Since most of these exciting methods are in the nature of concentrated sources, they will not in general excite purely one wave, but all waves that have field components in a favorable direction for the particular exciting source. That is, we see that one wave alone will not suffice to satisfy the boundary conditions of the guide complicated by the exciting source, so that many higher order waves must be added for this purpose. If the guide is large enough, several of these waves will then proceed to propagate. Most often, however, only one of the excited waves is above cutoff. This will propagate down the guide, and (if absorbed somewhere) will represent a resistive load on the source, comparable to the radiation resistance of antennas which we shall encounter further in Chapter 12. The higher order waves that are excited, if all below cutoff, will be localized in the neighborhood of the source and will represent purely reactive loads on the source. For practical application, it is then necessary to add, in the line that feeds the probe or loop or other exciting means, an arrangement for matching to the load that has a real part representing the propagating wave and an imaginary part representing the localized reactive waves.

_____ **Example 8.11** _____
Excitation of a Waveguide by a Coaxial Line

Let us look in more depth at the structure in Fig. 8.11b where a coaxial line is inserted in the center of the broad side of a waveguide of rectangular cross section in order to excite a TE_{10} mode. The waveguide is short circuited at a distance l from the probe to aid in matching the coaxial line to the waveguide. The fields associated with the probe excite both the desired TE_{10} mode and other higher order modes. The latter are cutoff and do not propagate, but they store reactive energy and therefore constitute a reactive component of the load on the coaxial line. Proper choice of the size and location of the probe for a given frequency and guide dimensions make the standing wave between the probe and the shorted end

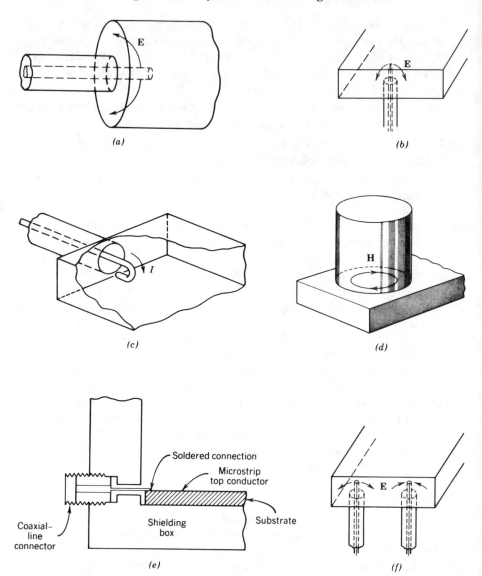

Fig. 8.11 (a) Antenna in end of circular guide for excitation of TM_{01} wave. (b) Antenna in bottom of rectangular guide for excitation of the TE_{10} wave. (c) Loop in end of rectangular guide for excitation of TE_{10} wave. (d) Junction between circular guide (TM_{01} wave) and rectangular guide (TE_{10} wave); large-aperture coupling. (e) Coaxial line coupling to microstrip. (f) Excitation of the TE_{20} wave in rectangular guide by two oppositely phased antennas.

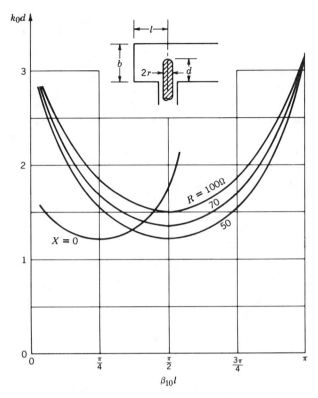

Fig. 8.11g Impedance parameters for coaxial line-to-waveguide transition in inset. $X = 0$ and R = constant contours in $k_0 d - \beta_{10} l$ plane for $a = 2.29$ cm, $b = 1.02$ cm, $r = 0.05a$, and $\lambda_0 = 3.14$ cm. From Ref. 7.

contain reactive energy of opposite sign and equal magnitude so that the net reactive component of the input impedance is zero. These adjustments are used to make the real part of the load impedance on the coaxial line equal to its characteristic impedance so that perfect matching is achieved and all the power is coupled into the guide. Figure 8.11g shows the calculated results for a probe of given radius in a guide of dimensions appropriate for use at about 10 GHz (X band).[7] The probe length and location required to make the reactive load X zero and for a selected resistive load R can be read from the graph. Similar graphs of R and X can be calculated using the formulas in Ref. 7 for other guide sizes and probe radii.

---

[7] R. E. Collin, *Field Theory of Guided Waves*, McGraw-Hill, New York, 1960, Sec. 7.1.

GENERAL PROPERTIES OF GUIDED WAVES

8.12   General Properties of *TEM* Waves on Multiconductor Lines

The classical two-conductor transmission system was studied extensively in Chapter 5, starting from a distributed circuits point of view. We used wave solutions to verify the results for the special case of *TEM* waves between parallel planes in Sec. 8.3. At this point we can show that *TEM* waves in any two-conductor cylindrical system with isotropic, homogeneous dielectric, and loss-free conductors are exactly those predicted by the transmission line equations.

The general relations between wave components as expressed by Eqs. 8.2(9)–(12) show that, with E_z and H_z zero, all other components must of necessity also be zero, unless $\gamma^2 + k^2$ is at the same time zero. Thus, a transverse electromagnetic wave must satisfy the condition

$$\gamma^2 + k^2 = 0$$

or

$$\gamma = \pm jk = \pm \frac{j\omega}{v} = \pm j\omega\sqrt{\mu\varepsilon} \tag{1}$$

For a perfect dielectric, the propagation constant γ is thus a purely imaginary quantity, signifying that any completely transverse electromagnetic wave must propagate unattenuated, and with velocity v, the velocity of light in the dielectric bounded by the guide.

With (1) satisfied, the wave equations, as written in the form of Eqs. 8.2(1) and 8.2(2), reduce to

$$\nabla^2_{xy}\mathbf{E} = 0; \qquad \nabla^2_{xy}\mathbf{H} = 0 \tag{2}$$

These are exactly the form of the two-dimensional Laplace's equation written for **E** and **H** in the transverse plane. Since E_z and H_z are zero, E and H lie entirely in the transverse plane. Since electric and magnetic fields both satisfy Laplace's equation under static conditions, the field distribution in the transverse plane is exactly a static distribution if boundary conditions to be applied to the fields in (2) are the same as those for a static field distribution. The boundary condition for the *TEM* wave on a perfect conducting guide is that electric field at the surface of the conductor can have a normal component only, which is the same as the condition at a conducting boundary in statics. The line integral of the electric field between conductors is the same for all paths lying in a given transverse plane, and may be thought of as corresponding to a potential difference between the conductors for that value of z.

To study the character of the magnetic field, note Eqs. 8.2(3) and 8.2(6) with zero E_z and H_z.

$$H_y = \frac{j\omega\varepsilon}{\gamma} E_x = \frac{E_x}{\eta} \tag{3}$$

and

$$H_x = -\frac{\gamma}{j\omega\mu} E_y = -\frac{E_y}{\eta} \tag{4}$$

[The signs of (3) and (4) are for a positively traveling wave; for a negatively traveling wave they are opposite.] Study shows that (3) and (4) are conditions that require that electric and magnetic field be everywhere normal to each other. In particular, magnetic field must be tangential to the conducting surfaces since electric field is normal to them. The magnetic field pattern in the transverse plane then corresponds exactly to that arising from static currents flowing entirely on the surfaces of the perfect conductors.

These characteristics show that a transverse electromagnetic wave may be guided by two or more conductors, or outside a single conductor, but not inside a closed conducting region, since it can have only the distribution of the corresponding two-dimensional static problem, and no electrostatic field can exist inside a source-free region completely closed by a conductor (see Prob. 8.12d).

We next may show an exact identity with the ordinary transmission line equations for *TEM* waves on the systems that support them. Consider a line consisting of two conductors *A* and *B* of any general shape, Fig. 8.12. The voltage between the two lines may be found by integrating electric field over any path between conductors, such as 1–0–2 of the figure. It will have the same value no matter which path is chosen, since **E** satisfies Laplace's equation in the transverse

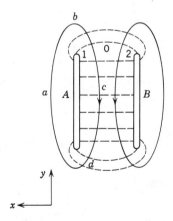

Fig. 8.12 Two-conductor transmission line with integration paths.

plane and so may be considered the gradient of a scalar potential in so far as variations in the transverse plane are concerned.

$$V = -\int_1^2 \mathbf{E} \cdot d\mathbf{l} = -\int_1^2 (E_x \, dx + E_y \, dy) \tag{5}$$

Differentiating the above equation with respect to z

$$\frac{\partial V}{\partial z} = -\int_1^2 \left(\frac{\partial E_x}{\partial z} \, dx + \frac{\partial E_y}{\partial z} \, dy \right) \tag{6}$$

But the curl relation,

$$\nabla \times \mathbf{E} = -\frac{\partial \mathbf{B}}{\partial t}$$

shows that, if E_z is zero,

$$\frac{\partial E_y}{\partial z} = \frac{\partial B_x}{\partial t} \quad \text{and} \quad \frac{\partial E_x}{\partial z} = -\frac{\partial B_y}{\partial t} \tag{7}$$

By substituting (7) in (6), we have

$$\frac{\partial V}{\partial z} = -\frac{\partial}{\partial t} \int_1^2 (-B_y \, dx + B_x \, dy) \tag{8}$$

A study of Fig. 8.12 reveals that the quantity inside the integral is the magnetic flux flowing across the path 1–0–2, per unit length in the z direction. According to the usual definition of inductance, this may be written as the product of inductance L per unit length and the current I so (8) becomes

$$\frac{\partial V}{\partial z} = -\frac{\partial}{\partial t}(LI) = -L\frac{\partial I}{\partial t} \tag{9}$$

Equation (9) is one of the differential equations used as a starting point for conventional transmission line analysis [Eq. 5.2(3)]. The other may be developed by starting with current in line A as the integral of magnetic field about a path a–b–c–d–a. (There is no contribution from displacement current since there is no E_z.)

$$I = \oint \mathbf{H} \cdot d\mathbf{l} = \oint (H_x \, dx + H_y \, dy) \tag{10}$$

Differentiating with respect to z,

$$\frac{\partial I}{\partial z} = \oint \left(\frac{\partial H_x}{\partial z} \, dx + \frac{\partial H_y}{\partial z} \, dy \right) \tag{11}$$

From the curl equation,

$$\nabla \times \mathbf{H} = \frac{\partial \mathbf{D}}{\partial t}$$

it follows that, if $H_z = 0$,

$$\frac{\partial H_y}{\partial z} = -\frac{\partial D_x}{\partial t} \quad \text{and} \quad \frac{\partial H_x}{\partial z} = \frac{\partial D_y}{\partial t} \tag{12}$$

Substituting (12) in (11), we have

$$\frac{\partial I}{\partial z} = -\frac{\partial}{\partial t} \oint (D_x \, dy - D_y \, dx) \tag{13}$$

Inspection of Fig. 8.12 shows that this must be the electric displacement flux per unit length of line crossing from one conductor to the other. Since it corresponds to the charge per unit length on the conductors, it may be written as the product of capacitance per unit length and the voltage between lines and (13) becomes

$$\frac{\partial I}{\partial z} = -C \frac{\partial V}{\partial t} \tag{14}$$

Equations (9) and (14) are exactly the equations used as a beginning for transmission-line analysis, if losses are neglected (Sec. 5.2). It is seen that these equations may be derived exactly from Maxwell's equations provided the conductors are perfect, and, since fields in the transverse plane satisfy Laplace's equation, the inductance and capacitance appearing in the equations are the same as those calculated in statics. This is of course not the case for the *TM* and *TE* waves met in the preceding sections. Since the transmission line equations correspond to a *TEM* field distribution and phase velocity in the former is $(LC)^{-1/2}$ and in the latter $(\mu\varepsilon)^{-1/2}$, then for any transmission line with uniform dielectric $LC = \mu\varepsilon$.

Transmission Lines with Losses   If the conductor of the transmission line has finite losses, the above argument does not apply exactly. There must be finite E_z at the conductor to force the axial currents through the imperfect conductors. In that case $\gamma^2 + k^2$ of Eqs. 8.2(9)–(12) cannot be zero, and Eqs. 8.2(1) and 8.2(2) do not reduce to Laplace's equation. But so long as the conductors are reasonably good, the axial component of electric field is small compared with the transverse component and corrections are small. The usual way of handling the losses through a series resistance in the transmission line equations can then be shown by perturbation arguments to be an excellent approximation.[8] Losses in the dielectric, however, do not in themselves disturb the *TEM* nature of the wave since these cause conduction currents to flow only in transverse directions. In this

[8] Ref. 7, Sec. 4.1.

case treatment by inclusion of shunt conductance computed from static concepts and by the wave method with ε replaced by $\varepsilon' - j\varepsilon''$ can be shown to be the same (Prob. 8.12c).

In addition to the principal *TEM* or *transmission-line* mode on the two-conductor system, there may propagate higher order modes as well. The higher order modes may be excited at discontinuities in the transmission line and may cause dispersion effects or radiation. It is difficult to work out the forms of the higher order modes in open structures such as the two-wire line, but is straight-forward to derive them for the coaxial line as was done in Sec. 8.10.

8.13　General Properties of *TM* Waves in Cylindrical Conducting Guides of Arbitrary Cross Section

In the earlier sections we have seen several specific examples of *TM* waves; it is the purpose of the present section to generalize the formulation for any cylindrical structure. The analysis can be done in a generalized coordinate system[9] and might appear more general, but for simplicity we will use rectangular coordinates with the understanding that boundaries may be of arbitrary shape.

The Differential Equation　With the assumed propagation constant $e^{j(\omega t - \beta z)}$, the finite axial component of electric field for the *TM* waves must satisfy the wave equation in the form of Eq. 8.2(17):

$$\nabla_{xy}^2 E_z = -k_c^2 E_z \tag{1}$$

$$k_c^2 = (\gamma^2 + k^2) = \gamma^2 + \omega^2 \mu\varepsilon \tag{2}$$

The value of k_c, which is a constant for a particular mode, is determined by the boundary condition to be applied to (1).

Boundary Condition for a Perfectly Conducting Guide　As in the examples, the first step in the solution of a practical waveguide problem is to assume that the waveguide boundaries are perfectly conducting. The appropriate boundary condition is $E_z = 0$. It is easily shown from the general relations for the transverse field components in Sec. 8.2 that

$$E_x = \mp\frac{\gamma}{k_c^2}\frac{\partial E_z}{\partial x} \qquad E_y = \mp\frac{\gamma}{k_c^2}\frac{\partial E_z}{\partial y} \tag{3}$$

$$H_x = \frac{j\omega\varepsilon}{k_c^2}\frac{\partial E_z}{\partial y} \qquad H_y = -\frac{j\omega\varepsilon}{k_c^2}\frac{\partial E_z}{\partial x} \tag{4}$$

[9] Ref. 7, Sec. 5.1

Relations (3) may be written in the vector form:

$$\mathbf{E}_t = \mp \frac{\gamma}{k_c^2} \nabla_t E_z \tag{5}$$

where $\mathbf{E}_t$ is the transverse part of the electric field vector, and ∇_t represents the transverse part of the gradient. By the nature of the gradient, the transverse electric vector $\mathbf{E}_t$ is normal to any line of constant E_z. It is then normal to the conducting boundary, as required, once the boundary is made a curve of constant $E_z = 0$. Thus $E_z = 0$ is the only required boundary condition for solutions of (1).

Cutoff Properties of *TM* Waves   Solution of the homogeneous differential equation (1) subject to the given boundary condition is possible only for discrete values of the constant k_c. These are the *characteristic values*, *allowed values*, or *eigenvalues* of the problem, any one of which determines a particular *TM* mode for the given guide. It can be shown (Prob. 8.13g) that, for any lossless dielectric region which is completely closed by perfect conductors, the allowed values of k_c must always be real. Hence the propagation constant from (2),

$$\gamma = \sqrt{k_c^2 - k^2} \tag{6}$$

always exhibits cutoff properties. That is, for a particular mode in a perfect dielectric, γ is real for the range of frequencies such that $k < k_c$, γ is zero for $k = k_c$, and γ is imaginary for $k > k_c$. The cutoff frequency of a given mode is then given by

$$2\pi f_c \sqrt{\mu\varepsilon} = \frac{2\pi}{\lambda_c} = k_c \tag{7}$$

and (6) may be written in terms of frequency f and cutoff frequency f_c:

$$\gamma = \alpha = k_c \sqrt{1 - \left(\frac{f}{f_c}\right)^2}, \qquad f < f_c \tag{8}$$

$$\gamma = j\beta = jk \sqrt{1 - \left(\frac{f_c}{f}\right)^2}, \qquad f > f_c \tag{9}$$

The phase velocity for all *TM* modes in an ideal guide then has the form

$$v_p = \frac{\omega}{\beta} = v \left[1 - \left(\frac{f_c}{f}\right)^2 \right]^{-1/2} \tag{10}$$

The group velocity is

$$v_g = \frac{d\omega}{d\beta} = v \left[1 - \left(\frac{f_c}{f}\right)^2 \right]^{1/2} \tag{11}$$

Universal curves for attenuation constant, phase velocity, and group velocity as functions of f/f_c are shown in Fig. 8.13. Phase velocity is infinite at cutoff

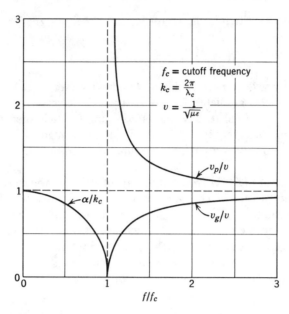

Fig. 8.13 Frequency characteristics of all TE and TM wave types.

frequency and is always greater than the velocity of light in the dielectric; group velocity is zero at cutoff and is always less than the velocity of light in the dielectric. As the frequency increases far beyond cutoff, phase and group velocities both approach the velocity of light in the dielectric.

Magnetic Fields of the Waves   Once the distribution of E_z is found by solution of the differential equation (1) subject to the boundary condition $E_z = 0$, the transverse electric field of a given mode may be found from relation (3) or (5). The transverse magnetic field may be found from relations (4). By comparing (3) and (4), we see that

$$\frac{E_x}{H_y} = -\frac{E_y}{H_x} = \pm \frac{\gamma}{j\omega\varepsilon} \qquad (12)$$

These relations show that transverse electric and magnetic fields are at right angles, and that their magnitudes are related by the quantity $\gamma/j\omega\varepsilon$, which may be thought of as the wave impedance or field impedance of the mode.

$$Z_{TM} = \frac{\gamma}{j\omega\varepsilon} = \eta\sqrt{1 - \left(\frac{f_c}{f}\right)^2} \qquad (13)$$

$$\eta = \sqrt{\frac{\mu}{\varepsilon}}$$

The wave impedance is imaginary (reactive) for frequencies less than the cutoff frequency, and purely real for frequencies above cutoff, approaching the intrinsic impedance of the dielectric at infinite frequency. This type of behavior is also found in the study of lumped-element filters, and it emphasizes that the wave can produce no average power transfer for frequencies below cutoff, where the impedance is imaginary.

The relations between electric and magnetic fields may also be given in the following vector form, which expresses the properties described above:

$$\mathbf{H} = \pm \frac{\hat{\mathbf{z}} \times \mathbf{E}_t}{Z_{TM}} \tag{14}$$

where $\hat{\mathbf{z}}$ is the unit vector in the z direction. The upper sign is for positively traveling waves, the lower sign for negatively traveling waves.

Power Transfer in the Waves   The power transfer down the guide is zero below cutoff if the conductor of the guide is perfect. Above cutoff it may be obtained in terms of the field components by integrating the axial component of the Poynting vector over the cross-sectional area. Since it has been shown that transverse components of electric and magnetic fields are in phase and normal to each other, the axial component of the average Poynting vector is one-half the product of the transverse field magnitudes. For a positively traveling wave,

$$W_T = \int_{cs} \frac{1}{2} \text{Re}[\mathbf{E} \times \mathbf{H}^*]_z \cdot d\mathbf{S} = \frac{1}{2} \int_{cs} |E_t| |H_t| \, dS = \frac{Z_{TM}}{2} \int_{cs} |H_t|^2 \, dS \tag{15}$$

Attenuation Due to Imperfectly Conducting Boundaries   When the conducting boundaries are imperfect, an exact solution would require solution of Maxwell's equations in both the dielectric and conducting regions. Because this procedure is impractical for most geometrical configurations, we take advantage of the fact that most practical conductors are good enough to cause only a slight modification of the ideal solution, and the expression $w_L/2W_T$ in formula 5.9(4) may be used. To compute the average power loss per unit length, we require the current flow in the guide walls, which is taken the same as that in the ideal guide. By the $\hat{n} \times \mathbf{H}$ rule, the current per unit width in the boundary is equal to the transverse magnetic field at the boundary, and flows in the axial direction since magnetic field is entirely transverse:

$$w_L = \oint_{\text{bound}} \frac{R_s}{2} |J_{sz}|^2 \, dl = \frac{R_s}{2} \oint_{\text{bound}} |H_t|^2 \, dl \tag{16}$$

The attenuation constant is then approximately

$$\alpha = \frac{w_L}{2W_T} = \frac{R_s \oint_{\text{bound}} |H_t|^2 \, dl}{2Z_{TM} \int_{cs} |H_t|^2 \, dS} \tag{17}$$

Attenuation Due to Imperfect Dielectric   It is noted that the general form for propagation constant (9) is exactly the same as that for the special case of the parallel-plane guide, Eq. 8.3(15). Hence, the modification caused by an imperfect dielectric, taken into account by replacing $j\omega\varepsilon$ by $j\omega(\varepsilon' - j\varepsilon'')$ or $\sigma + j\omega\varepsilon$, yields the same form for attenuation as Eq. 8.5(4):

$$\alpha_d = \frac{k\varepsilon''/\varepsilon'}{2\sqrt{1 - (f_c/f)^2}} = \frac{\sigma\eta}{2\sqrt{1 - (f_c/f)^2}} \quad \text{Np/m} \tag{18}$$

It is especially interesting to note that the form of the attenuation produced by an imperfect dielectric is the same for all modes and all shapes of guides, though of course the amount of attenuation is a function of the cutoff frequency, which does depend on the guide and the mode.

8.14   General Properties of *TE* Waves in Cylindrical Conducting Guides of Arbitrary Cross Section

Finally, we consider waves that have magnetic field but no electric field in the axial direction. Because of the similarity of treatment to that of *TM* waves in the preceding section, it will be given more briefly.

The Differential Equation   The finite H_z of the waves must satisfy the wave equation in the form of Eq. 8.2(18):

$$\nabla_t^2 H_z = -k_c^2 H_z \tag{1}$$

$$k_c^2 = \gamma^2 + k^2 \tag{2}$$

Boundary Conditions for a Perfectly Conducting Guide   Allowable solutions to (1) are determined by the single boundary condition that at perfect conductors the normal derivative of H_z must be zero:

$$\frac{\partial H_z}{\partial n} = 0 \quad \text{at boundary} \tag{3}$$

To show that (3) is the required boundary condition, write the transverse fields of the wave from Eqs. 8.2(9)–(12).

$$E_x = -\frac{j\omega\mu}{k_c^2}\frac{\partial H_z}{\partial y} \qquad E_y = \frac{j\omega\mu}{k_c^2}\frac{\partial H_z}{\partial x} \tag{4}$$

$$H_x = \mp\frac{\gamma}{k_c^2}\frac{\partial H_z}{\partial x} \qquad H_y = \mp\frac{\gamma}{k_c^2}\frac{\partial H_z}{\partial y} \tag{5}$$

Relation (5) may be written in the vector form,

$$\mathbf{H}_t = \mp\frac{\gamma}{k_c^2}\nabla_t H_z \tag{6}$$

If H_z has no normal derivative at the boundary, its transverse gradient has only a component tangential to the boundary, so, by (6), $\mathbf{H}_t$ does also. Comparison of (4) and (5) shows that transverse electric and magnetic field components are normal to one another, so electric field is normal to the conducting boundary as required.

Cutoff Properties of *TE* Waves   It was mentioned in the preceding section on *TM* waves that k_c is always real for dielectric regions completely closed by perfect conductors; the same can be shown for *TE* waves. By (2), γ then shows cutoff properties exactly the same as for *TM* waves:

$$\gamma = \sqrt{k_c^2 - k^2} \tag{7}$$

Formulas for attenuation constant below cutoff, phase constant, and phase and group velocities above cutoff then follow exactly as in Eqs. 8.13(8)–(11) and the universal curves of Fig. 8.13 apply.

Electric Field of the Wave   The electric field is everywhere transverse, and everywhere normal to the transverse magnetic field components. Transverse components of electric and magnetic field may again be related through a field or wave impedance:

$$\frac{E_x}{H_y} = -\frac{E_y}{H_x} = Z_{TE} \tag{8}$$

where, from (4) and (5),

$$Z_{TE} = \frac{j\omega\mu}{\gamma} = \eta\left[1 - \left(\frac{f_c}{f}\right)^2\right]^{-1/2} \tag{9}$$

This impedance is imaginary for frequencies below cutoff, infinite at cutoff, and purely real for frequencies above cutoff, approaching the intrinsic impedance η as f/f_c becomes large.

Electric field may also be written in the vector form

$$\mathbf{E} = \mp Z_{TE}(\hat{z} \times \mathbf{H}_t) \tag{10}$$

where $\hat{z}$ is the unit vector in the z direction, and the upper and lower signs apply respectively to positively and negatively traveling waves.

Power Transfer in *TE* Waves   Average power transfer in the propagating range is, as usual, obtained from the Poynting vector:

$$W_T = \frac{1}{2}\int_{cs} \text{Re}[\mathbf{E} \times \mathbf{H}^*] \cdot d\mathbf{S} = \frac{1}{2}\int_{cs} |E_t||H_t|\, dS$$

$$= \frac{Z_{TE}}{2}\int |H_t|^2\, dS \tag{11}$$

Attenuation Due to Imperfectly Conducting Boundaries　As with the *TEM* mode, there cannot be a true transverse electric wave in most guides with imperfect conductors, since most (but not all) of the *TE* modes have axial currents that require a certain finite axial electric field when conductivity is finite. This axial field is very small compared with the transverse field, however, so the waves are not renamed.

The axial component of current arises from the transverse component of magnetic field at the boundary:

$$|J_{sz}| = |H_t| = \frac{\beta}{k_c^2}|\nabla_t H_z| = \frac{\beta}{k_c^2}\frac{\partial H_z}{\partial l} \tag{12}$$

The last form follows since it has been shown that the transverse gradient of H_z has only a tangential component $\partial/\partial l$ at the boundary. There is in addition a transverse current arising from the axial magnetic field:

$$|J_{st}| = |H_z| \tag{13}$$

The power loss per unit length is then

$$w_L = \frac{R_s}{2}\oint [|H_z|^2 + |H_t|^2]\, dl \tag{14}$$

The attenuation caused by the conductor losses is

$$\alpha = \frac{R_s\oint [|H_z|^2 + |H_t|^2]\, dl}{2Z_{TE}\int |H_t|^2\, dS} \tag{15}$$

Attenuation Due to Imperfect Dielectrics　Since the propagation constant of the *TE* waves has the same form as for the *TM* waves, it follows that the form for attenuation due to an imperfect dielectric does also. For a reasonably good dielectric, the approximate form, Eq. 8.13(18), may be used.

8.15　Waves Below and Near Cutoff

The higher order waves that may exist in transmission lines and all waves that may exist in hollow-pipe waveguides are characterized by cutoff frequencies. If the waves are to be used for propagating energy, we are of course interested only in the behavior above cutoff. However, the behavior of these reactive or evanescent waves below cutoff is important in at least two practical cases:

1. Application to waveguide attenuators.
2. Effects of discontinuities in transmission systems.

The attenuation properties of these waves below cutoff have been developed in the previous analyses. It has been found that below the cutoff frequency there is an attenuation only and no phase shift in an ideal guide. The characteristic wave impedance is a purely imaginary quantity—a reemphasis of the fact that no energy can propagate down the guide. This is not a dissipative attenuation as is that due to resistance and conductance in transmission systems with propagating waves. It is a purely reactive attenuation, analogous to that in a filter section made of reactive elements, when this is in the cutoff region. The energy is not lost but is reflected back to the source so that the guide acts as a pure reactance to the source.

The expression for attenuation below cutoff in an ideal guide, Eq. 8.13(8), may be written as

$$\gamma = \alpha = k_c \sqrt{1 - \left(\frac{f}{f_c}\right)^2} = \frac{2\pi}{\lambda_c} \sqrt{1 - \left(\frac{f}{f_c}\right)^2} \tag{1}$$

As f is decreased below f_c, α increases from zero toward the constant value

$$\alpha = \frac{2\pi}{\lambda_c} \tag{2}$$

when $(f/f_c)^2 \ll 1$. This is an important point in the use of waveguide attenuators, since it shows that the amount of this attenuation is substantially independent of frequency if the operating frequency is far below the cutoff frequency.

Now let us look for a moment at the relations among the fields of both transverse magnetic and transverse electric waves below cutoff. If $\gamma = \alpha$ as given by (1) is substituted in the expressions for field components of transverse magnetic waves, Eqs. 8.13(3) and 8.13(4),

$$H_x = \frac{j}{\eta}\left(\frac{f}{f_c}\right)\frac{1}{k_c}\frac{\partial E_z}{\partial y} \qquad E_x = -\sqrt{1 - \left(\frac{f}{f_c}\right)^2}\frac{1}{k_c}\frac{\partial E_z}{\partial x}$$

$$H_y = -\frac{j}{\eta}\left(\frac{f}{f_c}\right)\frac{1}{k_c}\frac{\partial E_z}{\partial x} \qquad E_y = -\sqrt{1 - \left(\frac{f}{f_c}\right)^2}\frac{1}{k_c}\frac{\partial E_z}{\partial y} \tag{3}$$

For a given distribution of E_z across the guide section, which is determined once the guide shape and size and the wave type are specified, it is evident from relations (3) that, as frequency decreases, $f/f_c \to 0$, the components of magnetic field approach zero whereas the transverse components of electric field approach a constant value. We draw the conclusion that only electric fields are of importance in transverse magnetic or E waves far below cutoff. Similarly, only magnetic fields are of importance in transverse electric or H waves far below cutoff. If the waves are far below cutoff, the dimensions of the guide are small compared with wavelength. For any such region small compared with wavelength, the wave equation will reduce to Laplace's equation so that low-frequency analyses neglecting any tendency toward wave propagation are applicable.

The presence of losses in the guide below cutoff causes the phase constant to change from the zero value for an ideal guide to a small but finite value, and modifies slightly the formula for attenuation. These modifications are most important in the immediate vicinity of cutoff, for with losses there is no longer a sharp transition but a more gradual change from one region to another. It should be emphasized again that the approximate formulas developed in previous sections may become extremely inaccurate in this region. For example, the approximate formulas for attenuation caused by conductor or dielectric losses would yield an infinite value at $f = f_c$. The actual value is large compared with the minimum attenuation in the pass range since it is approaching the relatively larger magnitude of attenuation in the cutoff regime, but it is nevertheless finite. Previous formulas have also shown an infinite value of phase velocity at cutoff, and with losses it too will be finite.

8.16   Dispersion of Signals Along Transmission Lines and Waveguides

We have in several instances noted the dispersive properties of transmission systems when phase velocity or group velocity or both vary with frequency. In Chapter 5 we considered a simple two-frequency group in a dispersive system, but we now wish to be more general, using the Fourier integral of Sec. 7.11. There are two classes of problems of concern. One is that of a *base-band* signal, in which the detailed signal is of concern. Examples are audio or video signals, or electrical pulses from a computer, before being placed on other *carrier* frequencies. The other is that of modulated signals in which the base-band signal is placed on a high-frequency carrier. For the latter case we shall consider amplitude modulation and examine the distortion of the envelope.

Base-Band Signals   Given an audio signal, series of pulses, or similar electrical waveform, we can express it as a Fourier integral as in Eq. 7.11(15). For a time function $f(t)$, the transform pair may be written

$$f(t) = \frac{1}{2\pi} \int_{-\infty}^{\infty} g(\omega) e^{j\omega t}\, d\omega \tag{1}$$

$$g(\omega) = \int_{-\infty}^{\infty} f(t) e^{-j\omega t}\, dt \tag{2}$$

If each frequency component is delayed in phase by βz in propagating distance z along the transmission system, (1) gives the delayed function at z as

$$f(t, z) = \frac{1}{2\pi} \int_{-\infty}^{\infty} g(\omega) e^{j(\omega t - \beta z)}\, d\omega \tag{3}$$

Now if

$$\beta = \frac{\omega}{v_p} \tag{4}$$

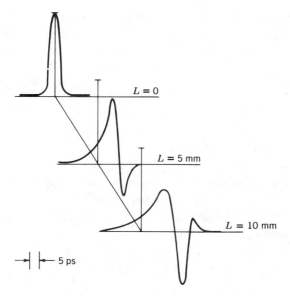

Fig. 8.16a Propagation of a 5-ps gaussian pulse along a microstrip line. Strip width $= 0.32$ mm, dielectric thickness $= 0.4$ mm and $\varepsilon_r = 6.9$. (© 1982 IEEE)

with v_p independent of ω, (3) is

$$f(t, z) = \frac{1}{2\pi} \int_{-\infty}^{\infty} g(\omega)e^{j\omega(t - z/v_p)}\, d\omega = f\left(t - \frac{z}{v_p}\right) \tag{5}$$

Thus the original function maintains its shape and propagates at the phase velocity, as we have assumed in many wave problems. But any dispersion in v_p modifies the function, at least to some degree.

Transmission lines are often used for base-band signals and have some dispersion through loss terms and internal inductance as affected by skin effect. Some lines, as the microstrip line of Sec. 8.6, have additional dispersion from the presence of multiple dielectrics. Figure 8.16a shows the result of a numerical calculation from (3), using the dispersion relation of Eq. 8.6(6), for the change in shape of a 5-ps gaussian pulse in propagating along a typical microstrip used with short electrical pulses.[10]

Modulated Signals   If the signal (1) is used to amplitude modulate a carrier of amplitude V_c and angular frequency ω_c, the resulting modulated wave may be written

$$v_m(t) = \text{Re}\{V_c e^{j\omega_c t}[1 + mf(t)]\} \tag{6}$$

---

[10] K. K. Li, G. Arjavalingam, A. Dienes, and J. R. Whinnery, *IEEE Trans.* **MTT-30**, 1270 (1982).

where m is a modulation coefficient. In substituting (1) in (6), we use ω_m for the frequency of the modulating (base-band) signal, and assume that its significant frequency components extend only over a band $-\omega_B \le \omega \le \omega_B$,

$$v_m(t, 0) = \mathrm{Re}\left\{ V_c e^{j\omega_c t}\left[1 + \frac{m}{2\pi} \int_{-\omega_B}^{\omega_B} g(\omega_m)e^{j\omega_m t}\, d\omega_m \right]\right\} \tag{7}$$

Or letting $\omega = \omega_c + \omega_m$,

$$v_m(t, 0) = \mathrm{Re}\left\{ V_c e^{j\omega_c t} + \frac{mV_c}{2\pi} \int_{\omega_c - \omega_B}^{\omega_c + \omega_B} g(\omega - \omega_c)e^{j\omega t}\, d\omega \right\} \tag{8}$$

Frequencies above ω_c in the integral in (8) correspond to upper sideband terms and those below ω_c to lower sideband terms. Each frequency component propagates according to its appropriate phase constant β. Let us expand β as a Taylor series about ω_c,

$$\beta(\omega) = \beta(\omega_c) + (\omega - \omega_c)\frac{d\beta}{d\omega}\bigg|_{\omega_c} + \frac{(\omega - \omega_c)^2}{2}\frac{d^2\beta}{d\omega^2}\bigg|_{\omega_c} + \cdots \tag{9}$$

So the modulated signal, after propagating a distance z, is

$$v_m(t, z) = \mathrm{Re}\left\{ V_c e^{j[\omega_c t - \beta(\omega_c)z]}\right.$$

$$\left. \times \left[1 + \frac{m}{2\pi}\int_{-\omega_B}^{\omega_B} g(\omega_m)e^{j[\omega_m(t - z/v_g) - (\omega_m^2/2)z(d^2\beta/d\omega^2) + \cdots]}\, d\omega_m \right]\right\} \tag{10}$$

where

$$\frac{1}{v_g} = \frac{d\beta}{d\omega}\bigg|_{\omega_c} \tag{11}$$

Now if $d^2\beta/d\omega^2$ and higher terms are negligible, (10) is interpreted as

$$v_m(t, z) = \mathrm{Re}\left\{ V_c e^{j[\omega_c t - \beta(\omega_c)z]}\left[1 + mf\left(t - \frac{z}{v_g}\right)\right]\right\} \tag{12}$$

so the envelope propagates without distortion at group velocity v_g (though the carrier inside moves at a generally different phase velocity). But if the higher order terms are not negligible, the envelope is distorted and there is said to be *group dispersion*. For a gaussian envelope,

$$f(t) = Ce^{-(2t/\tau)^2} \tag{13}$$

It can be shown (Prob. 8.16d) that the term $d^2\beta/d\omega^2$ causes the envelope to spread to a width τ' after propagating distance z, with τ' given by

$$\tau' = \tau\left[1 + \left(\frac{8z}{\tau^2}\frac{d^2\beta}{d\omega^2}\right)^2\right]^{1/2} \tag{14}$$

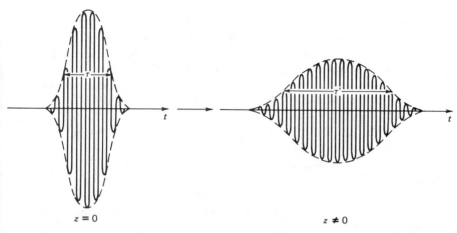

Fig. 8.16b Illustration of the spread of the modulated-envelope of a pulse as it travels down a system with group dispersion.

The spread of a gaussian envelope, illustrated in Fig. 8.16b, clearly limits data rates as pulses begin to overlap their neighbors. Although a factor in some waveguide problems (Prob. 8.16b) the limitation is most important for optical fibers and will be met again in Chapter 14.

PROBLEMS

8.2a As noted in Sec. 8.2, k^2 is complex if dielectric of the propagating system is lossy. In this case $\varepsilon' - j\varepsilon''$ is substituted for ε and propagation constant is generally complex, $\gamma = \alpha + j\beta$. Write the differential equations for E_z and H_z from 8.2(1) and 8.2(2), and the relations giving other field components 8.2(9)–(12) with these substitutions.

8.2b The division into TM and TE classes is not the only way of classifying guided waves, as noted in Sec. 8.2. Another frequently useful division employs longitudinal-section electric (LSE) with $E_x = 0$ but all other components present and longitudinal-section magnetic (LSM) with $H_x = 0$ but all other components present. Find the relations between E_z and H_z for each of these classes.

8.3a Add induced charges and current flows, with attention to sign, to the pictures of Figs. 8.3b and 8.3c for the positively traveling TM_1 and TE_1 waves. Repeat for negatively traveling waves.

8.3b Calculate cutoff frequency for TE_1, TE_2, TE_3, TM_1, TM_2, TM_3 waves between planes 1.5 cm apart with air dielectric. Repeat for a glass dielectric with $\varepsilon'/\varepsilon_0 = 4$.

8.3c The slope of an electric-field line in the x–z plane is $dx/dz = E_x/E_z$. Show that the curve for an electric field line of a TM_1 wave, obtained from the expressions for E_x and E_z of the wave, is defined by

$$\cos \beta z = [\cos \pi x_0/a][\cos(\pi x/a)]^{-1}$$

where x_0 is the value of x for a given curve at $z = 0$. Plot one or two lines to verify the form shown in Fig. 8.3b. [Hint: First express fields as real functions of z.)

8.3d Similarly to Prob. 8.3c, derive the expression defining magnetic field lines for a TE_1 wave and plot one or two lines to verify the form shown in Fig. 8.3c.

8.3e* Find the expression for electric field lines for a TM_2 wave, plot one or two lines, and sketch the remainder to give a plot similar to Fig. 8.3b. Similarly, plot and sketch magnetic field lines for a TE_2 wave.

8.3f Show that the expression for energy velocity as derived for TM_m waves [Eq. 8.3(37)] also applies to TE_m waves.

8.4a Calculate the angle θ as defined in Fig. 8.4a for ray directions of a TM_1 mode between planes 1.5 cm apart with glass dielectric, $\varepsilon'/\varepsilon_0 = 4$, for frequencies of 5, 6, 10, and 30 GHz.

8.4b Obtain the expressions for wave impedance of TM and TE waves, using the picture of uniform plane waves reflecting at an angle.

8.4c* By suitably changing coordinates as in Ex. 8.4, show that the expressions 6.09(18)–(20) for a wave polarized with electric field normal to the plane of incidence striking a conductor at an angle corresponds exactly to the field expressions for a TE_m wave.

8.5a Find average power transfer and conductor loss for a TE mode between parallel planes to verify the expression for attenuation, Eq. 8.5(12).

8.5b Calculate attenuation in decibels per meter for a TM_1 wave between copper planes 1.5 cm apart with air dielectric. Frequency is 12 GHz. For the same frequency and spacing, a glass dielectric with $\varepsilon'/\varepsilon_0 = 4$, $\varepsilon''/\varepsilon' = 2 \times 10^{-3}$ is introduced. Calculate attenuation from both dielectric and conductor losses.

8.5c Prove that the frequency of minimum attenuation for a TM_m mode, from conductor losses, is $\sqrt{3}f_c$, where f_c is cutoff frequency. Give the expression for the minimum attenuation and calculate for silver conductors 2 cm apart and air dielectric for the $m = 1$, 2, and 3 modes.

8.5d Show that the transmission line formula for attenuation constant, Eq. 5.9(7), gives precisely the same result as the approximate wave analysis of Sec. 8.5 for the TEM wave.

8.5e Derive the approximate formula for attenuation constant due to dielectric losses by using $\alpha = w_L/2W_T$.

8.5f* Since E_z is equal and opposite at top and bottom conductors for TEM wave in the parallel-plane line, it is reasonable to assume a linear variation between the two values.

$$E_z = (1 + j)\frac{R_s E_0}{\eta}\left(1 - \frac{2x}{a}\right)$$

Find the modification in the distribution for E_x to satisfy the divergence equation for **E**. Find the corresponding modification in H_y from Maxwell's equations. Describe qualitatively the average Poynting vector as a function of position in the guide.

8.6a* For a symmetrical stripline as in Fig. 8.6a with $w = 5$ mm, $t = 0.5$ mm, $d = 3$ mm, $\varepsilon_r = 2.7$, and copper conductors, calculate Z_0, phase velocity, and α_c. [Note that tables of elliptic integrals are required.]

8.6b For $\varepsilon_r = 1$, the lossless microstrip of Fig. 8.6b can propagate a true TEM wave at the velocity of light. Find inductance and capacitance per unit length for a 50-Ω line with such a dielectric, and the required w/d for this from Fig. 8.6c. Calculate the difference due to fringing fields between the capacitance per unit length found above and that given by the parallel-plane approximation, and express this as an equivalent extra width, $\Delta w/d$. Now maintaining w/d constant, assuming inductance is independent of ε_r, and transmission-line equations applicable, repeat for other values of ε_r and plot the extra equivalent $\Delta w/d$ due to fringing as a function of ε_r.

8.7a For a rectangular waveguide with inner dimensions 3 cm $\times$ 1.5 cm and air dielectric, calculate the cutoff frequencies of the TE_{10}, TE_{20}, TE_{11}, TE_{12}, TE_{21}, TE_{22}, TM_{11}, TM_{22} modes. Repeat for a glass dielectric with $\varepsilon'/\varepsilon_0 = 4$.

8.7b Derive the expression for magnetic field lines in the transverse plane of a TM_{11} wave and plot one or two such lines, comparing with Table 8.7. (See approach in Prob. 8.3c.)

8.7c Derive the expression for electric field lines in the transverse plane of a TE_{11} wave and plot one or two such lines, comparing with Table 8.7. (See approach in Prob. 8.3c.)

8.7d Show that the expression for attenuation because of conductor loss for a TM_{mn} mode in the rectangular guide is as given by Eq. 8.7(14).

8.7e Show that the expression for attenuation because of conductor loss for a TE_{mn} mode (neither m nor n zero) in the rectangular guide is as given by Eq. 8.7(26). Explain why this does not apply to $m = 0$ or $n = 0$ case.

8.7f Recalling that surface resistivity R_s is a function of frequency, find the frequency of minimum attenuation for a TM_{mn} mode. Show that the expression for attenuation of a TE_{mn} mode must also have a minimum.

8.7g* Of the wave types studied so far, those transverse magnetic to the axial direction were obtained by setting $H_z = 0$; those transverse electric to the axial direction were obtained by setting $E_z = 0$. For the rectangular waveguide, obtain the lowest order mode with $H_x = 0$ but all other components present. This may be called a wave transverse magnetic to the x direction. Show that it may also be obtained by superposing the TM and TE waves given previously of just sufficient amounts so that H_x from the two waves exactly cancel. Repeat for a wave transverse electric to the x direction. These wave types are also called *longitudinal section waves*. (See Prob. 8.2b.)

8.7h From the form of Eqs. 8.2(9)–(12), show that for a TM wave, imposition of the condition $E_z = 0$ on a perfectly conducting boundary of a cylindrical guide, causes the other tangential component of **E** also to be zero along that boundary.

8.8a For $f = 3$ GHz, design a rectangular waveguide with copper conductor and air dielectric so that the TE_{10} wave will propagate with a 30% safety factor ($f = 1.30f_c$) but also so that the wave type with next higher cutoff will be 30% below its cutoff frequency. Calculate the attenuation due to copper losses in decibels per meter.

8.8b For Prob. 8.8a, calculate the attenuation in decibels per meter of the three modes with cutoff frequencies closest to that of the TE_{10} mode, neglecting losses.

8.8c Design a guide for use at 3 GHz with the same requirements as in Prob. 8.8a except that the guide is to be filled with a dielectric having a permittivity four times that of air. Calculate the increase in attenuation due to copper losses alone, assuming that the dielectric is perfect. Calculate the additional attenuation due to the dielectric, if $\varepsilon''/\varepsilon' = 0.01$.

8.8d Find the maximum power that can be carried by a 6-GHz TE_{10} wave in an air-filled guide 4 cm wide and 2 cm high, taking the breakdown field in air at that frequency as 2×10^6 V/m.

8.8e The transmission-line analogy can be applied to the transverse field components, the ratios of which are constants over guide cross sections and are given by wave impedances, just as in the case of plane waves in Chapter 6. A rectangular waveguide of inside dimensions 4 cm × 2 cm is to propagate a TE_{10} mode of frequency 5 GHz. A dielectric of constant $\varepsilon_r = 3$ fills the guide for $z > 0$, with an air dielectric for $z < 0$. Assuming the dielectric-filled part to be matched, find the reflection coefficient at $z = 0$ and the standing wave ratio in the air-filled part.

8.8f Find the length and dielectric constant of a quarter-wave matching section to be placed between the air and given dielectric of Prob. 8.8e.

8.9a Derive the set of Eqs. 8.9(1)–(4) by utilizing Maxwell's equations in circular cylindrical coordinates and assuming propagation as $e^{-j\beta z}$.

8.9b What inner radius do you need for an air-filled round pipe to propagate the TE_{11} wave at 6 GHz with operating frequency 20% above the cutoff frequency? What is the guide wavelength for this mode? Find the attenuation in decibels per meter of the TM_{01} mode at this frequency, neglecting losses for that calculation.

8.9c Show that the expression for attenuation from conductor losses of a TM_{nl} mode is

$$\alpha_c = \frac{R_s}{a\eta\sqrt{1 - (\omega_c/\omega)^2}}$$

At what value of ω/ω_c is this a minimum?

8.9d* Show that the expression for attenuation from conductor losses of a TE_{nl} mode is

$$\alpha_c = \frac{R_s}{a\eta\sqrt{1 - (\omega_c/\omega)^2}}\left[\left(\frac{\omega_c}{\omega}\right)^2 + \frac{n^2}{p_{nl}'^2 - n^2}\right]$$

8.9e For a circular air-filled guide with copper conductor, select a radius so that the TE_{01} mode has attenuation of 3 dB/km for a frequency of 4 GHz. Estimate the number of modes (counting only the symmetric ones with $n = 0$) that have cutoff frequencies below the operating frequency.

8.10 Use the asymptotic forms of Bessel functions in Eqs. 8.10(1) and 8.10(2) for TM and TE waves, respectively, to show that for large $k_c r_i$ and r_0/r_i near unity, that the cutoff wavelength of the $n = 0, p = 1$ modes is approximately twice the spacing between conductors.

8.11 Sketch examples of mode couplings by each of the six methods described in Sec. 8.11 using for each a system different from the one utilized in Fig. 8.11 to illustrate it.

8.12a Demonstrate that, although in a TEM wave E does satisfy Laplace's equation in the transverse plane and so may be considered a gradient of a scalar in so far as variations in the transverse plane are concerned, E is not the gradient of a scalar when variations in all directions (x, y, and z) are included.

8.12b Two perfectly conducting cylinders of arbitrary cross-sectional shapes are parallel and separated by a dielectric of conductivity σ and permittivity ε. Show that the ratio of electrostatic capacitance per unit length to dc conductance per unit length is ε/σ.

8.12c If the conductors are perfect but the dielectric has conductivity σ as well as permittivity ε, show that γ must have the following value in order for a TEM wave to exist ($E_z = 0$, $H_z = 0$):

$$\gamma = \pm[j\omega\mu(\sigma + j\omega\varepsilon)]^{1/2}$$

Explain why the distribution of fields may be a static distribution as in the loss-free line, unlike the case for a lossy conducting boundary.

8.12d How many linearly independent *TEM* waves may exist on a three-conductor transmission line? Describe current relations for a basic set. Complete the proof that there can be no static field, and hence no *TEM* wave, inside a single infinite cylindrical conductor.

8.13a Show that the circuit of Fig. P8.13*a* may be used to represent the propagation characteristics of the transverse magnetic wave, if the characteristic wave impedance and propagation constant are written by analogy with transmission line results in terms of an impedance Z_1, and an admittance Y_1 per unit length, and the medium is μ_1, ε_1.

$$Z_{TM} = \sqrt{\frac{Z_1}{Y_1}}, \qquad \gamma = \sqrt{Z_1 Y_1}$$

Note the similarity between this and the circuits of conventional filter sections, remembering of course that all constants in this circuit are in reality distributed constants.

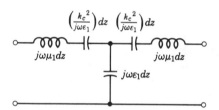

Fig. P8.13*a* Equivalent circuit for the transverse magnetic wave.

8.13b Show that all field components for a *TM* wave may be derived from the axial component of the vector potential **A**. Obtain the expressions relating E_x, H_x, and so on, to A_z, the differential equation for A_z, and the boundary conditions to be applied at a perfect conductor.

8.13c Repeat Prob. 8.13b, using the axial component of the Hertz potential as defined in Prob. 3.19b.

8.13d Show for a *TM* wave that the magnetic field distribution in the transverse plane can be derived from a scalar flux function, and relate this to E_z. With transverse electric field derivable from a scalar potential function and transverse magnetic field derivable from a scalar flux function, does it follow that both are static type distributions as in the *TEM* wave? Explain.

8.13e Show that the energy velocity equals the group velocity for the *TM* modes in a lossless waveguide of general cross section.

8.13f* Show that $E_z(x, y)$ for a general *TM* wave in a perfectly conducting guide satisfies the equation

$$k_c^2 = \left[\int_S (\nabla_t E_z)^2 \, dS \right] \left[\int_S E_z^2 \, dS \right]^{-1}$$

where $\mathbf{V}_t$ represents the transverse gradient and the integral is over the cross section of the guide. From this argue that k_c^2 is real and positive for waves in which phase is constant over the transverse plane.

8.13g* Numerical methods can be used to find the propagation constants for waveguides of arbitrary cross section. Apply the basic difference-equation relations 7.3(12)–(14) with the factor Ω set to 1.0 to make a numerical evaluation of k_c^2 for a TM_{11} mode in a rectangular waveguide, as an example. Assume a rectangular guide with side ratio $1:2$. The Helmholtz equation to be solved is Eq. 8.13(1). Divide the waveguide into a grid of 18 squares and number the interior points 1–10 left to right, top to bottom. A reasonable initial guess for the product $k_c^2 h^2 = u^2 h^2$ can be formed assuming a one-dimensional variation in the smallest dimension for Eq. 7.13(12); here take $k_c^2 h^2 = 1.1$. Start with E_z having the following values at the grid points as a first guess: for points 1, 5, 6, and 10, $E_z = 30$; for points 2, 4, 7, and 9, $E_z = 50$; for points 3 and 8, $E_z = 70$. Use simple relaxation twice to improve the values of E_z for the given $k_c^2 h^2$. Then calculate an improved value of $k_c^2 h^2$ using the following relation:

$$k_c^2 h^2 = \frac{\sum E_z(x, y)[E_{zN} + E_{zE} + E_{zS} + E_{zW} - 4E_z(x, y)]}{\sum E_z^2(x, y)}$$

where N, E, S, W indicate the points surrounding the grid point at (x, y) and the summations are over all grid points. Next make two more steps of relaxation to adjust the fields to the new $k_c^2 h^2$. Then use the above formula to get a second correction to $k_c^2 h^2$. Compare the result with the value of $k_c^2 h^2$ found using differential equations in Sec. 8.7.

8.14a As in Prob. 8.13a, show that the equivalent circuit for transverse electric waves in terms of distributed constants is as pictured in Fig. P8.14a.

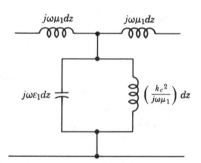

Fig. P8.14a Equivalent circuit for the transverse electric wave.

8.14b Show that fields satisfying Maxwell's equations in a homogeneous charge-free, current-free dielectric may be derived from a vector potential $\mathbf{F}$,

$$\mathbf{E} = -\frac{1}{\varepsilon}\mathbf{V} \times \mathbf{F}$$

$$\mathbf{H} = \frac{1}{j\omega\mu\varepsilon}\mathbf{V}(\mathbf{V} \cdot \mathbf{F}) - j\omega\mathbf{F}$$

$$(\mathbf{V}^2 + k^2)\mathbf{F} = 0$$

Obtain expressions for all field components of a TE wave from the axial component F_z of the above potential function, and give the differential equation and boundary conditions for F_z.

8.14c Show that, if one utilizes the potential function **A** instead of the **F** of Prob. 8.14b for derivation of a TE wave, more than one component is required.

8.14d Show for a TE mode that transverse distribution of electric field can be derived from a scalar flux function. How is this related to H_z?

8.14e Show that the energy velocity equals the group velocity for the TE modes in a lossless waveguide of general cross section.

8.14f Show, for a TM wave in any shape of guide passing from one dielectric material to another, that at one frequency the change in cutoff factor may cancel the change in η, and the wave may pass between the two media without reflection. Identify this condition with the case of incidence at polarizing angle in Sec. 6.13. Determine the requirement for a similar situation with TE waves, and show why it is not practical to obtain this.

8.15 A particular waveguide attenuator is circular in cross section with radius 1 cm. Plot attenuation in decibels per meter for the TE_{11} mode over the frequency range 1–4 GHz. Also plot attenuation of the mode with next nearest cutoff frequency.

8.16a For a hollow pipe waveguide, with β given by Eq. 8.13(9), find the group dispersion term $d^2\beta/d\omega^2$.

8.16b Using the result of Prob. 8.16a, find the waveguide length for which the width of a gaussian pulse with $\tau = 1$ ns is doubled if frequency is 10 GHz and $\omega_c/\omega = 0.7$.

8.16c Use transmission-line equations to set up the factors entering into group velocity v_g and $d^2\beta/d\omega^2$ for a transmission line (i) with series resistance R independent of frequency; (ii) with shunt conductance G independent of frequency; (iii) with no dielectric loss but resistance and internal inductance governed by skin effect. You need not carry out the differentiations and should start with fairly complete formulas.

8.16d* Start with a gaussian function $f(t)$ given by Eq. 8.16(13) and find its $g(\omega)$. Using this in Eq. 8.16(10), show that the envelope broadens with z as given by Eq. 8.16(14).

9

SPECIAL WAVEGUIDE TYPES

9.1 Introduction

The preceding chapter dealt with the important special case of waveguides with cylindrical conducting boundaries. In the present chapter, we examine several examples of waveguiding systems of different shapes and properties.

We start with dielectric guides which demonstrate that boundaries other than metal–dielectric boundaries can guide waves in cylindrical systems. Dielectric guides are now important for optical communication uses and are explored more in Chapter 14. There follows an examination of radial guiding, both in cylindrical coordinates and spherical coordinates. The former is important in certain classes of resonant systems and the latter in antenna theory. Waveguides of special cross section, used either for impedance matching or to lower the cutoff frequency for a given transverse dimension, are analyzed. Finally, classes of waves with phase velocity much slower than the velocity of light are studied both in uniform systems and periodic systems. These are important for such purposes as the interaction with electron beams in traveling-wave tubes and for confining electromagnetic energy near a surface.

The examples selected are not only important in themselves but illustrate a number of important principles. The principle of duality shows how the solution of one problem may sometimes be used for another by interchange of electric and magnetic fields. Cutoff in a waveguide is shown to correspond to a condition of transverse resonance, which leads to a variety of approximate and exact techniques for analyzing guides of irregular shape. Periodic systems show the importance of spatial harmonics for interpreting the behavior of such systems.

9.2 Dielectric Slab Guides

We shall see in this section that wave energy tends to be concentrated inside higher dielectric regions so that a rod or slab of high-dielectric material can serve to guide waves. Attention is focused on analysis of a flat slab configuration. Other configurations are studied in more detail in Chapter 14 in connection with their application in optics and so are discussed only briefly here. The plausibility of guidance can be based on the bouncing-plane-wave picture as in the case of the parallel-plane conducting guide in Sec. 8.4. The precise nature of the waves is best found by a procedure involving matching of field components at the boundary of the dielectric.

From the concept of total reflection in Sec. 6.12, we can argue that energy should be guided under certain conditions since, if a wave traveling in a dense dielectric strikes the boundary of a less-dense dielectric at an angle of incidence greater than a certain critical angle, all energy is reflected. This critical angle, Eq. 6.12(2), where 1 refers to the dense medium and 2 to the less-dense medium, is

$$\theta_c = \sin^{-1}\left(\sqrt{\frac{\mu_2 \varepsilon_2}{\mu_1 \varepsilon_1}}\right)$$

Thus in a dielectric slab as in Fig. 9.2a, which is assumed infinite in the direction normal to the paper, suppose that plane waves are excited inside the dielectric in some manner so that they travel as shown, striking the surface at an angle of incidence, θ. If $\theta > \theta_c$, the propagating energy is reflected from each side and will be retained in the slab, although there are evanescent fields in medium 2, as explained in Sec. 6.12.

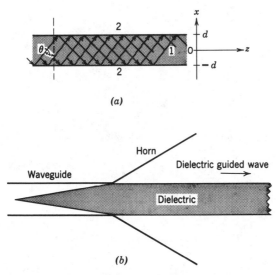

(a)

(b)

Fig. 9.2 *(a)* Plane-wave paths in a dielectric slab. *(b)* Excitation of waves in a dielectric slab.

This point of view provides a very useful physical picture and can be made to give exact quantitative relationships. However, it is more direct to go immediately to Maxwell's equations, writing appropriate wave solutions for the two dielectrics and matching tangential fields at the boundaries. If the propagating energy is to be confined to medium 1, the solutions in medium 2 must be "reactive," decaying exponentially away from the boundary. The field solutions in the slab may be broken into even and odd types, and there may be TM and TE mode types to consider. The appropriate fields for TM and TE waves having even and odd symmetries of their axial components follow, with subscripts 1 and 2 referring, respectively, to the bounded and unbounded regions of Fig. 9.2a. The equations may be checked by the planar Maxwell's equations for Eqs. 8.2(3)–(8) or the equivalent forms of Eqs. 8.2(9)–(12). The determinantal equation is obtained by equating tangential fields at $x = d$, resulting in the continuity relations,

$$TM: \left. \frac{E_{z1}}{H_{y1}} \right|_{x=d} = \left. \frac{E_{z2}}{H_{y2}} \right|_{x=d}, \qquad TE: \left. \frac{H_{z1}}{E_{y1}} \right|_{x=d} = \left. \frac{H_{z2}}{E_{y2}} \right|_{x=d}$$

The ratios apply also at the lower surface, $x = -d$, but by using even and odd solutions, this boundary need not be considered explicitly. The fields given below for region 1 apply to $-d < x < d$ and those for region 2 are for the upper region $x \geq d$. The relations for the TM modes with axial electric field having odd symmetry in the two regions are

$$E_{z1} = A \sin k_x x, \tag{1}$$

$$H_{y1} = \frac{\omega \varepsilon_1}{\beta} E_{x1} = \frac{-j\omega \varepsilon_1 A}{k_x} \cos k_x x \tag{2}$$

$$E_{z2} = C e^{-K_x(x-d)}, \tag{3}$$

$$H_{y2} = \frac{\omega \varepsilon_2}{\beta} E_{x2} = \frac{-j\omega \varepsilon_2 C}{K_x} e^{-K_x(x-d)}, \tag{4}$$

$$K_x = \frac{\varepsilon_2}{\varepsilon_1} k_x \tan k_x d \tag{5}$$

For TM modes with even symmetry of axial electric field,

$$E_{z1} = B \cos k_x x \tag{6}$$

$$H_{y1} = \frac{\omega \varepsilon_1}{\beta} E_{x1} = \frac{j\omega \varepsilon_1 B}{k_x} \sin k_x x \tag{7}$$

The fields have the same form as (3) and (4) for region 2 and

$$K_x = -\frac{\varepsilon_2}{\varepsilon_1} k_x \cot k_x d \tag{8}$$

For TE modes with odd symmetry of axial magnetic field,

$$H_{z1} = D \sin k_x x, \tag{9}$$

$$E_{y1} = \frac{-\omega\mu_1}{\beta} H_{x1} = \frac{j\omega\mu_1 D}{k_x} \cos k_x x, \tag{10}$$

$$H_{z2} = F e^{-K_x(x-d)}, \tag{11}$$

$$E_{y2} = -\frac{\omega\mu_2}{\beta} H_{x2} = \frac{j\omega\mu_2}{K_x} F e^{-K_x(x-d)} \tag{12}$$

$$K_x = \frac{\mu_2}{\mu_1} k_x \tan k_x d \tag{13}$$

For TE modes with even symmetry of axial magnetic field,

$$H_{z1} = E \cos k_x x \tag{14}$$

$$E_{y1} = \frac{-\omega\mu_1}{\beta} H_{x1} = \frac{-j\omega\mu_1 E}{k_x} \sin k_x x \tag{15}$$

The fields in region 2 have the same form as (11) and (12) so that

$$K_x = \frac{-\mu_2}{\mu_1} k_x \cot k_x d \tag{16}$$

In these relations, for guided waves,

$$k_x^2 = k_1^2 - \beta^2 > 0 \tag{17}$$

$$K_x^2 = \beta^2 - k_2^2 > 0 \tag{18}$$

There are several important points that may be found from the solutions in (1) to (18) whereby dielectric guides differ from guides closed by conducting pipes. The first is that the lowest order TM and TE modes with axial components having odd symmetry have no cutoff frequency but may exist for slab thickness very small compared with wavelength. Consider for example the odd TM mode with determinantal equation (5). If $k_x d \ll 1$, this reduces to

$$K_x \approx \frac{\varepsilon_2}{\varepsilon_1} k_x^2 d \tag{19}$$

Approximate solution of this equation combined with (17) and (18) yields

$$K_x \approx \frac{\varepsilon_2}{\varepsilon_1} (k_1^2 - k_2^2)d, \qquad \beta \approx k_2 \tag{20}$$

Thus as d becomes small, and/or $k_1 \to k_2$, the constant K_x becomes small, indicating that energy extends very far into the outer medium. The propagation constant in such cases becomes very close to the wave number in the outer region, since most of the energy is there. A similar solution exists for the lowest order TE mode defined by (13), but not for the even modes defined by (8) and (16). This is seen since no positive value of K_x is given by these equations until $k_x d$ is at least $\pi/2$.

A second point is that any of the Eqs. (5), (8), (13), and (16) have only a finite number of solutions (Prob. 9.2e) as compared with the infinite number in closed-pipe guides. Closely related is the fact that this finite number may not be sufficient to represent the total fields in the vicinity of an exciting source, as represented by the waveguide and horn excitation system of Fig. 9.2b. The difference is a set of waves reflected from and diffracted into the dielectric over a continuous range of angles. This continuous distribution of reflected waves behaves as a distribution, at $x = \pm\infty$, of plane waves, called *radiation modes*. The finite number of guided waves and the continuous spectrum of radiation modes are orthogonal, and form a complete set for expression of an arbitrary cross-sectional field distribution.[1,2]

Thin-film dielectric guides, in which the dielectric guiding region has different dielectrics above and below, and round dielectric rods or fibers are used primarily for guidance of optical signals. Detailed discussion of these cases is thus given in Chapter 14.

9.3 Parallel-Plane Radial Transmission Lines

Linearly propagating waves between parallel planes were studied in Secs. 8.3–8.5 and a connection was made between the TEM wave and a transmission-line wave, where the two plates are the conductors of the line. Here we analyze radially propagating TEM waves between parallel conducting planes and introduce a transmission-line type of formalism to facilitate impedance-matching calculations (Figs. 9.3a, 9.3b). Higher order modes are introduced in the following section. The wave under consideration has no field variations either circumferentially or axially. There are then no field components in the radial direction, but field components E_z and H_ϕ only. The component E_z, having no variations in the z direction, corresponds to a total voltage $E_z d$ between plates. The component H_ϕ corresponds to a total radial current $2\pi r H_\phi$, outward in one plate and inward in the other. This wave is then analogous to an ordinary transmission line wave and thus derives its name, radial transmission line.

For the simple wave described there are no radial field components, and analysis may be made by the nonuniform transmission line theory of Sec. 5.15, allowing L and C to vary with radius. However, the wave solution for fields may also be obtained directly from the results of Sec. 7.20. Since there are no ϕ or z variations, v and γ may be set equal to zero. Special linear combinations of Bessel functions

[1] R. E. Collin, *Field Theory of Guided Waves*, McGraw-Hill, New York, 1960, pp. 474–477.
[2] D. Marcuse, *Theory of Dielectric Optical Waveguides*, Academic Press, New York, 1974.

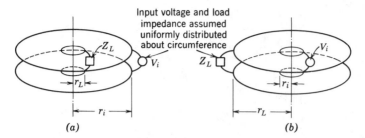

Fig. 9.3 (*a*) Radial transmission line with input at outer radius. (*b*) Radial transmission line with input at inner radius.

have been defined particularly for this problem[3] but for occasional solution of radial line problems, known forms of the Bessel functions are satisfactory. The form of Eq. 7.20(6) in the Hankel functions is particularly suitable since these can be shown to have the character of waves traveling radially inward or outward. Since $k_c = (\gamma^2 + k^2)^{1/2}$ by Eq. 8.2(19) and $\gamma = 0$ has been assumed, then $k_c = \omega\sqrt{\mu\varepsilon}$.

$$E_z = AH_0^{(1)}(kr) + BH_0^{(2)}(kr) \tag{1}$$

With v and γ zero, the only remaining field component in Eqs. 8.9(1)–(4) is H_ϕ.

$$H_\phi = \frac{1}{j\omega\mu}\frac{\partial E_z}{\partial r} = \frac{j}{\eta}[AH_1^{(1)}(kr) + BH_1^{(2)}(kr)] \tag{2}$$

The $H_n^{(1)}$ terms are identified as the negatively traveling wave and the $H_n^{(2)}$ terms as the positively traveling wave because of the asymptotic forms that approach complex exponentials (Sec. 7.15). It is convenient to utilize the magnitudes and phases of these functions,

$$H_0^{(1)}(r) = J_0(v) + jN_0(v) = G_0(v)e^{j\theta(v)} \tag{3}$$

$$H_0^{(2)}(v) = J_0(v) - jN_0(v) = G_0(v)e^{-j\theta(v)} \tag{4}$$

$$jH_1^{(1)}(v) = -N_1(v) + jJ_1(v) = G_1(v)e^{j\psi(v)} \tag{5}$$

$$jH_1^{(2)}(v) = N_1(v) + jJ_1(v) = -G_1(v)e^{-j\psi(v)} \tag{6}$$

where

$$G_0(v) = [J_0^2(v) + N_0^2(v)]^{1/2}, \qquad \theta(v) = \tan^{-1}\left[\frac{N_0(v)}{J_0(v)}\right] \tag{7}$$

$$G_1(v) = [J_1^2(v) + N_1^2(v)]^{1/2}, \qquad \psi(v) = \tan^{-1}\left[\frac{J_1(v)}{-N_1(v)}\right] \tag{8}$$

---

[3] N. Marcuvitz, in C. G. Montgomery, R. H. Dicke, and E. M. Purcell, *Principles of Microwave Circuits*, McGraw-Hill, New York, 1947, Chapter 8.

Expressions (1) and (2) then become

$$E_z = G_0(kr)[Ae^{j\theta(kr)} + Be^{-j\theta(kr)}] \tag{9}$$

$$H_\phi = \frac{G_0(kr)}{Z_0(kr)}[Ae^{j\psi(kr)} - Be^{-j\psi(kr)}] \tag{10}$$

where

$$Z_0(kr) = \eta \frac{G_0(kr)}{G_1(kr)} \tag{11}$$

is a radially dependent characteristic wave impedance.

Evaluation of the constants A and B follows from specification of two field values at given radii. For example, given E_a at r_a and H_b at r_b, the fields at any radius r are

$$E_z = E_a \frac{G_0 \cos(\theta - \psi_b)}{G_{0a} \cos(\theta_a - \psi_b)} + jZ_{0b}H_b \frac{G_0 \sin(\theta - \theta_a)}{G_{0b} \cos(\theta_a - \psi_b)} \tag{12}$$

$$H_\phi = H_b \frac{G_1 \cos(\psi - \theta_a)}{G_{1b} \cos(\theta_a - \psi_b)} + j \frac{E_a G_1 \sin(\psi - \psi_b)}{Z_{0a}G_{1a} \cos(\theta_a - \psi_b)} \tag{13}$$

These are similar in form to the ordinary transmission line equations except for the use of the special magnitudes and phases. A plot of these special radial line quantities versus kr is given in Fig. 9.3c. This is not very accurate for small values of kr, so the following approximations are then preferred:

$$G_0(v) \approx \frac{2}{\pi} \ln\left(\frac{\gamma v}{2}\right) \qquad \theta(v) \approx \tan^{-1}\left[\frac{2}{\pi} \ln\left(\frac{\gamma v}{2}\right)\right] \tag{14}$$

$$G_1(v) \approx \frac{2}{\pi v} \qquad \psi(v) \approx \tan^{-1}\left(\frac{1}{\pi v^2}\right) \tag{15}$$

where $\gamma = 0.5772\ldots$. More accurate values can be calculated from the definitions (7) and (8) utilizing tables of J and N, although some tables give magnitude and phase of $H_n^{(1)}$ and $H_n^{(2)}$ directly.

Forms similar to (12) and (13) can be derived for two values of E_z specified at different radii, or two values of H_ϕ. The most useful form, however, is that defining an input wave impedance $Z_i = E_{zi}/H_{\phi i}$ when load impedance $Z_L = E_{zL}/H_{\phi L}$ is given. This is

$$Z_i = Z_{0i}\left[\frac{Z_L \cos(\theta_i - \psi_L) + jZ_{0L} \sin(\theta_i - \theta_L)}{Z_{0L} \cos(\psi_i - \theta_L) + jZ_L \sin(\psi_i - \psi_L)}\right] \tag{16}$$

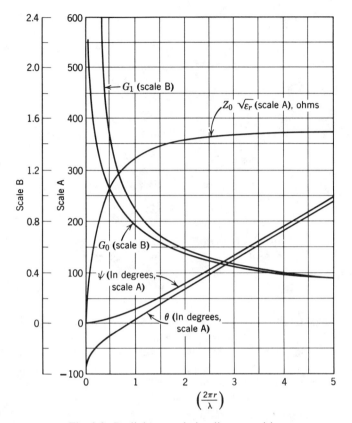

Fig. 9.3c Radial transmission line quantities.

Special forms of this for output shorted ($Z_L = 0$) or open ($Z_L = \infty$) are especially simple and are analogous to the corresponding forms for uniform transmission lines.

All of the foregoing relationships are given in terms of fields or wave impedances. Usually current, voltage, and total impedance are desired. The relations are

$$V = -E_z d; \qquad I = 2\pi r H_\phi \tag{17}$$

$$Z_{\text{total}} = \mp \frac{d}{2\pi r}\left(\frac{E_z}{H_\phi}\right) \tag{18}$$

The sign convention defines higher voltage in the upper plate and outward current in the upper plate as positive. The upper sign in (18) is for input radius less than that of the load, $r_i < r_L$, and the lower sign for the reverse, $r_i > r_L$, since in this case the convention for positive current would be the opposite of (17).

9.4   Circumferential Modes in Radial Lines; Sectoral Horns

Many higher order modes can exist in the radial transmission lines studied in the last section. All those with z variations require a spacing between plates greater than a half-wavelength for radial propagation of energy. For smaller spacing, modes are possible with circumferential variations but no z variations. The field components may be written

$$E_z = A_v Z_v(kr) \sin v\phi \tag{1}$$

$$H_\phi = -\frac{j}{\eta} A_v Z_v'(kr) \sin v\phi \tag{2}$$

$$H_r = \frac{jvA_v}{k\eta r} Z_v(kr) \cos v\phi \tag{3}$$

In the foregoing equations, Z_v denotes any solution of the ordinary vth order Bessel equation. For example, to stress the concept of radially propagating waves it may again be convenient to utilize Hankel functions:

$$Z_v(kr) = H_v^{(1)}(kr) + c_v H_v^{(2)}(kr) \tag{4}$$

These circumferential modes may be important as disturbing effects excited by asymmetries in radial lines intended for use with the symmetrical modes studied in the preceding section. In this case v must be an integer since the wave must have the same value at $\phi = 0$ and $\phi = 2\pi$. Waves of the form (1) to (3) may also be supported in a wedge-shaped guide with conducting planes at $\phi = 0$ and $\phi = \phi_0$ as well as at $z = 0$, d (Fig. 9.4a). The latter case is important as a sectoral electromagnetic horn[4] used for radiation. In this case, since E_z must be zero at $\phi = 0$, ϕ_0,

$$v = \frac{m\pi}{\phi_0} \tag{5}$$

The waves discussed here are interesting in one respect especially. If we think of the lowest order mode ($m = 1$) propagating radially inward in the pie-shaped guide for Fig. 9.4a, it would be quite similar to the TE_{10} mode of the rectangular guide, although modified by the convergence of the sides. We would consequently expect a cutoff phenomenon at such a radius r_c that the width $r_c\phi_0$ becomes a half-wavelength. Similarly, for the lowest order circumferential mode in the radial line of Figs. 9.3a, and 9.3b, we would expect a cutoff at such a radius that circumference is one wavelength:

$$2\pi r_c = \lambda \text{ for radial line}; \qquad \phi_0 r_c = \frac{\lambda}{2} \text{ for sectoral horn} \tag{6}$$

A casual inspection of (1) to (3) would not reveal this cutoff since there is no sudden change of mathematical form as there was in the rectangular guide at cutoff.

---

[4] W. L. Barrow and L. J. Chu, *Proc. I.R.E.* **27**, 51 (1939).

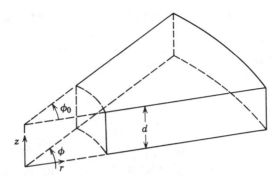

Fig. 9.4a Wedge-shaped guide or sectoral horn.

However, a more detailed study would reveal that there is a very effective cutoff phenomenon at about the radius predicted by (6) in that the reactive energy for given power transfer becomes very great for radii less than this. The radial field impedance for an inward traveling wave is

$$\left(\frac{E_z}{H_\phi}\right)_- = j\eta\,\frac{H_\nu^{(1)}(kr)}{H_\nu^{(1)\prime}(kr)} = R_\nu - jX_\nu \tag{7}$$

This impedance becomes predominantly reactive at a value $kr \approx \nu$, which is compatible with (6). Figure 9.4b shows real and imaginary parts of the wave impedances versus kr for $\nu = 1$.

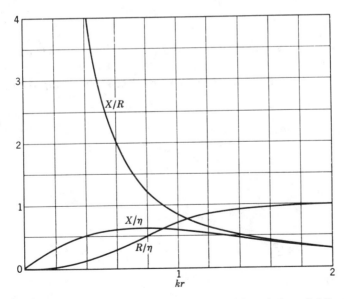

Fig. 9.4b Wave resistance and reactance for circumferential mode in radial line for $\nu = 1$.

9.5   Duality; Propagation Between Inclined Planes

Given certain solutions of Maxwell's equations, we may obtain other useful ones by making use of the simple but important principle of duality. This principle follows from the symmetry of the field equations for charge-free regions,

$$\mathbf{V} \times \mathbf{E} = -j\omega\mu\mathbf{H} \tag{1}$$

$$\mathbf{V} \times \mathbf{H} = j\omega\varepsilon\mathbf{E}. \tag{2}$$

It is evident that, if $\mathbf{E}$ is replaced by $\mathbf{H}$, $\mathbf{H}$ by $-\mathbf{E}$, μ by ε, and ε by μ, the original equations are again obtained. It follows that, if we are given any solution for such a dielectric, another may be obtained by interchanging components as stated. It may be difficult to supply appropriate boundary conditions for the new solution since the magnetic equivalent of the perfect conductor is not known at high frequencies, so the new solution is not always of practical importance.

One example in which the principle of duality may be utilized to save work in a practical problem is that of the principal mode in the wedge-shaped dielectric region between inclined plane conductors (Fig. 9.5). This mode has electric field E_ϕ and magnetic field H_z. If there are no variations with ϕ or z, it is evident that the field distributions can be obtained from those of the radial transmission line mode, Sec. 9.3, through the foregoing principle of duality. Replacing $\mathbf{E}$ by $\mathbf{H}$, $\mathbf{H}$ by $-\mathbf{E}$, μ by ε, and ε by μ in Eqs. 9.3(1) and 9.3(2),

$$H_z = AH_0^{(1)}(kr) + BH_0^{(2)}(kr) \tag{3}$$

$$E_\phi = -j\sqrt{\frac{\mu}{\varepsilon}} \left[AH_1^{(1)}(kr) + BH_1^{(2)}(kr)\right] \tag{4}$$

The real advantage is that all the derived expressions 9.3(9)–(16) may be used without rederivation, as well as the curves of Fig. 9.3c, with the interchange of quantities as above. Admittance should be read in place of impedance, and the

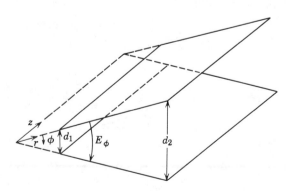

Fig. 9.5 Inclined-plane guide.

numerical scale of $Z_0\sqrt{\varepsilon_r}$ in ohms (Fig. 9.3c) should be divided by $(377)^2$ to give the characteristic admittance $Y_0/\sqrt{\varepsilon_r}$ in siemens. Total admittance is obtained from the field admittance as follows:

$$Y_{\text{total}} = \mp \frac{l}{r\phi_0}\left[-\frac{H_z}{E_\phi}\right] \tag{5}$$

where the upper sign is for $r_i < r_L$, the lower for $r_i > r_L$.

_____ **Example 9.5** _____
Use of Inclined Planes for Impedance Matching

One application of the inclined-plane transmission line might be in impedance matching between parallel-plate transmission lines of different spacings, d_1 and d_2 (Fig. 9.5). It is known from practical experience that such transitions, if gradual enough, supply a good impedance match over a wide band of frequencies (unlike schemes studied in Prob. 5.5c, which depend upon quarter-wavelengths of line). It is seen from Fig. 9.3c that for both kr_i and kr_L large (say, greater than 5) the characteristic admittance Y_0 is nearly $1/\eta$ and θ and ψ are nearly equal (i.e., $\theta_i \approx \psi_i$, $\theta_L \approx \psi_L$). If the parallel-plane line to the right is matched, its characteristic wave admittance is that of a plane wave, $1/\eta$. Equation 9.3(16) then shows that with the above approximations the input wave admittance is also approximately $1/\eta$, so that the parallel-plane line to the left is also nearly matched. This gives some quantitative support to the matching phenomenon mentioned.

_____

9.6 Waves Guided by Conical Systems

The problem of waves guided by conical systems (Fig. 9.6) is important to a basic understanding of waves along dipole antennas and in certain classes of cavity resonators. In particular, one very important wave propagates along the cones with the velocity of light and has no field components in the radial direction, and

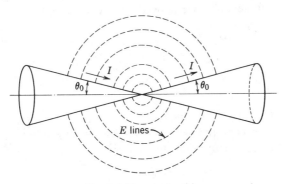

Fig. 9.6 Biconical guide.

so is analogous to the transmission line wave on cylindrical systems. This basic wave is symmetric about the axis of the guiding cones, so that if the two curl relations of Maxwell's equations are written in spherical coordinates with all ϕ variation eliminated, it is seen that there is one independent set containing E_θ, H_ϕ, and E_r only:

$$\frac{1}{r}\frac{\partial(rE_\theta)}{\partial r} - \frac{1}{r}\frac{\partial E_r}{\partial \theta} + j\omega\mu H_\phi = 0 \tag{1}$$

$$\frac{1}{r\sin\theta}\left[\frac{\partial}{\partial\theta}(\sin\theta H_\phi)\right] - j\omega\varepsilon E_r = 0 \tag{2}$$

$$-\frac{1}{r}\frac{\partial(rH_\phi)}{\partial r} - j\omega\varepsilon E_\theta = 0 \tag{3}$$

It can be checked by substitution that the following solution does satisfy the three equations.

$$E_r = 0 \tag{4}$$

$$rE_\theta = \frac{\eta}{\sin\theta}[Ae^{-jkr} + Be^{jkr}] \tag{5}$$

$$rH_\phi = \frac{1}{\sin\theta}[Ae^{-jkr} - Be^{jkr}] \tag{6}$$

These equations show the now-familiar propagation behavior, the first term representing a wave traveling radially outward with the velocity of light in the dielectric material surrounding the cones, the second term representing a radially inward traveling wave of the same velocity. The ratio of electric to magnetic field E_θ/H_ϕ is given by η for the positively traveling wave, by $-\eta$ for the negatively traveling wave. There is no field component in the radial direction, which is the direction of propagation.

The above wave looks much like the ordinary transmission line waves of uniform cylindrical systems. This resemblance is stressed if we note that the E_θ corresponds to a voltage difference between the two cones,

$$V = -\int_{\theta_0}^{\pi-\theta_0} E_\theta r\, d\theta = -\eta\int_{\theta_0}^{\pi-\theta_0}\frac{d\theta}{\sin\theta}[Ae^{-jkr} + Be^{jkr}]$$

$$= 2\eta\ln\cot\frac{\theta_0}{2}[Ae^{-jkr} + Be^{jkr}] \tag{7}$$

where the case treated is that of equal angle cones (Fig. 9.6). This is a voltage which is independent of r, except through the propagation term, $e^{\pm jkr}$. Similarly the azimuthal magnetic field corresponds to a current flow in the cones at $\theta = \theta_0$,

$$I = 2\pi r H_\phi \sin\theta_0$$
$$= 2\pi[Ae^{-jkr} - Be^{jkr}] \tag{8}$$

This current is also independent of radius, except through the propagation term. A study of the sign relations shows that it is in opposite radial directions in the two cones at any given radius.

The ratio of voltage to current in a single outward-traveling wave, a quanti.y which we call characteristic impedance in an ordinary transmission line, is obtained from (7) and (8) with $B = 0$:

$$Z_0 = \frac{\eta \ln \cot \theta_0/2}{\pi} \tag{9}$$

For a negatively traveling wave, the ratio of voltage to current is the negative of this quantity. This value of impedance is a constant, independent of radius, unlike those ratios defined for a parallel-plane radial transmission line in Sec. 9.3. We can also see this starting from the familiar concept of Z_0 as $\sqrt{L/C}$, since inductance and capacitance between cones per unit radial length are independent of radius. This comes about since surface area increases proportionally to radius, and distance separating the cone, along the path of the electric field, also increases proportionally to radius.

So far as this wave is concerned, the system arising from two ideal coaxial conical conductors can be considered a uniform transmission line. All the familiar formulas for input impedances and voltage and current along the line hold directly with Z_0 given by (9) and phase constant corresponding to velocity of light in the dielectric,

$$\beta = \frac{2\pi}{\lambda} = \omega\sqrt{\mu\varepsilon} \tag{10}$$

If the conducting cones have resistance, there is a departure from uniformity due to this resistance term, but this is usually not serious in any practical cases where such conical systems are used.

Of course, a large number of higher order waves may exist in this conical system and in similar systems. These will in general have field components in the radial direction and will not propagate at the velocity of light. We shall consider such general wave types for spherical coordinates in Sec. 10.7.

9.7 Ridge Waveguide

Of the miscellaneous shapes of cylindrical guides that have been utilized, one rather important one is the ridge waveguide, which has a central ridge added either to the top or bottom, or both, of a rectangular section as in Fig. 9.7a. It is interesting from an electromagnetic point of view since the cutoff frequency is lowered because of the capacitive effect at the center, and could in principle be made as low as desired by decreasing the gap width g sufficiently. Of course, the effective impedance of the guide also decreases as g is made smaller. One of the

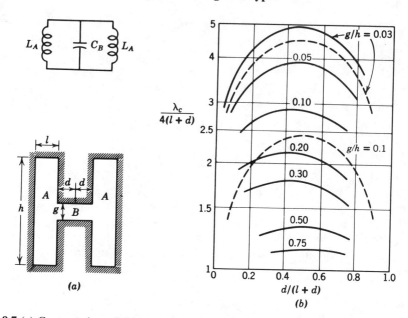

Fig. 9.7 (a) Cross section of ridge waveguide and approximate equivalent circuit for cutoff calculation. (b) Curves giving cutoff wavelength for a ridge waveguide as in (a). Solid curves: data from Cohn.[5] Dashed curves from (1).

important applications is as a nonuniform transmission system for matching purposes obtained by varying the depth of ridge along the guide, similar to the approach described in Sec. 9.5.

The calculation of cutoff frequency, which has been found to be one of the very important parameters for any shape of guide, also illustrates an interesting approach that may be applied to many guide shapes which cannot be solved exactly. At cutoff, there is no variation in the z direction ($\gamma = 0$), so one may think of this as a resonant condition for waves propagating only transversely in the given cross section, according to the desired mode. For example, the TE_{10} wave in a rectangular guide has a cutoff frequency equal to the resonant frequency for a plane wave propagating only in the x direction across the guide, thus corresponding to a half-wavelength in the x direction. A very approximate calculation of cutoff frequency for the ridge guide might then be made as in Fig. 9.7a by considering the gap a capacitance and the side sections one-turn solenoidal inductances, and writing the condition for resonance.

$$f_c = \frac{1}{2\pi}\left(C_B\frac{L_A}{2}\right)^{-1/2} = \frac{1}{2\pi}\left(\frac{2d\varepsilon}{g}\right)^{-1/2}\left(\frac{\mu lh}{2}\right)^{-1/2} = \frac{1}{2\pi}\left(\frac{g}{\mu\varepsilon lhd}\right)^{1/2} \qquad (1)$$

A better equivalent circuit for calculation of the transverse resonance is one in which the two sections A and B are considered parallel-plane transmission lines with a discontinuity capacitance C_d placed at the junction between them. (This junction effect will be discussed in Chapter 11.) Curves of cutoff frequency and a total impedance for the guide have been calculated in this manner by Cohn.[5] Some results are shown in Fig. 9.7*b* with comparisons of results from the lumped element approximation (1). As might be expected, agreement is better for the smaller gaps. Additional design data on practical ridge guides are available in handbooks.[6] The technique of finding cutoff by calculating transverse resonance frequencies—often by numerical means[7]—is valuable for a variety of irregularly shaped guides.

A conformal transformation of the Helmholtz equation, as illustrated in Ex. 7.8, may be used to simplify the boundaries of some guides. In fact, a numerical technique exists for transforming a waveguide of arbitrary cross section into one of rectangular form.[8] The trade-off is that there results a problem with a simpler boundary but an inhomogeneous dielectric.

The ridge guide illustrates that guides other than two-conductor transmission lines can be appreciably smaller than the half-wavelength measure found for the rectangular and circular guides if the boundaries concentrate energy appropriately. However, the boundary irregularities generally decrease power-handling capacity.

9.8 The Idealized Helix and Other Slow-Wave Structures

A wire wound in the form of a helix (Fig. 9.8*a*) makes a type of guide that has been found useful for antennas[9] and as slow-wave structures in traveling-wave tubes.[10] It is interesting as an example of a general class of structures that possess waves with a phase velocity along the axis much less than the velocity of light, as contrasted to most of the waves so far studied, which have phase velocities greater than the velocity of light. A rough picture suggests that the wave might follow the wire with about the velocity of light, so that its rate of progress along the axis would correspond to a phase velocity

$$v_p \approx c \sin \psi \tag{1}$$

where ψ is the pitch angle. It is rather surprising that this represents a good approximation over a wide range of parameters. It is also interesting to find that a useful analysis can be made by considering an idealization of the actual helix.

The idealization commonly analyzed,[10] referred to as the helical sheet, is a cylindrical surface in which the component of electric field along the direction of

[5] S. B. Cohn, *Proc. I.R.E.* **35**, 783–788 (August 1947).

[6] *ITT, Reference Data for Radio Engineers*, 6th ed., Howard W. Sams, Inc., New York, 1975.

[7] See, for example, R. Bulley, *IEEE Trans. Microwave Theory and Techniques* **MTT-18**, 1022 (1970).

[8] P. Silvester, *Modern Electromagnetic Fields*, Prentice-Hall, Englewood Cliffs, N.J., 1968, pp. 285–290.

[9] J. D. Kraus, *Antennas*, McGraw-Hill, New York, 1950, Chapter 7.

[10] J. R. Pierce, *Traveling-Wave Tubes*, Van Nostrand, Princeton, New Jersey, 1950, Chapter III and Appendix II.

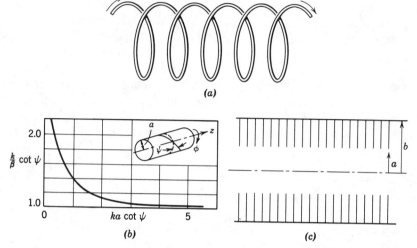

Fig. 9.8 (*a*) Wire helix. (*b*) Idealized conducting sheet and curve giving propagation constant.[10] (*c*) Section of disk-loaded waveguide.

ψ is assumed to be zero at all points of the sheet (Fig. 9.8*b*). Moreover, the component of electric field lying in the cylindrical surface normal to the direction of ψ is assumed to be continuous through the surface, as is the component of magnetic field along ψ (the latter because there is to be no current flow normal to the direction of ψ). Since the idealization takes these conditions to be the same over all the sheet, it would be expected to give best results for fine-wire helices of small pitch angle or for multifilar helices with fine wires close together.

To find the propagation constant, we set up the general solutions inside and outside the helical sheet and match them at the boundary according to continuity conditions stated above. The wave that satisfies the boundary conditions is a hybrid mode containing both TE and TM components. We assume an unattenuated wave so that $\gamma = j\beta$. Then the separation constant in the general solutions, Eqs. 7.20(5) is $k_c^2 = k^2 + \gamma^2 = k^2 - \beta^2$. Since $k = \omega/c$ and $\beta = \omega/v_p$, $\beta > k$ for waves with phase velocities below the speed of light. Therefore, $k_c^2 < 0$ and $k_c = j\tau$, where τ is a real quantity. Assuming axial symmetry and that the fields must be finite on the axis and zero at infinite radius, the general solutions can be expressed in terms of modified Bessel functions [Eqs. 7.14(16)–(17)]. The z variation $e^{-j\beta z}$ is understood in all of the following:

$$r < a \qquad\qquad\qquad r > a$$

$$E_{z1} = A_1 I_0(\tau r) \qquad\qquad E_{z2} = A_2 K_0(\tau r) \qquad\qquad (2)$$

$$E_{r1} = \frac{j\beta}{\tau} A_1 I_1(\tau r) \qquad\quad E_{r2} = -\frac{j\beta}{\tau} A_2 K_1(\tau r) \qquad (3)$$

$$H_{\phi 1} = \frac{j\omega\varepsilon}{\tau} A_1 I_1(\tau r) \qquad\quad H_{\phi 2} = -\frac{j\omega\varepsilon}{\tau} A_2 K_1(\tau r) \qquad (4)$$

$$H_{z1} = B_1 I_0(\tau r) \qquad\qquad H_{z2} = B_2 K_0(\tau r) \tag{5}$$

$$H_{r1} = \frac{j\beta}{\tau} B_1 I_1(\tau r) \qquad\qquad H_{r2} = -\frac{j\beta}{\tau} B_2 K_1(\tau r) \tag{6}$$

$$E_{\phi 1} = -\frac{j\omega\mu}{\tau} B_1 I_1(\tau r) \qquad E_{\phi 2} = \frac{j\omega\mu}{\tau} B_2 K_1(\tau r) \tag{7}$$

The idealized boundary conditions first described are

$$E_{z1} \sin\psi + E_{\phi 1} \cos\psi = 0 \tag{8}$$

$$E_{z2} \sin\psi + E_{\phi 2} \cos\psi = 0 \tag{9}$$

$$E_{z1} \cos\psi - E_{\phi 1} \sin\psi = E_{z2} \cos\psi - E_{\phi 2} \sin\psi \tag{10}$$

$$H_{z1} \sin\psi + H_{\phi 1} \cos\psi = H_{z2} \sin\psi + H_{\phi 2} \cos\psi \tag{11}$$

The fields (2)–(7) are substituted in (8)–(11). A nonvanishing solution for the field amplitude requires that the determinant of the coefficients be zero, which leads to

$$(\tau a)^2 \frac{I_0(\tau a) K_0(\tau a)}{I_1(\tau a) K_1(\tau a)} = (ka \cot\psi)^2 \tag{12}$$

A solution of this taken from Pierce[11] is shown in Fig. 9.8b. It is seen that for $ka \cot\psi > 4$, the approximation (1) gives good results.

Some general comments about slow-wave structures are in order. If it is desired to produce a wave with an electric field along the axis, propagating with a phase velocity less than that of light (as in a traveling-wave tube where the phase velocity should be approximately equal to the beam velocity for efficient interaction with the electrons) then, as discussed above, $k_c^2 < 0$:

$$\tau^2 = -k_c^2 = \beta^2 - k^2 \tag{13}$$

$$\tau = \beta \left(1 - \frac{v_p^2}{c^2} \right)^{1/2} \tag{14}$$

It is consequently necessary in a cylindrically symmetric system that the Bessel function solutions (Sec. 7.20) have imaginary arguments, and they may therefore be written as modified Bessel functions. For a TM wave,

$$E_z = A I_0(\tau r) \tag{15}$$

$$H_\phi = \frac{\omega\varepsilon}{\beta} E_r = \frac{j\omega\varepsilon}{\tau} I_1(\tau r) \tag{16}$$

[11] See Ref. 10, J. R. Pierce.

_____ **Example 9.8** _____
Cylindrical Reactance Wall to Produce Slow Waves

If we ask about the boundary conditions that might be supplied at a cylindrical surface $r = a$ in order to support slow waves, we see that, if uniform, it should be of the nature of a reactive sheet with

$$jX = -\frac{E_z}{H_\phi}\bigg|_{r=a} = jn\frac{\tau}{k}\frac{I_0(\tau a)}{I_1(\tau a)} \tag{17}$$

The helical sheet studied earlier may be considered as supplying this required reactance through the interaction with the TE waves and external fields caused by the helical cuts. The short-circuited sections of radial lines of a disk-loaded waveguide (Fig. 9.8c) may also be considered as supplying an approximation to the above required uniform reactance at $r = a$, and will therefore support a slow wave also. The approximate reactance supplied by this structure is

$$X = \eta\left[\frac{J_0(ka)N_0(kb) - J_0(kb)N_0(ka)}{J_1(ka)N_0(kb) - J_0(kb)N_1(ka)}\right] \tag{18}$$

Note that if $(v_p/c)^2 \ll 1$, τ is substantially equal to β. By the nature of the I_0 functions (Fig. 7.14c), the field on the axis of such slow-wave structures is much less than that on the boundary when βa is large. This is of course undesirable when it is the field on the axis that is to act on electrons, as in a traveling-wave tube. Of course, the presence of electron space charge will modify the forms of solution somewhat.

_____

9.9 Surface Guiding

The result of Sec. 9.8 suggests a localization of fields near a surface that possesses a reactive surface impedance. This concept appeared there in an interior region, but may be useful when the fields are in the external region also. Because of the surface-guiding principle, the energy is maintained near the surface so that it is not radiated or coupled seriously to nearby objects. Thus the external region corresponding to Fig. 9.8c would be as shown in Fig. 9.9a. The proper solutions for the external cylindrical region for a TM wave are the same as Eqs. 9.8(2)–(4) for $r > a$. For E_z and H_ϕ, with $e^{-j\beta z}$ understood,

$$E_z = AK_0(\tau r) \tag{1}$$

$$H_\phi = -\frac{j\omega\varepsilon}{\tau}AK_1(\tau r) \tag{2}$$

The field reactance at $r = a$ necessary to support such a wave is then

$$jX = \frac{E_z}{H_\phi}\bigg|_{r=a} = jn\frac{\tau}{k}\frac{K_0(\tau r)}{K_1(\tau r)} \tag{3}$$

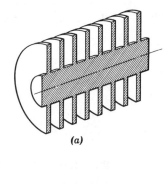

(a)

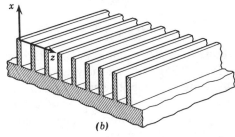

(b)

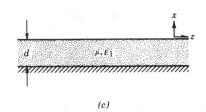

(c)

Fig. 9.9 (a) Rod with periodic radius variations capable of propagating a slow wave. (b) Planar equivalent of (a). (c) Dielectric-coated conducting surface.

Equation (3) again represents a positive, or inductive, reactance as did Eq. 9.8(17), and the solutions (1) and (2) die off for large r. Phase velocity is less than the velocity of light in the external dielectric in order to maintain τ real, as in Eq. 9.8(13),

$$\tau^2 = \beta^2 - k^2 \tag{4}$$

Example 9.9a
Surface Guiding by a Reactance Sheet

The above concept is perhaps more easily visualized for a plane sheet, as pictured in Fig. 9.9b. Substitution in Eqs. 8.2(13)–(16) easily verifies that the following are solutions of Maxwell's equations, using the definition (4), again with $e^{-j\beta z}$ understood:

$$E_z = Ce^{-\tau x} \tag{5}$$

$$H_y = -\frac{j\omega\varepsilon C}{\tau} e^{-\tau x} \tag{6}$$

$$E_x = -\frac{j\beta C}{\tau} e^{-\tau x} \tag{7}$$

A surface wave impedance defined for this example is then

$$jX = \frac{E_z}{H_y}\bigg|_{x=0} = \frac{j\tau}{\omega\varepsilon} \tag{8}$$

Again we see that the impedance sheet should be inductive for surface guiding of this TM wave (but see Prob. 9.9a), and that the value of β must be greater than k if the wave is to die away with increasing x. Therefore phase velocity is less than the velocity of light in the dielectric. The spaces between the conducting fins may be considered to be shorted parallel-plane transmission lines and the field impedance at the ends of the fins can be found from Sec. 8.3 as

$$\frac{E_z}{H_y} = j\eta \tan kd$$

where d is the height of the fins. It is clear that if $kd < \pi/2$, the surface appears as an inductive reactance.

Example 9.9b
Surface Guiding by a Dielectric Coating

Another important way of obtaining a surface impedance sheet is by coating a conductor with a thin layer of dielectric, as illustrated in Fig. 9.9c. The following TM solutions for the region 1 can be verified by substitution in Eqs. 8.2(13)–(16). The conductor is assumed perfect and the argument of the sine is selected to make E_z zero at the conductor surface, $x = -d$.

$$E_z = D \sin k_x(x + d) \tag{9}$$

$$H_y = -\frac{j\omega\varepsilon_1 D}{k_x} \cos k_x(x + d) \tag{10}$$

$$E_x = -\frac{j\beta D}{k_x} \cos k_x(x + d) \tag{11}$$

$$k_x^2 = k_1^2 - \beta^2 \tag{12}$$

From (9) and (10) a field impedance at $x = 0$ can be found as follows:

$$jX = \frac{E_z}{H_y}\bigg|_{x=0} = \frac{jk_x}{\omega\varepsilon_1}\tan k_x d \tag{13}$$

This is inductive as required for guiding TM waves, and if equated to (8) will define the β of the desired surface wave. For small thicknesses, $k_x d \ll 1$, (13) becomes

$$jX \approx \frac{jk_x^2 d}{\omega\varepsilon_1} \tag{14}$$

Using (4), (8), (12), and (14), we can show that the phase velocity is less than the velocity of light in the external dielectric and greater than that in the coating.

The principle is also applicable to lossy dielectrics and/or conductors, for which a finite but relatively small attenuation in the z direction is obtained. Zenneck[12] and Sommerfeld[13] provided the early classical analyses of this phenomenon; Goubau[14] illustrated its usefulness as a practical waveguiding means by using either thin dielectric coatings or corrugations on round wires; a thorough treatment is given by Collin.[15] It is clear that the problem treated in (9)–(12) (Fig. 9.9c) is just the upper half of the solution for the dielectric slab waveguide, Fig. 9.2a and Eqs. 9.2(1)–(5), so surface guiding and dielectric waveguiding are closely related phenomena.

9.10 Periodic Structures and Spatial Harmonics

The corrugated surfaces used as illustrations of reactance walls in the two preceding sections are special examples of periodic systems if the spacing between corrugations is uniform. Periodic systems have interesting properties and important applications, so will be examined more carefully in this section. We recognize that the treatment as a smooth reactance wall, used in Secs. 9.8 and 9.9, is only an approximation since the grooves will cause field disturbances not accounted for by this "smoothed out" approximation.

Let us begin by consideration of a specific example, the parallel-plane transmission line with periodic troughs in one plate, as illustrated in Fig. 9.10a. If the troughs are relatively narrow, they act to waves with z-directed currents in the bottom plate as shorted transmission lines in series with the conductor. These lines produce values of E_z and H_y at $x = 0$, which are essentially constant over the gap width w. Thus the boundary condition for E_z at $x = 0$ is as shown in Fig. 9.10b, a phase shift of $\beta_0 d$ being allowed over each period since we will stress

[12] J. Zenneck, *Ann. Phys.* **23**, 846 (1907).

[13] A. Sommerfeld, *Ann. Phys. u. Chemie* **67**, 233 (1899). Described in J. A. Stratton, *Electromagnetic Theory*, McGraw-Hill, New York, 1946, p. 527.

[14] G. Goubau, *Proc. I.R.E.* **39**, 619 (1951); *J. Appl. Phys.* **21**, 1119 (1950).

[15] See Ref. 1, R. E. Collin, Chapter 11.

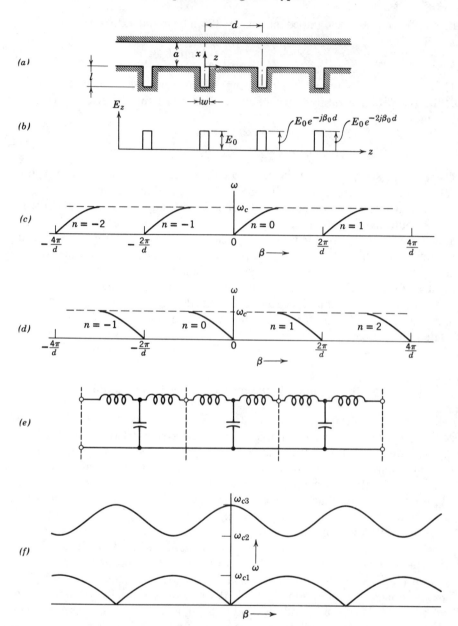

Fig. 9.10 (*a*) Parallel-plane transmission line with periodic short-circuited troughs. (*b*) Idealized variations of E_z along the lower plate in the structure of (*a*). (*c*) ω–β relation for spatial harmonics of wave with fundamental traveling in $+z$ direction. (*d*) ω–β relation for spatial harmonics of wave with fundamental traveling in $-z$ direction. (*e*) Low-frequency equivalent circuit for fundamental spatial harmonic of structure in (*a*). (*f*) Composite ω–β diagram including pass band at higher frequency.

propagating waves. The square waves shown neglect higher order fringing fields at the corners, but are better approximations than the complete smoothing out of the effect (as in Sec. 9.9) and are sufficient to illustrate the basic properties of periodic structures.

To fulfill the boundary condition at $x = 0$, we might expect to add solutions of Maxwell's equations just as we added solutions of Laplace's equation in Chapter 7 to satisfy boundary conditions not satisfied by a single solution.

For the present problem let us take $\partial/\partial y = 0$, and consider waves with E_x, E_z, and H_y only. Thus a sum of solutions satisfying Maxwell's equations and the boundary condition of $E_z = 0$ at $x = a$ may be written by adding waves of the TM form, the TEM component being included if the sum includes $n = 0$

$$E_z(x, z) = \sum_{n=-\infty}^{\infty} A_n \sin K_n(a - x)e^{-j\beta_n z} \tag{1}$$

$$E_x(x, z) = \sum_{n=-\infty}^{\infty} \frac{j\beta_n}{K_n} A_n \cos K_n(a - x)e^{-j\beta_n z} \tag{2}$$

$$H_y(x, z) = \sum_{n=-\infty}^{\infty} \frac{j\omega\varepsilon}{K_n} A_n \cos K_n(a - x)e^{-j\beta_n z} \tag{3}$$

where

$$K_n^2 = \omega^2\mu\varepsilon - \beta_n^2 = k^2 - \beta_n^2 \tag{4}$$

The boundary condition illustrated by Fig. 9.10b is a periodic function of z if the phase factor is taken out separately. If we expand the resulting periodic function in the complex form of the Fourier series (Prob. 7.11f), we may write for the boundary condition

$$E_z(0, z) = e^{-j\beta_0 z} \sum_{n=-\infty}^{\infty} C_n e^{-(j2\pi nz/d)} \tag{5}$$

where

$$C_n = \frac{1}{d} \int_{-d/2}^{d/2} E_z(0, z)e^{j\beta_0 z}e^{+j(2\pi nz/d)} \, dz = \frac{1}{d} \int_{-w/2}^{w/2} E_0 e^{+j(2\pi nz/d)} \, dz$$
$$= \frac{E_0}{\pi n} \sin\left(\frac{n\pi w}{d}\right) \tag{6}$$

By comparing (5) with (1) evaluated at $x = 0$, we can now identify A_n and β_n

$$A_n = \frac{C_n}{\sin K_n a} = \frac{E_0}{\pi n} \frac{\sin(n\pi w/d)}{\sin K_n a} \tag{7}$$

$$\beta_n = \beta_0 + \frac{2\pi n}{d} \tag{8}$$

A wave solution is thus determined for this problem with the approximations described.

We wish first to stress the different role of the TM solutions in this problem compared with that considered previously. In earlier sections TM solutions have been considered as "modes" with the inference that each may be excited independently. Here they are coupled by the periodic boundary condition and must exist in the proper relationship to each other to satisfy this boundary condition. In this capacity they are known as "spatial harmonics," a natural extension of the harmonic character of the Fourier series to the periodic system in space. We note especially from (8) that determination of β for any spatial harmonic automatically determines the value of all others. Thus the ω–β diagram (Sec. 5.11) is periodic in β with repetition at intervals of $2\pi/d$ as illustrated in Figs. 9.10c and 9.10d.

Determination of the shape of one period of this plot (say β_0) requires study of the boundary conditions on two field components. We have already considered E_z and now choose H_y as the second component for this example. If the solution for the troughs is well approximated by the shorted-transmission-line behavior, the value of H_y at $x = 0$, $z = 0$ for the shorted lines is given by

$$H_y = \frac{-jE_0}{\eta} \cot kl$$

Thus one might equate this to (3) evaluated at $x = 0$, with A_n substituted from (7):

$$-j\frac{1}{\eta}\cot kl = \sum_{n=-\infty}^{\infty} \frac{j\omega\varepsilon}{\pi n K_n} \sin\frac{\pi n w}{d} \cot K_n a \tag{9}$$

This with (4) and (8) determines β_0 in principle, although general solution may be difficult.

For the special case of $\beta_0 d \ll 1$, lumped element approximations may be used. The fundamental harmonic solution of the corresponding low-pass filter then gives the characteristics of the fundamental, and from that, the values of β for other spatial harmonics. The filter corresponding to the line used in the above example is shown in Fig. 9.10e. The shunt capacitance and a part of the series inductance represent the fields in the parallel-plate section between grooves and the shorted-transmission-line grooves contribute an additional series inductance. The capacitors in Fig. 9.10e each have the value $C = \epsilon(d - w)/a$ for a unit length in the y direction. The inductance from the section between grooves is $L_1 = \mu a(d - w)$ and that from the shorted grooves is $L_2 = (\eta w/\omega)\tan kl$ (assuming $kl < \pi/2$). Each inductor in Fig. 9.10e has the value $L = (L_1 + L_2)/2$.

Although illustrated for a particular example to provide concreteness, the periodic character of the ω–β diagram, and the relation (8) for phase constant of a spatial harmonic, apply to any structure of period d. Other important properties are as follows:

1. The group velocity of all spatial harmonics of a given wave are equal. This would be expected so that the wave would stay together, but may be noted either from the ω–β diagrams, or by differentiating (8),

$$\frac{1}{v_{gn}} = \frac{d\beta_n}{d\omega} = \frac{d\beta_0}{d\omega}$$

2. Of the infinite number of spatial harmonics, half are backward waves (Sec. 5.13) with phase and group velocities in the opposite directions. Thus the $n = 0, 1, 2$, and so on of Fig. 9.10c are forward waves whereas $n = -1, -2$, and so on are backward waves, as is seen by comparing the signs of ω/β and $d\omega/d\beta$ for the various portions of the figure. If the phase velocity of the fundamental spatial harmonic is negative, the harmonics are as shown in Fig. 9.10d. Here the $n = 0, -1, -2$, and so on have both phase and group velocities in the negative direction and are not backward waves, whereas the $n = 1, 2$, and so on are backward waves with positive phase velocities and negative group velocities. The structure of Fig. 9.10a has a fundamental forward wave, but other periodic structures may have fundamental backward waves.

3. The field distribution at any plane $z = md$ (m an integer) is the same as at $z = 0$ except for multiplication by the phase factor $e^{-j\beta_0 md}$. This important property is related to Floquet's theorem,[16] and is often taken as the fundamental starting point for the study of periodic systems. For the particular example used in this section, (1) and (8) yields

$$E_z(x, md) = \sum_{n=-\infty}^{\infty} A_n \sin K_n(a - x) e^{-j\beta_0 md} e^{-j2\pi mn}$$

$$= e^{-j\beta_0 md} \sum_{n=-\infty}^{\infty} A_n \sin K_n(a - x) = e^{-j\beta_0 md} E_z(x, 0) \qquad (10)$$

and similarly for the other field components.

4. The wave is cut off where group velocity becomes zero, shown as ω_c in Figs. 9.10c and 9.10d. Above this frequency there is a region of reactive attenuation, typical of filters in the attenuating region. As frequency is increased, however, other pass bands will be found with propagating waves, as illustrated in Fig. 9.10f. Each of these waves has spatial harmonics, just as the low-pass wave studied. The Pierce coupled-mode theory[17] is especially powerful in giving a picture of the entire ω–β diagram in a structure with relatively small periodic perturbations if the phase constants of the unperturbed system are known.

[16] L. Brillouin, *Wave Propagation in Periodic Structures*, Dover, New York, 1953.
[17] W. H. Louisell, *Coupled Mode and Parametric Electronics*, Wiley, New York, 1962.

PROBLEMS

9.2a For a dielectric slab guide with $\mu_1 = \mu_2$, plot the complement of the critical angle of Fig. 9.2a, $\psi = \pi/2 - \theta_c$, as a function of $\varepsilon_2/\varepsilon_1$. Many dielectric guides have small differences between the two permittivities, $\varepsilon_2 = \varepsilon_1(1 - \Delta)$ where $\Delta \ll 1$. For these, show that $\psi \approx \sqrt{\Delta}$.

9.2b Demonstrate that when $\theta = \theta_c$ (defined as cutoff for the dielectric slab guides) there is no decay in medium 2 (outside of slab) and propagation is at the velocity of light in medium 2. Show that when θ approaches $\pi/2$ (grazing incidence at the boundary) phase velocity approaches the velocity of light in medium 1.

9.2c Using the definition of cutoff given in Prob. 9.2b, show that k_x at cutoff is $(k_1^2 - k_2^2)^{1/2}$. Calculate the lowest cutoff frequencies (other than zero) for even and odd TM and even and odd TE modes in a dielectric slab guide with $d = 1$ mm, $\varepsilon_1 = 4.08$, $\varepsilon_2 = 4$, and $\mu_1 = \mu_2$.

9.2d Verify Eq. 9.2(20) for the stated approximations.

9.2e* The picture of "zigzag" rays shown in Fig. 9.2a can be made quantitative by suitably considering the reflections at the two surfaces. To illustrate this, consider a TE wave in a symmetrical slab. For an angle $\theta > \theta_c$, find the reflection coefficient (including phase) from Chapter 6. After two reflections and two traverses across the slab (with suitable transverse phase constant), the net phase change should be 2π or a multiple. Set up this relation and show how it determines β.

9.3a For a TM_{01} wave in a circular waveguide it is desired to insert a blocking impedance for a given frequency. To do this, a section of shorted radial line (Fig. P9.3a) is inserted in the guide, its outer radius a chosen so that with the guide radius b given, the impedance looking into the radial line is infinite at the given frequency. Suppose that the radius b is 1.25 times greater than cutoff radius at this frequency for the TM_{01} wave and find the radius a.

9.3b It is sometimes required to break the outer conductor of a coaxial line for insulation purposes, without interrupting the rf current flow. This may be accomplished by the radial line as shown (Fig. P9.3b) in which a is chosen so that with b and the operating wavelength specified, the radial line has zero input impedance seen from the line. Find the value of a, assuming that end effects are negligible, and that $2\pi b/\lambda = 1$.

9.3c Find the voltage at the radius a in terms of the coaxial line's current flowing into the radial line at radius b (Fig. P9.3b).

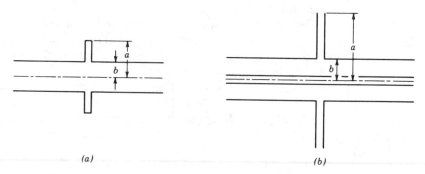

(a)   (b)

Fig. P9.3 (a) Circular guide with shorted radial line. (b) Circular guide with open-circuited radial line.

9.3d Taking the classical transmission line equations with distributed inductance and capacitance varying with radius as appropriate to the radial line,

$$L = \frac{\mu d}{2\pi r} \quad \text{and} \quad C = \frac{2\pi \varepsilon r}{d}$$

show that the equation for voltage as a function of radius is a Bessel equation, and that the solutions for voltage and current obtained in this manner are consistent with Eqs. 9.3(1) and 9.3(2).

9.3e Derive forms similar to Eqs. 9.3(12) and 9.3(13) if E_a is specified at r_a, E_b at r_b. Repeat for H_a specified at r_a, H_b at r_b.

9.4a Sketch lines of current flow for a circumferential mode with $v = 1$ in a radial line of Sec. 9.4. (Take $Z_n = J_n$ for this purpose.) Suggest methods of suppressing this mode by judicious cuts without disturbing the symmetrical mode.

9.4b A section of the wedge-shaped guide as in Fig. 9.4a may be used to join two waveguides of the same height but different width, both propagating the TE_{10} mode, and a good degree of match is obtained by the process so long as the transition is gradual. Discuss the transition between the fields and impedances of the waveguide and those of the sectoral guide in terms of Bessel functions.

9.5a In the use of the inclined plane line for matching as discussed in Sec. 9.5, suppose that $f = 3$ GHz, $d_1 = 1$ cm, $d_2 = 2$ cm, $kr_1 = 2.5$, and the dielectric is air. If the line to the right is perfectly matched, obtain the approximate standing-wave ratio in the line to the left. Compare with that which would exist with a sudden transition, considering only the discontinuity of characteristic impedance.

9.5b Apply the principle of duality to the TE_{11}, TE_{01}, and TM_{01} modes in circular cylindrical waveguides to obtain qualitatively the fields of the "dual" modes. For which of these might boundary conditions be supplied, allowing changes in the conductor position or shape from those of the original mode?

9.5c A wedge-shape dielectric region is bounded by conducting planes at $\phi = 0$ and ϕ_0, $z = 0$ and d. Find the field components of the lowest order mode with E_ϕ, H_r, and H_z. Is this the dual of the mode discussed for sectoral horns in Sec. 9.4?

9.5d Discuss the application of the mode of Prob. 9.5c to the matching between rectangular waveguides of different height, both propagating the TE_{10} mode, describing how to use the solution of Prob. 9.5c to estimate reflections and the approximations involved.

9.6a Derive the basic characteristics of the principal waves on a transmission line consisting of two coaxial, common-apex cones of unequal angles.

9.6b Write the dual of the mode studied in Sec. 9.6. Can it be supported by physical boundaries using perfect conductors?

9.6c One point of view toward a cylindrical conductor excited by a source across a gap at $z = 0$ (Fig. P9.6c) is that it is a nonuniform transmission line propagating effects away from the gap. (This point of view is used in one theory of cylindrical antennas to be presented in Sec. 12.24). That is, an elemental section at z may be considered as a biconical line passing through radius a at z. Write the expression for characteristic impedance as a function of z and plot versus z/a for air dielectric.

9.7a Demonstrate that the cutoff of a TM_{01} mode in circular guide may be found by considering transverse resonance of a radial line mode.

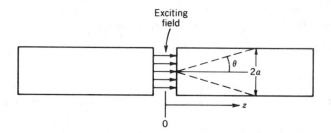

Fig. P9.6c Cylindrical conductor excited in gap.

9.7b For a TE mode in the small-gap ridge waveguide, the energy stored in electric fields is predominantly in the gap region. If electric field is assumed uniform at E_0 over this region, estimate the stored electric energy u_E per unit length for a single propagating wave in such a guide. If power transfer is $v_g(u_E + u_M)$, and magnetic energy is equal to electric energy, calculate power transfer for a guide with $g = 1$ mm, $2d = 1$ cm, $E_0 = 10^6$ V/m, air dielectric, and operating frequency 1.3 times cutoff frequency.

9.8a Assuming $(v_p/c)^2 \ll 1$, plot kaX/η versus βa for a slow-wave structure. State the requirements on the reactance in order that there may be any slow-wave solution of this type. What should X/η be for β_a large?

9.8b Show that a reactance sheet might be used as the boundary condition on fast waves of the TM_{01} type studied in Sec. 8.9. Plot the required value of kaX/η as a function of $k_c a$. Under what conditions might there be a slow wave and a series of fast waves in a given guide of this type?

9.8c** Imagine a parallel-plane transmission line of spacing $2a$ in the x direction, for which both upper and lower planes are cut with many fine cuts at angle ψ from the y direction. Assume no variations with y, and apply approximations as utilized in the helical-sheet analysis, obtaining the field components for propagation in the z direction, the complete equation determining β, and the approximate solution of this for $ka \cot \psi \gg 1$.

9.9a Analyze a TE surface wave over a plane with no variations in y and show that a capacitive reactance is necessary to produce an exponential decay with x.

9.9b For the TM surface wave established by the thin dielectric coating on a perfect conductor, find the average power transfer in the z direction.

9.9c Utilizing the result of Prob. 9.9b, find the approximate attenuation if the conductor at $x = -d$ has surface resistivity R_s.

9.9d Repeat Prob. 9.9c, but assume conductor perfect and dielectric with a small lossy part, $\varepsilon_1 = \varepsilon_1' - j\varepsilon_1''$ with $\varepsilon_1'' \ll \varepsilon_1'$.

9.9e It is desired to pass an electron beam 1 mm above the structure in Fig. 9.9b. To match the velocity of the electrons, a wave at 18 GHz should travel at 0.1 of the speed of light. Design the fins to produce the appropriate impedance at $x = 0$. How much weaker is the field at the beam than at $x = 0$?

9.10a* Making the assumptions that $kl \ll 1$ and $\beta_0 d \ll 1$, plot the lowest frequency pass band for the structure of Fig. 9.10a. Plot the curve for all β and state your reasoning.

9.10b* Take the same approximations as in Prob. 9.10a, but assume the "troughs" are open circuited. Find the equivalent lumped-element circuit, the propagation constant of the fundamental, and sketch the $\omega-\beta$ diagram of this wave. Note the fact that the cutoff region includes $\omega = 0$ and frequencies up to some lower cutoff frequency.

9.10c* Find the low-frequency portion of the $\omega-\beta$ diagram for a periodic transmission system composed of parallel $L-C$ circuits in both the series and shunt legs. Note that the fundamental spatial harmonic is a backward wave if resonant frequency of circuit in series leg is lower than that for shunt leg.

9.10d The example used in Sec. 9.10 was a closed line. Consider an open region, such as those studied in Sec. 9.9 for surface-wave propagation. What regions of the $\omega-\beta$ plot will represent nonradiating or guided surface waves?

9.10e* Illustrate Prob. 9.10d by solving a problem with the structure at $x = 0$ as in Fig. 9.10a but with the top plate removed so that the region extends to infinity.

10

RESONANT CAVITIES

10.1 Introduction

Lumped-element circuits, as was noted in Chapter 4, require components small compared with wavelength. With integrated-circuits technology, these elements may be made very small indeed, but the small elements have correspondingly low power-handling capabilities. There are thus needs for circuit elements of larger size. For microwave and millimeter-wave frequencies, these may be comparable with wavelength in size so that distributed effects must be considered. In fact the resonant properties of standing waves may be used to provide an important class of resonant elements for microwave and higher frequency oscillators and amplifiers.

The resonant properties of standing waves were first seen for transmission lines in Chapter 5. Similar resonances for plane waves were met in Chapter 6. Other wave-guiding systems, such as the guides with conducting boundaries or the dielectric guides, provide similar resonant systems when terminated to produce complete reflection and hence standing waves. Those with conducting boundaries are also self-shielding, with essentially no leakage or coupling to the exterior. Such closed systems are known as *cavity resonators*.

The analysis of cavity resonators requires solution of Maxwell's equations subject to the boundary conditions of the cavity walls, but for simple geometrical shapes such solutions may be found by superposing incident and reflected traveling waves for the corresponding guides to produce standing waves. For many applications, less regular shapes are required and solution of the boundary-value problem is more difficult, though approximate solutions are often adequate. For all types, circuit ideas such as those of impedance and Q, are useful. Methods of calculation will be provided for the simple shapes, and methods of measurement, for resonators of any shape. Also included is a study of resonances in dielectric bodies which are of importance in microwave integrated circuits.

10.2   Elemental Concepts of Cavity Resonators

Before beginning the detailed analysis, let us develop some physical pictures of the resonances occurring within cavities. The most direct is that of standing waves, referred to in the Introduction. Consider first a coaxial transmission line with both ends shorted, as pictured in Fig. 10.2a. From the transmission line analysis of Chapter 5, it is known that such a shorted line may support a standing wave if frequency is such that the length of line is exactly a half-wavelength. The line may be thought of as resonant at that frequency, since the standing-wave pattern set up has constant total energy in that section of line, with that energy oscillating between the electric and magnetic fields of the line. Thus, as in Fig. 10.2a, the standing wave of voltage has a zero at each end and a maximum at the center. The standing wave of current is 90 degrees out of phase with the voltage wave, has maxima at the two ends, and a zero at the center. These waves may exist inside this completely enclosed region without interference from, or radiation to, the outside. The shielding is complete if conductors are perfect, and practically so for any practical conductors at ultrahigh frequencies. This viewpoint is verified by the analyses of skin effect (Chapter 3) where it was found that depth of penetration at high frequencies is so small (of the order of 10^{-4} cm for copper at 3 GHz) that almost any practical thickness acts essentially as an infinite thickness.

Since the inside of the region is completely shielded from the outside, it will be necessary to excite the waves by some source, such as the small loop (Fig. 10.2a),

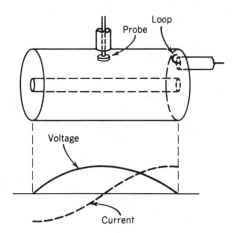

Fig. 10.2a Resonant coaxial system and standing waves of voltage and current.

designed to excite the magnetic field of the line at its maximum value, or the small probe introduced at the maximum of electric field. If one of these means is used to stimulate the line exactly at its resonant frequency, the oscillations may build up to a large value. In the steady-state limit, the exciting source need supply only the relatively small amount of energy lost to the finite conductivity of the walls, the relatively large stored energy being essentially constant and passing back and forth between electric and magnetic fields. If the source excites the line at a frequency somewhat off resonance, the energies in electric and magnetic fields do not balance. Some extra energy must be supplied over one part of the cycle which is given back to the source over another part of the cycle, and the line acts as a reactive load on the exciting source in addition to its small loss component. The similarity to ordinary tuned-circuit operation is evident, and as with such circuits, the concept of Q is useful in describing the effect of losses on bandwidth.

Similarly, one may consider the standing wave produced by a mode in any of the other waveguides studied as resonant if the length between conducting end plates (shorts) is a multiple of a half-guide wavelength for that mode. An example is given in Fig. 10.2b, in which fields are indicated for a circular cylindrical resonator which is one-half guide wavelength for the TM_{01} mode. It is evident from this picture that a particular cavity of fixed shape and size will have many different modes (actually an infinite number) corresponding to all the wave types that may exist in the corresponding waveguide, and to different numbers of half-waves between shorting ends. This picture is pursued in detail for cylindrical resonators of circular and square section in the following sections. We will see that coupling can be to either the electric or magnetic fields of the mode it is desired to excite.

Not all resonators are simple enough in shape to be considered as sections of a waveguiding system, and for these other methods of solving the boundary-value problem are required. We shall see a number of small-gap cavities that are particularly useful with electron devices. Some of these may be considered as capacitively loaded transmission lines. In others electric and magnetic energies are effectively separated so that they may be considered as lumped L–C circuits,

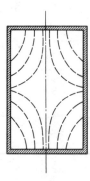

Fig. 10.2b Cylindrical cavity and electric field pattern on a longitudinal section plane.

with the one-turn inductance providing the self-shielding. These last provide a direct tie between cavity resonators and lumped-element resonant circuits. From still another point of view, the internal field patterns in any of these cavities may be thought of as interferences between plane waves multiply reflected from the several walls, just as with the acoustic resonances of a closed room or other enclosure. This is an extension of the idea of synthesizing the TE_{10} mode of rectangular guide from plane waves reflected from the side walls, Sec. 8.8. From all these points of view, resonances occur when energy interchanges between the electric and magnetic forms with only a small loss per cycle.

RESONATORS OF SIMPLE SHAPE

10.3 Fields of Simple Rectangular Resonator

For the first mode to be studied in some detail, we shall choose that mode in a rectangular conducting box which may be considered the standing wave pattern corresponding to the TE_{10} mode in a rectangular guide. As was done in the study of waveguides, the conducting walls will be taken as perfect, and losses in an actual resonator will be computed approximately by taking the current flow of the ideal mode as flowing in the walls of known conductivity.

In the rectangular conducting box of Fig. 10.3a, imagine a TE_{10} waveguide mode oriented with its electric field in the y direction and propagating in the z direction. The condition that E_y shall be zero at $z = 0$ and d, as required by the perfect conductors, is satisfied if the dimension d is a half-guide wavelength. Using Eq. 8.8(11),

$$d = \frac{\lambda_g}{2} = \frac{\lambda}{2\sqrt{1 - (\lambda/2a)^2}}$$

or

$$\lambda = \frac{2ad}{\sqrt{a^2 + d^2}} \tag{1}$$

To obtain the field distributions in the dielectric interior, we add positive and negative propagating waves of the form of Eqs. 8.8(4) and 8.8(5)

$$E_y = (E_+ e^{-j\beta z} + E_- e^{j\beta z}) \sin \frac{\pi x}{a} \tag{2}$$

$$H_x = -\frac{1}{Z_{TE}}(E_+ e^{-j\beta z} - E_- e^{j\beta z}) \sin \frac{\pi x}{a} \tag{3}$$

$$H_z = \frac{j}{\eta}\left(\frac{\lambda}{2a}\right)(E_+ e^{-j\beta z} + E_- e^{j\beta z}) \cos \frac{\pi x}{a} \tag{4}$$

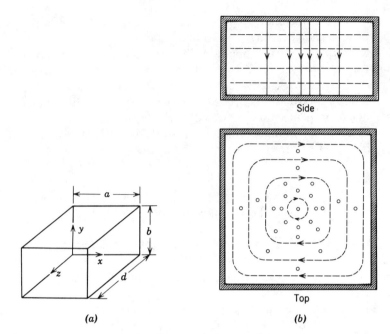

Fig. 10.3 (*a*) Rectangular cavity. (*b*) Electric and magnetic fields in rectangular resonator with TE_{101} mode. Solid lines represent electric field and dashed, magnetic field.

Since E_y must be zero at $z = 0$, $E_- = -E_+$, as we would expect, since the reflected wave from the perfectly conducting wall should be equal to the incident wave. E_y must also be zero at $z = d$, so that $\beta = \pi/d$. Then (2)–(4) may be simplified, letting $E_0 = -2jE_+$:

$$E_y = E_0 \sin \frac{\pi x}{a} \sin \frac{\pi z}{d} \tag{5}$$

$$H_x = -j \frac{E_0}{\eta} \frac{\lambda}{2d} \sin \frac{\pi x}{a} \cos \frac{\pi z}{d} \tag{6}$$

$$H_z = j \frac{E_0}{\eta} \frac{\lambda}{2a} \cos \frac{\pi x}{a} \sin \frac{\pi z}{d} \tag{7}$$

In studying the foregoing expressions, we find that electric field passes vertically from top to bottom, entering top and bottom normally and becoming zero at the side walls as required by the perfect conductors. The magnetic field lines lie in horizontal (*x–z*) planes and surround the vertical displacement current resulting from the time rate of change of E_y. Fields are sketched roughly in Fig. 10.3*b*. There are equal and opposite charges on top and bottom because of the normal electric field ending there. A current flows between top and bottom, becoming vertical in the side walls. Here we are reminded of a conventional resonant circuit with the top and bottom acting as capacitor plates and the side walls as the current path between them. But in the lumped-element circuit, electric and magnetic fields are separated whereas here they are intermingled.

Because the mode studied here has one half-sine variation in the x direction, none in the y direction, and one in the z direction, it is some times known as a TE_{101} mode. The coordinate system is of course arbitrary, but some choice must be made before the mode can be described in this manner.

10.4 Energy Storage, Losses, and Q of Simple Resonator

The energy storage and energy loss in the rectangular resonator of the preceding section are quantities of fundamental interest. Since the total energy passes between electric and magnetic fields, we may calculate it by finding the energy storage in electric fields at the instant when these are a maximum, for magnetic fields are then zero in the standing wave pattern of the resonator.

$$U = (U_E)_{max} = \frac{\varepsilon}{2} \int_0^d \int_0^b \int_0^a |E_y|^2 \, dx \, dy \, dz$$

Utilizing Eq. 10.3(5), we see that

$$U = \frac{\varepsilon}{2} \int_0^d \int_0^b \int_0^a E_0^2 \sin^2 \frac{\pi x}{a} \sin^2 \frac{\pi z}{d} \, dx \, dy \, dz$$

$$= \frac{\varepsilon E_0^2}{2} \cdot \frac{a}{2} \cdot b \cdot \frac{d}{2} = \frac{\varepsilon abd}{8} E_0^2 \qquad (1)$$

To obtain an approximation for power loss in the walls, we utilize the current flow in the ideal conductors as obtained from the tangential magnetic field at the surface. Referring to Fig. 10.3a

Front: $\quad J_{sy} = -H_x\|_{z=d}$		Back: $\quad J_{sy} = H_x\|_{z=0}$
Left side: $J_{sy} = -H_z\|_{x=0}$		Right side: $J_{sy} = H_z\|_{x=a}$
Top: $\quad J_{sx} = -H_z, J_{sz} = H_x$		Bottom: $\quad J_{sx} = H_z, J_{sz} = -H_x$

If the conducting walls have surface resistivity R_s, the foregoing currents will produce losses as follows:

$$W_L = \frac{R_s}{2} \left\{ 2 \int_0^b \int_0^a |H_x|_{z=0}^2 \, dx \, dy + 2 \int_0^d \int_0^b |H_z|_{x=0}^2 \, dy \, dz \right.$$

$$\left. + 2 \int_0^d \int_0^a [|H_x|^2 + |H_z|^2] \, dx \, dz \right\}$$

In this equation, the first term comes from the front and back, the second from left and right sides, and the third from top and bottom. Substituting from Eqs. 10.3(6) and 10.3(7) and evaluating the integrals,

$$W_L = \frac{R_s \lambda^2}{8\eta^2} E_0^2 \left[\frac{ab}{d^2} + \frac{bd}{a^2} + \frac{1}{2} \left(\frac{a}{d} + \frac{d}{a} \right) \right] \tag{2}$$

A Q of the resonator may be defined from the basic definition of Eq. 5.10(4),

$$Q = \frac{\omega_0 U}{W_L} \tag{3}$$

Substituting (1) and (2) with λ from Eq. 10.3(1), we have

$$Q = \frac{\pi \eta}{4R_s} \left[\frac{2b(a^2 + d^2)^{3/2}}{ad(a^2 + d^2) + 2b(a^3 + d^3)} \right] \tag{4}$$

Note that for a cube, $a = b = d$, this reduces to the expression

$$Q_{cube} = \frac{\sqrt{2}\pi}{6} \frac{\eta}{R_s} = 0.742 \frac{\eta}{R_s} \tag{5}$$

For an air dielectric, $\eta \approx 377 \ \Omega$, and a copper conductor at 10 GHz, $R_s \approx 0.0261 \ \Omega$, the Q is about 10,730. Thus we see the very large values of Q for such resonators as compared with those for lumped circuits (order of a few hundred) or even with resonant lines (order of a few thousand). In practice, some care must be used if Q's of the order of that calculated are to be obtained, since disturbances caused by the coupling system, surface irregularities, and other perturbations will act to increase the losses. Dielectric losses and radiation from small holes, when present, may be especially serious in lowering the Q.

It will be shown in Chapter 11 that a lumped circuit model applies to a cavity mode in the vicinity of resonance. From it we can deduce that the Q, defined in terms of stored energy and power loss, is also useful in estimating bandwidth of the cavity just as for the lumped resonant circuit. If Δf is the distance between points on the response curve for which amplitude response is down to $1/\sqrt{2}$ of its maximum value, (Sec. 5.10),

$$\frac{\Delta f}{f_0} \approx \frac{1}{Q} \tag{6}$$

Thus, for the foregoing, a Q of 10,000 in a cavity resonant at 10 GHz will yield a bandwidth between "half-power" points of 1 MHz.

10.5   Other Modes in the Rectangular Resonator

As has been noted, the particular mode studied for the rectangular box is only one of an infinite number of possible modes. If we adopt the point of view that a resonant mode is the standing wave pattern for incident and reflected waveguide modes, any one of the infinite number of possible waveguide waves might be used, with any integral number of half-waves between shorting ends. We recognize that this description of a particular field pattern is not unique, for it depends on the axis chosen to be the "direction of propagation" for the waveguide modes. Thus (see Prob. 10.3a) the simple mode studied in past sections would be a TE_{101} mode if the z axis or x axis were considered the direction of propagation, but it would be a TM_{110} mode if the vertical (y) axis were taken as the propagation direction. In the following, a coordinate system will be chosen as in Fig. 10.3a, and field patterns will be obtained by superposing incident and reflected waves for various waveguide modes propagating in the z direction.

The TE_{mnp} Mode   If we select the TE_{mn} mode of a rectangular waveguide (see Sec. 8.7), addition of positively and negatively traveling waves for H_z gives

$$H_z = (Ae^{-j\beta z} + Be^{j\beta z}) \cos \frac{m\pi x}{a} \cos \frac{n\pi y}{b}$$

Since the normal component of magnetic field, H_z, must be zero at $z = 0$ and $z = d$, then $B = -A$ and $\beta d = p\pi$ with p an integer. Let $C = -2jA$.

$$H_z = A(e^{-j\beta z} - e^{j\beta z}) \cos \frac{m\pi x}{a} \cos \frac{n\pi y}{b}$$

$$= C \cos \frac{m\pi x}{a} \cos \frac{n\pi y}{b} \sin \frac{p\pi z}{d} \tag{1}$$

Then, substituting (1) in Eqs. 8.2(13)–(16), remembering that for the negatively traveling waves all terms multiplied by β change sign,

$$H_x = -\frac{j\beta}{k_c^2} (Ae^{-j\beta z} - Be^{j\beta z})\left(-\frac{m\pi}{a}\right) \sin \frac{m\pi x}{a} \cos \frac{n\pi y}{b}$$

$$= -\frac{C}{k_c^2} \left(\frac{p\pi}{d}\right)\left(\frac{m\pi}{a}\right) \sin \frac{m\pi x}{a} \cos \frac{n\pi y}{b} \cos \frac{p\pi z}{d} \tag{2}$$

$$H_y = -\frac{C}{k_c^2} \left(\frac{p\pi}{d}\right)\left(\frac{n\pi}{b}\right) \cos \frac{m\pi x}{a} \sin \frac{n\pi y}{b} \cos \frac{p\pi z}{d} \tag{3}$$

$$E_x = \frac{j\omega\mu C}{k_c^2} \left(\frac{n\pi}{b}\right) \cos \frac{m\pi x}{a} \sin \frac{n\pi y}{b} \sin \frac{p\pi z}{d} \tag{4}$$

$$E_y = -\frac{j\omega\mu C}{k_c^2} \left(\frac{m\pi}{a}\right) \sin \frac{m\pi x}{a} \cos \frac{n\pi y}{b} \sin \frac{p\pi z}{d} \tag{5}$$

where

$$k_c^2 = \left(\frac{m\pi}{a}\right)^2 + \left(\frac{n\pi}{b}\right)^2 \tag{6}$$

$$\beta = \left[\left(\frac{2\pi}{\lambda}\right)^2 - k_c^2\right]^{1/2} = \frac{p\pi}{d}$$

so

$$k = \frac{2\pi}{\lambda} = \left[\left(\frac{m\pi}{a}\right)^2 + \left(\frac{n\pi}{b}\right)^2 + \left(\frac{p\pi}{d}\right)^2\right]^{1/2} \tag{7}$$

The TM_{mnp} Mode In a similar manner, positively and negatively traveling TM_{mn} modes in a rectangular waveguide may be combined to yield

$$E_z = D \sin \frac{m\pi x}{a} \sin \frac{n\pi y}{b} \cos \frac{p\pi z}{d} \tag{8}$$

$$E_x = -\frac{D}{k_c^2}\left(\frac{p\pi}{d}\right)\left(\frac{m\pi}{a}\right) \cos \frac{m\pi x}{a} \sin \frac{n\pi y}{b} \sin \frac{p\pi z}{d} \tag{9}$$

$$E_y = -\frac{D}{k_c^2}\left(\frac{p\pi}{d}\right)\left(\frac{n\pi}{b}\right) \sin \frac{m\pi x}{a} \cos \frac{n\pi y}{b} \sin \frac{p\pi z}{d} \tag{10}$$

$$H_x = \frac{j\omega\varepsilon D}{k_c^2}\left(\frac{n\pi}{b}\right) \sin \frac{m\pi x}{b} \cos \frac{n\pi y}{b} \cos \frac{p\pi z}{d} \tag{11}$$

$$H_y = -\frac{j\omega\varepsilon D}{k_c^2}\left(\frac{m\pi}{a}\right) \cos \frac{m\pi x}{b} \sin \frac{n\pi y}{b} \cos \frac{p\pi z}{d} \tag{12}$$

The quantity k_c^2 and resonant wavelength λ are as in (6) and (7).

General Comments We note first that TM and TE modes of the same order m, n, p have identical frequencies. Such modes with different field patterns but the same resonant frequency are known as *degenerate* modes. Other cases of degeneracy may exist as in a cube, $a = b = d$, where orders 112, 121, and 211 of both TM and TE types have the same resonant frequency.

It is also apparent from (7) that, as the order of a mode becomes higher, the wavelength decreases or resonant frequency increases. Put differently, it means that to be resonant at a given frequency, the box must be made bigger as the order increases. This is to be expected, since more half-sine waves are to fit in each dimension. It can be shown (Probs. 10.5b and 10.5c) that Q increases at a given frequency as one goes to higher mode orders. This too is logical, since the larger box has a greater volume-to-surface ratio, and energy is stored in the volume, whereas it is lost on the imperfectly conducting surface. The high-order modes are consequently useful in "echo boxes" where a high Q is desired so that the energy will decay at a very slow rate after being excited by a pulse. Because modes become very close together in frequency as the order increases, it may be difficult to excite one mode only in such applications.

10.6 Circular Cylindrical Resonator

For a circular cylindrical resonator, Fig. 10.6, there is a simple mode analogous to that first studied for the rectangular box (Sec. 10.3). The vertical electric field has a maximum at the center and dies off to zero at the conducting side walls. A circumferential magnetic field surrounds the displacement current represented by the time-varying electric field. Neither component varies in the axial or circumferential direction. Equal and opposite charges exist on the two end plates, and a vertical current flows in the side walls. The mode may be considered a TM_{01} mode in a circular waveguide operating at cutoff (to give the constancy with respect to z), or it may be thought of as the standing wave pattern produced by inward and outward radially propagating waves of the radial transmission line type, Sec. 9.3. From either point of view we obtain the field components

$$E_z = E_0 J_0(kr) \tag{1}$$

$$H_\phi = \frac{jE_0}{\eta} J_1(kr) \tag{2}$$

$$k = \frac{p_{01}}{a} = \frac{2.405}{a} \tag{3}$$

Then the resonant wavelength is

$$\lambda = \frac{2\pi}{k} = 2.61a \tag{4}$$

The energy stored in the cavity at resonance may be found from the energy in the electric fields at the instant these have their maximum value. Take a and d, respectively, as radius and length of the cavity.

$$U = d \int_0^a \frac{\varepsilon |E_z|^2}{2} 2\pi r \, dr = \pi \varepsilon \, d E_0^2 \int_0^a r J_0^2(kr) \, dr$$

This may be integrated by Eq. 7.15(22).

$$U = \pi \varepsilon \, d E_0^2 \frac{a^2}{2} J_1^2(ka) \tag{5}$$

If the walls are of imperfect conductors, the power loss may be calculated approximately:

$$W_L = 2\pi a d \frac{R_s}{2} |J_{sz}|^2 + 2 \int_0^a \frac{R_s}{2} |J_{sr}|^2 2\pi r \, dr$$

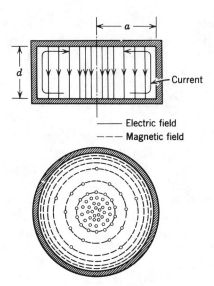

Fig. 10.6 Sections through a cylindrical cavity with fields of TM_{010} mode.

The first term represents losses on the side wall, the second on top and bottom. The current per unit width J_{sr} on top and bottom is $\pm H_\phi$, and J_{sz} on the side wall is the value of H_ϕ at $r = a$. Substituting from (2), we see that

$$W_L = \pi R_s \left[ad \frac{E_0^2}{\eta^2} J_1^2(ka) + 2 \int_0^a \frac{E_0^2}{\eta^2} r J_1^2(kr)\, dr \right]$$

This may also be integrated by Eq. 7.15(22), recalling that $J_0(ka) = 0$ is the condition for resonance.

$$W_L = \frac{\pi a R_s E_0^2}{\eta^2} J_1^2(ka)[d + a] \tag{6}$$

The Q of the mode may then be obtained as usual from power losses (6), and energy stored (5), using (3) for resonant frequency ω_0:

$$Q = \frac{\omega_0 U}{W_L} = \frac{\eta}{R_s} \frac{p_{01}}{2(a/d + 1)} \tag{7}$$

where

$$p_{01} \approx 2.405$$

An infinite number of additional modes may be obtained for the cylindrical resonator by considering others of the possible waveguide modes for circular cylindrical guides as propagating in the axial direction with an integral number of half guide wavelengths between end plates. In this manner the standing wave pattern formed by the superposition of incident and reflected waves fulfills the boundary conditions of the conducting ends. Table 10.6 shows a TE_{11} mode, a TM_{01} mode, and a TE_{01} mode, each with one-half guide wavelength between ends. The resonant wavelengths shown are obtained by solving the equation

$$d = \frac{p\lambda_g}{2} = \frac{p\lambda}{2}\left[1 - \left(\frac{\lambda}{\lambda_c}\right)^2\right]^{-1/2} \tag{8}$$

For these modes the integer p is unity in (8), and cutoff wavelength λ_c is obtained from Sec. 8.9. Note that in the designations TE_{111}, TM_{011}, TE_{011}, the order of subscripts is not in the cyclic order of coordinates, r, ϕ, z, since it is common in circular waveguides to designate the ϕ variation by the first subscript.

Of the foregoing modes, the TE_{011} is perhaps the most interesting since it has only circumferential currents in both the cylindrical wall and the end plates. Thus, if a resonator for such a wave is tuned by moving the end plate, one does not need a good contact between the ends and the cylindrical wall since no current flows between them. For both of the other modes shown (and in fact all except those of type TE_{0mp}) a finite current does flow between the cylinder and its ends so that any sliding contact must be good to prevent serious loss.

Table 10.6

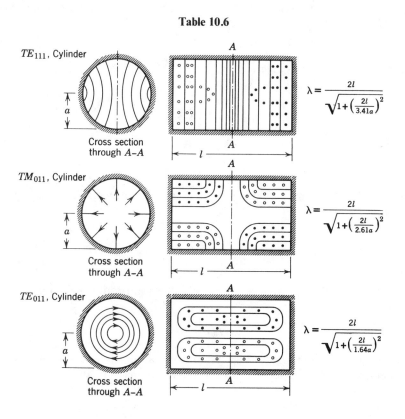

TE_{111}, Cylinder

Cross section through A–A

$$\lambda = \frac{2l}{\sqrt{1 + \left(\frac{2l}{3.41a}\right)^2}}$$

TM_{011}, Cylinder

Cross section through A–A

$$\lambda = \frac{2l}{\sqrt{1 + \left(\frac{2l}{2.61a}\right)^2}}$$

TE_{011}, Cylinder

Cross section through A–A

$$\lambda = \frac{2l}{\sqrt{1 + \left(\frac{2l}{1.64a}\right)^2}}$$

As with the rectangular resonator, it would be found that higher wave orders (those having more variations with any or all coordinates r, ϕ, z) would require larger resonators to be resonant at a given wavelength. The Q would become higher because of the increased volume-to-surface ratio, but the modes would become close together in frequency so that it might be difficult to excite one mode only.

10.7   Wave Solutions in Spherical Coordinates

Before considering the specific problem of a spherical cavity resonator, we shall look at the solutions of Maxwell's equations in spherical coordinates. We shall sketch here only those solutions with axial symmetry, $\partial/\partial\phi = 0$. The solutions with general ϕ variations are more involved, but have been given completely by Stratton.[1] It is found that with axial symmetry the solutions separate into waves with components E_r, E_θ, H_ϕ and those with components H_r, H_θ, E_ϕ. These are called TM and TE types, respectively, the spherical surface r constant serving here as the transverse surface.

Consider then TM spherical modes with axial symmetry by setting $\partial/\partial\phi = 0$ in Maxwell's equations in spherical coordinates. The three curl equations containing E_r, E_θ, H_ϕ are

$$\frac{\partial}{\partial r}(rE_\theta) - \frac{\partial E_r}{\partial \theta} = -j\omega\mu(rH_\phi) \tag{1}$$

$$\frac{1}{r\sin\theta}\frac{\partial}{\partial\theta}(H_\phi\sin\theta) = j\omega\varepsilon E_r \tag{2}$$

$$-\frac{\partial}{\partial r}(rH_\phi) = j\omega\varepsilon(rE_\theta) \tag{3}$$

Equations (2) and (3) may be differentiated and substituted in (1), leading to an equation in H_ϕ alone.

$$\frac{\partial^2}{\partial r^2}(rH_\phi) + \frac{1}{r^2}\frac{\partial}{\partial\theta}\left[\frac{1}{\sin\theta}\frac{\partial}{\partial\theta}(rH_\phi\sin\theta)\right] + k^2(rH_\phi) = 0 \tag{4}$$

To solve this partial differential equation, we follow the product solution technique. Assume

$$(rH_\phi) = R\Theta \tag{5}$$

where R is a function of r alone, Θ is a function of θ alone. If this is substituted in (4), the functions of r may be separated from the functions of θ, and these must then

---

[1] J. A. Stratton, *Electromagnetic Theory*, McGraw-Hill, New York, 1941, Chapter VII.

be separately equal to a constant if they are to equal each other for all values of r and θ. For a definitely ulterior motive, we label this constant $n(n + 1)$.

$$\frac{r^2 R''}{R} + k^2 r^2 = -\frac{1}{\Theta}\frac{d}{d\theta}\left[\frac{1}{\sin\theta}\frac{d}{d\theta}(\Theta \sin\theta)\right] = n(n + 1) \tag{6}$$

Thus there are two ordinary differential equations, one in r only, one in θ only. Let us consider that in θ first, making the substitutions

$$u = \cos\theta \qquad \sqrt{1 - u^2} = \sin\theta \qquad \frac{d}{d\theta} = -\sin\theta\,\frac{d}{du}$$

Then

$$(1 - u^2)\frac{d^2\Theta}{du^2} - 2u\frac{d\Theta}{du} + \left[n(n + 1) - \frac{1}{1 - u^2}\right]\Theta = 0 \tag{7}$$

The differential equation (7) is reminiscent of Legendre's equation (Sec. 7.18) and is in fact a standard form. This form is

$$(1 - x^2)\frac{d^2 y}{dx^2} - 2x\frac{dy}{dx} + \left[n(n + 1) - \frac{m^2}{1 - x^2}\right]y = 0 \tag{8}$$

One of the solutions is written

$$y = P_n^m(x)$$

and the function defined by this solution is called an associated Legendre function of the first kind, order n, degree m. These are related to the ordinary Legendre functions by the equation

$$P_n^m(x) = (1 - x^2)^{m/2}\frac{d^m P_n(x)}{dx^m} \tag{9}$$

As a matter of fact, (8) could be derived from the ordinary Legendre equation by this substitution. A solution to (7) may then be written

$$\Theta = P_n^1(u) = P_n^1(\cos\theta) \tag{10}$$

And, from (9),

$$P_n^1(\cos\theta) = -\frac{d}{d\theta}P_n(\cos\theta) \tag{11}$$

Thus for integral values of n these associated Legendre functions are also polynomials consisting of a finite number of terms. By differentiations according to (9) in Eq. 7.18(8), the polynomials of the first few orders are found to be

$$P_0^1(\cos\theta) = 0$$

$$P_1^1(\cos\theta) = \sin\theta$$

$$P_2^1(\cos\theta) = 3\sin\theta\cos\theta \tag{12}$$

$$P_3^1(\cos\theta) = \tfrac{3}{2}\sin\theta\,(5\cos^2\theta - 1)$$

$$P_4^1(\cos\theta) = \tfrac{5}{2}\sin\theta\,(7\cos^3\theta - 3\cos\theta)$$

Other properties of these functions that will be useful to us, and which may be found from a study of the above, are as follows:

1. All $P_n^1(\cos \theta)$ are zero at $\theta = 0$ and $\theta = \pi$.
2. $P_n^1(\cos \theta)$ are zero at $\theta = \pi/2$ if n is even.
3. $P_n^1(\cos \theta)$ are a maximum at $\theta = \pi/2$ if n is odd, and the value of this maximum is given by

$$P_n^1(0) = \frac{(-1)^{-(n-1)/2} n!}{2^{n-1} \left[\left(\frac{n-1}{2} \right)! \right]^2} \quad n \text{ odd} \tag{13}$$

4. The associated Legendre functions have orthogonality properties similar to those of the Legendre polynomials studied previously.

$$\int_0^\pi P_l^1(\cos \theta) P_n^1(\cos \theta) \sin \theta \, d\theta = 0, \quad l \neq n \tag{14}$$

$$\int_0^\pi [P_n^1(\cos \theta)]^2 \sin \theta \, d\theta = \frac{2n(n+1)}{2n+1} \tag{15}$$

5. The differentiation formula is

$$\frac{d}{d\theta} [P_n^1(\cos \theta)] = \frac{1}{\sin \theta} [n P_{n+1}^1(\cos \theta) - (n+1) \cos \theta P_n^1(\cos \theta)] \tag{16}$$

Note that only one solution for this second-order differential equation (7) has been considered. The other solution becomes infinite on the axis, and so will not be required in solutions valid on the axis, but will be needed in problems such as the biconical antenna analysis of Sec. 12.24.

To go back to the r differential equation obtainable from (6), substitute the variable $R_1 = R/\sqrt{r}$:

$$\frac{d^2 R_1}{dr^2} + \frac{1}{r} \frac{dR_1}{dr} + \left[k^2 - \frac{(n + \frac{1}{2})^2}{r^2} \right] R_1 = 0$$

By comparing with Eq. 7.14(3) it is seen that this is Bessel's differential equation of order $(n + \frac{1}{2})$. A complete solution may then be written

$$R_1 = A_n J_{n+1/2}(kr) + B_n N_{n+1/2}(kr) \tag{17}$$

and

$$R = \sqrt{r} R_1$$

If n is an integer, these half-integral order Bessel functions reduce simply to algebraic combinations of sinusoids.[2] For example, the first few orders are

$$J_{1/2}(x) = \sqrt{\frac{2}{\pi x}} \sin x \qquad\qquad N_{1/2}(x) = -\sqrt{\frac{2}{\pi x}} \cos x$$

$$J_{3/2}(x) = \sqrt{\frac{2}{\pi x}} \left[\frac{\sin x}{x} - \cos x \right] \qquad N_{3/2}(x) = -\sqrt{\frac{2}{\pi x}} \left[\sin x + \frac{\cos x}{x} \right]$$

$$J_{5/2}(x) = \sqrt{\frac{2}{\pi x}} \left[\left(\frac{3}{x^2} - 1 \right) \sin x \right. \qquad N_{5/2}(x) = -\sqrt{\frac{2}{\pi x}} \left[\frac{3}{x} \sin x \right.$$

$$\left. - \frac{3}{x} \cos x \right] \qquad\qquad \left. + \left(\frac{3}{x^2} - 1 \right) \cos x \right] \quad (18)$$

The linear combinations of the J and N functions into Hankel functions (Sec. 7.14) represent waves traveling radially inward or outward, and boundary conditions will be as found previously for other Bessel functions:

1. If the region of interest includes the origin, $N_{n+1/2}$ cannot be present since it is infinite at $r = 0$.
2. If the region of interest extends to infinity, the linear combination of J and N into the second Hankel function, $H^{(2)}_{n+1/2} = J_{n+1/2} - jN_{n+1/2}$, must be used to represent a radially outward traveling wave.

The particular combination of $J_{n+1/2}(kr)$ and $N_{n+1/2}(kr)$ required for any problem may be denoted as $Z_{n+1/2}(kr)$, and now by combining correctly (17), (10), and (5), H_ϕ is determined. E_r and E_θ follow from (2) and (3) respectively.

$$H_\phi = \frac{A_n}{\sqrt{r}} P_n^1(\cos \theta) Z_{n+1/2}(kr)$$

$$E_\theta = \frac{A_n P_n^1(\cos \theta)}{j\omega\varepsilon r^{3/2}} [nZ_{n+1/2}(kr) - krZ_{n-1/2}(kr)] \qquad (19)$$

$$E_r = -\frac{A_n n Z_{n+1/2}(kr)}{j\omega\varepsilon r^{3/2} \sin \theta} [\cos \theta P_n^1(\cos \theta) - P_{n+1}^1(\cos \theta)]$$

[2] Special notations for the spherical or half-integral order Bessel functions have been introduced and are useful if one has much to do with these functions. Thus Stratton, following Morse (*Vibration and Sound*, McGraw-Hill, New York, 1936, p. 246) uses $j_n(x)$ to denote $(\pi/2x)^{1/2} J_{n+1/2}(x)$, and similar small letters denote other spherical Bessel and Hankel functions. Schelkunoff follows the definitions of spherical Bessel functions given by Bateman (*Partial Differential Equations*, Dover, 1944, p. 386), although in a different notation, using $\hat{J}_n(x)$ to denote $(\pi x/2)^{1/2} J_{n+1/2}(x)$, and similarly for other Bessel and Hankel functions. Because of our limited need for spherical coordinates, we shall retain the original Bessel function forms so that standard recurrence formulas may be used.

The spherically symmetric TE modes may be obtained by the above and the principle of duality, Sec. 9.5. We then replace E_r and E_θ by H_r and H_θ, respectively, H_ϕ by $-E_\phi$, and ε by μ.

$$E_\phi = \frac{B_n}{\sqrt{r}} P_n^1(\cos\theta) Z_{n+1/2}(kr)$$

$$H_\theta = -\frac{B_n P_n^1(\cos\theta)}{j\omega\mu r^{3/2}} [n Z_{n+1/2}(kr) - kr Z_{n-1/2}(kr)] \tag{20}$$

$$H_r = \frac{B_n n Z_{n+1/2}(kr)}{j\omega\mu r^{3/2} \sin\theta} [\cos\theta P_n^1(\cos\theta) - P_{n+1}^1(\cos\theta)]$$

10.8 Spherical Resonators

The general discussion of spherical waves from the preceding section will now be applied to the study of some simple modes in a hollow conducting spherical resonator. Since the origin is included within the region of the solution, the Bessel functions can only be those of first kind, $J_{n+1/2}$. For the lowest order TM mode, let $n = 1$ in Eq. 10.7(19) and utilize the definitions of Eqs. 10.7(12) and 10.7(18). Letting $C = A_1(2k/\pi)^{1/2}$, we then have

$$H_\phi = \frac{C \sin\theta}{kr} \left(\frac{\sin kr}{kr} - \cos kr\right) \tag{1}$$

$$E_r = -\frac{2j\eta C \cos\theta}{k^2 r^2} \left(\frac{\sin kr}{kr} - \cos kr\right) \tag{2}$$

$$E_\theta = \frac{j\eta C \sin\theta}{k^2 r^2} \left[\frac{(kr)^2 - 1}{kr} \sin kr + \cos kr\right] \tag{3}$$

The mode may be designated TM_{101}, the subscripts here giving variations in the order r, ϕ, and θ. Electric and magnetic field lines are sketched in Fig. 10.8a.

To obtain the resonance condition, we know that E_θ must be zero at the radius of the perfectly conducting shell, $r = a$. From (3), this requires

$$\tan ka = \frac{ka}{1 - (ka)^2} \tag{4}$$

Roots of this transcendental equation may be determined numerically and the first is found at $ka \approx 2.74$, giving a resonant wavelength of

$$\lambda \approx 2.29a \tag{5}$$

The energy stored at resonance may be found from the peak energy in magnetic fields.

$$U = \int_0^a \int_0^\pi \frac{\mu}{2} |H_\phi|^2 2\pi r^2 \sin\theta \, d\theta \, dr$$

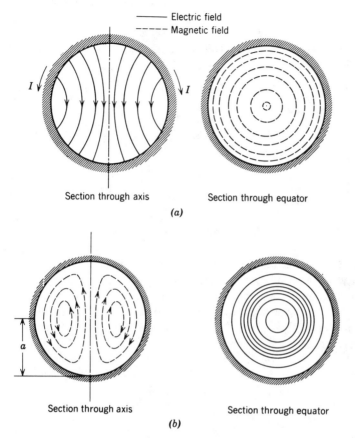

Fig. 10.8 (a) Field patterns for TM_{101} mode in spherical resonator. (b) Field patterns for TE_{101} mode in spherical resonator.

The value of H_ϕ is given by (1), and the result of the integration may be simplified by the resonance requirement (4):

$$U = \frac{2\pi\mu C^2}{3k^3}\left[ka - \frac{1 + (ka)^2}{ka}\sin^2 ka\right] \tag{6}$$

The approximate dissipation in conductors of finite conductivity is

$$W_L = \int_0^\pi \frac{R_s|H_\phi|^2}{2} 2\pi a^2 \sin\theta \, d\theta = \frac{4\pi R_s}{3} a^2 C^2 \sin^2 ka \tag{7}$$

So the Q of this mode is

$$Q = \frac{\eta}{2R_s(ka)^2}\left[\frac{ka}{\sin^2 ka} - \frac{1 + (ka)^2}{ka}\right] \approx \frac{\eta}{R_s} \tag{8}$$

The "dual" of the above mode is the TE_{101} mode, and its field components may be obtained by substituting in (1) to (3) E_ϕ for H_ϕ, $-H_r$ for E_r, and $-H_\theta$ for E_θ. The fields are sketched in Fig. 10.8b. Note that the resonance condition for this mode, obtained by setting $E_\phi = 0$ at $r = a$, requires

$$\tan ka = ka$$

Numerical solution of this yields $ka \approx 4.50$, or

$$\lambda \approx 1.395a \tag{9}$$

SMALL-GAP CAVITIES AND COUPLING

10.9 Small-Gap Cavities

Because of their shielded nature and high Q possibilities, resonant cavities are ideal for use in many high-frequency tubes such as klystrons, magnetrons, and microwave triodes. When they are used with an electron stream, it is essential for efficient energy transfer that the electron transit time across the active field region be as small as possible. If resonators such as those studied in preceding sections were used, very short cylinders or prisms would be required, and Q would be low and interaction weak. Certain special shapes are consequently employed which have a small gap in the region that is to interact with the electron stream. Several examples of useful small-gap cavities will follow.

_____ **Example 10.9a** _____
Foreshortened Coaxial Lines

Fig. 10.9a may be considered a coaxial line A terminated in the gap capacitance B (leading to the equivalent circuit of Fig. 10.9b) provided that the region B is small compared with wavelength. The method is particularly useful when the region B is not uniform, but contains dielectrics or discontinuities, so long as a reasonable estimate of capacitance can be made.

For resonance, the impedance at any plane should be equal and opposite, looking in opposite directions. Selecting the plane of the capacitance for this purpose,

$$jZ_0 \tan \beta l = -\left(\frac{1}{j\omega_0 C}\right)$$

or

$$\beta l = \tan^{-1}\left(\frac{1}{Z_0 \omega C}\right) \tag{1}$$

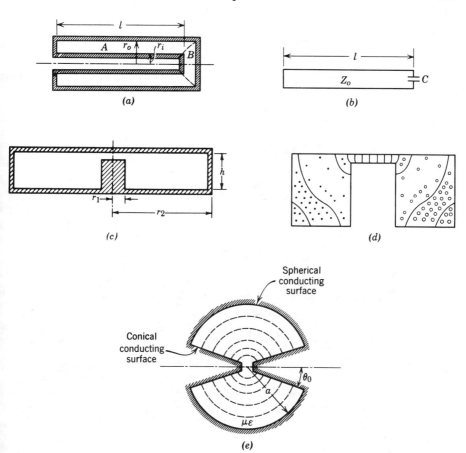

Fig. 10.9 (a) Foreshortened coaxial line resonator. (b) Approximate equivalent circuit for (a). (c) Foreshortened radial line resonator. (d) Resonator intermediate between foreshortened coaxial line and foreshortened radial line. (e) Conical line resonator.

If C is small ($Z_0 \omega C \ll 1$), the line is practically a quarter-wave in length. For larger values of C, the line is foreshortened from the quarter-wave value and would approach zero length if $Z_0 \omega C$ approached infinity.

_____ **Example 10.9b** _____
Foreshortened Radial Lines

If the proportions of the resonator are more as shown in Fig. 10.9c, it is preferable to look at the problem as one of a resonant radial transmission line (Sec. 9.3) loaded or foreshortened by the capacitance of the post or gap. Then, for resonance, the inductive reactance of the shorted radial line looking outward from radius

r_1 should be equal in magnitude to the capacitive reactance of the central post. Using the results and notation of Sec. 9.3 we see that

$$\frac{1}{\omega C} = -\frac{h}{2\pi r_1} Z_{01} \frac{\sin(\theta_1 - \theta_2)}{\cos(\psi_1 - \theta_2)}$$

or

$$\theta_2 = \tan^{-1}\left[\frac{\sin\theta_1 + (2\pi r_1/\omega C Z_{01}h)\cos\psi_1}{\cos\theta_1 - (2\pi r_1/\omega C Z_{01}h)\sin\psi_1}\right] \tag{2}$$

Once θ_2 is found, kr_2 is read from Fig. 9.3c.

_____ **Example 10.9c** _____
Resonators of Intermediate Shape

In the coaxial-line resonator of Fig. 10.9a the electric field lines would be substantially radial in the region far from the gap. In the radial line resonator of Fig. 10.9c the electric field lines would be substantially axial in the region far from the gap. For a resonator of the same general type, but with intermediate proportions, the field lines may be transitional between these extremes as indicated in Fig. 10.9d, and neither of these approximations may yield good results. Some useful design cuves for a range of proportions have been given in the literature.[3] Of course, if the capacitive loading at the center is great enough, the entire resonator will be relatively small compared with wavelength, and the outer portion may be considered a lumped inductance of value

$$L = \frac{\mu l}{2\pi} \ln\left(\frac{r_2}{r_1}\right) \tag{3}$$

Resonance is computed from this inductance and the known capacitance.

_____ **Example 10.9d** _____
Conical-Line Resonator

A somewhat different form of small-gap resonator, formed by placing a spherical short at radius a on a conical line as studied in Sec. 9.6, is shown in Fig. 10.9e. Since this is a uniform line, formula (1) applies to this case as well. For the conical line, $\beta = k$ and

$$Z_0 = \frac{\eta}{\pi} \ln \cot \frac{\theta_0}{2} \tag{4}$$

[3] T. Moreno, *Microwave Transmission Design Data*, McGraw-Hill, New York, 1948.

In the limit of zero capacitance (the two conical tips separated by an infinitesimal gap), the radius a becomes exactly a quarter-wavelength. The field components in this case, obtained by forming a standing wave from Eqs. 9.6(5) and 9.6(6), are

$$E_\theta = \frac{C}{\sin \theta} \frac{\cos kr}{r} \tag{5}$$

$$H_\phi = \frac{C}{j\eta \sin \theta} \frac{\sin kr}{r} \tag{6}$$

The Q of the resonator in this limiting case may be shown to be

$$Q \approx \frac{\eta\pi}{4R_s} \frac{\ln \cot (\theta_0/2)}{\ln \cot (\theta_0/2) + 0.825 \csc \theta_0} \tag{7}$$

10.10 Coupling to Cavities

The types of electromagnetic waves that may exist inside closed conducting cavities have been discussed without specifically analyzing ways of exciting these oscillations. Obviously they cannot be excited if the resonator is completely enclosed by conductors. Some means of coupling electromagnetic energy into and out of the resonator must be introduced from the outside. The most straightforward methods, similar to those discussed in Sec. 8.11 for exciting waves in waveguides, are:

1. Introduction of a conducting probe or antenna in the direction of the electric field lines, driven by an external transmission line.
2. Introduction of a conducting loop with plane normal to the magnetic field lines.
3. Introduction of a pulsating electron beam passing through a small gap in the resonator, in the direction of electric field lines, and similarly for carriers in solid-state devices.
4. Introduction of a hole or iris between the cavity and a driving wave guide, the hole being located so that some field component in the cavity mode has a common direction to one in the wave mode.

For example, in a velocity modulation device of the klystron type, as in Fig. 10.10a, the input cavity may be excited by a probe, the oscillations in this cavity producing a voltage across gap g_1 and causing a velocity modulation of the electron beam. The velocity modulation is converted to convection current modulation by a drifting action so that the electron beam may then excite electromagnetic oscillations in the second resonator by passing through the gap g_2. Power may be coupled out of this resonator by a coupling loop and a coaxial transmission line. Iris coupling between a TM_{010} mode in a cylindrical cavity and the TE_{10} mode in a rectangular waveguide is illustrated in Fig. 10.10b. Here the H_ϕ of the cavity and the H_x of the guide are in the same direction over the hole.

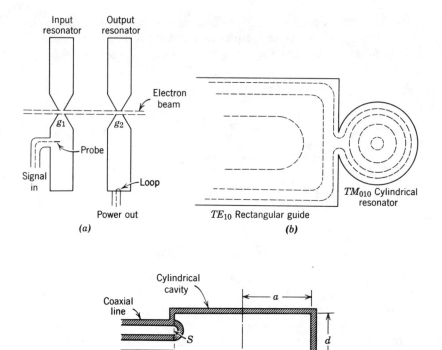

Fig. 10.10 (*a*) Couplings to the cavities of a velocity modulation tube amplifier. (*b*) Section showing approximate form of magnetic field lines in iris coupling between a guide and cavity. (*c*) Magnetic coupling to a cylindrical cavity.

The rigorous approach to a quantitative analysis of cavity coupling is given in the following chapter. Some comments and an approximate approach are, however, in order here.

_____ **Example 10.10** _____
Loop Coupling in a Cylindrical Cavity

Let us concentrate on the loop coupling to a TM_{010} cylindrical mode as sketched in Fig. 10.10c. If a current is made to flow in the loop, all wave types will be excited which have a magnetic field threading the loop. The simple TM_{010} mode is one of these, and, if it is near resonance, certainly it will be excited most. However, this wave is known to fit the boundary conditions imposed by the perfectly conducting box alone. Other waves will have to be superposed to make the electric field zero along the perfectly conducting loop, but these will in general be far from resonance and so will contribute only a reactive effect. In fact, they may be thought of as producing the self-inductive reactance of the loop, taking into account the presence of the cavity as a shield.

The voltage magnitude induced in the loop by the cavity mode is

$$|V| = \omega\mu S|H| \tag{1}$$

where $|H|$ is magnetic field from the TM_{010} mode, averaged over the loop, and S is loop area. Consider power loss as only that from the walls, Eq. 10.6(6). This can be written in terms of magnetic field at $r = a$ through Eq. 10.6(2):

$$W_L = \frac{\pi a(d + a)E_0^2 J_1^2(ka)R_s}{\eta^2} = \pi a(d + a)R_s|H|^2 \tag{2}$$

This loss requires an input resistance

$$R = \frac{|V|^2}{2W_L} = \frac{(\omega\mu S)^2}{2\pi a(d + a)R_s} \tag{3}$$

The reactive impedance from self-inductance of the loop, $j\omega L$, is added to this, but may be tuned out by a frequency shift in the cavity, or by a portion of the input line as will be discussed in the following section.

10.11 Cavity Q and Other Figures of Merit

The Q of a cavity has been defined in terms of power loss and energy storage and has been calculated for a number of ideal configurations. It has also been noted that the Q is useful in describing bandwidth of a cavity mode, just as for a lumped-element resonant system. The reason for this is that in the vicinity of resonance for a single mode, a lumped-element equivalent circuit such as that of Fig. 10.11a is a good representation. The excitation means may excite a number of modes, but in general only one is near resonance. The elements G, L, and C represent the mode near resonance, and jX the reactive effect of modes far from resonance. Such an equivalent circuit might be suspected to give correct qualitative results, but as will be shown in the next chapter, it actually gives useful quantitative results also. The equivalent circuit also permits one to devise ways of measuring Q when it is difficult or impossible to calculate.

Many methods of Q measurement are possible,[4] but we will describe only one simple technique by way of example. It is assumed that the cavity mode is coupled to a waveguide by a means which we will illustrate here by the ideal transformer of turns ratio $m:1$, Fig. 10.11b. The guide is assumed to have unity characteristic

---

[4] See, for example, E. L. Ginzton, *Microwave Measurements*, McGraw-Hill, New York, 1957, Chapter 9.

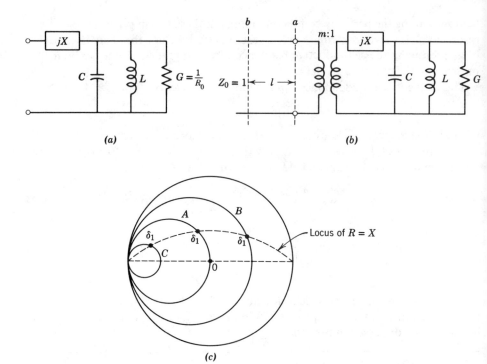

Fig. 10.11 (a) Cavity equivalent circuit. (b) Equivalent circuit for cavity with coupling to a waveguide. (c) Locus of impedance on Smith chart for Q measurement.

impedance for simplicity, so that terminating impedances are automatically normalized. The input impedance at reference a is then

$$Z_a = m^2\left[jX + \frac{1}{G + j(\omega C - 1/\omega L)}\right] \tag{1}$$

By defining $Q_0 = \omega_0 C/G$, $\omega_0^2 = 1/LC$, and $R_0 = 1/G$, this is

$$Z_a = m^2\left[jX + \frac{R_0}{1 + j(\omega/\omega_0 - \omega_0/\omega)Q_0)}\right] \tag{2}$$

In the vicinity of resonance, $\omega = \omega_0(1 + \delta')$ where δ' is small,

$$Z_a \approx jm^2 X + \frac{m^2 R_0}{1 + 2jQ_0\delta'} \tag{3}$$

The series reactance may be removed either by defining a new resonant frequency, or by referring input to a shifted point on the waveguide. The latter is common, and the new reference may be taken as the position where the impedance seen looking toward the cavity is zero when the cavity is tuned off resonance; this is called the "detuned short" position. The detuning is sufficient to make $Q_0 \delta' \gg 1$ either by detuning the cavity (by changing ω_0) or by changing frequency ω. By (3), $Z_a \approx j m^2 X$ and, from the impedance transformation formula, Eq. 5.5(13), which for $Z_0 = 1$ is

$$Z_b = \frac{Z_a + j \tan \beta l}{1 + j Z_a \tan \beta l} \tag{4}$$

we find that Z_b is zero when

$$\tan \beta l = -m^2 X \tag{5}$$

The impedance Z_b at arbitrary frequencies in the neighborhood of resonance is then

$$Z_b = \frac{m^2 R_{0b}}{1 + 2j Q_0 \delta} \tag{6}$$

where

$$R_{0b} = R_0 (1 + m^4 X^2)^{-1} \tag{7}$$

and

$$\delta = \delta' - \frac{m^4 X R_0}{2 Q_0 (1 + m^4 X^2)} \tag{8}$$

The locus of impedance is measured as δ' and, therefore, δ is varied either by changing frequency or detuning the cavity. As impedance is of the linear fraction form, Sec. 5.7, it will produce a circular locus when plotted on the Smith chart as illustrated by circles A, B, and C in Fig. 10.11c. Circle A, for which $m^2 R_{0b} = 1$, passes through the origin and is called the condition of *critical coupling* since it provides a perfect match to the guide at resonance; circle C with $m^2 R_{0b} < 1$ is said to be *undercoupled*, and circle B with $m^2 R_{0b} > 1$ is *overcoupled*. To match the last two, the coupling ratio m^2 would have to be changed. As with the lumped resonant system, the value of Q_0 can now be found from the specific value of δ which reduces impedance magnitude at reference b by $1/\sqrt{2}$ of its resonant value. On the Smith chart, this is the point $R = X$ and the corresponding δ may be denoted δ_1. At this point the known quantities are δ_1' corresponding to δ_1, $m^2 X$ found from (5), $m^2 R_{0b}$, which is the value of Z_b at resonance, and $2 Q_0 \delta_1 = 1$ in the denominator of (6). Then

$$Q_0 = \frac{1}{2 \delta_1} \tag{9}$$

and (8) may be solved to find Q_0 in terms of the known quantities.

The value of Q_0 thus determined is the "unloaded Q" since it does not account for loading by the guide. A loaded Q which accounts for this is also used, and may be found from Fig. 10.11b as

$$\frac{1}{Q_L} = \frac{G + m^2}{\omega_0 C} = \frac{1}{Q_0} + \frac{1}{Q_{ext}} \tag{10}$$

where "external Q," Q_{ext}, results from

$$Q_{ext} = \frac{\omega_0 C}{m^2} \tag{11}$$

It is assumed here that the generator is matched so that the impedance looking toward the guide is its characteristic impedance, taken here as unity.

In problems such as the klystron cavity described in Sec. 10.10, the gain is proportional to shunt resistance R_0, and bandwidth proportional to $1/Q$ by Eq. 10.4(6), so that R_0/Q may be a useful figure of merit in describing the effect of the cavity on *gain-bandwidth product*. This ratio can be recognized as equivalent to the ratio $(L/C)^{1/2}$ in lumped-element circuits. The quantity may be calculated for ideal cavities of simple shape of the type studied earlier in the chapter, although electron-beam loading might modify R_0 by adding losses in addition to those from the conductor. If R_0 of the cavity is defined in terms of a voltage across the gap, and electric field is approximately uniform,

$$W_L = \frac{V^2}{2R_0} = \frac{(E_0 d)^2}{2R_0} \tag{12}$$

and this figure of merit is

$$\frac{R_0}{Q} = \frac{(E_0 d)^2}{2W_L} \cdot \frac{W_L}{\omega_0 U} = \frac{(E_0 d)^2}{2\omega_0 U} \tag{13}$$

For the measurement of R_0/Q, we see by (13) that we need to measure the field E_0 along the axis and relate it to U, the energy stored. The field may be probed in a variety of ways, but the perturbation technique, to be described in the following section, is one of the most accurate.

10.12   Cavity Perturbations

Given a cavity at resonance, we know that average stored magnetic and electric energies are equal. If a small perturbation is made in one of the cavity walls, this will in general change one type of energy more than the other, and resonant frequency would then shift by an amount necessary to again equalize the energies. Slater[5] has given an important development for the amount of frequency

[5] J. C. Slater, *Microwave Electronics*, Van Nostrand, Princeton, N.J., 1950, p. 81 et seq.

shift when a small volume ΔV is removed from the cavity by pushing in the boundaries. This may be written

$$\frac{\Delta\omega}{\omega_0} = \frac{\int_{\Delta V} (\mu H^2 - \varepsilon E^2)\, dV}{\int_V (\mu H^2 + \varepsilon E^2)\, dV} = \frac{\Delta U_H - \Delta U_E}{U} \tag{1}$$

where ΔU_H is the magnetic energy removed, ΔU_E the electric energy removed, and U the total stored energy, all time averages. We illustrate with two examples.

_____ **Example 10.12a** _____
Perturbation of Resonant Parallel-Plane Line

We first take a simple case for which we can check the answer. Consider the parallel-plane transmission line of Fig. 10.12a, shorted at the two ends. The unperturbed resonance is at $l = \lambda/2$ or

$$\omega_0 = \frac{2\pi c}{\lambda} = \frac{\pi c}{l} \tag{2}$$

If we perturb by moving one end plate in by Δl, the new resonance is

$$(\omega_0 + \Delta\omega) = \frac{\pi c}{l - \Delta l} \approx \omega_0\left(1 + \frac{\Delta l}{l}\right) \tag{3}$$

In using the perturbation formula (1), only magnetic energy is removed. If unperturbed magnetic field is of the form

$$H_0(z) = H_0 \sin\frac{\pi z}{l} \tag{4}$$

Total stored energy (twice the average energy in magnetic fields) is

$$U = 2wd \int_{-l/2}^{l/2} \frac{\mu H_0^2}{4} \sin^2\frac{\pi z}{l}\, dz = w\, ld\, \frac{\mu H_0^2}{4} \tag{5}$$

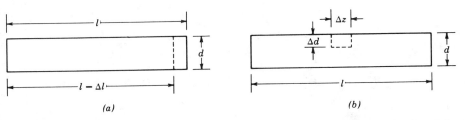

Fig. 10.12 Parallel-plane transmission line with perturbations: (*a*) by decreasing length by Δl; (*b*) by introducing a rectangular indentation at the center of the line.

where w is width of the line and d the spacing. The energy removed

$$\Delta U_H = \frac{\mu H_0^2}{4} w \, d\Delta l \tag{6}$$

so by (1)

$$\frac{\Delta \omega}{\omega_0} = \frac{\Delta l}{l} \tag{7}$$

which agrees with (3).

Now suppose the perturbation is at the center of the line where only electric field is removed, as shown by the dashed outline in Fig. 10.12b. If unperturbed electric field is

$$E_0(z) = E_0 \cos \frac{\pi z}{l} \tag{8}$$

Total energy stored is

$$U = 2U_E = 2wd \int_{-l/2}^{l/2} \frac{\varepsilon E_0^2}{4} \cos^2 \frac{\pi z}{l} \, dz = w \, ld \frac{\varepsilon E_0^2}{4} \tag{9}$$

And the electric energy removed

$$\Delta U_E = w \, \Delta d \, \Delta z \frac{\varepsilon E_0^2}{4} \tag{10}$$

so by (1)

$$\frac{\Delta \omega}{\omega_0} = -\frac{\Delta d \, \Delta z}{ld} \tag{11}$$

To check this, we may use the equivalent circuit of the capacitively loaded quarter-wave line shown in Fig. 10.12b. Resonance is given by

$$\omega \left(\frac{\Delta C}{2} \right) = Y_0 \cot \left(\frac{\omega l}{2c} \right) \tag{12}$$

where

$$\Delta C = \varepsilon w \, \Delta z \left(\frac{1}{d - \Delta d} - \frac{1}{d} \right) \approx \frac{\varepsilon w \, \Delta z \, \Delta d}{d^2} \tag{13}$$

If $\omega = \omega_0 + \Delta \omega$ and $\omega_0 l/c = \pi$, to first order (12) gives

$$\frac{\Delta \omega}{\omega_0} \approx -\frac{\Delta C c}{l Y_0} = -\left(\frac{\varepsilon w \, \Delta z \, \Delta d}{ld^2} \right) \left(\frac{1}{\mu \varepsilon} \right)^{1/2} \left(\frac{\mu}{\varepsilon} \right)^{1/2} \frac{d}{w} \tag{14}$$

This reduces to (11), including the check of sign.

_____ **Example 10.12b** _____
Perturbation on Bottom of Cylindrical Cavity

For a more practical example, imagine a small volume ΔV taken out of the pill-box resonator of volume V_0 in Fig. 10.12c, along the axis where electric field is maximum and magnetic field negligible. The change in energy stored is then

$$\Delta U_E = \approx \frac{\varepsilon E_0^2}{4} \Delta V \tag{15}$$

The total energy of the resonator is given by Eq. 10.6(5). Frequency shift from (1) for the lowest mode with $ka = 2.405$ is then

$$\frac{\Delta\omega}{\omega_0} = - \frac{\varepsilon E_0^2 \, \Delta V}{2\pi\varepsilon \, dE_0^2 a^2 J_1^2(ka)} = -1.85 \frac{\Delta V}{V_0} \tag{16}$$

The shift in resonant frequency determines the ratio E_0^2/U needed in Eq. 10.11(13) for determination of R_0/Q. Frequency shifts can be measured accurately, and the perturbation can be made in the form of a small conducting bead moved by an insulating thread along the axis. Field can be measured at all points on the axis, and thus its integral found even when field cannot be assumed to be uniform across the gap.

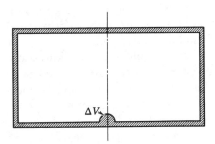

Fig. 10.12c Small perturbation in bottom of circular cylindrical cavity.

10.13   Dielectric Resonators

We saw in the earlier sections of this chapter that a section of a hollow metal wave-guide shorted at the ends constitutes a resonant structure with properties similar to resonant L–C circuits. Similarly, a section of dielectric waveguide (Sec. 9.2) exhibits resonances and can be used for the same purposes. Some materials have very high permittivities and wave energy is therefore strongly confined within the material. Wavelength is small so the dielectric resonator can be much smaller than an empty hollow metal structure. Early work[6] employed high-purity TiO_2 ceramic material with $\varepsilon_r \approx 100$ and $\varepsilon''/\varepsilon' \approx 10^{-4}$. Values of Q (not accounting for losses in supporting structures) then are high (about 10^4) as shown in Prob. 10.4d. The difficulty with TiO_2 is that its ε_r has intolerably strong temperature dependence of 10^3 parts per million (ppm) per degree Celsius, which leads to a resonant-frequency dependence of 500 ppm/°C. More recently, ceramics have been developed[7] that can be made with temperature coefficients selected to offset those of the supporting structures, giving a net zero temperature dependence. These have $\varepsilon_r \approx 37.5$ and $\varepsilon''/\varepsilon' \approx 2 \times 10^{-4}$ at about 10 GHz. Disks of these materials can conveniently be introduced as resonators into microwave-integrated circuits and use beyond 100 GHz is expected.

Exact analysis is possible for a sphere and a toroid, but shapes of greater technical interest such as rectangular prisms, disks, and rods must be treated approximately. It is seen that fields at the surface of a region of very high permittivity satisfy approximately the so-called *open-circuit* boundary condition for which the normal component of electric field and the tangential component of magnetic field are zero. This is made plausible by considering reflections of a plane wave in going from a dielectric of high permittivity (low intrinsic impedance) to one of low permittivity (high intrinsic impedance). One could calculate the resonant frequency of a dielectric resonator by surrounding it with a contiguous perfect magnetic conductor to impose the above stated conditions. However, since the open-circuit boundary conditions are only approximately satisfied for finite-permittivity resonators, modifications in the use of the perfect magnetic boundary have been found to give better results. The two lowest order modes for circular cylindrical disks and rods are one with zero electric field along the axis and one with zero magnetic field along the axis. The model for the former places the solid dielectric cylinder inside of a contiguous infinitely long, magnetic cylindrical waveguide so that the open-circuit conditions are imposed only on the cylindrical surface as in Fig. 10.13a. The model for the mode with zero magnetic field along the axis imposes the open-circuit condition only on the end faces of the dielectric cylinder by means of infinite parallel magnetic conducting plates. In either case the resonant fields and frequency are found by setting up propagating-type solutions of the wave equation inside and attenuating fields outside the dielectric and matching boundary conditions. The result is that the mode with zero axial electric field has the lowest resonant frequency for dielectric cylinders with length less than the diameter. The field distribution for this mode is shown in Fig. 10.13b. Some experimental results are shown in Fig. 10.13c. The curve labeled f_1 is the mode with zero axial electric field and f_2 has zero axial magnetic field. A general treatment of arbitrary shapes of dielectric resonators is given in Ref. 8.

[6] S. B. Cohn, *IEEE Trans. Microwave Theory Tech.* **MTT-16**, 218 (1968).

[7] M. R. Stiglitz, *Microwave J.* **24**, 19 (1981).

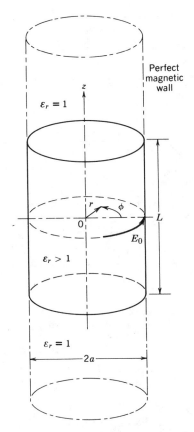

Fig. 10.13a Dielectric cylinder in magnetic-wall waveguide boundary.

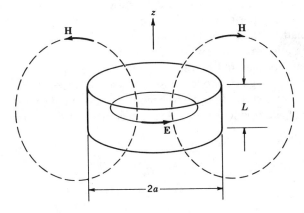

Fig. 10.13b Lowest-order mode where $L \lesssim 2a$ in dielectric resonator. Fields **H** outside $r = a$ are extensions of those given by model in Fig. 10.13a.

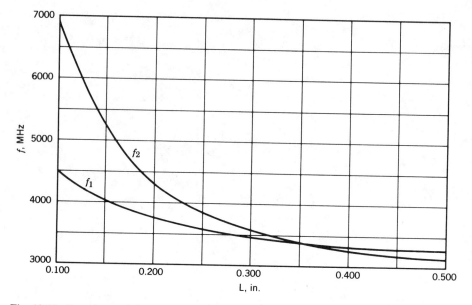

Fig. 10.13c Experimental data on the lowest resonant frequencies versus length of a dielectric cylinder of circular cross section. Radius $a = 0.162$ in. [After Ref. 6.] (© 1968 IEEE)

_____ **Example 10.13** _____
TE Mode in Dielectric Resonator

As an example we do the analysis for the case with the circular cylinder of dielectric enclosed in a perfectly conducting magnetic waveguide shown in Fig. 10.13a. Use is made of the concept of duality (Sec. 9.5) in which **E** is replaced by **H**, **H** by $-$**E**, μ by ε, and ε by μ in known field distributions to find another solution of Maxwell's equations that fits boundary conditions dual to those of the given field. The circular waveguide modes in Sec. 8.9 can be adapted. In particular, the TM mode with an electric boundary becomes a TE mode with the magnetic boundary. The H_ϕ component in Eq. 8.9(9) becomes the E_ϕ component and can be written as

$$E_\phi = E_0 \frac{J_1(k_c r)}{J_1(p_{01})} e^{-j\beta_d z}; \qquad |z| \le \frac{L}{2} \tag{1}$$

[8] J. Van Bladel, _IEEE Trans. Microwave Theory Tech._ **MTT-23**, 199 (1975).

where the propagation factor is included for a wave in the $+z$ direction and the subscript d signifies the dielectric region. Taking account of waves in both directions in the dielectric, one has

$$E_\phi = 2E_0 \frac{J_1(k_c r)}{J_1(p_{01})} \cos \beta_d z; \qquad |z| \le \frac{L}{2} \tag{2}$$

The frequency is assumed low enough that the waves outside the dielectric ($|z| > L/2$) are cut off. Choosing the coefficient to ensure continuity of E_ϕ across the end of the dielectric, we may write

$$E_\phi = 2E_0 \frac{J_1(k_c r)}{J_1(p_{01})} \cos \frac{\beta_d L}{2} e^{-\alpha_a(|z|-L/2)}; \qquad |z| \ge \frac{L}{2} \tag{3}$$

The magnetic field components tangential to the end faces of the dielectric are

$$H_r = j \frac{\beta_d}{\omega\mu} 2E_0 \frac{J_1(k_c r)}{J_1(p_{01})} \sin \beta_d z; \qquad |z| \le \frac{L}{2} \tag{4}$$

$$H_r = \pm \frac{j\alpha_a}{\omega\mu} 2E_0 \frac{J_1(k_c r)}{J_1(p_{01})} \cos \beta_d \frac{L}{2} e^{-\alpha_a(|z|-L/2)}; \qquad |z| \ge \frac{L}{2} \tag{5}$$

where we have used the relation dual to Eq. 8.9(9). The upper sign in (5) applies at $z = L/2$ and the lower at $z = -L/2$. Then equating H_r across the dielectric boundary at $z = \pm L$ we find the determinantal relation for the resonant frequency:

$$\beta_d \tan \frac{\beta_d L}{2} = \alpha_a \tag{6}$$

where

$$\beta_d = \sqrt{\frac{\omega^2 \varepsilon_r}{c^2} - \left(\frac{p_{01}}{a}\right)^2} \tag{7}$$

and

$$\alpha_a = \sqrt{\left(\frac{p_{01}}{a}\right)^2 - \left(\frac{\omega}{c}\right)^2} \tag{8}$$

These calculations give resonant frequencies about 10% lower than the experimental data in the range $0.24 < L/2a < 0.62$.

PROBLEMS

10.2 In Sec. 10.2, the analogy is made between acoustic resonances in a closed box with fixed walls and electromagnetic resonances in a box with walls of high conductivity. Develop the comparison further, stressing similarities and differences.

10.3a Show that the mode described in Sec. 10.3 (resonant condition and field expressions) would be obtained if one started with the point of view that it was a TE_{10} mode propagating in the x direction; similarly consider it a TM_{11} mode propagating in the y direction exactly at cutoff.

10.3b Find the total charge on top plate and bottom plate for the resonator of Sec. 10.3. Determine an equivalent capacitance that would give this charge with a voltage equal to that between top and bottom at the center of the box.

10.3c Find the total current in the side walls of the resonator of Sec. 10.3. Determine an equivalent inductance in terms of this current and the magnetic flux linking a vertical path at the center of the box. What resonant frequency would be given by this inductance and the equivalent capacitance of Prob. 10.3b? Compare with result of Eq. 10.3(1).

10.3d Suppose that in place of a perfect conductor at $z = d$, a "reactive wall" giving a wave reactance $E_y/H_x = jX$ is placed there. Obtain the condition for resonance. Discuss physical ways in which the reactance wall might be produced, at least as an approximation.

10.4a Calculate the maximum energy stored in magnetic fields for the simple mode of the rectangular resonator and show that it is the same as Eq. 10.4(1).

10.4b Convert the phasor forms of electric and magnetic fields of Sec. 10.3 to instantaneous forms and find stored electric and magnetic energy as functions of time. Show that the sum is a constant equal to the value of Eq. 10.4(1).

10.4c From the definition of Q in terms of energy storage and power loss, Eq. 10.4(3), show that the decay of energy in a natural oscillation after excitation is removed is of the form $\exp(-t/\tau)$ where $\tau = Q/\omega_0$.

10.4d For an imperfect dielectric, show that the Q for any resonant mode is just $\varepsilon'/\varepsilon''$, neglecting losses of the conducting walls. Give the value for the dielectrics of Table 6.4a at 10 GHz.

10.4e Modify the expression 10.4(2) for wall losses if front and back are of one material with surface resistivity R_{s1}, the two sides of another with R_{s2}, and top and bottom of a third material with R_{s3}. Give the special case of this for a cube.

10.4f Suppose that a perfect dielectric were available with $\varepsilon' = 5\varepsilon_0$. How would the Q of a dielectric-filled cube compare with that of an air-filled one for the simple mode studied? Why are they different? (Compare for the same resonant frequency in both cases.)

10.4g A square cavity resonator utilizing the simple TE_{101} mode is to be designed for a millimeter-wave electron device for operation at $f = 0.3$ THz. Height of the cavity must be kept small (0.1 mm) to minimize electron transit time. Take the cross section as square and calculate cavity size, the Q, bandwidth, and conductance $G = 2W_L/(E_0b)^2$ presented to the beam between top and bottom at the center. Dielectric in the copper cavity is vacuum.

10.5a For TE_{mnp} and TM_{mnp} modes in a cubic resonator, consider combinations of the integers with no integer higher than 2. How many *different* resonant frequencies do these represent? What is the degree of degeneracy (number of modes at a given frequency) for each frequency? (Note carefully which, if any, integers may be zero.)

10.5b* By combining incident and reflected waves as was done for the TE_{101} mode of Sec. 10.3, find electric and magnetic field expressions of the TE_{111} mode of the rectangular box. Sketch field distributions in the three coordinate planes. Find the expression for Q if conductors are imperfect.

10.5c Derive the expression for Q of a TE_{mmm} mode in a cube, $a = b = d$ and show that it increases as m increases for a given dielectric and resonant frequency. Explain the result.

10.6a Find expressions for resonant wavelength in terms of length and radius for the TM_{021} and TM_{111} modes of a circular cylindrical cavity.

10.6b A circular resonator has height h and radius a. Give the lowest frequency for which a degeneracy between two modes occurs and the designations of these modes. Make rough sketches of the field patterns. How might a small perturbation be added to change the frequency of one mode but not of the other?

10.6c Plot curves of d/λ versus a/d for all the significant modes in a circular cylindrical cavity over the range $0 < d/\lambda < 2$, $0 < a/d < 5$. Note especially ranges of operation where there is only one mode over a considerable region of operation.

10.6d* Give the field components and obtain expressions for energy storage, power loss, and Q for the TM_{011} mode of a circular cylindrical resonator.

10.6e* Repeat Prob. 10.6d for the TE_{011} mode.

10.7 Develop the TE set of spherical waves, Eq. 10.7(20), starting from Maxwell's equations with no ϕ variations, in parallel to the development for the TM set.

10.8a Determine an equivalent conductance for the TM_{101} mode of a spherical resonator in terms of the conductor losses and a voltage between poles taken along the axis.

10.8b By utilizing solutions and definitions of Sec. 10.7 write expressions for the components in a spherical TE mode with $n = 2$ in terms of sines and cosines.

10.8c* Determine the Q of the TE_{101} mode in the spherical resonator.

10.9a A coaxial line of radii 0.5 and 1.5 cm is loaded by a gap capacitance as in Fig. 10.9a of 1 pF. Find the length l for resonance at 3 GHz.

10.9b A radial line of spacing $h = 1$ cm has a central post as in Fig. 10.9c of radius 0.5 cm and capacitance 1 pF. Find the radius r_2 for resonance at 3 GHz.

10.9c Obtain expressions for the Q and the impedance referred to the gap ($V^2/2W_L$ where V is gap voltage) for the resonator of Fig. 10.9a, neglecting losses in region B. Calculate values for a copper conductor and the data of Prob. 10.9a.

10.9d Find Q and impedance if in addition to copper losses there are losses in region B representable by a shunt resistance R_0. Repeat the numerical calculation of Prob. 10.9c, taking $R_0 = 10,000 \,\Omega$.

10.9e For a cone angle θ_0 of 15 degrees in Fig. 10.9e, find radius a for resonance at 3 GHz if center capacitance is 1 pF.

10.9f For the conical resonator with no loading capacitance, show that there is a value of θ_0 which gives maximum Q. Calculate the value of Q for a copper resonator designed for $\lambda_0 = 15$ cm with this optimum angle.

10.10a For the simple mode in a rectangular resonator perform an approximate analysis as in Ex. 10.10 leading to an expression for input impedance of a loop introduced at the center of a side wall.

10.10b For the TM_{010} mode in the circular cylindrical resonator, suppose that the coupling to the line is by means of a small probe of length d' extending axially from the bottom center. Taking voltage induced in the probe as the probe length multiplied by electric field of the mode, find an expression for input admittance at resonance of the unperturbed mode, utilizing a procedure similar to that of Ex 10.10. The probe capacitance is C.

10.10c For a circular cylindrical cavity of radius 10 cm, height 10 cm, resonant in the TM_{010} mode, find the approximate resistance coupled into a transmission line by a loop of area 1 cm^2 introduced at the position of maximum magnetic field. Repeat for a probe of length 1 cm introduced at the position of maximum electric field.

10.11a For a cavity with $m^2 R_{ob} = 2$ and $Q_0 = 5000$, plot the locus of impedance on the Smith chart as δ is varied, showing selected values of δ on the locus. Modify m^2 to yield critical coupling and repeat.

10.11b Plot standing wave ratio in the guide versus δ for both parts of Prob. 10.11a. Describe how one might use the plot of SWR versus δ to determine Q as an alternate to the impedance function.

10.11c Find Q_L and Q_{ext} for Prob. 10.11a with $\beta\ell = 0.3$ radian.

10.11d Find R_0/Q for a pillbox resonator of the type studied in Sec. 10.6 operating in its lowest mode. Frequency is 10 GHz and $d = 0.5$ cm. The conductor is copper.

10.11e Derive the expression for R_0/Q for the small-gap coaxial line resonator pictured in Fig. 10.9a.

10.12a Obtain the approximate expression for frequency shift in a circular cylindrical cavity if small volume ΔV is taken from the side wall for the TM_{010} mode where magnetic field is large and electric field small.

10.12b* Discuss qualitatively the effect of a dielectric bead introduced along the axis of a resonator similarly to Fig. 10.12c; also a thin dielectric sheet along the bottom of the cavity; also a thin dielectric cylinder along the axis. (Note that the last two problems can be solved exactly.)

10.12c Consider a periodic circuit as in Fig. 9.10a with parallel perfectly conducting planes extending from $x = 0$ to $x = a$ and $z = d/2$ and $z = (n + \frac{1}{2})d$ where n is an integer. Discuss resonance for such a system. Will there be a resonance corresponding to each of the space harmonics defined in Sec. 9.10?

10.12d For Prob. 10.12c, show how measurement of the number of nodes (voltage minima) between planes as frequency is changed permits plotting of the $\omega-\beta$ diagram.

10.13 A dielectric resonator with zero axial electric field has radius $a = 0.162$ in., $\varepsilon_r = 100$, and is to be resonant at 3.5 GHz. Use Eqs. 10.13(6) and (7) to estimate length and compare with the experimental curve of Fig. 10.13c. What would be the estimated length if a magnetic short were used on the ends as well as on the side walls?

11

MICROWAVE NETWORKS

11.1 Introduction

We have so far considered individual components of a wave-type system, including transmission lines, waveguides, and cavity resonators. These are used in combination in practical systems, and it is found that many of the ideas from classical network theory, or an extension of these, are useful in handling such interconnections. In this chapter we consider some of these formulations and techniques. For convenience we refer to the subject as that of *microwave networks* since the techniques are most useful in the microwave frequency range, although they apply to any wave-type system from low-frequency transmission-line circuits to dielectric wave-guiding systems in the optical range.

One approach to the analysis of an interconnection of waveguides with other elements might be that of attempting to solve Maxwell's equations subject to boundary conditions for the entire system at once. This would be hopelessly complicated for most practical systems, and it would also give fields everywhere within the system, which is more information than is needed. One usually needs only the characteristics of each part of the system as a transducer or power-transfer element between units, or as a coupling element to adjacent units. A finite number of parameters (frequency-dependent, in general) may be defined to give that desired information. These parameters can be obtained by analysis in some cases, found from handbooks[1,2] for certain standard configurations, or determined from measurement if neither of the first two approaches work.

[1] ITT Handbook, *Reference Data for Radio Engineers*, 6th ed., Howard W. Sams & Co., Inc., New York, 1975.
[2] N. Marcuvitz, *Waveguide Handbook*, M.I.T. Rad. Lab. Series, Vol. 10, McGraw-Hill, New York, 1951.

Specifically, we shall mean by a microwave network a dielectric region of arbitrary shape having certain waveguide or transmission line inlets and outlets. The waveguides are assumed to support a finite number of noncutoff modes. Examples are the cavity resonator coupled to a single transmission line (Fig. 11.1*a*), the rectangular waveguide with change of height (Fig. 11.1*b*), the *E plane T* in rectangular waveguide (Fig. 11.1*c*), and the *magic T* or bridge (Fig. 11.1*d*). These may be said to be microwave networks with, respectively, one, two, three, and four waveguide terminal ports. (This assumes only one noncutoff mode per guide.) In considering the defined arrangements as microwave networks, it will also be assumed that we are interested only in the behaviors of the dominant modes in certain of the guides when various load conditions are placed on the remaining guides, and not in the detailed solution of the electromagnetic field in the vicinity of the discontinuities. Dielectric waveguides may also serve as the terminals for the guided modes with fields decaying properly away from the guide.

Although we may wish to excite only the dominant mode in any of the waveguide terminals, it is true that higher order modes are excited in the vicinity of the junctions, and, although these modes may be cut off, they have reactive energy which affects the transmission between the propagating dominant modes of the

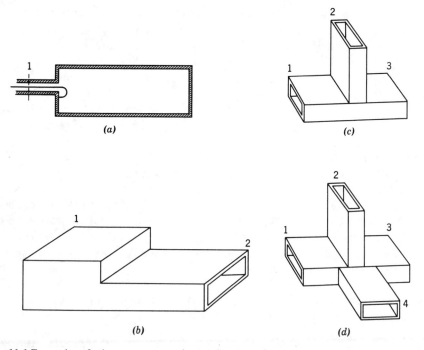

Fig. 11.1 Examples of microwave networks. (*a*) Coupling from a line to a cavity (one port). (*b*) Discontinuity in rectangular guide (two port). (*c*) *E*-plane T (three port). (*d*) Magic T or microwave bridge (four port).

various guides. But, if we are interested only in the manner in which such transmission is affected, it can be expressed in terms of certain coefficients or equivalent circuits, and the details of the higher mode fields need not be described. Thus the microwave two port of Fig. 11.1b may be represented by a T or π network in the same way as a lumped-element two port. It is interesting to note that Carson[3] recognized the validity of this representation as early as 1924, although the thorough development for distributed systems occurred much later.[4] Many formulations are possible, and since these are wave-type systems, some of the most useful relate incident and reflected waves in the various guides. Details of both types of formulations will be given in following sections.

Finally, a combination of elements such as those in the foregoing examples is also a microwave network, fitting the definition of the first paragraph. An important part of the study will be concerned with the finding of network parameters for an overall system when they are known for the individual components. The propagating media of this chapter will be considered to be linear and isotropic unless otherwise stated but not necessarily homogeneous.

11.2 The Network Formulation

As noted in the introductory section, microwave networks may be described either in terms of parameters relating incident and reflected waves at the terminals, or in terms of lumped-element equivalent circuits. Although the former approach may seem more natural, the latter is convenient in many cases and gives a tie to classical network theory, so will be considered first. The equivalent circuit approach does require definition of *voltage* and *current* for the microwave networks.

Voltage and current have been defined in usual ways for transmission lines propagating the *TEM* wave (Chapter 5 and Sec. 8.12). For the TE_{10} mode of rectangular guide (used in most of the examples of Fig. 11.1), one might think of a voltage as the line integral of electric field between top and bottom of the guide, but it is not clear if one should take the maximum value at the center or some sort of average. Axial current flows in the top and returns to the bottom much as in a transmission line, but it is not clear if one should worry about the transverse current flow in the sides (Sec. 8.8). For other modes it will become even more confusing if classical ideas of voltage and current are attempted. It is found, however, that a network formulation results if one follows these simple rules:

1. Voltage is defined as proportional to the transverse electric field of the mode and current is defined as proportional to the transverse magnetic field.

[3] J. R. Carson, *Proc. AIEE* **43**, 908 (1924).

[4] C. G. Montgomery, R. H. Dicke, and E. M. Purcell, *Principles of Microwave Circuits*, MIT Radiation Laboratory Series, Vol. 8, McGraw-Hill, New York, 1948. See also R. N. Ghose, *Microwave Circuit Theory and Design*, McGraw-Hill, New York, 1963; and A. F. Harvey, *Microwave Engineering*, Academic Press, New York, 1963.

2. One condition on the proportionality factors is that average power is given by Re[$VI*/2$] as in a circuit.
3. The second condition on the proportionality factors is that V/I of an incident wave should be a characteristic impedance of the mode of concern, often taken as unity to normalize automatically all impedances.

Concerning the last point, a characteristic impedance is clearly defined for *TEM* modes, but even there we found normalized values of impedance and admittance useful in applying the Smith chart (Chapter 5). For other waves the concept of a characteristic impedance is not so clear. The characteristic wave impedances (Secs. 8.13 and 8.14) are sometimes used, but it is usually better to normalize as suggested. In the following, however, Z_0 will first be retained in the expressions both for generality and for dimensional checks.

For a single traveling wave, using the first rule,

$$\mathbf{E}_t(x, y, z) = V_0 e^{-\gamma z} \mathbf{f}(x, y) \tag{1}$$

$$\mathbf{H}_t(x, y, z) = I_0 e^{-\gamma z} \mathbf{g}(x, y) \tag{2}$$

Applying the second and third rules

$$\mathrm{Re}(V_0 I_0^*) = 2W_T, \qquad \frac{V_0}{I_0} = Z_0 \tag{3a,b}$$

As an example, take the TE_{10} mode in loss-free rectangular guide:

$$E_y = E_0 \sin \frac{\pi x}{a} = V_0 f(x) \tag{4}$$

$$H_x = -\frac{E_0}{Z_z} \sin \frac{\pi x}{a} = I_0 g(x) \tag{5}$$

Utilizing (3a) we have

$$V_0 I_0^* = 2b \int_0^a \frac{E_0^2}{2Z_z} \sin^2 \frac{\pi x}{a} \, dx = \frac{abE_0^2}{2Z_z}$$

This result, combined with (3b), gives current and voltage:

$$V_0 = E_0 \left(\frac{abZ_0}{2Z_z}\right)^{1/2}; \qquad I_0 = \left(\frac{E_0}{Z_z}\right)\left(\frac{baZ_z}{2Z_0}\right)^{1/2} \tag{6}$$

and, by comparison with (4) and (5), the remaining functions are

$$f(x) = \left(\frac{2Z_z}{abZ_0}\right)^{1/2} \sin \frac{\pi x}{a}; \qquad g(x) = -\left(\frac{2Z_0}{baZ_z}\right)^{1/2} \sin \frac{\pi x}{a} \tag{7}$$

As noted in point 3, Z_0 can be made unity in order to normalize automatically all subsequent impedances.

The network formulation may now be obtained by making use of the specified linearity and a uniqueness argument. To be specific, consider the region with three

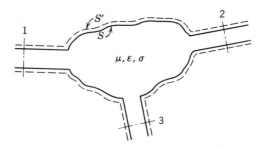

Fig. 11.2 General microwave network with three waveguide terminals.

waveguide terminals pictured in Fig. 11.1. Each waveguide is assumed to support one propagating mode only, and reference planes are at first chosen far enough from junctions so that all higher order (cutoff) modes have died out.[5] The forms of the propagating or dominant modes are assumed to be known, so that field is completely specified at each reference plane by giving two amplitudes, such as the voltage and current defined above. It is clear that it is not possible to specify independently all voltages and currents of the network and a uniqueness argument tells us how many of these may be specified to determine the problem. As was pointed out in Sec. 3.14 there is one and only one steady-state solution of Maxwell's equations within a region (except for possible undamped, uncoupled modes of no interest to us) if tangential electric field is specified over the closed boundary surrounding that region, or if tangential magnetic field is specified over the closed boundary, or if tangential electric field is specified over some of the boundary and tangential magnetic field over the remainder.

In Fig. 11.2 consider the closed region bounded by the conducting surface S and reference planes 1, 2, 3. If the conductor is first taken as perfectly conducting, the tangential electric field is known to be zero over the surface S. Then, if voltages are given for each of the reference terminals, tangential electric fields are known there, and, by the statement of uniqueness, one and only one solution of Maxwell's equations is possible. Then **E** and **H** are determinable for any point inside the region, including the reference planes, so that the currents (amplitudes of the tangential magnetic field distributions) may be found there. For linear media, the relations are linear ones and may therefore be written

$$I_1 = Y_{11} V_1 + Y_{12} V_2 + Y_{13} V_3$$
$$I_2 = Y_{21} V_1 + Y_{22} V_2 + Y_{23} V_3 \qquad (8)$$
$$I_3 = Y_{31} V_1 + Y_{32} V_2 + Y_{33} V_3$$

---

[5] The reference planes can actually be chosen for convenience at any place, but transmission-line measurements to determine the network should not be made in the region where local waves are of importance, nor will the calculations from the network give total fields in that region.

Similarly, if currents are given for all reference planes, tangential magnetic fields are known there, and, with the known zero tangential electric field over S, the uniqueness argument again applies so that tangential electric fields and hence voltages could be found at the reference planes. For linear media,

$$V_1 = Z_{11}I_1 + Z_{12}I_2 + Z_{13}I_3$$
$$V_2 = Z_{21}I_1 + Z_{22}I_2 + Z_{23}I_3 \qquad (9)$$
$$V_3 = Z_{31}I_1 + Z_{32}I_2 + Z_{33}I_3$$

Forms (8) and (9) are identical with the forms that would be found relating voltages and currents at the terminals of a three-port lumped-element network. Here also the coefficients Y_{ij} and Z_{ij} are functions of frequency and are known as the admittance parameters and impedance parameters, respectively. For an N port, in matrix form,

$$[I] = [Y][V], \qquad [V] = [Z][I] \qquad (10)$$

where $[I]$ and $[V]$ are column matrices of order N and $[Y]$ and $[Z]$ are $N \times N$ square matrices.

Although the argument has been given for a perfectly conducting surface S, the foregoing forms also apply to an imperfectly conducting boundary. A reasonably convincing way of seeing this comes from moving the bounding surface several depths of penetration within the conductor to S', Fig. 11.2. The electric field here is substantially zero, so that an imagined perfect conductor could be introduced along S' without changing the behavior of the system, and the argument would proceed as above. The conducting portion between S and S' will contribute to the parameters Y_{ij} or Z_{ij} since it is now part of the interior, and those coefficients will be complex because of the losses.

11.3 Conditions for Reciprocity

For most systems, the admittance and impedance matrices defined in the preceding article are symmetric. That is,

$$Y_{ij} = Y_{ji}, \qquad Z_{ij} = Z_{ji} \qquad (1)$$

This condition follows from a reciprocity theorem due to Lorentz, which states that fields $\mathbf{E}_a$, $\mathbf{H}_a$ and $\mathbf{E}_b$, $\mathbf{H}_b$ from two different sources at the same frequency satisfy the condition

$$\nabla \cdot (\mathbf{E}_a \times \mathbf{H}_b - \mathbf{E}_b \times \mathbf{H}_a) = 0 \qquad (2)$$

This theorem is easily verified for isotropic (but not necessarily homogeneous) media by substitution Maxwell's equations in complex form and can also be shown to hold for anisotropic media provided the permittivity and permeability matrices are symmetric. However, it does not hold if the matrices are asymmetric, thus explaining nonreciprocal properties of the gyrotropic media to be met in Secs. 13.15 and 13.17. If (2) is satisfied, a volume integral of (2), with application of the divergence theorem, gives

$$\oint_S (\mathbf{E}_a \times \mathbf{H}_b - \mathbf{E}_b \times \mathbf{H}_a) \cdot \mathbf{dS} = 0 \tag{3}$$

Consider Fig. 11.2 with all reference planes but 1 and 2 closed by perfect conductors (shorted). Fields at 1 and 2 may be written [Eqs. 11.2(1) and 11.2(2)]

$$\mathbf{E}_{t1} = V_1 \mathbf{f}_1(x_1, y_1) \qquad \mathbf{H}_{t1} = I_1 \mathbf{g}_1(x_1, y_1) \tag{4}$$

$$\mathbf{E}_{t2} = V_2 \mathbf{f}_2(x_2, y_2) \qquad \mathbf{H}_{t2} = I_2 \mathbf{g}_2(x_2, y_2) \tag{5}$$

By rule 2, voltage and current are defined to have the same relation to power flow in both guides, which requires that

$$\int_{S_1} (\mathbf{f}_1 \times \mathbf{g}_1) \cdot \mathbf{dS} = \int_{S_2} (\mathbf{f}_2 \times \mathbf{g}_2) \cdot \mathbf{dS} \tag{6}$$

The surface integral (3) is zero along the conducting surfaces S of Fig. 11.2 (or S', if imperfectly conducting), and along the shorted planes. For planes 1 and 2, substitution of (4) and (5) gives

$$(V_{1a}I_{1b} - V_{1b}I_{1a})\int_{S_1}(\mathbf{f}_1 \times \mathbf{g}_1)\cdot \mathbf{dS} + (V_{2a}I_{2b} - V_{2b}I_{2a})\int_{S_2}(\mathbf{f}_2 \times \mathbf{g}_2)\cdot \mathbf{dS} = 0$$

With (6), this reduces to

$$V_{1a}I_{1b} - V_{1b}I_{1a} + V_{2a}I_{2b} - V_{2b}I_{2a} = 0$$

Relations between current and voltage are introduced from Eq. 11.2(8):

$$V_{1a}(Y_{11}V_{1b} + Y_{12}V_{2b}) - V_{1b}(Y_{11}V_{1a} + Y_{12}V_{2a})$$
$$+ V_{2a}(Y_{21}V_{1b} + Y_{22}V_{2b}) - V_{2b}(Y_{21}V_{1a} + Y_{22}V_{2a}) = 0 \tag{7}$$
$$(V_{1a}V_{2b} - V_{1b}V_{2a})(Y_{12} - Y_{21}) = 0$$

In this argument the sources a and b are arbitrary so that the first factor need not be zero. Hence the second is zero. Then

$$Y_{21} = Y_{12} \tag{8}$$

The argument for the impedance coefficients may be supplied by placing "open circuits" at all but two of the terminals. This is done in the waveguides by placing a perfect short a quarter-wave in front of the reference planes. Moreover, since the numbering system is arbitrary, 1 and 2 may represent any two of the guides and the general relation (1) is valid.

TWO-PORT WAVEGUIDE JUNCTIONS

11.4   Equivalent Circuits for a Two Port

The microwave network with two waveguide terminals, as pictured in Fig. 11.1b, is of greatest importance since it includes the cases of discontinuities in a single guide or the coupling between two guides. Most filters, matching sections, phase-correction units, and many other components are of this type. There is a large body of literature on the lumped-element equivalents. The name "two port" is used for these, with a port denoting a single waveguide mode at a specific reference plane for a microwave network, and a terminal pair for a lumped-element network.

From Sec. 11.2 the equations for a two port may be written in terms of either impedance or admittance coefficients.

$$\begin{bmatrix} V_1 \\ V_2 \end{bmatrix} = \begin{bmatrix} Z_{11} & Z_{12} \\ Z_{21} & Z_{22} \end{bmatrix} \begin{bmatrix} I_1 \\ I_2 \end{bmatrix} \tag{1}$$

$$\begin{bmatrix} I_1 \\ I_2 \end{bmatrix} = \begin{bmatrix} Y_{11} & Y_{12} \\ Y_{21} & Y_{22} \end{bmatrix} \begin{bmatrix} V_1 \\ V_2 \end{bmatrix} \tag{2}$$

Another convenient form expresses input quantities in terms of output quantities.

$$\begin{bmatrix} V_1 \\ I_1 \end{bmatrix} = \begin{bmatrix} A & B \\ C & D \end{bmatrix} \begin{bmatrix} V_2 \\ -I_2 \end{bmatrix} \tag{3}$$

Algebraic elimination shows that relations among the above parameters are

$$Y_{11} = \frac{Z_{22}}{\Delta(Z)} = \frac{D}{B}$$

$$Y_{12} = \frac{-Z_{12}}{\Delta(Z)} = \frac{-(AD - BC)}{B} \tag{4}$$

$$Y_{21} = \frac{-Z_{21}}{\Delta(Z)} = \frac{-1}{B}$$

$$Y_{22} = \frac{Z_{11}}{\Delta(Z)} = \frac{A}{B}$$

where

$$\Delta(Z) = Z_{11}Z_{22} - Z_{12}Z_{21}$$

For a network satisfying reciprocity,

$$Z_{21} = Z_{12}; \qquad Y_{21} = Y_{12}; \qquad AD - BC = 1 \tag{5}$$

We shall assume such reciprocal networks in the remainder of this section.

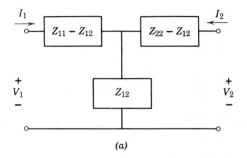

(a)

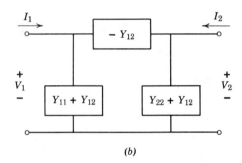

(b)

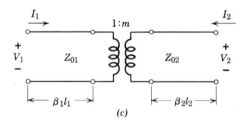

(c)

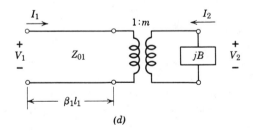

(d)

Fig. 11.4 (a) T equivalent circuit and (b) π equivalent circuit for a two-port satisfying reciprocity. (c) Equivalent circuit for a two port which satisfies reciprocity in terms of sections of transmission line and an ideal transformer. (d) Equivalent circuit in terms of section of transmission line, transformer, and shunt element.

An infinite number of equivalent circuits may be derived which yields the forms (1) to (5). Two important ones are the well-known T and π forms shown in Figs. 11.4a and 11.4b. They may be shown to be equivalent to (1) and (2), respectively, by setting down the circuit equations. Other interesting ones utilize ideal transformers and sections of transmission lines, two of which are pictured in Figs. 11.4c and 11.4d. These are of greatest importance for lossless microwave networks since the arbitrary reference planes in the input or output guides can be shifted in such a way that only an ideal transformer is left in the representation of Fig. 11.4c or an ideal transformer and shunt element in Fig. 11.4d. This will be explained in more detail when the measurement problem is discussed in the next section. The quantities of Fig. 11.4c are related to the impedance parameters as follows:

$$\tan \beta_1 l_1 = \left[\frac{1 + c^2 - a^2 - b^2}{2(bc - a)}\right] \pm \sqrt{\left[\frac{1 + c^2 - a^2 - b^2}{2(bc - a)}\right]^2 + 1}$$

$$\tan \beta_2 l_2 = \frac{1 + a \tan \beta_1 l_1}{b \tan \beta_1 l_1 - c} \tag{6}$$

$$\frac{m^2 Z_{01}}{Z_{02}} = \frac{1 + a \tan \beta_1 l_1}{b + c \tan \beta_1 l_1}$$

where

$$a = \frac{-j Z_{11}}{Z_{01}}$$

$$b = \frac{Z_{11} Z_{22} - Z_{12}^2}{Z_{01} Z_{02}} \tag{7}$$

$$c = \frac{-j Z_{22}}{Z_{02}}$$

11.5 Determination of Circuit Parameters by Measurement

As noted earlier, it is sometimes possible to find the circuit parameters representing a specific junction from handbooks, or by calculation, but it is often necessary to do this by measurement. There are many different ways to make the measurement. A few will be discussed here by way of example and more in a later section concerned with wave-type parameters. Techniques will be described for a single frequency, but since parameters are a function of frequency, measurements will need to be made for different frequencies over the range of interest.

It is straightforward to determine the parameters if loads and sources can be placed on either end of the network, and measurements of both voltage (electric field) and current (magnetic field), including phases, can be made. For example, consider the admittance parameters of Eq. 11.4(2). If V_2 is made zero by placing a short some multiple of a half-wavelength from junction 2, measurement of input admittance gives Y_{11}, and measurement of I_2/V_1 gives Y_{21}. The same procedure with source and short interchanged gives Y_{22} and Y_{12}. It is not always possible to reverse source and load in this way, and it is especially difficult in wave-type systems to make the measurements relating currents on one side to voltages on the other, including phase relations, so other techniques are usually necessary. In the remainder of the section we consider only networks satisfying reciprocity so that only three parameters need be determined at each frequency of interest.

Most often a two port is used to transform impedances from one side to the other. From Eq. 11.4(1), load impedance $Z_L = -V_2/I_2$ produces input impedance $Z_i = V_1/I_1$ as follows:

$$Z_i = Z_{11} - \frac{Z_{12}^2}{Z_{22} + Z_L} \tag{1}$$

Measurement of three different (Z_L, Z_i) pairs can then determine the three parameters Z_{11}, Z_{22}, and Z_{12}^2. A particularly simple way is to place a good short at three independent positions along the guide to produce the three known load impedances. Algebraic elimination from three equations of the form of (1) shows that, if Z_{L1} produces Z_{i1}, Z_{L2} produces Z_{i2}, and Z_{L3} produces Z_{i3}, the impedance parameters are

$$Z_{11} = \frac{(Z_{i1} - Z_{i3})(Z_{i1}Z_{L1} - Z_{i2}Z_{L2}) - (Z_{i1} - Z_{i2})(Z_{i1}Z_{L1} - Z_{i3}Z_{L3})}{(Z_{i1} - Z_{i3})(Z_{L1} - Z_{L2}) - (Z_{i1} - Z_{i2})(Z_{L1} - Z_{L3})} \tag{2}$$

$$Z_{22} = \frac{(Z_{i1}Z_{L1} - Z_{i2}Z_{L2}) - Z_{11}(Z_{L1} - Z_{L2})}{(Z_{i2} - Z_{i1})} \tag{3}$$

$$Z_{12}^2 = (Z_{11} - Z_{ip})(Z_{22} + Z_{Lp}), \qquad p = 1, 2, 3 \tag{4}$$

If the network is lossless and if purely reactive terminations are used, input impedances will also be pure reactances, and Z's may be replaced by X's everywhere in these equations. The form of (2) to (4) may also be shown to be valid for determination of admittance parameters Y_{11}, Y_{12}, and Y_{22} when pairs of input–output admittances $Y_{L1}Y_{i1}$, $Y_{L2}Y_{i2}$, $Y_{L3}Y_{i3}$ are measured. The Z's are then replaced by Y's in (2) to (4). Note that the sign of Z_{12} cannot be determined from impedance transformation measurements alone, since it does not enter into (1). For the same reason, it is of no interest if results are to be used only for impedance transformations by the network. For some purposes it may be necessary to know this sign, and, if so, an arrangement for measuring relative phase between input and output must be added. For simple configurations, the sign may sometimes be deduced by physical reasoning.

For regions that may be considered lossless, the representation of Fig. 11.4c is especially useful. This follows because a shift of the input reference plane from 1 to 1' (Fig. 11.5a) by a distance $\beta_1 y_0 = \beta_1 l_1 - \pi$ and a shift of output reference

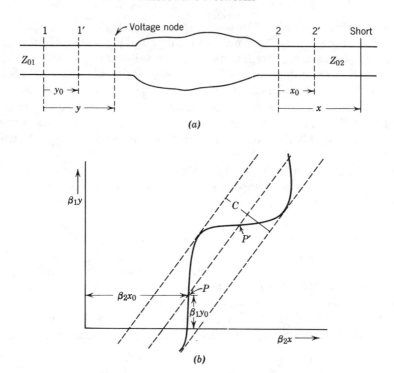

Fig. 11.5 (a) General two port. (b) Typical S curve obtained by measurement on (a).

plane from 2 to 2' by $\beta_2 x_0 = \pi - \beta_2 l_2$ gives as the equivalent circuit an ideal transformer with half-wave lines at input and output. But the half-wave lines give unity impedance transformation and so may be ignored, leaving only the ideal transformer representing the region between 2' and 1'. A load impedance referred to 2' is multiplied simply by $(1/m)^2$ to give the input impedance referred to 1'.

The parameters of the above representation may be determined as follows. The output guide is perfectly terminated ($Z_L = Z_{02}$); the position of the minimum impedance point on the input guide corresponds to 1', and the value of this minimum impedance gives m^2,

$$m^2 = \frac{Z_{02}}{Z_{min}} \tag{5}$$

Similarly, if the network is reversed, the input guide terminated, and like measurements made on the output guide, the location of reference plane 2' is obtained as well as a check on m^2.

An alternative procedure to the above has advantages in some cases. Weiss-floch[6] has shown that for a lossless junction a plot of position of voltage minimum

---

[6] N. Marcuvitz, Ref. 2, p. 122.

on the input guide as a function of position of a short on the output guide has the "S curve" form shown in Fig. 11.5b, where $\beta_1 y$ is the electrical distance of the minimum from the originally selected reference 1, and $\beta_2 x$ is the electrical distance of the short from 2. The form of the equation is easily found to be

$$\tan \beta_1(y - y_0) = \frac{Z_{02}}{m^2 Z_{01}} \tan \beta_2(x - x_0) \tag{6}$$

The new reference planes $1'$ and $2'$ are given by the positions x_0, y_0 of the maximum slope of the S curve, point P of Fig. 11.5b. The value of this maximum slope is $Z_{02}/m^2 Z_{01}$. The turns ratio may also be determined in terms of the distance C between the envelope tangents.

$$\frac{m^2 Z_{01}}{Z_{02}} = \tan^2 \left(\frac{\pi}{4} - \frac{\sqrt{2}C}{4} \right) \tag{7}$$

For the measurement, many points of input minimum are then measured as a short is moved along the output, and the curve determined. There is the advantage that the consistency of measurement and discrepancies caused by neglected losses may be told more easily than in the methods first described where only a few points are measured. Alternatively, one can use the crossover point P'. Turns ratio is the reciprocal of the above and reference planes are shifted by a quarter-wavelength on each side.

11.6   Scattering and Transmission Coefficients

The preceding discussions have been given in terms of the voltages, currents, and impedances defined for microwave networks. Definitions of these quantities are somewhat arbitrary. Moreover, the impedances are usually obtained by interpreting measured values of standing-wave ratios or reflection coefficients. It is then evident that for some problems it will be more convenient and direct to formulate the transformation properties of the two port in terms of waves. The two independent quantities required for each waveguide terminal are an incident and a reflected wave replacing the voltage and current. This section introduces two of the most useful forms based on wave quantities.

Suppose that incident and reflected voltage waves on the input guide are given in magnitude and phase at the chosen reference plane by V_{1+} and V_{1-} (Fig. 11.6). Similarly, incident and reflected waves looking toward the junction from reference plane 2 are V_{2+} and V_{2-}. It is common to normalize incident and reflected waves as follows,

$$a_n = \frac{V_{n+}}{\sqrt{Z_{0n}}}, \qquad b_n = \frac{V_{n-}}{\sqrt{Z_{0n}}} \tag{1}$$

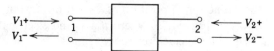

Fig. 11.6 Incident and reflected waves at ports of microwave network.

Thus voltage and current at reference plane n are related to these wave quantities as follows,

$$V_n = V_{n+} + V_{n-} = \sqrt{Z_{0n}}(a_n + b_n)$$

$$I_n = \frac{1}{Z_{0n}}(V_{n+} - V_{n-}) = \frac{1}{\sqrt{Z_{0n}}}(a_n - b_n)$$

(2)

The average power flowing into terminal n is

$$(W_n)_{\mathrm{av}} = \tfrac{1}{2}\,\mathrm{Re}(V_n I_n^*) = \tfrac{1}{2}\,\mathrm{Re}[(a_n a_n^* - b_n b_n^*) + (b_n a_n^* - b_n^* a_n)]$$

The first parenthesis is a purely real quantity, and the second purely imaginary. Thus

$$2(W_n)_{\mathrm{av}} = a_n a_n^* - b_n b_n^*$$

(3)

That is, $(W_n)_{\mathrm{av}}$ is the power carried into terminal n by the incident wave, less that reflected away.

In the first form to be used, we shall relate the two reflected waves to the two incident waves. For a linear medium,

$$\begin{bmatrix} b_1 \\ b_2 \end{bmatrix} = \begin{bmatrix} S_{11} & S_{12} \\ S_{21} & S_{22} \end{bmatrix} \begin{bmatrix} a_1 \\ a_2 \end{bmatrix}$$

(4)

or

$$[b] = [S][a]$$

(5)

with the $[S]$ array known as the *scattering matrix* and the coefficients S_{11}, and so on, known as *scattering coefficients*. For a physical interpretation, note that with the source applied to port 1, and the output guide matched so that $a_2 = 0$, then

$$b_1 = S_{11} a_1, \qquad b_2 = S_{21} a_1$$

(6)

Thus S_{11} is just the input reflection coefficient (in magnitude and phase) when the output is matched, and S_{21} is the ratio of waves to the right at output and input under this condition. The energy equation (3) for this matched condition becomes for the two terminals,

$$2(W_1)_{\mathrm{av}} = (1 - S_{11} S_{11}^*)a_1 a_1^*$$

$$2(W_2)_{\mathrm{av}} = -S_{21} S_{21}^* a_1 a_1^*$$

(7)

For a passive network with source at 1 as in this example, output power cannot be greater than that supplied at the input. Thus $(-W_2)_{av} \leq (W_1)_{av}$ or

$$S_{21}S_{21}^* \leq 1 - S_{11}S_{11}^* \tag{8}$$

The equality holds only when the network is loss free.

By substituting the definitions (2) into Eq. 11.4(1), and utilizing (4), we can relate the scattering coefficients to the impedance coefficients.[7] The results are

$$FS_{11} = (Z_{11} - Z_{01})(Z_{22} + Z_{02}) - Z_{12}Z_{21}$$
$$FS_{12} = 2\sqrt{Z_{01}Z_{02}}\,Z_{12}$$
$$FS_{21} = 2\sqrt{Z_{02}Z_{01}}\,Z_{21} \tag{9}$$
$$FS_{22} = (Z_{22} - Z_{02})(Z_{11} + Z_{01}) - Z_{21}Z_{12}$$

where

$$F = (Z_{11} + Z_{01})(Z_{22} + Z_{02}) - Z_{12}Z_{21} \tag{10}$$

From this we see that $S_{21} = S_{12}$ for a network satisfying reciprocity since then $Z_{21} = Z_{12}$. Through the relations of Eq. 11.4(4) the scattering coefficients may also be related to admittance coefficients or the transfer coefficients if necesary They can be obtained directly from reflection measurements, however, as is shown in a following section.

A second important linear transformation of (4) gives output wave quantities in terms of input quantities,

$$b_2 = T_{11}a_1 + T_{12}b_1$$
$$a_2 = T_{21}a_1 + T_{22}b_1 \tag{11}$$

The coefficients T_{11}, and so on, are known as the *transmission coefficients*, and are related to the scattering coefficients as follows,

$$T_{11} = S_{21} - \frac{S_{11}S_{22}}{S_{12}}, \qquad T_{12} = \frac{S_{22}}{S_{12}}$$

$$T_{21} = -\frac{S_{11}}{S_{12}}, \qquad T_{22} = \frac{1}{S_{12}} \tag{12}$$

This form is especially useful for cascaded networks, as will be illustrated in Sec. 11.8.

---

[7] C. G. Montgomery, R. H. Dicke, and E. M. Purcell, Ref. 4, pp. 146–148.

11.7   Scattering Coefficients by Measurement

From Eqs. 11.6(4), the reflection coefficient at the input terminal may be found in terms of the reflection coefficient at the output. We will define reflection coefficient for both guides looking toward the right,

$$\rho_i = \frac{b_1}{a_1}, \qquad \rho_L = \frac{a_2}{b_2} \tag{1}$$

Then

$$\rho_i = S_{11} - \frac{S_{12}^2}{S_{22} - 1/\rho_L} \tag{2}$$

As this is of the same form as Eq. 11.5(1), the equations 11.5(2)–(4) may be adapted to give S_{11}, S_{22} and S_{12}^2 if we measure three values of input ρ_i corresponding to three independent known values of output ρ_L (assuming satisfaction of reciprocity).

We may simplify this procedure by selecting special values of load terminations when practical. In particular, as shown in Sec. 11.6 if the output guide is matched so that $\rho_L = 0$, measurement of ρ_i gives S_{11} directly. If it is practical to place the source on guide 2 and match line 1, measurement of reflection coefficient looking into reference 2 similarly gives S_{22}. Then S_{12}^2 may be obtained by any other special termination, as a short on plane 2 giving $\rho_L = -1$ for (2). If the sign of S_{12} is required, a phase comparison measurement between input and output will be required. Note, however, that it will not be needed if the results are used only for impedance or reflection coefficient transfer through the network since only S_{12}^2 enters into (2).

Deschamps[8] has given a graphical method for determining the scattering coefficients utilizing plotted values of reflection coefficient directly on the Smith chart (Sec. 5.7). This method has the advantage that extra measurements can readily be utilized to improve accuracy.[8,9,10] Computer programs can also be devised to weight the available data.

11.8   Cascaded Two Ports

The *ABCD* transfer forms of Sec. 11.4 and the transmission coefficients of Sec. 11.6 are especially useful when two ports are connected in tandem or cascade, because the output quantities of one network become the input quantities of the following

---

[8] G. A. Deschamps, *J. Appl. Phys.* **24**, 1046 (1953).

[9] J. E. Storer, L. S. Sheingold, and S. Stein, *Proc. IRE*, **41**, 1004 (1953).

[10] E. L. Ginzton, *Microwave Measurements*, McGraw-Hill, New York, 1957, pp. 336–345.

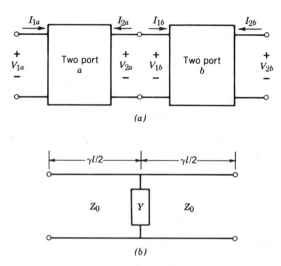

Fig. 11.8 (*a*) Cascade connection of two ports; (*b*) uniform transmission line with shunt element in center.

one. Thus, referring to Fig. 11.8*a*, we may successively apply the form of Eq. 11.4(3) to networks *a* and *b*:

$$\begin{bmatrix} V_{1a} \\ I_{1a} \end{bmatrix} = \begin{bmatrix} A_a & B_a \\ C_a & D_a \end{bmatrix} \begin{bmatrix} V_{2a} \\ -I_{2a} \end{bmatrix}, \qquad \begin{bmatrix} V_{1b} \\ I_{1b} \end{bmatrix} = \begin{bmatrix} A_b & B_b \\ C_b & D_b \end{bmatrix} \begin{bmatrix} V_{2b} \\ -I_{2b} \end{bmatrix}$$

but $V_{2a} = V_{1b}$ and $-I_{2a} = I_{1b}$, so we may combine the two to give

$$\begin{bmatrix} V_{1a} \\ I_{1a} \end{bmatrix} = \begin{bmatrix} A_a & B_a \\ C_a & D_a \end{bmatrix} \begin{bmatrix} A_b & B_b \\ C_b & D_b \end{bmatrix} \begin{bmatrix} V_{2b} \\ -I_{2b} \end{bmatrix} \tag{1}$$

Thus the transfer matrix for the two networks cascaded is

$$\begin{bmatrix} A & B \\ C & D \end{bmatrix} = \begin{bmatrix} A_a & B_a \\ C_a & D_a \end{bmatrix} \begin{bmatrix} A_b & B_b \\ C_b & D_b \end{bmatrix} \tag{2}$$

with obvious extension to more than two cascaded two ports. Similarly for the *T* matrices of Sec. 11.6, but here we start with the matrix for the last unit since these were defined to give output in terms of input quantities:

$$\begin{bmatrix} T_{11} & T_{12} \\ T_{21} & T_{22} \end{bmatrix} = \begin{bmatrix} (T_{11})_b & (T_{12})_b \\ (T_{21})_b & (T_{22})_b \end{bmatrix} \begin{bmatrix} (T_{11})_a & (T_{12})_a \\ (T_{21})_a & (T_{22})_a \end{bmatrix} \tag{3}$$

Table 11.8 gives the *ABCD* coefficients for some simple units.

Table 11.8

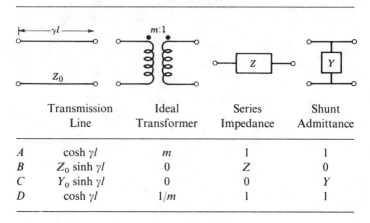

	Transmission Line	Ideal Transformer	Series Impedance	Shunt Admittance
A	$\cosh \gamma l$	m	1	1
B	$Z_0 \sinh \gamma l$	0	Z	0
C	$Y_0 \sinh \gamma l$	0	0	Y
D	$\cosh \gamma l$	$1/m$	1	1

―――――――――― **Example 11.8a** ――――――――――
Transmission Line with Shunt Element

To illustrate the use of the above, consider the circuit of Fig. 11.8*b* with a shunt admittance at the midpoint of a uniform transmission line. The matrix for the combination is

$$
\begin{bmatrix} A & B \\ C & D \end{bmatrix} = \begin{bmatrix} \cosh\left(\dfrac{\gamma l}{2}\right) & Z_0 \sinh\left(\dfrac{\gamma l}{2}\right) \\ Y_0 \sinh\left(\dfrac{\gamma l}{2}\right) & \cosh\left(\dfrac{\gamma l}{2}\right) \end{bmatrix} \begin{bmatrix} 1 & 0 \\ Y & 1 \end{bmatrix} \begin{bmatrix} \cosh\left(\dfrac{\gamma l}{2}\right) & Z_0 \sinh\left(\dfrac{\gamma l}{2}\right) \\ Y_0 \sinh\left(\dfrac{\gamma l}{2}\right) & \cosh\left(\dfrac{\gamma l}{2}\right) \end{bmatrix} \tag{4}
$$

Multiplication of the matrices and use of identities for hyperbolic functions gives

$$
A = D = \cosh \gamma l + \left(\frac{Y}{2Y_0}\right) \sinh \gamma l \tag{5}
$$

$$
B = Z_0 \left[\left(\frac{Y}{2Y_0}\right)(-1 + \cosh \gamma l) + \sinh \gamma l \right] \tag{6}
$$

$$
C = Y_0 \left[\left(\frac{Y}{2Y_0}\right)(1 + \cosh \gamma l) + \sinh \gamma l \right] \tag{7}
$$

_____ **Example 11.8b** _____
Periodic System as Cascade of N Two Ports

Periodic circuits, considered in Sec. 9.10 from a field point of view, may also be considered as a cascade of like networks. The overall transfer matrix for N like networks is just the Nth power of that for a single network,

$$\begin{bmatrix} A & B \\ C & D \end{bmatrix} = \begin{bmatrix} A_0 & B_0 \\ C_0 & D_0 \end{bmatrix}^N \tag{8}$$

For a network with reciprocity, this results in[11]

$$A = [A_0 \sinh N\Gamma - \sinh(N-1)\Gamma]/\sinh \Gamma \tag{9}$$

$$B = B_0 \sinh N\Gamma/\sinh \Gamma \tag{10}$$

$$C = C_0 \sinh N\Gamma/\sinh \Gamma \tag{11}$$

$$D = [D_0 \sinh N\Gamma - \sinh(N-1)\Gamma]/\sinh \Gamma \tag{12}$$

$$\cosh \Gamma = (A_0 + D_0)/2 \tag{13}$$

where $e^{\pm \Gamma}$ are the characteristic roots of the matrix for one section. As will be seen, Γ may be considered the propagation constant when the network is properly terminated.

In order to interpret the above, let us consider that each cell is symmetric, $(A_0 = D_0)$ in addition to being reciprocal $(A_0 D_0 - B_0 C_0 = 1)$ as already assumed. Let us also terminate the chain in a characteristic impedance Z_c defined so that each cell terminated in Z_c also gives impedance Z_c at its input.[12] That is,

$$Z_c = \frac{V_1}{I_1} = \frac{A_0 V_2 - B_0 I_2}{C_0 V_2 - A_0 I_2} = \frac{A_0 Z_c + B_0}{C_0 Z_c + A_0}$$

from which

$$Z_c = \left(\frac{B_0}{C_0}\right)^{1/2} \tag{14}$$

Equations (9)–(12) then reduce to

$$A = D = \cosh N\Gamma \tag{15}$$

$$B = Z_c \sinh N\Gamma \tag{16}$$

$$C = (Z_c)^{-1} \sinh N\Gamma \tag{17}$$

[11] See for example G. Birkhoff and S. Maclane, *A Survey of Modern Algebra*, Macmillan, New York, 1953.

[12] If the network is asymmetric, image impedances—one for each direction—are used. See, for example, Ghose, Ref. 4, p. 246.

By comparison with Table 11.8, we recognize that each cell now acts as a transmission line of characteristic impedance Z_c and overall propagation constant Γ. The cascaded N sections then just have overall propagation constant $N\Gamma$.

Note that from (13) if $|A_0 + D_0| \leq 2$, Γ is imaginary so that there is phase shift but no attenuation through the network. If $|A_0 + D_0| > 2$, Γ is real and there is attenuation. This filter-type behavior is most important in communication networks and will be illustrated more in the following section.

11.9 Examples of Microwave and Optical Filters

The filtering action described in the preceding section is extremely important for communication systems; filters pass desired frequencies with small attenuation while providing much more attenuation for noise or undesired signals outside the frequency range of interest. We shall give some examples in this section for different types of transmission systems. Many other configurations are useful, and a variety of techniques for design of the different types are discussed in treatises on this subject.[13]

_____ Example 11.9a _____
Filters with Periodic Shunt Elements in Transmission Lines

The first four forms to be considered, Figs. 11.9a–d, are specific cases of the symmetrical transmission system with periodic shunt elements analyzed in the preceding section. Figure 11.9a shows a coaxial line with capacitive diaphragms at intervals l along the line, and Fig. 11.9b is a similar arrangement in microstrip with the loading capacitors made as side taps. If the loading capacitors have value C_d and losses are neglected in the transmission line so that $\gamma = j\beta$, Eq. 11.8(5) and Eq. 11.8(13) give

$$\cosh \Gamma = \cos \beta l - \frac{\omega C_d}{2Y_0} \sin \beta l \tag{1}$$

These circuits would be expected to pass low frequencies since the periodic capacitors provide small shunting effects in that range. If $\beta l = \omega l \sqrt{LC} \ll 1$, (1) becomes

$$\cosh \Gamma \approx 1 - \frac{\omega^2 l C_d \sqrt{LC}}{2\sqrt{C/L}} \tag{2}$$

---

[13] G. L. Matthaei, L. Young, and E. M. T. Jones, _Microwave Filters, Impedance-Matching Networks and Coupling Structures_, McGraw-Hill, New York, 1964. Also, R. E. Collin, _Foundations of Microwave Engineering_, McGraw-Hill, New York, 1966.

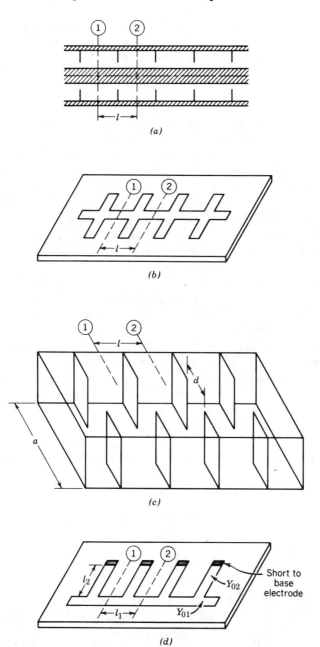

Fig. 11.9 (*a*) Coaxial line with capacitive disks at intervals *l*. (*b*) Microstrip with capacitive tabs at intervals *l*. (*c*) Rectangular waveguide with symmetrical inductive diaphragms introduced from sides at intervals *l*. (*d*) Microstrip with shorted microstrip stub lines in parallel at intervals *l*.

where L and C are inductance and capacitance of the microstrip per unit length. The region of imaginary Γ, which we call the *passband* of the filter, is from $-1 \le \cosh \Gamma \le 1$, which is from zero to ω_{c1} in angular frequency, where cutoff is

$$\omega_{c1} = 2\left(\frac{1}{C_d lL}\right)^{1/2} \tag{3}$$

In this approximation the transmission-line sections, short compared with wavelength, act as series inductors and with the shunt capacitors produce a classical lumped-element, low-pass filter. However, there are other passbands at higher frequencies. These occur in the vicinity of $\beta l = n\pi$ since $\sin \beta l$ becomes small there. Since $\omega C_d/2Y_0$ increases with frequency, the passbands become narrower as the order n increases (assuming of course that the capacitive representation for the discontinuities holds at these higher frequencies). In any event it is characteristic of the transmission-line circuits that they have multiple passbands, as was found from a different point of view in Sec. 9.10.

Example 11.9b

Filter with Periodic Inductive Diaphragms in Waveguide

Figure 11.9c pictures a rectangular waveguide with diaphragms at intervals l introduced symmetrically from the sides. Such diaphragms are found to act as inductive shunt susceptances to the TE_{10} mode, with admittance given approximately by[14]

$$\frac{Y}{Y_0} = -\frac{j\lambda_g}{a}\cot^2\left(\frac{\pi d}{2a}\right) \tag{4}$$

where a is guide width, d the aperture width, and λ_g guide wavelength,

$$\lambda_g = \lambda\left[1 - \left(\frac{\lambda}{2a}\right)^2\right]^{-1/2}$$

Equations 11.8(5) and 11.8(13) give for this case

$$\cosh \Gamma = \cos\left(\frac{2\pi l}{\lambda_g}\right) + \frac{\lambda_g}{2a}\cot^2\left(\frac{\pi d}{2a}\right)\sin\left(\frac{2\pi l}{\lambda_g}\right) \tag{5}$$

The waveguide is in itself a high-pass filter, but with typical values of λ_g/a and d/a, the added diaphragms cause it to remain attenuating to higher frequencies than the waveguide cutoff. But at least in the vicinity of $l = \lambda_g/2$ the last term in (5) is small and $|\cosh \Gamma| < 1$ so that there is a passband. Depending upon values of λ_g/a and d/a there may be additional attenuation bands and passbands at still higher frequencies, as in the first example.

[14] See Refs. 1, 2, or 4.

_____ **Example 11.9c** _____
Bandpass Filter in Microstrip

In Fig. 11.9d the shunt elements consist of shorted sections of microstrip of length l_2 and characteristic admittance Y_{02} connected in parallel with the main microstrip at intervals l_1. The shorted stub lines also short the main line at low frequencies (and at frequencies for which $\beta_2 l_2 = n\pi$), but present high impedances in parallel for frequency ranges near those for which the stub lengths are odd multiples of a quarter wavelength. The shunt admittances of the shorted stubs are

$$Y = -jY_{02} \cot \beta_2 l_2 \qquad (6)$$

so that Eqs. 11.8(5) and 11.8(13) yield

$$\cosh \Gamma = \cos \beta_1 l_1 + \frac{Y_{02}}{2Y_{01}} \cot \beta_2 l_2 \sin \beta_1 l_1 \qquad (7)$$

This clearly has passbands, in the sense defined, for $\beta_2 l_2$ near $(2m + 1)\pi/4$, as expected.

_____ **Example 11.9d** _____
Filtering by Coupled Microstrip Transmission Lines

A somewhat different approach to construction of a bandpass filter is illustrated by Fig. 11.9e. In this, one microstrip line is coupled to another through an intervening microstrip of finite length. For frequencies near those for which the coupling section is resonant, there is large coupling and nearly perfect transfer. For other frequencies, there is a small transfer of energy. Analysis of this, which requires consideration of the distributed couplings, is not given here but measured and calculated curves of insertion loss from a six-section filter of this type are shown in Fig. 11.8f.

_____ **Example 11.9e** _____
Optical Filter

As a final example we consider the optical filter made of alternating layers of dielectric with different refractive indices, as illustrated in Fig. 11.9g. (The equivalent of this in transmission-line form is made by alternating sections of transmission line of different characteristic impedance.) As explained in Sec. 11.1, microwave

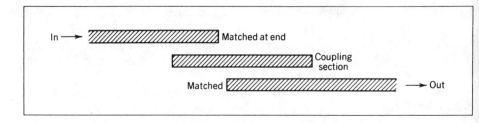

(e)

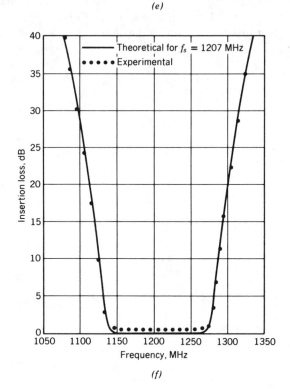

(f)

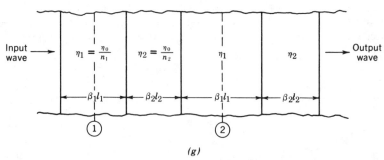

(g)

Fig. 11.9 (*e*) Top view of microstrip with coupling section between input and output transmission lines. (*f*) Insertion loss of parallel-coupled-resonator filter of six sections [S. B. Cohn, *IRE Trans. Microwave Theory Tech.* **MTT-6**, 223 (1958) © 1958 IRE (now IEEE)]. (*g*) Optical filter with periodic alternations of refractive index.

network concepts also apply at optical frequencies. The *ABCD* matrix for the symmetrical section between reference planes 1 and 2 is

$$
\begin{bmatrix} A & B \\ C & D \end{bmatrix} = \begin{bmatrix} \cos\dfrac{\beta_1 l_1}{2} & j\eta_1 \sin\dfrac{\beta_1 l_1}{2} \\[2ex] \dfrac{j}{\eta_1}\sin\dfrac{\beta_1 l_1}{2} & \cos\dfrac{\beta_1 l_1}{2} \end{bmatrix} \begin{bmatrix} \cos\beta_2 l_2 & j\eta_2 \sin\beta_2 l_2 \\[2ex] \dfrac{j}{\eta_2}\sin\beta_2 l_2 & \cos\beta_2 l_2 \end{bmatrix}
$$

$$
\times \begin{bmatrix} \cos\dfrac{\beta_1 l_1}{2} & j\eta_1 \sin\dfrac{\beta_1 l_1}{2} \\[2ex] \dfrac{j}{\eta_1}\sin\dfrac{\beta_1 l_1}{2} & \cos\dfrac{\beta_1 l_1}{2} \end{bmatrix} \tag{8}
$$

from which we obtain

$$
A = D = \cosh\Gamma = \cos\beta_1 l_1 \cos\beta_2 l_2 - \frac{1}{2}\left(\frac{\eta_1}{\eta_2} + \frac{\eta_2}{\eta_1}\right)\sin\beta_1 l_1 \sin\beta_2 l_2 \tag{9}
$$

This is propagating (imaginary Γ) for low frequencies, but eventually reaches an attenuating region with $|\cosh\Gamma| > 1$. There are then a series of attenuation bands and passbands as frequency is increased.

In this section we have considered only the propagation through one unit, and have looked for frequency ranges for which Γ is real (attenuation bands) or imaginary (passbands). As shown in the preceding section, if there are N such symmetric units in cascade, and the last is terminated in Z_c defined by Eq. 11.8(14), the overall attenuation is $N\Gamma$ for the attenuation bands, and N times the phase shift per unit in the pass bands. For a practical filter, the units need be neither symmetric, nor all the same. Moreover, Z_c is in general a function of frequency so that it is not usually possible to match its frequency characteristics exactly to that of the terminating impedance. Reflection losses thus also occur and the overall insertion loss—both in the passbands and attenuation bands—must be considered in the design of the filter. The several different techniques used for designing a filter which approximates a desired characteristic of insertion loss versus frequency are described in the references.[13]

N-PORT WAVEGUIDE JUNCTIONS

11.10   Circuit and *S*-Parameter Representation of *N* Ports

It was shown in Eq. 11.2(10) that the currents and voltages (as defined in Sec. 11.2) are related through impedance parameters,

$$
[V] = [Z][I] \tag{1}
$$

or through admittance parameters,

$$
[I] = [Y][V] \tag{2}
$$

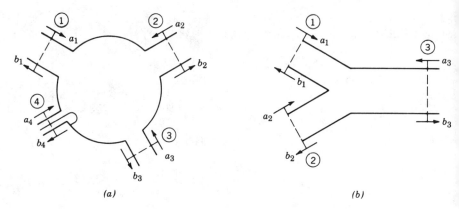

Fig. 11.10 (*a*) General four-port showing incident and reflected waves. (*b*) A Y junction (three port) for combining power from sources ① and ②.

where the admittance and impedance matrices are of order $N \times N$. The scattering coefficients, defined in Sec. 11.6 for a two port, may also be extended to the N port,

$$[b] = [S][a] \tag{3}$$

The column matrix $[b]$ represents the N waves leaving the junction, and $[a]$ the N waves incident upon the junction, as defined in Eq. 11.6(1). The representation is pictured in Fig. 11.10a for a four port. Other linear transformations are of course possible, and an indefinite number of equivalent circuits may be drawn, similar to those of Sec. 11.4, but the three formulations given above have proven the most generally useful. We now consider some specializations of the general N port.

Reciprocal Networks   For networks satisfying reciprocity, each of the matrices $[Z]$, $[Y]$, or $[S]$ is symmetric. That is,

$$Z_{ji} = Z_{ij}, \qquad Y_{ji} = Y_{ij}, \qquad S_{ji} = S_{ij} \tag{4}$$

The proof is easily accomplished by extending from that of the two port. For example, if all ports except for i and j are shorted at the reference planes, there remains a two port between i and j, for which the proof of Sec. 11.3 shows $Y_{ji} = Y_{ij}$ under the conditions of reciprocity, and similarly for the other relations of (4).

Loss-Free Networks   Internal losses may be neglected in many useful microwave junctions, so it is important to know the consequence of negligible loss. The complex Poynting theorem, Eq. 3.13(5), applied to power flow into the N ports gives

$$-\oint (\mathbf{E} \times \mathbf{H^*}) \cdot \mathbf{dS} = \sum_{m=1}^{N} V_m I_m^* = 2W_L + 4j\omega(U_H - U_E) \tag{5}$$

but

$$V_m = \sum_{n=1}^{N} Z_{mn} I_n \tag{6}$$

so

$$\sum_{n=1}^{N} \sum_{m=1}^{N} Z_{mn} I_n I_m^* = 2W_L + 4j\omega(U_H - U_E) \tag{7}$$

Now let all ports be open circuited except the ith,

$$Z_{ii} I_i I_i^* = 2W_L + 4j\omega(U_H - U_E) \tag{8}$$

So for a loss-free network with $W_L = 0$, Z_{ii} is imaginary since $I_i I_i^*$ is real. To study an off-diagonal term, let all ports but i and j be open-circuited.

$$Z_{ii} I_i I_i^* + Z_{jj} I_j I_j^* + Z_{ij} I_j I_i^* + Z_{ji} I_j^* I_i = 2W_L + 4j\omega(U_H - U_E) \tag{9}$$

The first two terms are imaginary, by the above, so if $W_L = 0$,

$$\text{Re}[Z_{ij} I_j I_i^* + Z_{ji} I_j^* I_i] = 0 \tag{10}$$

For a reciprocal network with $Z_{ji} = Z_{ij}$, Z_{ij} is then also imaginary. Similarly the admittance matrix is imaginary for a loss-free network satisfying reciprocity.

To show the properties of the scattering matrix for loss-free networks,

$$\sum_{m=1}^{N} V_m I_m^* = \sum_{m=1}^{N} (a_m + b_m)(a_m^* - b_m^*) = 2W_L + 4j\omega(U_H - U_E) \tag{11}$$

so for a loss-free network,

$$\sum_{m=1}^{N} b_m b_m^* = \sum_{m=1}^{N} a_m a_m^* \tag{12}$$

This equation can be written in matrix form,

$$[b]_t[b^*] = [a]_t[a^*] \tag{13}$$

where $[b]_t$ denotes the transpose of $[b]$, obtained by interchanging rows and columns. In particular the transpose of a column matrix is a row matrix, and it can be checked that the rule for matrix multiplication applied to (13) does give (12). Now substituting (3),

$$([S][a])_t([S][a])^* = [a]_t[a^*] \tag{14}$$

The transpose of a product is the product of transposes with order reversed, so

$$[a]_t[S]_t[S^*][a^*] = [a]_t[U][a^*] \tag{15}$$

where $[U]$ is the unit matrix. Thus

$$[S]_t[S^*] = [U] \tag{16}$$

From this we see that $[S]_t = [S^*]^{-1}$ so (16) may also be written

$$[S^*][S]_t = [U] \tag{17}$$

Matrices for which the transpose is the conjugate of the inverse matrix are called *unitary matrices*. Use of the product rule for matrices show that they have the following properties:

$$\sum_{n=1}^{N} S_{in} S_{in}^* = 1 \tag{18}$$

$$\sum_{n=1}^{N} S_{in} S_{jn}^* = 0, \qquad i \neq j \tag{19}$$

The above relations may be derived directly from conservation of energy, (18) by applying an incident wave to terminal i with all terminals matched, and (19) by applying incident waves to terminals i and j with all terminals matched. Note that reciprocity was not required in the derivation. The relations have important consequences as to what can or cannot be done with loss-free junctions, as will be illustrated with two examples.

─────────────── **Example 11.10a** ───────────────
Limitations on Loss-Free Three Ports

The three-port Y junction pictured in Fig. 11.10b is useful as a power divider or power combiner. It is assumed that it satisfies reciprocity. If sources are introduced at terminals 1 and 2 with the combined power obtained at 3, one might wish to have $S_{12} = 0$ in order to eliminate direct interaction between the two sources. But condition (19), with $i = 1, j = 2$, gives

$$S_{11} S_{21}^* + S_{12} S_{22}^* + S_{13} S_{23}^* = 0 \tag{20}$$

Thus if $S_{12} = 0$, either S_{13} or S_{23} is zero also, eliminating one of the two desired couplings. The junction will act as a power combiner, but there is interaction between the two sources, and if the two sources are not identical in magnitude and phase, one source will tend to feed power to the other.

─────────────── **Example 11.10b** ───────────────
Limitations on Ideal Isolating Networks

The ideal isolator would be a loss-free, one-way transmission line with $S_{12} = 0$ but $S_{21} \neq 0$. It was noted that the unitary property (17) applies to nonreciprocal as well as reciprocal networks. Equation (19) with $i = 1, j = 2$ gives

$$S_{11} S_{21}^* + S_{12} S_{22}^* = 0$$

so that if $S_{12} = 0$, either $S_{21} = 0$ or $S_{11} = 0$, so this ideal also is impossible. We shall see useful isolators employing nonreciprocal elements in Chapter 13, but because of the limitation shown here, they will have dissipative elements to absorb the reflected wave. (Note that $S_{11} = 0$ when $S_{12} = 0$ follows from (18).)

Some consequences of the unitary property to directional couplers and four-port hybrid networks are discussed in the following section.

11.11   Directional Couplers and Hybrid Networks

One of the most important four ports is the directional coupler, designed to couple in a separable fashion to the positively and negatively traveling waves in a guide. Figure 11.11a gives the simplest conception of this device. Imagine a main wave-guide with two small holes placed a quarter-wave apart coupling to an auxiliary guide terminated at each end by a matching resistance and meter as shown. If wave A progresses toward the right, coupled waves from the two holes at terminal 4 follow paths B and C of equal lengths, and the contributions add in that load, its meter indicating the strength of A. The couplings through the two holes cancel at terminal 3, however, since the paths E and D differ in length by a half-wavelength, and the couplings through the two holes are substantially the same in amount if the holes are small. By symmetry, a wave flowing to the left will register at terminal

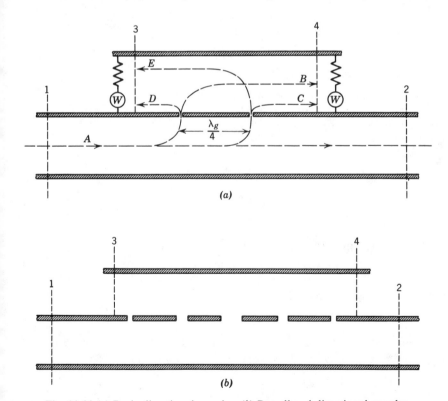

(a)

(b)

Fig. 11.11 (a) Basic directional coupler. (b) Broadband directional coupler.

3 but yield coupled waves which cancel at 4. Thus meter 4 reads the strength of the wave to the right, and meter 3 that to the left.

This simple coupler is frequency sensitive since it depends on the quarter-wave spacing of holes. A like effect with greater bandwidth may be obtained by supplying several holes with properly graded couplings, as illustrated in Fig. 11.11b. Still other embodiments are described in the references. All these couplers may be considered as four ports with the four reference planes as shown in Fig. 11.11a or Fig. 11.11b. Losses may normally be neglected. Several important general properties follow.

To study the properties of the coupler, it is most convenient to use the scattering matrix form of Eq. 11.10(3). It is desired not to couple between 1 and 3 with 2 and 4 matched, so $S_{13} = S_{31} = 0$. It is also desired to have no coupling between 2 and 4 with 1 and 3 matched, so $S_{24} = S_{42} = 0$. Moreover, the ideal directional coupler should be matched so that all the power entering at one terminal divides between the other two for which there is coupling, leaving no reflections at the input. Thus S_{11}, S_{22}, and so on, are zero. The network is also assumed to satisfy reciprocity so that $S_{12} = S_{21}$, $S_{14} = S_{41}$, and so on. Thus the scattering matrix for an ideal directional coupler has been specialized to

$$[S] = \begin{bmatrix} 0 & S_{12} & 0 & S_{14} \\ S_{12} & 0 & S_{23} & 0 \\ 0 & S_{23} & 0 & S_{34} \\ S_{14} & 0 & S_{34} & 0 \end{bmatrix} \tag{1}$$

For negligible loss within the network, power conservation leads to the unitary property of the scattering matrix as shown in the preceding section. Equation 11.10(18), for different values of the index i, gives

$$i = 1: \qquad S_{12}S_{12}^* + S_{14}S_{14}^* = 1 \tag{2}$$

$$i = 2: \qquad S_{12}S_{12}^* + S_{23}S_{23}^* = 1 \tag{3}$$

$$i = 3: \qquad S_{23}S_{23}^* + S_{34}S_{34}^* = 1 \tag{4}$$

$$i = 4: \qquad S_{14}S_{14}^* + S_{34}S_{34}^* = 1 \tag{5}$$

Comparison of (2) and (3) shows that $|S_{14}| = |S_{23}|$ and comparison of (2) and (5) shows that $|S_{12}| = |S_{34}|$. Moreover, reference plane 2 may be selected with respect to 1 so that S_{12} is real and positive, and similarly 4 with respect to 3 so that S_{34} is real and positive. Then

$$S_{12} = S_{34} \triangleq a \tag{6}$$

There remains the Eqs. 11.10(19), also following from the unitary nature of $[S]$. The specializations already made in (1) satisfy these identically except for

$$i = 1, \qquad j = 3: \qquad S_{12}S_{23}^* + S_{14}S_{34}^* = 0 \tag{7}$$

$$i = 2, \qquad j = 4: \qquad S_{12}S_{14}^* + S_{23}S_{34}^* = 0 \tag{8}$$

Use of (6) in either of the above requires $S_{14} = -S_{23}^*$. Reference plane 4 may then be selected with respect to 1 so that S_{14} is real and

$$S_{23} = -S_{14} \triangleq b \tag{9}$$

Thus we have reduced the scattering matrix to the very simple form

$$[S] = \begin{bmatrix} 0 & a & 0 & -b \\ a & 0 & b & 0 \\ 0 & b & 0 & a \\ -b & 0 & a & 0 \end{bmatrix} \tag{10}$$

Note that b gives the coupling from the main guide to the auxiliary guide and is known as the coupling factor (often expressed in decibels). The coefficient a may be called the transmission factor and the two are related by any of the energy relations (2) to (5)

$$a^2 + b^2 = 1 \tag{11}$$

Although a and b are the only two parameters of an ideal directional coupler, real units give some coupling to the terminal for which zero coupling is desired. That is, S_{13} and S_{24} will not be exactly zero for a real coupler. The coupling to the desired terminal in the auxiliary guide, as compared with the undesired terminal, is defined as the *front-to-back ratio* or the *directivity*, usually expressed in decibels.

Several important theorems may be proved for loss-free reciprocal four ports.

1. A four port with two pairs of noncoupling elements is completely matched. That is, the setting of S_{11}, S_{22}, and so on equal to zero was not a separate condition but followed from $S_{13} = 0$, $S_{24} = 0$ because of power relations resulting from the complete Eqs. 11.10(18) and 11.10(19).
2. Any completely matched junction of four waveguides is a directional coupler. (Note that this does not mean that an arbitrary four port may be made into a directional coupler by externally introducing matching transformers, since the adjustment of one of these in such a case disturbs matching for the other ports; it must be an internal property giving $S_{11} = S_{22} = S_{33} = S_{44} = 0$.)
3. A four port with two noncoupling terminals matched is a directional coupler. That is, if $S_{13} = 0$, $S_{11} = 0$, and $S_{33} = 0$, the other properties defined earlier follow.

The Magic T and Other Hybrid Networks   The special case of a directional coupler with $a^2 = b^2 = \frac{1}{2}$ is of particular interest in that it may be used as a bridge or "hybrid" network. (The latter name is taken from the properties of the classical hybrid coil.[15]) One of the most common configurations used for this

---

[15] C. G. Montgomery, R. H. Dicke, and E. M. Purcell, Ref. 4, p. 307.

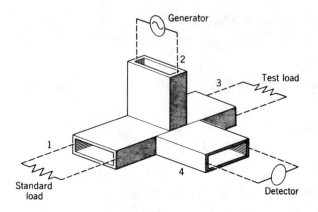

Fig. 11.11c Magic T network as a bridge.

purpose in rectangular guides for the TE_{10} mode is the magic T pictured in Fig. 11.11c. A wave introduced into the "E" arm, 2, will, from considerations of symmetry, divide equally between arms 1 and 3 but not couple to the "H" arm, 4. Conversely a wave introduced into 4 divides between arms 1 and 3 with no coupling to 2. Thus the scattering coefficient S_{24} is zero. By theorem III, above, this becomes a directional coupler if the unit is internally matched so that S_{22} and S_{44} are zero. This matching is normally accomplished by introducing pins or diaphragms or both within the guides near the junction. It then follows from the theorem that S_{11}, S_{33}, and S_{13} are also zero. By symmetry, the transmission and coupling coefficients are equal so that, as stated,

$$a^2 + b^2 = \tfrac{1}{2} \tag{12}$$

A typical use of one of these units is as a bridge. With the generator at 2 and detector at 4 in Fig. 11.11c, no output is observed if the loads on 1 and 3 are equal. Thus a standard load may be placed on 1 and test loads on 3 which are nominally the same. Any deviation from the standard will produce a reading in the detector at 4.

Other configurations accomplishing the same goal are known. One general type utilizes cancellation of waves around a "rat race" structure, as pictured for microstrip in Fig. 11.11d. The division of power and phase changes by the two paths to each outlet produce the cancellations or additions desired. Variations of this with greater bandwidth are possible.[16] Structures based on the same principles can also be made with coaxial lines.

---

[16] See, for example, Section II B of Jeffrey Frey (ed.), *Microwave Integrated Circuits*, Artech House, Inc., Dedham, Mass., 1975.

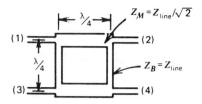

Fig. 11.11d Two-branch hybrid for microstrip.

FREQUENCY CHARACTERISTICS OF WAVEGUIDE NETWORKS

11.12   Properties of a One-Port Impedance

Let us consider a closed region with one waveguide terminal or one port, as in the cavity sketched in Fig. 11.1a or the probe-coupled cavity of Fig. 10.2a. A reference plane 1 is selected far enough from the junction so that only the dominant mode in the guide is important. Application of the complex Poynting theorem leads to Eq. 11.10(8), which for $i = 1$ is

$$Z = R + jX = \frac{2W_L + 4j\omega(U_H - U_E)}{II^*} \tag{1}$$

where W_L is the average power loss in the region, U_E and U_H are average stored energies in electric and magnetic fields, respectively. Similarly for input admittance Y,

$$Y = G + jB = \frac{2W_L + 4j\omega(U_E - U_H)}{VV^*} \tag{2}$$

Certain properties of these impedance and admittance functions will be discussed. Note that the comments apply to the function $Z_{ii}(\omega)$ of a general microwave network, since this would be the input impedance of a two port formed by shorting all but the ith terminal. Similarly, the theorems apply to Y_{ii} and to some other combinations of the impedance or admittance functions.

A study of (1) shows several simple results expected from physical reasoning. Impedance is purely imaginary (reactive) if power loss is zero. When power loss is finite, the real (resistance) part of Z is positive for a passive network. If stored electric and magnetic average energies are equal, reactance is zero and the network is said to be resonant. If average magnetic energy is greater than electric, the reactance is positive (inductive), and if electric energy is the greater, reactance is negative (capacitive). Since power and energy would be the same for positive and

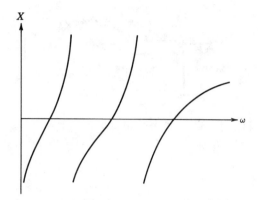

Fig. 11.12 Typical form of reactance versus frequency for a lossless one-port.

negative frequencies, (1) shows that $R(\omega)$ is an even function of frequency and $X(\omega)$ an odd function, if the range is extended to negative frequencies. Similar results can be deduced for the admittance function.

Loss-Free One Ports   When losses are negligible, (1) and (2) become

$$X = \frac{4\omega(U_H - U_E)}{II^*}, \qquad B = \frac{4\omega(U_E - U_H)}{VV^*} \tag{3}$$

Moreover, the variation of X and B with frequency may be related to stored energy through a variational form of the Poynting theorem (Prob. 11.12c). The results are

$$\frac{dX}{d\omega} = \frac{4(U_E + U_H)}{II^*}, \qquad \frac{dB}{d\omega} = \frac{4(U_E + U_H)}{VV^*} \tag{4}$$

It is evident that the average stored energy $(U_E + U_H)$ is positive, and II^* is positive, so the rate of change of reactance with frequency for the lossless one port will be positive. The reactance must then go through a succession of zeros and poles as sketched in Fig. 11.12. Similarly, the susceptance of the lossless one port has a positive slope for the same reason. It follows that the zeros and poles are all simple (first order) and that a zero must lie between two adjacent poles, and vice versa.

These results were first derived by Foster[17] for lumped-element networks, and so are known as Foster's reactance theorem.

Relations Between Real and Imaginary Parts of Impedance or Admittance Functions
If losses are finite, there remain some constraints between the frequency variation of resistance and of reactance, or of conductance and susceptance, just as in the

[17] R. M. Foster, *Bell Syst. Tech. J.* **3**, 259 (1924).

Kronig–Kramers relations between real and imaginary parts of permittivity (Sec. 13.2). This is because of the location of zeros and poles in the complex frequency plane, and the analytic character of the functions in a half-plane.

The complex "frequency" variable commonly utilized is $\alpha + j\omega$. Zeros or poles of Z (or Y) would mean that natural frequencies exist for such values, giving finite solutions without a driving source. For passive networks (those without internal sources of energy), such solutions can only decay from any transient initial state. Thus α must be negative for such natural frequencies of passive networks. In other words the impedance (or admittance) function can have no zeros or poles in the right half ($\alpha + j\omega$) plane. This analytic property permits the relation of real and imaginary parts.

Insight into the properties of circuits as functions of complex frequency can be obtained by considering an analogy with static electric potential and flux functions as complex variables (Sec. 7.5). In potential-function terms, specification of potential everywhere along the $j\omega$ axis determines potential everywhere in the right half-plane, if there are no sources there and potential dies off properly at infinity. Determination of potential at all points permits the finding of electric field, and this in turn permits the construction of a flux function. Conversely, a statement of flux along the $j\omega$ axis defines charge distribution there, and from it the potential is determined everywhere in the source-free region.

If the real part of a function $u + jv$ is defined along the imaginary axis, the imaginary part in a source-free right half-plane[18] (see also Prob. 11.12g) is

$$-v(\alpha, \omega) = \frac{-1}{\pi} \int_{-\infty}^{\infty} \frac{(\omega' - \omega)u(\omega') \, d\omega'}{\alpha^2 + (\omega' - \omega)^2} \tag{5}$$

Thus if $R(\omega)$ is given, this may be used directly to find $X(\omega)$. Several specializations may be used in applying to the network function. Since we wish X for real frequency, α is set equal to zero in (5). $R(\omega')$ is an even function of ω'. Finally we may add the reactance function of any lossless two port in series if we allow idealized elements in the circuit, since such a function is known to have no R. Use of these three points leads to

$$X(\omega) = \frac{1}{\pi} \int_0^{\infty} R(\omega') \left(\frac{1}{\omega' - \omega} + \frac{1}{-\omega' - \omega} \right) d\omega' + X_0(\omega)$$

$$X(\omega) = \frac{2\omega}{\pi} \int_0^{\infty} \frac{R(\omega')}{\omega'^2 - \omega^2} \, d\omega' + X_0(\omega) \tag{6}$$

where $X_0(\omega)$ denotes the reactance function for the lossless part.

Since jZ is also analytic in the complex plane, (5) may also be used with u denoting $-X$ and v denoting R. In this application we again set $\alpha = 0$, and

---

[18] H. Jeffreys and B. S. Jeffreys, *Methods of Mathematical Physics*, 3rd ed., Cambridge University Press, London, 1956.

utilize the fact that X is an odd function of ω. We can also add in series a constant resistance without changing X, so the result is

$$R(\omega) = \frac{1}{\pi} \int_0^\infty - X(\omega') \left(\frac{1}{\omega' - \omega} - \frac{1}{-\omega' - \omega} \right) d\omega' + R_0$$

$$R(\omega) = -\frac{2}{\pi} \int_0^\infty \frac{\omega' X(\omega')}{\omega'^2 - \omega^2} d\omega' + R_0 \tag{7}$$

Similar relations apply to admittance functions, and to relations between magnitude and phase. The analogy to potential and flux functions has allowed the interesting use of electrolytic tanks and resistance paper for the study of network functions.[19] For practical use, one should be able to consider resonances unimportant beyond some upper frequency (i.e., all sources in the potential analog should remain in an accessible part of the plane).

11.13 Equivalent Circuits Showing Frequency Characteristics of One Ports

The functional properties of the impedance (admittance) function developed in the preceding section allow one to develop several equivalent circuits showing the frequency characteristics. We first consider the impedance function, without losses, then the effect of small losses, and finally other forms.

First Foster Form for Loss-Free Case   It was shown that the reactance function of a loss-free one port is an odd function of frequency and must have an infinite number of simple poles. It is known from the theory of functions that such a function can be expanded in a series of "partial fractions" about the poles, provided that the following summation is convergent.[20]

$$X(\omega) = \sum_{n=1}^\infty \left(\frac{a_n}{\omega - \omega_n} + \frac{a_{-n}}{\omega - \omega_{-n}} \right) + \frac{a_0}{\omega} + f(\omega) \tag{1}$$

where a_0/ω represents the pole at zero frequency, if any is present, and $f(\omega)$ is an arbitrary *entire function* (one with no singularities in the finite plane). Since the function is odd, $\omega_{-n} = -\omega_n$ and $a_n = a_{-n}$. Moreover, $f(\omega)$ can have only odd powers of ω, and, since it must behave at most like a simple pole at infinity, it is known to be proportional to the first power of ω. With these specializations, (1) becomes

$$X(\omega) = \sum_{n=1}^\infty \frac{2\omega a_n}{\omega^2 - \omega_n^2} + \frac{a_0}{\omega} + \omega L_\infty \tag{2}$$

[19] W. W. Hansen and O. C. Lundstrom, *Proc. I.R.E.* **33**, 528 (1945).
[20] See, for example, K. Knopp, *Theory of Functions*, Part II, Dover, New York, 1947, Chapter 2.

In (2), a_n is known as the *residue* of the pole ω_n. It may be obtained in terms of the slope of the susceptance curve, which can in turn be related to energy storage. For, in the vicinity of ω_n, the nth term of (2) predominates and

$$B(\omega) = -\frac{1}{X(\omega)} \approx -\frac{\omega^2 - \omega_n^2}{2\omega a_n}$$

Differentiation shows that

$$\left.\frac{dB}{d\omega}\right|_{\omega=\omega_n} = -\frac{1}{a_n}$$

Then, utilizing Eq. 11.12(4)

$$a_n = -\frac{1}{(dB/d\omega)_{\omega=\omega_n}} = -\left[\frac{VV^*}{4(U_E + U_H)}\right]_{\omega=\omega_n} \tag{3}$$

The form of (2) suggests an equivalent circuit consisting of antiresonant LC circuits added in series as shown in Fig. 11.13a, since the nth component of this circuit yields a reactance

$$X_n = -\frac{1}{\omega C_n - 1/\omega L_n} = -\frac{\omega/C_n}{\omega^2 - 1/L_n C_n}$$

Comparison with the foregoing equation yields

$$a_n = -\frac{1}{2C_n} \qquad \omega_n^2 = \frac{1}{L_n C_n} \qquad a_0 = -\frac{1}{C_0} \tag{4}$$

or

$$C_n = -\frac{1}{2a_n} \qquad L_n = -\frac{2a_n}{\omega_n^2} \qquad C_0 = -\frac{1}{a_0} \tag{5}$$

This representation, known as the *first canonical form of Foster*,[21] is then applicable to any lossless one port for which the series in (2) is convergent. To find the circuit, we need to know the antiresonances, with energy storage quantities at those frequencies, both of which quantities were studied for cavity resonators in Chapter 10. The difficult part comes from the fact that the energy must be referred to the voltage in the input guide, see (3), and this requires some specific knowledge of the coupling network. The general representation may be useful for interpretation of measurements and for forming general conclusions even when this coupling problem cannot be solved.

Effect of Losses   The study of losses for practical cavity resonators in the last chapter was concerned with the calculation of a quality factor Q which expressed for a given mode the ratio of energy stored to energy lost per radian. For low-loss

---

[21] R. M. Foster, Ref. 17.

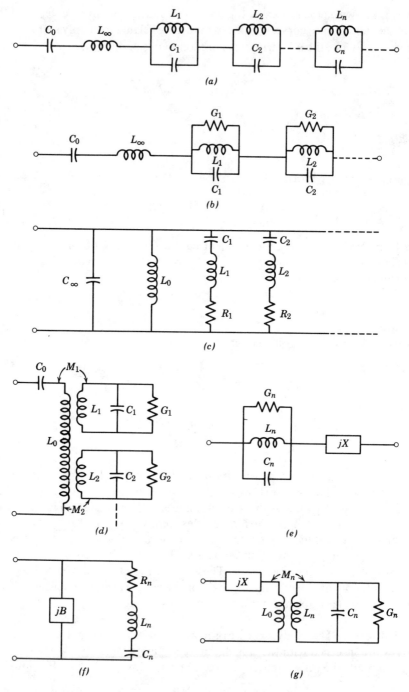

Fig. 11.13 (a)–(g) Various equivalent circuits for one ports.

cavities, it might be expected that the equivalent circuit of Fig. 11.13a would be modified by adding a shunt conductance to each antiresonant element, as shown in Fig. 11.13b. The value of a given conductance G_n would be adjusted so that the Q calculated from the nth antiresonant circuit would agree with the known Q_n of the mode which it represents. That is,

$$G_n = \frac{\omega_n C_n}{Q_n} \tag{6}$$

Justification for this procedure can be supplied by the theory of functions by making approximations appropriate to poles which are at a complex frequency near, but not exactly on, the real frequency axis.

If one accepts this modification of the lumped-circuit equivalent to account for losses, it is clear that the Q of a cavity, determined from energy calculations, is also useful for interpreting the frequency characteristics in the same manner as for a lumped circuit. This fact was stated without justification in Sec. 10.4.

Second Foster Form   An expansion of the susceptance function about its poles yields a form similar to (2):

$$B(\omega) = \sum_{m=1}^{\infty} \frac{2\omega b_m}{\omega^2 - \omega_m^2} + \frac{b_0}{\omega} + \omega C_\infty, \tag{7}$$

where the residues b_m are given by

$$b_m = -\frac{1}{[dX/d\omega]_{\omega=\omega_m}} = -\left[\frac{II^*}{4(U_E + U_H)}\right]_{\omega=\omega_m} \tag{8}$$

When the series is convergent, this equation has the equivalent circuit of Fig. 11.13c (known as the *second Foster canonical form*) with

$$L_m = -\frac{1}{2b_m} \qquad C_m = -\frac{2b_m}{\omega_m^2} \qquad L_0 = -\frac{1}{b_0} \tag{9}$$

Figure 11.13c also shows series resistance added to each resonant circuit to account for small losses, and, as in the foregoing discussion, these are selected to give the known Q for each mode.

$$R_m = \frac{\omega_m L_m}{Q_m} \tag{10}$$

Other Equivalent Circuits   Schelkunoff[22] has shown that other equivalent circuits may be derived by adding convergence factors to the series (2) or (7). These factors are necessary if the original series do not converge, the Mittag–Leffler theorem[23] from the theory of functions telling how they may be formed to insure

[22] S. A. Schelkunoff, *Proc. I.R.E.* **32**, 83 (1944).
[23] K. Knopp, Ref. 20.

convergence. They may also be desirable in other cases where the original series converge, but do so slowly. For example, Schelkunoff has shown that the form with one term of the convergence factor is

$$X(\omega) = \sum_{n=1}^{\infty} 2\omega a_n \left(\frac{1}{\omega^2 - \omega_n^2} + \frac{1}{\omega_n^2} \right) + \frac{a_0}{\omega} + \omega L_0 \qquad (11)$$

Note that, in addition to the convergence factor added in the series, the series inductance term has been modified, and inspection of (11) shows that L_0 is the entire series inductance of the circuit in the limit of zero frequency. The physical explanation of this procedure is then that this low-frequency inductance has been taken out as a separate term rather than being summed from its contributions from the various modes. It is reasonable to expect that this would often help convergence. A specific example for loop coupling to a cavity will be given in the following section.

The equivalent circuit of Fig. 11.13d gives the form of reactance function (11) (loss elements G_n being neglected at first), provided that

$$\frac{M_n^2}{L_n} = -\frac{2a_n}{\omega_n^2}, \qquad \frac{1}{L_n C_n} = \omega_n^2 \qquad (12)$$

Here one imagines the input guide coupled to the various natural modes of the resonator through transformers, which gives a very natural way of looking at a problem of loop coupling to a cavity. Note, however, that one cannot determine the elements of the circuit uniquely since there are three elements, L_n, C_n, M_n, to be determined from the two basic quantities a_n and ω_n for each mode. One of the three may be chosen arbitrarily—perhaps by reference to physical intuition, but any choice will give a circuit which properly duplicates the behavior with respect to impedance at input terminals. Small losses are again accounted for by adding conductances G_n, calculated from form (6), to the circuits as shown in Fig. 11.13d.

Approximations in the Vicinity of a Single Mode   Finally, we note that when we are interested in operation in the vicinity of the natural frequency for one mode, other resonances being well separated, the dominant factor will be the one representing that mode. Other terms will vary only slowly with frequency over this range and may be lumped together as a constant impedance or admittance (predominantly reactive). The equivalent circuits of Figs. 11.13b, c, d then reduce to the simplified representations of Fig. 11.13e, f, g, respectively. This is an important practical case, enabling one to use simplified lumped-element circuit analysis for the study of cavity resonator coupling problems.

11.14   Examples of Cavity Equivalent Circuits

Two examples will be given to clarify the calculation of element values in the equivalent circuits of Sec. 11.13. It should be stressed again that the difficult part comes in solving enough of the coupling problem to refer the energy quantities within the resonator to defined voltage or current in the guide. In the first example, a uniform line is considered so that energy can be expressed directly in terms of the input current. In the second example, a reasonable approximation to the coupling problem can be made.

_____ **Example 11.14a** _____
Open-Circuited Transmission Line

Let us consider a lossless open-circuited line of length l, inductance L per unit length, and capacitance C per unit length (Fig. 11.14a). We shall derive the second Foster form, Fig. 11.13c, starting from Eq. 11.13(7). For this calculation we need the natural modes having infinite susceptance at the input. Current is then a maximum at the input, zero at $z = l$, and length must be an odd multiple of a quarter-wavelength.

$$I_m(z) = I_{0m} \cos \omega_m z \sqrt{LC} \tag{1}$$

$$\omega_m = \frac{2\pi}{\lambda_m \sqrt{LC}} = \frac{2\pi m}{4l\sqrt{LC}} \quad (m \text{ odd}) \tag{2}$$

The sum of average U_E and U_H is equal to the total energy stored at resonance, which may be computed as maximum energy in magnetic fields.

$$U_E + U_H = (U_H)_{\max} = \int_0^l \frac{LI_{0m}^2}{2} \cos^2 \omega_m z \sqrt{LC}\, dz = \frac{lLI_{0m}^2}{4} \tag{3}$$

Substitution in Eq. 11.13(8) gives the residue for the mth mode.

$$b_m = -\frac{I_{0m}^2}{4(U_E + U_H)} = -\frac{1}{Ll} \tag{}$$

Inductance and capacitance for the mth circuit are found from Eq. 11.13(9)

$$L_m = -\frac{1}{2b_m} = \frac{Ll}{2} \tag{4}$$

$$C_m = -\frac{2b_m}{\omega_m^2} = \frac{8Cl}{\pi^2 m^2} \tag{5}$$

This leads to the equivalent circuit of Fig. 11.14b, which is valid for all frequencies, provided the equation for $B(\omega)$ obtained from Eq. 11.13(7) is convergent. The

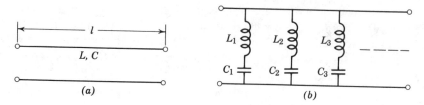

Fig. 11.14 (*a*) Open-circuited ideal line and (*b*) equivalent circuit.

series is convergent in this case, and in fact can be shown to be equivalent to the following closed form:

$$B(\omega) = - \sum_{m \text{ odd}} \frac{2\omega}{Ll[\omega^2 - m^2\pi^2/4l^2LC]} = \sqrt{\frac{C}{L}} \tan \omega l\sqrt{LC} \tag{6}$$

The last expression can be recognized as the input susceptance for an open-circuited ideal line obtained from simple transmission line theory, as it should be.

Example 11.14b
Loop-Coupled Cavity

For a second example, we shall return to the loop-coupled cylindrical cavity discussed in Sec. 10.10 from an energy point of view. In particular, we shall concern ourselves with behavior in the vicinity of resonance for the simple TM_{010} mode, all other resonances being well separated, so that one of the approximate forms of Fig. 11.13e, f, or g is appropriate, The form of Fig. 11.13g, arising from Eq. 11.13(11), is particularly useful because the self-inductance of the loop is separated out, and the remaining series may be thought of as representing more nearly the behavior of the unperturbed cavity. From the physical point of view, it is a natural equivalent circuit, since we picture the input line as being coupled to the cavity mode through a mutual inductance which represents the loop.

The voltage at the loop terminals (computed with no self-inductance drop, as is appropriate for the zero current of antiresonance) is found approximately by taking magnetic field of the unperturbed mode flowing through the small loop of area S, as in Sec. 10.10.

$$V = j\omega\mu HS \tag{7}$$

Energy stored in the mode from Sec. 10.6 may be written

$$(U_E + U_H) = (U_H)_{\max} = \tfrac{1}{2}\pi\mu \, dH^2 a^2 \tag{8}$$

Substitution in Eq. 11.13(3) gives the residue for the mode:

$$a_1 = -\frac{VV^*}{4(U_E + U_H)_{\omega=\omega_1}} = -\frac{\omega_1^2 \mu^2 S^2}{2\pi a^2 d} \tag{9}$$

Resonant frequency and Q are known from the analysis of Sec. 10.6.

$$\omega_1 = \frac{p_{01}}{a\sqrt{\mu\varepsilon}} \tag{10}$$

$$Q_1 = \frac{\eta p_{01} d}{2R_s(d + a)} \tag{11}$$

Here we meet the indeterminacy of the form selected, for we have three quantities, a_1, ω_1, Q_1 to determine four quantities, M_1, L_1, C_1, and G_1. As pointed out before, one of the four may be selected arbitrarily and the same input impedance will result. One choice is to calculate the conductance G_1 from power loss and voltage across the center. This makes sense, for example, when an electron beam is to be shot across the center, in which case a beam admittance, calculated on the same basis, can simply be placed in parallel with G_1 in the equivalent circuit. This can be shown to be

$$G_1 = \frac{R_s}{\eta^2} \frac{2\pi a(d + a)}{d^2} J_1^2(p_{01}) \tag{12}$$

Application of Eqs. 11.13(6) and 11.13(12) then yields

$$C_1 = \frac{Q_1 G_1}{\omega_1} = J_1^2(p_{01}) \left(\frac{\pi a^2 \varepsilon}{d}\right) \tag{13}$$

$$L_1 = \frac{1}{\omega_1^2 C_1} = \frac{\mu d}{\pi p_{01}^2 J_1^2(p_{01})} \tag{14}$$

$$M_1 = \left(-\frac{2a_1 L_1}{\omega_1^2}\right)^{1/2} = \frac{\mu S}{\pi a p_{01} J_1(p_{01})} \tag{15}$$

Input impedance, computed at resonance for the unperturbed mode $(\omega^2 L_1 C_1 = 1)$, can then be shown to yield the same result as was found in Eq. 10.10(3) by energy considerations.

$$Z = j\omega L_0 + \frac{\omega^2 M^2}{j\omega L_1 + 1/(G_1 + j\omega C_1)} \approx j\omega L_0 + \frac{(\omega\mu S)^2}{2\pi a R_s(d + a)} \tag{16}$$

11.15   Circuits Giving Frequency Characteristics of N Ports

In considering frequency characteristics of the type discussed in Sec. 11.12 for a one port, one notes that any coefficient Z_{ii} of an N port must satisfy the conditions for a one port, since this represents the input impedance for a one port formed by shorting all but the ith terminal. Similarly any Y_{ii} must satisfy conditions for the admittance function for a one port. The transfer coefficients, however, need not satisfy such conditions.

When the N port is a cavity resonator with more than one waveguide coupled to it, representation of frequency charcteristics by an equivalent circuit may be desirable, generalizing Sec. 11.13. Thus an extension of the form Fig. 11.13d would naturally lead to each waveguide terminal coupled to each of the normal modes by a mutual inductance as illustrated in Fig. 11.15a for a cavity with two ports.

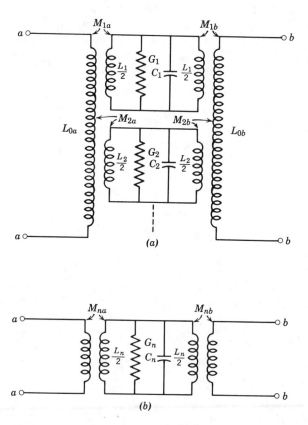

Fig. 11.15 (a) Equivalent circuit for a cavity coupled to two waveguide terminals, and (b) approximation in the vicinity of one resonant mode.

Justification for this procedure has been supplied by Schelkunoff[24] from complex function theory, and by Slater[25] by normal mode theory. In the vicinity of a resonance, the circuit simplifies to that of Fig. 11.15b.

If we wish to determine the characteristics of an N port by measurement, the simplest procedure is usually that of terminating all but two of the ports in known impedances, leaving a two port for which four of the parameters may be obtained by the methods of Sec. 11.5. Repetition of this procedure for different pairs of ports will eventually give all the coefficients.

JUNCTION PARAMETERS BY ANALYSIS

11.16   Quasistatic and Other Methods of Junction Analysis

It has been noted that we do not require the complete field solution for a network formulation. Nevertheless, solution of the boundary value problem for fields has been useful in obtaining the network representations for certain junctions. The results are largely tabulated in handbooks.[26, 27] A brief discussion of the approach to such problems may be helpful in the proper use of the tabulated results.

For junctions small in comparison with wavelength, it may be possible to set down a reasonably good equivalent circuit from quasistatic reasoning. The basis for this follows from the Helmholtz equation,

$$\frac{\partial^2 E}{\partial x^2} + \frac{\partial^2 E}{\partial y^2} + \frac{\partial^2 E}{\partial z^2} + \left(\frac{2\pi}{\lambda}\right)^2 E = 0 \tag{1}$$

Thus if variations arise from changes in dimensions x, y, or z that are small in comparison with wavelength, the first terms dominate over the last and the equation reduces to Laplace's equation giving static forms for the field solutions. As an example, consider the step in the parallel-plane line of Fig. 11.16a. The electrostatic solution of this problem is known (Sec. 7.7), and a "fringing" or "excess" capacitance may be found as the excess of total capacitance between electrodes over that which would exist if field lines were straight across. Letting $\alpha = a/b$, one has

$$C_d = \frac{\varepsilon}{\pi}\left[\left(\frac{\alpha^2 + 1}{\alpha}\right)\ln\left(\frac{1 + \alpha}{1 - \alpha}\right) - 2\ln\left(\frac{4\alpha}{1 - \alpha^2}\right)\right] \text{ F/m width} \tag{2}$$

[24] S. A. Schelkunoff, Ref. 22.
[25] J. C. Slater, *Microwave Electronics*, Van Nostrand, Princeton, N.J., 1950.
[26] T. Moreno, *Microwave Transmission Design Data*, McGraw-Hill, New York, 1948.
[27] N. Marcuvitz, Ref. 2.

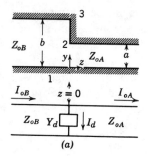

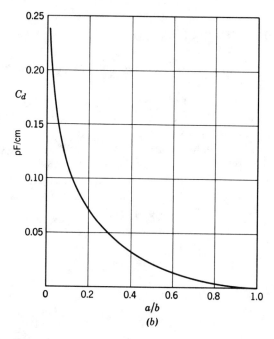

Fig. 11.16 (a) Step discontinuity in parallel-plane transmission line and exact equivalent circuit. (b) Curve of discontinuity capacitance per unit width for (a).

The curve for this is plotted in Fig. 11.16b. This capacitance is placed in shunt with the two transmission lines at the junction. As we shall see later, the shunt representation is exact. The use of

$$Y_d \approx j\omega C_d \tag{3}$$

is a good approximation if the transverse dimension b is less than about 0.2λ.

A second example in which quasistatic approximations are useful is the right-angle bend in a parallel-plane line illustrated in Fig. 11.16c. Physical reasoning

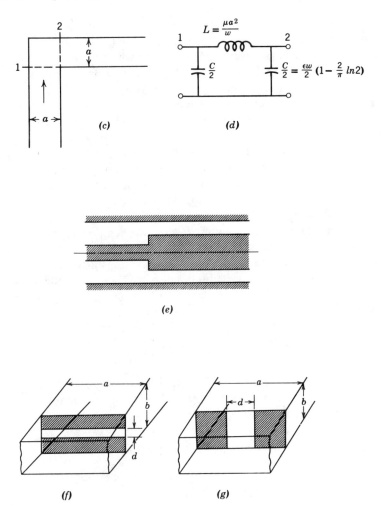

Fig. 11.16 (c), (d) Right-angle bend in parallel-plane transmission line and equivalent circuit. (e) Typical discontinuity in coaxial line. (f) Capacitive diaphragm in rectangular guide. (g) Inductive diaphragm in rectangular guide.

leads us to include an inductance to account for the Faraday's law difference in voltage between planes 1 and 2, and this may be estimated by assuming magnetic field uniform in the corner. There is also an excess capacitance as in the preceding example, and this may be divided equally because of the symmetry. There results the π equivalent circuit as shown in Fig. 11.16d with the element values there indicated and w, width into the page.

Let us return to Fig. 11.16a and argue more carefully the case for the shunt admittance as an exact representation of the junction effect. In an approximate transmission line treatment, it is common to consider this as two lines of different characteristic impedance joined at $z = 0$. Such a treatment, however, considers only the TEM or principal transmission-line waves which have E_y and H_x with no variations in y. The perfect conductor portion between points 2 and 3 requires that $E_y = 0$ here. If there were only principal waves, E_y would then have to be zero everywhere at $z = 0$ because of the lack of variations with y in the principal wave. There could then be no energy passing into the second line A regardless of its termination since the Poynting vector would then also be zero across the entire plane, $z = 0$. The difficulty is met by the higher order waves which are excited at the discontinuity, so that E_y in the principal waves is not generally zero at $z = 0$, but total E_y (sum of principal and higher order components) is zero from 2 to 3 but not from 1 to 2. For the example of Fig. 11.16a, the higher order waves excited are TM waves, since E_y, E_z, and H_x alone are required in the fringing fields. For spacings between planes small compared with wavelength, these waves are far below cutoff, so that their fields are localized in the region of the discontinuity.

To show that the effect of these local waves on the transmission of the principal waves may be expressed as a lumped admittance placed at $z = 0$ in the transmission line equivalent circuit, as in Fig. 11.16a, consider that current at any value of z may be expressed as one part $I_0(z)$ from the principal wave and a contribution $I'(z)$ from all local waves.

$$I(z) = I_0(z) + I'(z) \qquad (4)$$

Now *total* current must be continuous at the discontinuity $z = 0$, but current in the principal wave need not be, since the difference in principal-wave currents may be made up by the local-wave currents.

$$I_{0A}(0) + I'_A(0) = I_{0B}(0) + I'_B(0)$$

or

$$I_{0B}(0) - I_{0A}(0) = I'_A(0) - I'_B(0) \qquad (5)$$

Total voltage in the line as defined by $-\int \mathbf{E} \cdot \mathbf{dl}$ between planes, however, is only that in the principal wave, since a study of the local waves shows that their contribution is zero.

$$V(z) = V_0(z)$$

Continuity of total voltage across the discontinuity $z = 0$ then requires continuity of voltage in the principal wave.

$$V_{0A}(0) = V_{0B}(0) = V_0(0)$$

Now, if an equivalent circuit is drawn for the principal wave only, its continuity of voltage but discontinuity of current may be accounted for by a lumped discontinuity admittance at $z = 0$, the current through this admittance being

$$I_{0B}(0) - I_{0A}(0) = I_d = Y_d V_0(0)$$

Or, from (5),

$$Y_d = \frac{I'_A(0) - I'_B(0)}{V_0(0)} \tag{6}$$

The complete analysis[28] reveals that, when local-wave values are substituted in (6), numerical values of Y_d may be calculated which are independent of terminations so long as these are far enough removed from the discontinuity not to couple to the local-wave fields. For Fig. 11.16a the shunt admittance acts as a pure capacitance if transverse dimensions are negligible compared with wavelength, and the value is accurately given by Fig. 11.16b. Corrections are needed when transverse dimension is comparable with wavelength.[28]

Results are available[29] for several forms of coaxial discontinuity, carried out by an important series method as formulated by Hahn.[30] To a fair approximation, discontinuity capacitance for Fig. 11.16e may be found by multiplying values from Fig. 11.16b by outer circumference. If the step is in the outer conductor, values from Fig. 11.16b are multiplied by inner circumference.

Schwinger[31] and colleagues applied many of the powerful methods for boundary value problems to waveguide discontinuities including integral equation formulations and variational methods—probably the most powerful approximate methods for attacking wave problems of many types. The variational methods are described in several texts.[32,33] Approximate solution of the integral equations leads to approximate but useful forms for the two waveguide discontinuities shown in Figs. 11.16f and 11.16g. For the diaphragm extending from top and bottom of a rectangular guide propagating the TE_{10} mode (Fig. 11.16f), the energy of the higher order modes is predominantly capacitive. The susceptance for a symmetrical diaphragm of gap d in a guide of width a, height b, is approximately

$$\frac{B}{Y_0} = \frac{4b}{\lambda_g} \ln \csc \frac{\pi d}{2b} \tag{7}$$

[28] J. R. Whinnery and H. W. Jamieson, *Proc. IRE* **32**, 98 (1944).

[29] J. R. Whinnery, H. W. Jamieson, and T. E. Robbins, *Proc. IRE* **32**, 695 (1944).

[30] W. C. Hahn, *J. Appl. Phys.* **12**, 62 (1941).

[31] N. Marcuvitz and J. Schwinger, *J. Appl. Phys.* **22**, 806 (1951), and in unpublished work.

[32] R. E. Collin, *Field Theory of Guided Waves*, McGraw-Hill, New York, 1960.

[33] R. F. Harrington, *Time-Harmonic Electromagnetic Fields*, McGraw-Hill, New York, 1961.

For the diaphragm extending from the side walls as in Fig. 11.16g, higher order modes give a net stored magnetic energy and the corresponding inductive susceptance, with d the gap width in this case also, is

$$\frac{B}{Y_0} = -\frac{\lambda_g}{a}\cot^2\frac{\pi d}{2a} \tag{8}$$

PROBLEMS

11.2a Show that the functions **f** and **g** defined by Eqs. 11.2(1) and 11.2(2) are in general related as follows:

$$\mathbf{g}(x, y) = \left(\frac{Z_0}{Z_z}\right)\hat{z} \times \mathbf{f}(x, y)$$

Show that the values obtained for the rectangular guide satisfy this relation.

11.2b Use the principles of Sec. 11.2 to obtain definitions of voltage and current for the TE_{01} mode in a circular cylindrical guide.

11.2c* Repeat Prob. 11.2b for the TE_{11} mode in circular guide; for the TM_{11} mode in rectangular guide.

11.2d With voltage defined as the integral of maximum electric field between top and bottom of the TE_{10} mode in rectangular guide, and current as the total axial flow in the top surface, compare with the derived quantities of Sec. 11.2. How is the product VI related to power flow in this case?

11.2e Supply the proof of the uniqueness theorem cited in Sec. 3.14 and utilized in Sec. 11.2. To do this, assume that there are two possible solutions, $(\mathbf{E}_1, \mathbf{H}_1)$ and $(\mathbf{E}_2, \mathbf{H}_2)$, and apply the Poynting theorem to the difference field $(\mathbf{E}_1 - \mathbf{E}_2, \mathbf{H}_1 - \mathbf{H}_2)$. Note Sec. 1.17 for a typical uniqueness argument.

11.2f Suppose that an N port has a load impedance Z_L connected to the terminals 1, and voltage generators connected to the other $N - 1$ terminals. Show that the following Thévenin equivalent circuits are valid *so far as calculations of effects in the load are concerned*:

(i) A voltage generator V_0 connected to Z_L through a series impedance Z_g. V_0 is the voltage produced at terminals 1 with these terminals open circuited, and Z_g is the impedance seen looking into 1 with all voltage generators short circuited (and any current generators open circuited).

(ii) A current generator I_0 connected across Z_L with internal admittance Y_g in parallel. I_0 is the current that would flow at terminals 1 if these terminals were shorted, and $Y_g = 1/Z_g$.

11.3a Verify Eq. 11.3(2) for isotropic but not necessarily homogeneous media.

11.3b* Show that Eq. 11.3(2) does not apply for anisotropic media with the matrix representing permittivity or permeability asymmetric.

11.3c Complete a proof similar to that of Sec. 11.3 to show that $Z_{21} = Z_{12}$.

11.4a For $Z_{11} - Z_{12} = j2$, $Z_{22} - Z_{12} = j5$, $Z_{12} = j$, find the admittance coefficients, the π circuit, and the $ABCD$ constants.

11.4b For the numerical values of Prob. 11.4a, obtain the values in the equivalent circuit of Fig. 11.4c.

11.4c For a terminating impedance of 1 Ω, find input impedance using all the forms of Probs. 11.4a and 11.4b.

11.4d* Set up the relation between currents and voltages for Fig. 11.4d, and from these determine the impedance parameters in terms of Z_{01}, $\beta_1 l_1$, m, and B.

11.5a For a network in which it is possible to relate current and voltage on one side of the network to either quantity on the other side (including phase), describe simple measurements for determining the $ABCD$ constants of Eq. 11.4(3).

11.5b If one selects the point of minimum slope, P' of Fig. 11.5b to determine x_0, y_0 and equates this slope to $Z_{02}/m^2 Z_{01}$, a second correct representation results. Show that transformations calculated by the latter are equivalent to those from the representation described.

11.5c* For the numerical values of Prob. 11.4a, plot an "S" curve as in Fig. 11.5b. Show that the values of m^2, x_0, and y_0 agree with those calculated in Prob. 11.4b.

11.6a Relate the scattering coefficients to the admittance coefficients to obtain equations similar to Eqs. 11.6(9).

11.6b Imagine that the source of energy is introduced at reference 2 in Fig. 11.6, with guide 1 perfectly terminated so that $a_1 = 0$. Derive the condition corresponding to Eq. 11.6(8). (Note that although these conditions are derived for special terminations of the network, they relate the basic parameters and must hold, quite apart from the termination and driving conditions.)

11.6c Show that the condition for reciprocity in terms of the transmission coefficients is $T_{11}T_{22} - T_{12}T_{21} = 1$.

11.6d* The matrix equivalents of Eqs. 11.6(9) and Prob. 11.6a are

$$[S] = ([\bar{Z}] - [U])([\bar{Z}] + [U])^{-1} = ([U] - [\bar{Y}])([U] + [Y])^{-1}$$

where $\bar{Z}$ is normalized Z, i.e., $\bar{Z}_{nn} = Z_{nn}/Z_{0n}$, $\bar{Z}_{nm} = Z_{nm}/\sqrt{Z_{0n}Z_{0m}}$, and similarly for $\bar{Y}$. $[U]$ is the unit matrix. Derive these from matrix statements of the several formulations.

11.6e Find scattering and transmission coefficients for the network with numerical values given in Prob. 11.4a. Utilize the results to find input reflection coefficient if the output is matched.

11.8a Derive the transmission parameters, T_{11}, and so on, for each of the elements of Table 11.8.

11.8b In a transmission line with shunt element, as in Ex. 11.8a, let the line be lossless so that $\gamma = j\beta$ and the admittance be a pure reactance, $Y = jB$. Write the $ABCD$ parameters for one section and for N sections of such a line.

11.8c Repeat Prob. 11.8b for a lossless transmission line with series reactances replacing the shunt susceptances.

11.8d A dielectric window is formed by two adjacent dielectric slabs, ε_1 of length l_1, and ε_2 of length l_2, placed with air (ε_0) on each side. Find the transmission parameters for the overall unit. (Note that it is convenient to define all lines with unity Z_0, representing impedance changes by ideal transformers.) If there is only an outward propagating wave on the right, determine the conditions for no reflections to the left.

11.8e* Equation 11.8(13) for propagation constant may also be derived by a difference-equation approach. Assuming that voltage and current at the nth output are $\exp(-n\Gamma)$ times the values at the beginning, relate quantities at the nth, $(n + 1)$st, and $(n - 1)$st and derive the desired relation.

11.9a For the filter of Ex. 11.9a, take $C_d = 1$ pF, $Y_0 = 0.02$ S, $l = 1$ cm and find the upper cutoff frequency of the first passband, and upper and lower cutoff frequencies of the next higher passband.

11.9b A coaxial transmission line has periodic gaps, distance l apart, in the inner conductor. These act as series capacitors C. Find the expression for Γ and comment on the nature of the passbands and stop bands.

11.9c Derive Eq. 11.9(9) from 11.9(8). Note that for $\beta_1 l_1$ and $\beta_2 l_2$ each a multiple of π, one has real solutions for Γ. Explain in terms of the wave impedance transformation for such electrical lengths. For $\eta_2 = \eta_1(1 + \delta)$ show that to first order in δ, $\Gamma = j(\beta_1 l_1 + \beta_2 l_2)$. Explain.

11.10a Show that Eq. 11.10(12) is consistent with 11.10(13).

11.10b Show that Eqs. 11.10(18) and 11.10(19) follow from 11.10(16) or 11.10(17). Verify the statement that reciprocity is not required for these relations.

11.10c Derive Eqs. 11.10(18) and 11.10(19) from conservation of energy as suggested in the paragraph following 11.10(19).

11.10d For the Y junction of Fig. 11.10b, show that for symmetrical sources ($a_1 = a_2$) and port 3 matched so that $a_3 = 0$, the junction does act as a power combiner with all power incident on 1 and 2 exiting at 3. Assume symmetry ($S_{11} = S_{22}, S_{13} = S_{23}$) and that $S_{33} = 0$.

11.11a Prove theorem 1 [following Eq. 11.11(11)] for directional couplers.

11.11b Prove theorem 2 [following Eq. 11.11(11)] for directional couplers.

11.11c Prove theorem 3 [following Eq. 11.11(11)] for directional couplers.

11.11d For the "rat-race" hybrid of Fig. 11.11d, give physical arguments to explain why a signal introduced at port 2 gives no response at 4 when ports 1 and 3 are equally loaded.

11.12a It was noted in Sec 11.12 that diagonal terms of the impedance matrix for an N port have the properties of one-port impedance. To show that off-diagonal terms do not have all these properties, demonstrate that Z_{12} for length l of a loss-free transmission line does not satisfy Foster's reactance theorem.

11.12b Illustrate by example that $dX/d\omega$ is not always positive for lossy networks.

11.12c* Suppose that a small variation $\delta\omega$ in frequency produces variations $\delta \mathbf{E}$ and $\delta \mathbf{H}$ in fields. Starting from Maxwell's equations, show that

$$\oint_S (\mathbf{E} \times \delta\mathbf{H} - \delta\mathbf{E} \times \mathbf{H}) \cdot d\mathbf{S} = j\delta\omega \oint_V (\mu H^2 - \varepsilon E^2)\, dV$$

11.12d Apply the result of Prob. 11.12c to derive Eq. 11.12(4) for a lossless network with one waveguide input.

11.12e For a simple circuit with resistance R and capacitance C in parallel, show that $R(\omega)$ and $X(\omega)$ satisfy Eqs. 11.12(6) and 11.12(7).

11.12f If Z is analytic in the right half-plane, $\ln Z$ is also. Utilize this fact and the form of Eq. 11.12(5) to relate magnitude and phase of a passive network, obtaining equations analogous to Eqs. 11.12(6) and 11.12(7).

11.12g* Although Eq. 11.12(5) may be derived in a variety of ways, one interesting point of view is through potential theory. Let $u(\omega)$ correspond to a flux function given along the $j\omega$ axis. By the Cauchy–Riemann equations, $du/d\omega$ determines $dv/d\alpha$ for $\alpha = 0$, which is the normal electric field entering that plane. This may be interpreted to arise from a surface charge distribution $\rho_s = -2\, dv/d\alpha$ along the plane $\alpha = 0$. (Question: Why the factor of 2 when it is absent if the plane is a conductor?) An element of charge $\rho_s\, d\omega'$ at ω' may be thought of as a line charge perpendicular to the plane, for which the potential at point α, ω can be found. Write an expression for the potential and integrate by parts to obtain Eq. 11.12(5). What conditions did you have to assume at infinity for the result to be valid?

11.12h To what does scattering matrix reduce for a one port? Give the properties of this similar to those discussed in Sec. 11.12 for impedance or admittance.

11.13 Modify the form of $B(\omega)$, Eq. 11.13(7), to correspond to the form 11.13(11) with a convergence factor added, and find an equivalent circuit representation. *Hint*: Change the form of Fig. 11.13d by using T networks to replace the transformers, and look for the dual of this circuit.

11.14a Derive the equivalent circuit of Fig. 11.13a for a shorted length l of ideal line. Show that the series form for $X(\omega)$ converges to the usual expression for reactance of a shorted ideal line.

11.14b Derive the first Foster form for the open line and the second Foster form for the shorted line. Show that the series forms converge to proper expressions for reactance and susceptance, given that

$$\cot x = \frac{1}{x} + \sum_{n=1}^{\infty} \frac{2x}{x^2 + n^2\pi^2}$$

11.14c Show that the capacitance C_1 derived for the loop-coupled cavity of Ex. 11.14b is that which would be obtained by referring energy stored in electric fields to voltage at the center of the cavity. Compare the mutual M_1 to that which would be derived by referring induced voltage in the loop to total vertical current in the cavity wall.

11.14d Derive an equivalent circuit similar to Fig. 11.13e for coupling to the TM_{010} cylindrical mode by a small probe of length s extending axially from the top at the center. Assume that induced voltage is probe length times electric field of the unperturbed mode.

11.15a* Suppose that two well-separated loops, each of area 0.5 cm^2, are coupled to the TM_{010} cylindrical mode as for one loop in Sec. 11.14. Draw the equivalent circuit, and calculate element values (except L_0) for an air-filled cavity resonant at 4 GHz with $d = 1$ cm, $R_s = 0.02$ ohm.

11.15b* The Q of a cavity mode is sometimes measured by finding the curve of transmission versus frequency between two guides coupled to the cavity. Using the values of Prob. 11.15a, calculate from the equivalent circuit the transmission at resonance and at a frequency $f_0(1 + 1/Q)$, $Q \gg 1$, and compare with the ratio $\sqrt{2} : 1$. Take both lines of 50 ohms impedance and assume self-inductances of loops are tuned out. Use a series resonant circuit in the coupling loop.

11.16a Determine the form of the proper local waves in the example of Fig. 11.16a. Show that voltage between planes, $-\int \mathbf{E} \cdot \mathbf{dl}$, is zero for each of these.

11.16b Imagine a parallel-plane transmission line with two steps such as the one in Fig. 11.16a. The first is from spacing b to spacing a; the second is removed from the first by a half-wavelength and is from spacing a back to b. The line to the right of b is perfectly terminated by its characteristic impedance, Z_{0B}. If it were not for the discontinuity capacitances, the line to the left of the first discontinuity would also be perfectly terminated. Calculate reflection coefficient in this line, taking into account the discontinuity capacitances from Fig. 11.16b. Take $a = 1$ cm, $b = 2$ cm, $\lambda = 12$ cm. Assume air dielectric.

11.16c Using Fig. 11.16b, calculate an approximate discontinuity capacitance for the coaxial line of Fig. 11.16e. Take $r_1 = 0.5$ cm, $r_2 = 1$ cm, $r_3 = 1.2$ cm. Assume air dielectric.

11.16d A rectangular waveguide of dimensions 0.900 in. $\times$ 0.400 in. propagating the TE_{10} mode absent if the plane is a conductor?) An element of charge $\rho_s \, d\omega'$ at ω' may be thought of as a

minimum 0.55 cm in front of the horn entrance. Find the dimensions and placing of a capacitive diaphragm in order to produce a match of waves approaching from the left.

11.16e Repeat Prob. 11.16d, using an inductive diaphragm.

11.16f* Reason physically as to the field components required in higher order modes for the two waveguide discontinuities of Figs. 11.16*f* and 11.16*g*, and the type of energy storage predominant in each.

11.16g For $a \gg \lambda/2$, the capacitive waveguide discontinuity reduces to the corresponding capacitive diaphragm problem in a parallel-plane transmission line. Express Eq. 11.16(7) as a capacitance per unit width in this limit, and plot as a function of d/b. Compare with the step capacitance of Fig. 11.16*b*.

12

RADIATION

12.1 Introduction

Radiation of electromagnetic energy from a circuit, cavity resonator, or wave-guiding system may be important either as an undesired leakage phenomenon or as a desired process for exciting waves in space. In the former case one wishes to minimize the power lost by radiation, and this may be done by changing circuit configuration or by adding shielding. When radiation is desired, the goal is to excite waves from the given source in the required direction or directions, as efficiently as possible. The system that acts as the transition or matching unit between the source and waves in space is known as the *radiator* or *antenna*. The primary stress in this chapter will be on antenna principles, with radiation as the desired result.

In the design of antenna systems, some or all of the following information may be required:

1. The relative field strength for various directions (called the antenna pattern).
2. The total power radiated when the radiator is excited by a known voltage or current.
3. The input impedance of the radiator for matching purposes.
4. The bandwidth of the antenna with respect to any of the above properties.
5. The radiation efficiency, or ratio of power radiated to the total of radiated and dissipated power.
6. For high-power antennas, the maximum field strengths at key positions in air or dielectrics that may cause corona or dielectric breakdown.

577

The straightforward approach to any of the above questions might seem to be that of solving Maxwell's equations subject to the boundary conditions of the radiator, and at infinity. This is possible in a few cases (to be discussed in a later section), but most antenna configurations are too complicated for this direct approach, and approximations must be made. In order to make such approximations, it is important to have good physical pictures of radiation phenomena.

One physical picture of radiation is that found in Chapter 4 when we considered circuits that are comparable in size with wavelength. It was found that retardation effects from one part of the circuit to another cause a phase shift so that induced effects that are only reactive in a small circuit, now have a real part also. This real part represents power radiated from the circuit. From this point of view, an antenna is a circuit comparable with wavelength in size, purposely designed to maximize this retardation effect.

A picture closely related to the above but perhaps more direct follows from an examination of fields at a distance from the radiating system. If the system is small in size in comparison with wavelength, fields from positive and negative charges of the system (or oppositely directed current contributions) nearly cancel at large distances. For example, we found fields from a quasistatic electric dipole or small current loop falling off at least as fast as the inverse square of distance from the dipole or loop. But for larger circuits (in comparison with wavelength), there are phase differences from the distant point to various parts of the circuit so that cancellation is not so complete and we find terms inversely proportional to radius from the radiator.

For quantitative use of either of the above pictures, one needs the current distribution on the antenna. Exact determination of this would require solution of the difficult boundary-value problem, but for many antennas, reasonable assumptions of current distributions may be made, or they may be found by measurement.

We will examine a variety of radiators in the next section and find that several of these (horns, parabolic reflectors, etc.) utilize large apertures with stress on the field pattern excited in the aperture. For these, it is appropriate to use the concept of Huygen's principle, in which each elemental portion of the wave in the aperture can be considered a source of waves in space. This qualitative picture will be shown to lead to useful quantitative procedures in later sections. There, fields in the aperture are required, and they may often be approximated when not known exactly.

Finally, we may think of the antenna as a matching section between waves on whatever guiding system we use to excite the antenna, and the waves in space. The goal from this point of view is to design the unit for optimum matching over the desired frequency range, just as in matching units between different transmission lines or wave-guiding systems. An antenna which illustrates this and also shows the relation between the several points of view is the biconical antenna pictured in Fig. 12.1. This is the biconical transmission line of Sec. 9.6, open to space over the region shown dashed between regions I and II. A source is introduced at the origin, between apexes A and B, and this excites primarily the TEM

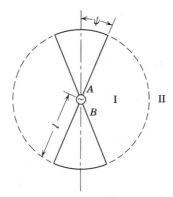

Fig. 12.1 Axial section through a biconical antenna.

or principal mode studied in Sec. 9.6. To a first approximation the surface at $r = l$ might be considered an open circuit and the principal wave reflection would then produce a sinusoidal standing wave on the cones. The requirement of field continuity on the dashed portion of the sphere requires that waves also be excited in space (region II), and these of course represent the desired radiation fields. It is found that continuity cannot be satisfied exactly between these and only the principal wave in I, so higher order waves on the biconical line are also required. This procedure can be made quantitative and some results from the process will be given later. Here we want to note that the process is like that from discontinuities in transmission lines or waveguides (Sec. 11.6), with the discontinuity here being the abrupt one between the biconical system and the spherical waves in space. Moreover, if the cone angle ψ is small, the cones may be thought of as wires, with the current to first order being the sinusoidal principal-wave current. We may calculate radiated fields from this approximation in either of the first two points of view. Alternatively, we can look at the principal-wave fields on the dashed part of the spherical surface between I and II as the source of radiation, with the system considered a biconical horn. So we see that the several pictures are related and the best for quantitative use may depend upon the specific configuration.

Most of the discussion in this chapter will relate directly to radiating or transmitting antennas but the results developed also can be applied to the same antenna when used for receiving applications. The radiating case is easier both for analysis and intuitive understanding and the relation to the receiving situation can be made rigorously using the principle of reciprocity that was introduced in Sec. 11.3.

12.2   Some Types of Practical Radiating Systems

To give point to comments and analyses that follow, let us look at some of the typical systems that have been used as radiators. No attempt will be made to

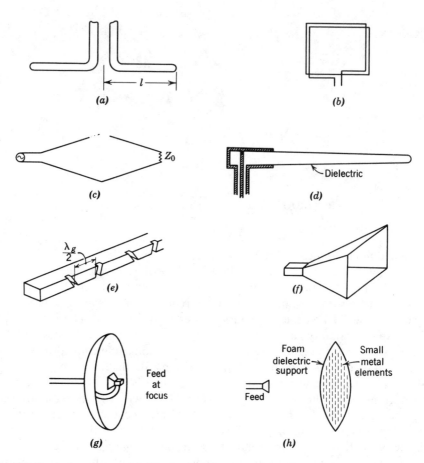

Fig. 12.2 Typical antennas: (*a*) Dipole. (*b*) Loop. (*c*) Rhombic. (*d*) Dielectric rod. (*e*) Slot array. (*f*) Pyramidal horn. (*g*) Parabolic reflector. (*h*) Artificial dielectric lens.

provide complete discussions of operation here, since the remainder of the chapter will be devoted to more thorough analyses of some of these systems. Nor does the list cover all types of radiators. The ones given are chosen as examples to make clearer the discussions of principles in sections to follow.

"Dipole" Antennas   Among the most common radiators is the dipole, which consists of a straight conductor (often a thin wire or circular cylinder of larger diameter) broken at some point where it is excited by a voltage derived from a transmission line, waveguide, or directly from a generator (Fig. 12.2*a*). In most cases, the exciting source is at the center, yielding a symmetrical dipole, although asymmetrical dipoles are also used. Resonant dipoles, and especially the half-wave dipole with 2*l* approximately equal to a half-wavelength, are common.

Loop Antennas   Radiation from a loop of wire excited by a generator has been discussed in Chapter 4. Such loop antennas are useful radiators, although often they have many turns, as in Fig. 12.2b. The field from a small loop is much like that from the small dipole, with electric and magnetic fields interchanged, as will be seen in the next section. Such a small loop is sometimes known for this reason as a *magnetic dipole*.

Traveling-Wave Antennas   If the radiator is made to have a traveling wave in one direction with a phase velocity about equal to the velocity of light, waves in space may be excited strongly in this direction as compared with other directions, thus yielding desired directivity. Figure 12.2c shows the important rhombic antenna, which utilizes waves along wires and is analyzed in more detail later. If only one-half of the rhombic is present, the result is the *V* antenna. These have been made in thin-film form for use at millimeter-wave frequencies. Figure 12.2d shows a version in which the traveling wave is guided by a dielectric as in Sec. 9.2. Near cutoff, phase velocity is approximately the velocity of light and the fields that extend outside a dielectric guide may excite appreciable radiation in space.

Slot and Aperture Antennas   As noted in Sec. 12.1, fields across an aperture may excite radiation in space. If the apertures are small, they must generally be resonant to excite appreciable amounts of power (although nonresonant holes in cavities may produce enough leakage to damage the Q of low-loss cavities). A narrow resonant half-wave slot has many similarities to the half-wave wire dipole, although electric and magnetic fields are interchanged, as will be seen. Figure 12.2e illustrates a series of half-wave slots in a rectangular guide to form a "leaky waveguide" array. When apertures are large, they need not be resonant to produce significant radiation. Electromagnetic horns are examples of radiators designed to match waves from a guiding system to a large radiating aperture by properly shaping the transition, much as in the acoustic horns used for sound waves. Figure 12.2f gives one example. Since the horns are not resonant, they, like the traveling-wave antennas, are especially useful for broadband signals.

Reflectors and Lenses   A parabolic reflector, as illustrated in Fig. 12.2g, is a most important device for microwave radiation. This may be considered as a mirror, serving to reflect the rays from the primary radiator at the focus. Geometrical optics would then predict an exactly parallel beam emerging from such a reflector if the primary radiator is infinitesimal. Alternatively, it may be considered that the primary source serves to illuminate the aperture, and radiation from this aperture produces the far field. From this point of view (which is also called "physical optics" or "diffraction theory"), there is always spreading of the beam. The spreading decreases as aperture size increases—an aperture 140 wavelengths in diameter producing a beam width of about 1 degree. Lenses serve a similar purpose in directing the rays from a primary radiator. These may be of solid dielectric, or artificial dielectrics as illustrated in Fig. 12.2h, or of metal waveguide paths to supply proper phase shifts to direct the beam. Lenses can also be considered as aperture radiators.

Arrays   Any of these radiators may be combined with like or different elements to form arrays that have particular directions in which phases add and radiation is concentrated. A most important use of arrays is in "electrical scanning" of the direction of concentration by control of phase shift from element to element. An example of the two-dimensional array of slot radiators is shown in Fig. 12.2*e*.

FIELD AND POWER CALCULATIONS WITH CURRENTS ASSUMED ON THE ANTENNA

12.3   Electric and Magnetic Dipole Radiators

In computing radiated power and the field distributions around an antenna when current distribution is assumed over the surface of the antenna's conductors, the simplest example is that of an ideal short linear element with current considered uniform over its length. Certain more complex antennas can be considered to be made up of a large number of such differential antennas with the proper magnitudes and phases of their currents. We shall consider only the case in which the current varies sinusoidally with time so we express it by its phasor I_0 that multiplies the factor $e^{j\omega t}$.

The current element is in the z direction with its location the origin of a set of spherical coordinates (Fig. 12.3*a*). Its length is h, with h very small compared with wavelength. By continuity, equal and opposite time-varying charges must exist on the two ends $\pm h/2$, so the element is frequently called a *Hertzian dipole*.

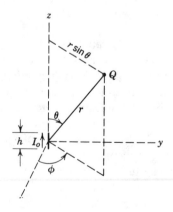

Fig. 12.3*a* Hertzian dipole.

One way of finding fields once current is given requires only the retarded potential $\mathbf{A}$ as seen in Sec. 3.21. For any point Q at radius r, $\mathbf{A}$ of Sec. 3.21 is in the z direction and becomes simply

$$A_z = \mu \frac{hI_0}{4\pi r} e^{-j(\omega r/v)} \tag{1}$$

Or, in the system of spherical coordinates,

$$\begin{aligned} A_r &= A_z \cos\theta = \mu \frac{hI_0}{4\pi r} e^{-jkr} \cos\theta \\ A_\theta &= -A_z \sin\theta = -\mu \frac{hI_0}{4\pi r} e^{-jkr} \sin\theta \end{aligned} \tag{2}$$

where $k = \omega/v = \omega\sqrt{\mu\varepsilon} = 2\pi/\lambda$. There is no ϕ component of $\mathbf{A}$, and there are no variations with ϕ in any expressions because of the symmetry of the structure about the axis. The electric and magnetic field components may be found directly from the components of $\mathbf{A}$ by use of the other equations listed in Sec. 3.21. Thus,

$$H_\phi = \frac{I_0 h}{4\pi} e^{-jkr} \left(\frac{jk}{r} + \frac{1}{r^2} \right) \sin\theta$$

$$E_r = \frac{I_0 h}{4\pi} e^{-jkr} \left(\frac{2\eta}{r^2} + \frac{2}{j\omega\varepsilon r^3} \right) \cos\theta \tag{3}$$

$$E_\theta = \frac{I_0 h}{4\pi} e^{-jkr} \left(\frac{j\omega\mu}{r} + \frac{1}{j\omega\varepsilon r^3} + \frac{\eta}{r^2} \right) \sin\theta$$

An interesting illustration of the fields produced by the dipole can be obtained by using E_r and E_θ components in (3) to show the patterns of field lines as functions of time, as was done in Sec. 8.3 for waveguide field patterns.[1] The expressions in (3) are converted into real functions of time in the usual way and plots are made of the electric field lines at various instants in time. At $t = 0$ the dipole current is maximum [Eq. 3.20(9)] so the charges on the ends of the dipole (Prob. 3.20b) are zero and the electric fields near the origin are zero. Previously generated patterns of fields have radiated away as shown in Fig. 12.3b. A quarter of a period later, the charge has built up on the ends of the dipole and electric field lines are produced in the neighborhood of the dipole as seen in Fig. 12.3c. The loops of the field lines close and "snap off" from the source as the charges return to zero.

For the region very near the element (r small) the most important term in H_ϕ is that varying as $1/r^2$. The important terms in E_r and E_θ are those varying as $1/r^3$. Thus, near the element, magnetic field is very nearly in phase with current, and H_ϕ may be identified as the usual induction field obtained from Ampère's law. Electric field in this region may be identified with that calculated for an electrostatic dipole. (By continuity, $I_0/j\omega$ represents the charge on one end of the dipole.)

[1] S. A. Schelkunoff and H. T. Friis, *Antennas: Theory and Practice*, Wiley, New York, 1952.

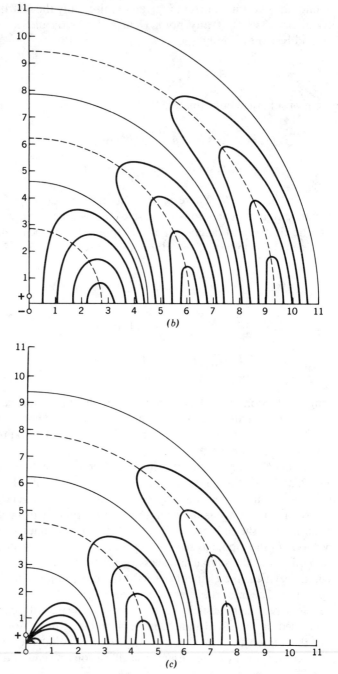

Fig. 12.3 (b) Fields near an oscillating electric dipole when the charges at its ends are zero. (c) Fields near an oscillating dipole when the charges at its ends have their maximum values. From S. A. Schelkunoff and H. T. Friis, Ref. 1, © 1952, John Wiley and Sons, New York.

As the important components of electric and magnetic field in this region are 90 degrees out of phase, these components represent no time average energy flow, according to the Poynting theorem.

At very great distances from the source, the only terms important in the expressions for E and H are those varying as $1/r$.

$$H_\phi = \frac{jkI_0 h}{4\pi r} \sin\theta \, e^{-jkr}$$

$$E_\theta = \frac{j\omega\mu I_0 h}{4\pi r} \sin\theta \, e^{-jkr} = \eta H_\phi \tag{4}$$

$$\eta = \sqrt{\mu/\varepsilon} \approx 120\pi \text{ ohms for space}$$

Equations (4) show the characteristics typical of uniform plane waves, as might be expected. E_θ and H_ϕ are in time phase, related by η, and at right angles to each other and the direction of propagation. The Poynting vector is then completely in the radial direction. The time-average P_r is

$$P_r = \frac{1}{2}\operatorname{Re}(E_\theta H_\phi^*) = \frac{\eta k^2 I_0^2 h^2}{32\pi^2 r^2}\sin^2\theta \qquad \text{W/m}^2 \tag{5}$$

The total power must be the integral of the Poynting vector over any surrounding surface. For simplicity this surface may be taken as a sphere of radius r:

$$W = \oint_S \mathbf{P}\cdot d\mathbf{S} = \int_0^\pi P_r 2\pi r^2 \sin\theta \, d\theta$$

$$= \frac{\eta k^2 I_0^2 h^2}{16\pi}\int_0^\pi \sin^3\theta \, d\theta$$

$$W = \frac{\eta\pi I_0^2}{3}\left(\frac{h}{\lambda}\right)^2 \cong 40\pi^2 I_0^2\left(\frac{h}{\lambda}\right)^2 \quad \text{W} \tag{6}$$

It is interesting and important to note that the field of the Hertzian dipole has the same form as the first-order TM spherical wave studied in Sec. 10.7. From Eqs. 10.7(19) with $n = 1$ and the Bessel function taken as the second Hankel form which is appropriate to a region extending to infinity,

$$H_\phi = A_1 r^{-1/2} P_1^1(\cos\theta) H_{3/2}^{(2)}(kr)$$

$$E_\theta = \frac{A_1}{j\omega\varepsilon r^{3/2}}[H_{3/2}^{(2)}(kr) - krH_{1/2}^{(2)}(kr)]P_1^1(\cos\theta) \tag{7}$$

$$E_r = -\frac{A_1 H_{3/2}^{(2)}(kr)}{j\omega\varepsilon r^{3/2}\sin\theta}[\cos\theta \, P_1^1(\cos\theta) - P_2^1(\cos\theta)]$$

But, from the relations of Sec. 10.7 it may be shown that

$$H_{3/2}^{(2)}(kr) = \sqrt{\frac{2}{\pi kr}}\, e^{-jkr}\left(\frac{j}{kr} - 1\right)$$

$$P_1^1(\cos\theta) = \sin\theta$$

With these substitutions and appropriate definition of constant A_1, (7) and (3) are identical. The electric field lines for this mode are the same as Figs. 12.3b or 12.3c.

Magnetic Dipole   The concept of duality was introduced in Sec. 9.5, where it was pointed out that exchanges **H** for **E**, $-\mathbf{E}$ for **H**, μ for ε, and ε for μ leave Maxwell's curl equations for source-free regions unchanged. Thus solutions for a problem with an electric source can be adapted to one with a magnetic source. For example, Sec. 2.10 showed that a small current loop of radius a and current I can be represented as a magnetic dipole $m = I\pi a^2$, and this gave the same form of magnetic field as the electric field given by the electric dipole $p = qh$. The charges q on the ends of the ac dipole are found from the ac current by the continuity equation which gives $j\omega q = I_0$ so that $p = I_0 h/j\omega$. Then replacing $I_0 h/j\omega$ by μ times the magnetic dipole in (3) and making the other above-stated duality interchanges, we find the fields of a magnetic ac dipole. The field components are

$$E_\phi = -\frac{j\omega\mu I a^2}{4} e^{-jkr}\left(\frac{jk}{r} + \frac{1}{r^2}\right)\sin\theta$$

$$H_r = \frac{j\omega\mu I a^2}{4} e^{-jkr}\left(\frac{2}{\eta r^2} + \frac{2}{j\omega\mu r^3}\right)\cos\theta \qquad (8)$$

$$H_\theta = \frac{j\omega\mu I a^2}{4} e^{-jkr}\left(\frac{j\omega\varepsilon}{r} + \frac{1}{j\omega\mu r^3} + \frac{1}{\eta r^2}\right)\sin\theta$$

In comparing the fields of the electric and magnetic dipole radiators, note that neither has azimuthal (ϕ) variations, there is a null along the z axis in the far zone, and they have orthogonal polarizations.

12.4   Systemization of Calculation of Radiating Fields and Power from Currents on an Antenna

For antennas with known current distributions, or with current distributions that can reasonably be assumed or measured, fields may be calculated from the retarded potentials as in Sec. 12.3, with integration over the current distributions. Or, equivalently, one can assume each infinitesimal element as a dipole and superpose the results of Sec. 12.3 by integration. We shall be able to simplify the procedures, however, whenever we are concerned only with the radiation or far-zone fields. The particular forms used here are useful ones introduced by Schelkunoff.[2,3]

---

[2] S. A. Schelkunoff, *Proc. I.R.E.* **27**, 660 (1939).
[3] S. A. Schelkunoff, *Electromagnetic Waves*, Van Nostrand, New York, 1943, Chapter IX.

The following assumptions are justified when field is calculated at a great distance from the radiator:

1. Differences in radius vector to different points of the radiator are unimportant in their effect on *magnitudes*.
2. All field components decreasing with distance faster than $1/r$ are negligible compared with those decreasing as $1/r$.
3. Differences in radius vector to different points on the radiator must be taken into account for phase considerations but may be approximated.

The vector potential at a point Q, a distance r from the origin of the coordinate system can be calculated using Eq. 3.20(4), where the integral is taken over the source region V' (Fig. 12.4):

$$\mathbf{A} = \mu \int_{V'} \frac{\mathbf{J} e^{-jkr''}}{4\pi r''} \, dV' \tag{1}$$

If the origin of coordinates is chosen close to the sources so that $r' \ll r$, the law of cosines may be used to find

$$r'' = \sqrt{r^2 + r'^2 - 2rr' \cos \psi} \cong r - r' \cos \psi \tag{2}$$

as suggested by Fig. 12.4. Using the first term of (2) for magnitudes (point 1 above) and both terms for phase (point 3), the vector potential (1) can be written

$$\mathbf{A} = \mu \frac{e^{-jkr}}{4\pi r} \int_{V'} \mathbf{J} \, e^{jkr' \cos \psi} \, dV' \tag{3}$$

The function of r is now outside the integral; *the integral itself is only a function of the assumed current distribution and the direction ψ between r and r'*. Define the integral as the *radiation vector* $\mathbf{N}$:

$$\mathbf{N} = \int_{V'} \mathbf{J} \, e^{jkr' \cos \psi} \, dV' \tag{4}$$

Then

$$\mathbf{A} = \mu \frac{e^{-jkr}}{4\pi r} \mathbf{N} \tag{5}$$

In the most general case, $\mathbf{A}$, and hence $\mathbf{N}$, may have components in any direction. In spherical coordinates, employing the unit vectors,

$$\mathbf{A} = \mu \frac{e^{-jkr}}{4\pi r} (\hat{\mathbf{r}} N_r + \hat{\boldsymbol{\theta}} N_\theta + \hat{\boldsymbol{\phi}} N_\phi)$$

A study of the equation $\mathbf{B} = \nabla \times \mathbf{A}$ in spherical coordinates (see inside back cover) shows that the only components that do not decrease faster than $1/r$ are

$$H_\theta = -\frac{1}{\mu r} \frac{\partial}{\partial r} (rA_\phi) = \frac{jk}{4\pi r} e^{-jkr} N_\phi$$

$$H_\phi = \frac{1}{\mu r} \frac{\partial}{\partial r} (rA_\theta) = -\frac{jk}{4\pi r} e^{-jkr} N_\theta \tag{6}$$

An examination of Eq. 3.21(7)

$$\mathbf{E} = -\frac{j\omega}{k^2} \nabla(\nabla \cdot \mathbf{A}) - j\omega\mathbf{A}$$

shows that the only components of $\mathbf{E}$ which do not decrease faster than $1/r$ are

$$E_\theta = -\frac{j\omega\mu}{4\pi r} e^{-jkr} N_\theta, \qquad E_\phi = -\frac{j\omega\mu}{4\pi r} e^{-jkr} N_\phi \qquad (7)$$

The Poynting vector has a time-average value

$$P_r = \frac{1}{2} \text{Re}[E_\theta H_\phi^* - E_\phi H_\theta^*] = \frac{\eta}{8\lambda^2 r^2} [|N_\theta|^2 + |N_\phi|^2] \qquad (8)$$

Total time average power radiated is

$$W = \int_0^\pi \int_0^{2\pi} P_r r^2 \sin\theta \, d\theta \, d\phi$$

$$= \frac{\eta}{8\lambda^2} \int_0^\pi \int_0^{2\pi} [|N_\theta|^2 + |N_\phi|^2] \sin\theta \, d\theta \, d\phi \qquad (9)$$

The expression is independent of r, as it should be.

The Poynting vector $\mathbf{P}$ gives the actual power density at any point. To obtain a quantity that does not depend on distance from the radiator, we define K, *radiation intensity*, as the power radiated in a given direction per unit solid angle. This is the time-average Poynting vector on a sphere of unit radius.

$$K = \frac{\eta}{8\lambda^2} [|N_\theta|^2 + |N_\phi|^2] \qquad (10)$$

and

$$W = \int_0^\pi \int_0^{2\pi} K \sin\theta \, d\theta \, d\phi \qquad (11)$$

Currents All in One Direction   If current in a radiating system flows all in one direction, this may be taken as the direction of the axis of a set of spherical coordinates. Vector $\mathbf{A}$ (hence $\mathbf{N}$) can have a z component only. Then

$$N_\phi = 0, \qquad N_\theta = -N_z \sin\theta$$

$$K = \frac{\eta}{8\lambda^2} |N_z|^2 \sin^2\theta \qquad (12)$$

Azimuthally Directed Currents   If all current in some radiating system is azimuthally directed about an axis, this axis may be taken as the axis of a set of spherical coordinates. Vector $\mathbf{A}$ (hence $\mathbf{N}$) can have a ϕ component only. Then

$$K = \frac{\eta}{8\lambda^2} |N_\phi|^2$$

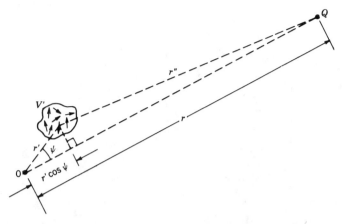

Fig. 12.4 Coordinates of a general current element in the volume V' and a distant point Q, where fields are to be evaluated, with respect to an origin of coordinates near the sources.

and

$$W = 2\pi \int_0^\pi K \sin \theta \, d\theta \tag{13}$$

Useful Relations for Spherical Coordinates   It sometimes may be desirable to calculate N_θ and N_ϕ from the Cartesian components N_x, N_y, N_z,

$$N_\theta = (N_x \cos \phi + N_y \sin \phi) \cos \theta - N_z \sin \theta$$
$$N_\phi = -N_x \sin \phi + N_y \cos \phi \tag{14}$$

The angle ψ appearing in the equation for radiation vector (4) may be found as follows, if θ, ϕ are the angular coordinates of the distant point Q, and θ', ϕ' are angular coordinates of the source point (Fig. 12.4):

$$\cos \psi = \cos \theta \cos \theta' + \sin \theta \sin \theta' \cos(\phi - \phi') \tag{15}$$

12.5   Long Straight Wire Antenna; Half-Wave Dipole

We argued in Sec. 12.1 that a very thin biconical antenna could serve as a model for a thin-wire antenna. It can be considered as a biconical transmission line with a very strong discontinuity at the end. The current along it should be essentially the same as that on an open-circuited transmission line and therefore nearly zero at the ends. Also, the propagation constant along the line is the same as for plane waves, $k = \omega\sqrt{\mu\varepsilon}$.

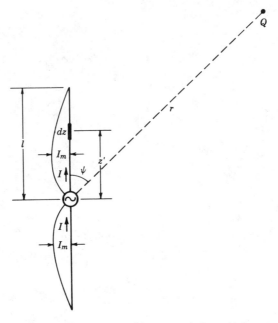

Fig. 12.5a Long straight wire antenna with assumed sinusoidal current distribution.

The long dipole of Fig. 12.5a with voltage applied at its midpoint is shown with an assumed sinusoidal distribution of current. The standing wave has zero current at the ends, and is selected with distance between zero and a maximum equal to a quarter free-space wavelength.

$$I = \begin{cases} I_m \sin[k(l - z)], & z > 0 \\ I_m \sin[k(l + z)], & z < 0 \end{cases} \qquad (1)$$

The fields and power radiated at a great distance from the antenna can be calculated using the general method of the preceding article. Since all current is z-directed, N and A have only z components also. In Eq. 12.4(4), the integral is converted to a line integral along the antenna wire with current density replaced by total current I and r' replaced by z'. Then since $\psi = \theta$ for all points along the wire,

$$N = \hat{z} \int_{-l}^{l} I e^{jkz' \cos\theta} \, dz'$$

$$= \hat{z} \int_{-l}^{0} I_m \sin[k(l + z')] e^{jkz' \cos\theta} \, dz'$$

$$+ \hat{z} \int_{0}^{l} I_m \sin[k(l - z')] e^{jkz' \cos\theta} \, dz' \qquad (2)$$

Using the integral formula

$$\int e^{ax} \sin(bx + c)\, dx = \frac{e^{ax}}{a^2 + b^2} [a \sin(bx + c) - b \cos(bx + c)]$$

(2) becomes

$$\mathbf{N} = \hat{z}\, \frac{2I_m}{k \sin^2 \theta} [\cos(kl \cos \theta) - \cos kl] \tag{3}$$

Since $N_\theta = -N_z \sin \theta$, Eq. 12.4(7) gives

$$E_\theta = \frac{j\eta I_m}{2\pi r} e^{-jkr} \left[\frac{\cos(kl \cos \theta) - \cos kl}{\sin \theta} \right] \tag{4}$$

and $H_\phi = E_\theta/\eta$.

The radiation intensity is found from (3) and Eq. 12.4(12) to be

$$K = \frac{\eta I_m^2}{8\pi^2} \left[\frac{\cos(kl \cos \theta) - \cos kl}{\sin \theta} \right]^2 \tag{5}$$

and the total power radiated is, from Eq. 12.4(11)

$$W = \frac{\eta I_m^2}{4\pi} \int_0^\pi \frac{[\cos(kl \cos \theta) - \cos kl]^2}{\sin \theta}\, d\theta \tag{6}$$

Example 12.5
The Half-Wave Dipole

The most important special case of the long center-fed antenna is that of the *half-wave dipole* in which $l = \lambda/4$. Field intensity and radiation intensity, from (4) and (5), with η for free space are respectively,

$$|E_\theta| = \frac{60 I_m}{r} \left| \frac{\cos[(\pi/2) \cos \theta]}{\sin \theta} \right| \qquad \text{V/m} \tag{7}$$

$$K = \frac{15 I_m^2}{\pi} \left\{ \frac{\cos[(\pi/2) \cos \theta]}{\sin \theta} \right\}^2 \qquad \text{W/steradian} \tag{8}$$

The half-wave antenna is a nearly resonant structure with its maximum of current near the drive point. It is not exactly resonant for any finite wire radius because treatment of the wire antennas as an open-circuited uniform transmission line is only an approximation. The half-length for resonance and the terminal resistance at resonance for different antenna radii a are given in Fig. 12.5b.[4]

---

[4] R. S. Elliott, *Antenna Theory and Design*, Prentice-Hall, Englewood Cliffs, N.J., 1981, p. 304.

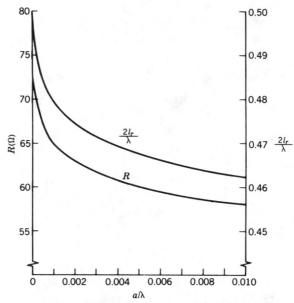

Fig. 12.5b Resonant length $2l_r$ and resistance at resonance for $\approx \lambda/2$ dipole as a function of radius of wire, normalized to wavelength. Robert S. Elliott, ANTENNA THEORY AND DESIGN, © 1981, p. 304. Reprinted by permission of Prentice-Hall, Inc., Englewood Cliffs, N.J.

12.6   Radiation Patterns and Antenna Gain

We pointed out earlier that the purposes of an antenna are to make an impedance match between a waveguide or transmission line and free space and to send the energy in the desired direction. The methods of characterizing the directivity of antennas will be elaborated in this Section.

Figure 12.6a shows a surface for which the distance from the origin to a given point on the surface is proportional to the radiation intensity in that direction from the antenna. The surface shown in Fig. 12.6a represents the half-wave dipole described by Eq. 12.5(8). The somewhat more complex pattern in Fig. 12.6b is for

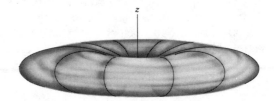

Fig. 12.6a Pattern of radiation intensity for a half-wave dipole.

Fig. 12.6b Pattern of radiation intensity for a dipole of length $3\lambda/2$.

a dipole of length $3\lambda/2$ (Problem 12.6c). Notice that both of these patterns have symmetry about the z axis since current is all z directed. Patterns of this type become quite complicated for asymmetric distributions.

Plots that are more easily made and read quantitatively are the intersections of surfaces like those in Figs. 12.6a and 12.6b with selected principal planes. Clearly, the intersections of planes perpendicular to the z axis with those surfaces are circles, reflecting the symmetry about the z axis. More interesting are the polar plots resulting from the intersection of the surface with a plane containing the z axis. This is shown for the surface of radiation intensity for the half-wave dipole (Fig. 12.6a) in Fig. 12.6c. Similar plots are shown for the magnitudes of electric field E_θ for the half-wave dipole and the Hertzian electric dipole, Eq. 12.3(4).

From the patterns in Figs. 12.6a–c, it is evident that radiated fields are stronger in some directions than in others. We say then that this antenna has a certain *directivity* as compared with an imagined isotropic radiator which radiates equally in all directions. This is, of course, an advantage if we desire to have the signal radiated in the direction of the maximum, since there is less power required to produce a given field in the desired direction than there would be for the isotropic radiator.

The ratio of the radiation intensity $K(\theta, \phi)$ from an antenna in a given direction, to the uniform radiation intensity for an isotropic radiator with the same total radiation power W is called the *directive gain*:

$$g_d(\theta, \phi) = \frac{K(\theta, \phi)}{(W/4\pi)} \tag{1}$$

where the factor 4π is the number of steradians in a full sphere. The value of directive gain in the direction for which it has its maximum value is called the *directivity*

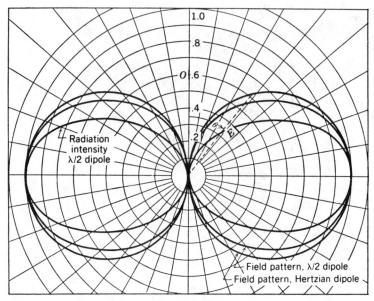

Fig. 12.6c Polar plots of field and radiation intensity for a half-wave dipole and of field for an Hertzian dipole in plane of dipole.

$(g_d)_{\max}$. For example, the directive gain for the half-wave dipole in free space can be found from Eqs. 12.5(5)–(6):

$$g_d(\theta, \phi) = 1.64 \left\{ \frac{\cos[(\pi/2)\cos\theta]}{\sin\theta} \right\}^2 \tag{2}$$

which has its maximum at $\theta = \pi/2$, as is clear from Figs. 12.6a or 12.6c, so 'he directivity is 1.64.

It is interesting to note that a similar calculation for the Hertzian dipole using Eqs. 12.3(5) and 12.3(6) with $\theta = \pi/2$ yields a directivity of 1.5, a value not much different from that of the half-wave dipole. In later parts of this chapter, we will see other antennas, especially antenna arrays, that have much higher directivities.

Another characterizing property of an antenna is the *power gain* $g_p(\theta, \phi)$. For this quantity, the radiation intensity is divided by the uniform radiation intensity that would exist if the total power *supplied to the antenna* by the connected wave-guide or transmission line were radiated isotropically:

$$g_p = \frac{K(\theta, \phi)}{W_T/4\pi} \tag{3}$$

Since the total supplied power W_T is the sum of that radiated W and the resistive losses W_l, then

$$g_p(\theta, \phi) = \frac{W}{W + W_l} g_d(\theta, \phi) \tag{4}$$

where the factor $W/(W + W_l)$ is called the *radiation efficiency* of the antenna. Although the directivity of an Hertzian dipole is almost equal to that of a half-wave dipole, as mentioned above, their radiation efficiencies are greatly different, assuming they are of the same material. This is because the power radiated by the dipole decreases as l^2 whereas the losses decrease as l. (The Hertzian dipole, having uniform current, models end-loaded short dipoles.) A similar result obtains for unloaded short antennas with $I = 0$ at the end. For other antennas, radiation efficiency is near unity and there is little difference between directive gain and power gain.

For a given length in comparison with wavelength, antenna diameter also affects radiation pattern and directivity.[5]

12.7   Radiation Resistance

The input impedance of an antenna can be represented by a series frequency-dependent impedance $Z = R + jX$ and is of great importance for optimum matching into the antenna. Methods for calculating R and X for certain lossless antennas are discussed in Secs. 12.24–12.25. The resistive component, in general, is a measure of the sum of radiated power W and ohmic losses W_l, as discussed in connection with power gain in the preceding section. It could be represented by two resistors in series, one for each component,

$$R_r = \frac{2W}{I^2} \tag{1}$$

where I is the magnitude of the terminal current, and

$$R_l = \frac{2W_l}{I^2} \tag{2}$$

The quantity R_r is called *radiation resistance*; for a small dipole current element, it can be found from (1) and Eq. 12.3(6) to be

$$R_r = 80\pi^2 \left(\frac{l}{\lambda}\right)^2 \quad \Omega \tag{3}$$

where we use l in place of h for the length. The power loss in the same element is $W_l = \frac{1}{2} R_s |J_s|^2 A$ where $J_s = I/2\pi a$, A is the surface area of the wire, a is its radius, and R_s is surface resistance.

$$W_l = \frac{R_s l}{4\pi a} I^2 \tag{4}$$

and

$$R_l = \frac{l R_s}{2\pi a} \tag{5}$$

J. D. Kraus, *Antennas*, McGraw-Hill, New York, 1950, p. 246.

Then the efficiency is

$$\frac{W}{W + W_l} = \frac{R_r}{R_r + R_l} = \frac{80(l/\lambda)}{80(l/\lambda) + (\lambda R_s/2\pi^3 a)} \tag{6}$$

Thus, it can be seen that the efficiency decreases as the length l decreases, becoming proportional to l/λ in the limit.

For an antenna with $l \ll \lambda$ and no end loading, the current decreases linearly from the source to the end (linear portion of the sinusoidal distribution in Fig. 12.5a). In this case the average I^2 along the antenna is one-quarter the terminal value I_m^2 so the radiated power is also one-quarter the value in the uniform-current case and radiation resistance is

$$R_r = \frac{2W}{I_m^2} = 20\pi^2 \left(\frac{l}{\lambda}\right)^2 \tag{7}$$

The average losses are also one-quarter of the uniform-current case so the conclusion about efficiency is again that it is proportional to the length of the element.

The radiation resistance of a lossless filamentary half-wave dipole can be found using Eq. 12.5(6)

$$R_r = \frac{2W}{I_m^2} = \frac{\eta}{2\pi} \int_0^\pi \frac{\cos^2((\pi/2)\cos\theta)}{\sin\theta} \, d\theta \tag{8}$$

which can be shown (Problem 12.7b) to have the value

$$R_r = 73.09 \, \Omega \tag{9}$$

It is seen in Fig. 12.5b that this equals the terminal resistance of the half-wave dipole as wire radius goes to zero.

12.8   Antennas Above Earth or Conducting Plane

Many antennas are placed near plane conductors. If the latter are large enough to be considered effectively infinite, and of high enough conductivity to be considered perfect conductors, it is possible to account for the conductor by imaging the antenna in it. For example, given a single cone with axis vertical above a perfectly conducting plane (Fig. 12.8a), the boundary condition of zero electric field tangential to the plane may be satisfied by removing the plane and utilizing a second cone as an image of the first. The problem then reduces to that of the biconical antenna discussed in Sec. 12.1. Note that current is in the *same vertical* direction at any instant in the two cones. Given a single wire above a plane conductor and parallel to it, as in Fig. 12.8b, our knowledge of symmetry in the transmission line problem tells us that the condition of electric field lines normal to the earth is met by removing the plane and placing the image with current in the *opposite horizontal direction*. Generalizing from these two cases, we see that current direction in the

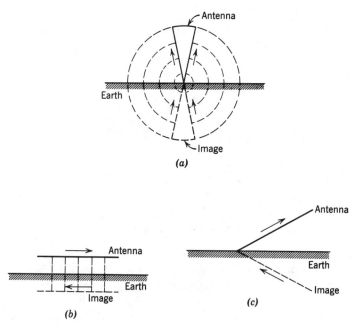

Fig. 12.8 (*a*) Cone above plane conducting earth and image cone. (*b*) Horizontal wire above plane conducting earth and image wire. (*c*) Inclined wire above plane conducting earth and image wire.

image will be selected so that vertical components are in the same direction, horizontal components in opposite directions at any instant. An example is shown in Fig. 12.8c.

The technique of replacing the conducting plane by the antenna image, of course, gives the proper value of field only above the plane. The proper value below the perfectly conducting plane should be zero. For example, given a long straight vertical antenna above a plane conductor, excited at the base, the image reduces the problem to that solved in Sec. 12.5. Field strength for maximum current I_m in the antenna is given exactly by Eq. 12.5(4) for all points above the earth $(0 < \theta < \pi/2)$, but is zero for all points below $(\pi/2 < \theta < \pi)$. Thus, for power integration, the integral of Eq. 12.5(5) extends only from 0 to $\pi/2$, and the power radiated is just half that for the corresponding complete dipole.

$$W = \frac{\eta I_m^2}{4\pi} \int_0^{\pi/2} \frac{[\cos(kl\cos\theta) - \cos kl]^2}{\sin\theta} \, d\theta \quad \text{W}$$

Calculations of radiation from antennas above an imperfectly conducting earth are much more complicated. The radiation from an electric dipole over a plane, finite-conductivity earth has been calculated. The fields are conveniently divided into a *space wave* and a *surface wave*. The space-wave fields decay as $1/r$ and the

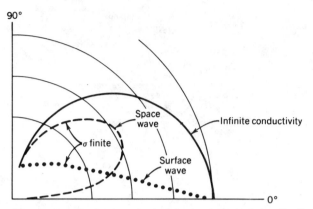

Fig. 12.8d Radiation pattern for field E_θ for infinitesimal vertical electric dipole over planes of finite-conductivity and infinite-conductivity. (From Ref. 6.)

surface-wave fields as $1/r^2$ so the important one at large distances is the space wave. As an example, consider the fields of an Hertzian dipole vertically oriented and at the level of a finite-conductivity earth. The electric field for the case of infinitely conducting earth is shown by the solid line for reference in Fig. 12.8d. The space-wave field magnitude ($1/r$ dependence) for a frequency at which the conduction current and displacement current in the earth are equal (typically, 5 MHz) is indicated by the dashed line. The surface-wave component of the electric field is shown by the dotted line; it decays as $1/r^2$. Each component is separately normalized; their relative magnitudes depend on the distance from the antenna because of the different dependences on r. An important conclusion is that the space-wave fields are greatly reduced for angles near the surface of the earth of finite conductivity.[6] Other properties of the real earth, such as curvature and the presence of the troposphere or ionosphere, are important in some frequency ranges and add still more complication.[7]

12.9 Traveling Wave on a Straight Wire

Some important types of antennas involve traveling waves of current on the wires that comprise them rather than the standing waves seen in Sec. 12.5 for the long wire antenna. As preparation, we consider the radiation from a traveling wave of current on an isolated wire. Let us consider a straight wire extending from $z = 0$ to $z = l$, excited by a single traveling wave of current, assumed to be unattenuated and with phase velocity equal to $1/\sqrt{\mu\varepsilon}$ (Fig. 12.9a). Since all current is in the z

[6] E. C. Jordan and K. G. Balmain, *Electromagnetic Waves and Radiating Systems*, 2nd ed., (Prentice-Hall, Englewood Cliffs, N.J. 1968, Fig. 16.7.
[7] J. R. Wait, *Electromagnetic Waves in Stratified Media*, Pergamon Press, New York, 1962.

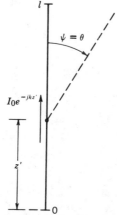

Fig. 12.9a Thin wire of length l supporting a progressive wave.

direction, the radiation vector of Eq. 12.4(4) has only a z component. The special forms of Eqs. 12.4(12) then apply.

$$N_z = I_0 \int_0^l e^{-jkz'} e^{jkz' \cos\theta} \, dz'$$

$$= \frac{I_0[1 - e^{-jkl(1 - \cos\theta)}]}{jk(1 - \cos\theta)}$$

$$|N_z| = \frac{2I_0 \sin[(kl/2)(1 - \cos\theta)]}{k(1 - \cos\theta)} \tag{1}$$

$$K = \frac{\eta|N_z|^2}{8\lambda^2}\sin^2\theta = \frac{I_0^2\eta}{2\lambda^2}\frac{\sin^2[(kl/2)(1 - \cos\theta)]}{k^2(1 - \cos\theta)^2}\sin^2\theta \tag{2}$$

In addition, since there is symmetry about the axis,

$$W = 2\pi \int_0^\pi K \sin\theta \, d\theta = 2\pi \int_0^\pi \frac{I_0^2\eta}{2\lambda^2}\frac{\sin^2[(kl/2)(1 - \cos\theta)]}{k^2(1 - \cos\theta)^2}\sin^3\theta \, d\theta$$

$$W = 30I_0^2 \int_0^\pi \frac{\sin^3\theta \sin^2[(kl/2)(1 - \cos\theta)]}{(1 - \cos\theta)^2} \, d\theta$$

If the foregoing integral is evaluated,[8]

$$W = 30I_0^2\left[1.415 + \ln\frac{kl}{\pi} - \text{Ci}(2kl) + \frac{\sin 2kl}{2kl}\right] \tag{3}$$

where

$$\text{Ci}(x) = -\int_x^\infty \frac{\cos x}{x} \, dx$$

[8] J. A. Stratton, *Electromagnetic Theory*, McGraw-Hill, New York, 1941, p. 445.

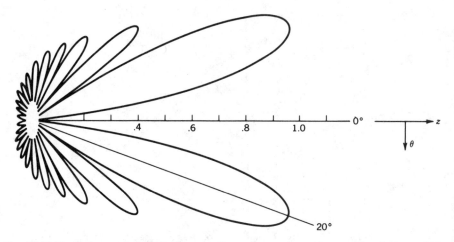

Fig. 12.9b Radiation pattern (E_θ) of a traveling-wave antenna six wavelengths long. (After Jordan and Balmain, Ref. 6, p. 357.)

This single traveling wave might be expected to produce a maximum of radiation in the direction of propagation. Actually the radiation is zero at $\theta = 0$, as seen from (2), but only because the radiation from each element of current is zero in that direction. A complete radiation pattern of a traveling-wave wire radiator six wavelengths long is shown in Fig. 12.9b, where it is clear that the lobes near $\theta = 0$ are largest and those near $\theta = \pi$ are smallest.

It should be pointed out that a single traveling-wave wire does not make a very desirable antenna because of the large amount of energy in the side lobes. The V-type and rhombic antennas discussed in the following section combine traveling-wave wire lines in advantageous ways.

12.10   V and Rhombic Antennas

In this section we will study two kinds of antenna, V and rhombic, based on the radiation from a traveling wave given in the preceding article. If the wires of the V are terminated in an appropriate resistance (for example, by connection of the ends of the V through resistors to ground), there will be a negligible reflected wave. The results of the preceding section can then be used. For example, if two wires carrying traveling-wave currents are oriented in a V with an angle between them equal to twice the angular displacement of the main lobe from the wire, the result is a unidirectional pattern with a maximum in the plane normal to that of the V and containing its bisector. The maximum would also be in the plane of the V if the effect of the ground were not significant; the effect of the image in a perfectly conducting ground is discussed at the end of this section.

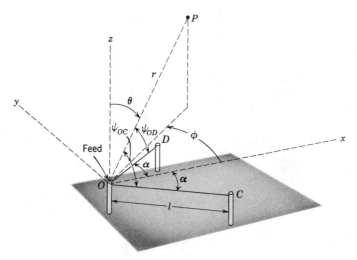

Fig. 12.10a Coordinates for calculation of radiation from a traveling-wave V-type antenna.

We can make a direct calculation of the radiation from the two arms of the V being driven from point 0 in Fig. 12.10a, assuming the ends at C and D are matched so there are only waves traveling away from 0. We also neglect, for now, the effect of the earth. The analysis is based on the radiation of a single wire given in the preceding section. The currents in the two arms are in different directions so addition of the effects is by orthogonal components.

The radiation vector for a single wire, with energy traveling at the velocity of light in only one direction, has only the direction of the wire. From 12.9(1),

$$N_s = \frac{I_0[1 - e^{-jkl(1 - \cos \psi)}]}{jk(1 - \cos \psi)} = \frac{I_0}{jk} f(\psi) \tag{1}$$

The subscript s denotes the direction of the wire, and ψ is the angle between the wire and the radius vector to the distant point (r, θ, ϕ) at which the radiation field quantities are desired. The angles for the elements, in terms of the coordinates in Fig. 12.10a are

$$\begin{aligned} \cos \psi_{OC} &= \sin \theta \cos(\phi + \alpha) \\ \cos \psi_{OD} &= \sin \theta \cos(\phi - \alpha) \end{aligned} \tag{2}$$

The currents at O for OC and OD are 180 degrees out of phase; they may be taken as I_0 and $-I_0$, respectively. Components of the radiation vectors of the two wires may be added:

$$N_x = \frac{I_0}{jk} \cos \alpha [f(\psi_{OC}) - f(\psi_{OD})] \tag{3}$$

$$N_y = -\frac{I_0}{jk} \sin \alpha [f(\psi_{OC}) + f(\psi_{OD})] \tag{4}$$

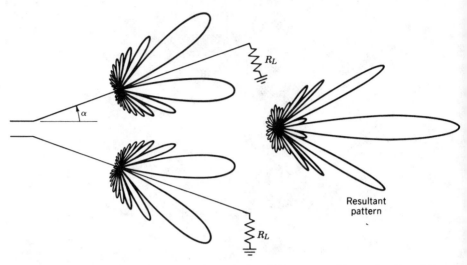

Fig. 12.10b Radiation intensity pattern for an isolated V antenna with arm lengths of 6λ and 32 degrees between arms. From W. L. Stutzman and G. A. Thiele, Ref. 9. Reprinted by permission. © John Wiley and Sons, 1981, New York.

To simplify expressions, we define

$$A \equiv f(\psi_{OC}), \qquad B \equiv f(\psi_{OD}) \tag{5}$$

The components of the radiation vector in spherical coordinates may be written in terms of the Cartesian components N_x and N_y from Eq. 12.4(14):

$$N_\theta = \frac{I_0}{jk} [A \cos(\alpha + \phi) - B \cos(\alpha - \phi)] \cos \theta \tag{6}$$

$$N_\phi = \frac{I_0}{jk} [-A \sin(\phi + \alpha) + B \sin(\phi - \alpha)] \tag{7}$$

Then the radiation intensity Eq. 12.4(10) is

$$K = \frac{\eta}{8\lambda^2} [|N_\theta|^2 + |N_\phi|^2]$$

$$= \frac{\eta I_0^2}{32\pi^2} \{ |A \cos(\alpha + \phi) - B \cos(\alpha - \phi)|^2 \cos^2 \theta$$

$$+ |-A \sin(\phi + \alpha) + B \sin(\phi - \alpha)|^2 \} \tag{8}$$

Figure 12.10b shows the radiation intensity pattern in the plane of the V for a structure with arm lengths of 6λ and with an angle of 32 degrees between arms.[9]

[9] W. L. Stutzman and G. A. Thiele, *Antenna Theory and Design*, Wiley, New York, 1981, p. 242.

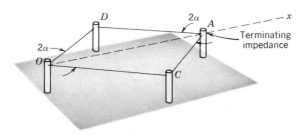

Fig. 12.10c Rhombic antenna.

If the V antenna described here is located above earth, as would be required for installing terminating resistors, the effect of the earth must be taken into account. Assuming the earth to be plane and perfectly conducting, the image effect can be accounted for by using the results to be worked out in Prob. 12.10a. The total radiation intensity K_T is

$$K_T = 4K \sin^2(kh \cos \theta) \qquad (9)$$

where K is the radiation intensity of a single V given by (8) and h is the height of the antenna above the earth. Although derived for a special case, (9) is a general relation for horizontal antennas above a perfectly conducting earth.

A similar analysis is used for the *rhombic antenna*, which consists of two oppositely directed V antennas with the ends of their arms joined as in Fig. 12.10c. It can be demonstrated that the V-type antenna has the same pattern (in the direction of the traveling waves) for either direction of use (Prob. 12.10b) so the radiations from the two Vs reinforce each other. Alternatively, the analysis given above can be extended to treat the rhombic antenna. The termination impedance is attached to the apex of the rhombus opposite from the drive point. Operation can be over about a 2:1 frequency range without large changes of the performance. Typical lengths of the arms are 2–7 wavelengths and the rhombus is usually located 1–2 wavelengths above the earth. The acute angles 2α are typically 35–60 degrees.

The form of the radiation pattern can be altered significantly by changing the angle α and the lengths of the arms. Moreover, for antennas used with long distance communication, it is desired to have the beam make a certain angle with the earth for optimum reflection from the ionosphere. This is adjusted by the distance of antenna above earth and calculated as for the V antenna, using (9). Figures 12.10d and 12.10e show, respectively, the calculated pattern in a vertical plane containing the x axis and the pattern in an azimuthal plane tilted 10 degrees above the x axis. The arms of the rhombus are 6λ long, $\alpha = 20$ degrees, and the rhombus is 1.1λ above the earth,[10] which is assumed perfectly conducting.

[9] E. A. Laport, in *Antenna Engineering Handbook*, H. Jasik, Ed., McGraw-Hill, New York, 1961, Fig. 4-14.

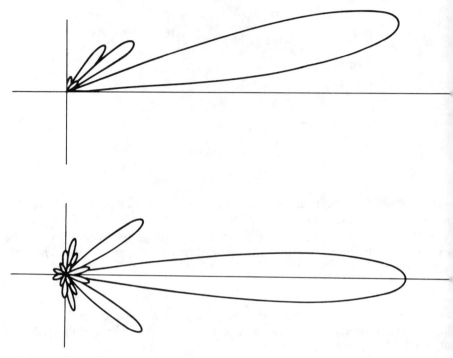

Fig. 12.10 (*d*) Radiation intensity pattern in the vertical plane containing the *x* axis for a typic rhombic antenna (*c*) over a perfectly conducting earth. (*e*) Radiation intensity pattern in t aximuthal plane containing the maximum intensity for the antenna of (*c*). From J. D. Kra Ref. 5, p. 408. © McGraw-Hill, 1950, New York.

12.11   Methods of Feeding Wire Antennas

In this section we consider three aspects of the problem of driving wire antenna The simplest excitation problem is that of the wire perpendicular to a conductir plane. Another common situation is the feeding of an isolated dipole by a balance two-wire transmission line, where special techniques must be used to achieve satisfactory impedance match. Finally, we discuss circuit arrangements to perm the feeding of an antenna, which is balanced with respect to ground, by a tran: mission line having one side at ground potential (usually the outer conductor a coaxial line).

Monopole Perpendicular to a Conducting Plane   Consider first a wire monopo that is effectively an extension of the inner conductor of a coaxial drive line, who: outer conductor is connected to an infinite ground plane, as shown in Fig. 12.11. The load on the coaxial line is resistive when the antenna is resonant. With $l \cong \lambda$ the resonant resistance is about 37.5 Ω, with only weak deviations caused b

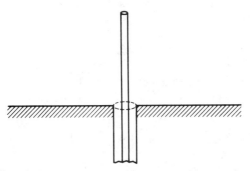

Fig. 12.11a Coaxial line feeding a monopole through a ground plane.

changes of the gap between inner and outer coaxial conductors at the connection to the ground plane.[11] To make the line impedance also 37.5 Ω, the radii ratio should be $b/a = 1.868$. With available experimental data on the admittance of such structures, it can be deduced that the antenna will have an impedance that is purely resistive and of value ≈ 36.8 Ω at resonance if the radius and length of the antenna are given by $a/\lambda = 0.00159$ and $l/\lambda = 0.236$. Thus, for an antenna operating at 100 MHz, the diameter and length should be 0.95 cm and 0.71 m, respectively. The power reflected at the feed point with this design is less than 0.01 %. The data in Ref. 11 can be used to find the bandwidth of feed to keep power transfer above some required level.

Folded Dipole   Antennas for television and FM broadcast applications are commonly coupled to a two-wire balanced transmission line. Such lines typically have characteristic impedances of 300–600 Ω. If such a line were to drive a half-wave dipole, which has a characteristic impedance on the order of 75 Ω, a serious mismatch would exist. A folded dipole one-half wavelength long with parallel conductors of the same diameter, shorted at the ends as shown in Fig. 12.11b, has a characteristic impedance four times that of the single half-wave dipole for a mode with currents in the parallel wires in the same direction. The currents are of equal magnitude and in the same direction in the two closely spaced equal-diameter wires. If a terminal current I_t is supplied, the effective radiating current is $2I_t$ so the power is four times that of the single half-wave dipole with current I_t. At resonance, the power is related to the input resistance R as $P = (\frac{1}{2})I_t^2 R$ so the four-fold increase of power for a given I_t implies that R is four times that of the half-wave dipole, or about 300 Ω. If another equal-diameter shorted wire is added in parallel to the folded dipole in Fig. 12.11b, the resonant input impedance for the mode with parallel currents in the same direction becomes nine times that of the single half-wave dipole. Another advantage of the folded dipole is that its input

[11] R. S. Elliot, Ref. 4, pp. 352–355.

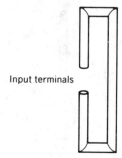

Input terminals

Fig. 12.11b Folded-dipole antenna.

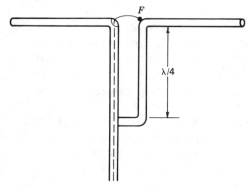

Fig. 12.11c One type of balun used for feeding balanced antenna from unbalanced transmission line (coaxial line in this case). Edward C. Jordan, Keith G. Balmain, ELECTROMAGNETIC WAVES AND RADIATING SYSTEMS, 2nd Ed., © 1968, p. 406. Reprinted by permission of Prentice-Hall, Inc., Englewood Cliffs, N. J.

impedance is acceptably constant over a larger frequency range than is that of the single dipole. There are many derived forms of the folded dipole giving a wide range of impedances.[12]

Baluns An isolated dipole or one oriented parallel to a ground plane is balanced with respect to ground. To avoid unbalancing the antenna, it should be fed by a balanced line, such as a two-wire line. For some situations (for frequencies higher than about 200 MHz or for a dipole mounted parallel to a ground plane), it is preferred to drive the antenna by a coaxial line rather than a two-wire line. Since the outer conductor of the coaxial line is grounded, its direct connection to a balanced dipole would lead to a distortion of the radiation pattern. The connection arrangement that avoids unbalancing a dipole is called a *balun*, a contraction for *balanced/unbalanced*. Baluns take a variety of forms; one is shown in Fig. 12.11c for a dipole parallel to a ground plane. At the feed point (marked *F*) the coaxial

[12] J. D. Kraus, Ref. 5, pp. 415–419.

line drives the dipole and, in parallel electrically, a two-wire transmission line consisting of its own outer conductor and the parallel support for the dipole. By making the distance to the shorting termination $\lambda/4$, the impedance of the two-wire line, as seen at the drive point, is infinite, so the dipole has equal currents in its two arms and is thus balanced with respect to ground. More complicated matching circuits[13] may be necessary to maintain matching over the bandwidth of broadband systems.

RADIATION FROM FIELDS OVER AN APERTURE

12.12   Fields as Sources of Radiation

For wire antennas, it is fairly natural to assume a current distribution over the antenna, and to consider the current elements as the sources of radiation. For other antennas, such as the electromagnetic horn, the slot antennas, the parabolic reflectors, the lens directors, and all optical systems, it is more natural to think in terms of the fields as sources. Huygen's principle states that any wave front can be considered the source of secondary waves that add to produce distant wave fronts. Thus the knowledge (or assumption) of field distribution over an aperture should yield the distant field. We wish now to make a quantitative statement of this general principle. There are several possible approaches, but we shall start with one that considers the fields as arising from equivalent current sheets in the aperture. This approach provides good physical pictures, and builds directly on the formulations already developed in this chapter.

Consider the aperture in a plane, Fig. 12.12, with sources to the left and the field desired in the region to the right. For present purposes the plane may be considered absorbing, so that it has no fields or currents except in the aperture. The exact boundary-value problem can be solved in only a few cases, but it is often possible to make a reasonable estimate of the aperture fields, just as was done for antenna currents in the radiators considered previously. We thus assume that tangential components of fields $E_t(x', y')$ and $H_t(x', y')$ are known in the aperture. Although these fields arise from sources to the left, they may be considered to be produced by equivalent sources located in the aperture plane. In particular, we have seen on many occasions that the relation between tangential magnetic field and a surface current is

$$\mathbf{J}_s = \hat{\mathbf{n}} \times \mathbf{H} \tag{1}$$

Similarly the tangential electric field can be related to a term which we can interpret as a surface magnetic current $\mathbf{M}_s$,

$$\mathbf{M}_s = -\hat{\mathbf{n}} \times \mathbf{E} \tag{2}$$

---

[13] D. F. Bowman, in *Antenna Engineering Handbook*, H. Jasik, Ed., McGraw-Hill, New York, 1961.

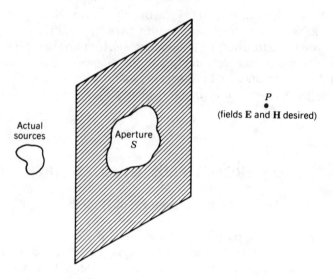

Fig. 12.12 Radiation through an aperature.

Note that (1) and (2) individually assume that the tangential fields H_t and E_t on the left side of the aperture plane are zero. Jointly, $\mathbf{J}_s$ and $\mathbf{M}_s$ produce fields that satisfy that assumption while producing the required fields on the right side. Whether or not true magnetic charges and currents are found in nature, the concept of such equivalent sources is a useful one in a variety of circumstances including this one. Maxwell's equations, augmented by magnetic charge density ρ_m and magnetic surface current $\mathbf{J}_m$, become

$$\nabla \cdot \mathbf{D} = \rho_e$$

$$\nabla \cdot \mathbf{B} = \rho_m$$

$$\nabla \times \mathbf{E} = -\mathbf{J}_m - \frac{\partial \mathbf{B}}{\partial t} \tag{3}$$

$$\nabla \times \mathbf{H} = \mathbf{J}_e + \frac{\partial \mathbf{D}}{\partial t}$$

A second retarded potential $\mathbf{F}$ can be defined in terms of the magnetic sources as $\mathbf{A}$ was related to electric sources. For a homogeneous isotropic medium, the relations in terms of surface currents are

$$\mathbf{A} = \mu \int_{S'} \frac{\mathbf{J}_s e^{-jkr}}{4\pi r} \, dS', \qquad \mathbf{F} = \varepsilon \int_{S'} \frac{\mathbf{M}_s e^{-jkr}}{4\pi r} \, dS' \tag{4}$$

and the fields in terms of these potentials are

$$E = -j\omega A - \frac{j\omega}{k^2} \nabla(\nabla \cdot A) - \frac{1}{\varepsilon} \nabla \times F \tag{5}$$

$$H = -j\omega F - \frac{j\omega}{k^2} \nabla(\nabla \cdot F) + \frac{1}{\mu} \nabla \times A \tag{6}$$

In order to calculate fields at any point in the region to the right, the actual sources in the left-hand region are replaced by equivalent sources in the aperture defined by (1) and (2) and the calculation made through the set (4) to (6). If the plane is conducting and not absorbing, any actual currents flowing on the right side of the plane must be added in the calculation of A in addition to the equivalent sources in the aperture, although this correction is often negligible.

If we are concerned only with the radiation field, the usual approximations appropriate to great distances can be employed, and the general formulation of Sec. 12.4 extended. A magnetic radiation vector L may be related to vector potential F as N was to A. Thus, consistent with the assumptions listed previously,

$$A = \mu \frac{e^{-jkr}}{4\pi r} N, \qquad F = \varepsilon \frac{e^{-jkr}}{4\pi r} L \tag{7}$$

where

$$N = \int_{S'} J_s e^{jkr' \cos \psi} \, dS', \qquad L = \int_{S'} M_s e^{jkr' \cos \psi} \, dS' \tag{8}$$

Here r' and ψ are as defined in Sec. 12.4.

If electric and magnetic field components are now written in the usual way in terms of these two vector potentials, the only components not decreasing faster than $(1/r)$ are

$$E_\theta = \eta H_\phi = -j \frac{e^{-jkr}}{2\lambda r} (\eta N_\theta + L_\phi) \tag{9}$$

$$E_\phi = -\eta H_\theta = j \frac{e^{-jkr}}{2\lambda r} (-\eta N_\phi + L_\theta) \tag{10}$$

So the radiation intensity, or power per unit solid angle, is

$$K = \frac{\eta}{8\lambda^2} \left[\left| N_\theta + \frac{L_\phi}{\eta} \right|^2 + \left| N_\phi - \frac{L_\theta}{\eta} \right|^2 \right] \tag{11}$$

In optics, this far-zone field is called the region of *Fraunhofer diffraction*. For the near zone, or region of *Fresnel diffraction*, better approximations to r are necessary.[14]

[14] M. Born and E. Wolf, *Principles of Optics*, 6th ed., Pergamon Press, Oxford, 1980, p. 660.

One does not need to use the concept of magnetic currents explicitly since the expressions (1) and (2) may be substituted in (4) and then one works with the aperture fields directly. In fact Stratton and Chu derived such expressions by direct integration of the field equations.[15] For many purposes, however, the physical pictures inherent in the formulation with magnetic currents is helpful, because of the duality with fields from electric currents.

12.13   Plane Wave Sources

The radiation vector and radiation intensity may be calculated for a differential surface element on a uniform plane wave. Such an element might be considered the elemental radiating source in radiation calculations from field distributions, as was the differential current element for radiation calculations from current distributions.

The plane wave source, that is, one that produces **E** and **H** of constant direction, normal to each other, and in the ratio of magnitudes η over the area of interest may be replaced by equivalent electric and magnetic current sheets over that area, Fig. 12.13a.

If

$$\mathbf{E} = \hat{x}E_{x0} \quad \text{and} \quad \mathbf{H} = \hat{y}H_{y0} = \frac{\hat{y}E_{x0}}{\eta} \tag{1}$$

the equivalent current sheets are

$$J_x = -H_{y0} = -E_{x0}/\eta, \qquad M_y = -E_{x0} \tag{2}$$

If this is a source of infinitesimal area dS (actually it need only be small compared with wavelength for following results to hold true), the radiation vectors **N** and **L** given by Eqs. 12.12(8) are

$$N_x = -\frac{E_{x0}\, dS}{\eta}, \qquad L_y = -E_{x0}\, dS$$

since the source element is at the origin of coordinates. The components in spherical coordinates are from Eq. 12.4(14):

$$N_\theta = -\frac{E_{x0}\, dS}{\eta}\cos\phi\cos\theta, \qquad N_\phi = \frac{E_{x0}\, dS}{\eta}\sin\phi \tag{3}$$

$$L_\theta = -E_{x0}\, dS\sin\phi\cos\theta, \qquad L_\phi = -E_{x0}\, dS\cos\phi$$

from which far-zone fields, by Eqs. 12.12(9)–(10) are

$$E_\theta = \frac{jE_{x0}\, dS\, e^{-jkr}}{2\lambda r}(1 + \cos\theta)\cos\phi \tag{4}$$

$$E_\phi = -j\frac{E_{x0}\, dS\, e^{-jkr}}{2\lambda r}(1 + \cos\theta)\sin\phi \tag{5}$$

[15] J. A. Stratton, Ref. 8, Sec. 8.14.

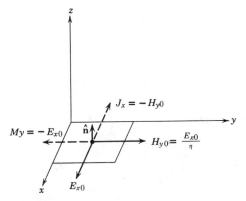

Fig. 12.13a Small plane wave source and equivalent current sheets.

and radiation intensity, by Eq. 12.12(11) is

$$K = \frac{E_{x0}^2 (dS)^2}{8\eta\lambda^2} [(-\cos\phi\cos\theta - \cos\phi)^2 + (\sin\phi + \sin\phi\cos\theta)^2]$$

$$K = \frac{E_{x0}^2 (dS)^2}{2\eta\lambda^2} \cos^4\frac{\theta}{2} \tag{6}$$

Paraxial Approximation for Large Apertures   In most cases of radiation from large apertures and horns, we are interested in small values of θ (*paraxial approximation*) so that $\cos\theta$ may be replaced by unity. Let us also convert to rectangular components in the radiation field,

$$E_x = E_\theta \cos\theta \cos\phi - E_\phi \sin\phi \approx E_\theta \cos\phi - E_\phi \sin\phi \tag{7}$$

$$E_y = E_\theta \cos\theta \sin\phi + E_\phi \cos\phi \approx E_\theta \sin\phi + E_\phi \cos\phi \tag{8}$$

Substituting (4) and (5) with $\cos\theta = 1$,

$$E_x \approx j\frac{E_{x0}\, dS\, e^{-jkr}}{\lambda r}, \qquad E_y \approx 0 \tag{9}$$

Thus in the paraxial approximation, the radiation field is in the direction of the source field and we may drop the subscript on E. In order to integrate over the aperture shown in Fig. 12.13b, we must first adapt the above results to an element at an arbitrary location x', y' in the aperture. This can be done using Eqs. 12.12(8) but a direct approach is more convenient here. Within the paraxial approximation, (9) gives the field produced by the element at x', y' if E_{x0} is replaced by $E(x', y')$ and r by r''. Now integrating over the given aperture distribution,

$$E(x, y, z) = \frac{j}{\lambda}\int_{S'} \frac{E(x', y')e^{-jkr''}}{r''}\, dx'\, dy' \tag{10}$$

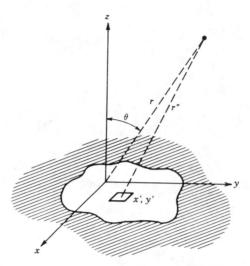

Fig. 12.13b Plane wave element as part of a plane aperature located at $z = 0$.

where r'' is distance between the element and the field point and may be approximated for the far zone by the binomial expansion,

$$r'' = [(x - x')^2 + (y - y')^2 + z^2]^{1/2} \approx r - (xx' + yy')/r \qquad (11)$$

Also, as in other radiation problems, the difference between r and r'' is important only to phase and not to amplitude. Thus (10) becomes

$$E(x, y, z) = \frac{je^{-jkr}}{\lambda r} \int_{S'} E(x', y')e^{jk(xx' + yy')/r} \, dx' \, dy' \qquad (12)$$

This is a standard form of diffraction for the Fraunhofer region. Moreover, (12) can be recognized as a two-dimensional Fourier integral (Sec. 7.11), so in the paraxial approximation, the *far-zone field is the Fourier transform of the aperture field*.

12.14 Examples of Radiating Apertures Excited by Plane Waves

The expressions developed in the preceding section will now be applied to several important examples. It should be remembered that **E** and **H** at any point in the aperture are assumed to be related as in a plane wave (though strength may vary over the aperture), that we are ignoring contributions from any induced currents outside of the aperture, and that we are restricting ourselves to angles near the polar axis so the paraxial approximation can be used. These assumptions are best satisfied for apertures large compared to wavelength.

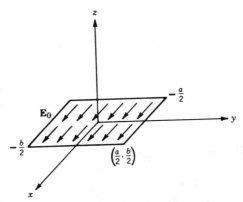

Fig. 12.14a Uniformly illuminated aperture.

Rectangular Aperture with Uniform Illumination   Consider first a rectangular aperture as in Fig. 12.14a with uniform illumination E_0 over the aperture. Equation 12.13(12) becomes

$$E(x, y, z) = \frac{je^{-jkr}}{\lambda r} \int_{-a/2}^{a/2} \int_{-b/2}^{b/2} E_0 \, e^{jkxx'/r} \, e^{jkyy'/r} \, dx' \, dy' \tag{1}$$

The integrations are readily performed to give

$$E(x, y, z) = \frac{je^{-jkr}}{\lambda r} E_0 ab \, \text{sinc}\left(\frac{kax}{2r}\right) \text{sinc}\left(\frac{kby}{2r}\right) \tag{2}$$

where $\text{sinc } X \triangleq (\sin X)/X$. The pattern in the plane $y = 0$ has nulls of the field at

$$(\theta)_{\text{null}} \approx \left(\frac{x}{r}\right)_{\text{null}} = \frac{m\lambda}{a}, \qquad m = 1, 2, 3, \ldots \tag{3}$$

The pattern in the $x = 0$ plane is of the same form with y replacing x and b replacing a. Thus as expected, the primary radiating lobe becomes narrower as the aperture dimensions become larger in comparison with wavelength.

Directivity of this aperture is readily calculated since the maximum field is that at $x = 0$, $y = 0$ so that power radiated by an isotropic radiator of this field strength is

$$(W)_{\text{isotropic}} = 4\pi r^2 \frac{|E_{\text{max}}|^2}{2\eta} = \frac{2\pi E_0^2 a^2 b^2}{\lambda^2 \eta} \tag{4}$$

Actual power radiated may be calculated from the fields in the aperture neglecting currents in the surrounding aperture plane:

$$W = \frac{E_0^2}{2\eta} ab \tag{5}$$

Thus directivity is

$$(g_d)_{max} = \frac{W_{isotropic}}{W} = \frac{4\pi ab}{\lambda^2} = \frac{4\pi(area)}{\lambda^2} \tag{6}$$

which is most accurate for large apertures. The relation between directivity and area is very important and will be found for other aperture radiators.

Long Slit   A long slit is often used to demonstrate diffraction effects with visible light, and if uniformly illuminated, is just a special case of the above. With b very long in comparison with wavelength, one would expect an extremely thin pattern in the plane $x = 0$, and this is the case if the light used is coherent (in phase) over the length of the slit. In many demonstrations this is not the case, so that there are no appreciable interferences and, therefore, no variations in the long y direction, but only over the shorter x direction. The resulting radiation intensity can be found as proportional to $|E|^2$ using only the variable x in (2). Since this problem has only two-dimensional symmetry, it is advantageous to express it in cylindrical coordinates with ϕ measured about the y axis and $\phi = 0$ along the x axis. Then in the approximation $x/r \approx \cos \phi$ radiation intensity is

$$K = A \left\{ \frac{\sin[(ka/2)\cos \phi]}{(ka/2) \cos \phi} \right\}^2 \tag{7}$$

where A is a constant related to the illumination of the aperture. The resulting pattern is shown in Fig. 12.14b. We will return to this point, showing patterns using visible light when we consider multiple slits in Sec. 12.20.

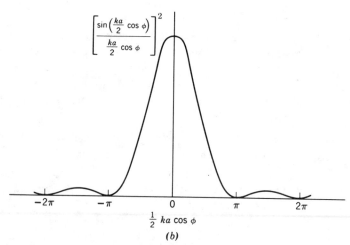

Fig. 12.14b Radiation intensity variation in one coordinate for a rectangular aperture.

Circular Aperture with Uniform Illumination   In considering the circular aperture, let us first transform to polar coordinates. Let

$$x = r \sin \theta \cos \phi, \qquad y = r \sin \theta \sin \phi, \qquad x' = r' \cos \phi', \qquad y' = r' \sin \phi'$$

Then Eq. 12.13(12) becomes

$$E(r, \theta, \phi) = \frac{je^{-jkr}}{\lambda r} \int_0^{2\pi} \int_0^a E(r', \phi')e^{jkr' \sin \theta \cos(\phi - \phi')}r' \, dr' \, d\phi' \tag{8}$$

If $E(r', \phi')$ is independent of ϕ', $E(r, \theta, \phi)$ is independent of ϕ, so we may take $\phi = 0$ and for the ϕ' integration may use the integral[16]

$$\int_0^{2\pi} e^{jq \cos \psi} \, d\psi = 2\pi J_0(q) \tag{9}$$

where $J_0(q)$ is a zeroth order Bessel function. Then

$$E(r, \theta) = \frac{2\pi je^{-jkr}}{\lambda r} \int_0^a E(r')J_0(kr' \sin \theta)r' \, dr' \tag{10}$$

If $E(r')$ is a constant E_0, Eq. 7.15(20) may be utilized to give

$$E(r, \theta) = \frac{2\pi je^{-jkr}}{\lambda r} E_0 a^2 \frac{J_1(ka \sin \theta)}{ka \sin \theta} \tag{11}$$

The magnitude of the last term in $E(r, \theta)$ is plotted as a function of $ka \sin \theta$ in Fig. 12.14c. The first null is reached at an angle

$$\theta_0 = \sin^{-1}\left(\frac{3.83\lambda}{2\pi a}\right) \approx \frac{0.61\lambda}{a} \tag{12}$$

Directivity of the circular aperture may be calculated in the same fashion as for the rectangular aperture, taking the ratio of power from an isotropic radiator having intensity the same as that at $\theta = 0$, to that of the plane wave power through the actual aperture,

$$(g_d)_{max} = \frac{(4\pi r^2/2\eta)(2\pi E_0 a^2/2\lambda r)^2}{(\pi a^2 E_0^2/2\eta)} = \left(\frac{4\pi}{\lambda^2}\right)(\pi a^2) \tag{13}$$

Directivity is thus related to aperture area exactly as for the rectangular aperture in (6).

Circular Aperture with Gaussian Illumination   In order to compare the effects of a tapered illumination with a uniform illumination, we next consider a gaussian distribution of field over the circular aperture,

$$E(r', \phi') = E_0 e^{-(r'/w)^2} = E_0 e^{-(x'^2 + y'^2)/w^2} \tag{14}$$

---

[16] I. S. Gradshteyn and I. M. Ryzhik, *Table of Integrals, Series and Products*, 4th ed., Academic Press, New York, 1965, 8.411(7).

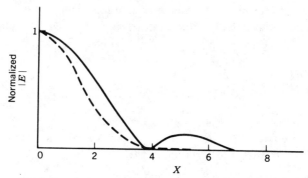

Fig. 12.14c Normalized electric field magnitude for circular aperture with uniform illumination (solid) and gaussian illumination (dashed). $X = ka \sin \theta$ for uniform case and $X = kw \sin \theta$ for gaussian illumination, with a and w defined in text.

but assume that **E** and **H** are related as in a plane wave. The quantity w is beam radius to the $1/e$ point in field. We assume that w is small compared with aperture radius a so that the latter may be assumed effectively infinite. Since field is a function of r' only, we could use (10) directly, but it is simpler to return to the rectangular form, Eq. 12.13(12):

$$E(x, y, z) = \frac{je^{-jkr}}{\lambda r} \int_{-\infty}^{\infty} \int_{-\infty}^{\infty} E_0 e^{-(x'^2/w^2 - jkxx'/r)} e^{-(y'^2/w^2 - jkyy'/r)} \, dx' \, dy' \quad (15)$$

The imaginary parts of the complex exponentials are odd functions and integrate to zero, so

$$E(x, y, z) = \frac{4je^{-jkr}}{\lambda r} \int_{0}^{\infty} \int_{0}^{\infty} E_0 \left[e^{(-x'^2/w^2)} \cos \frac{kxx'}{r} \right] \left[e^{(-y'^2/w^2)} \cos \frac{kyy'}{r} \right] dx' \, dy' \quad (16)$$

This integral is tabulated[17] and gives

$$E(x, y, z) = \frac{je^{-jkr}}{\lambda r} E_0 \pi w^2 e^{(-k^2(x^2 + y^2)w^2/4r^2)} = \frac{je^{-jkr}}{\lambda r} E_0 \pi w^2 e^{-(k^2 w^2 \sin^2 \theta)/4} \quad (17)$$

The angle θ_0 in the direction where the radiation field is e^{-1} of its value at $\theta = 0$ is

$$\theta_0 = \sin^{-1}\left(\frac{2}{kw}\right) \approx \frac{2}{kw} = \frac{\lambda}{\pi w} \quad (18)$$

The pattern as a function of $kw \sin \theta$ is compared with that for the case of uniform illumination in Fig. 12.14c. It is seen that the tapered illumination eliminates the side lobes present in the uniformly illuminated example.

---

[17] Gradshteyn and Ryzhik, Ref. 16; 3.896(4).

Practical Parabolic Reflectors   The examples above give some information about the important parabolic reflector with circular aperture. If illumination were uniform, (11) would apply. In general it is not uniform, both because of the practical difficulties in making it so and also because decreased illumination near the edge has the desirable effect of decreasing side lobes, as illustrated by the gaussian example above. For most primary sources at the focus, such as a small horn or dipole radiator, there is a difference in illumination of the reflector in two orthogonal planes, resulting in a difference of radiation patterns between those planes. Taking into account the above practical problems, a typical design value for beam angle between half-power directions, for a paraboloid of radius a, is[18]

$$(2\theta_0)_{\text{half-power}} \approx (70°)(\lambda/2a)$$

12.15   Electromagnetic Horns

The electromagnetic horns discussed qualitatively in Sec. 12.2 are of interest both as directive radiators in themselves, and also as feed systems for reflectors or directive lens systems. In the horn, there is a gradual flare from the waveguide or transmission line to a larger aperture. This large aperture is desired to obtain directivity, and also to produce more efficient radiation by providing a better match to space. Usually, a fair approximation to aperture field may be made by studying the fields in the feeding system and the possible modes in the horn structure. This then makes possible approximate radiation calculations starting from these fields, by the methods discussed in preceding sections. The idealized examples of Sec. 12.14 give some ideas of the patterns and directivity for certain practical cases. In general, however, there is not only a variation of fields across the aperture, but also a relation between electric and magnetic fields in the aperture somewhat different from that of a plane wave and some effect of currents induced on the exterior surfaces or supporting members.

We consider first the effect of variations across the aperture assuming **E** and **H** related as in a plane wave and neglecting currents on exterior surfaces and supports. For a rectangular horn excited by the TE_{10} mode of rectangular guide, Fig. 12.15, a reasonable approximation to aperture field is

$$E(x', y') = E_0 \sin\left(\frac{\pi x'}{a}\right) \tag{1}$$

Use of Eq. 12.13(12) for the radiation field gives the integral

$$E(x, y, z) = \frac{je^{-jkr}}{\lambda r} \int_{-b/2}^{b/2} \int_{-a/2}^{a/2} E_0 \sin\left(\frac{\pi x'}{a}\right) e^{jk(xx' + yy')/r} \, dx' \, dy' \tag{2}$$

---

[18] H. Jasik, Ed., *Antenna Engineering Handbook*, McGraw-Hill, New York, 1961, p. 2–30.

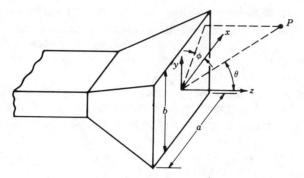

Fig. 12.15 Waveguide horn.

These integrals are standard and yield

$$E(x, y, z) = \frac{2E_0\,ab\,e^{-jkr}}{\pi}\,\frac{1}{\lambda r}\left[\frac{\cos(kax/2r)}{1 - (kax/\pi r)^2}\right]\mathrm{sinc}\left(\frac{kby}{2r}\right) \tag{3}$$

The dependence on y is of the same form as with the uniform distribution (since E is not a function of y), but that for x will be found to be somewhat broader than that of Eq. 12.14(2). Directivity for this case is

$$(g_d)_{\max} = \frac{(4\pi r^2/2\eta)(2E_0\,ab/\pi\lambda r)^2}{(E_0^2 ab/4\eta)} = \frac{4\pi}{\lambda^2}\left(\frac{8ab}{\pi^2}\right) \tag{4}$$

The quantity $8ab/\pi^2$ can be considered an equivalent area [by comparing with Eqs. 12.14(6) or (13)], and is slightly less than the actual area.

Generalization to a horn with a TE_{mn} distribution in the aperture, with β_{mn} the phase constant of the mode for that plane, and ρ the complex reflection coefficient referred to the aperture, can be carried out by using both electric and magnetic current sheets as explained in Sec. 12.12. The result is[19]

$$E_\theta = -\frac{\eta(\pi ab)^2 \sin\theta}{2\lambda^3 r k_{cmn}^2}\left[1 + \frac{\beta_{mn}}{k}\cos\theta + \rho\left(1 - \frac{\beta_{mn}}{k}\cos\theta\right)\right]$$

$$\times \left[\left(\frac{m\pi}{a}\sin\phi\right)^2 - \left(\frac{n\pi}{b}\cos\phi\right)^2\right]\Psi_{mn}(\theta, \phi) \tag{5}$$

$$E_\phi = -\frac{\eta(\pi ab)^2 \sin\theta \sin\phi \cos\phi}{2\lambda^3 r}$$

$$\times \left[\cos\theta + \frac{\beta_{mn}}{k} + \rho\left(\cos\theta - \frac{\beta_{mn}}{k}\right)\right]\Psi_{mn}(\theta, \phi) \tag{6}$$

[19] S. Silver, *Microwave Antenna Theory and Design*, McGraw-Hill, New York, 1949, Chapter 10.

where

$$\Psi_{mn}(\theta, \phi) = \left[\frac{\sin(\frac{1}{2}ka\sin\theta\cos\phi + m\pi/2)}{(\frac{1}{2}ka\sin\theta\cos\phi)^2 - (m\pi/2)^2}\right]\left[\frac{\sin(\frac{1}{2}kb\sin\theta\sin\phi + n\pi/2)}{(\frac{1}{2}kb\sin\theta\sin\phi)^2 - (n\pi/2)^2}\right] \quad (7)$$

12.16   Resonant Slot Antenna

Another important class of radiators in which the emphasis is on the field in the aperture is that of the slot antenna mentioned qualitatively in Sec. 12.2. Let us consider the resonant slot antenna (approximately a half-wave long) in an infinite plane conductor,[20] as shown in Fig. 12.16a. For apertures in perfectly conducting planes, it is only necessary to consider the electric field in the aperture in order to specify completely the boundary conditions since tangential **E** is zero along the conductor and dies off at infinity (see uniqueness argument, Sec. 3.14). We will take the electric field in the aperture to be uniform in x and to have a half-sine distribution in z with its maximum at the center. By reference to Eq. 12.12(2), the magnetic current sheet equivalent to this would be found to lie in the z direction and to be equal to E_x:

$$M_z = E_x = E_m\cos kz \quad (1)$$

Since E_z is assumed to be zero behind M_z, the gap can be considered closed with an electric conductor. It is convenient to account for the infinite conductor by replacing it with the image of M_z. As shown in Prob. 12.16a, M_z and its image are in the same direction so the value in (1) must be doubled in using free-space field expressions.

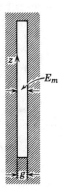

Fig. 12.16a Resonant half-wave slot.

[20] R. T. Compton, Jr. and R. E. Collin, Chapter 14 of R. E. Collin and F. J. Zucker, *Antenna Theory*, Part 1, McGraw-Hill, New York, 1969.

If gap width g is taken as small, the equation for magnetic radiation vector, Eq. 12.12(8), becomes

$$L_z = \int_{-\lambda/4}^{\lambda/4} 2gE_m \cos kz' \, e^{jkz' \cos \theta} \, dz' \qquad (2)$$

The integral is evaluated to give

$$L_z = \frac{4gE_m \cos[(\pi/2) \cos \theta]}{k \sin^2 \theta} \qquad (3)$$

Then, utilizing Eq. 12.12(10), we see that the fields are

$$E_\phi = -\eta H_\theta = \frac{je^{-jkr} gE_m}{\pi r} \left\{ \frac{\cos[(\pi/2) \cos \theta]}{\sin \theta} \right\} \qquad (4)$$

We should note here that this is of the same form as the expression for fields about a half-wave dipole antenna (Sec. 12.5) except for the interchange in electric and magnetic fields.

The power radiated corresponding to (4), assuming radiation only into one side, is

$$W = \frac{\pi(gE_m)^2}{2\pi^2\eta} \int_0^\pi \frac{\cos^2[(\pi/2) \cos \theta]}{\sin \theta} \, d\theta \qquad (5)$$

This may be interpreted in terms of a radiation conductance defined in terms of the maximum gap voltage:

$$(G_r)_{\text{slot}} = \frac{2W}{(gE_m)^2} = \frac{1}{\pi\eta} \int_0^\pi \frac{\cos^2[(\pi/2) \cos \theta]}{\sin \theta} \, d\theta \qquad (6)$$

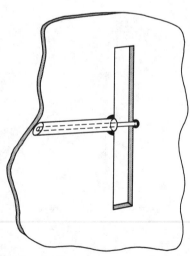

Fig. 12.16b Coaxial-line feed for resonant-slot antenna.

By comparing with the expression for radiation resistance of the half-wave dipole, Eqs. 12.7(8)–(9), we find

$$(G_r)_{\text{slot}} = \frac{2(R_r)_{\text{dipole}}}{\eta^2} \approx 0.00103 \text{ S} \tag{7}$$

If radiation is to both sides, conductance is double the above value. The reciprocity between results for the slot and dipole can also be shown to follow from *Babinet's principle*,[21] which is an extension of the principle of duality discussed in Sec. 9.5. A possible means of exciting the slot is shown in Fig. 12.16*b*.

12.17   Lenses for Directing Radiation

The lenses used in radiating systems, like the parabolic reflector, act to focus the radiation from primary sources into desired directions. Most lenses are conceived from the point of view of geometrical optics, and in this approximation would produce a plane parallel beam of radiation from an ideal point radiator at the focus. The actual lens behavior is disturbed from this ideal by the finite character of the primary radiator, and by diffraction arising from the finite aperture of the lens. The latter effect causes the beam to spread in a manner predicted by the aperture analysis of Sec. 12.14. Thus we will leave this aspect of the matter and consider the various first-order designs suggested by geometrical optics.

The most direct conception is that of a primary source placed at the focus of a converging lens so that the beam emerges parallel. The technique is well developed at optical frequencies, including the coating of the surfaces to eliminate reflections (Ex. 6.8c), The solid lenses most often used are small enough for convenient handling. The use of solid dielectric lenses can be extended to submillimeter and millimeter wavelengths. However, for microwaves the lenses are large and heavy if the aperture is made large enough to produce a reasonably narrow beam. Thus one of the two variants to be described are commonly used.

A fairly direct extension of the solid lens concept to microwaves is possible with reasonable weights through the use of "artificial dielectrics." These lenses are made of metal spheres, disks, strips, or rods imbedded in a light material such as polyfoam.[22] If the metal particles are small in comparison with wavelength, they act very much as individual molecules in a solid dielectric. That is, the metal particles are "polarized" by the applied field (Fig. 12.17*a*), with the positive and negative charges displaced from each other. Each particle then acts as a small dipole, contributing to total displacement and thus to an effective dielectric constant. In general both electric and magnetic field components contribute to the polarization of the metallic particles. In spherical particles, for example, the electric and magnetic effects tend to cancel. Particles having a form of thin sheets (disks

---

[21] See, for example, R. S. Elliott, Ref. 4, Appendix F.

[22] W. E. Kock, *Bell Syst. Tech. J.* **27**, 58 (1948).

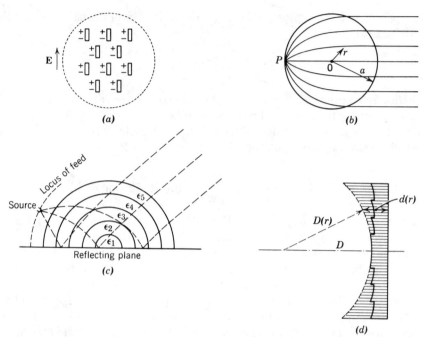

Fig. 12.17 (*a*) Artificial-dielectric lens. (*b*) Luneberg lens. (*c*) Practical form of Luneberg lens. (*d*) Metal lens antenna.

or strips) parallel to the electric and magnetic fields, as in Fig. 12.2*h*, have negligible magnetically induced polarization and therefore give a larger *n*. They are also lighter than the other shapes. For example, the effective relative permittivity for disks which are small compared with wavelength has been shown to be $\varepsilon_r = 1 + 5.3Na^3$ where *a* is the disk radius and *N* is the number per cubic meter. Effective permittivities in the hundreds have been obtained with usefully light structures. The use of disks or cylinders leads to anisotropic properties, but the structure can be designed for use with the desired ray direction and direction of polarization. Inhomogeneous properties are also possible by varying spacing or size of particles; this technique may be used instead of shaping to produce focusing action.

One of the most important lenses that utilizes an inhomogeneous effective dielectric constant is the Luneberg lens.[23] This is made in the form of a sphere with effective dielectric constant varying with radius as

$$\varepsilon_r(r) = 2 - (r/a)^2, \qquad r \le a \tag{1}$$

Techniques for analyzing rays in inhomogeneous media will be given in Chapter 14. Use of these can show that all rays originating from a point on the surface will emerge from the other side as a parallel beam in the direction of the diameter

[23] R. F. Rinehart. *J. Appl. Phys.* **19**. 860 (1948)

through the source point (Fig. 12.17b). This lens is important because small-distance movements of the source along the surface can provide substantial angular scanning of the beam. It is more practical to have the source point a small distance away from the surface of the lens, to use a reflection plane to eliminate half the sphere, and to use steps in effective dielectric constant, as illustrated in Fig. 12.17c. The required gradation of ε_r for these modifications has been calculated.[24]

An interesting and different means of realizing lens-type focusing action is with the *metal lens antenna*[25] which utilizes sections of parallel-plane waveguides, as illustrated in Fig. 12.17d. Polarization is such that the mode between the plates is a *TE* mode, normally with one half-sine variation in field distribution. The ratio of spacing to wavelength is such that phase velocity v_p is appreciably greater than that in free space. The concept used is one in which all paths from focus to the lens plane yield the same phase,

$$\frac{D}{c} + \frac{d(0)}{v_p} = \frac{D(r)}{c} + \frac{d(r)}{v_p} \tag{2}$$

where v_p is given by Eq. 8.3(18) and r is distance from the axis in cylindrical coordinates. Since the waveguide sections advance rather than delay phase, the outer parts of the lens are longer than the center parts for a converging lens, as illustrated in Fig. 12.17d. Whole wavelengths contribute nothing from the point of view of phase, so these may be cut out to save weight, resulting in the final stepped configuration shown by the solid lines of the figure.

ARRAYS OF ELEMENTS

12.18   Radiation Intensity with Superposition of Effects

If there are several complete radiators operating together, currents or fields might be assumed over the entire group, and a complete calculation made for potentials or fields at any point in the radiation field. If the radiation pattern of the individual elements is known, however, some labor may be saved by superposing these with proper attention to phase, direction, and magnitude of the fields. This is actually only an extension of the procedure we have already used to add up the effects of infinitesimal elements. As a practical matter, the synthesis of desired antenna patterns by the addition of individual radiators is a most important tool, so that efficient methods for the analysis of arrays are essential. A few of these, with examples, are set down in this and several following sections.

[24] E. A. Wolff, *Antenna Analysis*, Wiley, New York, 1966.
[25] W. E. Kock, "Metal Lens Antenna," *Proc. I. R. E.* **34**, 828–836 (November 1946). Also see Elliott, Ref. 4, pp. 542–545.

The usual problem in array theory is that of identical radiators with similar current or field distributions (although magnitudes and phases of currents or fields in individual radiators may differ). The radiation vectors for one of these alone may be calculated as $\mathbf{N}_0$ and $\mathbf{L}_0$. ($\mathbf{N}_0$ alone is sufficient if we consider contributions only from electric current elements.) In adding contributions from the individual radiators for the far-zone field, the usual approximation is made in that differences in distances to the individual radiators need be taken into account only with respect to phase; also, the approximate form for distance differences may be calculated by assuming the field point far removed. The form of the radiation vectors can be found by evaluating the distance r'' from an element of radiating current to the field point in terms of the parameters of the array. We can see from Fig. 12.18a that for an element on the nth radiator, r'' can be expressed in terms of the distance r from the arbitrary reference point in the array to the field point as

$$r'' \approx r - r'_n \cos \psi_n - r'_0 \cos \psi_0 \tag{1}$$

where r'_n and ψ_n locate the nth radiator relative to the reference point and r'_0 and ψ_0 determine the location of the differential radiating element relative to the reference point on the radiator. Since all radiators have the same orientation, r'_0

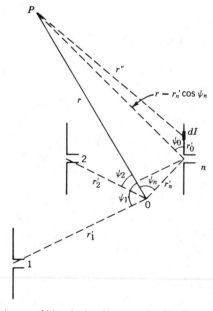

Fig. 12.18a Array of identical radiators, all with the same orientation.

and ψ_0 apply equally to all n radiators. Substituting (1) in Eq. 12.4(1), we get the far-field vector potential at P for the nth radiator:

$$\mathbf{A}_n = \frac{\mu e^{-jkr}}{4\pi r} \left[e^{jkr'_n \cos \psi_n} \int_{V'_n} \mathbf{J}_n e^{jkr'_0 \cos \psi_0} \, dV'_n \right]$$

$$= \frac{\mu e^{-jkr}}{4\pi r} [C_n e^{jkr'_n \cos \psi_n} \mathbf{N}_0] \tag{2}$$

where C_n is a complex constant and $\mathbf{N}_0$ is the normalized radiation vector for one of the identical radiators. Thus the radiation vector for a system of radiators with electric current is

$$\mathbf{N} = \mathbf{N}_0 (C_1 e^{jkr'_1 \cos \psi_1} + C_2 e^{jkr'_2 \cos \psi_2} + \cdots) \tag{3}$$

Likewise, for radiators with magnetic currents,

$$\mathbf{L} = \mathbf{L}_0 (C_1 e^{jkr'_1 \cos \psi_1} + C_2 e^{jkr'_2 \cos \psi_2} + \cdots) \tag{4}$$

It follows that the total radiation intensity may be written in terms of the radiation intensity K_0 for one radiator alone:

$$K = K_0 |C_1 e^{jkr'_1 \cos \psi_1} + C_2 e^{jkr'_2 \cos \psi_2} + \cdots|^2 \tag{5}$$

Examples will follow to make the use of these forms more specific. We should note one important point, however, that will not be touched on specifically in the examples. Mutual couplings between the elements, including effects from the near-zone fields, may affect the amount and phase of current that a given array element receives from the driving source, so that the excitation problem may become quite difficult. A discussion of the coupling problems is given for one type of array in Sec. 12.22.

_____ **Example 12.18** _____
Array of Two Half-Wave Dipoles

Consider two half-wave dipoles separated by a quarter-wavelength and fed by currents equal in magnitude and 90 degrees out of time phase, as in Fig. 12.18b. For a single dipole, we have Eq. 12.5(8):

$$K_0 = \frac{15}{\pi} I_m^2 \frac{\cos^2[(\pi/2) \cos \theta]}{\sin^2 \theta}$$

For the two dipoles with the origin as shown in Fig. 12.18b,

$$r'_1 = 0, \qquad r'_2 = \frac{\lambda}{4}, \qquad \theta'_2 = \frac{\pi}{2}, \qquad \phi'_2 = 0$$

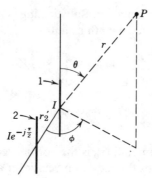

Fig. 12.18b Combination of two half-wave dipoles.

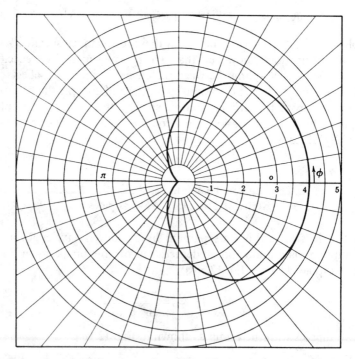

Fig. 12.18c Polar plot of relative power intensity radiation for array in Fig. 12.18b in the plane $\theta = \pi/2$.

Using the general formula, Eq. 12.4(15)

$$\cos \psi_2 = \sin \theta \cos \phi$$

If

$$I_2 = I_1 e^{-j(\pi/2)}$$

then (5) gives

$$K = K_0 |1 + e^{-j(\pi/2)} e^{j(\pi/2)(\sin \theta \cos \phi)}|^2$$

$$= 4K_0 \cos^2 \left[\frac{\pi}{4} (\sin \theta \cos \phi - 1)\right]$$

A horizontal radiation intensity pattern is plotted in Fig. 12.18c.

12.19   Linear Arrays

An especially important class of arrays is that in which the elements are arranged along a straight line, usually with equal spacing between them, as indicated in Fig. 12.19a. Let the line be the z axis, with the basic spacing d and coefficients $a_0, a_1, \ldots, a_{N-1}$ representing the relative currents in elements at $z = 0$, $d, \ldots, (N - 1)d$. (Note that any of the elements can be missing, in which case the coefficient is zero, so the elements need only be of commensurate spacing instead of equal spacing.) If the elements have a radiation vector $\mathbf{N}_0$, Eq. 12.18(3) becomes for this case

$$\mathbf{N} = \mathbf{N}_0[a_0 + a_1 e^{jkd \cos \theta} + \cdots + a_{N-1} e^{j(N-1)kd \cos \theta}] = \mathbf{N}_0 S(\theta) \qquad (1)$$

where $S(\theta)$ may be called the *space factor* of the array,

$$S(\theta) = \sum_{n=0}^{N-1} a_n e^{jnkd \cos \theta} \qquad (2)$$

The radiation intensity from Eq. 12.18(5) is then

$$K = K_0 |S|^2 \qquad (3)$$

Broadside Array   If all currents in the linear array are equal in magnitude and phase, it is evident from physical reasoning that the contributions to radiation will

Fig. 12.19a Coordinate system for a linear array.

add in phase in the plane perpendicular to the axis of the array ($\theta = \pi/2$). For this reason, the array is called a *broadside array*. Moreover, it is evident that, if the total length $l = (N - 1)d$ is long compared with wavelength, the phase of contributions from various elements will change rapidly as angle is changed slightly from the maximum, so that maximum in this case would be expected to be sharp. To see this from (2), let all $a_n = a_0$:

$$S(\theta) = a_0 \sum_{n=0}^{N-1} e^{jnkd \cos \theta} = a_0 \frac{1 - e^{jNkd \cos \theta}}{1 - e^{jkd \cos \theta}} \tag{4}$$

Summation (4) is effected by the rule for a geometric progression. Then

$$|S| = a_0 \left| \frac{\sin \frac{1}{2}(Nkd \cos \theta)}{\sin \frac{1}{2}(kd \cos \theta)} \right| \tag{5}$$

Relation (5) is plotted as a function of $kd \cos \theta$ in Fig. 12.19b for $N = 10$. Note that the peak of the main lobe occurs at $kd \cos \theta = 0$ (or $\theta = \pi/2$) as expected. The width of the main lobe may be described by giving the angles at which radiation goes to zero. If we set these angles as $\theta = \pi/2 \pm \Delta/2$,

$$Nkd \cos\left(\frac{\pi}{2} \pm \frac{\Delta}{2}\right) = \mp 2\pi$$

$$\Delta = 2 \sin^{-1}\left(\frac{\lambda}{Nd}\right) \approx \frac{2\lambda}{l} \tag{6}$$

The last approximation is for large N. So we see that the beam becomes narrow as l/λ becomes large, as predicted.

If N is large the denominator of (5) remains small over several of the lobes near the main lobe. Over this region it is then a good approximation to set the sine equal to the angle in the denominator.

$$|S| \approx Na_0 \left| \frac{\sin \frac{1}{2}(Nkd \cos \theta)}{\frac{1}{2} Nkd \cos \theta} \right| \tag{7}$$

This approximation is compared as a dotted curve with the accurate curve for $N = 10$ in Fig. 12.19b, and is found to agree well over several maxima. Thus, for large N, this universal form applies near $\theta = \pi/2$ and the first secondary maximum is observed to be about 0.045 of the absolute maximum, or 13.5 dB below.

The directivity of an array is usually expressed as though the elements were isotropic radiators. It is of course modified if actual elements having some directivity are employed, but, for high-gain arrays, the modification is small. (Note that it is not correct to multiply directivity of the array by directivity of a single element.) For the array,

$$(g_d)_{max} = \frac{4\pi |S_{max}|^2}{2\pi \int_0^\pi |S|^2 \sin \theta \, d\theta} \tag{8}$$

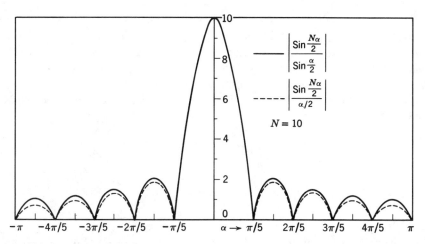

Fig. 12.19*b* Plot of magnitude of space factor and approximation with 10 elements. α is $kd\cos\theta$ for a broadside array and $kd(1 - \cos\theta)$ for an end-fire array.

The high-gain broadside array gives most of its contribution to the integral in the denominator near $\theta = \pi/2$, where the approximate expression (7) applies

$$\int_0^\pi |S|^2 \sin\theta \, d\theta = \frac{2N^2 a_0^2}{Nkd} \int_0^{Nkd} \frac{\sin^2 \alpha/2}{(\alpha/2)^2} \, d\alpha \approx \frac{4N^2 a_0^2}{Nkd} \cdot \frac{\pi}{2}$$

where Nkd is effectively infinite. From (8),

$$(g_d)_{\max} \approx \frac{2l}{\lambda} \qquad \left(\text{large } \frac{l}{\lambda}\right) \tag{9}$$

End-Fire Arrays   If the elements of the array are progressively delayed in phase just enough to make up for the retardation of the waves, it would be expected that the radiation from all elements of the array could be made to add in the direction of the array axis. Such an array is called an *end-fire array*. To accomplish this, let

$$a_n = a_0 e^{-jnkd} \tag{10}$$

Then in (2),

$$S = a_0 \sum_{n=0}^{N-1} e^{-jnkd(1-\cos\theta)} = a_0 \frac{1 - e^{-jNkd(1-\cos\theta)}}{1 - e^{-jkd(1-\cos\theta)}} \tag{11}$$

$$|S| = \frac{\sin\frac{1}{2}[Nkd(1-\cos\theta)]}{\sin\frac{1}{2}[kd(1-\cos\theta)]} a_0 \tag{12}$$

By comparison with (5), it is recognized that the plot of Fig. 12.19*b* made for a broadside array may be utilized for the end fire also if the abscissa is interpreted as $kd(1 - \cos \theta)$. The pattern as a function of θ of course looks different, but the ratio of secondary to primary maxima is the same. It may also be shown to follow that the formula for directivity (9) applies also to an end-fire array with large l/λ. To obtain the angular width of the main lobe, let $\Delta/2$ be the angle at which S goes to zero.

$$Nkd\left(1 - \cos \frac{\Delta}{2}\right) = 2\pi$$

$$\Delta \approx 2\sqrt{\frac{2\lambda}{l}} \tag{13}$$

Phase Scanning of Arrays By interpretation of the broadside and end-fire examples, it is clear that a phase delay between elements somewhere between the extremes for the two cases should give a maximum lobe between these extremes. Moreover, if this phase delay is controllable by any means (usually with ferrite or semiconductor devices) the direction of the lobe can be scanned without physical motion of the antenna. This may be most important in large installations that must continually scan a large range of directions in short times, as in airport surveillance radar.[26]

Quantitatively, for the linear array, we see the scanning if we allow the phase delay to be $\Delta\varphi$ between elements. Then replacing (10) for the end fire,

$$a_n = a_0 e^{-jn\Delta\varphi} \tag{14}$$

Following procedures as for (5) or (12), the space factor here is

$$|S| = \frac{\sin[(N/2)(kd \cos \theta - \Delta\varphi)]}{\sin \frac{1}{2}(kd \cos \theta - \Delta\varphi)} a_0 \tag{15}$$

The direction of the principal maximum is then given by

$$kd \cos \theta_m = \Delta\varphi \tag{16}$$

and so can be varied if $\Delta\varphi$ is varied.

This principle can of course be extended to two-dimensional as well as linear arrays, although the number of phases to control then increases.

Unequally Spaced Arrays Although the formulation above has been given only for arrays of equal element spacing, unequal spacing of elements provides an ad-

---

[26] R. J. Mailloux, *Proc. IEEE* **70**, 246 (1982).

ditional degree of freedom which may sometimes be used to advantage. Unequally spaced arrays have been used to give greater gain and lower side lobes than an equally spaced array with the same number of elements.[27] The amplitude of excitation of the elements may be retained more nearly constant in the array of unequal spacing. Analysis can be carried out by digital computation once element spacings and excitations are specified.

12.20   Radiation from Diffraction Gratings

In Sec. 6.10 we saw the basic idea of the diffraction grating, which has become of increased importance since coherent light has become available from lasers. Here we make use of the array factors developed in Sec. 12.19 to explain the patterns observed for diffraction gratings. The grating could take the form of a parallel array of slits of the type discussed in Sec. 12.14, with each slit forming a radiation source. For practical reasons, these are often engraved lines ("rulings") on a solid reflector rather than open slits, but they act very similarly with regard to radiation pattern.

Consider first the radiation pattern of two parallel slits, each of width a, separated by distance d between central axes. Assuming the slits are illuminated with the same phase, Eq. 12.19(5) for the broadside array may be adapted, with $N = 2$:

$$|S|^2 = a_0^2 \frac{\sin^2 \frac{1}{2}(2kd \cos \phi)}{\sin^2 \frac{1}{2}(kd \cos \phi)} = 4a_0^2 \cos^2\left(\frac{kd}{2} \cos \phi\right) \tag{1}$$

Here, because the slits are infinitely long, the array factor is put in terms of ϕ in cylindrical coordinates. Then the radiation intensity is given by Eq. 12.19(3) with K_0 representing the radiation intensity for a single slit, Eq. 12.14(7). We let $4a_0^2 A = B$ and find

$$K = B\left[\frac{\sin(ka/2 \cos \phi)}{ka/2 \cos \phi}\right]^2 \cos^2\left(\frac{kd}{2} \cos \phi\right) \tag{2}$$

which is shown in Fig. 12.20a for $d = 2a$.

Array factors for larger number of radiators have shapes like that in Fig. 12.19b. Shown there is a principal maximum with subsidiary maxima (side lobes). If the spacing d between slits exceeds one wavelength, there will be repeated principal maxima, as was seen in Sec. 6.10. using the idea of constructive interference. Photographs of diffraction patterns for different sets of slits are shown in Fig. 12.20b. The broad maximum of the single slit is seen with its subsidiary maxima for $N = 1$. With three slits the pattern becomes sharper and the subsidiary maxima

[27] D. D. King, R. F. Packard, and R. K. Thomas, *IRE Trans. Antennas and Propagation* **AP-8**, 380 (1960).

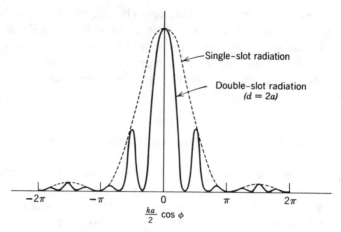

Fig. 12.20a Fraunhofer intensity pattern for two slits compared with that of one of the slits alone. Slit width is a and spacing d.

may be seen only vaguely. In the case of $N = 20$, the larger number of slits leads to very distinct principal maxima and the intensity of the subsidiary maxima is low enough that they do not register in the photograph. The clear separation of the lines is of utmost importance in the use of diffraction gratings for spectroscopy or other applications where it is desired to separate patterns of different frequency. In practical gratings there are hundreds, or even thousands, of slits (or rulings) so that the principal maxima are very sharp and the subsidiary maxima are very near the principal maxima.

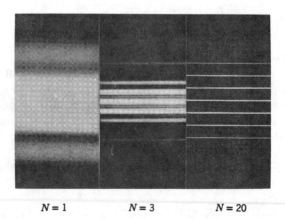

$N = 1$ $\qquad$ $N = 3$ $\qquad$ $N = 20$

Fig. 12.20b Diffraction patterns showing the increased sharpness resulting from increasing the number of slits.

12.21 Polynomial Formulation of Arrays and Limitations on Directivity

Schelkunoff[28] has shown that linear arrays with equal element spacings may be formulated as polynomials, with useful results obtained by interpretations of the zeros of the polynomials in the complex plane. If we define

$$\zeta = e^{j\psi} = e^{jkd\cos\theta} \tag{1}$$

then Eq. 12.19(2) may be written as a polynomial in ζ,

$$S = \sum_{n=0}^{N-1} a_n \zeta^n \tag{2}$$

Concentration is now on the properties of S in the complex ζ plane.

Note first that real θ corresponds to values of ζ on the unit circle with phase angles between $-kd$ and kd. All, a part, or none of the $N-1$ zeros of S may occur in this part of the unit circle. When they do so occur, they correspond to true zeros of the pattern, or "cones of silence." The broadside array, Eq. 12.19(5), has its zeros spread out uniformly over the entire unit circle except for the missing one at $\psi = 0$ (Fig. 12.21a), where the very large main lobe builds up. One approach to the synthesis of arrays is then that of positioning zeros on this picture so that they are close together where the pattern is to be of small amplitude, and farther apart where it is to build up to a relatively large value.

If we limit ourselves to linear arrays with currents in phase, it can be shown that the uniform array (one with all currents of equal amplitude) gives more directivity than arrays with nonuniform excitations. Still greater directivities are possible in principle, however, if one goes to excitations of other phases. Arrays having more directivity than the uniform array are known as *supergain arrays*.

In terms of the picture in the ζ plane, we see that for an element spacing less than a half-wavelength ($kd = \pi$) the range of real θ covers only a part of the unit circle. The uniform array, however, has its zeros spread over all the unit circle, so some are "invisible." Schelkunoff has shown that arbitrarily high directivity for a given antenna size is possible in principle by moving the zeros into the range of real θ, properly distributed to give the desired directivity (Fig. 12.21b). The trouble here is that a monstrous lobe builds up in the "invisible" range from which the zeros have been eliminated, which might seem to be of no concern, but it turns out to represent reactive energy and so is of importance. It is surprising to find the rapidity with which this limitation takes over. For high-gain broadside arrays, no significant increase in gain is possible over that of the uniform array before the reactive energy becomes impossibly large.[29] For end-fire arrays, a modest increase is possible, and has been utilized in practice.[30] Another way of stating

---

[28] S. A. Schelkunoff, *Bell Syst. Tech. J.* **22**, 80 (1943).

[29] L. J. Chu, *J. Appl. Phys.* **19**, 1163 (1948).

[30] W. W. Hansen and J. R. Woodyard, *Proc. I.R.E.* **26**, 333 (1938).

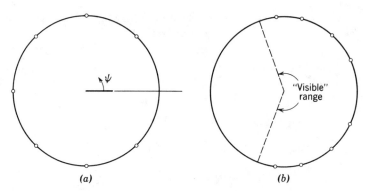

Fig. 12.21 (*a*) Location of zeros of polynomial representing a uniform broadside array. (*b*) Positioning of zeros of polynomial to produce a supergain array.

this limitation is that currents of the elements become huge for a given power radiated and fluctuate in phase from one element to the next so that it would be impossible to feed such an array.

A physical picture is provided by looking at the problem from a wave point of view. Imagine that we are attempting to produce a high-gain broadside array with a thin pancake pattern near the equator. We may imagine the distant fields (H_ϕ and E_θ) of this pattern expanded in a series of the spherically symmetrical *TM* modes of the type studied in Sec. 10.7. If the pattern is to be sharp, it is clear that waves of very high order are needed to represent this pattern (order of $2\pi/\Delta$ for a narrow beam of angle Δ). A study of the Hankel functions shows that these functions change character at a radius such that n is of the order kr, becoming rapidly reactive for radii less than this value. Hence, the antenna boundary must extend approximately to this radius if excessive reactive power is to be avoided. That is,

$$\frac{2r}{\lambda} \approx \frac{n}{\pi} \approx \frac{2}{\Delta} \tag{3}$$

which gives a relation between angle and length equivalent to that for a uniform broadside array, Eq. 12.19(6). The phenomenon is a cutoff of the type found in sectoral horns, Sec. 9.4, where it was found that reactive effects caused an effective cutoff when the cross section became too small to support the required number of half-wave variations in the pattern. The rapidity with which the limitation takes over must again be stressed. For an array 50 wavelengths long, a halving in size from that of the uniform array would require reactive power 10^{59} times the radiated power.

It should not be inferred from the foregoing discussion that uniform arrays are always best. Often the side-lobe level (13.5 dB) is higher than can be tolerated.

Side lobes can be reduced with a sacrifice in directivity. Dolph[31] has given the procedure for finding the array of a given number of elements which gives the lowest side lobes for a prescribed antenna directivity, or highest directivity for a prescribed side-lobe level. The polynomial $S(\zeta)$ in (2) has in this case the form of a Chebyshev polynomial.

12.22   Yagi–Uda Arrays

One difficulty in using multielement arrays to achieve directivity is the need to provide controlled feeds to the various elements. The so-called Yagi–Uda array avoids this problem by feeding only one element and having other elements with currents induced by the first; correct phasing is achieved by adjusting the size and positions of the other elements.

Consider the situation shown in Fig. 12.22a where there is one driven element and one parallel "parasitic" element. The linearity of Maxwell's equations makes it possible to write a set of equations relating the voltages at the center of the antennas to the currents at the feed points:

$$V_1 = Z_{11}I_1 + Z_{12}I_2$$
$$V_2 = Z_{21}I_1 + Z_{22}I_2 \tag{1}$$

where the Z_{ij} are constants that depend upon the lengths l_1 and l_2 and the separation d of the elements. Using the fact that the voltage at the drive point of element 2 is zero, one finds from (1) that I_2 is determined (by induction) by the current I_1 and the coefficients in (1):

$$I_2 = -\frac{Z_{21}}{Z_{22}}I_1 \tag{2}$$

The array factor in Eq. 12.19(2) can be written for this array as

$$S(\theta) = a_0 + a_1 e^{jkd\cos\theta} = I_1\left(1 + \frac{I_2}{I_1} e^{jkd\cos\theta}\right) \tag{3}$$

Therefore, we see that the pattern depends upon the spacing d and upon the coefficients Z_{12} and Z_{22}.

The methods to be studied in Sec. 12.26 can be used to show that the mutual impedance Z_{21} is rather insensitive to the lengths $2l_1$ and $2l_2$ when they are nearly $\lambda/2$. Therefore, the phase of the current I_2 depends, according to (2), mainly upon

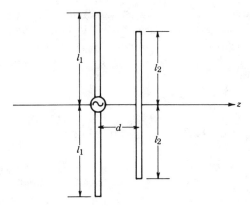

Fig. 12.22a Driven antenna element with nearby, parallel, parasitic element.

the self-impedance Z_{22} of element 2. The direction of maximum radiation of course depends upon relative phases of currents I_1 and I_2.

If the arrangement is such as to maximize radiation in the $-z$ direction, element 2 is called a *reflector*. The maximum directivity is achieved if the spacing d is about 0.16λ. If the driven element has length $2l_1 = \lambda/2$, the length of the reflector should be slightly greater: $0.51 < 2l_2/\lambda < 0.52$.

If the radiation is maximized in the $+z$ direction, element 2 is called a *director*. In this case, the maximum directivity is achieved with a spacing of about 0.11λ. With $2l_1 = \lambda/2$, the director length should be in the range $0.38 < 2l_2/\lambda < 0.48$, that is, somewhat smaller than the driven element.

The simplest array that is called a Yagi–Uda type has both a reflector and a director. The situation is then considerably more complicated because of the interaction of the three elements, but the same conclusions as above are roughly applicable. The reflector is still larger, and the director smaller, than the driven element. The directivity turns out to not depend critically upon the spacings of the director and reflector from the driven element. The input impedance is given by

$$Z_{in} = Z_{22} + \left(\frac{I_1}{I_2}\right) Z_{21} + \left(\frac{I_3}{I_2}\right) Z_{23} \qquad (4)$$

Because of the phasing of the currents, Z_{in} tends to be low. It can be raised if the driven element is made a folded dipole (Sec. 12.11) for which Z_{22} is four times that of the single dipole.

The addition of other, parallel reflectors has little effect since the field behind the first is small, but additional directors can be added advantageously. Typical antennas for television reception have several directors and have a radiation pattern as shown in Fig. 12.22b. Directivities between 10 and 100 are obtainable depending on the number of director dipoles.

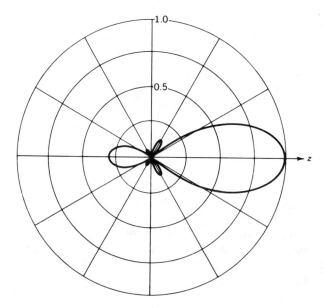

Fig. 12.22b Intensity pattern of a typical Yagi–Uda type television receiving antenna, having four directors and one reflector. (After W. L. Stutzman and G. A. Thiele, Ref. 9.)

12.23 Frequency-Independent Antennas: Logarithmically Periodic Arrays

The determining factor for the radiation pattern and the impedance of a lossless antenna is the ratio of the dimensions to wavelength. Thus, if an antenna is to be scaled for use at another frequency, all dimensions must be multiplied by the ratio of the wavelengths. If the structural form of an antenna could be defined such that the multiplication of all dimensions by the wavelength ratio for two frequencies would leave the antenna unchanged, that antenna would behave identically at the two frequencies. The earliest work[32] on structures having this property considered an equiangular spiral on either a plane or conical surface with arms having the form $r = r_0 e^{a\phi}$. It is easy to see that a magnification of a structure having this form is identical to the original except for a rotation (which can be an undesirable feature). Frequency independence assumes the unrealizable condition that the spirals start at $r = 0$ (where the feed must be connected) and extend to infinity. Therefore, the properties of such structures are, in practice, only approximately frequency independent. A number of derived forms that eliminate some of the disadvantages of the equiangular spiral have been studied.

[32] V. H. Rumsey, *IRE National Convention Record*, Part I, 114 (1957).

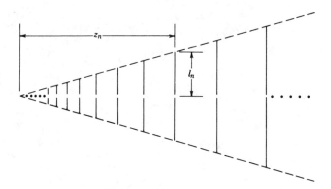

Fig. 12.23*a* Logarithmically periodic antenna array.

In this section we concentrate on a more convenient type of (nearly) frequency independent antenna.[33] The idealized form is shown in Fig. 12.23*a*, in which the locations of the parallel-wire elements and their lengths increase from one to the next in the fixed ratio τ:

$$z_n = \tau z_{n-1}$$
$$l_n = \tau l_{n-1} \tag{1}$$

If the wavelength is multiplied by τ, then all arm lengths and positions should be multiplied by τ so that everywhere the ratio of length to wavelength is the same. The result is a structure that is identical to the original. Therefore, the single structure appears the same at λ_1, $\tau\lambda_1$, $\tau^2\lambda_1$, and so on. Expressed in terms of logarithms,

$$\log \lambda_n = \log \lambda_1 + (n-1)\log \tau \tag{2}$$

and correspondingly for frequency

$$\log f_n = \log f_1 - (n-1)\log \tau \tag{3}$$

The structure is called *logarithmically periodic*.

An understanding of the radiation properties can be aided by noting the qualitative relationships between three adjacent elements in this antenna and the three-element Yagi–Uda array discussed in the preceding section. It is found from detailed calculations that maximum radiation along the array occurs when the element has a length of about $\lambda/2$. This could be expected from the fact that the element at resonance will have the maximum current for a given drive voltage. Let us consider the resonant element and its two neighbors. We saw in the Yagi–Uda array that the larger acts as a reflector and the smaller as a director, with radiation, therefore, toward the small end of the array.

---

[33] D. E. Isbell, *IRE Trans. Antennas Propagation* **AP-8**, 260 (1960).

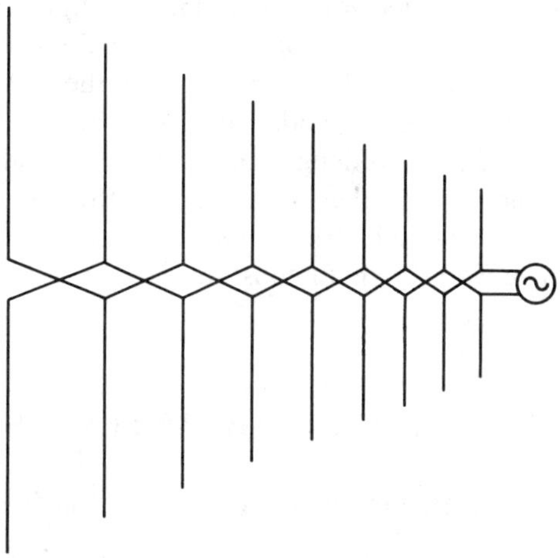

Fig. 12.23b Manner of connecting antenna elements to the driving transmission line for a logarithmically periodic antenna.

The input to the transmission line feeding the array is at the small end for a reason to be discussed below. By using alternating connections to the line as shown in Fig. 12.23b, and spacing the elements by one-half of the element length at that point in the array, the phasing of adjacent elements in the region of maximum radiation is such that the current in the $n + 1$th element leads that in the nth by $\pi/2$ rad. Since the nth element is resonant ($l_n = \lambda/4$), the spacing to the $n + 1$th element is $\lambda/4$ and its radiation in the direction toward the small end of the array is in phase with that from the nth element.

Detailed calculations[34] show that the phase of the transmission-line voltage provided to the dipoles advances uniformly away from the feedpoint up to the region of the strongly radiating dipoles. The amplitude decays slowly because the short elements are radiating inefficiently up to the region of strong radiation. Over a range of a few elements, the voltage level drops by a factor of 10 and then slowly decays over the remainder of the array. The point of maximum radiation moves toward the small end as frequency is increased and vice versa. The pattern and input impedance was earlier said to repeat at wavelengths $\tau^n \lambda_1$; and detailed calculations show that there is little variation between those wavelengths. Because any real antenna is of limited length and there is a minimum practical size at the small end, the frequency (or wavelength) range of constant pattern and input impedances is limited. Roughly speaking, the range of operation is bounded by wavelengths equal to twice the lengths of the dipoles at the ends of the array. In fact, since several dipoles are involved in the region of large radiation, the bandwidth is somewhat smaller.

[34] W. L. Weeks, *Antenna Engineering*, McGraw-Hill, New York, 1968, Section 7.2.

As pointed out above, the need for alternating the dipole connections to the feeding transmission line arises because of feeding from the small end of the array. The reason for not feeding from the large end is that if the $\lambda/2$ resonance for some given frequency were near the small end, the power flowing from the large end would have encountered dipoles of lengths $n\lambda/2$, which would radiate and excite the other resonant dipoles toward the small end with the result of multiple radiation points. Their locations would depend upon frequency so neither the pattern nor the input impedance would remain constant.

FIELD ANALYSIS OF ANTENNAS

12.24   The Antenna as a Boundary-Value Problem

In our discussion of antennas, we have thus far made assumptions about currents along conductors, or fields in the apertures of the system. This approach has proved useful, especially for calculating radiation patterns, directivity, and other aspects of the far-zone solution. One would like, however, to check the assumed current or field distribution, and to have better knowledge of the near-zone fields. These last are especially important in calculating antenna impedances, or the couplings between nearby radiating elements. In principle, one has only to solve Maxwell's equations subject to boundary conditions, much as we did for waveguides and cavity resonators. There are only a few shapes for which this has been found possible in analytic form.[35] Even though practical antennas are often more complicated, solutions for these idealized shapes have proved useful in developing a fundamental understanding of the radiation problem, and in showing several important features of real antennas. We thus present briefly in this section the boundary-value approach to three of the important idealized shapes that have contributed to the understanding of radiation theory.

Spherical and Spheriodal Antennas   One straightforward approach is that of adding wave solutions to fit the boundary conditions, much as we did in static problems and in waveguides. The simplest configuration to illustrate the procedure is the spherical one, illustrated in Fig. 12.24a. Although not terribly important in itself, a straightforward extension to spheroidal shapes does lead to more practical results. Field solutions in spherical coordinates have been given in Sec. 10.7. The antenna shown is excited with an azimuthally symmetric E_θ at the equator, so the symmetric TM set with E_θ, E_r, and H_ϕ is appropriate. We have seen in Sec. 12.3 that the lowest order of these represents radiation from an infinitesimal dipole,

---

[35] Numerical methods are becoming increasingly important. See, for example, W. L. Stutzman and G. A. Thiele. Ref. 9. Chapter 7.

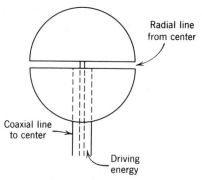

Fig. 12.24a Spherical antenna and possible driving system.

but now we form a series. From Eqs. 10.7(19), using the second Hankel function for the Bessel function in the region extending to infinity,

$$H_\phi(r, \theta) = \sum_{n=1}^{\infty} A_n r^{-1/2} P_n^1(\cos \theta) H_{n+1/2}^{(2)}(kr) \tag{1}$$

$$E_\theta(r, \theta) = \sum_{n=1}^{\infty} \frac{jA_n}{\omega\varepsilon} r^{-3/2} P_n^1(\cos \theta)[kr H_{n-1/2}^{(2)}(kr) - n H_{n+1/2}^{(2)}(kr)] \tag{2}$$

If the exact excitation field in the gap were known, the coefficients A_n could be determined from the orthogonality relations for associated Legendre polynomials [Eqs. 10.7(14)–(15)]. Then the boundary values at $r = a$ are

$$E_\theta(a, \theta) = \sum_{n=1}^{\infty} b_n P_n^1(\cos \theta) \tag{3}$$

where

$$b_n = \frac{2n+1}{2n(n+1)} \int_0^\pi E_\theta(a, \theta) P_n^1(\cos \theta) \sin \theta \, d\theta \tag{4}$$

If the conductor is good, E_θ may be made zero everywhere except in the gap, but field is usually not known exactly there. One might assume a form (say a uniform field over a finite gap), but for our purposes we take the gap as infinitesimal, given only that the integral of field is equal to applied voltage. This is equivalent to assuming an impulse or delta function $V_0 \delta(\cos \theta)$ in the gap. Then b_n from (4) is readily obtained,

$$b_n = \frac{(2n+1)P_n^1(0)V_0}{2n(n+1)a} \tag{5}$$

and by comparison with (2), with $r = a$,

$$A_n = \frac{\omega\varepsilon a^{3/2} b_n}{j[ka H_{n-1/2}^{(2)}(ka) - n H_{n+1/2}^{(2)}(ka)]} \tag{6}$$

Substitution of (6) in (1) and (2) gives the complete solution for the fields around the sphere. In order to find input admittance, we first calculate current at the gap from the θ-directed surface current $J_{s\theta}$:

$$I = -2\pi a J_{s\theta}(\pi/2) = 2\pi a H_\phi(a, \pi/2) \tag{7}$$

Input admittance is

$$Y = I/V_0 = \sum_{n=1}^{\infty} Y_n \tag{8}$$

and with (1), (5), and (6),

$$Y_n = \frac{j\pi(2n + 1)[P_n^1(0)]^2}{n(n + 1)\eta} \left[\frac{n}{ka} - \frac{H_{n-1/2}^{(2)}(ka)}{H_{n+1/2}^{(2)}(ka)}\right]^{-1} \tag{9}$$

It is found that the conductive or real part converges and gives a useful value for this antenna shape. The imaginary or susceptive part does not converge however, since there is infinite capacitance for the infinitesimal gap. One can make an estimate of the finite number of terms to retain for a finite gap size, but in a practical case the connection to the feed line has much to do with the actual input capacitance. In any event the principles of finding exact field solutions are well illustrated by this example.

Stratton and Chu[36] used a similar approach but with prolate spheroidal wave functions for the more-practical, spheroidal antennas illustrated in Fig. 12.24b. Curves of input resistance versus L/λ are shown in Fig. 12.24c and of reactance versus L/λ for a selected gap size in Fig. 12.24d. Several features associated with past antenna knowledge are shown. Resonance (zero reactance) is near the half-wave condition, but not exactly there. Resistance at the resonant condition is near the 72-Ω value known for filamentary half-wave dipoles (see Sec. 12.5). Fatter antennas gave broader bandwidth (i.e., less variation of input impedance about the resonant point). For thin antennas $(L/D \gg 1)$ the analysis showed current distribution well approximated by a sinusoid, as assumed in our earlier calculations for wire antennas.

The Biconical Antenna and Its Extensions   Another shape for which the boundary-value problem may be formulated in terms of a series of spherical wave solutions is that of the biconical antenna used as an example for qualitative discussion in Sec. 12.1. This has been especially important in giving the picture of the antenna as a transducer between waves on the input transmission system and waves in space. Referring to Fig. 12.1, the solutions for the exterior region, $r > l$, are of the same form as (1) and (2). For the interior region, $r < l$, the Bessel function selected is $J_{n+1/2}(kr)$ and a second solution[37] for the associated Legendre function is

[36] J. A. Stratton and L. J. Chu, *J. Appl. Phys.* **12**, 236, 241 (1941).

[37] The second solution is often designated $Q_n^1(\cos \theta)$ but for general angles ψ, n is not an integer so that $P_\nu^1(-\cos \theta)$ may then be shown to be a linearly independent second solution. See, for example, S. A. Schelkunoff, *Proc. I.R.E.* **29**, 493 (1941); S. A. Schelkunoff and C. B. Feldman, *Proc. I.R.E.* **30**, 512 (1942).

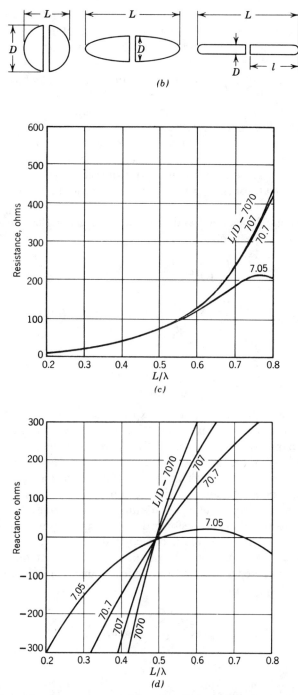

Fig. 12.24 (*b*) Transition from sphere to thin spheroidal wire dipole. (*c*) Resistance and (*d*) reactance of spheroidal antennas fed at the center. From Stratton and Chu, Ref. 36.

required to satisfy the two boundary conditions of $E_r = 0$ at $\theta = \psi$ and $\pi - \psi$. The continuity condition at $r = l$ requires that E_θ and H_ϕ be continuous for $\psi \leq \theta \leq \pi - \psi$ and E_θ be zero over the conducting caps $0 < \theta < \psi$, $\pi - \psi < \theta < \pi$. Tai[38] and Smith[39] carried out the matching exactly for certain angles, but probably most useful are some approximate matchings carried out by Schelkunoff for relatively small-angle cones. One of his methods expands the far-zone fields calculated from a sinusoidal distribution of currents along the cones in spherical TM modes for the exterior region, then shows that they are nearly equal term by term to the modes of the interior region. All these higher order modes, inside and outside, are shown to act exactly as a terminating imped-ance for the TEM mode on the biconical transmission line system, and this ter-mination may be transformed to the source at the origin by transmission-line theory.

The characteristic impedance and load admittance of the transmission-line representation are, respectively,

$$Z_0 = \left(\frac{\eta}{\pi}\right) \ln \cot\left(\frac{\psi}{2}\right) \tag{10}$$

$$Y_L = \frac{1}{Z_0^2} \sum_{m=0}^{\infty} b_m J_{2m+3/2}(kl) H^{(2)}_{2m+3/2}(kl) \tag{11}$$

where

$$b_m = \frac{30\pi k l(4m+3)}{(m+1)(2m+1)} \tag{12}$$

The input impedance plotted from these expressions as functions of l/λ show the same general features as those given above for the spheriodal antenna (with some differences for shape) for lengths $2l \approx \lambda/2$, but were also extended to larger values of l/λ. In the vicinity of the antiresonance, $2l \approx \lambda$, input resistance is high (order of thousands of ohms) and very sensitive to antenna diameter.

Schelkunoff extended his point of view to dipole antennas of other shapes by considering them to have essentially the same load impedance as the biconical antenna, with this impedance transformed to the input by a nonuniform trans-mission-line theory depending upon the configuration of the antenna. In the cylindrical dipole of Fig. 12.24e, for example, an element at radius r from the origin may be considered a section of biconical transmission line of inductance and capacitance (found from (10))

$$L = \frac{\mu\varepsilon}{C} \approx \frac{\mu}{\pi} \ln\left(\frac{2r}{a}\right) \tag{13}$$

[38] C. T. Tai, *J. Appl. Phys.* **20**, 1076 (1949).
[39] P. D. Smith, *J. Appl. Phys.* **19**, 11 (1948).

Fig. 12.24e Cylindrical dipole interpreted as a nonuniform transmission line.

As r varies, L and C vary, but nonuniform transmission-line theory may be used to transform Y_L of (11) to the input. Curves of input resistance and reactance calculated by Schelkunoff in this way are shown in Figs. 12.24f and 12.24g. The values of Z_0 shown are the average values over the length of the dipole.

The biconical model is also useful in telling something about the current distribution along dipole antennas. The current in the dominant TEM wave of the loss-free biconical line is exactly sinusoidal, so departures from a sinusoidal distribution are due either to losses, the higher order modes near the end of the antenna, or the perturbations arising from nonuniform line effects if the shape is other than biconical. For low-loss, long, thin antennas, these effects are small so that the sinusoidal assumptions made earlier are reasonable.

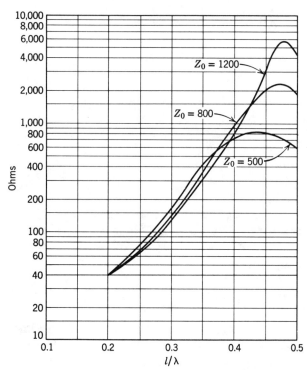

Fig. 12.24f Input resistance for cylindrical dipole antenna. From Schelkunoff. [Ref. 37.]

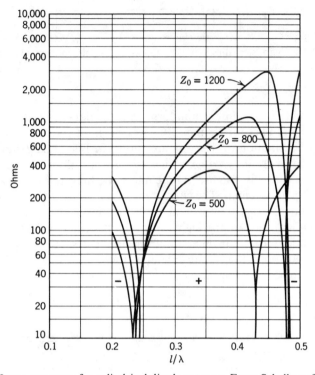

Fig. 12.24g Input reactance for cylindrical dipole antenna. From Schelkunoff. [Ref. 37.]

Integral Equation Approach to the Cylindrical Antenna   A different procedure which has contributed to the theory of antennas is illustrated by the hollow cylindrical antenna pictured in Fig. 12.24h. A surface current element J_{sz} on the surface of the cylinder contributes to the axial component of electric field at some point r, ϕ, z by the relation

$$dE_z(r, \phi, z) = G(r, \phi, z, a, \phi', z')J_{sz}(a, \phi', z') \, dz'a \, d\phi' \tag{14}$$

The function G is called the *Green's function* of this particular problem and may be calculated from the retarded potentials. By integrating over all surface current elements, total E_z is found. This is independent of ϕ because of symmetry, so at the surface of the cylinder may be written

$$E_z(a, z) = \int_{-l}^{l} \int_{0}^{2\pi} G(a, \phi = 0, z, a, \phi', z')J_{sz}(a, \phi', z')a \, d\phi' \, dz' \tag{15}$$

E_z may be taken as zero along the conducting portion, and if it is known or assumed in the exciting gap, the only unknown is the current, which is within the integral. An equation with the unknown within the integral is called an integral equation (this one is a *Fredholm equation of first kind*) and is generally difficult to solve

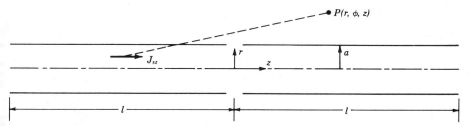

Fig. 12.24h Cylindrical tubular antenna showing current element and field point P.

exactly. Many useful approximate solutions have been given, however, and have verified that for thin-wire antennas the current distribution is very nearly sinusoidal. Moreover, the solutions have made clear the importance of assumptions concerning the exciting region at the center. Thus the junction from the exciting transmission line or waveguide to the input gap is of major importance to the impedance matching problem. For details of the integral equation method see certain of the advanced texts.[40]

12.25   Direct Calculation of Input Impedance for Wire Antennas

We saw in the preceding section that input impedance can be found from the fields produced by antenna structures and currents flowing on their surfaces. Here we look at the direct calculation of input impedance. The integral-equation approach discussed briefly in the preceding section can be adapted to direct calculation of terminal impedance without finding the current distribution. Another way is the so-called *EMF method*, which was introduced in Sec. 4.12 and will be extended below. Refinements of these two approaches give comparable values for wire antennas.[41]

Consider a circular cylindrical radiator of length $2l$ and radius a with an infinitesimal gap in the center. (See Fig. 12.25.) The cylinder is a perfectly conducting tube with negligibly thin walls so end currents can be neglected. The electric field $E_{az}(z)$ is zero on the surface at $r = a$ except in the gap at the center where we assume it can be represented by a delta function so that

$$- \int_{-l}^{l} E_{az}(z)\, dz = V \quad \text{and} \quad E_{az}(z) = 0 \quad \text{for} \quad z \neq 0 \tag{1}$$

---

[40] For example, R. S. Elliott, Ref. 4, Chapter 7; or R. W. P. King in R. E. Collin and F. J. Zucker, *Antenna Theory*, Part I, McGraw-Hill, New York, 1969.

[41] R. S. Elliott, Ref. 4, p. 315.

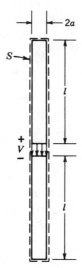

Fig. 12.25 Circular cylindrical dipole radiator used in calculation of terminal impedance by EMF method. The spacing between S and the cylinder is infinitesimal.

where V is the applied voltage. Let us construct a surface S just outside the radiator and have the current $J_{az}(z)$ that flows on the surface of the radiator flow, instead, on the surface S. Then the conducting cylinder can be removed and the fields outside S are the same as in the original situation. Additionally, we place an arbitrary (but later to be specified) current $I_b(z)$ along the axis. Since both currents are inside S, we can apply the reciprocity theorem quoted in Eq. 11.3(3),

$$\oint_S (\mathbf{E}_a \times \mathbf{H}_b - \mathbf{E}_b \times \mathbf{H}_a) \cdot d\mathbf{S} = 0 \tag{2}$$

where $\mathbf{E}_a$ and $\mathbf{H}_a$ are fields of current J_a and fields $\mathbf{E}_b$ and $\mathbf{H}_b$ are those of current I_b.

Assuming very small radius a, the fields on the end surfaces of S can be neglected. Also making use of symmetry in ϕ, (2) becomes

$$\int_{-l}^{l} 2\pi a [E_{az}(a, z) H_{b\phi}(a, z) - E_{bz}(a, z) H_{a\phi}(a, z)] \, dz = 0 \tag{3}$$

The field $H_{a\phi}(a, z)$ can be put in terms of current since

$$2\pi a H_{a\phi}(a, z) = 2\pi a J_{sa}(z) = I_a(z) \tag{4}$$

Then from (3) with (1) and (4), we get

$$2\pi a H_{b\phi}(a, 0)V = -\int_{-l}^{l} E_{bz}(a, z) I_a(z) \, dz \tag{5}$$

Assuming radius a small, we can make the approximation

$$2\pi a H_{b\phi}(a, 0) \approx I_b(0) \tag{6}$$

Since $I_b(z)$ is arbitrary, we can take it to be identical with $I_a(z)$. Using these arguments and dropping the now-superfluous subscripts a, we have

$$I(0)V = -\int_{-l}^{l} E_z(a, z)I(z)\,dz \tag{7}$$

and the induced *EMF* formula for input impedance becomes

$$Z = \frac{V}{I(0)} = -\frac{1}{I^2(0)}\int_{-l}^{l} E_z(a, z)I(z)\,dz \tag{8}$$

To make use of (8), we first assume a current distribution $I(z)$ and calculate the field $E_z(z)$ that $I(z)$ would produce along the surface of S at $r = a$ and then perform the integration. Using the sinusoidal current distribution in Eqs. 12.5(1) that has been shown to be reasonably accurate for thin antennas and the methods in Secs. 3.20–3.21, the electric field E_z along the surface S can be found to be

$$E_z = -\frac{j\eta}{4\pi}I_m\left(\frac{e^{-jkr_1}}{r_1} + \frac{e^{-jkr_2}}{r_2} - 2\cos kl\,\frac{e^{-jkr}}{r}\right) \tag{9}$$

where I_m is the peak value of the standing current wave and the other quantities are

$$r_1 = [(l - z)^2 + a^2]^{1/2}$$
$$r_2 = [(l + z)^2 + a^2]^{1/2} \tag{10}$$
$$r = (a^2 + z^2)^{1/2}$$

Substituting (9) in (8) gives the impedance

$$Z = j30\left(\frac{I_m}{I(0)}\right)^2 \int_{-l}^{l} \sin k(l - |z|)\left(\frac{e^{-jkr_1}}{r_1} + \frac{e^{-jkr_2}}{r_2} - 2\cos kl\,\frac{e^{-jkr}}{r}\right)dz \tag{11}$$

Equation (11) can be evaluated numerically and or can be put in terms of tabulated functions, *sine integrals*, $Si(x)$ and *modified cosine integrals* $Cin(x)$.[42] The results indicate that in this approximation the resistive part of Z is independent of radius a but the reactance is not.

The impedance formula (8) can be recast into a form that is *stationary* with respect to the choice of $I(z)$; that is, no first-order error is made in calculation of Z when the guess for $I(z)$ contains first-order errors.[43]

---

[42] E. C. Jordan and K. G. Balmain, Ref. 6, pp. 540–547.
[43] R. S. Eliott, Ref. 4, p. 305.

The approach of this section is clearly related to the circuit approach of Sec. 4.12, and the self-impedance determination a generalization of the method of calculating inductance by considering induced fields from an equivalent current on the axis of the wire (Sec. 4.7).

12.26   Mutual Impedance Between Thin Dipoles

When arrays of dipoles are used, interactions between them affect the impedances seen by the feed system at their drive points. Also, a dipole with a reflector is equivalent to a pair of dipoles so the terminal impedance is similarly affected. Circuit equations relating terminal voltages and currents can be written for an array of n dipoles:

$$V_1 = Z_{11}I_1 + Z_{12}I_2 + \cdots + Z_{1n}I_n$$
$$V_2 = Z_{21}I_1 + Z_{22}I_2 + \cdots + Z_{2n}I_n$$
$$\vdots$$
$$V_n = Z_{n1}I_1 + Z_{n2}I_2 + \cdots + Z_{nn}I_n$$

(1)

The input impedance for dipole 1 is

$$Z = \frac{V_1}{I_1} = Z_{11} + Z_{12}\frac{I_2}{I_1} + \cdots + Z_{1n}\frac{I_n}{I_1}$$

(2)

In the absence of terminal currents on the other dipoles $Z = Z_{11}$, which is usually assumed to be the terminal impedance of the isolated dipole calculated in the preceding section. (The currents induced on the open-circuited neighboring dipoles produce an important reaction on the one excited only if they are extremely close or the length is near $n\lambda/2$.) The off-diagonal terms represent the mutual effects and are called *mutual impedances*. It is further assumed that mutual impedances can be calculated between any two of the dipoles with the others open-circuited and not producing significant effect on the pair being considered.

Let us consider two dipoles as shown in Fig. 12.26a where we have made them parallel but not necessarily of the same length or having their centers in the same plane. The mutual impedance is defined as the ratio of the open-circuit voltage in dipole 1 produced by a current in dipole 2 to the terminal current of dipole 2:

$$Z_{12} = \frac{(V_1)_{oc}}{I_2(0)}$$

(3)

This is recognized as an extension of mutual inductance ideas with retardation included. It can be argued from the general reciprocity theorem[44] that

$$Z_{12} = -\frac{1}{I_1(0)I_2(0)} \int_{-l_2}^{l_2} E_z^1(z_2)I_2(z_2)\,dz_2$$

(4)

[44] R. S. Elliott, Ref. 4, Section 7.13.

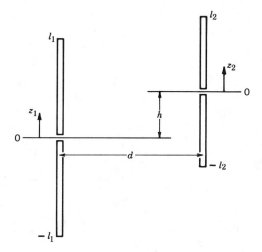

Fig. 12.26a Model for mutual impedance calculation for a pair of thin dipoles.

Here $I_2(z_2)$ is the assumed current distribution in dipole 2 with terminal value $I_2(0)$, and $E_z^1(z_2)$ is the electric field along the surface of dipole 2 that would exist in its absence if a current were impressed on dipole 1 with terminal value $I_1(0)$. The electric field $E_z^1(z_2)$ can be expressed by Eq. 12.25(9) with l being l_1 in the present problem, if we make the usual assumption of sinusoidal current distribution for I_1. In this case,

$$r = [d^2 + (h + z_2)^2]^{1/2}$$
$$r_1 = [d^2 + (h + z_2 - l_1)^2]^{1/2} \tag{5}$$
$$r_2 = [d^2 + (h + z_2 + l_1)^2]^{1/2}$$

where the displacements d and h are shown in Fig. 12.26a. Then making $I_2(z_2)$ also to be sinusoidal $[I_2(z_2) = I_{m2} \sin k(1_2 - |z_2|)]$ one finds

$$Z_{12} = \frac{j30}{(\sin kl_1)(\sin kl_2)} \int_{-l_2}^{l_2} \left(\frac{e^{-jkr_1}}{r_1} + \frac{e^{-jkr_2}}{r_2} - 2 \cos kl_1 \frac{e^{-jkr}}{r} \right) \sin k(l_2 - |z_2|)\, dz_2$$

$$\tag{6}$$

A typical result is shown in Fig. 12.26b for two half-wave dipoles with no vertical displacement ($h = 0$) as a function of spacing d.

It should be noted that the result (6) is not dependent on the radii of the dipoles as long as they are small compared with wavelength and, by the same reasoning, the dipoles need not be of circular cross section for this result to apply.

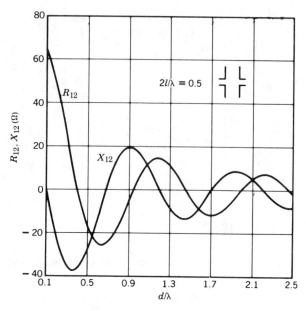

Fig. 12.26b Real and imaginary parts of mutual impedance between half-wave dipoles with zero vertical separation [$(h = 0$ in 12.26a)] as a function of spacing. Robert S. Elliott, AN-TENNA THEORY AND DESIGN, © 1981, p. 333. Reprinted by permission of Prentice-Hall, Inc., Englewood Cliffs, N.J.

RECEIVING ANTENNAS AND RECIPROCITY

12.27   A Transmitting–Receiving System

The discussion in previous sections has generally implied that the radiating system was to be used as a transmitting antenna, exciting waves in space from some source of high-frequency energy. The same devices useful for transmission are also useful for reception, and it will be seen that the quantities already calculated (such as the pattern, directivity, and input impedance for transmission) are the useful parameters in the design of a receiving system also. This might at first seem surprising, since the two problems have some noticeable differences. In the transmitting antenna, a generator is generally applied at localized terminals, and waves that are set up go out in space approximately as spherical wave fronts. In the receiving antenna, a wave coming in from a distant transmitter approximates a portion of a uniform plane wave, and so sets up an applied electric field on the antenna system quite different from that associated with the localized sources in the transmitting case. As a consequence, the induced fields must be different so that total field meets the boundary conditions of the antenna, and the current distribution in general

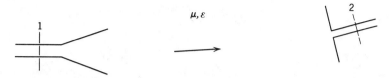

Fig. 12.27a A system of transmitting and receiving antennas.

is different for the same antenna on transmission or reception. The currents set up on the receiving antenna system by the plane wave will convey useful power to the load (probably through a transmission line or guide), but will also produce reradiation or scattering of some of the energy back into space. The mechanism of this scattering is exactly the same as that discussed for radiation from a transmitting antenna, but the form may be different for a given antenna because of the different current distribution.

Thus, we seem to have somewhat different pictures of the mechanism of transmitting and receiving electromagnetic radiation. Reciprocity theorems related to those already discussed (Sec. 11.3) provide ties between the two phenomena such as the following:

1. The antenna pattern for reception is identical with that for transmission.
2. The input impedance of the antenna on transmission is the internal impedance of the equivalent generator representing a receiving system.
3. An effective area for the receiving antenna can be defined and by reciprocity is related to the directivity (Sec. 12.6).

These points will be discussed in this and following sections. An excellent treatment in more detail has been given by Silver.[45]

We wish to begin the discussion by considering the transmitting and receiving antennas with intermediate space (Fig. 12.27a) as a system in which energy is to be transferred from the first to the second. We select terminals in the feeding guide where voltage and current may be defined in the manner explained in Sec. 11.2, and similarly select a reference in the guide from the receiving antenna. The region between, including both antennas, the space, and any intermediate conductors and dielectric (assumed linear) may be represented as a two port or transducer as indicated in Fig. 12.27b. That is,

$$V_1 = Z_{11}I_1 + Z_{12}I_2 \tag{1}$$

$$V_2 = Z_{21}I_1 + Z_{22}I_2 \tag{2}$$

The systems discussed in Chapter 11, for which proofs of (1) and (2) were given, were assumed to be closed by conducting boundaries, whereas the present system

---

[45] S. Silver, Ref. 19, Chapter 2.

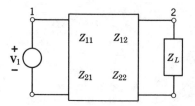

Fig. 12.27b Equivalent representation for Fig. 12.27a.

extends to infinity. The theorems given there, however, can be extended to regions extending to infinity[46] because of the manner in which fields die off there.

The present system is specialized in another respect in that the coupling impedance Z_{12} in (1) is very small for a large separation between transmitter and receiver. It may then be neglected in (1), and the impedance coefficient Z_{11} is just the input impedance of the transmitting antenna calculated by itself.

$$V_1 \approx Z_{11}I_1 \approx Z_A I_1 \tag{3}$$

The coupling term in (2) cannot be neglected, since this coupling is the effect being studied. Equation (2), however, can be represented by the usual equivalent circuit of Thévenin's theorem (Prob. 11.2f) in which an equivalent voltage generator $I_1 Z_{21}$ is connected to the load impedance Z_L through an antenna impedance Z_{22} (which is essentially the input impedance of antenna 2 if driven as a transmitter). Thus, because of the small coupling, the reaction of the receiving antenna on the transmitting antenna can be neglected and the equivalent circuit separated as in Fig. 12.27c.

The equivalent circuit will be discussed again later. For the moment, we shall discuss transmission over the system from another point of view. For this purpose, an *effective area* of the receiving antenna is defined so that the useful power removed by the receiving antenna is given by this area multiplied by the average Poynting vector (power density) in the oncoming wave.

$$W_r = A_{er} P_{av} \tag{4}$$

Like directive gain defined previously, A_{er} is in general a function of direction about the antenna, and of the condition of match in the guide. When not otherwise specified, it will be assumed to be the value for a matched load. The power density at the receiver is the power density of an isotropic radiator ($W_t/4\pi r^2$) multiplied by directive gain of the transmitting antenna in the given direction. Therefore, from (4),

$$W_r = W_t \frac{A_{er}}{4\pi r^2} g_{dt} \tag{5}$$

[46] W. K. Saunders, *Proc. Nat. Acad. Sci.* **38**, 342 (1952).

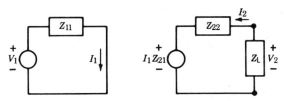

Fig. 12.27c Approximate equivalent circuit neglecting reaction of receiver back on transmitting system.

where W_t is the power transmitted, r is the distance between transmitter and receiver, g_{dt} is the directive gain of the transmitting antenna, and A_{er} is the effective area of the receiving antenna. We will see in Sec. 12.28 that effective area always is related to directive gain by the same constant; the constant may be found for any specific antenna and any direction and then applied to all other cases. It was seen in Sec. 12.14, for an aperture with an area A large enough to ignore source currents on the surfaces surrounding it, that

$$(g_d)_{max} = \frac{4\pi}{\lambda^2} A \qquad (6)$$

The last term in (6) follows from recognition that the power W_r received through a large aperture oriented perpendicular to $\mathbf{P}$ is $P_{av} A$. Making use of the above-mentioned general applicability of the constant $4\pi/\lambda^2$, (5) may be written in either of the following forms given by Friis.[47] Subscripts r and t refer to receiving and transmitting antennas, respectively.

$$\frac{W_r}{W_t} = \frac{\lambda^2 g_{dr} g_{dt}}{(4\pi r)^2} = \frac{A_{er} A_{et}}{\lambda^2 r^2} \qquad (7)$$

12.28 Reciprocity Relations

From the equivalent circuit of Fig. 12.27c, we can obtain a different form for the power delivered to the receiving antenna. For this purpose, let us assume that there is a conjugate match

$$Z_L = Z_{22}^* = R_{r2} - jX_{r2} \qquad (1)$$

which is known to be the condition for maximum power transfer from the equivalent generator to the load. The power delivered to the load under this condition is

$$W_r = \frac{|I_1 Z_{21}|^2}{8R_{r2}} \qquad (2)$$

[47] H. T. Friis, *Proc. I.R.E.* **34** 254 (1946).

If the transmitting antenna has radiation resistance R_{r1}, transmitted power is

$$W_t = \tfrac{1}{2}|I_1|^2 R_{r1} \tag{3}$$

so

$$\frac{W_r}{W_t} = \frac{|Z_{21}|^2}{4R_{r1}R_{r2}} \tag{4}$$

By comparing (4) with Eq. 12.27(5), we see that

$$|Z_{21}|^2 = \frac{R_{r1}R_{r2}g_{d1}A_{e2}}{\pi r^2} \tag{5}$$

If we now reverse the roles of transmitting and receiving antennas, we find for the transfer impedance in the reverse direction

$$|Z_{12}|^2 = \frac{R_{r2}R_{r1}g_{d2}A_{e1}}{\pi r^2} \tag{6}$$

By the reciprocity argument of Sec. 11.3 (modified so that it applies to a region extending to infinity), Z_{12} and Z_{21} are equal, so we conclude

$$\frac{A_{e1}}{g_{d1}} = \frac{A_{e2}}{g_{d2}} \tag{7}$$

The antennas in the foregoing argument were arbitrary, so it follows from (7) that the ratio of effective area to directive gain for all antennas is the same. As discussed in Sec. 12.27, the constant of proportionality is known from large-aperture theory to be $\lambda^2/4\pi$. It could also be shown for other specific antennas, such as the Hertzian dipole[48] or for the small loop antenna.

Although, for large apertures effective area equals actual area, this is not true for small antennas. In fact, for the Hertzian dipole, we found the directivity to be 1.5, so the maximum effective area is

$$(A_e)_{max} = \frac{\lambda^2}{4\pi}(g_d)_{max} = \frac{3}{8\pi}\lambda^2 \qquad \text{(for dipole)} \tag{8}$$

which is finite and sizable even though antenna size is infinitesimal.

Another relation following from reciprocity is that the pattern of a given antenna is the same for transmission or reception. This is useful because the pattern may then be calculated or measured in the easiest way and then be used for both transmission or reception designs. To show this, imagine, as in Fig. 12.28a, that antenna 2 is moved about the arc of a circle to measure the pattern of antenna 1. Let $\theta = 0$ be the angle of maximum response (position a) and angle θ (position b) be a general

---

[48] D. O. North, *RCA Review* **6**, 332 (1942).

position. Then, if 1 is transmitting and 2 receiving, the power received in position b, as compared with position a, by (4) is

$$\frac{W_{2b}}{W_{2a}} = \frac{|Z_{21}|_b^2}{|Z_{21}|_a^2} \tag{9}$$

If 2 is transmitting and 1 receiving, the powers received for the two positions are related by

$$\frac{W_{1b}}{W_{1a}} = \frac{|Z_{12}|_b^2}{|Z_{12}|_a^2} \tag{10}$$

Because of reciprocity $|Z_{12}|_a = |Z_{21}|_a$, and similarly for b, so the ratios (9) and (10) are the same. Thus, the same relative power pattern will be measured with antenna 1 transmitting or receiving. The same result can be argued from the arrangement in Fig. 12.28b.

It is, of course, important to remember that the reciprocity relation may be violated if the transmission path contains a medium such as the ionosphere, which may not have strictly bilaterial properties. It is obvious that frequency must be kept constant when receiver and transmitter are interchanged. Also, if there are obstacles or other secondary radiators in the field, they must keep their same position relative to the system when the interchange is made. (See Prob. 12.28c.)

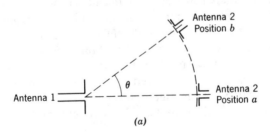

(a)

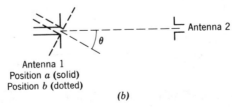

(b)

Fig. 12.28 Possible system for pattern measurement.

12.29   Equivalent Circuit of the Receiving Antenna

For a study of the circuit problem in matching the antenna to a receiver, the second part of the equivalent circuit of Fig. 12.27c is useful and is repeated in Fig. 12.29a. In this the internal impedance of the generator Z_{22} is essentially the input impedance of the same antenna if driven at the same terminals.

$$Z_{22} \approx Z_{i2} \tag{1}$$

This follows by the same argument given for antenna 1 in Sec. 12.27, meaning that reaction back through Z_{21} is negligible *when the antenna is driven.*

The voltage generator in Fig. 12.29a is given from Eq. 12.27(2) as $I_1 Z_{21}$, but transmitter current and transfer impedance are not convenient parameters for most calculations, so other forms in terms of the power density of the oncoming wave are preferable. By substitution from Eq. 12.27(4) and Eq. 12.28(2), we can find this voltage in terms of the average Poynting vector or power density of the oncoming wave P_{av}, the radiation resistance of the receiving antenna R_{r2}, and the effective area A_{e2}. (The effective area is calculated on the assumption of a matched load, but may be a function of the orientation of the antenna with respect to the oncoming wave.)

$$|V_a| = |I_1 Z_{21}| = (8R_{r2} A_{e2} P_{av})^{1/2} \tag{2}$$

The equivalent circuit is useful, for example, in computing the transfer of power from the antenna to the useful load through a transmission line which may have discontinuities, matching sections, or filter elements. All these may be lumped together as a transducer, as explained in the preceding chapter (Fig. 12.29b), and the problem from here on is a standard circuit calculation. It must be emphasized, as in any Thévenin equivalent circuit, that the equivalent circuit was derived to tell what happens in the load under different load conditions, and significance cannot be automatically attached to a calculation of power loss in the internal impedance of the equivalent circuit. In the present case, it is tempting to interpret this as the power reradiated or scattered by currents on the receiving antenna, and one would conclude that as much power is scattered under a condition of perfect match as is absorbed in the load. This conclusion is not true, except in

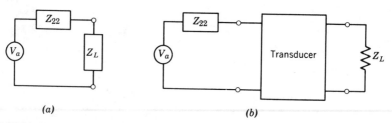

(a)   (b)

Fig. 12.29 (a) Equivalent circuit of receiving antenna for power transfer calculations. (b) Circuit for receiving antenna coupled to the load through a transducer.

special cases where the current distribution may be the same for reception as for transmission.

PROBLEMS

12.3a An inspection of Eqs. 12.3(3) shows that there are other in-phase parts than those considered in forming the power flow. Taking into account all terms, show that the result (6) is correct for average power radiated.

12.3b Study the $n = 2$ TM wave and show that it corresponds to a quadrupole field, that is, field from two small current elements at right angles.

12.3c Show that the fields of Eqs. 12.3(8) are those of a first-order spherical TE wave of Sec. 10.7.

12.3d Plot $|H_\phi|$ of the Hertzian dipole (appropriately normalized) versus kr for $\theta = \pi/2$ and note (approximately) near-zone, far-zone, and transition regions. Do similarly for $|E_\theta|$.

12.3e Plot $|H_\theta|$ of the magnetic dipole (normalized) versus kr for $\theta = \pi/2$ and identify near-zone, far-zone, and transition regions.

12.3f What current is required to radiate 100 W from a loop of circumference equal to 0.1 wavelength?

12.3g Show that a circularly polarized wave may be formed in the far zone if a Hertzian dipole is placed at the center of a small current-carrying loop, the dipole being directed along the axis of the loop. Give the relation of currents in the dipole and the loop.

12.4a Verify the relations of Eq. 12.4(6) as the only field components of **H** remaining if components with higher power of r^{-1} are neglected.

12.4b As in Prob. 12.4a, verify Eqs. 12.4(7).

12.4c Verify Eqs. 12.4(14) and 12.4(15).

12.4d* For the small loop antenna of radius $a \ll \lambda$, assume current distribution uniform around the loop, $I = \hat{\phi}I_0$. Take the z axis perpendicular to the plane of the loop and find A_ϕ in the far zone. From this, obtain fields and compare with those of the magnetic dipole of Sec. 12.3. Also calculate radiation intensity and total power radiated.

12.5a* Although Eq. 12.5(6) may be readily integrated numerically, it can also be evaluated in terms of tabulated functions. Let $u = \cos \theta$, separate denominator by partial fractions, and show that for $\eta = 120\pi$,

$$W = 30I_m^2\{C + \ln 2kl - \text{Ci}(2kl) + \tfrac{1}{2}(\sin 2kl)[\text{Si}(4kl) - 2\,\text{Si}(2kl)]$$
$$+ \tfrac{1}{2}(\cos 2kl)[C + \ln(kl) + \text{Ci}(4kl) - 2\text{Ci}(2kl)]\}$$

where

$$\text{Si}(x) = \int_0^x \frac{\sin x}{x}\,dx, \qquad \text{Ci}(x) = -\int_x^\infty \frac{\cos x}{x}\,dx, \qquad C = 0.5772\ldots.$$

12.5b Following the procedure in Sec. 12.5, find the far-zone fields and radiated power if the antenna is loaded in such a way that current I is substantially constant along the antenna, even though kl is not negligibly small.

12.5c For antennas that are short in comparison with wavelength, and not end-loaded, current-distribution is nearly linear from a maximum at the center to zero at the ends. For such a short, symmetrical dipole, give the mathematical form of currents, far-zone fields, and radiated power.

12.5d Make the calculations required in Prob. 12.5c except that here the current is assumed to have a quadratic distribution with z, $I(z) = I_m[1 - (z/l)^2]$.

12.5e* Obtain Eq. 12.5(4) by integrating far-zone contributions to E_θ from small current elements (Hertzian dipoles) along the antenna, assuming current distribution as in (1). Make the far-zone approximations used in Sec. 12.4.

12.6a Plot polar patterns of field and intensity for a full-wave antenna with $2l = \lambda$.

12.6b Find the directivity of a loop antenna with diameter $\ll \lambda$.

12.6c Find the direction of maximum directive gain and directivity of a dipole of length $3\lambda/2$, for which the pattern is pictured in Fig. 12.6b. Use W from Prob. 12.5a.

12.6d* Find the directivity for the antennas with assumed current distributions as in Probs. 12.5b, 12.5c, and 12.5d.

12.7a Compare the currents that would be required in a half-wave dipole and a small dipole of height 0.05λ to produce 100 W of radiated power from each.

12.7b* Calculate and plot radiation resistance as defined by Eq. 12.7(1) with $0 < kl < \pi$ for thin-wire antennas with sinusoidal distribution. (Note that this problem requires tables of Si and Ci functions, as described in Prob. 12.5a.)

12.7c Simpson's rule is useful for evaluation of radiation integrals. If the area to be evaluated is divided into $2m$ even-numbered portions by $(2m + 1)$ lines spaced an equal distance Δ apart, and values of the function at these lines are $f_0, f_1, \ldots, f_{2m+1}$, the area under the curve is approximately

$$I = \frac{\Delta}{3}[(f_0 + f_{2m}) + 4(f_1 + f_3 + \cdots + f_{2m-1}) + 2(f_2 + f_4 + \cdots + f_{2m-2})]$$

Use this to find the radiation resistance of a half-wave dipole, taking $m = 3$. Compare with the value 73.09 Ω found by more precise methods (Prob. 12.7b).

12.7d* Find expressions for the radiation resistance of antennas with assumed current distributions as in Probs. 12.5b, 12.5c, and 12.5d.

12.7e When current maximum occurs at a position other than the input terminals, radiation resistance is sometimes calculated by referring radiated power to the maximum rather than the input current. (This is particularly useful when assumed current distribution has a zero at the input.) Although not the input resistance in this case, it does allow one to estimate the level of antenna current necessary to radiate a given power. Define such a radiation resistance for general kl in the long antenna with sinusoidal distribution in Sec. 12.5 and make a numerical estimate for the case of $kl = \pi$, for which $I(z = 0) = 0$ from Eq. 12.5(1).

12.8 Prove by a study of the resulting vector potential the *same vertical direction*, opposite *horizontal direction* rule for image currents given in Sec. 12.8.

12.9a Plot the form of radiation intensity as a function of θ of a wire with traveling wave for $kl = \pi, 2\pi, 4\pi$.

12.9b Obtain the value of θ for maximum radiation intensity, for the wire with traveling wave, as a function of kl. Plot θ_{max} versus kl. Find directivity for $kl = \pi$.

12.9c Suppose the traveling wave on the wire has a different phase constant β than the free space k. Find radiation intensity for this case. For a slow wave on the wire ($\beta > k$) describe qualitatively the main effects on the pattern.

12.9d As in Prob. 12.9c, but assume attenuation so that current along the wire is $I_0 \exp[-(\alpha + j\beta)z']$.

12.10a Derive Eq. 12.10(9) for V-antenna distance h above perfectly conducting earth.

12.10b Analysis has been given for a V-antenna with traveling wave from the vertex of the V to the wide end. Show that the same pattern results (for a given kl) if the V is reversed in direction with the traveling waves propagating from the wide end to the vertex.

12.10c V antennas may also be made with standing waves in the two arms in place of the traveling waves considered in Sec. 12.10. For example, the radiation pattern of the dipole of length $3\lambda/2$ sketched in Fig. 12.6b has maximum radiation at an angle 42.5 degrees from the wire axis. If two such wires form a V with angle 85 degrees between them, one would then expect maxima in the forward and reverse directions. Sketch the expected pattern.

12.10d For the V antenna with standing waves, it is found that interaction between arms modifies the simple superposition assumed in Prob. 12.10c. For $0.5 < l/\lambda < 3$, an empirical formula gives optimum angle $2\alpha = 60(\lambda/l) + 38$ degrees with resulting directivity $(g_d)_{max} = 1.2 + 2.3(l/\lambda)$. [W. L. Weeks, *Antenna Engineering*, McGraw-Hill, New York, 1968, p. 141.] Plot 2α and $(g_d)_{max}$ versus l/λ over the range of validity.

12.11 For the folded dipole pictured in Fig. 12.11b, one obvious mode that might seem to be excited is the parallel-wire transmission line mode with equal and opposite currents in the parallel conductors. With the length from input terminals to shorted ends approximately a quarter-wavelength, estimate the input impedance for such a mode and explain why it is not much excited by an input line of a few hundred ohms characteristic impedance.

12.12 Demonstrate that Eqs. 12.12(5) and 12.12(6) do satisfy 12.12(3), with **A** and **F** satisfying inhomogeneous Helmholtz equations.

12.13a* From Eq. 12.13(2) we see that the elemental plane wave source is equivalent to crossed electric and magnetic infinitesimal dipoles. Superpose directly the far-zone fields of these from Sec. 12.3 to obtain Eqs. 12.13(4) and 12.13(5). (Note the required transformation of coordinate axes for this demonstration.)

12.13b Suppose **E** and **H** in an aperture are not related as in a plane wave, but by a general wave impedance Z_w, so that $E_x/H_y = -E_y/H_x = Z_w$. Give the modified form of Eq. 12.13(6).

12.13c* Using the result of Prob. 12.13b, how is the paraxial approximation, Eq. 12.13(10), modified if **E** and **H** in the aperture are related by Z_w?

12.14a* Analyze the rectangular aperture with uniform illumination without making the paraxial approximation (i.e., for θ not necessarily small) and compare with Eq. 12.14(2).

12.14b Analyze, using the paraxial approximation, the far-zone field from a rectangular aperture with uniform variation in y but a quadratic variation in x,

$$E = E_0\left[1 - \left(\frac{2x}{a}\right)^2\right]; \qquad -\frac{a}{2} \le x \le \frac{a}{2}$$

12.14c* In calculating radiated power for the rectangular aperture, we have utilized power conservation and found the flow through the aperture. Check this, using approximations appropriate to small θ, from the far-zone field, Eq. 12.14(2).

12.14d For the circular aperture with uniform illumination, suppose that all radiation was confined to a cone of angle θ_0 given by Eq. 12.14(12) and was of constant amplitude over this cone. Give the directivity and compare with Eq. 12.14(13).

12.14e For $ka = 10$, plot K versus θ for the circular aperture with uniform illumination. Find beam width and directivity.

12.14f* As in Prob. 12.14c, use the far-zone field and paraxial approximations to check radiated power from the circular aperture.

12.14g Find far-zone fields from an "elliptic" gaussian distribution,

$$E(x', y') = E_0 e^{-(x'/w_1)^2} e^{-(y'/w_2)^2}.$$

As in the text, aperture size is large compared with either w_1 or w_2.

12.15 Show that Eq. 12.15(3) is a special case of Eqs. 12.15(5) and 12.15(6), making clear the specializations necessary. Hint: Refer to Prob. 12.13b.

12.16a Prove that the image of a magnetic current in a perfectly conducting plane requires horizontal currents in the same direction for source and image. In what sense is the field for the resonant slot the dual of the field for the half-wave dipole?

12.16b* For the open end of a coaxial line of radii a and b, set up the problem of determining power radiated, assuming the electric field of the *TEM* mode as the only important field in the aperture. Make approximations appropriate to a line small in radius compared with wavelength, and show that the equivalent radiation resistance referred to the open end is

$$R = \frac{3\eta}{2\pi^3} \left[\frac{\lambda^2 \ln (b/a)}{\pi(b^2 - a^2)} \right]^2$$

Calculate the value for $a = 0.3$ cm, $b = 1$ cm, $\lambda = 30$ cm. What standing-wave ratio would this produce in the line?

12.16c* In Prob. 12.16b assume the magnetic field of the *TEM* mode at the open end of sufficient magnitude to account for the power calculated there. Show that a recalculation of power radiated including effects from this magnetic field leads to only a small correction to the first calculation. Take radii small compared with wavelength.

12.16d* Read one of the references on the Babinet principle and supply the derivation of the relation between slot conductance and dipole resistance found in Eq. 12.16(7).

12.17a Utilizing the relation for effective permittivity of disks given in Sec. 12.17, estimate the radius of disks and the spacing (assumed uniform) if $\varepsilon_r = 100$ is desired for use at a frequency of 10 GHz.

12.17b Utilize Eq. 12.17(2) to solve for $d(r)$ in terms of $d(0)$, D, r, and v_p/c. Plot $d(r)$ versus r if $d(0) = 0.1$ m, $D = 1$ m, and the waves between plates act as TE_1 modes with $\lambda = 0.1$ m and plate spacing 0.075 m.

12.18a Find K and sketch the radiation pattern versus ϕ at $\theta = \pi/2$ if phase is as in the Fig. 12.18b but $I_2 = I_1/2$.

12.18b As in Prob. 12.18a but $I_2 = I_1$ in magnitude and phase. Then repeat for $I_2 = -I_1$.

12.18c A section of parallel-wire transmission line when properly terminated may be approximately considered as two wires along which waves are propagating, the currents being opposite in phase at any point along the line. Neglect the radiation from the termination and the mutual effect between the termination and the lines, and obtain an expression for the radiated power from a line of length l and spacing d between thin wires by the method described in Sec. 12.18. Use results for a single wire with traveling wave from Sec. 12.9.

12.18d Consider a radiator with only vertical currents and a radiation intensity K. It is placed a distance h above a plane, perfectly conducting earth. Find an expression for total radiation intensity K_T corresponding to Eq. 12.10(9).

12.19a Explain the statement of the text that directivity of an array of elements is not the product of array directivity and element directivity. Illustrate with a high-directivity array made up of low-directivity elements.

12.19b Considering only the array factor, plot the direction of maximum radiation θ_m for a linear array versus the normalized phase delay $\Delta\varphi/kd$ between elements for element spacings $kd = 0.1, 0.25, 0.4, 0.5$. What can you say about the relative sensitivities of θ_m to changes in $\Delta\varphi$ for nearly endfire versus nearly broadside arrays?

12.20 Plot patterns similar to Fig. 12.20a for $d = 4a$, $d = 6a$.

12.21a The factored form of the polynomial of Eq. 12.21(2) is

$$S = C(\zeta - \zeta_1) \cdots (\zeta - \zeta_{N-1}).$$

Find the locations of the zeros of a broadside array with $N = 4$.

12.21b Consider making a five-element linear array to be "supergain" by choosing $kd = \pi/2$ and placing the zeros at $\zeta = e^{\pm j\pi/6}$, $e^{\pm j\pi/3}$. Find the coefficients of the resulting polynomial and interpret in terms of required current excitation.

12.21c A potential analog of the linear array problem can be set up and has been utilized in array synthesis [T. T. Taylor and J. R. Whinnery, *J. Appl. Phys.* **22**, 19 (1951)] although numerical methods are now preferable. Show that if the logarithm of the factored form of Prob. 12.21a is taken, the result may be interpreted in terms of potential and flux about line charges.

12.22a For a three-element Yagi–Uda array with element 1 a passive reflector, element 2 driven by voltage V_2, and element 3 a passive director, write the equivalent of Eq. 12.22(1) and solve for I_1, I_2, I_3 in terms of V_2.

12.22b Estimate the directivity for the array with pattern sketched in Fig. 12.22b, assuming, for simplicity, that the given pattern also applies in planes normal to that shown.

12.23 Sketch a log-periodic array with seven elements, starting with $l_0 = 0.1$ m, $z_0 = 0.05$ m, and $\tau = 1.1$. Using the physical arguments of Sec. 12.23, estimate the frequency range over which this might be expected to radiate effectively.

12.24a The value of Z_0 shown in Figs. 12.24f and 12.24g for the cylindrical antenna is the average of the variable $Z_0 = (L/C)^{1/2}$ with variables L and C defined by Eq. 12.24(13):

$$Z_0 = \frac{1}{l}\int_0^l Z_0(r)\, dr$$

Show that for the cylindrical antennas this is $Z_0 = 120[\ln(2l/a) - 1]$.

12.24b Using the result of Prob. 12.24a, interpret Z_0 in terms of l/D where $2l = $ total length and $D = $ antenna diameter. Then compare Figs. 12.24f and g with those for spheriodal antennas, Figs. 12.24c and 12.24d.

12.25a Verify Eq. 12.25(9).

12.25b Find the input impedance formula corresponding to Eq. 12.25(11) for a short antenna so loaded that the current is uniform along its length.

12.25c An approximate solution of Eq. 12.25(11), valid where $1.3 \le kl \le 1.7$ and $0.0016 \lesssim a/\lambda \lesssim 0.0095$, is given by [Ref. 4, p. 302]

$$Z = [122.65 - 204.1kl + 110(kl)^2]$$

$$- j\left[120\left(\ln\frac{2l}{a} - 1\right)\cot kl - 162.5 + 140kl - 40(kl)^2\right]$$

Calculate the input impedance for a dipole with $kl = \pi/2$ and $a/\lambda = 0.005$. Compare with the result for the $\lambda/2$ filamentary dipole.

12.25d* It can be shown that Eq. 12.25(11) can be integrated with the assumption that $a \ll \lambda$ and $a \ll l$ to give

$$Z = \frac{j60}{\sin^2 kl} \{[4 \cos^2 kl][S(kl)] - [\cos 2kl][S(2kl)] $$
$$ - [\sin 2kl][2C(kl) - C(2kl)]\}$$

where

$$C(kl) = \ln \frac{2l}{a} - \frac{1}{2} \text{Cin}(2kl) - \frac{j}{2} \text{Si}(2kl)$$

and

$$S(kl) = \frac{1}{2} \text{Si}(2kl) - \frac{j}{2} \text{Cin}(2kl) - ka$$

Use tables of $\text{Si}(x)$ and $\text{Cin}(x)$ integrals to calculate the input impedance for a dipole with $kl = \pi/2$ and $a/\lambda = 0.005$ and confirm the result found by the simpler formula in Prob. 12.25c.

12.26 Give the proof of Eq. 12.26(4).

12.27a Calculate power received corresponding to a transmitted power of 100 W and a distance between transmitter and receiver of 10^3 m under the following conditions:

(i) Directivity of transmitting antennas = 1.5; effective area of receiver = 0.40 m^2.
(ii) Directivity of both antennas = 2; wavelength = 0.10 m.
(iii) Effective area of both antennas = 1 m^2; wavelength = 0.03 m.

12.27b Synchronous satellites orbit at about 4×10^4 km from the earth. About what diameter would you need for a uniformly illuminated circular aperture antenna on such a satellite to give full earth coverage if frequency is 7.5 GHz? (Earth radius $\approx$ 6370 km.) What power would have to be transmitted from the satellite if the earth stations have antennas 4 m in diameter and the receivers are capable of responding with acceptable error to signals of 10^{-10} W? Repeat for a satellite with earth coverage reduced to a region of diameter 100 km.

12.28a Calculate the effective area for a half-wave dipole antenna. Compare with that for the infinitesimal dipole.

12.28b Calculate the effective area for the small loop antenna.

12.28c If the antennas are in free space, the pattern of 1 may be measured, as was described in Fig. 12.28a, by moving antenna 2 about the arc of a circle or as in Fig. 12.28b by rotating 1 about an axis to give the same relative position. Explain why the same results might not be obtained by the two methods if there are fixed obstacles in the transmission path.

12.28d For the data of Prob. 12.27a(ii) and the radiation resistance of both antennas equal to 50 Ω, calculate $|Z_{12}|$. Now, allowing separation r to vary, find the separation for which magnitude of Z_{12} becomes comparable to real part of Z_{11}.

12.29a To demonstrate that a power calculation in the internal impedance of a Thevenin equivalent circuit does not necessarily represent the power lost internally, consider the source of constant voltage 100 V coupled to a resistance load of 2 Ω through the series-parallel resistances shown in Fig. P12.29a. Calculate the actual power lost in the generator, and compare with that calculated by taking load current flowing through the internal impedance of a Thévenin equivalent circuit.

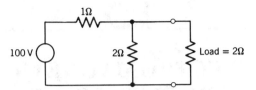

Fig. P.12.29 a Circuit for Prob. 12.29a.

12.29b An antenna having an input impedance on transmission of $70 + j30$ Ω is used on reception by connecting directly to a 50-Ω transmission line which is perfectly matched to a pure resistance load. The effective area is 0.40 m². Find the power transfer to the load if the antenna is in a plane wave field of 100 μV/m. Compare this with the power that could be obtained with a conjugate match to the antenna impedance.

13

ELECTROMAGNETIC PROPERTIES OF MATERIALS

13.1 Introduction

Materials react to applied electromagnetic fields in a variety of ways including displacements of both free and bound electrons by electric fields and the orientation of atomic moments by magnetic fields. In most cases, these responses can be treated as *linear* (proportional to the applied fields) over useful ranges of field magnitudes. Often the response is independent of the direction of the applied field; the material is called *isotropic*. The responses of these linear, isotropic materials to time-varying fields may depend to a significant extent upon the frequency of the fields. The materials considered heretofore in this text have been isotropic and linear and we have been able to represent them by scalar values of ε and μ for analysis at a given frequency. Common dielectrics such as glass, fused quartz, and polystyrene and common conductors such as copper, aluminum, and brass behave this way in a vast majority of applications.

Some comment should be made here about terminology. We use the terms *dielectric, conductor*, and *magnetic material* to indicate the character of the dominant response. A study would show that in all solids and liquids the permittivity differs appreciably from that of free space as a result of the behavior of the bound electrons in the constituent atoms. In a conductor, however, the movement of the free electrons in response to the electric field produces an electromagnetic response which overwhelms that of the bound electrons. Similarly, ferromagnetic materials are mostly rather highly conductive but we refer to them as magnetic materials as that property is the most significant in their application. All materials have some response to magnetic fields but, except for ferromagnetic and ferrimagnetic types, this is usually small, and differs from μ_0 by a negligible fraction.

The first part of the chapter deals with isotropic linear media, giving physical explanations for the electric and magnetic responses. These are necessarily qualitative or semiquantitative and in many cases incomplete, since detailed analysis requires quantum mechanics, which is beyond the scope of this book. Conductive materials are reasonably well represented by a free-electron model; we introduce it as a simple vehicle to show how the complex permittivity comes out of first principles. A section is devoted to clarifying the meaning we have given to a *perfect conductor* and to distinguishing it from a *superconductor*.

Important device applications depend upon the nonlinear behavior of certain materials. These include conversion of energy from one frequency to another, for example, frequency multiplers for the mixing of two signals of different frequencies to give a third. The nonlinear, hysteretic behavior and extremely high permeability of ferromagnetic materials significantly determine their use.

The final portion of the chapter treats anisotropic materials, both from the view of the physical sources of the anisotropy and with regard to calculating wave propagation in such media. Dielectric crystals with permittivity that depends upon the orientation of the electric field can be described succinctly by a so-called *dielectric ellipsoid*. This is introduced and wave propagation in such crystals is discussed. Plasma (ionized gases with equal positive and negative charge densities) and ferrites (ferrimagnetic solids) situated in a steady magnetic field exhibit a type of anisotropy in which one set of characteristic propagating waves are circularly polarized. The fact that the behavior depends on the propagation direction relative to the magnetic field leads to the possibility of nonreciprocal microwave and optical devices.

LINEAR ISOTROPIC MEDIA

13.2 Characteristics of Dielectrics

In this section we consider linear, isotropic dielectrics to explain the connection between microscopic effects and their representation by a permittivity. It is of special importance to discuss the frequency dependence of permittivity. We consider here usual dielectrics with $\mu = \mu_0$.

We saw in connection with the Poynting theorem in Sec. 3.12 that currents in phase with the electric field (there considered to be conduction currents) lead to an energy-loss term. In the general characterization of lossy dielectrics we introduce a complex permittivity

$$\varepsilon = \varepsilon' - j\varepsilon'' \tag{1}$$

which enters Maxwell's equations as the current $j\omega(\varepsilon' - j\varepsilon'')\mathbf{E}$, the second term of which is in phase with electric field. Both ε' and ε'' are in general, functions of frequency. The various physical phenomena contributing to ε' and ε'' differ for solids, liquids, and gases and are too complicated to summarize here in any detail.

Several excellent books and survey articles give thorough treatments.[1-8] Nevertheless, discussion of some of the properties and the simple models for these help to give insight into the most important characteristics.

It can be shown that, if the atoms or molecules are polarized (or their natural dipole moments have a nonzero average alignment), a dipole moment density can be defined by

$$\mathbf{P}_d = \lim_{\Delta V \to 0} \frac{\sum_i \mathbf{p}_i}{\Delta V}$$

and that this, neglecting higher-order multipoles is identical to the so-called *polarization*[9] that enters the relation between $\mathbf{D}$ and $\mathbf{E}$:

$$\mathbf{D} = \varepsilon_0 \mathbf{E} + \mathbf{P} = \varepsilon_0(1 + \chi_e)\mathbf{E} \qquad (2)$$

The term χ_e is the dielectric susceptibility and we have assumed that the polarization is linearly dependent on the field $\mathbf{E}$. A few dielectrics, such as electrets,[10] have a permanent polarization, but we are chiefly concerned with dielectrics in which $\mathbf{P}$ is induced by the applied electric field $\mathbf{E}$ and shall consider only these in the following. If there are N like molecules per unit volume, the induced polarization may be written

$$\mathbf{P} = \varepsilon_0 \chi_e \mathbf{E} = N\alpha_T \mathbf{E}_{\mathrm{loc}} = N\alpha_T g\mathbf{E} \qquad (3)$$

where α_T is the *molecular polarizability* and g the ratio between local field $\mathbf{E}_{\mathrm{loc}}$ acting on the molecule and the applied field $\mathbf{E}$. The local field differs from applied field because of the effect of surrounding molecules. We also write electric flux density as

$$\mathbf{D} = \varepsilon\mathbf{E} = \varepsilon_0 \varepsilon_r \mathbf{E} \qquad (4)$$

so by comparing with (2) and (3) we may write relative permittivity

$$\varepsilon_r = 1 + \frac{N\alpha_T g}{\varepsilon_0} = 1 + \chi_e \qquad (5)$$

[1] C. Kittel, *Introduction to Solid State Physics*, 5th ed., Wiley, New York, 1976.

[2] N. W. Ashcroft and N. D. Mermin, *Solid State Physics*, Holt, Rinehart and Winston, New York, 1976.

[3] J. S. Blakemore, *Solid State Physics*, W. B. Saunders Co., Philadelphia, 1974.

[4] W. F. Brown, in *Encyclopedia of Physics*, Vol. XVII, Springer-Verlag, Berlin, 1956.

[5] G. Wyllie, in *Progress in Dielectrics*, Vol. 2, Wiley, New York, 1960.

[6] A. R. von Hippel, *Dielectrics and Waves*, Wiley, New York, 1954.

[7] A. R. von Hippel, *Dielectric Materials and Applications*, M.I.T. Press and Wiley, New York, 1954.

[8] *American Institute of Physics Handbook*, 3rd ed., McGraw-Hill, New York, 1972.

[9] Not to be confused with polarization of a wave in the sense of Sec. 6.3. See footnote to Sec. 6.3.

[10] Electrets are readily formed in the laboratory by applying a dc electric field across certain types of molten wax and allowing the resulting dipoles to "freeze in" as the wax solidifies. For a bibliography on electrets see B. Gross, *Charge Storage in Solid Dielectrics*, Elsevier Publishing Co., Amsterdam, 1964.

If the surrounding molecules act in a spherically symmetric fashion on the molecule for which E_{loc} is being calculated, g can be shown to be $(2 + \varepsilon_r)/3$ and (5) may then be written as

$$\frac{\varepsilon_r - 1}{\varepsilon_r + 2} = \frac{N\alpha_T}{3\varepsilon_0} \tag{6}$$

This expression is known as the Clausius–Mossotti relation, or when frequency effects in α_T are included, the Debye equation. It is most accurate for gases but does give qualitative behavior for liquids and solids.

The molecular polarizability α_T, defined above, has contributions from several different atomic or molecular effects. One part, called electronic, arises from the shift of the electron cloud in each atom relative to its positive nucleus, much as was pictured in Sec. 1.3. Another part, called ionic, comes from the displacement of positive and negative ions from their neutral positions. Still another part may arise if the individual molecules have permanent dipole moments. Application of the electric field tends to align these permanent dipoles against the randomizing forces of molecular collision, and since random motion is a function of temperature, this last effect is clearly temperature dependent. The three effects together constitute the total molecular polarizability,

$$\alpha_T = \alpha_e + \alpha_i + \alpha_d \tag{7}$$

where α_e, α_i, and α_d are electronic, ionic, and permanent dipole contributions, respectively.

Frequency Effects　　In the classical model of the electronic polarization, any displacement of the charge cloud from its central ion produces a restoring force and its interaction with the inertia of the moving charge cloud produces a resonance as in a mechanical spring-mass system. Similarly, the displacement of one ion from another produces resonances in the ionic polarizability but at lower frequencies than for the electronic contribution because of the larger masses in motion. There are also losses or damping in each of the resonances, pictured as arising from radiation or interaction with other charges. The Lorentz model of an atom in which damping is proportional to the velocity of the oscillating charge cloud, is expressed by the equation of motion for displacement:

$$\frac{d^2r}{dt^2} + \Gamma \frac{dr}{dt} + \omega_0^2 r = -\frac{e}{m} E_{loc} \tag{8}$$

where Γ is a damping constant and ω_0 is related to the restoring force. Assuming the field and displacement in phasor form, one finds

$$r = \frac{-(e/m)E_{loc}}{(\omega_0^2 - \omega^2) + j\omega\Gamma} \tag{9}$$

The electronic polarizability is the ratio of the dipole moment $p = -er$ to the local field:

$$\alpha_e = \frac{(e^2/m)}{(\omega_0^2 - \omega^2) + j\omega\Gamma} \tag{10}$$

We can generalize this to represent all electronic and ionic resonant responses:

$$\alpha_j = \frac{F_j}{(\omega_j^2 - \omega^2) + j\omega\Gamma_j} \tag{11}$$

where F_j measures the strength of the jth resonance. This has the form of the impedance of a parallel resonant circuit. Real and imaginary parts of the expression contribute to ε' and ε'', respectively, in a manner shown by the electronic and ionic resonances pictured for a hypothetical dielectric in Fig. 13.2a. Near a resonance the lossy part goes through a peak. The contribution to ε' from a given resonance, like the reactance of the tuned circuit, has peaks of opposite sign on either side of the resonance.

The dynamic response of the permanent dipole contribution to permittivity is different in that the force opposing complete alignment of the dipoles in the direction of the applied field is that from thermal effects. It acts as a viscous force and the dynamic response is "overdamped." The frequency response of such an overdamped system is of the form

$$\alpha_d = \frac{p^2}{3k_B T(1 + j\omega\tau)} \tag{12}$$

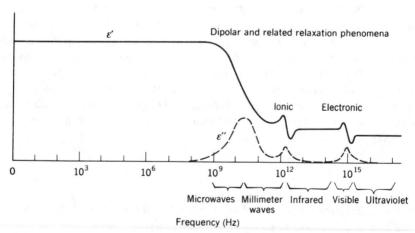

Fig. 13.2a Frequency response of permittivity and loss factor for a hypothetical dielectric showing various contributing phenomena.

where T is temperature, k_B is Boltzmann's constant, p is the permanent dipole moment, and τ is a relaxation time for the effect (i.e., the time for polarization to fall to $1/e$ of the original value if orienting fields are removed). Dipole contributions are also illustrated for the hypothetical dielectric of Fig. 13.2a and produce smoother decreases in ε as one goes through the range $\omega\tau \sim 1$, along with a peak of absorption. The picture of dipole orientation and relaxation as given applies best to gases and liquids possessing such dipoles, but there are a variety of similar relaxation effects in solids.

Relation Between Real and Imaginary Parts of Permittivity   The analytic properties of the complex permittivity defined by (1) provide certain relationships between $\varepsilon'(\omega)$ and $\varepsilon''(\omega)$ so that the frequency behavior of one part is not independent of the frequency behavior of the other. These interesting and important relationships, known as the Kramers–Kronig relations[11] may be written

$$\varepsilon'(\omega) = \varepsilon_0 + \frac{2}{\pi} \int_0^\infty \frac{\omega' \varepsilon''(\omega')\, d\omega'}{(\omega'^2 - \omega^2)} \tag{13}$$

$$\varepsilon''(\omega) = -\frac{2\omega}{\pi} \int_0^\infty \frac{[\varepsilon'(\omega') - \varepsilon_0]\, d\omega'}{\omega'^2 - \omega^2} \tag{14}$$

They are similar to relations between resistance and reactance functions of frequency in circuit theory (see Sec. 11.12).

At optical frequencies it is common to use the index of refraction defined by Eq. 6.2(28) rather than permittivity. This also is complex for a material with losses (absorption). With $\mu = \mu_0$,

$$n_c(\omega) = n_r(\omega) - jn_i(\omega) = \{[\varepsilon'(\omega) - j\varepsilon''(\omega)]/\varepsilon_0\}^{1/2} \tag{15}$$

(n_r and n_i are often listed as n and k, respectively, in optical tables but, to avoid confusion with wavenumber, we use the self-defining n_i for the *extinction coefficient k*.)

Values of $\varepsilon'/\varepsilon_0$ and $\varepsilon''/\varepsilon'$ for some representative materials at radio and microwave frequencies were given in Table 6.4a. Values of refractive index n_r are plotted versus wavelength for several optical materials in Fig. 13.2b. The imaginary part n_i is generally small except in or near absorption bands, where it is a strong function of wavelength. Reference 8 gives curves for numerous materials in their absorption regimes.

---

[11] J. D. Jackson, *Classical Electrodynamics*, 2nd ed., Wiley, New York, 1975, p. 311.

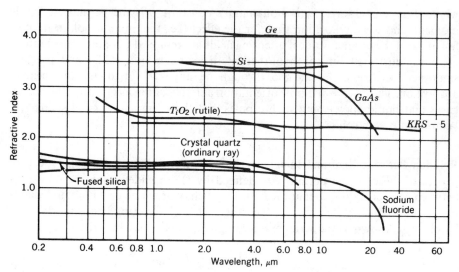

Fig. 13.2b Refractive index versus wavelength for several materials with useful values of optical and infrared transparency. Data from *American Institute of Physics Handbook*, D. E. Gray, (ed) McGraw-Hill, New York, 1972.

13.3   Imperfect Conductors and Semiconductors

A *good conductor* was defined in Chapter 3 as one that follows Ohm's law and for which displacement current is negligible compared with conduction current. In that case the current and electric field are in phase. There are a number of important materials and situations in which the currents can have significant components at phase quadrature to the electric field. At higher frequencies the displacement current becomes increasingly important; for example, at microwave frequencies semiconductors can have comparable conduction and displacement currents. Also, at optical frequencies the charge motion in metals can give current components both in phase with the electric field and at phase quadrature. The latter has the same effect as a displacement current and is included in the real part of the permittivity.

Let us consider a medium consisting of a density n_e of free electrons with a background of fixed positive ions of the same density. This can be considered to be a model of a conductor, semiconductor, or an ionized gas (plasma); only the values of the parameters are different. For most cases of interest, the derived complex permittivity will be used to study the behavior of *TEM* or quasi-*TEM* waves; we will use that fact to justify certain simplifications of the analysis.

The force on charges arising from interactions with the magnetic field in a wave is typically several orders of magnitude smaller than the electric forces (Prob.

13.3a), so we will neglect the former. In this analysis we bring in the effect of collisions by introducing a term representing the resulting average rate of loss of momentum. The equation of motion for an average electron is then

$$m\frac{d\mathbf{v}}{dt} = -e\mathbf{E} - mv\mathbf{v} \tag{1}$$

If it is assumed that each collision causes a total loss of momentum of the electron, then v is the collision frequency.

The total time derivative is taken as we move with the particle and must in general be written as

$$\frac{d\mathbf{v}}{dt} = \frac{\partial\mathbf{v}}{\partial t} + \frac{\partial\mathbf{v}}{\partial x}\frac{dx}{dt} + \frac{\partial\mathbf{v}}{\partial y}\frac{dy}{dt} + \frac{\partial\mathbf{v}}{\partial z}\frac{dz}{dt} \tag{2}$$

To simplify, consider a uniform plane wave propagating in the z direction so that the field is transverse and all velocity components are also in transverse directions. Thus the last term of (2), which contains the z-directed velocity dz/dt, vanishes. By virtue of the assumed uniformity of the fields in the transverse plane, the variations of velocity with respect to x and y are zero. Hence the total and partial derivatives with respect to time are equal here. Since we are considering the steady-state behavior of the system we may let $\mathbf{E} = \mathbf{E}e^{j\omega t}$ and $\mathbf{v} = \mathbf{v}e^{j\omega t}$. Equation (1) may then be arranged to give the velocity in terms of the electric field as

$$\mathbf{v} = \frac{-e\mathbf{E}}{m(v + j\omega)} \tag{3}$$

The electron density n_e is undisturbed by the motions of the electrons since all paths lie in the transverse planes and are parallel with each other. Then the convection current density is

$$\mathbf{J} = -n_e e\mathbf{v} = \frac{n_e e^2 \mathbf{E}}{m(v + j\omega)} \tag{4}$$

This can be inserted in Maxwell's curl equation, $\nabla \times \mathbf{H} = j\omega\varepsilon_1\mathbf{E} + \mathbf{J}$:

$$\nabla \times \mathbf{H} = j\omega\varepsilon_1\mathbf{E} + \frac{n_e e^2 \mathbf{E}}{m(v + j\omega)}$$

$$= j\omega\left[\left(\varepsilon_1 - \frac{n_e e^2}{m(v^2 + \omega^2)}\right) - j\frac{n_e e^2 v}{\omega m(v^2 + \omega^2)}\right]\mathbf{E} \tag{5}$$

where ε_1 is used in the displacement current term to represent the effect of the bound electrons of the positive-ion background. It is assumed for conductive materials to have an insignificant imaginary part compared with the effect of $\mathbf{J}$. Separating the real and imaginary parts of the right-hand side, we can identify the components of the complex permittivity $\varepsilon = \varepsilon' - j\varepsilon''$. Thus,

$$\varepsilon' = \varepsilon_1 - \frac{n_e e^2}{m(v^2 + \omega^2)} \tag{6}$$

and

$$\varepsilon'' = \frac{n_e e^2 v}{\omega m(v^2 + \omega^2)} \tag{7}$$

Note that $n_e e^2 / mv$ is the low-frequency conductivity σ.

One important application of the above results is to materials with moderate-to-low conductivity (say, $\sigma \lesssim 1$ S/m) such as semiconductors with moderate doping. Then for microwave and millimeter-wave frequencies where $\omega^2 \ll v^2$, the second term in (6) is negligible and ε becomes

$$\varepsilon = \varepsilon_1 - j\frac{\sigma}{\omega} \tag{8}$$

which is the form used in Eq. 6.4(11).

At the other extreme is the ionized gas with very low density and negligible collision frequency. In that case we can take $\varepsilon_1 = \varepsilon_0$ and $\varepsilon'' = 0$ so that

$$\varepsilon = \varepsilon_0\left(1 - \frac{\omega_p^2}{\omega^2}\right) \tag{9}$$

where ω_p is the so-called *plasma frequency*:

$$\omega_p = \sqrt{\frac{n_e e^2}{\varepsilon_0 m}} \tag{10}$$

The physical meaning of the plasma frequency is as follows: if alternate plane layers of compression and rarefaction of the electrons in a uniform positive-ion background were set as initial conditions, the charge would subsequently oscillate at the plasma frequency because of the interaction of forces between charges and the inertial effects.

An immediate observation about the form (9) for permittivity is that it is negative for all frequencies below the plasma frequency. Therefore, the intrinsic wave impedance $\eta = \sqrt{\mu/\varepsilon}$ is imaginary, so that a wave with $\omega < \omega_p$ incident from free space on a region of ionized gas is reflected. This is also seen from the propagation constant, Eq. 6.2(25), which becomes, for the plasma described by (9),

$$k^2 = \omega^2 \mu \varepsilon_0\left(1 - \frac{\omega_p^2}{\omega^2}\right) \tag{11}$$

If $\omega < \omega_p$, k^2 is negative so $jk = \alpha$ and $\beta = 0$. Then

$$E_x = E_+ e^{-\alpha z} + E_- e^{\alpha z} \tag{12}$$

That is, purely attenuated waves exist for $\omega < \omega_p$. For frequencies higher than ω_p, k^2 is positive and unattenuated waves propagate in the gas. The ω–β diagram for *TEM* waves in an ionized gas is shown in Fig. 13.3a.

The effect described above for a plasma can also be observed in some metals (e.g., aluminum and the alkali metals) at optical frequencies. Very high fractional reflection is observed for frequencies below the plasma frequency, which is typically

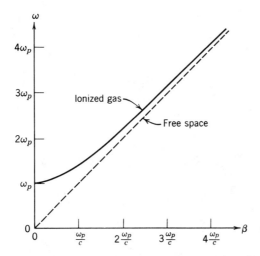

Fig. 13.3a The ω–β diagram for plane wave propagating through an ionized gas.

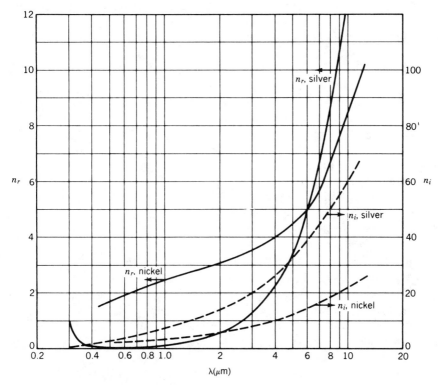

Fig. 13.3b Real and imaginary parts of refractive index for silver and nickel over a part of the optical range. Data from *American Institute of Physics Handbook*, D. E. Gray, (ed) McGraw-Hill, New York, 1972. (Measured values depend strongly on surface conditions.)

between the visible and ultraviolet ranges. The effects in metals are actually more complex than the simple model above would suggest and a complete treatment requires quantum mechanics.[12,13] As noted in Eq. 13.2(15), the properties at optical frequencies are usually expressed by a complex index of refraction n_c. Both parts of n_c may vary appreciably with frequency, the variation being a characteristic of the particular metal. Curves for silver and nickel for a part of the visible and infrared ranges are shown in Fig. 13.3b.

13.4 Perfect Conductors and Superconductors

At numerous points in this text—and very commonly in electromagnetic theory—we refer to *perfect conductors*. Here we wish to comment on what is, and what is not, meant by that term and to see how it relates to *superconductors*.

A perfect conductor is usually understood to be a material in which there is no electric field at any frequency. Maxwell's equations assure that there is then also no time-varying magnetic field in the perfect conductor. On the other hand, a truly static magnetic field should be unaffected by conductivity of any value, including infinity. In all time-invariant electric problems, the use of the perfect-conductor approximation for an electrode assures uniform potential over its surface. This is often an excellent approximation to a metal electrode even in problems with direct current flowing (for example, a metal electrode on a semiconductor). In rf problems the fields are essentially zero a few skin depths inside the surface of a metal electrode. If the spacing between electrodes in the system being analyzed is much greater than the skin depth, the perfect-conductor assumption that $E = 0$ at the surface is a good one for finding field distributions, though the penetration of the fields cannot be neglected if losses are to be found.

The real approximation to the perfect conductor would be a conductor in which the collision frequency approaches zero. This was believed to be the correct model for a superconductor in the period between its discovery in 1911 by H. Kamerlingh Onnes and the experiment by W. Meissner and R. Ochsenfeld in 1933.[14] A collisionless conductor can be modeled as a dense plasma and the results of the preceding section may be applied with $v = 0$. For frequencies below the plasma frequency, Eq. 13.3(12) indicates that the fields inside the conductor (Fig. 13.4) can only exist near the surface where they are excited. This phenomenon is like the skin effect (Sec. 3.16) except that here the behavior of the shielding currents is determined by inertia of the electrons rather than collisions. For frequencies a factor of about 10 or more below the plasma frequency, the attenuation factor $\alpha = (\mu n_e e^2/m)$ is independent of frequency and has values typically on the order of $(0.1 \ \mu m)^{-1}$ so the

[12] F. Abeles, *Optical Properties of Metals*, North Holland Publ., Amsterdam, 1972.
[13] F. Wooten, *Optical Properties of Solids*, Academic Press, New York, 1972.
[14] F. London, *Superfluids*, Vol. 1, Dover, New York, 1961.

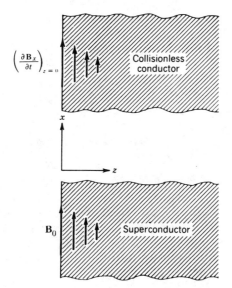

Fig. 13.4 Coordinates for analysis of fields at surface of a half-space of collisionless conductor or superconductor.

interior of bulk samples can be considered essentially free of time-varying fields. The results of the plasma wave analysis appear to be valid to zero frequency; however in static fields in the absence of collisions, the electrons would be accelerated to infinite velocities and the analysis is inapplicable.

Extremely pure crystals of metal can have very low collision frequencies at temperatures near the absolute zero, but such materials are not of much interest in practice since it is difficult to construct useful systems of single metal crystals. On the other hand, the superconductor is a candidate. The collisionless conductor excludes time-varying fields but the experiment of Meissner and Ochsenfeld on a superconductor showed that constant magnetic fields are also excluded from the interior. The London theory (1935) showed that this result would be obtained if

$$\nabla \times \mathbf{J} = -\Lambda \mathbf{B} \tag{1}$$

where $\Lambda = n_e e^2/m$ at $T = 0$. Substituting (1) in Maxwell's curl $\mathbf{H}$ equation and neglecting displacement current, one finds

$$\nabla \times \nabla \times \mathbf{B} = -\mu\Lambda \mathbf{B} \tag{2}$$

Using a vector identity and $\nabla \cdot \mathbf{B} = 0$ in (2) yields

$$\nabla^2 \mathbf{B} - \mu\Lambda \mathbf{B} = 0 \tag{3}$$

This can be applied to the fields inside the surface of a superconductor by considering a half-space of superconductor filling $z \geq 0$ as in Fig. 13.4. We assume that only an x component of **B** exists and it varies only in the z direction. Then (3) becomes

$$\frac{d^2 B_x}{dz^2} - \frac{B_x}{\lambda_L^2} = 0 \qquad (4)$$

where $\lambda_L = (m/\mu n_e e^2)^{1/2}$ at $T = 0$ and is called the *London penetration depth*.[15] The solution that remains bounded is

$$B_x = B_{x0} e^{-z/\lambda_L} \qquad (5)$$

which is the same as the form in Eq. 13.3(12) since $\lambda_L = 1/\alpha$. Some typical London penetration depths are 0.037 μm for lead, 0.039 μm for niobium, 0.016 μm for aluminum. The important difference in the case of the superconductor is that (5) applies even for static magnetic field. Superconductors must be held at temperatures within several degrees of $T = 0$; the values of the above expressions for Λ and λ_L are only approximately correct as other factors must be taken into account, including temperature. At any given temperature, fields at microwave and higher frequencies penetrate into a superconductor less than the dc penetration. It should also be pointed out that at any nonzero temperature, rf fields within the penetration region at the surface of a superconductor produce some losses (although very significantly smaller than in a real normal conductor), unlike in the ideal collisionless conductor. The physical phenomena for the behavior of superconductors are discussed in Ref. 16.

In summary, the commonly used concept of a perfect conductor (**E** = 0 inside) is a convenient artifice for calculation of the field distributions outside when the actual conductor has a field penetration that is small compared with the space between conductors. This is true for either a highly conductive electrode with a small skin depth or for a superconductor. However, neither a conductor nor a superconductor has exactly zero **E** inside, near the surface. To find field distributions where the spacing between electrodes is comparable with or smaller than the depth of field penetration, the latter must be taken into account. For example, a tunnel junction consisting of parallel films of normal metal or superconductor has an insulating layer that is typically 10–100 times thinner than the depth of field penetration so that waves propagating in this parallel-plane waveguide are strongly affected by the field penetration.

---

[15] Since $\mu \approx \mu_0$ for almost all superconductive materials the definition of λ_L is usually given in terms of μ_0.

[16] T. Van Duzer and C. W. Turner, *Principles of Superconductive Devices and Circuits*, Elsevier, New York, 1981.

13.5 Diamagnetic and Paramagnetic Responses

As noted in Sec. 13.1, all materials have some response to magnetic fields. Except for certain categories which we will consider separately (ferromagnetic and ferrimagnetic), the magnetic response is very weak. Magnetic response can either decrease or increase the flux density for a given **H**. If **B** is decreased, the material is said to be *diamagnetic*; if increased, it is called *paramagnetic*.

Atoms have a diamagnetic response that arises from changes of the electron orbits when the magnetic field is applied. As we saw in our study of Faraday's law, an electric field is produced by a changing magnetic field. That field induces currents which, in turn, produce a magnetic field that opposes the change. The response of the electron orbits in an atom is of this kind. The effect produces fractional changes of μ from μ_0 by only 10^{-8} to 10^{-5}.

The diamagnetic response is usually obscured in materials having natural net magnetic dipole moment. Such materials give a paramagnetic response which can arise from dipole moments in atoms, molecules, crystal defects, and conduction electrons. The dipole moments tend to align with the magnetic field but are deflected from complete alignment by their thermal activity. The fields resulting from the partially aligned dipoles add to the applied field.

It can be shown that, if there are induced magnetic dipoles in the atoms or there is a nonzero average alignment of natural dipoles, a *dipole density* can be defined by

$$\mathbf{M} \triangleq \lim_{\Delta V \to 0} \frac{\sum_i \mathbf{m}_{0i}}{\Delta V} \tag{1}$$

and this is identical to the so-called *magnetization* that enters the relation between **B** and **H**[17]:

$$\mathbf{B} = \mu_0(\mathbf{H} + \mathbf{M}) \tag{2}$$

It can also be shown[18] that the magnitude of the magnetization can be expressed in terms of the magnetic field $\mathbf{H}_i$ acting on the atomic dipoles and the absolute temperature T by

$$M = Nm_0\left(\coth \frac{m_0 \mu_0 H_i}{k_B T} - \frac{k_B T}{m_0 \mu_0 H_i}\right) \tag{3}$$

where N is the number of dipoles per unit volume, m_0 is the natural dipole moment, and k_B is Boltzmann's constant. Equation (3) is derived from considerations of the statistical distribution of the dipole orientations. The direction of **M** is parallel with $\mathbf{H}_i$ in materials of this type and the molecular field $\mathbf{H}_i$ is negligibly different

[17] R. Plonsey and R. E. Collin, *Principles and Application of Electromagnetic Fields*, McGraw-Hill, New York, 1961, Chapter 7.

[18] J. R. Reitz and F. J. Milford, *Foundations of Electromagnetic Theory*, Addison-Wesley, Reading, Mass., 1960, Chapter 11.

from the macroscopic field. For magnetic fields that are not too strong and temperatures not too low, the parentheses in (3) may be approximated by the first term in its series expansion. In this case the magnetization is linearly dependent on the applied field:

$$\mathbf{M} = \chi_m \mathbf{H} \tag{4}$$

and

$$\chi_m \cong \frac{N\mu_0 m_0^2}{3k_B T} \tag{5}$$

In a variety of materials at room temperature χ_m has values on the order of 10^{-5}.

Some of the most important of magnetic properties are those that involve coupling effects between atomic magnetic moments; these are discussed in the following section.

NONLINEAR ISOTROPIC MEDIA

13.6 Materials with Residual Magnetization

If the temperature of a paramagnetic substance is reduced below a certain value, which depends on the material, the magnetization $\mathbf{M}$ may be sufficient to produce the field necessary to hold the dipoles aligned even when the external field is not present. The molecular field produced by the magnetization is[19]

$$\mathbf{H}_i = \kappa \mathbf{M} \tag{1}$$

But from Eq. 13.5(3) it is seen that a second relation between $\mathbf{H}_i$ and $\mathbf{M}$ exists. Equating M of the two expressions, the conditions for spontaneous magnetization are obtained. The temperature below which a material exhibits spontaneous magnetization is called its *Curie temperature*. From experimental knowledge of the Curie temperatures for various materials, it has been found that κ, the factor measuring the interaction of neighboring dipoles, must be on the order of 1000. On the basis of purely magnetic interactions, it would be expected to be about one-third, as in the corresponding factor for electric dipoles. The values are explained using quantum-mechanical considerations on the basis of electric interaction between molecules resulting from distortions of the charge distributions in the molecules. In some materials it is energetically favorable for dipoles to align parallel to their neighbors; these are called *ferromagnetic* substances. In some other materials it is energetically favorable for neighboring dipoles to assume an antiparallel orientation, as indicated in Fig. 13.6a; these are called *antiferromagnetic* substances. In *ferrimagnetic* materials, also called *ferrites*, the neighboring dipoles are aligned in

[19] J. R. Reitz and F. J. Milford, Ref. 18.

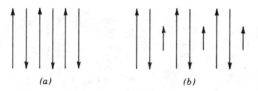

Fig. 13.6 (*a*) Arrangement of dipoles in antiferromagnetic material. (*b*) Unbalanced antiparallel dipoles in ferrites.

an antiparallel arrangement but different types of atoms are present and the dipoles do not cancel. This is illustrated schematically in Fig. 13.6*b*.

From the foregoing discussion, one would expect that a single crystal of ferromagnetic or ferrite material would act as a permanent magnet with a common alignment of molecular dipoles. However, except in the case of very small crystals, it is energetically favorable for there to occur a subdivision into regions, called *domains*, within which the dipoles are all parallel.[20] The subdivision is the one that minimizes the total energy of the crystal, which includes energy in the walls between domains and stored magnetic energy both within and outside the crystal. Figure 13.6*c* shows a photograph made by a special technique that reveals the domains in a single crystal of nickel. Notice that the arrows indicating the magnetization vectors of the domains tend to cancel each other when averaged over the crystal. There are factors such as crystal shape, energetically preferred crystal directions, and the magnetic history of the sample that leads to nonzero net magnetization. (The existence of domains in thin films in the presence of an applied field is exploited for the so-called *bubble memory* devices for computer memories.[21])

Practical materials ordinarily contain large numbers of crystallites with random orientations so that the macroscopic effects are isotropic. Within each crystallite is a domain structure of the type shown in Fig. 13.6*c*. When a magnetic field is applied, the walls of the domains move in such a way as to enlarge those having components of their dipole vectors in the direction of the magnetic field. If the magnetic field intensity is further increased, the dipole direction for each domain as a whole rotates toward the direction of the applied magnetic field. This process is nonlinear in that the magnetization and hence the flux density $\mathbf{B}$ is not simply proportional to the applied field. The magnetization tends to lag the field intensity with the result that the relation between $\mathbf{B}$ and $\mathbf{H}$ has the pattern shown in Fig. 13.6*d*, which is called a *hysteresis loop*. The linear parts of the loop at high values of magnetic field correspond to the condition in which essentially all of the dipole moments are aligned, and ΔB and ΔH are therefore related by μ_0 (neglecting diamagnetic effects).

[20] C. Kittel, Ref. 1, pp. 484–493.

[21] H. Chang, *Magnetic Bubble Technology: Integrated Circuit Magnetics for Digital Storage and Processing*, IEEE Press, New York, 1975.

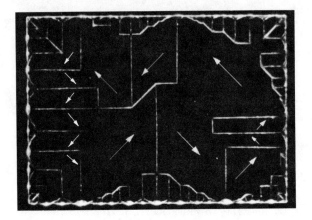

Fig. 13.6c Ferromagnetic domains in a single-crystal platelet of nickel. (From C. Kittel, Ref. 1, © John Wiley and Sons, 1976, New York. Original photograph from R. W. de Blois, *J. Appl. Phys.* **36**, 1647 (1965).)

The processes of wall movement and rotation of the domains require the expenditure of energy. The energy required for the traversal of the loop by varying **H** from a large negative value to a large positive value and back again can be found from the expression for the stored magnetic energy, Eq. 2.16(1). For nonlinear materials, the energy required to change **B** by a differential amount can be shown to be

$$dU = \int_V \mathbf{H} \cdot d\mathbf{B} \, dV \tag{2}$$

where the integration is over the volume of the material. This differential energy is shown as a shaded bar on the hysteresis loop in Fig. 13.6d. When the field is

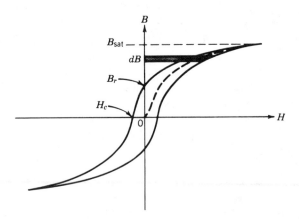

Fig. 13.6d Typical hysteretic *B–H* curve for a ferromagnetic material.

Table 13.6a
Hard Magnetic Materials

Material	Comment or Percent Composition	Remanence B_r (T)	Coercive Force H_c (A-m^{-1})
Iron	Bonded powder	0.6	0.0765
Iron–cobalt	Bonded powder	1.08	0.0980
Alnico V	14 Ni, 8 Al, 24 Co, 3 Cu, 51 Fe	1.27	0.0650
Ticonol II	15 Ni, 7 Al, 34 Co, 4 Cu, 5 Ti, 35 Fe	1.18	0.1315

decreased, a portion of the energy indicated by the part of the bar outside the loop is returned to the field. The result of integrating around the loop is that the total expended energy per unit volume is equal to the area of the loop.

Some important applications of ferromagnetic materials include permanent magnets, signal recording tapes and disks, magnetic shielding, and cores for electromagnets, transformers, electric motors, and generators. These materials can be roughly divided into categories called *hard* and *soft*. The designations refer to permanence of magnetization when the applied field is removed, with hard materials retaining a strong magnetization.

Hard materials generally have a wide, rectangular hysteresis loop. Two important points on the hysteresis loop for hard materials are the *remanence B_r* and the *coercive force H_c* (see Fig. 13.6d). The remanence is the flux density that remains in the material when H is reduced to zero and is the value existing in a permanent magnet. The coercive force is the magnetic field that must be applied in the opposite direction to reduce the flux density to zero. Its importance lies in the fact that the retention of magnetization in the presence of disturbances (thermal, mechanical, etc.) improves with increasing coercive force. The properties of some technically important materials for permanent magnets are listed in Table 13.6a.[22]

Soft materials should usually have large *saturation induction* (B_{sat} in Fig. 13.6d), small coercive force, large *initial permeability* (dB/dH at $B = H = 0$), large *maximum permeability* (maximum B/H), and small hysteresis losses. The specific application may make some characteristics more important than others; for example, a material to be used to shield something from a weak magnetic field would have high initial permeability as its most important criterion. Properties of some important soft materials are listed in Table 13.6b.[22,23] When ferromagnetic metals are used in situations where the magnetic field is time varying, as in a generator or a transformer, so-called *eddy currents* are induced by electric fields prescribed by

[22] A. H. Morrish, *The Physical Principles of Magnetism*, Robert E. Krieger, Malabar, Florida, 1982.
[23] R. Boll, Ed., *Soft Magnetic Materials*, Heyden, London, Philadelphia, Rheine, 1979.

Table 13.6b

Soft Magnetic Materials

Material	Comment and Percent Composition	Initial Permeability $(\mu_r)_0$	Maximum Permeability $(\mu_r)_{max}$	Coercive Force $\times 10^4$ (A-m^{-1})	Saturation Induction (T)
Iron	Commercial (99 Fe)	2×10^2	6×10^3	0.9	2.16
Iron	Pure (99.9 Fe)	2.5×10^4	3.5×10^5	0.01	2.16
Hypersil	97 Fe, 3 Si	9×10^3	4×10^4	0.15	2.01
78 permalloy	78 Ni, 22 Fe	4×10^3	1×10^5	0.05	1.05
Mumetal	18 Fe, 75 Ni, 5 Cu, 2 Cr	2×10^4	1×10^5	0.05	0.75
Supermalloy	15 Fe, 79 Ni, 5 Mo, 0.5 Mn	9×10^4	1×10^6	0.004	0.8
Cryoperm[a]	Usable at cryogenic temperatures	6.5×10^4			

[a] Reference 23. Permeability given for $T = 4.2$ K.

Faraday's law (Sec. 3.2). The currents are large because the metals have rather high conductivity; the resulting Joule (I^2R) losses can be unacceptable. The currents and resulting losses are minimized by making the cores laminated with the induced current directions perpendicular to the laminations. The losses in metallic ferromagnetic materials become intolerable for radio and microwave frequencies. Ferrites are important because they typically have very low conductivities [$\ll 10^{-4}$ S/m]. They are used extensively for purposes such as cores in transformers and filters and antenna rods for frequencies throughout the radio frequency ranges.[24] Several microwave applications utilize an anisotropic property of ferrites; these are discussed in Sec. 13.19.

For the materials discussed here, the permeability as defined by $\mu = B/H$ is not a constant. In solving field problems, however, one is often concerned with small variations of the field about some large average value. If the variations are small enough, any part of the hysteresis loop can be considered a straight line, the slope of which depends upon the point on the loop. The slope is called the *incremental permeability* and is the ratio $\Delta B/\Delta H$.

An effect similar to that discussed for magnetization is found for electric polarization in some materials (for example, barium titanate). These materials, called

[24] J. Verweel, in *Magnetic Properties of Materials*, J. Smit, Ed., McGraw-Hill, New York, 1971.

ferroelectric because of the behavior analogous to ferromagnetic materials, exhibit a hysteresis loop in the relationship between the flux density D and the electric field E. As in ferromagnetic materials, a domain structure is found in ferroelectric materials. The details of the interactions between neighboring dipoles are discussed in texts on solid-state physics.[25]

13.7 Nonlinear Optics

Most of this text is concerned with linear materials. We have, however, seen in the preceding section that the nonlinear characteristics of ferromagnetic materials must usually be considered. For these, magnetization **M** is a nonlinear function of magnetic field **H**. The polarization **P** of dielectrics may similarly become a nonlinear function of electric field **E** for sufficiently high fields. It is less common to have to consider this nonlinearity than that of ferromagnetic materials, but there is one area in which it is central and has led to a number of useful effects. This is in the area of *nonlinear optics* made possible by the high fields of laser beams. This subject is too specialized to treat in detail, but we give here the basis for nonlinear polarizations and applications to heterodyning and harmonic generation. Other important applications such as parametric amplification and oscillation and four-wave mixing to produce conjugate waves (which can be used, for example, to correct wavefront distortion arising from material inhomogeneities[26]) are included in some of the special texts on this subject.[27,28,29]

A complete theoretical treatment of this subject is complicated by the use in practice, of anisotropic crystals, in which case the material response is in a different direction from that of the applied field and directions of propagation of various components can differ. The analysis here avoids those complications while giving the essentials for understanding the most important phenomena. With proper interpretation the analysis given below can be adapted to anisotropic crystals.

In Sec. 13.2, we discussed the Lorentz model for the dynamics of the charge cloud in an atom when an oscillating field is applied. For the calculation of the nonlinear polarizability or susceptibility, a modified form of the Lorentz equation is used.

$$\frac{d^2 r}{dt^2} + \Gamma \frac{dr}{dt} + \omega_0^2 r - \xi r^2 = -\frac{e}{m} E_{\text{loc}} \tag{1}$$

The nonlinearity becomes significant when the field is large enough to make the displacement r a significant fraction of the radius of the equilibrium electron

[5] For example, see C. Kittel, Ref. 1.

[6] A. Yariv, *IEEE J. Quantum Electronics* **QE-14**, 650 (1978).

[7] N. Bloembergen, *Nonlinear Optics*, W. A. Benjamin, New York, 1954.

[8] F. Zernike and J. E. Midwinter, *Applied Nonlinear Optics*, Wiley, New York, 1973.

[9] M. D. Levensen, *Introduction to Nonlinear Laser Spectroscopy*, Academic Press, New York, 1982.

[.] Yariv and P. Yeh, *Optical Waves in Crystals*, Wiley, New York, 1984, Chap. 12. H. A. Haus, *Waves and Fields in Optoelectronics*, Prentice-Hall, Englewood Cliffs, N.J., 1984, Chap. 13.

orbit. Such atomic fields are on the order of 3×10^{10} V/m.[30] One can begin to observe nonlinearities with fields typically 10^{-5} to 10^{-4} of this value so that the perturbation theoretical approach discussed below can be used to calculate r.

We will examine a typical case of interest for applications in which the electric field is the sum of the fields in several waves of different frequencies $E = E(\omega_1) + E(\omega_2) + \cdots$. It is possible to work through the algebra for this case to find expressions for the frequency dependences of the nonlinear polarizations of various orders. The total polarization can be written as

$$P = P_1 + P_2 + \cdots \tag{2}$$

where

$$P_i = -Ner_i \tag{3}$$

in which N is the volume density of atoms, e is the electronic charge, and the displacements r_i are related to the total electric field by

$$r_i = a_i(gE)^i \tag{4}$$

where g is the factor relating the macroscopic field to that acting on the atom. Then the total polarization is

$$P = (-Nea_1 g)E + (-Nea_2 g^2)E^2 + \cdots$$
$$= \varepsilon_0 \chi_e E + \varepsilon_0 \chi_e^{(2)} E^2 + \cdots \tag{5}$$

The coefficients a_i, and therefore the susceptibilities $\chi_e^{(i)}$ are frequency dependent. For example, the form of the frequency dependence of a_1 was given in Eq. 13.2(9). The frequency dependences in the higher order terms are quite complicated when E is the sum of the fields of several frequencies.

When all frequencies of applied fields are in the optical transmission region of a crystal, it is a good approximation to neglect the frequency dependence of the susceptibility $\chi_e^{(2)}$. Making this assumption we consider now the polarization due to the nonlinearity, making the often-useful approximation that P_3, P_4, and so on, are negligibly small. Thus

$$P_2 = \varepsilon_0 \chi_e^{(2)} E^2 = 2\varepsilon_0 d\, E^2 \tag{6}$$

where $2d = \chi_e^{(2)}$. (For materials having inversion symmetry, including isotropic materials, the lowest nonlinear susceptibility is $\chi_e^{(3)}$.)

Let us find the nonlinear polarization when two plane waves with parallel propagation and different frequencies coexist in a crystal. We write the two waves as[31]

$$E_1(z, t) = E_1 \cos(\omega_1 t - k_1 z)$$
$$E_2(z, t) = E_2 \cos(\omega_2 t - k_2 z) \tag{7}$$

---

[30] N. Bloembergen, Ref. 27, p. 7.

[31] Note that if complex phasors are used, full equivalence to (7) must be included, $E_i e^{j\omega t} + E_i^* e^{-j\omega t}$ since real and imaginary parts do not remain separated in nonlinear problems.

The electric field in (6) is the sum of the fields of the two waves in (7); making the substitution gives

$$P_2 = 2\varepsilon_0 d[E_1^2 \cos^2(\omega_1 t - k_1 z) + E_2^2 \cos^2(\omega_2 t - k_2 z)$$
$$+ 2E_1 E_2 \cos(\omega_1 t - k_1 z) \cos(\omega_2 t - k_2 z)] \tag{8}$$

Use of trigonometric identities on (8) reveals that it contains components at several different frequencies. There are components at the second harmonic of the two applied frequencies:

$$P_{2\omega_1} = d\varepsilon_0 E_1^2 \cos[2(\omega_1 t - k_1 z)] \tag{9}$$

$$P_{2\omega_2} = d\varepsilon_0 E_2^2 \cos[2(\omega_2 t - k_2 z)] \tag{10}$$

Also, there are components at the sum and difference frequencies,

$$P_{\omega_1 + \omega_2} = 2 d\varepsilon_0 E_1 E_2 \cos[(\omega_1 + \omega_2)t - (k_1 + k_2)z] \tag{11}$$

$$P_{\omega_1 - \omega_2} = 2 d\varepsilon_0 E_1 E_2 \cos[(\omega_1 - \omega_2)t - (k_1 - k_2)z] \tag{12}$$

and a time-independent term.

Thus at any instant of time there is a complicated polarization pattern in the crystal. The situation is even more complicated than portrayed since E_1, E_2, and P are in general not in the same direction. Thus, for instance, if a crystal is used, the z direction referred to above is not necessarily along a crystal axis. For the essential features these polarization aspects can be ignored; however, without changing the notation one may consider E_1 and E_2 as well as P as representing components in different directions if desired. The constant d depends on the relative directions.

It is of use to visualize the polarization for a simple case. Consider the situation where $E_2 = 0$; only the constant polarization and that at $2\omega_1$ remain from the nonlinear terms. Figure 13.7 shows the spatial variation of the applied field and the polarization at $2\omega_1$. The polarization has a phase constant of $2k_1$. In order for all parts of this pattern of dipoles to produce fields that add constructively, the phase constant of these fields k_2 must equal $2k_1$. Since $k_2 = n(2\omega_1)\omega_2/c$, $2k_1 = n(\omega_1)2\omega_1/c$, and $\omega_2 = 2\omega_1$, one must have

$$n(2\omega_1) = n(\omega_1) \tag{13}$$

The same phase-synchronism condition is true for the other components when both sources are present. For example, for $P_{\omega_1 - \omega_2}$, the phase constant of the polarization is $k_1 - k_2$. Then for there to be constructive interference so that a net output is created, the index of refraction at $\omega_1 - \omega_2$ must equal that at ω_1,

$$n(\omega_1 - \omega_2) = n(\omega_1) \tag{14}$$

The technique for getting contributions to add constructively is called *phase matching*; it requires control of the index of refraction at the frequencies of the fields

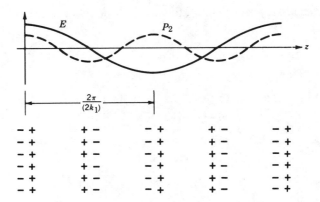

Fig. 13.7 Nonlinear polarization resulting from a single applied electric field of wave number k_1.

involved in generating a particular polarization component and at the frequency of the polarization source term. It is generally possible to do this with only one nonlinear interaction process at a time. Techniques for accomplishing this include temperature control and crystal orientation.

Two important applications are illustrated by the components discussed above. Efficient generation of the second harmonic makes possible signal sources in frequency ranges not otherwise accessible. For example, useful amounts of light at $\lambda_0 = 0.53$ μm can be generated by focusing an intense laser beam from a Nd^{3+}:YAG laser at $\lambda_0 = 1.06$ μm on a crystal of $BaNaNb_5O_{15}$ which has a relatively large value of the constant d. Also, nonlinear polarization can be used to frequency-shift a signal say, at ω_2 to a frequency for which more convenient detectors may be available. Suitable choice of ω_1 can give the desired difference frequency component at $\omega_3 = \omega_2 - \omega_1$.

Power in nonlinear effects of this kind (lossless) is governed by the important Manley–Rowe relations.[32] The changes in power ΔW_{Ti} in the various components are related by

$$\frac{\Delta W_{T1}}{\omega_1} = \frac{\Delta W_{T2}}{\omega_2} = -\frac{\Delta W_{T3}}{\omega_3} \tag{15}$$

where $\omega_3 = \omega_1 + \omega_2$. For example, these show that in the sum-frequency generation, the two source waves at ω_1 and ω_2 lose power to the sum in amounts related to their relative frequencies.

[32] A. Yariv, *Introduction to Optical Electronics*, 2nd. ed., Holt, Rinehart and Winston, New York, 1976, pp. 210–221.

ANISOTROPIC MEDIA

13.8 Representation of Anisotropic Dielectric Crystals

The materials studied up to this point have been representable by scalar permittivities and permeabilities, which may be frequency dependent and nonlinear. In a number of technically important materials, responses to fields with different orientations can differ; that is, they are anisotropic. This anisotropy may be either in the response to the electric field or to the magnetic field. In the former case, the permittivity must be represented by a matrix, an array of nine scalar quantities that may be frequency-dependent and/or nonlinear. The anisotropy in the magnetic field response is represented by a matrix permeability.[33] It is rare that both the permeability and the permittivity are significantly anisotropic and we will not consider that possibility here.[34] We first treat anisotropic dielectric crystals, then plasmas having anisotropy of the permittivity caused by the application of a constant magnetic field, and finally ferrites with an applied magnetic field. In addition to analyzing the representation of these anisotropic materials, we investigate the types of waves that can exist in the various media and comment on applications.

The relation between $\mathbf{D}$ and $\mathbf{E}$ for an anisotropic dielectric is given by[35]

$$D_x = \varepsilon_{11}E_x + \varepsilon_{12}E_y + \varepsilon_{13}E_z$$
$$D_y = \varepsilon_{21}E_x + \varepsilon_{22}E_y + \varepsilon_{23}E_z \tag{1}$$
$$D_z = \varepsilon_{31}E_x + \varepsilon_{32}E_y + \varepsilon_{33}E_z$$

or, in matrix form,

$$\begin{bmatrix} D_x \\ D_y \\ D_z \end{bmatrix} = \begin{bmatrix} \varepsilon_{11} & \varepsilon_{12} & \varepsilon_{13} \\ \varepsilon_{21} & \varepsilon_{22} & \varepsilon_{23} \\ \varepsilon_{31} & \varepsilon_{32} & \varepsilon_{33} \end{bmatrix} \begin{bmatrix} E_x \\ E_y \\ E_z \end{bmatrix} \tag{2}$$

or still more compactly,

$$[D] = [\varepsilon][E] \tag{3}$$

---

[33] It is sometimes convenient to use either *tensor* or *dyadic* representation for the anisotropic permittivity and permeability, for example, in equations involving curl operators. For an introduction to dyadic notation, see R. E. Collin, *Field Theory of Guided Waves*, McGraw-Hill, New York, 1960, pp. 568–570 and for tensors see J. A. Stratton, *Electromagnetic Theory*, McGraw-Hill, New York, 1941, Sec. 1.20. For our purposes, the more common matrix representation suffices.

[34] A treatment of the case where both permittivity and permeability are significantly anisotropic, as occurs in yttrium–iron–garnet (YIG) at infrared frequencies, is given by E. E. Bergmann, *Bell System Tech. J.* **61**, 935 (1983).

[35] More details may be found in M. Born and E. Wolf, *Principles of Optics*. 6th ed., Pergamon Press, 1980, Chapter XIV. See also Yariv and Yeh, Ref. 29, Chaps. 4 and 5 and Haus, Ref. 29, Chap. 11.

It can be shown that, for physically real materials, the permittivity operator (represented here by a matrix) is Hermitian, a property defined by

$$\varepsilon_{ij} = \varepsilon_{ji}^* \tag{4}$$

We will first consider loss-free crystals so (4) implies real and symmetric matrices.[36]

It is a property of symmetrical matrices that they can be diagonalized by a rotation of coordinates so that (1) reduces to

$$D_x = \varepsilon_{11} E_x; \qquad D_y = \varepsilon_{22} E_y; \qquad D_z = \varepsilon_{33} E_z \tag{5}$$

where ε_{11}, ε_{22}, and ε_{33} are called the *principal permittivities*. The properties of the matrix can be studied with no loss of generality and with considerable algebraic simplification by using the *principal coordinates*, in which the matrix is diagonalized.

Now we will introduce a geometrical formalism for describing crystal permittivity that aids in the understanding of wave propagation. The electric energy density can be found in the same form, Eq. 1.21(7), as for the isotropic case:[37]

$$U = \tfrac{1}{2} \mathbf{D} \cdot \mathbf{E} \tag{6}$$

Therefore, from (5) and (6):

$$U = \frac{1}{2} \left(\frac{D_x^2}{\varepsilon_{11}} + \frac{D_y^2}{\varepsilon_{22}} + \frac{D_z^2}{\varepsilon_{33}} \right) \tag{7}$$

If we define quantities X, Y, and Z measured along the three principal spatial axes by

$$X = \frac{D_x}{\sqrt{2\varepsilon_0 U}}; \qquad Y = \frac{D_y}{\sqrt{2\varepsilon_0 U}}; \qquad Z = \frac{D_z}{\sqrt{2\varepsilon_0 U}} \tag{8}$$

then (7) becomes

$$\frac{X^2}{(\varepsilon_{11}/\varepsilon_0)} + \frac{Y^2}{(\varepsilon_{22}/\varepsilon_0)} + \frac{Z^2}{(\varepsilon_{33}/\varepsilon_0)} = 1 \tag{9}$$

This is the equation of an ellipsoid; it is known of as the *index ellipsoid* (also called the *index of wave normals, optical indicatrix,* or *reciprocal ellipsoid*) and has as its semiaxes the square roots of the relative permittivities in the three principal directions, as shown in Fig. 13.8. The name *index ellipsoid* derives from the fact that the index of refraction equals the square root of relative permittivity when $\mu = \mu_0$.

The shape of the ellipsoid depends on the crystalline properties of the materials. Crystals can be divided into three optical classifications. Type I materials have

---

[36] For magnetically anisotropic real materials, the permeability is Hermitian so that $\mu_{ij} = \mu_{ij}^*$. For loss-free ferrites and plasmas with applied magnetic field and, to a lesser extent in dielectrics, the $[\mu]$ matrices are antisymmetric, so the off-diagonal terms must be imaginary.

[37] L. D. Landau and E. M. Lifshitz, *Electrodynamics of Continuous Media*, Pergamon Press, Oxford 1960, p. 313.

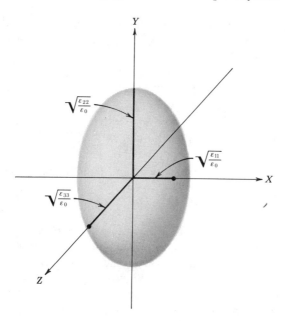

Fig. 13.8 The index ellipsoid.

ubic symmetry, for which there are three equivalent directions. In this case $\varepsilon_{11} = \varepsilon_{22} = \varepsilon_{33}$, the ellipsoid reduces to a sphere, and the properties are isotropic. Silicon, gallium arsenide, and cadmium telluride are examples of cubic crystals. Type II materials include trigonal, tetragonal, and hexagonal crystals, which all have one axis of symmetry. This axis is one of the principal axes of the permittivity ellipsoid. The symmetry of the crystal about this axis requires corresponding symmetry of the ellipsoid. It is, therefore, an ellipsoid of revolution, or spheroid. Crystals of this type are called *uniaxial* and include such important materials as calcium carbonate, quartz, lithium niobate, and cadmium sulfide. Type III or *biaxial* materials are crystals having no axes of symmetry; these are the orthorhombic, monoclinic, and triclinic systems. All three principal axes of the ellipsoid are different, as shown for the general ellipsoid in Fig. 13.8.

13.9   Plane-Wave Propagation in Anisotropic Crystals

The general properties of plane-wave propagation in anisotropic crystals are developed in this section. We will see that $\mathbf{E}$, $\mathbf{D}$, the wave normal $\boldsymbol{\beta}$, and the Poynting vector $\mathbf{P}$ all lie in a common plane perpendicular to $\mathbf{H}$. It is then shown that there are always two linearly polarized waves (except in certain degenerate situations) with orthogonal orientations of the $\mathbf{D}$ vectors for each direction of propagation.

It is convenient to introduce here the expression for a wave with a plane phase front oriented in an arbitrary direction. The physical quantities of the wave may be assumed to vary as $e^{-j\boldsymbol{\beta}\cdot\mathbf{r}}$ where

$$\boldsymbol{\beta} = \hat{\mathbf{x}}\beta_x + \hat{\mathbf{y}}\beta_y + \hat{\mathbf{z}}\beta_z$$

is a vector phase factor and

$$\mathbf{r} = \hat{\mathbf{x}}x + \hat{\mathbf{y}}y + \hat{\mathbf{z}}z \qquad (1$$

is the vector distance from an arbitrary origin. Thus we may write

$$\mathbf{E} = \mathbf{E}e^{-j\boldsymbol{\beta}\cdot\mathbf{r}}; \qquad \mathbf{D} = \mathbf{D}e^{-j\boldsymbol{\beta}\cdot\mathbf{r}}; \qquad \mathbf{H} = \mathbf{H}e^{-j\boldsymbol{\beta}\cdot\mathbf{r}} \qquad (2$$

where the coefficients **E**, **D**, and **H** are constant complex vectors for the electric field intensity, electric flux density, and magnetic field intensity with all spatial dependence removed. Constant phase surfaces are those for which $\boldsymbol{\beta}\cdot\mathbf{r} = \beta r \cos\theta =$ constant, where θ is the angle between $\boldsymbol{\beta}$ and $\mathbf{r}$. It is evident from Fig. 13.9a that the constant phase surfaces are planes.

Let us substitute **D** and **H** in the form (2) into Maxwell's equation

$$\nabla \times \mathbf{H} = j\omega\mathbf{D}$$

and apply the vector identity

$$\nabla \times (\Phi\mathbf{A}) = \nabla\Phi \times \mathbf{A} + \Phi\nabla \times \mathbf{A}$$

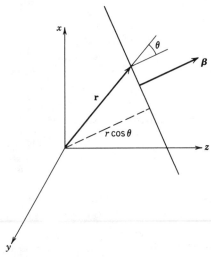

Fig. 13.9a Coordinates for a plane wave propagating in an arbitrary direction $\boldsymbol{\beta}$.

where Φ is a scalar and **A** is an arbitrary vector. The result is

$$(-j\boldsymbol{\beta} \times \mathbf{H})e^{-j\boldsymbol{\beta}\cdot\mathbf{r}} + e^{-j\boldsymbol{\beta}\cdot\mathbf{r}}\nabla \times \mathbf{H} = j\omega\mathbf{D}e^{-j\boldsymbol{\beta}\cdot\mathbf{r}}$$

Noting that $\nabla \times \mathbf{H} = 0$ since **H** is spatially invariant, we obtain

$$-\boldsymbol{\beta} \times \mathbf{H} = \omega\mathbf{D} \tag{3}$$

Application of the same operations to the other of Maxwell's curl equations, $\nabla \times \mathbf{E} = -j\omega\mu\mathbf{H}$, yields

$$\boldsymbol{\beta} \times \mathbf{E} = \omega\mu\mathbf{H} \tag{4}$$

Since the vector product is normal to both terms in the product, (3) shows that $\boldsymbol{\beta}$ is perpendicular to **D**. Similarly, (4) shows that $\boldsymbol{\beta}$ is also normal to **H**. Therefore $\boldsymbol{\beta}$ is normal to the direction defined by **D** and **H**, which means that constant phase surfaces contain **D** and **H**.

Another important direction of wave propagation is the direction of power flow. This is defined by the Poynting vector

$$\mathbf{P} = \mathbf{E} \times \mathbf{H} \tag{5}$$

In optics the unit vector in the direction of **P** is called the *ray vector*.

We see from (3)–(5) that **D**, **E**, $\boldsymbol{\beta}$, and **P** all lie in the plane perpendicular to **H**. It is also seen that the ray, or power flow, direction is different from that of the wave normal; the angle separating them is that between **E** and **D**, as seen in Fig. 3.9b where **H** is normal to the page.

Let us now derive the *Fresnel equation of wave normals* with which it is possible to make some important conclusions about the nature of the allowed waves. Combining (3) and (4) gives

$$\mathbf{D} = -\frac{n^2}{c^2\mu}[\hat{\mathbf{b}} \times (\hat{\mathbf{b}} \times \mathbf{E})] \tag{6}$$

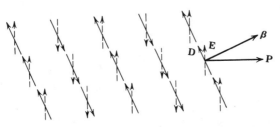

Fig. 13.9b Relative directions of the field vectors **D** and **E**, the wave vector $\boldsymbol{\beta}$, and the Poynting vector **P** in an anisotropic crystal. Magnetic field is normal to the plane of the other vectors.

where $\hat{\mathbf{b}}$ is the unit vector in the direction of $\boldsymbol{\beta}$, c is the velocity of light in vacuum, and n is the unknown index of refraction c/v_p in the direction of propagation defined by $\hat{\mathbf{b}}$. Using a vector identity, (6) can be converted to the form

$$\mathbf{D} = \frac{n^2}{c^2\mu}[\mathbf{E} - \hat{\mathbf{b}}(\hat{\mathbf{b}} \cdot \mathbf{E})] \tag{7}$$

Considering the field components lying along the principal axes of the index ellipsoid in order to simplify the analysis (without any loss of generality) and using Eqs. 13.8(5), (7) becomes

$$c^2\mu\varepsilon_{11}E_x = n^2[E_x - b_x(\hat{\mathbf{b}} \cdot \mathbf{E})]$$
$$c^2\mu\varepsilon_{22}E_y = n^2[E_y - b_y(\hat{\mathbf{b}} \cdot \mathbf{E})] \tag{8}$$
$$c^2\mu\varepsilon_{33}E_z = n^2[E_z - b_z(\hat{\mathbf{b}} \cdot \mathbf{E})]$$

Solving any one of the equations (8) gives

$$E_k = \frac{n^2 b_k(\hat{\mathbf{b}} \cdot \mathbf{E})}{n^2 - c^2\mu\varepsilon_{ii}} \tag{9}$$

where $k = x$, y, or z and i is the corresponding value 1, 2, or 3. Multiplying E_k by b_k, summing the three expressions $b_k E_k$, and dividing by $\hat{\mathbf{b}} \cdot \mathbf{E}$ gives

$$\frac{b_x^2}{n^2 - c^2\mu\varepsilon_{11}} + \frac{b_y^2}{n^2 - c^2\mu\varepsilon_{22}} + \frac{b_z^2}{n^2 - c^2\mu\varepsilon_{33}} = \frac{1}{n^2} \tag{10}$$

If we define the principal velocities of propagation by

$$v_x = \frac{1}{\sqrt{\mu\varepsilon_{11}}}, \qquad v_y = \frac{1}{\sqrt{\mu\varepsilon_{22}}}, \qquad v_z = \frac{1}{\sqrt{\mu\varepsilon_{33}}} \tag{11}$$

Then substituting (11) and $n = c/v_p$ in (10) and subtracting $b_x^2 + b_y^2 + b_z^2 = 1$ from both sides, we get

$$\frac{b_x^2}{v_p^2 - v_x^2} + \frac{b_y^2}{v_p^2 - v_y^2} + \frac{b_z^2}{v_p^2 - v_z^2} = 0 \tag{12}$$

Also (9) can be converted using $n = c/v_p$ and (11) to

$$E_k = \frac{v_k^2}{v_k^2 - v_p^2} b_k(\hat{\mathbf{b}} \cdot \mathbf{E}) \tag{13}$$

Equation (12) is *Fresnel's equation of wave normals*. It is easily seen to be quadratic in v_p^2. Therefore, for each set (b_x, b_y, b_z) that defines the direction of propagation, there are two values of phase velocity.

Using (13) for each value of phase velocity, we can find E_x, E_y, and E_z and, therefore, their ratios, which define the direction of $\mathbf{E}$ corresponding to that v_p. Likewise, the direction of $\mathbf{D}$ for each value of v_p can be found from $\mathbf{E}$ using the

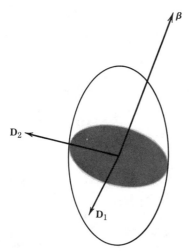

Fig. 13.9c Index ellipsoid with intersecting plane passing through its center. The lengths of the semiaxes of the resulting ellipse give the indices of refraction for the two waves and the flux-density vectors lie along the semiaxes.

relations in Eqs. 13.8(5). If this procedure is carried through, one finds that the ratios of the components of both **D** and **E** are real, which signals that the two waves are linearly polarized.

There is a helpful construction that can be used to give the phase velocities for an arbitrary direction of propagation.[38] It can also be used to prove that the **D** vectors of the two waves are mutually perpendicular and to determine their directions. A plane passed through the center of the index ellipsoid is oriented so its normal lies in the direction of the wave vector. The ellipse formed by the intersection of the plane and the ellipsoid is shown shaded in Fig. 13.9c. It can be shown (Prob. 13.9d) that the lengths of the two semiaxes give the indices of refraction (or, equivalently, the phase velocities) of the two waves. The same analysis shows that the flux density vectors **D** for the two waves lie along the minor and major axes of the shaded ellipse as shown.

13.10   Plane-Wave Propagation in Uniaxial Crystals

It was pointed out in Sec. 13.8 that the index ellipsoid for a uniaxial crystal is an ellipsoid of revolution; its axis of symmetry is called the *optic axis*. The Fresnel equation, Eq. 13.9(12) can be specialized to this case by writing $v_z = v_e$ and $v_x = v_y = v_o$:

$$(v_p^2 - v_o^2)[(b_x^2 + b_y^2)(v_p^2 - v_e^2) + b_z^2(v_p^2 - v_o^2)] = 0 \qquad (1)$$

[38] M. Born and E. Wolf, Ref. 35, Section 14.2.3.

If θ is the angle between $\boldsymbol{\beta}$ and the z axis, we can write

$$b_x^2 + b_y^2 = \sin^2 \theta; \qquad b_z^2 = \cos^2 \theta \tag{2}$$

With (2), the Fresnel equation (1) can be expressed as

$$(v_p^2 - v_o^2)[(v_p^2 - v_e^2) \sin^2 \theta + (v_p^2 - v_o^2) \cos^2 \theta] = 0 \tag{3}$$

The two roots of (3) are

$$v_{p1}^2 = v_o^2$$
$$v_{p2}^2 = v_o^2 \cos^2 \theta + v_e^2 \sin^2 \theta \tag{4}$$

Notice that one wave propagates like a wave in an isotropic medium in that its phase velocity is independent of the direction of propagation. This is called the *ordinary wave*, which is the reason for the subscript o. The other wave has a phase velocity that depends upon the direction of propagation and is called the *extra-ordinary wave* (subscript e). The two phase velocities are equal only when $\boldsymbol{\beta}$ is along the optic axis (z), as can be seen from (4) with $\theta = 0$.

The polarizations of the flux density vectors of the two waves are along the principal axes of the intersection ellipse in Fig. 13.10a. Since the ellipsoid is axially symmetric for the uniaxial case being considered now, the minor axis of the intersection ellipse is unchanged as the direction of propagation is changed. Therefore, the polarization with $\mathbf{D}$ along the minor axis is that of the ordinary wave and the index measured along that axis corresponds to the velocity v_o. Power flow P_o is parallel to $\boldsymbol{\beta}$ for the ordinary wave as predicted by Eqs. 13.9(3) and 13.9(5) since $\mathbf{D}$ and $\mathbf{E}$ are in the same direction ($\mathbf{D} = \hat{\mathbf{y}} D_y = \hat{\mathbf{y}} \varepsilon_{22} E_y$). The extraordinary wave has $\mathbf{D}_e$ along the major axis of the intersection ellipse. Since the angle between $\boldsymbol{\beta}$

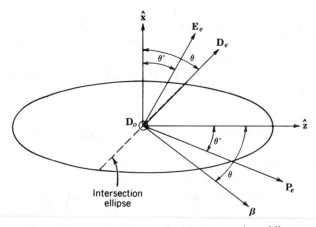

Fig. 13.10a Index ellipsoid for a uniaxial crystal with propagation oblique to the optic axis. The intersection ellipse contains the broken line along $\mathbf{D}_e$ and also the $\mathbf{D}_o$ vector normal to the page.

and **P** is the same as that between **D** and **E**, we can find the angle between the ray and wave-vector directions by using the permittivities. It is seen in Fig. 13.10a that the angle θ between $\boldsymbol{\beta}$ and the z axis is the same as that between $\mathbf{D}_e$ and the x axis so that $\theta = \tan^{-1}(D_z/D_x)$. The **E** vector lies in the plane of $\mathbf{D}_e$, $\boldsymbol{\beta}$, and the z axis (called the *principal section*) and its angle θ' from the x axis is the same as that between **P** and the z axis. Therefore,

$$\tan \theta' = \frac{E_z}{E_x} = \frac{\varepsilon_o}{\varepsilon_e} \frac{D_z}{D_x} = \frac{\varepsilon_o}{\varepsilon_e} \tan \theta \tag{5}$$

where $\varepsilon_o = \varepsilon_{11} = \varepsilon_{22}$ and $\varepsilon_e = \varepsilon_{33}$.

Although the details of reflection and refraction of waves at surfaces of anisotropic materials can become quite complicated, the principles are straightforward. Tangential components of **E** and **H** remain continuous across the boundary. Consider, for example, a crystal cut at an angle oblique to the optic axis, as shown in Fig. 13.10b. The y axis is the outward normal to the page. It is assumed that the incident plane wave has electric field vectors both in the plane of the page and normal to it. The situation inside the crystal is described by the index ellipsoid in Fig. 13.10a. As we saw from Eq. 13.9(3), the **D** vectors lie in the planes of the phase fronts and therefore must lie in the plane of the surface of the crystal since the incoming wave has constant phase on that surface. The dashed line across the ellipsoid in Fig. 13.10a is the edge of the intersection ellipse. The flux density $\mathbf{D}_o$ of the ordinary wave is in the y direction out of the page and $\mathbf{D}_e$ of the extraordinary wave lies in the plane of the page on the major axis of the intersection ellipse. We see from (5) that $\theta' < \theta$ for the ellipsoid illustrated since $\varepsilon_e > \varepsilon_o$. The path of the ray in real space, corresponding to the $\mathbf{P}_e$ direction in the ellipsoid, is shown in Fig. 13.10b. Notice that $\boldsymbol{\beta}$ is the same for the two waves, so the phase fronts are parallel even though the extraordinary ray follows a different path from the ordinary ray. The splitting of the refracted wave is called *double refraction* or *birefringence*. The two rays are again parallel after emergence from the planar slab. This is because the continuity conditions depend upon the phase variations along the surface and

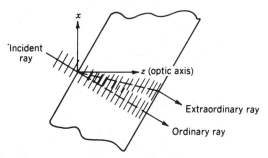

Fig. 13.10b Excitation of ordinary and extraordinary waves in a uniaxial crystal with parallel surfaces cut at an angle oblique to the optic axis.

these are zero for both waves. A common experiment shows this effect. A calcium carbonate crystal is placed on a spot on a piece of paper and two spots are observed, one from the ordinary wave and one from the extraordinary wave, provided the incident wave is unpolarized. If a polarizer is introduced into the incident beam, it can be rotated to eliminate one or the other of the spots.

13.11   Electro-Optic Effects

For certain materials the permittivity can be changed through application of an electric field. For liquids such as carbon disulfide and nitrobenzene, and for centrosymmetric solids, the effect is proportional to the square of electric field and is known as the *Kerr effect*. For certain noncentrosymmetric solids,[39] such as lithium niobate and gallium arsenide, the change may be directly proportional to applied electric field and is known as the *Pockels effect*. Both effects are relatively small for practical values of electric field, but the latter especially is important in a number of modulators, switches, and deflectors, so that the remainder of this section is devoted to this linear effect in crystals. Its description becomes annoyingly complicated because each component of the permittivity matrix is affected differently depending upon the orientation of the applied field with respect to the crystal axes.

Although one could describe the electro-optic effect by giving the change in permittivity element ε_{ij} when field component E_k is applied, it is customary to define the effect in terms of the reciprocal matrix. Define the matrix

$$\left[\frac{1}{n^2}\right] = \left[\frac{\varepsilon}{\varepsilon_0}\right]^{-1} \tag{1}$$

Then if electric field is applied, the change in an element of this matrix is

$$\Delta\left(\frac{1}{n^2}\right)_{ij} = \sum_{k=1}^{3} r_{ijk} E_k \tag{2}$$

where i, j, k each range over the three spatial coordinates (1, 2, 3 denote x, y, z, respectively). It is also usual to take advantage of the symmetry of the matrices $[\varepsilon]$ and hence $[1/n^2]$ to utilize a contracted notation: $11 \to 1$, $22 \to 2$, $33 \to 3$, $23 = 32 \to 4$, $13 = 31 \to 5$, $12 = 21 \to 6$ so that (2) may be written

$$\Delta\left(\frac{1}{n^2}\right)_{p} = \sum_{k=1}^{3} r_{pk} E_k \tag{3}$$

[39] A. Yariv, Ref. 32, Sec. 9.1. Also see Yariv and Yeh, Ref. 29, Chap. 7 and Haus, Ref. 29, Chap. 12.

with k ranging from 1 to 3 and p from 1 to 6. In matrix form,

$$\begin{bmatrix} \Delta\left(\dfrac{1}{n^2}\right)_1 \\[2ex] \Delta\left(\dfrac{1}{n^2}\right)_2 \\[2ex] \Delta\left(\dfrac{1}{n^2}\right)_3 \\[2ex] \Delta\left(\dfrac{1}{n^2}\right)_4 \\[2ex] \Delta\left(\dfrac{1}{n^2}\right)_5 \\[2ex] \Delta\left(\dfrac{1}{n^2}\right)_6 \end{bmatrix} = \begin{bmatrix} r_{11} & r_{12} & r_{13} \\ r_{21} & r_{22} & r_{23} \\ r_{31} & r_{32} & r_{33} \\ r_{41} & r_{42} & r_{43} \\ r_{51} & r_{52} & r_{53} \\ r_{61} & r_{62} & r_{63} \end{bmatrix} \begin{bmatrix} E_1 \\ E_2 \\ E_3 \end{bmatrix} \tag{4}$$

Values of r_{pk} for a few crystals are given in Table 13.11.

_____ **Example 13.11a** _____
Electro-Optic Phase Modulation

Consider a crystal of lithium niobate as shown in Fig. 13.11a, with a wave propagating in the y direction, polarized with electric field in the z direction, and with electrodes placed so that the modulating field is also in the z direction (the optical axis of this crystal). From Table 13.11 we see that applied field $E_z = E_3$, through r_3, changes $(1/n^2)_1$ and $(1/n^2)_3$ and no other elements so that the matrix remains diagonalized even after application of the modulating field. (This is not true in general.) The change of $(1/n^2)_1$ cannot affect a wave with $E_z = E_3$ only but the change of $(1/n^2)_3$ is important. From (3),

$$\left(\frac{1}{n^2}\right)_3 = \left(\frac{1}{n_e^2}\right) + r_{33}E_3 \tag{5}$$

where n_e is the index of refraction for extraordinary waves defined in the preceding article. Because of the small value of r_{33},

$$n_3 = \left[\left(\frac{1}{n^2}\right)_3\right]^{-1/2} \approx n_e - \frac{r_{33}n_e^3}{2}E_3 \tag{6}$$

So if the modulating field is

$$E_3 = E_m \cos \omega_m t \tag{7}$$

Table 13.11[a]
Electro-Optic Properties

Material	$n_0(\lambda_0 = 550 \text{ nm})$	$n_e(\lambda_0 = 550 \text{ nm})$	r_{pk}^b in units of 10^{-12} m/V
KDP	1.51	1.47	$r_{41} = 8.6$
(KH$_2$PO$_4$)			$r_{63} = 8.6$
GaAs(10.6 μm)	3.34		$r_{41} = 1.6$
LiNbO$_3$	2.29	2.20	$r_{33} = 30.8$
			$r_{13} = 8.6$
			$r_{22} = 3.4$
			$r_{42} = 28$
BaTiO$_3$	2.437	2.365	$r_{33} = 23$
			$r_{13} = 8.0$
			$r_{42} = 820$

[a] From A. Yariv, *Introduction to Optical Electronics*, Ref. 32, p. 251.
[b] All elements not listed are zero.

the phase shift after propagating distance l is

$$\phi(l, t) = \frac{\omega l}{c} n_3 = \frac{\omega l}{c} n_e + \Delta\phi_m \cos \omega_m t \tag{8}$$

where the *modulation index* $\Delta\phi_m$ is

$$\Delta\phi_m = -\frac{\omega l r_{33} n_e^3 E_m}{2c} \tag{9}$$

To make this equal to π (a useful value for the applications to be discussed next) for a signal with free-space wavelength $\lambda_0 = 550$ nm with $l = 1$ cm, E_m would need to be 1.68×10^5 V/m. Although a high field, the required applied voltages are reasonable in thin crystals or thin-film modulators using the optical guiding described in Sec. 14.7.

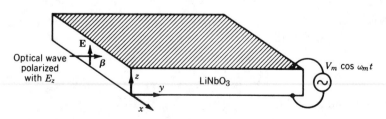

Fig. 13.11a Electro-optic phase modulation in LiNbO$_3$ crystal.

_____ **Example 13.11b** _____
Conversion of Phase Modulation to Amplitude
Modulation by Interferometry

Although phase modulation may be used directly for some purposes, it is often desirable to convert this to amplitude modulation. One of the simplest methods is by means of a Mach–Zehnder interferometer sketched in Fig. 13.11b.[40] An optical wave is divided into two equal parts by the first Y junction, electro-optic phase shift is introduced into one arm, and then the two waves are combined in the second Y junction. With zero phase shift the two add, with π phase shift they cancel, and with intermediate phases there is a value between so that 100% modulation is possible. The device is thus also useful as a switch.

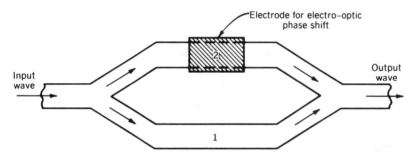

Fig. 13.11b Top view of optical-waveguide Mach–Zehnder interferometer for converting electro-optic phase modulation to amplitude modulation.

_____ **Example 13.11c** _____
Conversion of Phase Modulation to Amplitude
Modulation by Polarization Shifts

A somewhat more complicated case is that in which phase shifts are converted to a change in direction of polarization, and this in turn is converted to amplitude modulation through use of an analyzer. It is the classic electro-optic modulator, and also illustrates that new principal axes are in general required after application of the electric field. A sketch of the arrangement is shown in Fig. 13.11c. The crystal used for the example is KDP and the modulating field is applied in the z (i.e., 3) direction, which is also the optic axis and the direction of wave propagation. From Table 13.11 the pertinent electro-optic coefficient is then r_{63}, telling us

[40] T. R. Ranganath and S. Wang, _IEEE J. Quantum Electronics_ **QE-13**, 290 (1977).

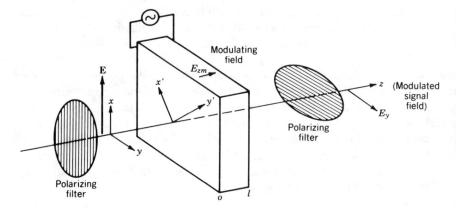

Fig. 13.11c Electro-optic modulator that employs conversion of phase modulation to amplitude modulation.

that application of electric field E_z adds an off-diagonal element $(1/n^2)_6 = (1/n^2)_{12}$ to the matrix. New principal axes x' and y' after application of E_z are rotated by 45 degrees relative to the crystal axes x and y (Prob. 13.11a). For these

$$\frac{1}{n_{x'}^2} = \frac{1}{n_o^2} + r_{63} E_z, \qquad \frac{1}{n_{y'}^2} = \frac{1}{n_o^2} - r_{63} E_z \tag{10}$$

Since $r_{63} E_z$ is small, these lead to

$$n_{x'} \approx n_o - \frac{n_o^3}{2} r_{63} E_z, \qquad n_{y'} \approx n_o + \frac{n_o^3}{2} r_{63} E_z \tag{11}$$

The polarizer of Fig. 13.11c polarizes with field E_x, which gives equal components in the x' and y' directions,

$$\mathbf{E}(0) = \frac{E_x}{\sqrt{2}} [\hat{\mathbf{x}}' + \hat{\mathbf{y}}'] \tag{12}$$

After propagating distance l,

$$\mathbf{E}(l) = \frac{E_x}{\sqrt{2}} e^{-j(\omega l n_o/c)} [\hat{\mathbf{x}}' e^{j(\omega l n_o^3 r_{63} E_z/2c)} + \hat{\mathbf{y}}' e^{-j(\omega l n_o^3 r_{63} E_z/2c)}] \tag{13}$$

The analyzer selects the y component, which is

$$E_y(l) = \frac{1}{\sqrt{2}} [-E_{x'}(l) + E_{y'}(l)]$$

$$= \frac{E_x}{2} e^{-j(\omega l n_o/c)} [-e^{j(\omega l n_o^3 r_{63} E_z/2c)} + e^{-j(\omega l n_o^3 r_{63} E_z/2c)}] \tag{14}$$

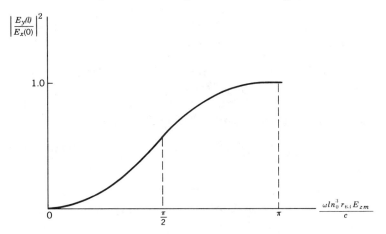

Fig. 13.11d Output signal field intensity as a function of modulating field E_{zm}, for Fig. 13.11c.

The square of magnitude of this is

$$|E_y(l)|^2 = |E_x|^2 \sin^2 \frac{\omega l n_o^3 r_{63} E_z}{2c} \tag{15}$$

So that it is seen to vary with modulating field E_z according to the curve of Fig. 13.11d. A "quarter-wave plate" can be added to shift the operating point with zero modulation to the linear, center part of the curve (Prob. 13.11c). Such a plate produces a phase difference of $\pi/2$ between the two polarizations.

13.12   Permittivity of a Stationary Plasma in a Magnetic Field

In this section we develop explicit, frequency-dependent, expressions for the elements of the permittivity matrix of a plasma with an applied steady magnetic field. The results are important in understanding electromagnetic wave propagation through the ionosphere in the presence of the earth's magnetic field, the magnetically confined plasma used in fusion research, and certain laser discharges. Later extensions, with an average electron velocity added, will apply to traveling-wave tubes and other electron-beam amplifiers.

Because of the low density of the plasma in comparison with liquids and solids, the primary dielectric behavior is given by analysis of the electron motion. The applied magnetic field in general causes the permittivity matrix to have off-diagonal elements, since if magnetic field is axial, electron motion in one transverse direction produces forces in the orthogonal transverse direction. The matrix can be diagonalized, but not by a simple rotation of coordinates; the characteristic waves in this case turn out to be circularly polarized. We shall later see very similar behavior for the permeability matrix of a ferrite with applied magnetic field.

In order to bring out the main physical effects, several idealizations are made in the model. The plasma is considered uniform and stationary in this analysis, although later a drift velocity for electrons will be added (Sec. 13.14). Magnetic field is stationary, uniform, and in the z direction. Because of the heavy mass of the ions, their motion is neglected; they serve only to neutralize the dc space-charge fields of electrons. The effects of thermal velocities and collisions are neglected, as are the forces arising from the magnetic fields of the electromagnetic waves (see Prob. 13.3a).

The first step is to find the convection current in the plasma. The dc plus ac current density is

$$\mathbf{J}_0 + \mathbf{J}_1 = (\rho_0 + \rho_1)(\mathbf{v}_0 + \mathbf{v}_1) \tag{1}$$

Here $\mathbf{v}_0$ and $\mathbf{J}_0$ are equal to zero since the plasma is assumed stationary. Thus, neglecting products of ac terms since they are of second-order smallness, we see that

$$\mathbf{J}_1 = \rho_0 \mathbf{v}_1 \tag{2}$$

The velocity $\mathbf{v}_1$ is found from the Lorentz equation [Eq. 3.6(5)]

$$\frac{d\mathbf{v}_1}{dt} = -\frac{e}{m}(\mathbf{E} + \mathbf{v}_1 \times \mathbf{B}_0) \tag{3}$$

where $\mathbf{B}_0 = \hat{z}B_{0z}$ is the applied steady magnetic field.

To find the conditions required for linearization of the equations, let us consider a uniform plane wave having arbitrary direction of propagation, such that the ac spatial variations may be expressed by

$$e^{-j\boldsymbol{\beta}\cdot\mathbf{r}} = e^{-j(\beta_x x + \beta_y y + \beta_z z)} \tag{4}$$

The left side of (3), expanded in partial derivatives and evaluated using (4), becomes

$$\frac{d\mathbf{v}_1}{dt} = j(\omega - \beta_x v_{1x} - \beta_y v_{1y} - \beta_z v_{1z})\mathbf{v}_1 e^{j\omega t} \tag{5}$$

If the ac velocity is assumed to be small compared with any possible phase velocities, then $\beta_i v_{1i} = \omega v_{1i}/v_{pi} \ll \omega$ and (5) reduces to

$$\frac{d\mathbf{v}_1}{dt} \cong j\omega \mathbf{v}_1 e^{j\omega t} \tag{6}$$

Writing (3) in component phasor form using (6), we have

$$j\omega v_{1x} = -\frac{e}{m}E_{1x} - \frac{e}{m}B_{0z}v_{1y} \tag{7a}$$

$$j\omega v_{1y} = -\frac{e}{m}E_{1y} + \frac{e}{m}B_{0z}v_{1x} \tag{7b}$$

$$j\omega v_{1z} = -\frac{e}{m}E_{1z} \tag{7c}$$

Equations (7) may be solved for velocity components in terms of the fields with the result

$$v_{1x} = \frac{-j\omega(e/m)E_{1x} + (e/m)\omega_c E_{1y}}{\omega_c^2 - \omega^2} \tag{8a}$$

$$v_{1y} = \frac{-(e/m)\omega_c E_{1x} - j\omega(e/m)E_{1y}}{\omega_c^2 - \omega^2} \tag{8b}$$

$$v_{1z} = \frac{j(e/m)}{\omega} E_{1z} \tag{8c}$$

where $\omega_c = (e/m)B_{0z}$ is called the angular cyclotron frequency.

It is convenient to define an equivalent flux density **D** that includes the effect of convection currents. In matrix notation, the equivalent displacement current is

$$j\omega[D] = j\omega\varepsilon_0[E] + [J]$$
$$= j\omega[\varepsilon][E] \tag{9}$$

where $[\varepsilon]$ is the equivalent-permittivity matrix. Using (2) and (8):

$$[\varepsilon] = \begin{bmatrix} \varepsilon_{11} & \varepsilon_{12} & 0 \\ \varepsilon_{21} & \varepsilon_{22} & 0 \\ 0 & 0 & \varepsilon_{33} \end{bmatrix} \tag{10}$$

where

$$\varepsilon_{11} = \varepsilon_{22} = \varepsilon_0\left[1 + \frac{\omega_p^2}{\omega_c^2 - \omega^2}\right] \tag{11a}$$

$$\varepsilon_{12} = \varepsilon_{21}^* = \frac{j\omega_p^2(\omega_c/\omega)\varepsilon_0}{\omega_c^2 - \omega^2} \tag{11b}$$

$$\varepsilon_{33} = \varepsilon_0\left[1 - \frac{\omega_p^2}{\omega^2}\right] \tag{11c}$$

Here ω_p is the plasma frequency, Eq. 13.3(10).

For plane waves in a plasma without a magnetic field (Sec. 13.3), ω_p was observed to be a characteristic frequency in that no propagation occurs for $\omega < \omega_p$. It is clear from the form of (11a)–(11b) that the cyclotron frequency ω_c is also a characteristic frequency in this case. The singularity at ω_c is seen from (8a)–(8b) to result from the singularity in the velocities produced by the wave. A moving electron in a magnetic field without an electric field rotates at an angular frequency ω_c. If an applied alternating electric field oscillates at ω_c, it is so phased on each cycle that it continually pumps the electron to higher and higher velocities. That leads to the infinite response for the continuous-wave case in this model, although collisions would limit excursions in practice.

13.13   *TEM* Waves on a Plasma with Infinite Magnetic Field

In the use of the plasma permittivity matrix, we consider first the simple special case of *TEM* waves on a plasma with a magnetic field strong enough that no significant electron motion is induced perpendicular to the magnetic field. This is modeled by setting $B_{0z} = \infty$. The effect of this assumption is to make the off-diagonal terms zero.

Taking the curl of Maxwell's curl **E** equation and substituting the curl **H** equation with a scalar permeability (taking $\mu = \mu_0$) we get[41]

$$\nabla \times \nabla \times \mathbf{E} = \omega^2 \mu_0 \mathbf{D} \tag{1}$$

Using a common vector identity, (1) becomes

$$\nabla^2 \mathbf{E} - \nabla(\nabla \cdot \mathbf{E}) + \omega^2 \mu_0 \mathbf{D} = 0 \tag{2}$$

For a *TEM* wave propagating either parallel or perpendicular to the steady magnetic field, either the electric field component or its derivative is zero in all directions, so $\nabla \cdot \mathbf{E} = 0$. (The more complicated problem of *TEM* waves propagating at arbitrary angles involves a nonvanishing $\nabla \cdot \mathbf{E}$ and will not be treated here.) Then

$$\nabla^2 \mathbf{E} + \omega^2 \mu_0 \mathbf{D} = 0 \tag{3}$$

If the *TEM* wave is propagating in the z direction as $e^{-j\beta z}$, the field **E** may have x and y components only and $\mathbf{D} = \varepsilon_0 \mathbf{E}$. Separating (3) and using the assumed $e^{-j\beta z}$ variation and the dielectric matrix Eqs. 13.12(10)–(11), we obtain

$$(\omega^2 \mu_0 \varepsilon_0 - \beta^2)E_x = 0 \tag{4}$$

and the same form for E_y. Thus we see that such a wave propagates as though it were in free space, since no electron motion can be induced normal to the infinite magnetic field.

Now consider a *TEM* wave propagating in the x direction. Note that here $\beta_z = 0$. For a wave with $\mathbf{E} = \hat{y}E_y$, (3) may be separated to give

$$(\omega^2 \mu_0 \varepsilon_0 - \beta^2)E_y = 0 \tag{5}$$

which is again a wave propagating as though in free space. But, for a wave with $\mathbf{E} = \hat{z}E_z$, $\mathbf{D} = \varepsilon_{33}\mathbf{E}$ and (3) gives

$$(\omega^2 \mu_0 \varepsilon_{33} - \beta^2)E_z = 0 \tag{6}$$

where

$$\varepsilon_{33} = \varepsilon_0 \left[1 - \frac{\omega_p^2}{\omega^2} \right] \tag{7}$$

---

[41] To get the right side of vector equations like (1) in terms of **E** would require tensor or dyadic notation (see Ref. 33). We choose to avoid that less-familiar notation by using **D**.

We see that if a wave propagates normal to an infinite magnetic field with the electric field directed along the dc magnetic field, the propagation is the same as for a transverse wave propagating through a plasma without a magnetic field. That is, the phase constant is

$$\beta = \omega\sqrt{\mu_0 \varepsilon_{33}} \tag{8}$$

This goes to zero as the wave frequency approaches the plasma frequency, and for lower frequencies the wave does not propagate. (See Fig. 13.3a.)

A plane wave having both E_z and E_y components propagating in the x direction can be treated as two separate waves having different phase velocities, as was done with the anisotropic crystal. Thus the plasma will act to convert linear polarization to elliptic polarization above the cutoff frequency $\omega = \omega_p$.

13.14 Space-Charge Waves on a Moving Plasma with Infinite Magnetic Field

We consider now waves propagating in a plasma parallel to an infinite magnetic field and having a component of electric field in the direction of propagation, as shown in Fig. 13.14. An important application of the results is as a model for waves on electron beams that have an average velocity parallel to the magnetic field. For the purpose of analyzing the space-charge waves, we will first show the effect of v_{0z} on the permittivity matrix developed in Sec. 13.12.

The ac current density, Eq. 13.12(1), when linearized is

$$\mathbf{J}_1 = \rho_0 \mathbf{v}_1 + \mathbf{v}_0 \rho_1 \tag{1}$$

To find ac charge density ρ_1 which arises from the bunching of electrons, we make use of the continuity equation for the ac current density

$$\nabla \cdot \mathbf{J}_1 = \frac{\partial J_{1x}}{\partial x} + \frac{\partial J_{1y}}{\partial y} + \frac{\partial J_{1z}}{\partial z} = -\frac{\partial \rho_1}{\partial t} \tag{2}$$

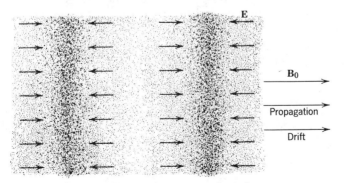

Fig. 13.14 Space-charge waves in a plasma.

With an infinite magnetic field, $J_{1x} = J_{1y} = 0$. Then with all ac quantities having wave form, $\exp j(\omega t - \beta z)$, (2) gives

$$\rho_1 = \frac{\beta J_{1z}}{\omega} \tag{3}$$

To get the value of v_{1z} needed in (1), we return to the force equation Eq. 13.12(3) and write the z component of the left side, taking account of the average velocity v_{0z}:

$$\frac{dv_{1z}}{dt} = \frac{\partial v_{1z}}{\partial t} + v_{0z} \frac{\partial v_{1z}}{\partial z} \tag{4}$$

Noting that $\mathbf{v}_1 \times \mathbf{B} = 0$, the force equation becomes, in phasor form,

$$j(\omega - \beta v_{0z})v_{1z} = -\frac{e}{m} E_{1z} \tag{5}$$

or

$$v_{1z} = \frac{j(e/m)}{\omega - \beta v_{0z}} E_{1z} \tag{6}$$

Substituting (3) and (6) into (1) and using the definition of the plasma frequency ω_p in Eq. 13.3(10), we get

$$J_{1z} = \frac{-j\omega\varepsilon_0 \omega_p^2}{(\omega - \beta v_{0z})^2} E_{1z} \tag{7}$$

Proceeding as in Eqs. 13.12(9)–(11) yields

$$\varepsilon_{33} = \varepsilon_0 \left[1 - \frac{\omega_p^2}{(\omega - \beta v_{0z})^2} \right] \tag{8}$$

Note that the difference in ε_{33} between (8) and that of the stationary plasma, Eq. 13.12(11c), can be considered a Doppler shift in the frequency seen by the electrons. In like manner, the other components of the permittivity matrix can be found by using a Doppler-shifted frequency $\omega - \beta v_{0z}$ wherever ω appears in Eqs. 13.12(11a) and (11b).

Using Gauss's law and assumed $e^{-j\beta z}$ form of variations, the z component of Eq. 13.13(2) may be written as

$$\nabla_t^2 E_z - \beta^2 E_z + j\beta \frac{\rho_1}{\varepsilon_0} + \omega^2 \mu_0 \varepsilon_{33} E_z = 0 \tag{9}$$

where $\nabla_t^2 E_z$ contains partial derivatives transverse to the z direction. Combining (3) and (7) we find

$$\rho_1 = -j \frac{\varepsilon_0 \omega_p^2 \beta}{(\omega - \beta v_{0z})^2} E_z \tag{10}$$

Then combining (9) and (10) and using the value of ε_{33} from (8), we obtain the desired wave equation

$$\nabla_t^2 E_z + (\omega^2 \mu_0 \varepsilon_0 - \beta^2)\left[1 - \frac{\omega_p^2}{(\omega - \beta v_{0z})^2}\right] E_z = 0 \tag{11}$$

We now restrict our attention to uniform plane longitudinal waves; that is, waves with electric fields along the direction of propagation and no transverse variation. Therefore the first term of (11) vanishes. Equation (11) may then be satisfied if either $\omega^2 \mu_0 \varepsilon_0 - \beta^2$ or the term in brackets is zero. The first alternative gives the usual *TEM field waves*, with zero E_z. More interesting is the second alternative,

$$1 - \frac{\omega_p^2}{(\omega - \beta v_{0z})^2} = 0 \tag{12}$$

This leads to two waves with phase constants

$$\beta_{1,2} = \frac{\omega \pm \omega_p}{v_{0z}} \tag{13}$$

These waves involve interaction with the electrons and are called *space-charge waves*.[42] Typically, $\omega_p \ll \omega$, and we may use the binomial expansion to obtain the phase velocities in the form

$$(v_p)_{1,2} = \frac{\omega}{\beta_{1,2}} \approx v_{0z}\left(1 \mp \frac{\omega_p}{\omega}\right) \tag{14}$$

the values of which are seen to be somewhat greater and somewhat less than the drift velocity v_{0z}. The group velocity is

$$v_g = \frac{\partial \omega}{\partial \beta} = v_{0z} \tag{15}$$

for both space-charge waves. With the plasma drift v_{0z} reduced to zero, a perturbation of the electron charge density leads to a local oscillation at a frequency $\omega = \omega_p$ but the effect does not propagate.

13.15   *TEM* Waves on a Stationary Plasma in a Finite Magnetic Field

The general study of waves on a plasma in a finite magnetic field reveals many interesting wave types. Analysis reveals that a *TEM* wave can propagate (in an unbounded region), but that no pure *TE* or *TM* modes can exist. The actual

---

[42] S. Ramo, *Phys. Rev.* **56**, 276 (1939); the concept was first usefully applied to electron devices by W. C. Hahn, *Gen. Elect. Rev.* **42**, 258; 497 (1939).

hybrid modes have both E_z and H_z. We will examine here only the simple TEM wave. The reader is referred to books on plasma waves for the details of the other types.[43]

We restrict our attention here to the very important case of TEM waves propagating in the z direction parallel to the applied steady magnetic field. Since we will use the phasor form of the wave equation, the appropriate form of z dependence is $e^{\pm j\beta z}$. Also, the TEM wave has no E_z component and no variations of fields with respect to transverse coordinates so the $\mathbf{V} \cdot \mathbf{E} = 0$ and Eq. 13.13(2) may be written in matrix form

$$\frac{d^2}{dz^2} \begin{bmatrix} E_x \\ E_y \\ 0 \end{bmatrix} + \omega^2 \mu_0 \begin{bmatrix} \varepsilon_{11} & \varepsilon_{12} & 0 \\ \varepsilon_{21} & \varepsilon_{22} & 0 \\ 0 & 0 & \varepsilon_{33} \end{bmatrix} \begin{bmatrix} E_x \\ E_y \\ 0 \end{bmatrix} = 0 \tag{1}$$

Here we have made use of the form of the permittivity in Eq. 13.12(10).

Previously, our wave solutions were taken to be linearly polarized waves which, by suitable orientation of the coordinate system, might have one component only (e.g., $\mathbf{E} = \hat{\mathbf{x}} E_x$). If we substitute such an assumed form of solution in (1), we find a vector with only an x component in the first term whereas the vector in the second term has both x and y components. For the sum of two vectors to be zero, their components must individually add to zero. Therefore, the magnitude of the y component must be zero. This is possible only if either ε_{21} or E_x is zero. We know from Sec. 13.12 that ε_{21} is not zero, and if E_x is zero, there is no wave. Thus we must conclude that a linearly polarized wave is not a possible solution; that is, it is not a *normal mode* of propagation for the medium.

Let us now consider the possibility of a solution in the form of a circularly polarized wave. For a clockwise polarized wave (field vectors rotate clockwise as a function of time at a fixed plane in space, looking in the direction of propagation), the field has the form

$$\mathbf{E}_+^{cw} = E_+^{cw}(\hat{\mathbf{x}} - j\hat{\mathbf{y}}) \tag{2}$$

Substituting (2) in (1) and making use of Eqs. 13.12(10) and (11), we find

$$(\beta_+^{cw})^2 = \omega^2 \mu_0 (\varepsilon_{11} - j\varepsilon_{12})$$

$$= \omega^2 \mu_0 \varepsilon_0 \left(1 + \frac{\omega_p^2/\omega}{\omega_c - \omega}\right) \tag{3}$$

We see that the circularly polarized wave is a possible solution, or normal mode of propagation; this is called *gyrotropic* behavior. For the forward wave we use the form $e^{-j\beta z}$ so we choose the positive square root of (3):

$$\beta_+^{cw} = \omega\sqrt{\mu_0 \varepsilon_0} \left(1 + \frac{\omega_p^2/\omega}{\omega_c - \omega}\right)^{1/2} \tag{4}$$

[43] T. Stix, *Theory of Plasma Waves*, McGraw-Hill, New York, 1962; or W. P. Allis, S. T. Buchsbaum, and A. Bers, *Waves in Anisotropic Plasmas*, MIT Press, Cambridge, Mass., 1963.

For a wave propagating in the positive z direction with counterclockwise polarization, the electric field may be represented by

$$\mathbf{E}_+^{\text{ccw}} = E_+^{\text{ccw}}(\hat{\mathbf{x}} + j\hat{\mathbf{y}}) \tag{5}$$

Using the assumed form $e^{-j\beta z}$ and proceeding as above, we find

$$\beta_+^{\text{ccw}} = \omega\sqrt{\mu_0 \varepsilon_0}\left(1 - \frac{\omega_p^2/\omega}{\omega_c + \omega}\right)^{1/2} \tag{6}$$

For waves traveling in the $-z$ direction, clockwise and counterclockwise have meaning opposite from those in the wave traveling in the $+z$ direction. Therefore

$$\mathbf{E}_-^{\text{cw}} = E_-^{\text{cw}}(\hat{\mathbf{x}} + j\hat{\mathbf{y}}) \tag{7}$$

$$\mathbf{E}_-^{\text{ccw}} = E_-^{\text{ccw}}(\hat{\mathbf{x}} - j\hat{\mathbf{y}}) \tag{8}$$

and it is easily seen that

$$\beta_-^{\text{ccw}} = \beta_+^{\text{cw}} \quad \text{and} \quad \beta_-^{\text{cw}} = \beta_+^{\text{ccw}} \tag{9}$$

Transmission-Line Analogy   We can make use of the transmission-line analogy to solve problems of reflection and transmission with considerable advantage of simplification. To assure continuity of electric and magnetic fields at a boundary for circularly polarized waves, forward- and reverse-traveling waves having the same absolute direction of rotation must be combined. Thus we sum the fields of the positive-traveling clockwise wave and the negative-traveling counterclockwise wave. The other pair of waves (positive-traveling counterclockwise and negative-traveling clockwise) are handled as a separate problem in the same manner. The two pairs of waves are illustrated in Figs. 13.15a and 13.15b.

Let us consider the pair of waves in Fig. 13.15a. The electric fields for both waves are described by (2) and the magnetic fields by

$$\mathbf{H}_+^{\text{cw}} = H_+^{\text{cw}}(j\hat{\mathbf{x}} + \hat{\mathbf{y}}) \tag{10}$$

and

$$\mathbf{H}_-^{\text{ccw}} = -H_-^{\text{ccw}}(j\hat{\mathbf{x}} + \hat{\mathbf{y}}) \tag{11}$$

If (2) and (10) or (2) and (11) are substituted in Maxwell's curl E equation, one finds the ratio of magnitudes, which can be called the characteristic wave impedance:

$$\eta_a = \frac{E_+^{\text{cw}}}{H_+^{\text{cw}}} = \frac{E_-^{\text{ccw}}}{H_-^{\text{ccw}}} = \frac{\omega\mu_0}{\beta_a} = \sqrt{\frac{\mu_0}{\varepsilon_{11} - j\varepsilon_{12}}} \tag{12}$$

where the subscript a denotes the case in Fig. 13.15a.

The total fields for case a can then be written as

$$E_a(z) = E_+^{\text{cw}}e^{-j\beta_a z} + E_-^{\text{ccw}}e^{j\beta_a z} \tag{13}$$

and

$$H_a(z) = \frac{1}{\eta_a}(E_+^{\text{cw}}e^{-j\beta_a z} - E_-^{\text{ccw}}e^{j\beta_a z}) \tag{14}$$

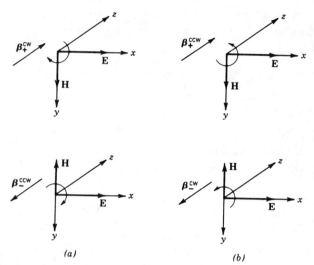

(a)   (b)

Fig 13.15 (a) Pair of waves having rotation of fields clockwise relative to the $+z$ axis. (b) Pair of waves having rotation counterclockwise relative to the $+z$ axis.

where β_a is given by (4). These are in the form of the transmission-line equations and the analogy is completed by defining a wave impedance at any plane z as

$$Z_a(z) = \frac{E_a(z)}{H_a(z)} \tag{15}$$

Then all the apparatus of Sec. 6.6 can be adapted to solve problems of reflection and transmission for a circularly polarized wave.

The pair of waves in Fig. 13.15b are handled in the same way. In this case the electric fields of the counterclockwise wave in the $+z$ direction and the clockwise wave in the $-z$ direction both have the form in (5) and the magnetic fields are

$$\mathbf{H}_+^{ccw} = H_+^{ccw}(-j\hat{\mathbf{x}} + \hat{\mathbf{y}}) \tag{16}$$

and

$$\mathbf{H}_-^{cw} = -H_-^{cw}(-j\hat{\mathbf{x}} + \hat{\mathbf{y}}) \tag{17}$$

Substitution in Maxwell's equations gives the characteristic wave impedance:

$$\eta_b = \frac{E_+^{ccw}}{H_+^{ccw}} = \frac{E_-^{cw}}{H_-^{cw}} = \sqrt{\frac{\mu_0}{\varepsilon_{11} + j\varepsilon_{12}}} \tag{18}$$

and the total fields are

$$E_b(z) = E_+^{ccw}e^{-j\beta_b z} + E_-^{cw}e^{j\beta_b z} \tag{19}$$

$$H_b(z) = \frac{1}{\eta_b}(E_+^{ccw}e^{-j\beta_b z} - E_-^{cw}e^{j\beta_b z}) \tag{20}$$

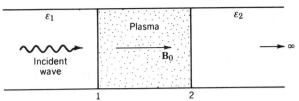

Fig. 13.15c Example of a problem utilizing transmission-line methods for circularly polarized waves.

with β_b given by (6). For this case the total impedance is

$$Z_b(z) = \frac{E_b(z)}{H_b(z)} \tag{21}$$

An example of the use of these tools in a calculation of the propagation of waves through a layer of plasma between two dielectric regions with the propagation direction parallel to the dc magnetic field as shown in Fig. 13.15c. If the incident wave is not circularly polarized, it can be decomposed into two oppositely rotating circularly polarized waves. Each of these can be handled separately since we are considering the media to be linear. The load impedance Z_L at plane 2 for each of type a and b waves in Figs. 13.15a and 13.15b is the characteristic wave impedance $\eta = \sqrt{\mu_0/\varepsilon_2}$. The impedance transformation Eq. 6.6(10) with k replaced by either β_a or β_b gives the wave impedance at plane 1. Then the reflection coefficient Eq. 6.6(11) gives the relative amplitude of the reflected wave in region 1 for each direction of polarization. Knowing the amplitudes of the incident waves of each circular polarization, the total reflected wave is constructed as the sum of those in the two polarizations.

13.16 Faraday Rotation

In this section we shall see how the difference of propagation constants between clockwise and counterclockwise polarized waves can be used to show that a linearly polarized wave is rotated when it passes through a gyrotropic medium. Let us first consider a linearly polarized wave propagating in the $+z$ direction with **E** vector in the x direction at the plane $z = 0$ in the plasma of Sec. 13.15. This field may be decomposed into circularly polarized modes of propagation

$$\mathbf{E}^{\mathrm{cw}} = (\hat{\mathbf{x}} - j\hat{\mathbf{y}})\frac{E}{2} e^{-j\beta^{\mathrm{cw}}z} \tag{1}$$

and

$$\mathbf{E}^{\mathrm{ccw}} = (\hat{\mathbf{x}} + j\hat{\mathbf{y}})\frac{E}{2} e^{-j\beta^{\mathrm{ccw}}z} \tag{2}$$

since the sum of (1) and (2) at $z = 0$ is

$$\mathbf{E} = \hat{\mathbf{x}}E$$

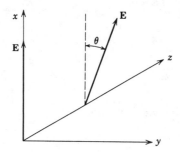

Fig. 13.16 Rotation of a linearly polarized wave in a gyrotropic medium.

This can be thought of as finding magnitudes of the natural modes of the system necessary to match the given boundary condition at $z = 0$.

The field at any other plane, $z = $ constant, is found by adding (1) and (2):

$$\mathbf{E} = \frac{E}{2}[\hat{\mathbf{x}}(e^{-j\beta^{ccw}z} + e^{-j\beta^{cw}z}) + j\hat{\mathbf{y}}(e^{-j\beta^{ccw}z} - e^{-j\beta^{cw}z})] \tag{3}$$

which can be cast into the form

$$\mathbf{E} = Ee^{-(j/2)(\beta^{cw} + \beta^{ccw})z}\left[\hat{\mathbf{x}}\cos\left(\frac{\beta^{cw} - \beta^{ccw}}{2}\right)z - \hat{\mathbf{y}}\sin\left(\frac{\beta^{cw} - \beta^{ccw}}{2}\right)z\right] \tag{4}$$

The sum wave (4) is seen to have field vectors in a fixed direction at each value of z since there is no phase difference between the x and y components. The direction of the field vector relative to that at $z = 0$ is, as seen in Fig. 13.16, given by

$$\tan\theta = \frac{E_y}{E_x} = -\tan\left(\frac{\beta^{cw} - \beta^{ccw}}{2}\right)z \tag{5}$$

or

$$\theta = \left(\frac{\beta^{ccw} - \beta^{cw}}{2}\right)z \tag{6}$$

The sign of the angle θ is positive for clockwise rotation about the positive z axis. Thus we see that the linearly polarized wave rotates upon passing along the gyrotropic axis (z axis) and has a propagation constant which is the average of those of the clockwise and counterclockwise modes.

It is of special interest to note that the direction of rotation about the positive z axis is the same for waves traveling in the positive and negative z directions. In Sec. 13.19 we see applications of this phenomenon.

The rotation of plane of polarization of an optical wave in passing through a dielectric with applied magnetic field was first observed by Michael Faraday around 1845—hence, the name *Faraday rotation*. The angle of rotation (6) can be written as

$$\theta = VHl \tag{7}$$

where H is magnetic field strength, l the interaction length, and V a constant known as the *Verdet constant*. The effect is small for ordinary dielectrics. For example, V is about 0.014 min/Oe-cm in silica at $\lambda_0 = 600$ nm, decreasing at longer wavelengths approximately as λ^{-2}. With the long propagation lengths obtainable with optical-fiber guides, useful nonreciprocal devices can nevertheless be made using this effect, but ingenious design is necessary to maintain magnetic field axial in the multiple passes through the magnet.[44]

13.17　Permeability Matrix for Ferrites

A group of materials, called ferrites, have the important characteristics of low loss and strong magnetic effects at microwave frequencies. These are solids with a particular type of crystal structure made up of oxygen, iron, and another element which might be lithium, magnesium, zinc, or any of a number of others.[45,46] The important factors in the study of wave propagation in ferrites are that the losses at microwave frequencies are small, the dielectric constants are relatively high (within a factor of about two of $\varepsilon_r = 15$), and anisotropic behavior of permeability results when the ferrite is subjected to a steady magnetic field.

A complete description of the energy state of an atom requires specification of both the orbits and spins of the electrons. In the language of quantum mechanics, there are orbital and spin quantum numbers, both of which must be given to define the energy state of an atom. A strong magnetic moment is associated with the electron spin. In paramagnetic substances these magnetic moments are randomly oriented with respect to those in neighboring atoms, but in ferromagnetic, antiferromagnetic, and ferrimagnetic (ferrite) materials, there exists a strong coupling between spin magnetic moments of neighboring atoms, causing parallel or antiparallel alignment as discussed in Sec. 13.6. We will study the situation in which all the domains are aligned in one direction by a strong applied steady field; the material is saturated. An equation of motion for the spin magnetic moments will be found. By assuming small perturbations of the magnetic field quantities about the large steady values, a permeability matrix for the perturbations will be derived.

The model of the spinning electron used in the derivation is shown diagrammatically in Fig. 13.17a. The analogy between the spinning electron and a gyroscope

[44] K. Kobayashi and M. Seki, *IEEE J. Quantum Electronics* **QE-16**, 11 (1980).

[45] B. Lax and K. J. Button, *Microwave Ferrites and Ferrimagnetics*, McGraw-Hill, New York, 1962.

[46] R. F. Soohoo, *Theory and Applications of Ferrites*, Prentice-Hall, Englewood Cliffs, N.J., 1960.

is evident. For any rotating body the rate of change of the angular momentum **J** equals the applied torque **T**:

$$\frac{d\mathbf{J}}{dt} = \mathbf{T} \tag{1}$$

Note as an example the gyroscope shown in Fig. 13.17*b*. The earth's gravitational attraction applies a force or torque to the gyroscope and the angular momentum vector along the axis of the gyroscope rotates slowly about a vertical line through

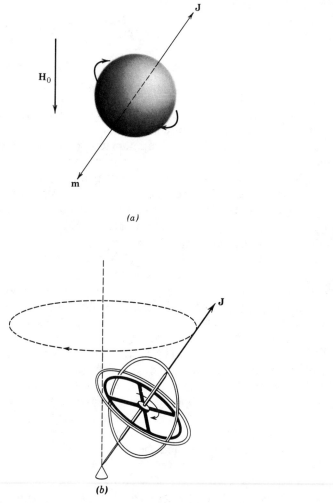

(a)

(b)

Fig. 13.17 (*a*) Spherical model of spinning electron in magnetic field. (*b*) Precession of a gyroscope.

the pivot. This rotation, called *precession*, is at a rate sufficient to conserve the initial angular momentum. For the electron, we relate the magnetic moment **m** to the applied torque since

$$\mathbf{m} = \gamma \mathbf{J} \qquad (2)$$

where γ is called the *gyromagnetic ratio*.[47] If the electron is considered to be a uniform charge distribution in a spherical volume and γ is found by classical calculations of **m** and **J**, the result is in error by a factor of almost exactly 2. The correct value of γ which is approximately -1.759×10^{11} m^2/Wb-s must be found from quantum mechanics.

The torque resulting from subjecting a magnetic moment **m** to a magnetic field $\mathbf{B}_i$ is

$$\mathbf{T} = \mathbf{m} \times \mathbf{B}_i \qquad (3)$$

There are also torques resulting from loss mechanisms but these will be neglected here. We may combine (1) to (3) to get

$$\frac{d\mathbf{m}}{dt} = \gamma(\mathbf{m} \times \mathbf{B}_i) \qquad (4)$$

where B_i is the total field to which the spinning charges are subjected. The total magnetic field acting on a molecule in a magnetic material can be adequately represented by

$$\mathbf{B}_i = \mu_0 \mathbf{H} + (\text{constant})(\mathbf{M}) \qquad (5)$$

The magnetization vector (or magnetic dipole density) **M** is $N_0 \mathbf{m}$ where N_0 is the effective number density since all spins in a saturated material act in concert. The magnetic intensity **H** is the value averaged over the space of many molecules within the material. The relation between **H** and the external applied field depends upon the shape of the ferrite body.[48]

Then combining (4) and (5),

$$\frac{d\mathbf{M}}{dt} = \gamma\mu_0(\mathbf{M} \times \mathbf{H}) \qquad (6)$$

Now consider the applied magnetic field in the form of a sum of dc and ac terms

$$\mathbf{H} = \hat{\mathbf{z}}H_0 + \mathbf{H}_1 e^{j\omega t} \qquad (7)$$

The material is saturated, so the magnetization vector must have the form

$$\mathbf{M} = \hat{\mathbf{z}}M_0 + \mathbf{M}_1 e^{j\omega t} \qquad (8)$$

---

[47] The γ of this section is not to be confused with propagation constant, which has the same symbol in other parts of the book.

[48] B. Lax and K. J. Button, Ref. 45, Chapter 4.

where M_0 is the saturation magnetization and M_1 has only x and y components. The expressions (7) and (8) are substituted, in phasor component form, in (6), and all products of ac terms are considered to be negligible compared with products involving one steady term and one alternating term. The result is

$$j\omega M_x = \gamma\mu_0(M_y H_0 - M_0 H_y)$$
$$j\omega M_y = \gamma\mu_0(M_0 H_x - M_x H_0) \tag{9}$$
$$j\omega M_z = 0$$

where the subscript 1 is deleted from the ac term for simplicity. Equations (9) may be solved to give **M** in terms of **H** and the result substituted in

$$\mathbf{B} = \mu_0(\mathbf{H} + \mathbf{M}) \tag{10}$$

to obtain the result

$$[B] = [\mu][H] \tag{11}$$

or

$$\begin{bmatrix} B_x \\ B_y \\ B_z \end{bmatrix} = \begin{bmatrix} \mu_{11} & \mu_{12} & 0 \\ \mu_{21} & \mu_{22} & 0 \\ 0 & 0 & \mu_0 \end{bmatrix} \begin{bmatrix} H_x \\ H_y \\ H_z \end{bmatrix} \tag{12}$$

where

$$\mu_{11} = \mu_{22} = \mu_0\left[1 + \frac{\omega_0 \omega_{M_c}}{\omega_0^2 - \omega^2}\right] \tag{13}$$

$$\mu_{12} = \mu_{21}^* = j\frac{\mu_0 \omega\omega_M}{\omega_0^2 - \omega^2} \tag{14}$$

where

$$\omega_M = -\mu_0\gamma M_0, \quad \omega_0 = -\mu_0\gamma H_0 \tag{15}$$

We see that the permeability of a ferrite with a finite dc magnetic field has the same form as the permittivity of the plasma with a finite magnetic field [Eqs. 13.12(10) and 13.12(11)]. Note that a resonance occurs at the frequency $\omega_0 = -\gamma\mu_0 H_0 = (e/m)\mu_0 H_0$. The resonance frequency of the stationary plasma is $\omega_c = (e/m)\mu_0 H_{0z}$. The difference between these frequencies is that the magnetic field H_0 within the ferrite differs from the applied field, as mentioned after (5), whereas the H_{0z} within the plasma is essentially the same as the applied field. The singularity at resonance vanishes if the mechanisms responsible for damping the precession are considered in the analysis.

13.18   *TEM* Wave Propagation in Ferrites

We consider here *TEM* wave propagation in a ferrite parallel to the dc magnetic field. Taking the curl of Maxwell's curl **H** equation, substituting the curl **E** equation, and making use of a vector identity, we have

$$\nabla^2 \mathbf{H} - \nabla(\nabla \cdot \mathbf{H}) + \omega^2 \varepsilon \mathbf{B} = 0 \tag{1}$$

Since there is no field in the direction of variations (z) and no variation in the direction of the fields (x and y), $\nabla \cdot \mathbf{H} = 0$. The Laplacian reduces to $d^2 H/dz^2$ and we use Eq. 13.17(12) to transform (1) to

$$\frac{d^2}{dz^2} \begin{bmatrix} H_x \\ H_y \\ 0 \end{bmatrix} + \omega^2 \varepsilon \begin{bmatrix} \mu_{11} & \mu_{12} & 0 \\ \mu_{21} & \mu_{22} & 0 \\ 0 & 0 & \mu_0 \end{bmatrix} \begin{bmatrix} H_x \\ H_y \\ 0 \end{bmatrix} = 0 \tag{2}$$

This matrix has the same form as Eq. 13.15(1), the wave equation for a plasma in a finite magnetic field. Therefore the phase constants found by substituting

$$\mathbf{H}^{cw}_+ = H^{cw}_+(\hat{\mathbf{x}} - j\hat{\mathbf{y}})$$

and

$$\mathbf{H}^{ccw}_+ = H^{ccw}_+(\hat{\mathbf{x}} + j\hat{\mathbf{y}}) \tag{3}$$

in (2), have the same form as those in Sec. 13.15. Similar procedure applies for the negative propagating waves. We may summarize the results as

$$(\beta^{cw}_+)^2 = (\beta^{ccw}_-)^2 = \omega^2 \varepsilon(\mu_{11} - j\mu_{12}) \tag{4}$$

$$(\beta^{cw}_-)^2 = (\beta^{ccw}_+)^2 = \omega^2 \varepsilon(\mu_{11} + j\mu_{12}) \tag{5}$$

where μ_{11} and μ_{12} are given by Eqs. 13.17(13) and 13.17(14). It is common to consider $\mu_{11} - j\mu_{12}$ and $\mu_{11} + j\mu_{12}$ as the equivalent permeabilities. Then it is clear that if $\mu_{11} - j\mu_{12}$ is positive, the phase constant for the clockwise forward wave (and counterclockwise reverse wave) is real

$$\beta^{cw}_+ = \omega\sqrt{\varepsilon}(\mu_{11} - j\mu_{12})^{1/2} \tag{6}$$

If $\mu_{11} - j\mu_{12}$ is negative, we have a purely attenuated (evanescent) wave. If losses are taken into account in the derivation of the permeability components, there are neither purely attenuating nor purely propagating regimes. We will restrict our attention to the simpler lossless case. From the forms of the matrix elements in Eqs. 13.17(13) and 13.17(14), it may be observed that the signs of μ_{11} and μ_{12} depend upon frequency. In Figs. 13.18a and 13.18b variations of equivalent permeability are illustrated for the clockwise and counterclockwise waves, respectively. The frequency ranges of propagation and attenuation are delineated.

It is evident that the difference of phase constants provides for Faraday rotation of linearly polarized waves. Several devices based on this principle have become important in microwave systems and are reviewed in Sec. 13.19.

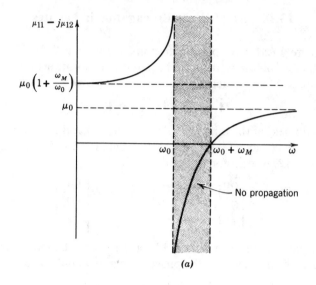

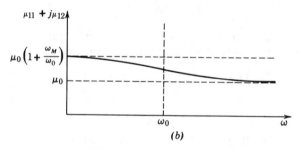

Fig. 13.18 (*a*) Equivalent permeability for clockwise circularly polarized wave propagating in $+z$ direction. (*b*) Equivalent permeability for counterclockwise circularly polarized wave propagating in $+z$ direction.

Intrinsic impedances may be defined for a ferrite as they were for the gyrotropic plasma in Sec. 13.15. Substitution of (3) and corresponding expressions for **E** into Maxwell's curl **H** equation yields

$$\eta_+^{cw} = \eta_-^{ccw} = \sqrt{(\mu_{11} - j\mu_{12})/\varepsilon}$$

and   (7)

$$\eta_+^{ccw} = \eta_-^{cw} = \sqrt{(\mu_{11} + j\mu_{12})/\varepsilon}$$

With these definitions, the transmission-line formalism can be used to solve propagation problems.

13.19 Ferrite Devices

Ferrite devices have found use of microwave applications for both waveguides and striplines. One family operates on the basis of the Faraday effect described in Sec. 13.16 and the others on different principles, to be discussed below, that also depend upon the asymmetry caused by an applied steady magnetic field.

Waveguide Devices The group of devices that employ Faraday rotation take advantage of the fact that the rotation of a linearly polarized wave is in the same direction relative to the direction of the dc magnetic field for both the positive and negative traveling waves. (See Fig. 13.19a.) The first and simplest of the microwave devices using the Faraday rotation effect was the *gyrator*,[49] shown in Fig. 13.19b. By virtue of a twist of the waveguide, a wave coming from the left is rotated counterclockwise by 90 degrees before reaching the ferrite. The magnetic field and the dimensions of the ferrite are chosen so an additional 90 degrees counterclockwise rotation of the wave takes place in the ferrite region. Thus, the wave progressing toward the right is inverted, or equivalently, shifted in phase by 180 degrees. A wave passing from right to left is rotated by 90 degrees counterclockwise with respect to the direction of the magnetic field and this is cancelled by the rotation in the twisted portion of the guide to the left of the ferrite. The gyrator, therefore, serves to produce a phase shift of 180 degrees in one direction and no shift in the opposite direction.

Another important Faraday rotation device is the *absorption isolator* shown in Fig. 13.19c. Here the 45 degree twist to the left of the ferrite rotates the wave coming from the left so that its electric field vectors are perpendicular to the thin resistance card just after the twist. With this orientation, the field suffers a minimum of conduction losses in the card. The field is then rotated back to its original orientation when it passes through the ferrite and leaves the isolator essentially unmodified. A wave traveling to the left is rotated in the ferrite in the same direction relative to the magnetic field as the wave moving to the right and is therefore oriented with its electric field vectors along the resistance card. In this way the wave traveling to the left is appreciably attenuated (typically 30 dB) and the wave moving to the right suffers little loss (usually less than 0.5 dB).

Another application of Faraday rotation is the *circulator* shown in Fig. 13.19d. The function of a circulator is to transmit a wave from guide 1 to guide 2, a wave from guide 2 to 3, 3 to 4, and 4 to 1, with all other couplings prohibited. The ferrite rod and magnetic field are chosen to give 45 degree rotation. The TE_{10} mode in guide 1 excites the TE_{11} mode in the circular guide. The symmetry of the TE_{11} circular mode (see Table 8.9) is such that it does not excite propagating waves in guide 3. The right half of the structure is rotated by 45 degrees relative to guides 1 and 3 as seen in Fig. 13.19d. A 45 degree rotation of the wave coming from guide 1 places it in the same orientation with respect to 2 and 4 as it had with respect to 1

---

[49] C. L. Hogan, *Bell System Tech. Jour.* **31**, 1 (1952).

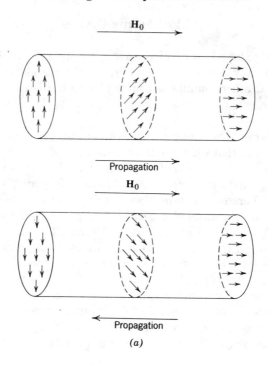

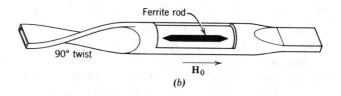

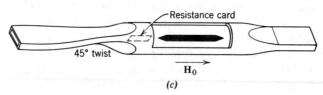

Fig. 13.19 (*a*) Wave rotation of waves in two directions in magnetized ferrite rod. (*b*) Microwave gyrator. (*c*) Microwave isolator.

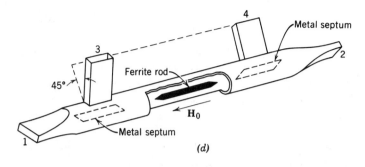

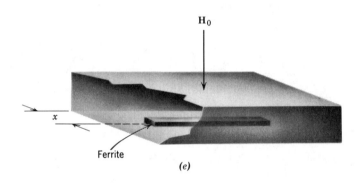

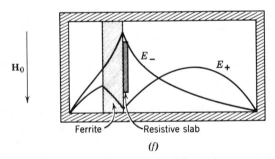

Fig. 13.19 (*d*) Microwave circulator. (*e*) Ferrite location in resonance isolator. (*f*) Field patterns in field-displacement isolator.

and 3 so it passes out through guide 2. A TE_{10} wave entering guide 3 excites a TE_{11} mode in the circular guide, but it is oriented at 90 degrees from the mode excited by a wave from guide 1. It cannot excite propagating modes in 1. When rotated by 45 degrees in the ferrite, the wave is oriented to excite waves in guide 4 but not 2. The analyses of waves entering guides 2 and 4 follow similar reasoning. The metal septums shown in the figure aid in preventing unwanted couplings. The circulator can also be used as a switch or for modulation by controlling the field that magnetizes the ferrite.

The choice of diameter and shape of the ferrite rods used in Faraday rotation devices is made to minimize reflections and maximize power-handling capability while achieving the required rotation with reasonable magnetic fields. Because power is dissipated in the small rod isolated from the walls, a Faraday rotation device is useful only for low power. Some of the following devices have the ferrite connected to the metal wall and can carry higher power fields.

The functions just described can also be achieved in *resonance* and *field-displacement* devices which employ magnetic fields transverse to the guide. Let us consider, for a rectangular waveguide with TE_{10} mode, the rf magnetic fields in the planes perpendicular to the dc magnetization. The vector sum of H_x and H_z of Eqs. 8.8(4) and 8.8(5) is

$$\mathbf{H}_{xz} = \left[\hat{\mathbf{x}} - j\hat{\mathbf{z}}\left(\frac{\lambda}{2a}\right)\frac{Z_{TE}}{\eta} \cot \frac{\pi x}{a} \right] |H| \sin \frac{\pi x}{a} \tag{1}$$

where Z_{TE} is positive for a wave in the $+z$ direction and negative for the reverse wave. It is clear from (1) that at the value of x in the guide such that

$$\left| \left(\frac{\lambda}{2a}\right)\frac{Z_{TE}}{\eta} \cot \frac{\pi x}{a} \right| = 1 \tag{2}$$

the field $\mathbf{H}_{xz}$ is circularly polarized. The direction of polarization depends upon the sign of Z_{TE}.

If a piece of ferrite is placed on the top or bottom of the guide at the value of x given by (2) as shown in Fig. 13.19e the precessing electron-spin moments will be subjected to a circularly polarized field which either enhances the precession or is very little coupled to it, depending upon the direction of the circular polarization. By appropriate choice of polarity of the dc magnetic field, the direction of spin precession can be made opposite to the direction of rotation of the field vectors of the forward wave. In this case the forward wave is little affected. The backward wave, having an opposite polarization of $\mathbf{H}_{xz}$, pumps the precessing spins and loses energy in the process. This energy is transferred from the electrons to the lattice of the ferrite by microscopic damping mechanisms. If the magnetic field is adjusted to set the resonance frequency equal to the field frequency ($\omega = -\gamma\mu_0 H_0$), the reverse loss is maximum. This device, used as an isolator, requires more dc field than the Faraday rotation isolator described previously, since it must operate at resonance. It has the advantage, however, of allowing convenient cooling of the ferrite. The same mechanism can be used to modulate a wave.

There is a third class of nonreciprocal devices utilizing magnetized ferrites. These also use transverse magnetization but make use of the fact that forward and reverse waves with the ferrite present may have different transverse field distributions. They are called *field-displacement* devices. If a slab of ferrite is

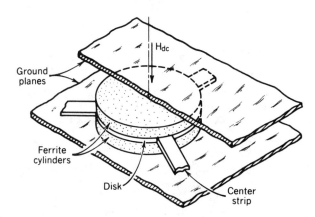

Fig. 13.19g Stripline Y-junction circulator. (After Fay and Comstock, Ref. 50, © 1965, IEEE.)

approximately located and magnetized, an approximate null of electric field can be made to exist at its edge for the forward wave but not for the reverse wave. This is shown schematically in Fig. 13.19*f.* A resistive strip attached to the side of the ferrite then attenuates the reverse wave but does not affect the forward wave. As with Faraday rotation devices, the field-displacement scheme is limited to low-power application.

Stripline Devices   A device called the *stripline Y-junction circulator* shown in Fig. 13.19*g* can also be used as a switch or isolator. A dc magnetic field is applied normal to the ferrite disks. Operation can be described in terms of standing mode patterns in the disks, similarly to the modes in dielectric resonators (Sec. 10.13).[50] The lowest resonant mode is shown in Fig. 13.19*h* for the case with no applied magnetic field; the rf electric field is perpendicular to the plane disk surfaces and magnetic field is transverse. The fields shown are excited by the input in line 1 and lines 2 and 3 receive excitation of magnitude one-half that of the input. In the presence of a magnetic field, the standing pattern comprises two counterrotating field patterns. This new pattern is rotated about the disk axis by an amount that depends upon the magnetic field strength. Figure 13.19*i* shows the pattern rotated to a position in which line 2 receives an excitation equal to the input and line 3 is isolated. The complete symmetry of the device implies that the roles of the lines can be rotated so that excitation in line 2 produces an output in line 3 and nothing in line 1 and similarly for excitation in line 3. A computer-aided design procedure is given in Ref. 51.

[50] C. E. Fay and R. L. Comstock, *IEEE Trans. Microwave Theory Tech.* **MTT-13**, 15 (1965).
[51] Z. Uzdy, *IEEE Trans. Microwave Theory Tech.* **MTT-28**, 1134 (1980).

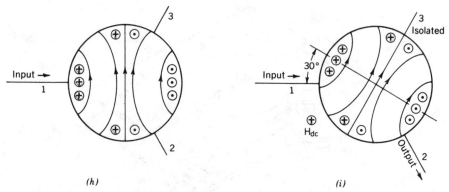

Fig. 13.19 (*h*) Standing-wave pattern in ferrite disk in absence of H_{dc}. (*i*) With selected value of H_{dc}, standing wave pattern is rotated to isolate line 3. (After Fay and Comstock, Ref. 50, © 1965, IEEE.)

PROBLEMS

13.2a The susceptibility χ_e of nitrogen at 20° C and atmospheric pressure is about 5.5×10^{-4}. What molecular polarizability α_T does this correspond to? Assuming the ideal gas law, and α_T independent of temperature, give susceptibility as a function of temperature.

13.2b Find the resonant frequency for an electronic resonance (charge cloud in an atom) in the Lorentz model, assuming the atom to have 10 electrons distributed uniformly in the cloud and acting as a group. Check your result by comparison with Fig. 13.2*a*. The maximum displacement r of the charge cloud from the equilibrium position is less than its radius $a = 0.1$ nm.

13.2c Consider a liquid with $\varepsilon_r = 4.0$ at room temperature for dc fields. Take the polarization **P** to be proportional to **E**. Assume, for simplicity, that the polarizability results only from orientation of permanent dipoles and that there is spherical symmetry about each molecule. The density of molecules is 10^{28} m^{-3} and the relaxation time 10^{-11} s. Find the dc polarizability and the field frequency at which $\varepsilon' = 2.0\varepsilon_0$.

13.2d* Find from Eq. 13.2(11) the corresponding permittivity with g and Γ taken as real and frequency independent. Show that ε' and ε'' satisfy the Kramers–Kronig relations, Eqs. 13.2(13)–(14). Assume small damping; use limits $-\infty$ to ∞ as in Eq. 11.12(5).

13.2e* Repeat Prob. 13.2d for a permittivity based solely on Eq. 13.2(12).

13.3a Show that the ratio of magnetic to electric forces on the electrons in an ionized gas resulting from the fields of a uniform plane wave is v/v_p, where v is electron velocity and v_p is the phase velocity of the wave. Assume a 10-MHz wave carrying a power of 1 W/m^2 and estimate the maximum electron velocity, assuming negligible collision frequency. Give an upper limit for v/v_p in this case. Assume $w \gg w_p$.

13.3b Assume a neutral plasma in which all electrons in the region x to $x + \Delta x$ are given a displacement Δx. Considering the forces acting to restore neutrality, show that the displaced sheet will oscillate at the "plasma frequency" ω_p given by Eq. 13.3(10).

13.3c A sample of silicon ($\varepsilon_1 = 11.7\varepsilon_0$) uniformly doped *n*-type with a dopant concentration of 5×10^{17} cm^{-3} is found to have a dc conductivity of 2.5×10^3 S at $T = 300$ K. The electron effective mass is $0.26m_0$ where m_0 is the electron rest mass. Determine the collision frequency, assuming that the electron density equals the dopant density. Find the frequency at which ε'' differs from the value in Eq. 13.3(8) by 10%. Evaluate the ratio of second to first terms on the right side of Eq. 13.3(6) where $\omega \to 0$ and $\omega = v$. Comment on the restrictions on the use of Eq. 13.3(8).

13.3d A simple model of the ionosphere may be formulated by assuming that the number density of free electrons increases linearly from zero at height z_0 to 10^{12} per cubic meter at a height of z_1 and then decreases linearly to zero at a height of $2z_1 - z_0$. The collisions may be neglected. A uniform plane wave traveling directly upward encounters the ionosphere. What is the highest frequency for which this wave will be totally reflected? Plot the maximum the real and imaginary components of total current.

13.3e Find the real and imaginary components of the dielectric constant, ε' and ε'', respectively, of silver and nickel for waves from a CO_2 laser (10.6-μm wavelength) using Fig. 13.3b assuming the complex behavior can be cast in that form. Determine the ratio of the magnitudes of the real and imaginary components of total current.

13.3f Plot attenuation in nepers per meter versus wavelength for nickel and silver over the wavelength range of Fig. 13.3b, using data of that figure.

13.3g Plot curves of n_r and n_i versus wavelength for silver assuming it a good conductor with the conductivity given in Table 3.17, and compare with Fig. 13b over the wavelength range shown. Comment on reasons for differences.

13.4a As discussed in Sec. 13.4, time-dependent fields decay into the surface of a collisionless conductor if $\omega < \omega_p$. Show this from Maxwell's equations in time-dependent form using Eq. 13.3(1) with $v = 0$ and $dv/dt \approx \partial v/\partial t$. Specifically, show that

$$\left(\frac{\partial B_x}{\partial t}\right)_z = \left(\frac{\partial B_x}{\partial t}\right)_{z=0} \exp(-\alpha z)$$

where $a^2 = \mu n_e e^2/m$ and B_x is tangential to the plane surface of a half-space as in Fig. 13.4.

13.4b* The *two-fluid model* of a superconductor assumes that a portion n_n of the total electron concentration is in the normal state described by Eq. 13.3(1) and the remainder n_s is in the superconducting state for which $m \, dv_s/dt = -e\mathbf{E}$. Calculate current density for each component and add to get $\mathbf{J}_{\text{total}} = \sigma\mathbf{E}$. Show that for $(\omega/v)^2 \ll 1$ and $n = n_s + n_n$, $\sigma = \sigma_n(n_n/n) - j(1/\omega\mu\lambda_L^2)$ where $\sigma_n = ne^2/vm$. Use the usual expression for skin depth to show that

$$\delta = \sqrt{2}\lambda_L\left[\left(\frac{\omega}{v}\right)\left(\frac{n_n}{n_s}\right) - j\right]^{-1/2}$$

and that in the low-frequency limit $\exp[-(1 + j)z/\delta]$ becomes $\exp(-z/\lambda_L)$.

13.4c Bulk superconductors (dimensions very large compared with penetration depth) are often considered to be perfectly diamagnetic ($\mu = 0$). Use the analysis in Sec. 7.18 to show that a sphere of superconducting material in an otherwise uniform magnetic field will have $B = 0$ inside and $H = \frac{3}{2}H_0$ at the equator at $r = a$.

13.4d Taking account of the penetration of tangential magnetic field into superconductor surfaces, show that the inductance per unit length of a parallel plate structure, assuming plates are identical and much thicker than λ_L, is $L = (\mu_0/w)(2\lambda_L + d)$, where w is width and d, spacing.

13.4e Suppose a quasi-TEM wave propagates in a parallel-plate structure in which thick superconducting electrodes are spaced by (3 nm) of niobium oxide having a relative permittivity of 30.

(i) Take account of the penetration of magnetic field into the superconductors as in Prob. 13.4d to find an expression for the phase velocity of this transmission line.

(ii) Find v_p for the given structure assuming $\lambda_L = 0.039$ μm for niobium and compare with the velocity of a plane wave in the same dielectric.

(iii) How would the result differ if niobium were a collisionless conductor?

13.5a* The potential energy of a dipole in a magnetic field H_i is $-m_0 \mu_0 H_i \cos \theta$, where θ is the angle between m_0 and H_i. Boltzmann theory states that the relative probability of finding a dipole with an energy U is given by $\exp(-U/kT)$. Write an expression for the average value of the component of dipole moment in the direction of H_i, $\langle m_0 \cos \theta \rangle$, and do the integrations to show the validity of Eq. 13.5(3).

13.5b Magnetic dipole moment of a small current loop of radius a is $m_0 = \pi a^2 I$ (Sec. 2.10). Verify that the argument of the hyperbolic cotangent in Eq. 13.5(3) and χ_m are dimensionless.

13.6a Spontaneous magnetization can only occur if there is some value of H_i for which Eqs. 13.5(3) and 13.6(1) are simultaneously satisfied. Sketch roughly the forms of these two expressions as functions of H_i, indicating a point of intersection. As temperature is raised, the point of intersection moves to lower H_i. The highest temperature at which there is a simultaneous solution is the Curie temperature T_c. At this temperature the initial slopes of the two curves are equal. Use this fact to find a relation between κ and T_c.

13.6b Find κ for iron using the result of Prob. 13.6a and Curie temperature of 1043 K. There are 8.6×10^{22} atoms/cm^3. Assume that the magnetic moment of the iron atom is 2.22 Bohr magnetons, where a Bohr magneton is $eh/4\pi m$. Here e is the electronic charge, h is Planck's constant (6.63×10^{-34} J-s), and m is the mass of an electron.

13.6c For the material depicted by Fig. 13.6d assume $B_{sat} = 0.1$ T. Deduce the scale on the H axis and estimate the energy loss per cycle if H is swept through the range shown. Find values for coercive force, remanence, initial permeability, and maximum permeability. (Note that since Fig. 13.6d is only diagrammatic, numbers obtained will not be representative of real materials.)

13.6d Assume that a rod of circular cross section of radius a and length l has been uniformly magnetized parallel to its axis. Find the current per meter in a solenoid of the same length to get the same flux density at the ends on the axis. Write an expression for the magnetic flux density on the axis at the end of the rod.

13.6e When an ellipsoidal sample of isotropic magnetic material is placed in an otherwise uniform magnetic field H_{app}, a complex field distribution results outside the sample but it is uniform inside. If the field is along a principal axis of the ellipsoid the field inside is parallel with H_{app} and its magnitude is given by a *demagnetization factor* $\mathscr{D}$ and the *constitutive relation* between B and H,

$$H = \frac{H_{app} - \mathscr{D}B/\mu_0}{(1 - \mathscr{D})}$$

(i) A rod of circular cross section is modeled by a long prolate ellipsoid of revolution with H perpendicular to the long axis, $\mathscr{D} = \frac{1}{2}$. Suppose a rod of $\mu = 100$ material in this arrangement and find the ratio of internal B to applied $B = \mu_0 H_{app}$.

(ii) Show that the result for a sphere in Eq. 7.18(25) can be found from the above relation with $\mathscr{D} = \frac{1}{3}$.

(iii) Verify the result for a superconducting sphere given in Prob. 13.4c.

(iv) Argue from boundary conditions why $\mathscr{D} = 0$ for field parallel with major axis of a long prolate ellipsoid of revolution.

13.7 Suppose two coparallel waves with the same electric field polarization pass through a 1-cm-long crystal. The fields are intense and produce nonlinear effects. The wavelengths (in free space) of the waves are 632.8 nm and 592.1 nm.

(a) What is the maximum difference in the index n between $\omega_1 - \omega_2$ and ω_1 [assume $n(\omega_1) = n(\omega_2)$] if negligible cancellation is to result from differences of the polarization at the two ends of the crystal? Use as a criterion that the phase of the polarization at the ends should differ by no more than $\pi/2$ from the ideal.

(b) An output difference-frequency wave of 1.0 mW is observed. What is the power lost or gained by each of the two driving waves?

13.8a As a simple example of the idea of diagonalizing matrices by coordinate transformations, consider a two-dimensional system where

$$\begin{bmatrix} D_x \\ D_y \end{bmatrix} = \varepsilon_0 \begin{bmatrix} 1.75 & 0.433 \\ 0.433 & 1.25 \end{bmatrix} \begin{bmatrix} E_x \\ E_y \end{bmatrix}$$

Write expressions for $\mathbf{D}$ and $\mathbf{E}$ in a primed coordinate system rotated by θ from the unprimed system. By combining these with the given relation between $\mathbf{D}$ and $\mathbf{E}$, find the amount of rotation necessary to put $[\varepsilon]$ in the diagonal form

$$[\varepsilon] = \varepsilon_0 \begin{bmatrix} 1 & 0 \\ 0 & 2 \end{bmatrix}$$

13.8b Barium titanate (BaTiO$_3$) has $\epsilon_{11} = \epsilon_{22} = 5.94\epsilon_0$ and $\epsilon_{33} = 5.59\epsilon_0$. Sketch (showing dimensions) and describe the index ellipsoid. What are the values of X, Y, and Z that pertain to a plane wave propagating in the y direction with its $\mathbf{D}$ vector in the z direction?

13.9a Derive Eq. 13.9(4).

13.9b Verify the conversion of Eq. 13.9(10) to Eq. 13.9(12), the "Fresnel equation of wave normals."

13.9c* Assume waves propagating with direction such that $b_x = 0.900$, $b_y = 0.100$, and $b_z = 0.424$ in a crystal with $\mu = \mu_0$ and $\epsilon_{11} = 12\epsilon_0$, $\epsilon_{22} = 14\epsilon_0$, and $\epsilon_{33} = 16\epsilon_0$. Determine the values of phase velocity for the two allowed waves. Find for each the vector expressions for $\mathbf{D}$ and show that $\mathbf{D}_1 \perp \mathbf{D}_2$.

13.9d* For the index ellipsoid shown in Fig. 13.9c with an arbitrary propagation direction $\boldsymbol{\beta}$, show that if an ellipse is defined by the intersection of the ellipsoid and a plane normal to $\boldsymbol{\beta}$, major and minor axes of the ellipse define two basic solutions for the direction of $\mathbf{D}$, and corresponding indices of refraction. [Hint: Find the extrema of $R^2 = X^2 + Y^2 + Z^2$ subject to the two constraints on X, Y, Z, on the boundary of the intersection ellipse using Lagrange multipliers. This will lead to three equations of the form $X(1 - R^2/n_x^2) + b_x R^2[(Xb_x/n_x^2) + (Yb_y/n_y^2) + (Zb_z/n_z^2)]$ where $n_x^2 = \epsilon_{11}/\epsilon_0$, etc., which can be recast into the form of Eqs. 13.9(8).]

13.10a A beam of light is normally incident on a surface of a calcite (CaCO$_3$) crystal as shown in Fig. 13.10b. The optical axis is at an angle of 29 degrees from the inward surface normal. Take $\varepsilon_{11} = \varepsilon_{22} = 2.7\varepsilon_0$ and $\varepsilon_{33} = 2.2\varepsilon_0$. Treat the incident light as unpolarized plane waves. Find the spatial separation of ordinary and extraordinary rays at the opposite (parallel) crystal surface 1 cm away.

13.10b For a beam of light incident at an oblique angle on a planar slab of double-refracting (birefringent) material, show that the two emergent rays are parallel.

13.10c A wide beam of linearly polarized light with $\lambda_0 = 550$ nm is normally incident on the surface of a plane slab of $LiNbO_3$ cut with the z axis at 45 degrees to the surface normal. The beam is wide enough that there is appreciable overlap of ordinary and extraordinary beams at the second surface. Find a thickness that would lead to total cancellation in the overlap region, taking $n_0 = 2.29$ and $n_e = 2.20$ and assuming that input excitation is adjusted so that the exciting polarization components are of equal amplitude.

13.10d A linearly polarized wave at $\lambda_0 = 550$ nm is incident with its electric field at 45 degrees to the y axis on the surface of a $1.0\,\mu m$ slab of $LiNbO_3$ cut so the y-z plane lies in the surface ("x-cut"). Indices are $n_0 = 2.29$ and $n_e = 2.20$. Find the wave propagated beyond the slab.

13.11a Show that the new principal axes x' and y' are rotated by 45 degrees from the crystal axes x and y when a crystal of KDP is subjected to an electric field E_z. Prove that the 11 and 22 elements of the $[1/n^2]$ matrix in the primed coordinate system are given by Eqs. 13.11(10).

13.11b A $BaTiO_3$ crystal is to be used as a modulator as shown for $LiNbO_3$ in Ex. 13.11a. Assume the light is polarized with $\mathbf{E} = \hat{x}E_x$ and find the magnitude of field required for π rad of phase shift. Note the large value of r_{42} for $BaTiO_3$. How would you orient wave propagation direction, polarization, and direction of modulating field with respect to crystal axes to make use of this coefficient? Take $\lambda = 550$ nm and $l = 1$ cm.

13.11c Design a "quarter-wave" plate of KDP (choose length and crystal orientations) with no applied steady electric field that can be placed parallel with the modulation plate shown in Fig. 13.11c so that the ratio $|E_z(l)/E_x(0)|^2$ is at the most linear part of its variation in Fig. 13.11d. Take $\lambda_0 = 550$ nm. Odd multiples may be used for convenience.

13.12a Assume plane waves propagating in the z direction in a plasma so that $E_z = 0$, in the case where there is an average drift velocity v_0 of the electrons in the z direction parallel to the steady magnetic field. Find ε_{11}, ε_{12}, ε_{21}, and ε_{22}. Note in the result that the electrons see a field with Doppler shifted frequency.

13.12b Repeat the derivation of Sec. 13.12 to find the permittivity matrix taking account of the motion of ions as well as electrons.

13.13a A linearly polarized plane wave with equal y and z components of electric field is launched in the x direction in a neutral plasma with infinite magnetic field in the z direction. The frequency is 4 GHz and the number density of electrons is 10^{17} per cubic meter. Find the distance in which the linearly polarized wave becomes circularly polarized.

13.13b* A beam of millimeter waves is normally incident on a 0.06-mm-thick planar metal membrane in air. The temperature of the membrane is low enough to neglect collisions of the electrons. An extremely strong steady magnetic field $\mathbf{B}_0$ is applied to the membrane with $\mathbf{B}_0$ parallel to the surface. Take electron density to be 10^{23} cm^{-3}, permittivity of the ions as $30\varepsilon_0$, and $\mathbf{E}$ in the incident wave to be 45 degrees from $\mathbf{B}_0$. Find the wave that exists beyond the membrane. Take $f = 100$ GHz.

13.14a Evaluate the propagation constants in Eq. 13.14(13) for a beam of electrons which has been accelerated to an energy of 2000 eV. The current density is 10^5 A/m^2. Plot an ω-β diagram for the space-charge waves. For convenience use $\omega\sqrt{\mu_0\varepsilon_0}$ as the ordinate.

13.14b Consider a plasma with electrons drifting along an infinite magnetic field as in Sec. 13.14 but located inside a circular cylindrical waveguide with perfect-conductor walls. Use the differential equation 13.14(11) in the appropriate form to find an expression for the cutoff frequencies of the allowed electromagnetic modes.

13.15a Verify Eqs. 13.15(6) and 13.15(9).

13.15b Suppose a stationary plasma filling the half-space $z > 0$ in a magnetic field in the $+z$ direction. The other half-space $z < 0$ has only vacuum. Take the number density of electrons to be 10^{17} m^{-3} and the magnetic field to be 0.1 T. A clockwise polarized plane wave of frequency 2 GHz propagating in the $+z$ direction is incident on the plasma. What is the value of the field of the wave propagating in the plasma as a fraction of the field in the incident wave?

13.15c Find conditions for propagation or attenuation of clockwise and counterclockwise polarized waves propagating along the $+z$ axis in Sec. 13.15.

13.15d* A linearly polarized plane wave is incident on a region of plasma as in Fig. 13.15c, with $\varepsilon_1 = \varepsilon_2 = \varepsilon_0$. Assume electron density, magnetic field, and frequency as in Prob. 13.15b. The thickness of plasma region is 10 cm. Find the reflected wave in region 1 using transmission-line methods.

13.16a Verify Eq. 13.16(4) for the rotating linearly polarized wave and draw a sketch of the instantaneous field distribution at $t = 0$ for various values of z.

13.16b For the plasma with electron density of 10^{16} m^{-3}, plot the angle of rotation per meter of a linearly polarized wave at $f = 3$ GHz as a function of magnetic field from $B = 0$ to $B = 0.1$ T. Discuss the result.

13.16c Show that Eq. 13.11(13) represents an elliptically polarized wave and contrast with the Faraday effect.

13.16d Show that Faraday rotation for a wave traveling in the $-z$ direction is in the same direction relative to the magnetic field as for a wave in $+z$ direction. Sketch the orientation of **E** for various z at $t = 0$ in the wave in $-z$ direction.

13.16e Some materials possess a property called *natural optical rotation* whereby the plane of polarization of a wave passing through them is rotated without the application of electric or magnetic fields. This is explained by the "screw-like" nature of individual molecules. For example, a 10-cm column of a cane sugar solution (0.1 g/cm^3) produces about 6.7 degrees of rotation. Assuming the individual molecules to be represented crudely by right-hand screws, explain why there remains a net effect in a solution where these molecules are randomly oriented. Explain also why the reflected wave returns to its original polarization, in contradistinction to the Faraday effect where the reflected wave rotates through an additional angle. What should be the rotation per meter of a 0.05 g/cm^3 solution of sugar?

13.17a Assume a model of an electron as a spinning sphere of uniform mass and charge and calculate its magnetic moment and angular momentum. Show that γ in Eq. 13.17(2) differs by a factor of 2 from the value given in the text.

13.17b Verify the matrix coefficients in Eq. 13.17(13)–(14) starting with 13.17(9).

13.18a* Consider a 10-cm thick plane slab of ferrite of infinite cross section normal to the z axis. Assume a dc z-directed magnetic field H_0 of 10^5 A/m inside the ferrite. The saturation magnetization M_0 of the ferrite is 1.5×10^5 A/m and the relative permittivity is 13. A linearly polarized plane wave of frequency 10^9 Hz moving in the $+z$ direction is incident on the slab. Find the wave on the other side of the slab in terms of the incident wave using transmission-line methods.

13.18b The literature on ferrite properties often gives saturation magnetization as $4\pi M_0$ and magnetic field H_0 in gaussian units. Calculate the gaussian values of $4\pi M_0$ and H_0 from the MKS values of M_0 and H_0 given in Prob. 13.18a. Also equate values of ω_0 for the two systems of units to show that γ in the expression $\omega_0 = -\gamma H_0$ using gaussian units has the value -1.76×10^7 rad/O-s.

13.19 Consider a resonance isolator in a rectangular waveguide. Operating frequency is 18 GHz in TE_{10} mode at 1.3 times its cutoff frequency. Absorbing strip is square cross section; approximate as a circular rod and use information in Prob. 13.6e. Assume B saturated at 0.2 T. Determine location of strip and magnitude and sign of the applied field H_{app}.

14

OPTICS

14.1 Introduction

Maxwell's theory demonstrated that light is an electromagnetic phenomenon, as was explained in Chapter 3. We have consequently used optical examples extensively to this point. But there are enough special considerations in the generation of coherent light by means of lasers, and in the use of this radiation, that a chapter for specific discussion of these is justified. We consequently concentrate on optical waveguides, optical resonators, and some related optical components in this chapter. In referring to "optical frequencies," we generally mean the visible range, the near infrared, and the near ultraviolet (wavelengths, say from around 10 μm to 100 nm), but the key point is the relation of size and curvature of wavefronts to wavelength. Thus most of the concepts apply to the far infrared (wavelengths ~ 100 μm) and many of the analyses to millimeter waves and microwave radio frequencies when relation of size to wavelength is appropriate.

For a number of practical applications, the propagation of light may be described by the behavior of the *rays*, which are the normals to the wave fronts. These are straight lines in a homogeneous medium, change direction according to Snell's law at the boundary between dielectric media, and are in general curved for inhomogeneous media. The basis for these rules and some important applications will be presented in the first part of this chapter.

The guiding of optical waves is generally accomplished by dielectric waveguides of the type studied in Sec. 9.2. Hollow-pipe guides with metal boundaries are not only difficult to fabricate in the small sizes one would find at these wavelengths, but the metal boundaries would produce much greater losses at the optical frequencies than can be obtained with good dielectric guides. The principles

732

of such dielectric guiding have been presented in the earlier section but there are a number of special considerations worth developing in more detail than was done in that introduction. In particular a class of inhomogeneous dielectric guides, known as *graded-index* guides, has unique and useful properties and leads to waves with gaussian form in the transverse plane. Similar but diverging *gaussian beams* are found in space (or other homogeneous media) and their transformation by lenses and other optical components is especially important.

Optical resonators are likewise not different in principle from other resonators we have studied in Chapter 10. But as with the guides, the questions of size in relation to wavelength and the properties of materials at optical frequencies produce some practical differences. In particular, there is much use of resonant systems which are open to space in the transverse direction, with the waves reflected longitudinally between plane or spherical mirrors placed normally to an axis. Gaussian beams again play a role in such resonators.

Other important topics in the electromagnetics of optics include the propagation in anisotropic materials such as crystals, and the diffraction of light from apertures. It has seemed better to present the former of these subjects in the chapter on materials, and the latter in relation to the material on radiation. But we will see eventually that there is a very close relationship between diffraction and the spreading of gaussian beams in a homogeneous medium. And the diffraction laws together with the transforming properties of lenses leads to important information-processing applications. Many other important optical effects, such as the generation of coherent light by stimulated emission (the laser principle) and the many interaction phenomena of light with semiconductors, are beyond the scope of the text but are covered in books on quantum electronics.[1]

RAY OR GEOMETRICAL OPTICS

14.2 Geometrical Optics Through Applications of Laws of Reflection and Refraction

We have already applied the results of the plane wave analysis to many optical problems. The results of that analysis (Chapter 6) are strictly applicable only to plane waves which are uniform over an infinite wave front, and which fall upon infinite plane boundaries between media. Nevertheless, one would expect the results to be useful whenever the wave front extends and is uniform over many wavelengths, and when the boundaries are large in comparison with wavelength. Both boundaries and wavefront may be nonplanar so long as radii of curvature

---

[1] See, for example, A. Yariv, *Quantum Electronics*, 2nd. ed., Wiley, New York, 1975. Much of the material in this chapter is covered in more detail in the excellent books, H. A. Haus, *Waves and Fields in Optoelectronics*, Prentice-Hall, Englewood Cliffs, N. J., 1984, and A. Yariv and P. Yeh, *Optical Waves in Crystals*, Wiley, New York, 1984.

are also large in comparison with wavelength. In such cases one obtains a great deal of information about the waves by tracing the *rays* that represent the normals to the wave fronts, using locally the law of reflection and Snell's law of refraction. These were developed in Chapter 6 for uniform plane waves by considering phase requirements on continuity conditions at a boundary. These laws were actually first developed by observing the behavior of rays of light falling upon various materials before the wave nature of light was clearly established, and formed the basis for an extensive art and science of optics well before Maxwell. It is called *geometrical optics* or *ray optics*. We shall show the formal development in the next section; here we use only the physical justification discussed above and give some useful examples. In this section we consider only homogeneous media, except for discontinuities at boundaries.

_____ **Example 14.2a** _____
Spherical Mirror

Consider first a reflecting surface of spherical form with rays (normals to the wave fronts of the plane waves) parallel to the axis and falling on the mirror. As shown in Fig. 14.2a, a ray at radius r from the axis makes angle θ with the radius vector from P to the center of the sphere C. Then if the law of reflection applies locally to the region around P, the reflected ray PF also makes angle θ with radius PC and crosses the axis at F. We tentatively call this the *focus F* and its distance from the sphere, *focal length f*. The reflected ray makes angle 2θ with respect to the axis. Then, from trigonometric relationships,

$$\frac{r}{R - z} = \tan \theta \tag{1}$$

$$\frac{r}{f - z} = \tan 2\theta \equiv \frac{2 \tan \theta}{1 - \tan^2 \theta} \tag{2}$$

The equation for the sphere is

$$r^2 + (R - z)^2 = R^2 \tag{3}$$

By substituting (1) in (2) and utilizing (3) we can solve for f,

$$f = R\left\{1 - \frac{1}{2}\left[1 - \left(\frac{r}{R}\right)^2\right]^{-1/2}\right\} \tag{4}$$

We see that f is a function of radius r at which rays enter, so not all rays are focused to the same point. There is spherical *aberration* as shown in Fig. 14.2b. However, for rays very near the axis, with r negligible in comparison with R, the expression (4) reduces to

$$f \approx \frac{R}{2} \tag{5}$$

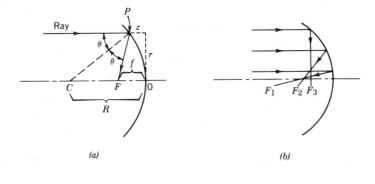

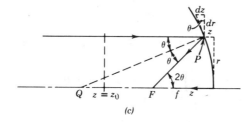

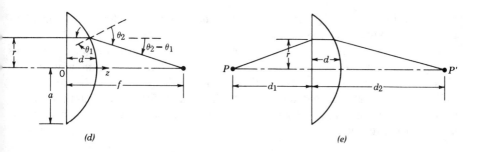

Fig. 14.2 (*a*) Ray entering spherical mirror parallel to axis. (*b*) Rays entering spherical mirror at various radii. (*c*) Parabolic mirror and ray reflected to focus *F*. (*d*) Thin lens showing focusing of entering parallel rays to a point. (*e*) Imaging of point *P* to *P'* by thin lens.

so that in this approximation the focal length is independent of r and is just half the radius of curvature of the mirror.

Spherical mirrors are used in forming the resonant systems for various lasers. They typically have radii of curvature of a few meters, with rays entering a few millimeters from the axis so that the spherical aberration is small and the approximation (5) for focal length well justified.

———————————————— **Example 14.2b** ————————————————
Parabolic Mirror

It is well known that to eliminate the spherical aberration just shown, the reflecting surface should be a paraboloid. The paraboloid (Fig. 14.2c) may be defined by

$$z = \frac{r^2}{4g} \tag{6}$$

where g is a constant shown below to be focal length. Slope of the mirror at radius r is then

$$\frac{dr}{dz} = \left(\frac{dz}{dr}\right)^{-1} = \frac{2g}{r} \tag{7}$$

The slope of the normal to the mirror, PQ, is just the negative reciprocal of (7),

$$\left(\frac{dr}{dz}\right)_{QP} = -\frac{r}{2g} = -\tan\theta \tag{8}$$

From the trigonometry of the figure,

$$\tan 2\theta = \frac{r}{f - z} \tag{9}$$

Using the identity in (2), it can be shown that f is independent of the incident radius r and equals the constant g in (6).

We have here used only the law of reflection but it is interesting to use this example to show the relation to wave optics. In the latter, one can show that all rays entering parallel to the axis and reflecting to the focus have the same phase delay from some reference plane $z = z_0$ to the focus F. Since the medium is homogeneous, this means that all such paths are of equal length, as is known to be the case for a paraboloid.

———————————————— **Example 14.2c** ————————————————
Thin Lens

We next consider the thin lens, the requirement of thinness implying that lens thickness is small in comparison with focal length. For simplicity we make one surface a plane as illustrated in Fig. 14.2d. The lens material is considered transparent, with index of refraction n, the surrounding medium being air or space with refractive index unity. Consider first the derivation of the law for the curved surface from Snell's law of refraction. A ray coming in parallel to the axis at radius r from the axis experiences no refraction at the plane surface, but does refract at the right-hand curved surface, as shown in Fig. 14.2d. If θ_1 and θ_2 are the angles on the two sides from the normal to the curved surface, Snell's law gives

$$\frac{\sin \theta_2}{\sin \theta_1} = \frac{n_1}{n_2} = n \tag{10}$$

If rays are "paraxial," angles will be small so that sines may be approximated by angles themselves,

$$\theta_2 \approx n\theta_1 \tag{11}$$

Also r/f is small so that

$$\frac{r}{f} = \tan(\theta_2 - \theta_1) \approx \theta_2 - \theta_1 \approx (n - 1)\theta_1 \tag{12}$$

The angle θ_1 can be related to the slope of the curved surface,

$$\frac{dz}{dr} = -\tan \theta_1 \approx -\theta_1 \approx -\frac{r}{(n - 1)f} \tag{13}$$

Upon integrating we find

$$z = -\frac{r^2}{2(n - 1)f} + d \tag{14}$$

where the constant d is the value of z at $r = 0$, the maximum thickness of the lens. Thus the lens should be parabolic in order to have a constant focal length f.

To show the same result by the constancy of phase paths for the different rays, let us require the same phase for the ray at distance r from the axis and that on the axis, we must consider the different phase delay within the lens medium, so we will write phase terms completely. Equality of phases for the two specified paths for a sinusoidal wave of angular frequency ω requires

$$\frac{n\omega}{c}z + \frac{\omega}{c}\sqrt{(f - z)^2 + r^2} = \frac{n\omega}{c}d + \frac{\omega}{c}(f - d) \tag{15}$$

or

$$(f - z)^2 + r^2 = [(f - d) + n(d - z)]^2 \tag{16}$$

If we neglect terms in z^2, zd, and d^2 in comparison with f^2 according to the thin-lens approximation, we finally obtain

$$z \approx d - \frac{r^2}{2(n - 1)f} \tag{17}$$

which is the equation for the parabolic surface found in (14). In practice, this surface is frequently formed as a spherical surface having a radius equal to the radius of curvature of the parabolic surface at the origin,

$$R \approx (n - 1)f \tag{18}$$

Note that the extra phase delay introduced by the lens to a plane wave incident upon it is

$$\Delta\phi(r) = \frac{\omega}{c}(n-1)z = \frac{k_0}{2f}[2f(n-1)d - r^2]$$

Or in terms of maximum lens radius a where $z = 0$,

$$\Delta\phi(r) = \frac{k_0}{2f}(a^2 - r^2) \tag{19}$$

_____ **Example 14.2d** _____
Relation of Object and Image Distances in Thin Lens

Either Snell's law or the requirement on constancy of phase for different paths may be used to derive the relationship for object and image distances in a thin lens. Let us use the latter method for the simple lens analyzed above. This is redrawn in Fig. 14.2e to show a ray path going from object point P, distance d_1 from the lens, to image point P', distance d_2 on the other side. If the phase delay for the path going through the lens at radius r is set equal to that along the axis so that all paths add contributions in phase, we have

$$\frac{\omega}{c}\sqrt{d_1^2 + r^2} + \frac{n\omega z}{c} + \frac{\omega}{c}\sqrt{r^2 + (d_2 - z)^2} = [d_1 + nd + (d_2 - d)]\frac{\omega}{c} \tag{20}$$

If r^2 is small in comparison with d_1^2 and $(d_2 - z)^2$, the binomial approximation to the square roots may be taken so that

$$d_1\left(1 + \frac{r^2}{2d_1^2}\right) + nz + (d_2 - z)\left[1 + \frac{r^2}{2(d_2 - z)^2}\right] \approx d_1 + d_2 + (n-1)d \tag{21}$$

If (17) is used for $(d - z)$,

$$\frac{r^2}{2d_1} + \frac{r^2}{2(d_2 - z)} = (n-1)(d - z) = \frac{(n-1)r^2}{2(n-1)f} \tag{22}$$

Neglect of z in comparison with d_2 in the second term leads to the geometrical optics relationship between object distance d_1, image distance d_2, and focal length f,

$$\frac{1}{d_1} + \frac{1}{d_2} = \frac{1}{f} \tag{23}$$

14.3   Geometrical Optics as Limiting Case of Wave Optics

In the preceding section we argued the case for geometrical optics on physical grounds. Now we show formally the terms that are neglected in this approximation. Let us assume fields of the following form:

$$\mathbf{E} = \mathbf{e}(x, y, z)e^{-jk_0 S(x, y, z)} \tag{1}$$

$$\mathbf{H} = \mathbf{h}(x, y, z)e^{-jk_0 S(x, y, z)} \tag{2}$$

where $k_0 = \omega\sqrt{\mu_0 \varepsilon_0}$. We next substitute (1) and (2) in Maxwell's equations for a linear, source-free, isotropic medium, making use of the expansion of the curl of a vector multiplied by a scalar:

$$\nabla \times \mathbf{E} = [\nabla \times \mathbf{e} - jk_0 \nabla S \times \mathbf{e}]e^{-jk_0 S} = -j\omega\mu\mathbf{h}e^{-jk_0 S} \tag{3}$$

$$\nabla \times \mathbf{H} = [\nabla \times \mathbf{h} - jk_0 \nabla S \times \mathbf{h}]e^{-jk_0 S} = j\omega\varepsilon\mathbf{e}e^{-jk_0 S} \tag{4}$$

Rearranging, we have

$$\nabla S \times \mathbf{e} - \frac{\omega\mu}{k_0}\mathbf{h} = \frac{1}{jk_0}\nabla \times \mathbf{e} \tag{5}$$

$$\nabla S \times \mathbf{h} + \frac{\omega\varepsilon}{k_0}\mathbf{e} = \frac{1}{jk_0}\nabla \times \mathbf{h} \tag{6}$$

At this point we make the approximation that leads to the geometrical optics formulation, assuming that the multipliers $\mathbf{e}$ and $\mathbf{h}$ vary little in a wavelength. The curl of $\mathbf{e}$ represents a derivative with distance, and $k_0 = 2\pi/\lambda_0$, so the right side of (5) is small if $\mathbf{e}$ varies only a small amount in distance λ_0. Similarly, the right side of (6) can be neglected if $\mathbf{h}$ likewise varies a small amount in a wavelength. Equations (5) and (6) then lead to

$$\mathbf{h} \approx \frac{1}{\mu_r \eta_0}\nabla S \times \mathbf{e} \tag{7}$$

$$\mathbf{e} \approx -\frac{\eta_0}{\varepsilon_r}\nabla S \times \mathbf{h} \tag{8}$$

Let us now interpret the expressions. If S is real, the function $k_0 S$ may be considered a phase function with surfaces $S = $ constant being surfaces of constant phase and the gradient of S giving the local "direction of propagation." We see from (7) and (8) that $\mathbf{e}$ and $\mathbf{h}$ are then in phase, normal to each other and to the local direction of propagation, as would be expected if the local behavior is that of a plane wave [Eqs. 13.9(3)–(4)]. If S is complex, (1) and (2) show that there is attenuation of the wave, and then (7) and (8) yield a phase shift between $\mathbf{e}$ and $\mathbf{h}$, as is expected for attenuating media.

We next obtain a rather important equation relating the gradient of S to the local refractive index. To do this, substitute (7) in (8) and expand the triple vector product to obtain

$$\mathbf{e} = -\frac{1}{\mu_r \varepsilon_r} [\nabla S(\mathbf{e} \cdot \nabla S) - \mathbf{e}(\nabla S \cdot \nabla S)] \tag{9}$$

Since $\mathbf{e}$ and ∇S are at right angles, the first term on the right is zero and

$$|\nabla S|^2 = n^2(x, y, z) \tag{10}$$

where n is refractive index $= (\mu_r \varepsilon_r)^{1/2}$. This equation is known as the *eikonal* equation of geometrical optics.[2] In rectangular coordinates,

$$\left(\frac{\partial S}{\partial x}\right)^2 + \left(\frac{\partial S}{\partial y}\right)^2 + \left(\frac{\partial S}{\partial z}\right)^2 = n^2(x, y, z) \tag{11}$$

Note that for a homogeneous medium with n a constant this is satisfied by

$$S = n(x \cos \alpha + y \cos \beta + z \cos \gamma) \tag{12}$$

with

$$\cos^2 \alpha + \cos^2 \beta + \cos^2 \gamma = 1 \tag{13}$$

which is the phase variation one would expect for a uniform plane wave propagating in an oblique direction defined by direction cosines $\cos \alpha$, $\cos \beta$, and $\cos \gamma$. Substitution in (7) and (8) would lead to the expected relative orientation of $\mathbf{e}$ and $\mathbf{h}$ for such a wave (Prob. 14.3c).

Let us next look at power and energy relations for the rays in this formulation. The average Poynting vector is

$$\mathbf{P}_{av} = \tfrac{1}{2} \operatorname{Re}[\mathbf{E} \times \mathbf{H}^*] = \tfrac{1}{2} \operatorname{Re}[\mathbf{e} \times \mathbf{h}^*] \tag{14}$$

Substitution of (7) and expansion of the triple vector product gives

$$\mathbf{P}_{av} = \frac{1}{2\mu_r \eta_0} [\nabla S(\mathbf{e} \cdot \mathbf{e}^*) - \mathbf{e}^*(\mathbf{e} \cdot \nabla S)] \tag{15}$$

The last term is zero since $\mathbf{e}$ is normal to ∇S. The first term may be written in terms of average energy density in electric fields, $u_e = \varepsilon(\mathbf{e} \cdot \mathbf{e}^*)/4$:

$$\mathbf{P}_{av} = \frac{4u_e}{2\mu_r \eta_0 \varepsilon_r \varepsilon_0} \nabla S = \frac{c}{n} (2u_e)\left(\frac{\nabla S}{n}\right) \tag{16}$$

---

[2] M. Born and E. Wolf, *Principles of Optics*, 6th ed., Pergamon Press, New York, 1980, p. 112.

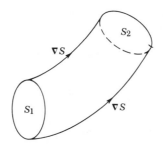

Fig. 14.3 Tube of constant power flow.

Note that the average stored energy densities in electric and magnetic fields are equal, as can be shown from (7) and (8):

$$\frac{\varepsilon}{4}(\mathbf{e} \cdot \mathbf{e}^*) = -\frac{1}{4c}[\mathbf{e} \cdot (\nabla S \times \mathbf{h}^*)] = \frac{\mu}{4}(\mathbf{h} \cdot \mathbf{h}^*) \tag{17}$$

Thus the term $2u_e$ is total energy density and we see from (16) that the Poynting vector is equal in magnitude to energy density multiplied by c/n, as would be expected if local behavior is as in plane waves. It also has the ray direction given by ∇S so that a tube defined by ray directions (Fig. 14.3) has, for a lossless medium, the same power traversing each cross section. This is analogous to the continuity of flux in a flux tube as met in static fields.

Corrections to the geometrical optics formulation have been made[3] by expanding $\mathbf{e}$ and $\mathbf{h}$ in asymptotic series in terms of wavelength and including contributions of various orders from the right-hand sides of (5) and (6).

14.4   Rays in Inhomogeneous Media

The general formulation for ray propagation given in the preceding section gave the expected but almost trivial result for homogenous media, but is of much more use with materials for which the refractive index varies with position. In such media the rays are curved and the geometrical optics approach can often give a great deal of information about the nature of the waves. Important practical examples of inhomogeneous media include the graded-index fibers used for guiding of optical waves, the surface guides of integrated optics made by diffusion or ion implantation, and the lower-frequency example of radio waves in the ionosphere.

Before applying the general formulation, we can obtain an expression for curvature of the ray and a physical understanding of the bending process, by assuming that the change of index occurs in a series of steps of differential size. If Snell's law

M. Kline and I. W. Kay, *Electromagnetic Theory and Geometrical Optics*, Interscience, New York, 1965

is applied to the surface separating a differential region with index n from one with $n + dn$ as in Fig. 14.4a, there results

$$\frac{n}{n + dn} = \frac{\sin(\theta + d\theta)}{\sin \theta} = \frac{\sin \theta + d\theta \cos \theta}{\sin \theta} \tag{1}$$

from which we find

$$\cot \theta \, d\theta = -\frac{dn}{n} \tag{2}$$

The radius of curvature is found from the intersection of the lines erected perpendicular to the ray, and can be related to the length ds of the ray segment in the layer with index $n + dn$ and the angle $d\theta$, or alternatively in terms of the ray segment dR along the radius of curvature:

$$R = \frac{ds}{d\theta} = \frac{dR}{d\theta \cot \theta} \tag{3}$$

Using (2), this may be written as

$$\frac{1}{R} = -\frac{1}{n}\frac{dn}{dR}$$

or

$$\frac{1}{R} = -\hat{\mathbf{R}} \cdot \nabla(\ln n) \tag{4}$$

In graphical ray tracing, the radius of curvature can be calculated from point to point as the index changes, and the ray traced by joining the corresponding arcs at suitably small intervals.

   With this physical understanding of the basis for ray curvature, let us apply the formal development of Sec. 14.3. As shown in Fig. 14.4b, let $\mathbf{r}'$ be the vector position of a point on the ray from an arbitrary reference. Its differential change is

$$d\mathbf{r}' = \hat{\mathbf{s}} ds \tag{5}$$

where s is distance along the ray and $\hat{\mathbf{s}}$ is a unit vector in the ray direction. But from Sec. 14.3, gradient of S has a magnitude equal to refractive index and the direction of $\hat{\mathbf{s}}$,

$$\nabla S = n\hat{\mathbf{s}} \tag{6}$$

Thus

$$n\frac{d\mathbf{r}'}{ds} = \nabla S \tag{7}$$

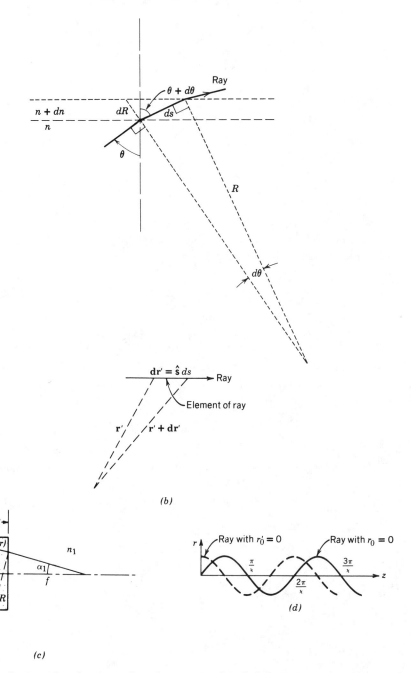

Fig. 14.4 (*a*) Construction showing radius of curvature of ray in inhomogeneous medium. (*b*) Element of ray and definitions used in general formulation. (*c*) A cell filled with an inhomo-geneous material, acting as a lens. (*d*) Ray paths through a long quadratic-index medium.

Taking an additional derivative,

$$\frac{d}{ds}\left(n\frac{d\mathbf{r'}}{ds}\right) = \frac{d}{ds}(\nabla S) \tag{8}$$

But because of (6)

$$\frac{d}{ds}(\nabla S) = \nabla n \tag{9}$$

so that (8) becomes

$$\frac{d}{ds}\left(n\frac{d\mathbf{r'}}{ds}\right) = \nabla n \tag{10}$$

Still other useful forms, including the relation (4), may be derived from this (Prob. 14.4c). We now illustrate the use of the above through two practical examples.

_____ **Example 14.4a** _____
Thin Cell with Quadratic Index Variation

As a simple but useful example, consider a thin cell, Fig. 14.4c, with refractive index varying quadratically with radius r from the axis,

$$n(r) = n_0\left[1 - \Delta\left(\frac{r}{a}\right)^2\right] \tag{11}$$

Such a variation is approximated when a laser beam passes through certain materials, either because of heating or other nonlinear effects. Self-focusing or self-defocusing of the laser beam may result. A ray entering from the left, parallel to the axis and at radius r from it, is curved, with radius of curvature given by (4),

$$\frac{1}{R} = -\frac{1}{n(r)}\frac{dn(r)}{dr} = \frac{2n_0 r \Delta}{a^2 n(r)} \tag{12}$$

If cell thickness d is small, this curvature applies throughout the cell and the angle at exit, measured inside the cell, is approximately

$$\alpha_0 \approx \frac{d}{R} \tag{13}$$

There is a further deflection by Snell's law at exit from the cell, and if angles are small and the external medium one with index n_1,

$$\alpha_1 \approx \frac{n(r)\alpha_0}{n_1} = \frac{2n_0 r \Delta d}{a^2 n_1} \tag{14}$$

The focal length of this lens is then

$$f \approx \frac{r}{\alpha_1} = \frac{a^2 n_1}{2 n_0 \, \Delta \, d} \tag{15}$$

We thus see that the cell with quadratic index variation acts as a thin lens, converging if Δ is positive (index decreasing with increasing radius) and diverging for the opposite variation.

_____ **Example 14.4b** _____

Rays in Graded-Index Fiber with Quadratic Variation of n

Suppose the dielectric with quadratic index variation with radius is not a thin slice, as in the first example, but a long cylinder, as in graded-index fibers used for optical guiding. Assume that all rays have small slopes so that $ds \approx dz$. Then (10) becomes

$$\frac{d}{dz}\left[n(r) \frac{d\mathbf{r}'}{dz} \right] = \hat{\mathbf{r}} \frac{dn(r)}{dr} \tag{16}$$

The radius vector $\mathbf{r}'$ from origin 0 to a general point on the ray is

$$\mathbf{r}' = \hat{\mathbf{r}} r + \hat{\mathbf{z}} z \tag{17}$$

Substitution in (16) leads to the simpler relation

$$\frac{d}{dz}\left[n(r) \frac{dr}{dz} \right] = \frac{dn(r)}{dr} \tag{18}$$

If $n(r)$ varies quadratically with r, as in (11),

$$\frac{d^2 r}{dz^2} = \frac{-2 n_0 r \Delta}{a^2 n_0 [1 - \Delta (r/a)^2]} \approx -\left(\frac{2\Delta}{a^2} \right) r \tag{19}$$

The last approximation assumes Δ small. The solution to (19) is

$$r(z) = r_0 \cos \kappa z + \frac{r_0'}{\kappa} \sin \kappa z \tag{20}$$

where

$$\kappa = \frac{\sqrt{2\Delta}}{a} \tag{21}$$

and r_0 and r_0' are, respectively, radius and slope of the ray at $z = 0$. Two special ray paths are shown in Fig. 14.4d.

_____

14.5 Paraxial Ray Optics—Ray Matrices

The rays in many optical instruments and laser resonators or guiding systems have only small slopes with respect to an axis of symmetry, as in the example of the quadratic-index rod or fiber analyzed in the preceding section. In such cases the rays are called *paraxial rays* and the optics deriving from this condition is called *gaussian optics*.[4] It is convenient for this class of problems (and sometimes more generally) to describe the effect of various optical components by *ray matrices* that relate output radius and slope of a ray to the input radius and slope,

$$\begin{bmatrix} r_{out} \\ r'_{out} \end{bmatrix} = \begin{bmatrix} A & B \\ C & D \end{bmatrix} \begin{bmatrix} r_{in} \\ r'_{in} \end{bmatrix} \tag{1}$$

We illustrate with some simple examples.

_____ **Example 14.5a** _____
Homogeneous Medium

This is the simplest case, and although almost trivial, nevertheless an important one. As shown in Fig. 14.5a, the ray in a homogeneous medium is a straight line so that for a distance d,

$$\begin{matrix} r_{out} = r_{in} + d r'_{in} \\ r'_{out} = r'_{in} \end{matrix} \quad \text{or} \quad \begin{bmatrix} A & B \\ C & D \end{bmatrix} = \begin{bmatrix} 1 & d \\ 0 & 1 \end{bmatrix} \tag{2}$$

_____ **Example 14.5b** _____
Thin Lens

Interpretation of Eq. 14.2(23) shows that a lens acts to change the slope of an incoming ray, but if it is thin enough, the change in radius is negligible. Thus

$$\begin{matrix} r_{out} = r_{in} \\ r'_{out} = r'_{in} - \dfrac{r_{in}}{f} \end{matrix} \quad \text{or} \quad \begin{bmatrix} A & B \\ C & D \end{bmatrix} = \begin{bmatrix} 1 & 0 \\ -\dfrac{1}{f} & 1 \end{bmatrix} \tag{3}$$

_____ **Example 14.5c** _____
Parabolic or Spherical Mirror

A curved mirror is similar to the lens except that the reflected ray returns rather than passing through to the other side. For a fixed coordinate system, this would

_____

[4] W. Born and E. Wolf, Ref. 2, Sec. 6.4, for uses with lenses; see A. Yariv, Ref. 1, for use with laser resonators.

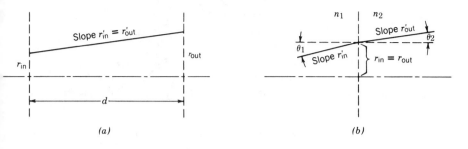

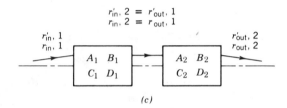

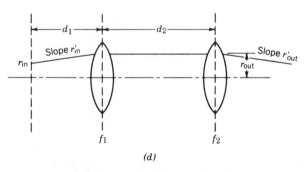

Fig. 14.5 (*a*) Ray in homogeneous medium. (*b*) Ray passing between media of different refractive index at small angle. (*c*) Cascade arrangement of two general optical elements. (*d*) Cascade arrangement of two thin lenses with homogeneous regions between and on each side.

result in opposite signs for the slopes of the two cases, and thus reverse the sign of the $1/f$ term in (3). However, because of the use of these in cascaded systems to be described below, it is more convenient to take the direction of the ray as determining the positive coordinate. That is, for the mirror, positive z is toward the mirror for the incident wave and away from the mirror for the reflected wave. Thus for a parabolic mirror of focal length f, the matrix is the same as in (3). For a spherical mirror, spherical aberration is negligible for paraxial rays so that $f = R/2$.

Example 14.5d
Plane Discontinuity Between Dielectrics

If a plane perpendicular to the axis separates a medium with refractive index n_1 from a second medium with n_2, Fig. 14.5b, the ray is not changed in radius at the boundary, but slope is changed by Snell's law. For paraxial rays, sines of the angles may be replaced by slopes.

$$r_{out} = r_{in}$$
$$r'_{out} = \frac{n_1}{n_2} r'_{in} \quad \text{or} \quad \begin{bmatrix} A & B \\ C & D \end{bmatrix} = \begin{bmatrix} 1 & 0 \\ 0 & \dfrac{n_1}{n_2} \end{bmatrix} \tag{4}$$

Example 14.5e
Rod with Quadratic Index Variation

For a rod or fiber of length d with refractive index varying quadratically with radius, Eqs. 14.4(20) and 14.4(21) give the ray matrix:

$$\begin{bmatrix} r_{out} \\ r'_{out} \end{bmatrix} = \begin{bmatrix} \cos \kappa d & \dfrac{1}{\kappa} \sin \kappa d \\ -\kappa \sin \kappa d & \cos \kappa d \end{bmatrix} \begin{bmatrix} r_{in} \\ r'_{in} \end{bmatrix}; \quad \kappa = \frac{\sqrt{2\Delta}}{a} \tag{5}$$

Example 14.5f
Elements in Tandem

The advantage of the matrix formulation is primarily in its ease of handling combinations of elements. Thus if one element is followed by another, as sketched in Fig. 14.5c, $r_{out, 1} = r_{in, 2}$ and $r'_{out, 1} = r'_{in, 2}$ so that

$$\begin{bmatrix} r_{out, 2} \\ r'_{out, 2} \end{bmatrix} = \begin{bmatrix} A_2 & B_2 \\ C_2 & D_2 \end{bmatrix} \begin{bmatrix} r_{in, 2} \\ r'_{in, 2} \end{bmatrix} = \begin{bmatrix} A_2 & B_2 \\ C_2 & D_2 \end{bmatrix} \begin{bmatrix} A_1 & B_1 \\ C_1 & D_1 \end{bmatrix} \begin{bmatrix} r_{in, 1} \\ r'_{in, 1} \end{bmatrix} \tag{6}$$

The ray matrix of the combination is thus the product of the individual ray matrices starting from the one nearest the output. The procedure is extended in a straightforward manner to any number of elements in cascade, with the final element first in the product chain. As an example, consider the combination of two lenses and two homogeneous regions sketched in Fig. 14.5d. The ray matrix for the combination is

$$\begin{bmatrix} A & B \\ C & D \end{bmatrix} = \begin{bmatrix} 1 & 0 \\ -\dfrac{1}{f_2} & 1 \end{bmatrix} \begin{bmatrix} 1 & d_2 \\ 0 & 1 \end{bmatrix} \begin{bmatrix} 1 & 0 \\ -\dfrac{1}{f_1} & 1 \end{bmatrix} \begin{bmatrix} 1 & d_1 \\ 0 & 1 \end{bmatrix} \tag{7}$$

from which

$$A = 1 - \frac{d_2}{f_1}$$

$$B = d_1 + d_2\left(1 - \frac{d_1}{f_1}\right)$$

$$C = -\frac{1}{f_1} - \frac{1}{f_2}\left(1 - \frac{d_2}{f_1}\right)$$

$$D = -\frac{1}{f_2}\left[d_1 + d_2\left(1 - \frac{d_1}{f_1}\right)\right] + \left(1 - \frac{d_1}{f_1}\right)$$

(8)

We shall interpret these expressions in the following section to show that this combination is useful in forming optical guiding systems or optical resonators.

14.6 Guiding of Rays by a Periodic Lens System or in Spherical Mirror Resonators

If the lens pair of Fig. 14.5d is repeated, as in Fig. 14.6a, a periodic system results. It will be seen that for such systems, rays may be guided or confined for certain combinations of parameters, but may diverge for other combinations. Let us apply the ray matrix between reference planes m and $m + 1$ with A, B, C, D defined for one unit of the periodic system:

$$r_{m+1} = Ar_m + Br'_m \tag{1}$$

$$r'_{m+1} = Cr_m + Dr'_m \tag{2}$$

Equation (1) may be solved for r'_m.

$$r'_m = \frac{1}{B}(r_{m+1} - Ar_m) \tag{3}$$

For this periodic system, a similar equation applies between references $m + 1$ and $m + 2$:

$$r'_{m+1} = \frac{1}{B}(r_{m+2} - Ar_{m+1}) \tag{4}$$

If (3) and (4) are substituted in (2) there results

$$r_{m+2} - (A + D)r_{m+1} + (AD - BC)r_m = 0 \tag{5}$$

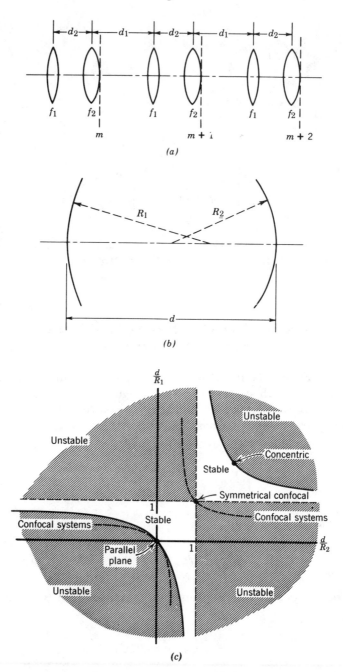

Fig. 14.6 (*a*) A periodic lens system. (*b*) Spherical mirror resonator showing two coaxial mirrors separated by distance *d*. (*c*) Diagram showing stable (confined) conditions for spherical mirror resonator of (*b*).

For optical systems of concern to us, it can be shown that $AD - BC = 1$. (This can be checked from Eqs. 14.5(7)–(8) for the system of Fig. 14.6a.) Equation (5) then simplifies to

$$r_{m+2} - (A + D)r_{m+1} + r_m = 0 \tag{6}$$

This linear difference equation may be solved by assuming solutions of exponential form[5], much as with linear differential equations with constant coefficients. Let

$$r_m = r_0 e^{\pm jm\theta} \tag{7}$$

Substitution of (7) in (6) yields

$$e^{\pm 2j\theta} - (A + D)e^{\pm j\theta} + 1 = 0 \tag{8}$$

This is a quadratic in $e^{\pm j\theta}$ so that the solution is

$$e^{\pm j\theta} = \left(\frac{A + D}{2}\right) \pm \left[\left(\frac{A + D}{2}\right)^2 - 1\right]^{1/2} \tag{9}$$

But since $e^{j\theta} = \cos\theta + j\sin\theta$, this is consistent with

$$\cos\theta = \left(\frac{A + D}{2}\right) \tag{10}$$

Thus there are real solutions for θ with r_m bounded if

$$-1 \le \left(\frac{A + D}{2}\right) \le 1 \tag{11}$$

But if (11) is not satisfied, θ is imaginary and one of the solutions (7) is a growing exponential as m increases so that the ray is not bounded. The two cases are frequently called *stable* and *unstable*, respectively.

For the system of Fig. 14.6a, Eqs. 14.5(8) apply and the condition for a stable or confined solution is

$$-1 \le 1 - (d_1 + d_2)\left(\frac{1}{2f_1} + \frac{1}{2f_2}\right) + \frac{d_1 d_2}{2f_1 f_2} \le 1 \tag{12}$$

If all spacings are the same, $d_1 = d_2 = d$, (12) may be rearranged so that the stability condition is

$$0 \le \left(1 - \frac{d}{2f_1}\right)\left(1 - \frac{d}{2f_2}\right) \le 1 \tag{13}$$

---

[5] Floquet's theorem shows that periodic systems always have solutions of this form. See, for example, L. Brillouin, *Wave Propagation in Periodic Structures*, 2nd ed., Dover, New York, 1953.

The above discussion for the periodic lens system may be applied directly to the spherical mirror resonator by the relations between mirrors and lenses discussed in Sec. 14.5. A typical optical resonator used in laser systems consists of two spherical mirrors with radii of curvature R_1 and R_2, aligned with a common axis (Fig. 14.6b). Rays bounce back and forth between the two mirrors (in the geometrical optics picture), and if the system is stable, they are confined within some maximum radius. From Sec. 14.5, this system is equivalent to the periodic lens systems of Fig. 14.6a with $d_1 = d_2$ so that (13) applies directly, with $f_1 = R_1/2$ and $f_2 = R_2/2$. The stability condition for Fig. 14.6b is thus

$$0 \le \left(1 - \frac{d}{R_1}\right)\left(1 - \frac{d}{R_2}\right) \le 1 \tag{14}$$

It will be shown later that this condition is confirmed by a wave analysis, so it is a very important relation for such resonators. Figure 14.6c shows the stable (unshaded) and unstable (shaded) regions in a plane with d/R_1 as ordinate and d/R_2 as abscissa. The special cases shown are the parallel plane ($R_1 = R_2 = \infty$), the concentric ($R_1 = R_2 = d/2$), and the confocal ($R_1/2 + R_2/2 = d$). These will be discussed more following the wave analysis.

DIELECTRIC OPTICAL WAVEGUIDES

14.7   Dielectric Guides of Planar Form

The principle of guiding electromagnetic waves by dielectric guides was shown in Sec. 9.2. Such guides have become very useful for optical communication devices. We will consider the optical fibers used for transmission of optical communication signals in following sections. Here we want to consider the simpler planar forms, which have been ultilized in the thin-film devices of integrated optics.

Figure 14.7a shows three dielectrics separated by parallel-plane interfaces. If the refractive index of the intermediate region 2 is higher than that of the materials on either side, there is the possibility of having guided modes through the phenomenon of total reflection at the boundaries, as explained in Sec. 9.2. Modification of the analysis for the symmetrical example of that article is straightforward. For a wave propagating as $e^{-j\beta z}$ and no variations with y, the field components of the TE wave are as follows:

$$\left.\begin{aligned} E_{y1} &= Ae^{-qx} = -\left(\frac{\omega\mu}{\beta}\right)H_{x1} \\[2mm] H_{z1} &= -\frac{1}{j\omega\mu}\frac{\partial E_y}{\partial x} = \frac{q}{j\omega\mu}Ae^{-qx} \end{aligned}\right\} x > 0 \tag{1}$$

$$E_{y2} = B \cos hx + C \sin hx = -\left(\frac{\omega\mu}{\beta}\right)H_{x2} \Bigg|$$

$$H_{z2} = \frac{h}{j\omega\mu}[B \sin hx - C \cos hx] \Bigg\} \quad -d \le x \le 0 \quad (2)$$

$$E_{y3} = De^{px} = -\left(\frac{\omega\mu}{\beta}\right)H_{x3} \Bigg|$$

$$H_{z3} = -\frac{p}{j\omega\mu}De^{px} \Bigg\} \quad x < -d \quad (3)$$

where

$$q^2 = \beta^2 - k_1^2, \qquad h^2 = k_2^2 - \beta^2, \qquad p^2 = \beta^2 - k_3^2 \quad (4)$$

Continuity of tangential field components requires that E_y and H_z be continuous at $x = 0$, and again at $x = -d$. The four resulting equations allow reduction from the four arbitrary constants A, B, C, and D to a single one, and development of the determinantal equation

$$\tan hd = \frac{h(q + p)}{h^2 - pq} \quad (5)$$

Although this can be solved graphically[6] for the symmetrical case, and for the useful case in which $n_3 - n_1 \gg n_2 - n_3$ (Prob. 14.7d), it has been solved numerically[7] and the important results are shown in Fig. 14.7b with these definitions:

$$v = \frac{2\pi d}{\lambda_0}\sqrt{n_2^2 - n_3^2} \quad (6)$$

$$b = \frac{n_{eff}^2 - n_3^2}{n_2^2 - n_3^2} \quad (7)$$

$$a = \frac{n_3^2 - n_1^2}{n_2^2 - n_3^2} \quad (8)$$

where

$$n_{eff} = \frac{\beta}{k_0} \quad (9)$$

It is seen that the parameter v is proportional to the ratio of thickness to wavelength, but also depends upon the differences in refractive index between guiding region and substrate. Parameter b determines the value of β in terms of an effective

[6] R. E. Collin, *Field Theory of Guided Waves*, McGraw-Hill, New York, 1960, pp. 471–473.
[7] H. Kogelnik and V. Ramaswamy, *Appl. Opt.* **13**, 1857 (1974).

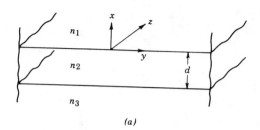

(a)

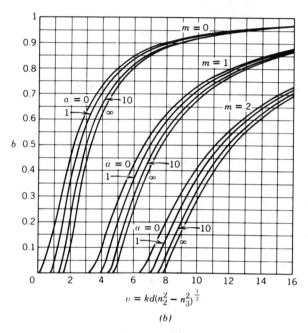

$$v = kd(n_2^2 - n_3^2)^{\frac{1}{2}}$$

(b)

Fig. 14.7 (a) Section of parallel-plane dielectric guide. Propagation is in the z direction. (b) diagram showing normalized phase constant for first few modes of planar slab dielectric waveguides versus normalized frequency, with a range of parameters. (Taken from Ref. 7.)

refractive index n_{eff}. Note that when $b = 0$, the wave travels with the velocity of light in the substrate material, and when b is unity, it travels with the velocity in the film material. The parameter a describes the degree of asymmetry. For $a = 0$, $n_1 = n_3$ and for $a = \infty$, $n_3 - n_1 \gg n_2 - n_3$. As an example of use of the figure, take a glass film on silica substrate with air above, $n_1 = 1, n_2 = 1.55, n_3 = 1.50$. Thickness $d = 1.46\ \mu m$ and $\lambda_0 = 0.6\ \mu m$. Then $v = 6.0$ from (6) and $a = 8.2$ from (8). From Fig. 14.7b we find that two modes, the $m = 0$ and $m = 1$ modes, are guided. Their b values are read as 0.82 and 0.31, respectively. Using (7), this gives $n_{eff} = 1.541$ and 1.516, respectively. Both have velocities between light velocities of film and substrate, but the higher order wave is nearer cutoff and so has more energy in the substrate and travels with a velocity near c/n_3. The lower order mode is further from cutoff, has more of its energy in the film, and so has velocity nearer c/n_2.

For TM waves the analysis is essentially the dual of the above except that we must take into account the different ε's (μ was assumed the same for all regions). The resulting determinantal equation is

$$\tan hd = \frac{h[pn_2^2/n_3^2 + qn_2^2/n_1^2]}{h^2 - pqn_2^4/n_1^2n_3^2} \tag{10}$$

Figure 14.7b may also be used for this case when $n_2 \approx n_3$ if a is taken as

$$a_{TM} = \frac{n_2^4 (n_3^2 - n_1^2)}{n_1^4 (n_2^2 - n_3^2)} \tag{11}$$

Note from the figure that for all modes except the $m = 0$ mode for the symmetric case ($n_1 = n_3$), there is a cutoff condition. For lower frequencies or lesser thicknesses, the modes change from guided modes to radiating modes as the condition for total reflection from the interfaces can no longer be satisfied. The ranges are (assuming $n_2 > n_3 > n_1$)

Guided waves: $k_0 n_3 < \beta < k_0 n_2$
Substrate radiation modes: $k_0 n_1 < \beta < k_0 n_3$
Modes with radiation above and below: $0 < \beta < k_0 n_1$
Physically unrealizable (growing in both regions): $\beta > k_0 n_2$

Although there is only a discrete set of guided modes, there is a continuous spectrum of radiation modes. It can be shown that the modes (including the radiation modes) are orthogonal over the interval $-\infty < x < \infty$ and that the totality of guided and radiation modes form a complete set. However, the set is awkward to use for expansion of arbitrary functions because of the infinite extent of the radiation modes, and their continuous spectrum. Excellent discussions of these points have been given by Marcuse.[8,9]

Finally we note that although we have developed the determinantal equation from a direct solution of Maxwell's equations, it can also be developed rigorously from the picture of wave reflections at an angle, which we have given as a physical picture of the phenomenon. For this, one finds the phase of wave reflection at the boundary for each interface. The sum of these angles, plus the delay up and back in the x direction, should be a multiple of 2π. Thus if ϕ_{2-1} is the phase shift upon reflection from the upper interface at angle θ and ϕ_{2-3} that from the lower interface, the condition is

$$2k_2 d \cos \theta + \phi_{2-1} + \phi_{2-3} = 2m\pi, \qquad m = 0, 1, 2, \ldots \tag{12}$$

[8] D. Marcuse, *Light Transmission Optics*, 2nd ed. Van Nostrand Reinhold, New York, 1982.
[9] D. Marcuse, *Theory of Dielectric Optical Waveguides*, Academic Press, New York, 1974.

where m is the mode order as used in Fig. 14.7b. The phase angles may be found from Sec. 6.11 and are, for TE and TM waves, respectively,

$$(\phi_{2-i})_{TE} = -2 \tan^{-1} \left[\frac{\sqrt{n_2^2 \sin^2 \theta - n_i^2}}{(n_2 \cos \theta)} \right] \tag{13}$$

$$(\phi_{2-i})_{TM} = -2 \tan^{-1} \left[\frac{n_2 \sqrt{n_2^2 \sin^2 \theta - n_i^2}}{(n_i^2 \cos \theta)} \right] \tag{14}$$

Tien[10] has pioneered this "zig-zag wave" approach to analysis of the modes.

14.8   Dielectric Guides of Rectangular Form

The study of planar guides in the preceding section established the basic principles of dielectric guiding but for most practical purposes it is desirable to confine the wave laterally as well as in depth. One useful configuration is that of a dielectric of rectangular cross section imbedded in a substrate of lower index, as indicated in Fig. 14.8a. Often the higher dielectric region is formed by diffusion or ion implantation to make an inhomogeneous guiding region, as illustrated in Fig. 14.8b. The latter case can be analyzed only numerically once the distribution of index is known (although analytic solutions for certain index variations have been given[11]), but analysis of the rectangular approximation to Fig. 14.8b may still be useful. Even the rectangular configuration is hard to analyze, but approximate solutions are available which give a good idea of the behavior.

A numerical analysis of a dielectric guide of rectangular cross section surrounded by a dielectric of lower index, was given by Goell.[12] This utilized expansions of the wave in circular harmonics. Most often, one is concerned with operation well above cutoff so that most of the energy is concentrated in the guiding region, and Marcatili has supplied a very useful approximate method for this range.[13] Consider the rectangular dielectric 1 of Fig. 14.8c. In the general case, different dielectrics 2, 3, 4, and 5 surround the guiding region on the four sides. Except near cutoff, the evanescent fields in the external regions die off rapidly so that one need not worry about the difficult problem of matching fields in the shaded corner regions. Moreover, if index differences are not too great, the waves are nearly transverse electromagnetic and are found to break into two classes: E_{pq}^x with

[10] P. K. Tien, *Appl. Opt.* **10**, 2395 (1971).

[11] H. Kogelnik, in T. Tamir, Ed., *Integrated Optics*, Springer-Verlag, New York, 1975.

[12] J. E. Goell, *Bell Syst. Tech. J.* **48**, 2133 (1969).

[13] E. A. J. Marcatili, *Bell Syst. Tech. J.* **48**, 2071 (1969).

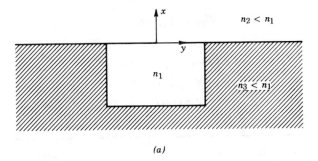

(a)

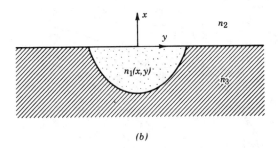

(b)

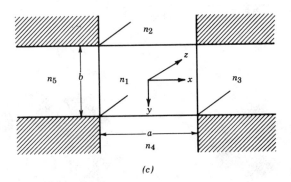

(c)

Fig. 14.8 (a) Dielectric guide with guiding confined in the y direction. (b) Similar inhomogeneous guide made by diffusion or ion implantation techniques. (c) Model for Marcatili analysis.

negligible E_y, and E_{pq}^y with negligible E_x. For the latter class, with propagation as $e^{-j\beta z}$,

$$E_{y1} = C_1 \cos(k_x x + \phi_1) \cos(k_y y + \phi_2) \tag{1}$$

$$E_{y2} = C_2 \cos(k_x x + \phi_1)e^{\alpha_2 y} \tag{2}$$

$$E_{y3} = C_3 e^{-\alpha_3 x} \cos(k_y y + \phi_2) \tag{3}$$

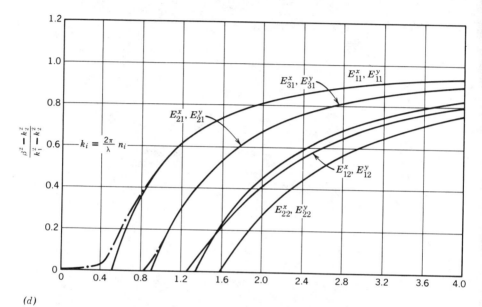

(d)

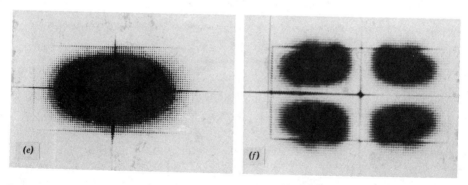

(e)   (f)

Fig. 14.8 (d) Curves giving normalized phase constant versus normalized frequency of several modes in rectangular dielectric guide surrounded by a common dielectric. Solid curves are by Marcatili's approximate method (Ref. 13) and ——·— curves by Goell's computer solutions (Ref. 12). (e) Intensity picture of E_{11}^y mode for $a/b = 2$ from Goell (Ref. 12). (f) Similar picture for E_{22}^y mode (Ref. 12). Figs. d, e, and f reprinted with permission of the Bell System Technical Journal, copyright 1969, A.T.&T.

where

$$\alpha_2 = [\beta^2 + k_x^2 - k_2^2]^{1/2} \tag{4}$$

$$\alpha_3 = [\beta^2 + k_y^2 - k_3^2]^{1/2} \tag{5}$$

The solution for region 4 is similar to 2, and 5 to 3, but with exponentials of opposite sign. Remaining field components are obtained from Maxwell's equations. Components E_x and H_y are negligible for this class; continuity of other tangential components at the four boundaries $x = \pm a/2$, $y = \pm b/2$, relate all constants to C_1 and give the determinantal equation for β. Figure 14.8d gives results when all surrounding dielectrics are the same, $n_5 = n_4 = n_3 = n_2$, with $n_1/1.05 < n_2 < n_1$ and $b/a = \frac{1}{2}$. Also plotted for comparison are results from Goell's computer calculations for these parameters showing that there is good agreement except near cutoff. Figures 14.8e and f show a picture of two E_{mn}^y modes in a dielectric with $b/a = \frac{1}{2}$, $n_1 = 1.02$, $n_2 = 1$, and $(2b/\lambda_0)(n_1^2 - n_2^2)^{1/2} = 2$.

14.9 Dielectric Guides of Circular Cross Section

Many practical dielectric guides, including the fibers used for optical communications, are of circular cross section. We consider in this section a dielectric with uniform permittivity ε_1 extending to $r = a$, with a second dielectric of lower permittivity ε_2 surrounding it, and permeability of both materials taken as μ_0 (Fig. 14.9a). (This is called a *step-index fiber* in optical communications; the useful *graded-index fiber* is discussed in a following section.)

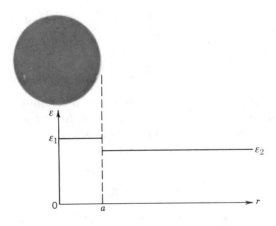

Fig. 14.9a Step-index optical fiber.

Any rectangular component of field, such as E_z, satisfies the scalar Helmholtz equation, which for a wave propagating as $e^{\mp j\beta z}$, in circular cylindrical coordinates, is

$$\frac{\partial^2 E_z}{\partial r^2} + \frac{1}{r}\frac{\partial E_z}{\partial r} + \frac{1}{r^2}\frac{\partial^2 E_z}{\partial \phi^2} + (k^2 - \beta^2)E_z = 0$$

For guided modes, $k_1^2 - \beta^2 > 0$ in the core, so ordinary Bessel functions result there. Only the first kind is used, to maintain fields finite on the axis. For the outer material or cladding, $\beta^2 - k_2^2 > 0$ for guided modes so that modified Bessel functions are utilized. Only the second solution is retained, so that fields die off properly at infinity. Axial field H_z satisfies a similar equation and we may write

$$r < a \qquad\qquad\qquad r > a$$

$$E_{z1} = AJ_l\left(\frac{ur}{a}\right)\begin{cases}\cos l\phi \\ \sin l\phi\end{cases} \qquad E_{z2} = CK_l\left(\frac{wr}{a}\right)\begin{cases}\cos l\phi \\ \sin l\phi\end{cases}$$

$$H_{z1} = BJ_l\left(\frac{ur}{a}\right)\begin{cases}\sin l\phi \\ \cos l\phi\end{cases} \qquad H_{z2} = DK_l\left(\frac{wr}{a}\right)\begin{cases}\sin l\phi \\ \cos l\phi\end{cases} \qquad (1)$$

$$u^2 = (k_1^2 - \beta^2)a^2 \qquad\qquad w^2 = (\beta^2 - k_2^2)a^2$$

The transverse field components may be obtained from these through Eqs. 8.9(1)–(4), where $k_{c1}^2 = u^2/a^2$ for region 1 and $k_{c2}^2 = -w^2/a^2$ for region 2. The results for E_ϕ and H_ϕ (needed in applying continuity) are

$$E_{\phi 1} = \left[\pm\frac{j\beta a^2 l}{ru^2}AJ_l\left(\frac{ur}{a}\right) + \frac{j\omega\mu a}{u}BJ_l'\left(\frac{ur}{a}\right)\right]\begin{cases}\sin l\phi \\ \cos l\phi\end{cases} \qquad (2)$$

$$E_{\phi 2} = \left[\mp\frac{j\beta a^2 l}{rw^2}CK_l\left(\frac{wr}{a}\right) - \frac{j\omega\mu a}{w}DK_l'\left(\frac{wr}{a}\right)\right]\begin{cases}\sin l\phi \\ \cos l\phi\end{cases} \qquad (3)$$

$$H_{\phi 1} = \left[-\frac{j\omega\varepsilon_1 a}{u}AJ_l'\left(\frac{ur}{a}\right) \mp \frac{j\beta a^2 l}{u^2 r}BJ_l\left(\frac{ur}{a}\right)\right]\begin{cases}\cos l\phi \\ \sin l\phi\end{cases} \qquad (4)$$

$$H_{\phi 2} = \left[\frac{j\omega\varepsilon_2 a}{w}CK_l'\left(\frac{wr}{a}\right) \pm \frac{j\beta a^2 l}{w^2 r}DK_l\left(\frac{wr}{a}\right)\right]\begin{cases}\cos l\phi \\ \sin l\phi\end{cases} \qquad (5)$$

For the symmetric case with $l = 0$, the solutions break into separate TM and TE sets, the former with E_z, H_ϕ and E_r (expression for the last component not shown) and the TE set with H_z, E_ϕ and H_r. The continuity condition of $E_{z1} = E_{z2}$ and $H_{\phi 1} = H_{\phi 2}$ at $r = a$ gives for the TM set

$$\frac{J_1(u)}{J_0(u)} = -\frac{\varepsilon_2 u}{\varepsilon_1 w}\frac{K_1(w)}{K_0(w)} \qquad (6)$$

The continuity condition of $H_{z1} = H_{z2}$, $E_{\phi 1} = E_{\phi 2}$ at $r = a$ for the TE set gives the same equation with the factor $\varepsilon_2/\varepsilon_1$ missing. Cutoff for the dielectric guide

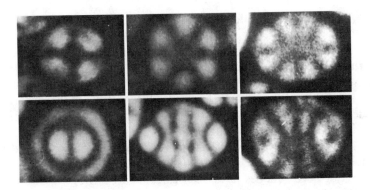

Fig. 14.9b Photographs of modes in the step-index optical fiber. (From Ref. 15.)

may be considered to be the condition for which fields in the outer guide extend to infinity, which happens if $w = 0$. For $w = 0$, $K_1/K_0 = \infty$, so that $J_0(u_c) = 0$. So at cutoff,

$$u_c = a(k_1^2 - k_2^2)^{1/2} = 2.405, 5.52, \ldots \tag{7}$$

If $l \neq 0$, the fields do not separate into TM and TE types, but all fields become coupled through the continuity conditions. Applying continuity of E_z, H_ϕ, H_z, and E_ϕ at $r = a$, the four arbitrary constants of (1)–(5) reduce to a single constant and there results the determinantal equation

$$\left[\frac{k_1 J_l'(u)}{u J_l(u)}\right]^2 + \left[\frac{k_2 K_l'(w)}{w K_l(w)}\right]^2 + \frac{(k_1^2 + k_2^2)}{uw}\left[\frac{J_l'(u)K_l'(w)}{J_l(u)K_l(w)}\right] = \frac{\beta^2 l^2 v^4}{u^4 w^4} \tag{8}$$

where

$$v^2 = u^2 + w^2 = (k_1^2 - k_2^2)a^2 \tag{9}$$

Solution of this transcendental equation leads to *hybrid* modes. Although not purely TM or TE, H_z is dominant in one set of solutions, designated HE_{lp} modes, whereas E_z is dominant in a set designated EH_{lp} modes. The HE_{11} mode is special in that it has no cutoff frequency and so is often called the *dominant mode*. Although it has no strict cutoff, energy is primarily in the guiding core only when the core size is appreciable in comparison with wavelength. This mode has been used in dielectric radiators,[14] and is also important in optical fibers.[15] Some pictures showing the light distribution in various modes are shown in Fig. 14.9b.

[14] G. E. Mueller and W. A. Tyrrell, *Bell Syst. Tech. J.* **26**, 837 (1947).
[15] N. S. Kapany, *J. Opt. Soc. of America* **51**, 1067 (1961).

14.10   Weakly Guiding Fibers: Dispersion

Most fibers used in optical communications have only a small index difference between the inner core and the outer cladding. It has been shown[16,17] that in this case the determinantal equation, Eq. 14.09(8) greatly simplifies, and that modes of order $l + 1$ and $l - 1$ have nearly the same values of β. When these nearly degenerate modes are superposed, it is found that fields combine in such a way that they are linearly polarized—one set with E_x and H_y, another with E_y and H_x as the dominant fields. Thus a wave with E_x constructed from the $l + 1$, p and $l - 1$, p hybrid modes is designated $LP_{l,p}^x$. Curves of phase constant (normalized) for these "weakly guided" modes have been calculated by Gloge[17] and are shown in Fig. 14.10a. The variable v is given by Eq. 14.9(9) while b is similar to that used with planar guides and is defined as

$$b = \frac{[\beta^2/k_0^2 - n_2^2]}{[n_1^2 - n_2^2]} = \frac{[\beta/k_0 - n_2][\beta/k_0 + n_2]}{[n_1 - n_2][n_1 + n_2]} \tag{1}$$

Since n_1, n_2 and β/k_0 are all near one another, the factors with positive sign essentially cancel and

$$b \approx \frac{[\beta/k_0 - n_2]}{[n_1 - n_2]} \tag{2}$$

As with the planar guides, most of the power flow near cutoff is in the cladding and wave velocity is determined by the index of the cladding ($b = 0$). Well above cutoff, most of the power is in the core and wave velocity is determined by n_1 ($b = 1$).

In an information transmission system, the optical wave is modulated (often digitally), and by Fourier analysis there must be a band of frequencies transmitted to represent the modulated wave. If group velocity varies over this frequency band, the envelope will be distorted as the signal propagates down the fiber as shown in Sec. 8.16. Such *group dispersion* limits the useful transmission distance for a given information rate. There are three possible sources of such group dispersion; intermode dispersion, modal dispersion, and material dispersion. These are discussed below.

1. If the fiber has a large enough v value to guide more than one mode, the different modes have different group velocities and so propagate the information (which we will think of here as pulses) at different rates. If there are many modes, some will be near cutoff with most of the energy in the cladding, and other modes will be far above cutoff with most of the energy in the core. Thus

[16] A. W. Snyder and J. D. Love, *Optical Waveguide Theory*, Chapman and Hall, London, 1983.
[17] D. Gloge. *Appl. Opt.* **10**, 2252 (1971).

an estimate of the *inter-mode group delay* for a length L of this multimode fiber is

$$\Delta T_g \approx \frac{L}{c}(n_1 - n_2) \tag{3}$$

2. There is also a contribution to group dispersion from any single mode. Group delay, inversely proportional to group velocity, is proportional to $d\beta/d\omega$.

$$T_g = \frac{L}{v_g} = L\frac{d\beta}{d\omega} \tag{4}$$

In terms of b, defined by (2), and the v of Eq. 14.9(9), group delay (4) from the wave guiding factors can be shown to be

$$T_g \approx \frac{L}{c}(n_1 - n_2)\frac{d(vb)}{dv} \tag{5}$$

Gloge[17] has calculated curves of $d(vb)/dv$ for the LP modes, as shown in Fig. 14.10b. Slopes of these curves are proportional to $dT_g/d\omega$ and hence to the dispersion of group delays, called *modal dispersion*.

3. The curves of Fig. 14.10b assume that refractive index is constant with frequency and take into account only the redistribution of energy with frequency between core and cladding, but there is also a *material dispersion* contribution resulting from the actual change of index with frequency. This is usually expressed as change in index per unit wavelength and for silica is of the order $-3 \times 10^{-5}(\text{nm})^{-1}$ at $\lambda_0 = 0.6\ \mu\text{m}$, but becomes zero and changes sign near $\lambda_0 = 1.3\ \mu\text{m}$. For small dispersions, the material dispersion and the modal dispersion due to geometrical factors add. The contribution from material dispersion is

$$\frac{dT_g}{d\omega} = \frac{L}{c}\frac{d^2(n\omega)}{d\omega^2} \tag{6}$$

For a coherent source, such as a good laser, the frequency spectrum of the modulating pulse determines the spread. A gaussian pulse of initial width τ, after propagation through a dispersive medium of length L, then spreads according to Eq. 8.16(14).

For moderately broad pulses from incoherent sources, such as light-emitting diodes (LEDs) or imperfect lasers, the bandwidth $\Delta\omega$ of the source may determine the spread in group delay since it may exceed the frequency spread required to form the pulse. Then

$$\Delta T_g \approx \Delta\omega L\frac{d^2\beta}{d\omega^2} \tag{7}$$

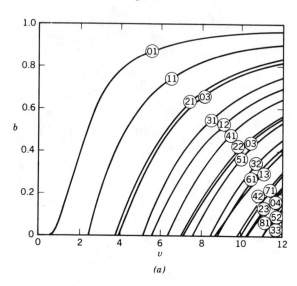

(a)

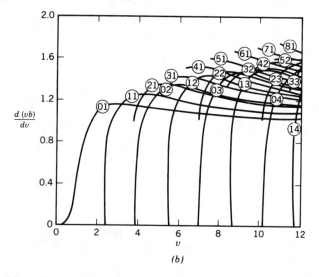

(b)

Fig. 14.10 (*a*) Normalized propagation constant versus normalized frequency for weakly guided modes from Gloge (Ref. 17). (*b*) Normalized group delay from Gloge for the modes of (*a*).

14.11   Propagation of Gaussian Beams in Graded-Index Fibers

Dielectric rods or fibers with a quadratic variation of index with radius (or nearly so) are found to be very useful for guiding optical waves. From a ray-optics point of view each section appears as a section of a lens, with a consequent confining of rays by this distributed lens as shown in Ex. 14.4b. We now wish to study the guided modes of such a rod by a field analysis and will find that the fundamental mode has a gaussian distribution over the cross section. Later we will see similar gaussian modes in space, but spreading in the absence of a guiding medium. Gaussian modes, and their higher order extensions, are consequently of major importance in work with coherent optics and we will find ways of studying their transformation by lenses, mirrors, and other optical components.

Let μ be the permeability of space but ε a function of radius. Maxwell's equations are

$$\mathbf{V} \cdot [\varepsilon(r)\mathbf{E}] = \varepsilon \mathbf{V} \cdot \mathbf{E} + \mathbf{E} \cdot \mathbf{V}\varepsilon = 0 \tag{1}$$

$$\mathbf{V} \times \mathbf{H} = j\omega\varepsilon(r)\mathbf{E} \tag{2}$$

$$\mathbf{V} \times \mathbf{E} = -j\omega\mu_0 \mathbf{H} \tag{3}$$

As usual, we take the curl of (3):

$$\mathbf{V} \times \mathbf{V} \times \mathbf{E} = -j\omega\mu_0 \mathbf{V} \times \mathbf{H} \tag{4}$$

The left side is expanded and (2) is substituted on the right:

$$-\mathbf{V}^2\mathbf{E} + \mathbf{V}(\mathbf{V} \cdot \mathbf{E}) = \omega^2\mu_0\varepsilon(r)\mathbf{E} \tag{5}$$

Now $\mathbf{V} \cdot \mathbf{E}$ is substituted from (1) and the terms rearranged to give

$$\mathbf{V}^2\mathbf{E} + \omega^2\mu_0\varepsilon(r)\mathbf{E} = -\mathbf{V}\left(\frac{\mathbf{E} \cdot \mathbf{V}\varepsilon}{\varepsilon}\right) \tag{6}$$

This is the general equation for $\mathbf{E}$ in a region with inhomogeneous ε. However, if the change of ε is small in a wavelength, as is usual, it can be shown that the term on the right is small in comparison with the second term on the left. We consequently neglect the right side:

$$\mathbf{V}^2\mathbf{E} + \omega^2\mu_0\varepsilon(r)\mathbf{E} \approx 0 \tag{7}$$

We take the variation of ε as follows:

$$\varepsilon(r) = \varepsilon_0 n^2(r) = \varepsilon_0 n^2(0)\left[1 - \Delta\left(\frac{r}{a}\right)^2\right]^2 \tag{8}$$

where $n(r)$ is the corresponding index variation and a is some reference radius. For one transverse rectangular component of **E**, equation (7), neglecting the term in Δ^2, becomes

$$\nabla^2 E + k^2(0)\left[1 - 2\Delta\left(\frac{r}{a}\right)^2\right]E = 0 \tag{9}$$

where

$$k^2(0) = \omega^2\mu_0\varepsilon(0) = \frac{\omega^2}{c^2}n^2(0) = \left(\frac{2\pi}{\lambda_0}\right)^2 n^2(0) \tag{10}$$

Let us first consider circular cylindrical coordinates with no ϕ variations. Equation (9) then becomes

$$\frac{\partial^2 E}{\partial r^2} + \frac{1}{r}\frac{\partial E}{\partial r} + \frac{\partial^2 E}{\partial z^2} + k^2(0)\left[1 - 2\Delta\left(\frac{r}{a}\right)^2\right]E = 0 \tag{11}$$

It can be shown by substitution that this is solved by

$$E(r, z) = Ae^{-(r/w)^2}e^{-j\beta z} \tag{12}$$

where

$$w = \left(\frac{2a^2}{\Delta k^2(0)}\right)^{1/4} = \left[\frac{a\lambda_0}{\pi n(0)}\right]^{1/2}\left(\frac{1}{2\Delta}\right)^{1/4} \tag{13}$$

and

$$\beta = k(0) - \frac{2}{w^2 k(0)} = k(0) - \frac{2}{a}\left(\frac{\Delta}{2}\right)^{1/2} \tag{14}$$

So that the variation of E with radius is of gaussian form with a radius w to the $1/e$ value of field dependent upon a, λ_0, $n(0)$, and Δ. This radius is usually called *beam radius* although fields do extend beyond. As a numerical example, if $\Delta = 0.01$, $a = 50\ \mu\text{m}$, $\lambda_0 = 1\ \mu\text{m}$, $n(0) = 1.5$, w is found to be 8.67 μm.

When w is large compared with wavelength, transverse variation of E is small in a wavelength, and the mode is nearly transverse electromagnetic with axial components of **E** and **H** negligible and the transverse components normal to each other and related by $[\mu_0/\varepsilon(0)]^{1/2}$. This fundamental mode is consequently designated TEM_{00}.

Higher order modes may be found by returning to (9). First expanding ∇^2 in rectangular coordinates,

$$\frac{\partial^2 E}{\partial x^2} + \frac{\partial^2 E}{\partial y^2} + \frac{\partial^2 E}{\partial z^2} + k^2(0)\left[1 - 2\Delta\left(\frac{r}{a}\right)^2\right]E = 0 \tag{15}$$

the solution may be shown to be

$$E = A_{mp} H_m\left(\frac{\sqrt{2}x}{w}\right) H_p\left(\frac{\sqrt{2}y}{w}\right) e^{-(x^2+y^2)/w^2} e^{-j\beta_{mp}z} \tag{16}$$

where $H_m(\xi)$ are Hermite polynomials of order m satisfying the differential equation[18]

$$\frac{d^2 H_m(\xi)}{d\xi^2} - 2\xi \frac{dH_m(\xi)}{d\xi} + 2mH_m(\xi) = 0 \tag{17}$$

and defined by

$$H_m(\xi) = (-1)^m e^{\xi^2} \frac{d^m}{d\xi^m} e^{-\xi^2} \tag{18}$$

In these higher order modes, w is the same as in (13) but β_{mp} is given by

$$\beta_{mp}^2 = k^2(0) - 2\frac{\sqrt{2\Delta}}{a} k(0)(m + p + 1) \tag{19}$$

Or with Δ small,

$$\beta_{mp} \approx k(0) - \frac{\sqrt{2\Delta}}{a}(m + p + 1) \tag{20}$$

Similarly if ∇^2 is expanded in circular cylindrical coordinates, (9) is

$$\frac{\partial^2 E}{\partial r^2} + \frac{1}{r}\frac{\partial E}{\partial r} + \frac{1}{r^2}\frac{\partial^2 E}{\partial \phi^2} + \frac{\partial^2 E}{\partial z^2} + k^2(0)\left[1 - 2\Delta\left(\frac{r}{a}\right)^2\right]E = 0 \tag{21}$$

The solution may be shown to be

$$E = B_{mp}\left(\frac{\sqrt{2}r}{w}\right)^m L_p^m\left(\frac{2r^2}{w^2}\right) e^{-r^2/w^2} e^{\pm jm\phi} e^{-j\beta_{mp}z} \tag{22}$$

where $L_p^m(\xi)$ are associated Laguerre polynomials[19] of order p and degree m, satisfying the differential equation

$$\xi \frac{d^2 L_p^m(\xi)}{d\xi^2} + (m + 1 - \xi)\frac{dL_p^m(\xi)}{d\xi} + (p - m)L_p^m(\xi) = 0 \tag{23}$$

and defined by

$$L_p^m(\xi) = \frac{d^m}{d\xi^m}\left[e^\xi \frac{d^p}{d\xi^p}(\xi^p e^{-\xi})\right] \tag{24}$$

[18] M. R. Spiegel. *Mathematical Handbook*, Schaum's Outline Series, McGraw-Hill, New York, 1968, p. 151.
[19] Ref. 18, p. 153.

Here also w is as in (13) but phase constant β is

$$\beta \approx k(0) - \frac{\sqrt{2\Delta}}{a}(2m + p + 1) \tag{25}$$

Since the fiber is cylindrical, it might seem that we would only be interested in the latter set, but the Hermite forms are not only simpler, but are often generated by asymmetrical excitations. Each of the sets is complete, so an arbitrary distribution can be expanded in either of the two sets.

In looking at the phase constants for the higher order modes, from (20) or (25), we find one of the great advantages of the graded-index fiber. The time delay of a pulse or other modulation envelope on the wave for a length L is given approximately by

$$T_d = \frac{L}{v_g} \tag{26}$$

where v_g is group velocity given by $d\omega/d\beta$. Thus, using (20),

$$T_d = L\frac{d\beta_{mp}}{d\omega} = L\frac{dk(0)}{d\omega} = \frac{L}{c}\left[n(0) + \frac{\omega\,dn(0)}{d\omega}\right] \tag{27}$$

and similarly for (25). Thus to this degree of approximation group delay is the same for all the modes. The more accurate expression, (19), would show some dispersion, but probably more important is the material dispersion and the discontinuity in slope of $n(r)$ when the quadratic variation stops. Nevertheless the graded-index fiber is better than the step-index fiber for multimode use in moderately high-data-rate systems.

GAUSSIAN BEAMS IN SPACE AND IN OPTICAL RESONATORS

14.12   Propagation of Gaussian Beams in a Homogeneous Medium

The modes of a graded-index fiber, studied in the preceding section, propagate with a constant pattern in the fiber if the modes are properly started. Any tendency to diffract is just countered by the distributed focusing effect of the lens-like medium. Let us imagine that they come to the end of the fiber, as in Fig. 14.12, and excite similar modes in space or other homogeneous material. It seems clear that they will spread by diffraction in the external region. Gaussian modes, and their higher order extensions, thus become fundamental forms in homogeneous regions and may be excited in a variety of ways by lasers or other coherent sources.

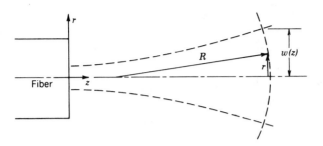

Fig. 14.12 A gaussian beam, after leaving the graded-index fiber in which it is guided, will spread by diffraction effects.

We shall return to the reasons for this after seeing some of their properties.

The main propagation variation in the homogeneous region is still expected as e^{-jkz} but there are other variations with z so that we may write

$$E(r, \phi, z) = \psi(r, \phi, z)e^{-jkz} \tag{1}$$

We make use of Eq. 14.11(9) with $\Delta = 0$ and $k(0) = k$ and assume that ψ is slowly varying in a wavelength so that ψ'' may be neglected in comparison with ψ', where primes denote derivatives with z. There results

$$\nabla_t^2 \psi - 2jk\psi' = 0 \tag{2}$$

Considering again the fundamental mode with $\partial/\partial\phi = 0$, it has been found that a useful form[20] for ψ is

$$\psi(r, z) = A \exp\left\{-j\left[P(z) + \frac{kr^2}{2q(z)}\right]\right\} \tag{3}$$

where $P(z)$ and $q(z)$ are complex. In particular,

$$\frac{1}{q(z)} = \frac{1}{R(z)} - \frac{2j}{kw^2(z)} \tag{4}$$

where $R(z)$ is a radius of curvature of the approximately spherical wave fronts, Fig. 14.12, and $w(z)$ is a measure of the gaussian beam radius at position z. Thus the imaginary part of $1/q$ gives the gaussian variation with radius and the real part gives the phase related to the curvature of wave fronts. The latter may be seen for paraxial rays with $r \ll z$ since, by Fig. 14.12,

$$kz = k\sqrt{R^2 - r^2} \approx kR\left(1 - \frac{r^2}{2R^2}\right) = k\left(R - \frac{r^2}{2R}\right) \tag{5}$$

[20] H. Kogelnik, *Applied Optics* **4**, 1562 (1965).

If (3) is substituted in (2) with ∇_t^2 in cylindrical coordinates with $\partial/\partial\phi = 0$,

$$-\left(\frac{k}{q}\right)^2 r^2 - 2j\left(\frac{k}{q}\right) + \frac{k^2 r^2 q'}{q^2} - 2kP' = 0 \tag{6}$$

Since the equation is to hold for all r, coefficients of like powers of r must separately be zero:

$$\left(\frac{k}{q}\right)^2 = \left(\frac{k}{q}\right)^2 q' \tag{7}$$

$$P' = \frac{-j}{q} \tag{8}$$

Equation (7) has a solution of the form

$$q(z) = q_0 + z \tag{9}$$

which, substituted in (8) leads to

$$P(z) = -j \ln\left(1 + \frac{z}{q_0}\right) \tag{10}$$

At $z = 0$, curvature of the wave fronts is zero (Fig. 14.12), so from (4)

$$q_0 = \frac{jkw_0^2}{2} \tag{11}$$

Substitution of (9) and (10) in (3), and the result in (1), after some manipulation with the complex quantities q and P, leads to the following expression for E:

$$E(r, z) = A \frac{w_0}{w(z)} e^{-r^2/w^2(z)} e^{-j[kz + kr^2/2R(z) - \eta(z)]} \tag{12}$$

where

$$w(z) = w_0 \left[1 + \frac{z^2}{z_0^2}\right]^{1/2} \tag{13}$$

$$R(z) = z + \frac{z_0^2}{z} \tag{14}$$

$$\eta(z) = \tan^{-1}\left(\frac{z}{z_0}\right) \tag{15}$$

$$z_0 = \frac{kw_0^2}{2} \tag{16}$$

As expected, the gaussian beam spreads out, while maintaining its gaussian form for each cross section. For large z, the angle of divergence is

$$\theta \approx \frac{w(z)}{z} \approx \frac{w_0}{z_0} = \frac{2w_0}{kw_0^2} = \frac{2}{kw_0} \tag{17}$$

This is consistent with a diffraction analysis for the distribution at the waist, $z = 0$, as seen in Eq. 12.14(18).

Higher order modes exist, as with the graded-index fiber. In rectangular coordinates, these are

$$E_{m,p}(x, y, z) = A_{mp} \frac{w_0}{w(z)} H_m\left(\frac{\sqrt{2}x}{w(z)}\right) H_p\left(\frac{\sqrt{2}y}{w(z)}\right)$$

$$\times e^{-(x^2+y^2)/w_0^2 e - j[kz + k(x^2+y^2)/2R(z) - (m+p+1)\eta(z)]} \tag{18}$$

where the Hermite polynomials H_m and H_p are as defined in Sec. 14.11, and $w(z)$, $R(z)$, and $\eta(z)$ are as in (13), (14), and (15), respectively. For circular cylindrical coordinates, it can be shown that

$$E_{mp}(r, \phi, z) = A_{mp}\left[\frac{\sqrt{2}r}{w(z)}\right] L_p^m\left(\frac{2r^2}{w^2(z)}\right) e^{-r^2/w^2(z)} e^{\pm jm\phi} e^{-j[kz + kr^2/2R(z) + (2m+p+1)\eta(z)]}$$

$$\tag{19}$$

where $L_p^l(\xi)$ is the associated Laguerre polynominal defined in Sec. 14.11 and other quantities are as defined above.

Other uses and properties of the gaussian beams will be studied in the following two sections.

14.13   Transformation of Gaussian Beams by Ray Matrix

The transformation of gaussian beams, passing through a combination of optical elements, including lenses, homogeneous regions, and dielectric discontinuities, has been shown[20] to follow from the q parameter of Eq. 14.12(4) and the A, B, C, D parameters of the ray matrix defined in Sec. 14.5. The gaussian beams are assumed to be coaxial with the optical elements, and the paraxial (small slope) approximations satisfied. Thus, within these approximations, the value of q_2 at the output of a region with defined parameters A, B, C, D in terms of input q_1 is

$$q_2 = \frac{Aq_1 + B}{Cq_1 + D} \tag{1}$$

where

$$\frac{1}{q_i} = \frac{1}{R_i} - \frac{2j}{kw_i^2}, \quad i = 1, 2 \tag{2}$$

R_i is the radius of curvature of the wave front and w_i is the beam radius to e^{-1} value of field, as in Sec. 14.11. We will first show this for special elements and then argue the more general case.

Homogeneous Region   For a homogeneous region of length z, from Eq. 14.5(2), $A = 1, B = z, C = 0$, and $D = 1$ so (1) gives

$$q_2 = q_1 + z \tag{3}$$

which is consistent with the solution found in Eq. 14.12(9). In particular, if z is measured from the waist so that $q_1 = q_0 = jkw_0^2/2$, values of R and w at plane z are

$$\frac{1}{R(z)} - \frac{2j}{kw^2(z)} = \frac{1}{q_2} = \frac{1}{z + jkw_0^2/2} \tag{4}$$

Rationalization of this gives

$$R(z) = z + \frac{k^2 w_0^4}{4z} \tag{5}$$

$$w^2(z) = w_0^2 \left[1 + \frac{4z^2}{k^2 w_0^4} \right] \tag{6}$$

which are the forms found in Eqs. 14.12(14) and 14.12(13).

Thin Lens   A thin lens would be expected to leave beam radius w unchanged, but modify radius of curvature R by $-1/f$ (Fig. 14.13a). Thus the transformation expected is

$$\frac{1}{q_2} = \frac{1}{q_1} - \frac{1}{f} \tag{7}$$

Since $A = 1$, $B = 0$, $C = -1/f$, and $D = 1$ from Eq. 14.5(3), (7) is also consistent with (1).

Combination of Elements   If one element follows another, as a lens following a homogeneous region (Fig. 14.13b), then one can extend the form (1) to following elements, so that

$$q_3 = \frac{A_2 q_2 + B_2}{C_2 q_2 + D_2} \tag{8}$$

But this bilinear form is consistent with matrix multiplication so that we can define total A_t, B_t, C_t, D_t parameters as in Eq. 14.5(6)

$$\begin{bmatrix} A_t & B_t \\ C_t & D_t \end{bmatrix} = \begin{bmatrix} A_2 & B_2 \\ C_2 & D_2 \end{bmatrix} \begin{bmatrix} A_1 & B_1 \\ C_1 & D_1 \end{bmatrix} \tag{9}$$

and then

$$q_3 = \frac{A_t q_1 + B_t}{C_t q_1 + D_t} \tag{10}$$

In addition to the lens and homogeneous region, the transformation can be shown to apply to a plane dielectric discontinuity with the gaussian beam at normal incidence (Prob. 14.13a) and, of course, to a spherical mirror since the

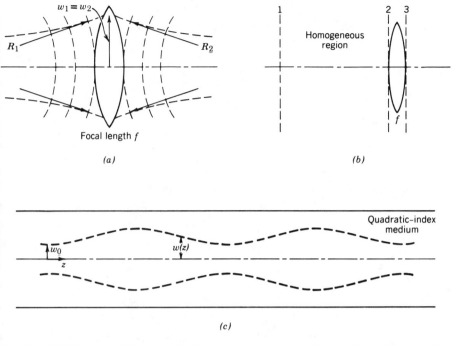

Fig. 14.13 (a) Thin lens of focal length f leaves gaussian beam radius unchanged but modifies radius of curvature by Eq. 14.13(7). (b) Combination of regions transforms gaussian beam according to overall ray matrix for the regions. (c) Gaussian beam in a graded-index medium has variable radius $w(z)$ if not started with zero slope at the equilibrium radius.

relationship of this to a lens has already been shown. Since most useful optical systems are composed of combinations of these basic elements, the $ABCD$ law for gaussian beams is very general and most useful.

_____ **Example 14.13** _____

Rod with Quadratic Index Variation

For the rod with quadratic index variation, which has been shown to act as a continuous lens to rays and which can support gaussian modes of constant radius if excited properly, we may now utilize the transformations of this section to see what happens if the gaussian beam is not properly started. We use (1) with $ABCD$ values from Eq. 14.5(5). Let $q_1 = q_0$ and $q_2 = q(z)$.

$$\frac{1}{q(z)} = \frac{C + D/q_0}{A + B/q_0} = \frac{-\kappa \sin \kappa z + (\cos \kappa z)/q_0}{\cos \kappa z + (\sin \kappa z)/\kappa q_0} \tag{11}$$

This equation tells how $R(z)$ and $w(z)$ vary with distance for a general incident R_0 and w_0. To interpret, let us assume plane wave fronts, $R_0 = \infty$ at the input. Then (11) gives

$$\frac{1}{R(z)} - \frac{2j}{kw^2(z)} = \frac{-\kappa \sin \kappa z - (2j/kw_0^2) \cos \kappa z}{\cos \kappa z - (2j/kw_0^2\kappa) \sin \kappa z} \tag{12}$$

from which beam radius is

$$w^2(z) = w_0^2\left[\cos^2 \kappa z + \left(\frac{2}{kw_0^2\kappa}\right)^2 \sin^2 \kappa z\right] \tag{13}$$

or using trigonometric identities,

$$w^2(z) = w_0^2\left[\left(1 + \frac{4}{k^2w_0^2\kappa^2}\right) + \left(1 - \frac{4}{k^2w_0^2\kappa^2}\right)\cos 2\kappa z\right] \tag{14}$$

so that beam radius is constant at w_0 if

$$\kappa^2 = \frac{4}{k^2w_0^2} \tag{15}$$

which is consistent with the condition Eq. 14.11(13), but if this condition is not satisfied, the beam varies periodically with z as sketched in Fig. 14.13c. A similar variation will occur if the beam is introduced with nonzero slope.

14.14   Gaussian Modes in Optical Resonators

It was noted in Sec. 14.6 that a typical laser resonator consists of the region between two coaxial spherical mirrors, as pictured in Fig. 14.14a. The active laser material may fill all or part of the internal region. Such arrangements (including the limiting case of plane mirrors) have long been used as interferometers. Schawlow and Townes[21] suggested that these would also be useful as the resonant structures for laser action. In contrast to the closed cavity resonators studied in Chapter 10, radiation from the open sides will damp out all but the modes with primarily axial propagation. It is not entirely obvious that any low-loss modes will remain, but it seems that if the diffraction pattern from mirror M_1 is largely contained within the diameter of mirror M_2 and vice versa, diffraction losses should

---

[21] A. L. Schawlow and C. H. Townes, *Phys. Rev.* **112**, 1940 (1958).

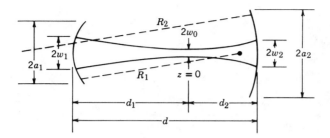

Fig. 14.14a Typical spherical mirror resonator with mirrors matching the curved wave fronts of the gaussian beam.

be small. From diffraction theory (Sec. 12.12) the condition requires that the Fresnel number N be larger than unity,

$$N = \frac{a_1 a_2}{\lambda d} > 1 \tag{1}$$

where a_1 and a_2 are radii of the mirrors and d is the spacing.

The above general conclusions were verified through an important computer analysis of Fox and Li.[22] They assumed an initial field distribution over one of the mirrors, and from diffraction theory calculated the resulting pattern at the second mirror. Using this, diffraction theory was used to give the new distribution over mirror 1 and so on through many iterations. There was eventual convergence to stable field patterns showing different modal forms, and with low diffraction loss when condition (1) is satisfied.

Once it is known that stable, low-loss modes exist in the open structure, it is reasonable to look for approximate solutions of Maxwell's equations to represent these. Through work of Pierce,[23] Boyd and Gordon,[24] Goubau and Schwering,[25] and Kogelnik,[26] there developed the understanding of the guassian, Hermite-gaussian and Lagurerre–gaussian modes discussed in the preceding sections. The remainder of this discussion will consequently apply these to the resonator of Fig. 14.14a. See Fig. 14.14b for some modal patterns.

In considering the gaussian modes within spherical-mirror resonators, let us consider first the fundamental gaussian mode. It has been shown in Sec. 14.12 that these have nearly spherical wave fronts, so the approach is to match the

[22] A. G. Fox and T. Li, *Bell Syst. Tech. J.* **40**, 453 (1961).

[23] J. R. Pierce, *Proc. Nat. Acad. Sci.* **47**, 1808 (1961).

[24] G. D. Boyd and J. P. Gordon, *Bell Syst. Tech. J.* **40**, 489 (1961).

[25] G. Goubau and F. Schwering, *IRE Trans. on Antennas and Propagation* **AP-9**, 248 (1961).

[26] H. Kogelnik, *Applied Optics* **5**, 1550 (1966).

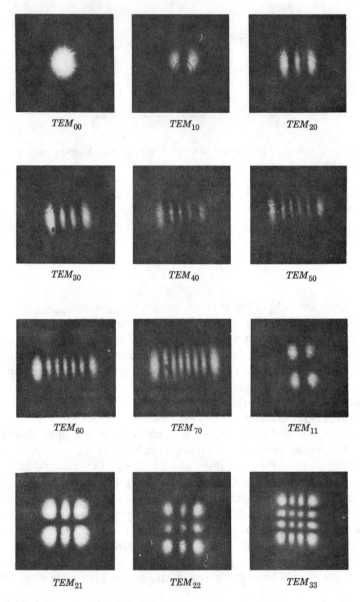

Fig. 14.14b Modal patterns in optical resonators. From H. Kogelnik and W. W. Rigrod, "Visual Display of Isolated Optical Modes," *Proc. IRE* **50**, 220 (1962).

curvature of these to the mirrors. Using the forms for radius of curvature given by Eq. 14.12(14), we may then write[27]

$$R_1 = d_1 + \frac{z_0^2}{d_1} \tag{2}$$

$$R_2 = d_2 + \frac{z_0^2}{d_2} \tag{3}$$

where

$$z_0 = \frac{k w_0^2}{2} = \frac{\pi w_0^2 n}{\lambda_0} \tag{4}$$

and n is the refractive index. One kind of design problem is that in which wavelength and position of the beam waist (position of minimum radius) are specified, and w_0 selected to fill as much of the active region as possible. Then mirror curvatures are immediately calculable from (2)–(4). As example, for a CO_2 laser with $\lambda_0 = 10.6 \ \mu m$, $n = 1$, $d_1 = d_2 = 0.2$ m, suppose w_0 is selected as 3 mm. Then $R_1 = R_2 = 35.8$ m.

A second type of design problem for laser resonators is one in which R_1 and R_2 are given together with wavelength and spacing d and it is desired to find position of the waist, its radius, and the beam radii at the two mirrors. Equations (2) and (3) are first solved for d_1 and d_2:

$$d_1 = \frac{R_1}{2} \pm \sqrt{\left(\frac{R_1}{2}\right)^2 - z_0^2} \tag{5}$$

$$d_2 = \frac{R_2}{2} \pm \sqrt{\left(\frac{R_2}{2}\right)^2 - z_0^2} \tag{6}$$

and

$$d_1 + d_2 = d \tag{7}$$

Substitution of (5) and (6) in (7), after some algebra, gives

$$z_0^2 = \frac{d(R_1 - d)(R_2 - d)(R_1 + R_2 - d)}{(R_1 + R_2 - 2d)^2} \tag{8}$$

With z_0 known, minimum spot size is determined from (4) and spot size at the mirrors, by using Eq. 14.12(13), to give

$$w_i = w_0 \left[1 + \left(\frac{d_i}{z_0}\right)^2\right]^{1/2}, \qquad i = 1, 2 \tag{9}$$

[27] It is more logical from a geometric point of view to consider radii of curvature negative when center of curvature is to the right of the mirror, but from a practical point of view the sign for a given mirror should not depend upon its orientation, so we make the practical choice and consider radii of curvature positive when the reflecting surface is concave.

As an example, if $d = 1$ m, $R_1 = 2$, $R_2 = 4$, $z_0^2 = 0.938$ from (8). Then $d_1 = 1.5$ and $d_2 = 0.5$ from (2) and (3). If $\lambda_0 = 1$ μm, minimum spot size w_0 is 0.554 mm from (4) and $w_1 = 0.70$ mm, $w_2 = 0.57$ mm from (9). Several special cases are worth considering individually.

Plane Mirrors If $R_1 = R_2 = \infty$, (8) shows that z_0 is infinite, and from (4) and (9), $w_0 = w_1 = \infty$. From a practical point of view, the mode spreads out until limited by the edges of the mirrors. Although there are then appreciable diffraction effects, the numerical analysis of Fox and Li demonstrated that under certain conditions such systems are still useful. Plane reflectors are most often met in solid-state lasers such as ruby or neodymium–YAG. for which the dielectric discontinuity at $r = a$ provides additional confinement.

The Concentric Resonator Consider a symmetrical resonator with $R_1 = R_2 = d/2$ so that centers of curvature coincide in the midplane. From (8), $z_0 = 0$, which would make minimum spot size zero from (4) and spot size at the mirrors infinite from (9). This unrealistic situation is also limited by diffraction of the finite mirrors, yielding a finite minimum spot size at the center in a practical case. Like the plane example, diffraction losses are high, but usable in some cases.

The Symmetric Confocal Resonator If $R_1 = R_2 = d$ so that the foci of the two mirrors coincide at the midpoint of the resonator, expression (8) becomes indeterminant. We consequently look at the gaussian beam transformation of the preceding section. The ray matrix elements can be obtained from the equivalent periodic lens system and Eq. 14.5(8) with $d_1 = d_2 = d = 2f_1 = 2f_2$. Then $A = -1$, $B = 0$, $C = 0$, and $D = -1$ for a round trip. Thus a beam introduced with any beam radius and radius of curvature reproduces itself after a round trip by the transformation of Eq. 14.13(1).

$$q_1 = \frac{Aq_2 + B}{Cq_2 + D} = q_2 \tag{10}$$

This interesting case is highly degenerate. However, if we select the symmetrical mode with waist at the midplane, then the proper solution of (8) gives $z_0 = d/2$ and

$$w_0 = \left(\frac{d\lambda_0}{2n}\right)^{1/2} \tag{11}$$

Since z_0 is half the cavity length in this case, it is often interpreted as the confocal parameter. Spot sizes at the mirrors are then $\sqrt{2}w_0$ for this symmetrical mode.

It is to be noticed that the above three special cases are on the edge of the stability diagram derived from ray concepts in Sec. 14.6. The applicability of the diagram for gaussian modes, and the consequences of operation outside the "stable" range, will be discussed in the following section.

14.15   Stability and Resonant Frequencies of Optical-Resonator Modes

From the ray optics point of view, it was found in Sec. 14.6 that rays are confined within some maximum radius in a periodic system if

$$\left| \frac{A + D}{2} \right| \leq 1 \tag{1}$$

where A and D are ray matrix elements for the system. For the resonator with spacing d and coaxial mirrors having curvature R_1 and R_2, this led to

$$0 \leq \left(1 - \frac{d}{R_1}\right)\left(1 - \frac{d}{R_2}\right) \leq 1 \tag{2}$$

Resonators satisfying this condition are said to be *stable*, and the diagram of Fig. 14.6c showing regions for which the condition is satisfied is called a *stability diagram*. To show that the condition applies to gaussian modes of the type studied in the preceding section, let A, B, C, D be the ray matrix parameters for a round trip in the resonator, and q the gaussian beam parameter defined by Eq. 14.13(2). Then q_2 at the end of a round trip is related to q_1 at the beginning by Eq. 14.13(1). But if the mode remains of the same form, $q_1 = q_2$ and

$$\frac{1}{q_1} = \frac{Cq_1 + D}{Aq_1 + B} \tag{3}$$

This leads to a quadratic equation in $1/q$ and has the solution

$$\frac{1}{q_1} = \left(\frac{D - A}{2B}\right) \pm \left[\left(\frac{D - A}{2B}\right)^2 + \frac{C}{B}\right]^{1/2} \tag{4}$$

As noted in Sec. 14.5, the ray parameters are real for the elements we have considered, and also $AD - BC = 1$, so (4) reduces to

$$\frac{1}{q_1} = \frac{1}{R_1} - \frac{2j}{kw_1^2} = \left(\frac{D - A}{2B}\right) \pm \left[\left(\frac{D + A}{2}\right)^2 - 1\right]^{1/2} \tag{5}$$

Thus for a real value of beam radius w_1, $(A + D)/2$ must not be greater than unity so that (1) applies in general, and (2) for the specific system of two mirrors separated by a homogeneous region of length d.

As has been noted, stability of an optical resonator means, in a ray-optics point of view, that rays are confined within some maximum radius; from the wave point of view, it means that the field energy is essentially all within this maximum radius. Unstable resonators have the opposite property, but Siegman has shown that they may nevertheless be useful.[28] The rays, or diffraction field, escaping outside the mirror may be utilized as the output, and this has proved to be the preferred means of output coupling for many high-power lasers.

[28] A. E. Siegman, *Proc. IEEE* **53**, 277 (1965); also see summary in A. Yariv, *Quantum Electronics*, Ref. 1, (Sec. 7.5).

The resonant frequencies are determined by the requirement that the phase shift after a round trip is a multiple of 2π, or of π for one way. Thus from Eq. 14.12(18),

$$kz_2 - kz_1 + (m + p + 1)[\eta(z_2) - \eta(z_1)] = l\pi \tag{6}$$

where m, p, and l are integers. Using Eq. 14.12(15) and the sign conventions of the preceding section, $z_2 = d_2$ and $z_1 = -d_1$,

$$k(d_1 + d_2) + (m + p + 1)\left[\tan^{-1}\left(\frac{d_2}{z_0}\right) + \tan^{-1}\left(\frac{d_1}{z_0}\right)\right] = l\pi \tag{7}$$

Longitudinal modes are defined by the integer l. Thus, the frequency spacing between longitudal modes is determined by maintaining m and p constant and changing l by unity,

$$\frac{(\omega_{l+1} - \omega_l)n}{c}(d_1 + d_2) = \frac{2\pi \Delta f_l nd}{c} = \pi \tag{8}$$

or

$$\Delta f_l = \frac{c}{2nd} \tag{9}$$

As example, if $d = 0.10$ m and $n = 1.5$, $\Delta f_l \approx 1$ GHz.

Transverse modes are defined by the integers m and p. Frequency spacing between these is determined by maintaining l constant and changing either m or p by unity,

$$\frac{(\omega_{m+1} - \omega_m)n(d_1 + d_2)}{c} + \tan^{-1}\left(\frac{d_2}{z_0}\right) + \tan^{-1}\left(\frac{d_1}{z_0}\right) = 0 \tag{10}$$

or

$$\Delta f_t = \frac{(\omega_{m+1} - \omega_m)}{2\pi} = -\frac{c}{2\pi nd}\left[\tan^{-1}\left(\frac{d_2}{z_0}\right) + \tan^{-1}\left(\frac{d_1}{z_0}\right)\right] \tag{11}$$

Note that as the mirrors approach planes, $z_0 \to \infty$ as shown in the preceding section, and transverse modes become very close together. For a concentric resonator, $z_0 \to 0$ and $\Delta f_t \to c/2nd$. That is, transverse mode spacing is the same as longitudinal mode spacing for concentric resonators. For a symmetrical confocal resonator, $d_1 = d_2 = z_0$, $\Delta f_t = c/4nd$ showing that transverse mode spacing is half the longitudinal mode spacing in this case.

BASIS FOR OPTICAL INFORMATION PROCESSING

14.16   Fourier Transforming Properties of Lenses

Combinations of lenses have been found useful for information processing by optical means because of the Fourier transforming properties of these systems[29,30,31] which follow from diffraction theory, Sec. 12.13. Here we will use the Fresnel form in the paraxial approximation, Eq. 12.13(10),

$$\psi(x, y, z) = \frac{je^{-jkL}}{\lambda L} \int_{-\infty}^{\infty} \int_{-\infty}^{\infty} \psi(x', y', z') \exp\left\{ -\frac{jk}{2L} [(x - x')^2 + (y - y')^2] \right\} dx'\, dy'$$

(1)

where $\psi(x', y', z')$ is the given distribution of some scalar field component (such as E_x) in the plane z' and $\psi(x, y, z)$ is the resulting distribution of that component in a plane distance L away, where $L = |z - z'|$, Fig. 14.16a.

First note that for L large enough, terms kx'^2/L and ky'^2/L in (1) are negligible,

$$\psi(x, y, z) = \frac{je^{-jk[L + (x^2 + y^2)/2L]}}{\lambda L} \int_{-\infty}^{\infty} \int_{-\infty}^{\infty} \psi(x', y', z') \exp\left[\frac{jk}{L} (xx' + yy') \right] dx'\, dy'$$

(2)

which is the Fraunhofer diffraction studied earlier. The integral in the above is a two-dimensional Fourier transform (Sec. 7.12). The phase factor in front is often unimportant in optical problems, but note that it can be maintained constant if ψ is observed on a spherical surface $L + (x^2 + y^2)/L = $ constant. Thus with these understandings of the phase term, the diffracted distribution in the Fraunhofer approximation can be considered the Fourier transform of the given distribution.

The next point to be made is that the restriction to large distances can be eliminated if a lens is placed just beyond the input plane to correct for the phase factors involving x'^2 and y'^2. An ideal thin lens does not change amplitude, but modifies phase quadratically with distance from the axis, as in Eq. 14.2(19). Thus referring to Fig. 14.16b,

$$\psi_2(x', y', 0+) = \psi_1(x', y', 0-)e^{-(jk/2f)[a^2 - x'^2 - y'^2]}$$

(3)

where a is the outer radius of the lens and f is focal distance. We assume that a is large enough to intercept all of the important optical field so that we can continue to integrate over infinite limits. If (3) is then substituted in (1), the phase terms in x'^2

[9] J. W. Goodman, *Introduction to Fourier Optics*, McGraw-Hill, New York, 1968.
[10] S. H. Lee, Ed., *Optical Information Processing—Fundamentals*, Springer-Verlag, Berlin, 1981.
[11] D. Marcuse, Ref. 8.

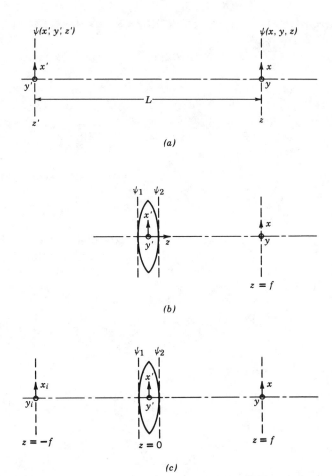

Fig. 14.16 (a) Diffraction of a scalar field component between planes z' and z. (b) Diffraction with a lens introduced to correct spherical wave fronts. (c) Diffraction from input focal plane of lens to output focal plane, yielding Fourier-transforming properties.

and y'^2 are canceled, provided $L = f$. Thus at the focal plane of the lens, within the Fresnel approximation,

$$\psi(x, y, f) = \frac{je^{-jk[f + (x^2 + y^2 + a^2)/2f]}}{\lambda f}$$

$$\times \int_{-\infty}^{\infty} \int_{-\infty}^{\infty} \psi_1(x', y', 0-) \exp\left[\frac{jk}{f}(xx' + yy')\right] dx' \, dy' \qquad (4)$$

So Fourier-transforming properties are again seen, with the multiplying phase factor again a function of x and y. As above, this phase factor can be made constant by observing ψ on a nearly spherical surface passing through the focus.

Finally, the variable phase term of (4) can be eliminated if one expresses the field $\psi_1(x', y', 0)$ just before the lens as a transform of its value at the focal plane before the lens at $z = -f$ (see Fig. 14.16c). Transformation from the input at $z = -f$ to the plane just in front of the lens is by the Fresnel diffraction integral (1),

$$\psi_1(x', y', 0-) = \frac{je^{-jkf}}{\lambda f} \int_{-\infty}^{\infty} \int_{-\infty}^{\infty} \psi_i(x_i, y_i, -f)$$

$$\times \exp\left\{-\frac{jk}{2f}[(x' - x_i)^2 + (y' - y_i)^2]\right\} dx_i\, dy_i \qquad (5)$$

When (5) is substituted in (4) there results

$$\psi_0(x, y, f) = \frac{je^{-jk(2f + a^2/2f)}}{\lambda f} \int_{-\infty}^{\infty} \int_{-\infty}^{\infty} \psi_i(x_i, y_i, -f)$$

$$\times \exp\left[\frac{jk}{f}(xx_i + yy_i)\right] dx_i\, dy_i \qquad (6)$$

The above made use of the integral

$$\int_{-\infty}^{\infty} e^{j(\alpha x^2 + \beta x)}\, dx = \sqrt{\frac{j\pi}{\alpha}}\, e^{-j(\beta^2/4\alpha)} \qquad (7)$$

Here the phase factor before the integral represents a constant phase delay so that the distribution in the output focal plane is directly proportional to the Fourier transform of field distribution in the input focal plane. This may be seen more clearly by defining

$$p = x_i\sqrt{\frac{k}{f}}, \qquad q = y_i\sqrt{\frac{k}{f}}, \qquad u = x\sqrt{\frac{k}{f}}, \qquad v = y\sqrt{\frac{k}{f}} \qquad (8)$$

$$f(u, v) = j\psi_0(x, y, f)e^{jk(2f + a^2/f)}, \qquad g(p, q) = \psi_i(x_i, y_i, -f)$$

in which case (6) is in the exact form of the two-dimensional Fourier integral,

$$f(u, v) = \frac{1}{2\pi} \int_{-\infty}^{\infty} \int_{-\infty}^{\infty} g(p, q)e^{j(up + vq)}\, dp\, dq \qquad (9)$$

This important property has made possible Fourier transforming, convolution, correlation and filtering of spatial functions by lenses and combinations of lenses.[29,30]

The Fourier transforming properties noted above also give us some physical insight as to why gaussian modal forms are found for spherical mirror resonators or periodic lens systems in preceding sections. A gaussian function is known to Fourier transform to a gaussian, so this form would be expected to persist as a mode in the periodic lens system. The resonator with two spherical mirrors has already been shown to be equivalent to the periodic lens system. Hermite–gaussian and Laguerre–gaussian forms likewise transform into themselves, so these would be expected to be the higher order mode forms. In fact an alternate way of deriving such modes is by setting up the Fresnel diffraction integrals and requiring that the function repeat in a period. Solution of the resulting integral equation leads to modal solutions of the form already found.[32]

PROBLEMS

14.2a Obtain from Eq. 4.2(4) the approximate spread of focal length Δf from $r = 0$ to r_{max} when $(r_{max}/R) \ll 1$. For a spherical mirror of radius of curvature 1 m, used with a laser of wavelength $\lambda_0 = 1$ μm, give the maximum radius of rays from the axis if spread in focal length is not to be more than a wavelength. What is the f number of the mirror if its diameter is limited to meet this limitation on spread of focal length? (f number = focal length/diameter of aperture.) This standard use of the symbol f is not to be confused with equally standard use as focal length.

14.2b Carry out the derivation corresponding to that of Ex. 14.2c for a diverging thin lens with plane surface at $z = 0$ and a parabolic surface $z = d + r^2/4g$, $0 < r < a$, (i) by ray analysis; (ii) by phase considerations.

14.2c Derive the shapes for converging and diverging thin cylindrical lenses for which the plane wave is focused to a line in the focal plane.

14.2d Carry out the derivation corresponding to that of Ex. 14.2c except that the lens is symmetrical, with a paraboloidal surface on the left side also. Obtain an expression for focal length.

14.4e Consider the imaging for a spherical mirror by ray reflection. That is, for an object point on the axis distance d_1 from a mirror with radius of curvature R, find the image distance d_2 from the mirror at which a ray reflected at radius r crosses the axis. Under what approximations does Eq. 14.2(23) apply?

14.3a For a uniform plane wave, polarized with e_z only, propagating at angle α from the x axis with $\cos \gamma = 0$, give all quantities $\mathbf{e}$, $\mathbf{h}$, S, u_c, and u_h used in the general formulation. Material has refractive index n.

14.3b We will find later that many useful beams have gaussian forms. Assume one with form in the transverse plane, $\mathbf{e} = \hat{x} \exp[-(x^2 + y^2)/w_0^2]$ propagating substantially in the z direction, $S \approx nz$. Compare the neglected term on the right side of Eq. 14.3(5) with the term $\nabla S \times \mathbf{e}$ in magnitude and direction and state under what conditions it is negligible.

14.3c Utilizing Eq. 14.3(12) for the eikonal of a plane wave, show that $\mathbf{E}$ and $\mathbf{H}$ are related as in an arbitrarily propagating plane wave in a homogeneous medium.

14.4a Find the required form of refractive index variation to maintain a ray in a circular path of constant radius R.

14.4b Very low absorptions can be measured by using the thermal lens effect resulting from passing a beam through a cell containing the sample, as in Ex. 14.4a. It has been shown [J. R.

---

[32] See, for example, A. E. Siegman, *An Introduction to Lasers and Masers*, McGraw-Hill, New York, 1971, Chapter 8.

Whinnery, *Accounts of Chemical Research* **7**, 225 (1974)] that the heating effect results, in steady state in a quadratic index variation near the axis as in Eq. 14.4(11), with

$$\Delta = - \frac{\alpha P(dn/dT)}{4\pi k n_0}$$

where α is absorption coefficient, P power in the laser beam, dn/dT the variation of refractive index with temperature, k the thermal conductivity, and n_0 the refractive index on the axis. Find the expected focal length of a cell 1 cm long filled with carbon disulfide ($n_0 = 1.63$, $dn/dT = 7.9 \times 10^{-4}\,\mathrm{K}^{-1}$, $k = 6.82 \times 10^{-4}\,\mathrm{J/(cm\text{-}s\text{-}K)}$, $\alpha = 6 \times 10^{-4}\,\mathrm{cm}^{-1}$) for a laser beam with 0.5 W power and a beam radius $a = 0.5$ mm.

14.4c Derive Eq. 14.4(4) from Eq. 14.4(10).

14.4d A typical graded-index fiber used in an optical communication system with moderate data rates has $n_0 = 1.5$, $\Delta = 0.005$, and $a = 25\ \mu m$. For a ray crossing the axis at an angle r_0' find the distance at which it returns to the axis. What is the maximum angle of crossing to maintain r_{max} less than a?

14.5a Derive the ray matrix for a ray of small slope passing from medium of refractive index n_1 to one with refractive index n_2 if the surface separating the media is a sphere with center on the axis in material 1.

14.5b Use the result of Prob. 14.5a, with that of a plane dielectric interface, Eq. 14.5(4), to obtain the ray matrix for a diverging lens as shown in Fig. P14.5b. Assume $n_2 > n_1$ and $R \gg d$. Interpret the result in terms of the lens formula (3).

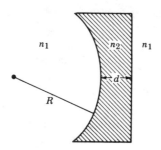

n_1 n_2 n_1

d

R

Fig. P14.5b Diverging lens.

14.5c A ray passes from a medium of index n_1 into a slab of thickness d and index n_2, then exits into n_1. Compare radius and slope at the output from the ray matrix and an exact Snell's law calculation if $n_1 = 1$, $n_2 = 2$, $d = 1$ cm, and angle of input ray is (i) 20 degrees; (ii) 40 degrees from the perpendicular to the slab.

14.5d Slabs of dielectric with index n_1 and thickness d_1 alternate with slabs having index n_2 and thickness d_2. Using the paraxial (small-angle) approximation, find the ray matrix for one period of this combination, then two periods and induce the result for N periods. Interpret the result.

14.6a An alternate form of the solution in Eq. 14.6(7) to the difference equation for the periodic lens system is

$$r_m = C_1 \cos m\theta + C_2 \sin m\theta$$

Relate C_1 and C_2 to the values of initial radius and slope of the rays. The ray matrix for one period is assumed known.

14.6b A periodic lens system has $d_1 = d_2$ and $f_1 = f_2 = d_1/5$. Check stability and sketch the ray paths through a few lenses starting with zero slope and $r = 0.1 f_1$ at the first lens.

14.6c Repeat Prob. 14.6b for $d_1 = d_2, f_1 = f_2 = d_1$.

14.6d Repeat Prob. 14.6b for the case of alternate converging and diverging lenses, $d_1 = d_2$, $f_1 = -f_2 = d_1$.

14.7a In the example of the glass film guide given in Sec. 14.7, thickness d is increased to 2 μm. Use Fig. 14.7b to find n_{eff} for all of the guided TE and TM modes.

14.7b In a certain semiconductor laser, the active region consists of a layer of GaAs 0.3 μm thick, with $n_2 = 3.35$. This is surrounded above and below with GaAlAs with $n_1 = n_3 = 3.23$. Use Fig. 14.7b to find the modes that can be guided by this active layer, and their values of n_{eff}. Wavelength λ_0 is 0.85 μm.

14.7c* To illustrate the graphical method of solution for Eq. 14.7(5), consider the symmetric case with $n_1 = n_3$ so that $q = p$. Show that (5) is then satisfied either by $pd = hd \tan(hd/2)$ or $pd = -hd \cot(hd/2)$. Sketch some curves of pd versus hd from these equations. Then show from (4) that the loci for constant v [defined by (6)] in the pd versus hd plane are circles and sketch these for $v = 2, 5, 8$. Solutions (modes) are given by points of intersection between the circles and the first sets of curves plotted. (Note that p must be positive for guided modes.) Check the modes predicted for the three values of v from Fig. 14.7b.

14.7d* Find the approximation to Eq. 14.7(5) when $n_3 - n_1 \gg n_2 - n_3$, as is the case for many guides with dielectric substrates but air above. Show that a graphical solution similar to that of Prob. 14.7c may be utilized for this case also.

14.9 For a step-index fiber of silica ($n = 1.500$) cladding on a glass ($n = 1.505$) core of 6.0 μm diameter, find the cutoff wavelengths of the two lowest axially symmetric TE modes.

14.10a A fiber with $a = 10$ μm has $n_1 = 1.51$ and $n_2 = 1.50$. Use Fig. 14.10a to estimate the number of modes that may propagate at $\lambda_0 = 1.3$ μm. What radius would be required for only one propagating mode?

14.10b A multimode fiber has $n_1 = 1.51$ and $n_2 = 1.50$. About what maximum data rate could be used with this fiber over a distance of 7 km?

14.10c A single-mode fiber with refractive indices as in Prob. 14.10b is designed with $v = 2$. Use Fig. 14.10b to estimate $dT_g/d\omega$ (neglecting material dispersion) for $\lambda_0 = 1.3$ μm. About what maximum data rate could be used with this dispersion over a distance of 7 km if the source is (i) a light-emitting diode with a spectral bandwidth $\Delta\lambda_0 = 10$ nm; (ii) a laser with $\Delta\lambda_0 = 0.2$ nm?

14.10d For the single-mode fiber of Prob. 14.10c, the laser source may be considered coherent. Over what distance will a 1-ns gaussian pulse just double in width?

14.10e For a GaAlAs laser source at $\lambda_0 = 0.85$ μm with a single-mode fiber, material dispersion is likely to be dominant. If $d^2n/d\lambda^2 \approx 3.2 \times 10^{10}$ m^{-2} at this wavelength, what approximate data rate is usable over a length of 20 km?

14.11a For a fiber with quadratic index variation having $n(0) = 1.5, a = 10$ μm, and $\lambda_0 = 1$ μm, what Δ is required for a beam radius $w = 3$ μm? Find the percentage difference of phase velocity from that of a plane wave in material with index $n(0)$.

14.11b* For the numerical value given in Prob. 14.11a, estimate the term on the right of Eq. 14.11(6) in comparison with the second term on the left.

14.11c If one uses the expression 14.11(19) rather than the approximation (20), group velocity does depend upon mode order. Find group velocity from (19) and the intermode dispersion

ased upon this result. (In a practical case, the boundary effect at $r = a$, and the departure
rom the ideal profile may be more important, but this is at least one component of dispersion.)

4.11d Using Eq. 14.11(17) show that (16) does satisfy (15) with the conditions on w and β_{mp}
s given.

4.11e* Using Eq. 14.11(23) show that (22) satisfies (21) with conditions on w and β_{mp} as
iven.

4.12a Verify the form in Eq. 14.12(12).

4.12b A typical helium–neon laser with $\lambda_0 = 633$ nm has a beam radius of about 0.5 mm.
'aking this as w_0, find beam radius at the other side of a bay 10 km away. Similarly find beam
adius on the moon of a Nd–YAG laser ($\lambda_0 = 1.06 \ \mu m$) 3.84×10^3 km from the earth where
. starts with $w_0 = 5$ mm.

4.12c For the examples of Prob. 14.12b, there is an optimum w_0 to produce minimum beam
adius at the receiver a given distance away. Find the optimum w_0, and the corresponding
(z) at the targets, for the two examples of that problem.

4.12d** Show that Eq. 14.12(18) is a solution of (2) in rectangular coordinates.

4.12e** Show that Eq. 14.12(19) is a solution of (2) in circular cylindrical coordinates.

4.13a A gaussian beam of beam radius w_1 and radius of curvature R_1 passes from dielectric
/ith index n_1 to one with index n_2, the plane interface being normal to the beam axis. Find
aussian beam properties in medium 2.

4.13b One practical problem is that of focusing the output of a laser, assumed to be of funda-
nental gaussian beam form, onto a fiber distance L away. If beam radius of the laser (assumed
waist) is w_1 and that at the fiber (also a waist) is w_2, find position d from the laser and focal
ength f of a thin lens for the desired focusing. (Hint: work forward from the laser and backward
:om the fiber until gaussian beams intersect.)

4.13c This is a variation of Prob. 14.13b in which there is a specific lens with given focal length
', but its placing and the laser-fiber spacing L are variable. Find d and L for the given w_1,
$_2$, and f in this case.

4.13d In the rod with quadratic index variation described in Sec. 14.11 [$n(0) = 1.5$, $\Delta = 0.01$,
$_0 = 1 \ \mu m$, $a = 50 \ \mu m$], a gaussian beam is introduced with zero slope but $w(0) = 12 \ \mu m$.
)escribe the beam propagation for $z > 0$.

4.13e This is similar to Prob. 14.13d except that the gaussian beam is introduced into the
:raded-index fiber at the equilibrium radius but with slope $dw/dz = 0.2$.

4.14a Derive Eq. 14.14(8) from the equations (5), (6), and (7).

4.14b For a helium–neon laser, the discharge tube has a diameter of 5 mm and length of
0 cm. A plane mirror is placed at one end and it is desired to keep gaussian beam diameter
lot more than 3 mm at the other end to minimize wall losses. Find radius of curvature of a
nirror to be placed close to the second end.

4.14c For a typical solid-state laser such as ruby, Nd–YAG, or alexandrite, $a_1 = a_2 \approx 5$ mm,
≈ 10 cm, and $\lambda_0 \approx 1 \ \mu m$ with refractive index ≈ 1.7. Check the condition on Fresnel number
V for stability of such resonators. (Note that dielectric discontinuity at the crystal side bound-
ry further confines the beam.)

4.14d The resonator for a CO_2 laser with $\lambda_0 = 10.6 \ \mu m$ has radii of curvature $R_1 = 10$ m,
$R_2 = 20$ m (sign convention as in Fig. 14.14a) and spacing $d = 1.5$ m. Show location on the
:ability diagram in Fig. 14.6c and find radius and position of the beam waist, and the beam
adii at the two mirrors.

14.14e A half-confocal resonator is made by inserting a plane mirror at the midplane of a symmetric confocal resonator. The resulting resonator has $R_1 = 2d$, $R_2 = \infty$, and by an image argument, would seem to be equivalent to the original symmetric resonator. Check the $ABCD$ matrix for a round trip of this resonator and comment on the differences from the symmetric confocal resonator.

14.15a Show location on a stability diagram of the resonators of Probs. 14.14b, 14.14d, and the half-confocal resonator of Prob. 14.14e.

14.15b A ruby rod 10 cm long has refractive index $n = 1.77$ at free-space wavelength $\lambda_0 = 0.6943\ \mu m$. If the plane ends form the resonant reflectors, find the longitudinal mode number nearest to the given wavelength, and the frequency separation between longitudinal modes.

14.15c The ruby rod of Prob. 14.15b now has its ends ground to form a confocal resonator. Give the radius of curvature needed and calculate minimum spot size and spot size at the ends. For a given mode number l, how much is frequency shifted from the value for the plane mirrors?

14.15d Find longitudinal mode separation and transverse mode separation for the resonators of Probs. 14.14b and 14.14d.

14.15e Show that there are stable configurations in which the mirror curvatures are in the same direction. That is, the mode exists between a concave and convex surface.

14.16a Supply the details of the derivation of Eq. 14.16(9) with the definitions noted in the text.

14.16b Show specifically by use of Eq. 14.16(9) that a gaussian beam at the input focal plane does transform to a gaussian at the output focal plane, as stated in the text.

14.16c Convert the integral of Eq. 14.16(6) to polar coordinates (r, ϕ) and find the form of ψ at the output focal plane if that at the input focal plane corresponds to a uniformly illuminated circle of radius a, centered on the axis.

_____ Appendix 1 _____

CONVERSION FACTORS BETWEEN SYSTEMS OF UNITS

The system of units in this text, and in much of applied electromagnetics, is the SI system,[1] which is a rationalized meter-kilogram-second (MKS) system. Since the additional basic electric unit is selected as the ampere, it is also designated the MKSA system. The classical or gaussian CGS system is still used in much of the scientific literature. In this, electrical quantities are based upon the electrostatic system of units (ESU) and designated statcoulombs, statvolts, and so on. Current may be in this system (statampere) or in the electromagnetic system of units (EMU) system (abamperes), as may be some of the circuit quantities, resistance, inductance, and so on. Conversions are consequently given in the following table to both systems for certain selected quantities.

The factor given is the number of gaussian units required to equal one SI unit. A result in SI units would thus be multiplied by this number to obtain the equivalent answer in gaussian units. For example, if length is l_m meters, it will be $100 l_m$ centimeters. For simplicity, velocity of light is taken as 3×10^8 m/s; for more accurate work, all multipliers of 3 should be replaced by 2.997925. Additional details and conversions are given in the references.[1,2]

E. A. Mechtly, "The International System of Units—Physical Constants and Conversion Factors," National Aeronautics and Space Administration, Publication SP-7012, 1973). Available from Superintendent of Documents, U.S. Government Printing Office, Washington, DC 20402.

ASTM Standard Metric Practice Guide, Designation E 380-70, American Society for Testing and Materials, 1916 Race St., Philadelphia, PA 19103, 1970.

Physical Quantity	Symbol	SI Unit	Factor	Gaussian
Length	l	meter (m)	10^2	centimeter (cm)
Mass	m	kilogram (kg)	10^3	gram
Time	t	second (s)	1	second
Frequency	f	hertz (Hz)	1	hertz
Force	F	newton (N)	10^5	dyne
Energy	U	joule (J)	10^7	erg
Power	W	watt (W)	10^7	erg/s
Charge	Q	coulomb (C)	3×10^9	statcoulomb
Charge density	ρ	C/m^3	3×10^3	statcoul/cm^3
Current	I	ampere (A)	3×10^9	statampere
Current	I	ampere (A)	$1/10$	abampere
Potential, voltage	Φ, V	volt (V)	$(1/3) \times 10^{-2}$	statvolt
Potential, voltage	Φ, V	volt (V)	10^8	abvolt
Electric field	E	V/m	$(1/3) \times 10^{-4}$	statvolt/cm
Electric flux density	D	C/m^2	$4\pi(3 \times 10^5)$	statcoul/cm^2
Electric polarization	P	C/m^2	3×10^5	dipole moment/cm^3
Magnetic field	H	A/m	$4\pi \times 10^{-3}$	oersted
Magnetization	M	A/m	10^{-3}	magnetic moment/cm^3
Magnetic flux	ψ_m	weber (Wb)	10^8	maxwell
Magnetic flux density	B	tesla (T)	10^4	gauss
Capacitance	C	farad (F)	$(3)^2 \times 10^9$	statfarad
Inductance	L	henry (H)	$(3)^{-2} \times 10^{-11}$	stathenry
Inductance	L	henry (H)	10^9	abhenry
Resistance	R	ohm (Ω)	$(3)^{-2} \times 10^{-11}$	statohm
Resistance	R	ohm (Ω)	10^9	abohm
Conductivity	σ	siemens/meter (S/m)	$(3)^2 \times 10^9$	$(\text{statohm cm})^{-1}$
Conductivity	σ	siemens/meter (S/m)	10^{-11}	$(\text{abohm cm})^{-1}$

_____Appendix 2_____

COORDINATE SYSTEMS AND VECTOR RELATIONS

The Rectangular, Cylindrical, and Spherical Coordinate Systems

The three systems utilized in this book are rectangular coordinates, circular cylindrical coordinates, and spherical coordinates. These are defined briefly before generalizing.

The intersection of two surfaces is a line; the intersection of three surfaces is a point; thus the coordinates of a point may be given by stating three parameters, each of which defines a coordinate surface. In rectangular coordinates, the three planes $x = x_1$, $y = y_1$, $z = z_1$ intersect at a point designated by the coordinates x_1, y_1, z_1. The elements of length in the three coordinate directions are dx, dy, and dz, the elements of area are $dx\,dy$, $dy\,dz$, and $dz\,dx$, and the element of volume is $dx\,dy\,dz$.

In the circular cylindrical coordinate system, the coordinate surfaces are (a) a set of circular cylinders ($r = $ constant), (b) a set of planes all passing through the axis ($\phi = $ constant), (c) a set of planes normal to the axis ($z = $ constant). Coordinates of a particular point may then be given as r_1, ϕ_1, z_1 (Fig. 2a). The r, ϕ, and z coordinates are known respectively as the radius, the azimuthal angle, and the distance along the axis. Elements of length are dr, $r\,d\phi$, dz, and the element of volume is $r\,dr\,d\phi\,dz$. The system shown is a right-hand system in the order of writing r, ϕ, z.

In spherical coordinates the surfaces are (a) a set of spheres (radius r from the origin $=$ constant), (b) a set of cones about the axis ($\theta = $ constant), (c) a set of planes passing through the polar axis ($\phi = $ constant). The intersection of sphere $r = r_1$, cone $\theta = \theta_1$, and plane $\phi = \phi_1$ gives a point whose coordinates are said to be r_1, θ_1, ϕ_1 (Fig. 2b). r is the radius, θ the polar angle or co-latitude, and ϕ is

791

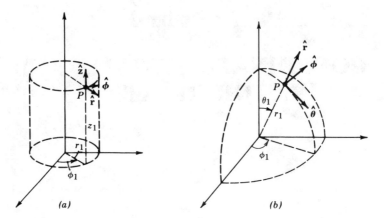

Fig. 2 (*a*) System of circular cylindrical coordinates. (*b*) System of spherical coordinates.

the azimuthal angle or longitude. Elements of distance are dr, $r\,d\theta$, and $r\sin\theta\,d\phi$, elements of area are $r\,dr\,d\theta$, $r^2\sin\theta\,d\theta\,d\phi$, and $r\sin\theta\,d\phi\,dr$, and the element of volume is $r^2\sin\theta\,dr\,d\theta\,d\phi$.

General Curvilinear Coordinates

Each of the three systems above, and many others utilized in mathematical physics, are orthogonal coordinate systems in that the lines of intersection of the coordinate surfaces are at right angles to one another at any given point. It is possible to develop general expressions for divergence, curl, and other vector operations for such systems which make it unnecessary to begin at the beginning each time a new system is met.

Suppose that a point in space is thus defined in any orthogonal system by the coordinate surfaces q_1, q_2, q_3. These then intersect at right angles and a set of three unit vectors, $\hat{1}, \hat{2}, \hat{3}$ may be placed at this point. These should point in the direction of increasing coordinates. (See Fig. 2*c*). The three coordinates need not necessarily

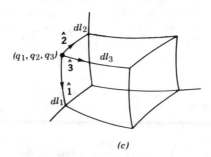

(c)

Fig. 2 (*c*) Element in arbitrary orthogonal curvilinear coordinates.

express directly a distance (consider, for example, the angles of spherical coordinates) so that the differential elements of distance must be expressed:

$$dl_1 = h_1\, dq_1; \qquad dl_2 = h_2\, dq_2, \qquad dl_3 = h_3\, dq_3, \tag{1}$$

where h_1, h_2, h_3 in the most general case may each be functions of all three coordinates, q_1, q_2, q_3.

Scalar and Vector Products A reference to the fundamental definitions of the two vector multiplications will show that these do not change in form in orthogonal curvilinear coordinates. Thus, for scalar or dot product,

$$\mathbf{A} \cdot \mathbf{B} = A_1 B_1 + A_2 B_2 + A_3 B_3, \tag{2}$$

and, for the vector or cross product,

$$\mathbf{A} \times \mathbf{B} = \begin{vmatrix} \hat{1} & \hat{2} & \hat{3} \\ A_1 & A_2 & A_3 \\ B_1 & B_2 & B_3 \end{vmatrix} \tag{3}$$

When one of these vectors is replaced by the operator ∇, the foregoing expressions do not hold, as will be shown below.

Gradient According to previous definitions, the gradient of any scalar Φ will be a vector whose component in any direction is given by the change of Φ for a change in distance along that direction. Thus

$$\nabla\Phi = \hat{1}\,\frac{\partial\Phi}{h_1\partial q_1} + \hat{2}\,\frac{\partial\Phi}{h_2\partial q_2} + \hat{3}\,\frac{\partial\Phi}{h_3\partial q_3} \tag{4}$$

Divergence In forming the divergence, it is necessary to account for the variations in surface elements as well as the vector components when one changes a coordinate. If the product of surface element by the appropriate component is first formed and then differentiated, both of these changes are taken into account:

$$\nabla \cdot \mathbf{D} = \frac{1}{h_1 h_2 h_3\, dq_1\, dq_2\, dq_3}\left[dq_1\,\frac{\partial}{\partial q_1}(D_1 h_2 h_3\, dq_2\, dq_3) \right.$$

$$\left. + dq_2\,\frac{\partial}{\partial q_2}(D_2 h_1 h_3\, dq_1\, dq_3) + dq_3\,\frac{\partial}{\partial q_3}(D_3 h_2 h_1\, dq_2\, dq_1)\right]$$

$$\nabla \cdot \mathbf{D} = \frac{1}{h_1 h_2 h_3}\left[\frac{\partial}{\partial q_1}(h_2 h_3 D_1) + \frac{\partial}{\partial q_2}(h_1 h_3 D_2) + \frac{\partial}{\partial q_3}(h_2 h_1 D_3)\right] \tag{5}$$

Note that for the spherical coordinate system $dl_1 = dr$, $dl_2 \quad r\, d\theta$, and $dl_3 = r \sin\theta\, d\phi$, so that $h_1 = 1$, $h_2 = r$, and $h_3 = r \sin\theta$. A substitution of these in (5) leads directly to Eq. 1.11(9).

Curl   In forming the curl, it is necessary to account for the variations in length elements with changes in coordinates as one integrates about an elemental path. This may again be done by forming the product of length element and proper vector component and then differentiating. The result may be written

$$
\nabla \times \mathbf{H} = \begin{vmatrix} \dfrac{\hat{\mathbf{1}}}{h_2 h_3} & \dfrac{\hat{\mathbf{2}}}{h_3 h_1} & \dfrac{\hat{\mathbf{3}}}{h_1 h_2} \\[2mm] \dfrac{\partial}{\partial q_1} & \dfrac{\partial}{\partial q_2} & \dfrac{\partial}{\partial q_3} \\[2mm] h_1 H_1 & h_2 H_2 & h_3 H_3 \end{vmatrix} \tag{6}
$$

Laplacian   The Laplacian of a scalar, which is defined as the divergence of the gradient of that scalar, may be found by combining (4) and (5):

$$
\nabla^2 \Phi = \nabla \cdot \nabla \Phi
$$

$$
= \frac{1}{h_1 h_2 h_3} \left[\frac{\partial}{\partial q_1} \left(\frac{h_2 h_3}{h_1} \frac{\partial \Phi}{\partial q_1} \right) + \frac{\partial}{\partial q_2} \left(\frac{h_3 h_1}{h_2} \frac{\partial \Phi}{\partial q_2} \right) + \frac{\partial}{\partial q_3} \left(\frac{h_1 h_2}{h_3} \frac{\partial \Phi}{\partial q_3} \right) \right] \tag{7}
$$

Laplacian of Vectors   For the Laplacian of a vector in a system of coordinates other than rectangular, it is convenient to use the vector identity

$$
\nabla^2 \mathbf{F} = \nabla(\nabla \cdot \mathbf{F}) - \nabla \times \nabla \times \mathbf{F} \tag{8}
$$

Each of the operations on the right has been defined earlier.

Differentiation of Vectors   The derivative of a vector is sometimes required as in Newton's law for the motion of particles.

$$
\mathbf{F} = m \frac{d\mathbf{v}}{dt} = m \frac{d}{dt} (\hat{\mathbf{1}} v_1 + \hat{\mathbf{2}} v_2 + \hat{\mathbf{3}} v_3) \tag{9}
$$

If this is expanded, we have

$$
\frac{\mathbf{F}}{m} = \hat{\mathbf{1}} \frac{dv_1}{dt} + \hat{\mathbf{2}} \frac{dv_2}{dt} + \hat{\mathbf{3}} \frac{dv_3}{dt} + v_1 \frac{d\hat{\mathbf{1}}}{dt} + v_2 \frac{d\hat{\mathbf{2}}}{dt} + v_3 \frac{d\hat{\mathbf{3}}}{dt}
$$

The last three terms involve changes in the unit vectors, which by definition cannot change magnitude, but may change in direction as one moves along the coordinate system. Consider for example the fourth term:

$$
v_1 \frac{d\hat{\mathbf{1}}}{dt} = v_1 \left(\frac{\partial \hat{\mathbf{1}}}{\partial q_1} \frac{dq_1}{dt} + \frac{\partial \hat{\mathbf{1}}}{\partial q_2} \frac{dq_2}{dt} + \frac{\partial \hat{\mathbf{1}}}{\partial q_3} \frac{dq_3}{\partial t} \right)
$$

Partials of the form $\partial \hat{\mathbf{1}}/\partial q_1$, and so on, may not be zero. As an example, consider the term $\partial \hat{\boldsymbol{\theta}}/\partial \theta$ in spherical coordinates. From Fig. 2d the vector $d\theta \, \partial \hat{\boldsymbol{\theta}}/\partial \theta$ is seen to have magnitude $d\theta$ and has direction given by $-\hat{\mathbf{r}}$. Thus $\partial \hat{\boldsymbol{\theta}}/\partial \theta = -\hat{\mathbf{r}}$. Other

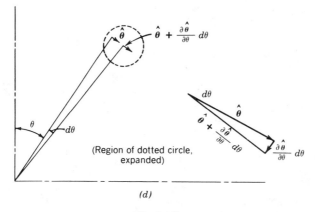

(d)

Fig. 2 *(d)*

partials of unit vectors in this and the cylindrical coordinate system are listed below.

Coordinates and Derivatives of Unit Vectors for Various Coordinate Systems

Rectangular Coordinates

$$q_1 = x \qquad q_2 = y \qquad q_3 = z$$
$$h_1 = 1 \qquad h_2 = 1 \qquad h_3 = 1$$

All partials of unit vectors ($\partial\hat{x}/\partial x$, $\partial\hat{x}/\partial y$, etc.) are zero.

Cylindrical Coordinates

$$q_1 = r \qquad q_2 = \phi \qquad q_3 = z$$
$$h_3 = 1 \qquad h_2 = r \qquad h_3 = 1$$

All partials of unit vectors are zero except

$$\frac{\partial\hat{r}}{\partial\phi} = \hat{r} \qquad \frac{\partial\hat{\phi}}{\partial\phi} = -\hat{r}$$

Spherical Coordinates

$$q_1 = r \qquad q_2 = \theta \qquad q_3 = \phi$$
$$h_1 = 1 \qquad h_2 = r \qquad h_3 = r\sin\theta$$

All partials of unit vectors are zero except

$$\frac{\partial \hat{\mathbf{r}}}{\partial \theta} = \hat{\boldsymbol{\theta}}; \qquad \frac{\partial \hat{\boldsymbol{\theta}}}{\partial \theta} = -\hat{\mathbf{r}}$$

$$\frac{\partial \hat{\mathbf{r}}}{\partial \phi} = \hat{\boldsymbol{\phi}} \sin \theta; \qquad \frac{\partial \hat{\boldsymbol{\theta}}}{\partial \phi} = \hat{\boldsymbol{\phi}} \cos \theta; \qquad \frac{\partial \hat{\boldsymbol{\phi}}}{\partial \phi} = -(\hat{\mathbf{r}} \sin \theta + \hat{\boldsymbol{\theta}} \cos \theta)$$

Appendix 3

SKETCH OF THE DERIVATION OF MAGNETIC FIELD LAWS

In presenting various forms for the laws of static magnetic fields, many of the less obvious steps were left out, and the order was chosen as that most convenient for presentation of the laws rather than that of the logical development. This appendix sketches the omitted steps.

We will start from Ampère's law, Eq. 2.3(2), which we may write

$$\mathbf{H}(\mathbf{r}) = \int \frac{I'(\mathbf{r}')\, \mathbf{dl}' \times \mathbf{R}}{4\pi R^3} \tag{1}$$

by assuming the origin of the coordinate system to be at the current element at each point through the integration. $I'(\mathbf{r}')$ is the current in a contributing element $\mathbf{dl}'$ at point (x', y', z') and $\mathbf{R}$ is the vector running from $\mathbf{dl}'$ to point (x, y, z) at which $\mathbf{H}(\mathbf{r})$ is to be computed.

$$\mathbf{R} = \mathbf{r} - \mathbf{r}' = \hat{\mathbf{x}}(x - x') + \hat{\mathbf{y}}(y - y') + \hat{\mathbf{z}}(z - z')$$

We wish to find $\mathbf{B} = \mu\mathbf{H}$ in terms of the derivatives at the point of observation (x, y, z) of the vector potential $\mathbf{A}$. It may be shown that

$$\frac{\mathbf{dl}' \times \mathbf{R}}{R^3} = \nabla\left(\frac{1}{R}\right) \times \mathbf{dl}' \tag{2}$$

where ∇ denotes derivatives with respect to $x, y,$ and z. Also, using the vector identity of Prob. 2.6b:

$$\nabla\left(\frac{1}{R}\right) \times \mathbf{dl}' = \nabla \times \left(\frac{\mathbf{dl}'}{R}\right) - \frac{1}{R} \nabla \times \mathbf{dl}' \tag{3}$$

797

Since ∇ represents derivatives with respect to x, y, and z, which are not involved with $\mathbf{dl}'$, the last term is zero. Therefore

$$\mathbf{B}(\mathbf{r}) = \mu \int \frac{I'(\mathbf{r}')}{4\pi} \nabla \times \left(\frac{\mathbf{dl}'}{R}\right) = \nabla \times \mathbf{A}(\mathbf{r}) \tag{4}$$

where

$$\mathbf{A}(\mathbf{r}) = \mu \int \frac{I'(\mathbf{r}')\mathbf{dl}'}{4\pi R} \tag{5}$$

The curl operation in (4) could be taken outside of the integral since it is with respect to $\mathbf{r}$, and the integration is with respect to $\mathbf{r}'$. Thus, the vector potential forms of Sec. 2.9 have been derived from Ampère's law for which the arbitrary coordinate origin was chosen at the current element for simplicity.

For the next step, let us note the x component of (5):

$$A_x(\mathbf{r}) = \mu \int \frac{I'_x(\mathbf{r}')\,dx'}{4\pi R} = \mu \int_{V'} \frac{J'_x(\mathbf{r}')\,dV'}{4\pi R} \tag{6}$$

This may be compared with Poisson's equation and the integral expression for electrostatic potential:

$$\nabla^2 \Phi = -\frac{\rho}{\varepsilon}; \qquad \Phi(\mathbf{r}) = \int_{V'} \frac{\rho(\mathbf{r}')\,dV'}{4\pi\varepsilon R} \tag{7}$$

Although these equations were obtained from a consideration of the properties of electrostatic fields, the second of the two equations (7) may be considered a solution in integral form for the first, for any continuous scalar functions Φ and ρ/ε. Consequently, by direct analogy between (6) and (7), we write

$$\nabla^2 A_x = -\mu J_x \tag{8}$$

$$\nabla^2 \mathbf{A} = -\mu \mathbf{J} \tag{9}$$

This was the differential equation relating $\mathbf{A}$ to current density discussed in Sec. 2.12. By reversing the steps of that section, the differential equation for magnetic field may be derived from it:

$$\nabla \times \mathbf{H} = \mathbf{J}. \tag{10}$$

And as was shown in Sec. 2.8, the integral form may be derived from this by use of Stokes's theorem:

$$\oint \mathbf{H} \cdot \mathbf{dl} = I \tag{11}$$

There remains the argument for $\nabla \cdot \mathbf{A} = 0$ used in Sec. 2.12. In this it is necessary to make use of the del operator with respect to both $\mathbf{r}$ and $\mathbf{r}'$. The former will be denoted V, and the latter V'. Recall that the integration is with respect to the primed variables. Then

$$\frac{1}{\mu} \nabla \cdot \mathbf{A} = \nabla \cdot \int_{V'} \frac{\mathbf{J}'(\mathbf{r}') \, dV'}{4\pi R} = \int_{V'} \nabla \cdot \left(\frac{\mathbf{J}'(\mathbf{r}')}{R} \right) \frac{dV'}{4\pi} \tag{12}$$

Using the vector equivalence of Prob. 1.11a,

$$\frac{1}{\mu} \nabla \cdot \mathbf{A} = - \int_{V'} \left[\frac{\nabla \cdot \mathbf{J}'(\mathbf{r}')}{R} + \mathbf{J}'(\mathbf{r}') \cdot \nabla \left(\frac{1}{R} \right) \right] \frac{dV'}{4\pi}$$

The first term is zero since $\mathbf{J}'$ is not a function of $\mathbf{r}$. In the latter term, from the definition of R, we can write

$$\nabla \left(\frac{1}{R} \right) = - \nabla' \left(\frac{1}{R} \right)$$

Then

$$\frac{1}{\mu} \nabla \cdot \mathbf{A} = - \int_{V'} \mathbf{J}'(\mathbf{r}') \cdot \nabla' \left(\frac{1}{R} \right) \frac{dV'}{4\pi}$$

$$= \int_{V'} \left[\frac{1}{R} \nabla' \cdot \mathbf{J}'(\mathbf{r}') - \nabla' \cdot \frac{\mathbf{J}'(\mathbf{r}')}{R} \right] \frac{dV'}{4\pi}$$

In the last step we have again used the vector equivalence of Prob. 1.11a. The first term is zero because we are concerned with direct currents, which, by continuity, give $\nabla' \cdot \mathbf{J}' = 0$. The second term is transformable to a surface integral by the divergence theorem. Thus

$$\frac{1}{\mu} \nabla \cdot \mathbf{A} = \oint_{S'} \frac{\mathbf{J}'(\mathbf{r}')}{4\pi R} \cdot d\mathbf{S}' \tag{13}$$

But, if the surface encloses all the current, as it must, there can be no current flow through the surface and the result in (13) is seen to be zero. Thus all the major laws given have been shown to follow from the original experimental law of Ampère. It should be noted that the argument given is for a homogeneous medium (permeability not a function of position). For an inhomogeneous or anisotropic medium, the equations forming the fundamental starting point are

$$\nabla \times \mathbf{H} = \mathbf{J}; \qquad \nabla \cdot \mathbf{B} = 0 \tag{14}$$

and for an anisotropic medium, $\mathbf{B}$ and $\mathbf{H}$ are not related by a simple constant.

Appendix 4

COMPLEX PHASORS AS USED IN ELECTRICAL CIRCUITS

If a voltage $V \cos \omega t$ is applied to a linear, time-invariant circuit containing R, L, and C in series, the equation to be solved is

$$L \frac{dI}{dt} + RI + \frac{1}{C} \int I \, dt = V_m \cos \omega t \tag{1}$$

But

$$\cos \omega t = \frac{e^{j\omega t} + e^{-j\omega t}}{2} \tag{2}$$

If we assume that the current has the steady-state solution

$$I = A e^{j\omega t} + B e^{-j\omega t} \tag{3}$$

the result of substituting in (1) is

$$j\omega L(A e^{j\omega t} - B e^{-j\omega t}) + R(A e^{j\omega t} + B e^{-j\omega t}) + \frac{1}{j\omega C}(A e^{j\omega t} - B e^{-j\omega t})$$

$$= \frac{V_m}{2}(e^{j\omega t} + e^{-j\omega t}) \tag{4}$$

This equation can be true for all values of time only if coefficients of $e^{j\omega t}$ are the same on both sides of the equation, and similarly for $e^{-j\omega t}$.

$$A\left[R + j\left(\omega L - \frac{1}{\omega C}\right)\right] = \frac{V_m}{2} \tag{5a}$$

$$B\left[R - j\left(\omega L - \frac{1}{\omega C}\right)\right] = \frac{V_m}{2} \tag{5b}$$

The complex quantity in the bracket of (5a) may be called Z and written in its equivalent form

$$Z = R + j\left(\omega L - \frac{1}{\omega C}\right) = |Z|e^{j\psi}$$

where

$$|Z| = \sqrt{R^2 + \left(\omega L - \frac{1}{\omega C}\right)^2} \tag{6}$$

and

$$\psi = \tan^{-1}\frac{(\omega L - 1/\omega C)}{R} \tag{7}$$

Similarly,

$$R - j\left(\omega L - \frac{1}{\omega C}\right) = |Z|e^{-j\psi}$$

Then

$$A = \frac{V_m}{2|Z|}e^{-j\psi}$$

$$B = \frac{V_m}{2|Z|}e^{j\psi}$$

(A and B are conjugates: they have the same real parts and equal and opposite imaginary parts.) Substituting in (3),

$$I = \frac{V_m}{|Z|}\left[\frac{e^{j(\omega t - \psi)} + e^{-j(\omega t - \psi)}}{2}\right] \tag{8}$$

By comparing with (2),

$$I = \frac{V_m}{|Z|}\cos(\omega t - \psi) \tag{9}$$

This final result gives the desired magnitude and phase angle of the current with respect to the applied voltage. That information is contained in either constant A or constant B, and no information is given in one which is not in the other. Constant B is of necessity the conjugate of A, since this is the only way in which the two may add up to a real current, and the final exact answer for current must be real. It follows that half of the work was unnecessary. We could have started only with $V_m e^{j\omega t}$ in place of the two-term expression which is exactly equivalent to $V_m \cos \omega t$. For current, there would then be only

$$I = \frac{V_m}{|Z|}e^{j(\omega t - \psi)} \tag{10}$$

Although this cannot actually be the expression for current, since it is a complex and not a real quantity, it contains all the information we wish to know: magnitude of current, $V/|Z|$, and its phase with respect to applied voltage, ψ. This procedure may be made exact by writing

$$V(t) = \text{Re}[V_m e^{j\omega t}] \tag{11}$$

$$I(t) = \left[\text{Re} \frac{V_m}{|Z|} e^{j(\omega t - \psi)} \right] \tag{12}$$

where Re denotes, "the real part of." Because of the inconvenience of this notation, it is usually not written explicitly but it is understood. That is, *if any single frequency sinusoid $f(t)$ is expressed by its magnitude and phase, $Me^{j\theta}$, or by its real (in-phase) and imaginary (out-of-phase) parts $A + jB$, the instantaneous expression may be found by multiplying by $e^{j\omega t}$ and taking the real part:*

$$f(t) = \text{Re}[Me^{j(\omega t + \theta)}] = \text{Re}[(A + jB)e^{j\omega t}] \tag{13}$$

So to summarize, voltage and current can be written[1]

$$V(t) = \text{Re}[V_c e^{j\omega t}] \tag{14}$$

$$I(t) = \text{Re}[I_c e^{j\omega t}] \tag{15}$$

where V_c and I_c are the complex or *phasor* representations of voltage and current respectively,

$$V_c = |V|e^{j\theta_V}, \qquad I_c = |I|e^{j\theta_I} \tag{16}$$

The differential equation (1) becomes an algebraic equation relating V_c and I_c:

$$j\omega L I_c + R I_c + \frac{I_c}{j\omega C} = V_c \tag{17}$$

from which

$$I_c = \frac{V_c}{Z} = \left(\frac{V_c}{|Z|} \right) e^{-j\psi} \tag{18}$$

so that

$$|I| = \frac{|V|}{|Z|}, \qquad \theta_I = \theta_V - \psi \tag{19}$$

where magnitude and phase of impedance Z are given by (6) and (7), respectively.

Although the subscript c is used to denote complex phasors in this development to stress their nature, the complex nature is understood from the context in most circuit analyses, without the special designation and similarly it is understood in the phasor representation of fields used in much of this text.

---

[1] An alternative form much used in the physics literature is $V(t) = \frac{1}{2}[V_c e^{j\omega t} + \text{c.c.}]$ where c.c. stands for complex conjugate of the first term in brackets.

For nonlinear or time-varying circuit elements, the complete equivalence (2) must be used if excitation is by a sinusoidal voltage, since frequency components other than ω are generated.

Power Calculations   Given a sinusoidal voltage $V(t) = V_m \cos(\omega t + \phi_1)$ and a corresponding sinusoidal current $I(t) = I_m \cos(\omega t + \phi_2)$, the expression for instantaneous power is

$$W(t) = V(t)I(t) = V_m I_m \cos(\omega t + \phi_1)\cos(\omega t + \phi_2)$$

By trigonometric identities, this may be written

$$W(t) = \frac{V_m I_m}{2} [\cos(\phi_1 - \phi_2) + \cos(2\omega t + \phi_1 + \phi_2) \tag{20}$$

The equivalent in terms of phasor current and voltage is

$$W(t) = \tfrac{1}{2} \operatorname{Re}[V_c I_c^* + V_c I_c e^{2j\omega t}] \tag{21}$$

where I_c^* denotes the complex conjugate of I_c. Often we are interested only in average power, which is given by the first terms of (20) and (21).

In another useful power calculation with complex voltage and current, $V_c I_c^*$ is found to give average power and the difference between stored energy in magnetic and electric fields. As an example, consider a simple series RLC circuit

$$V_c I_c^* = ZI_c I_c^* = \left[R + j\left(\omega L - \frac{1}{\omega C}\right)\right] I_c I_c^*$$

$$= 2\left(\frac{RI_c I_c^*}{2}\right) + 4j\omega\left[\frac{LI_c I_c^*}{4} - \frac{I_c I_c^* C}{4\omega^2 C^2}\right] \tag{22}$$

The real term is recognized as twice the average power dissipated in the resistor. The first term in brackets is average power stored in the inductor. (One factor of 2 comes from the form $(\tfrac{1}{2})LI^2$, the other from the average of squared sinusoids.) The second term in brackets is the average energy stored in the capacitor since $|I_c/\omega C|$ is magnitude of capacitor voltage. Thus (22) may be written

$$V_c I_c^* = 2W_L + 4j\omega(U_M - U_E) \tag{23}$$

where W_L is average power loss, U_M average energy in the magnetic fields of the inductor, and U_E average energy in electric fields of the capacitor. This is a special case of the complex Poynting theorem discussed in Sec. 3.13 that is applicable to any passive circuit.

INDEX

ABCD network parameters, 530
ABCD optics parameters, 746
Abeles, F., 676
Aberration, spherical, 734–735
Abramowitz, M., 93, 365
Acoustic resonances, 520
Adler, R.B., 296
Admittance diagram, 235
Admittance parameters, 528–530, 547, 555
Allis, W.P., 710
Ampère, A.M., 68–71
Ampère's law, 71–76, 102, 144, 165
Analytic functions, 327, 557
Angle:
  Brewster, 309
  critical, 307, 457
  of reflection, 298
Anisotropic media, 572, 689–726
Anomalous dispersion, 256
Antenna, 177, 203–206, 577–665
  above earth, 596–598
  aperture, 581, 607–623
  arrays, 623–640
  as boundary-value problem, 640–647
  dipole, 582–586
  directivity, 593
  effective area, 654
  frequency independent, 637–640
  impedance, 640–652

  loop, 203, 586
  receiving, 652–658
Antennas, mutual couplings of, 625, 650
Antiferromagnetic materials, 680–681
Antireflective coating for lens, 294
Aperture antenna, see Antenna, aperture
Arjavalingam, G., 447
Artificial dielectrics, 621–623
Ashcroft, N.W., 668
Associated Legendre function, 499–500
Atsuki, K., 409
Attenuation constant, 241
  from conductor losses, 406, 413, 417, 420,
    427, 441, 444
  from dielectric losses, 405, 442, 444
  of plane waves, 280–281
  reactive (filter-type), 253, 398, 413, 439,
    445, 547
  in transmission lines, 241–243, 248–253
  in waveguides, 405, 406, 413, 417, 420, 427
Attenuation bands, 547

Babinet's principle, 621
Backward wave, 257, 481, 485
Bahl, I.J., 410
Balmain, K.G., 598, 649
Balun, 606
Bandwidth:
  of antenna, 623–640, 642

Bandwidth (*Continued*)
  of cavity resonator, 492
  relation to Q, 246
Barrow, W.L., 464
Base-band signals, 446
Bateman, H., 501
Ber and bei functions, 179, 182, 207
Bergmann, E.E., 689
Berkey, R.F., 341
Bers, A., 710
Bessel equation, 178, 361, 500
Bessel functions, 181, 207, 362–373
  asymptotic forms, 367–369
  expansion in terms of, 371–373
  of imaginary arguments, 367
  of real arguments, 364
  spherical, 501
  zeros of, 369
Betatron, 111, 112, 114, 161
Biaxial materials, 691
Biconical antenna, 500, 578, 642
Biconical resonator, 506, 507
Biconical transmission line, 467–469
Bilinear transformation, 387
Biot, Jean-Baptiste, 68
Biot and Savart law, 71
Birefringence, 697, 729
Birkhoff, G., 541
Blakemore, J.S., 668
Bloembergen, N., 685, 686
Blumlein pulse generator, 269
Boll, R., 683
Boltzmann's constant, 47, 679
Borgnis, F.E., 343
Born, M., 277, 609, 689, 695, 740, 746
Boundary conditions, 38, 143–147
  for conduction problem, 40
  between dielectrics, 286
  in electrostatics, 38–44, 49, 319, 324
  in magnetostatics, 99–100
  at perfect conductor, 146
  for waveguide modes, 438
Bowman, D.F., 607
Boyd, G.D., 775
Bracewell, R., 354
Branch point, 385
Breakdown, 115
Brewster angle, 309
Bridge-type microwave circuit, 554
Brillouin, L., 256, 481, 751
Broadside arrays, 627–628
Brown, J.W., 341
Brown, W.F., 668

Bubble memory, 681
Buchsbaum, S.T., 710
Bulley, R., 471
Button, K.J., 715, 717

Cable for pulse transmission, 220
Capacitance, 27–28, 56, 64, 65, 66
  per unit length, 28, 216, 252
Capacitive waveguide diaphragm, 569, 576
Capacitor, 27, 121, 163, 172, 201
Carson, J.R., 198, 525
Cascaded transmission lines, 236
Cascaded two-ports, 537–542
Cauchy-Riemann equations, 327, 344, 385, 574
Cavity resonators, 486–522, 555–565
  circular cylindrical, 495–498
  coupling to, 507–509, 564
  dielectric, 515–519
  equivalent circuits, 562–567
  perturbations, 512–515
  Q of, 492, 496, 509–512
  rectangular, 489–495
  small-gap, 504–507
  spherical, 502–504
Chang, H., 681
Characteristic impedance, 214–218, 248, 408, 541–542
Characteristic roots, 541
Characteristic wave impedance, 286, 399, 401, 440, 443
Chebyshev method, 323, 384
Chu, L.J., 296, 464, 633, 642, 643
Churchill, R.V., 341
Circuits, 168–209
  elements of, 184–196
  equations of, 169–177
  microwave, 523–576
  with radiation, 203–206
  skin effect in, 178–183
Circular harmonics, 387
Circular polarization, 277, 279, 712
Circulator, 721, 725
Circumferential mode, 464, 483
Clausius-Mossotti relation, 669
Coaxial cylinders:
  capacitance, 28, 67
  with different dielectrics, 42–44, 65
  displacement current, 163
  electrostatic field, 10, 24, 54, 62
  inductance, 81, 106, 207
  magnetic field, 76
Coaxial loops, 107
Coaxial transmission line:

attenuation, 267
characteristic impedance, 214, 264
discontinuity, 571
higher order modes, 428–430
internal impedance, 153
resonator, 504
wave velocity, 214
see also Coaxial cylinder
Coercive force, 683
Cohn, S.B., 471, 516
Collin, R.E., 157, 283, 408, 433, 460, 477,
542, 571, 619, 647, 679, 689, 753
Collision frequency, 673, 676
Complete set, 755
Complex Fourier series, 388
Complex permittivity, 673
Complex phasors in electrical circuits, 800–803
in field problems, 127–129
Complex Poynting vector, 142–143, 555
Complex variables, 326–346
Compton, R.T., Jr., 619
Computer interconnection, 220–221
Comstock, R.L., 725
Conductance, 56, 57, 66
Conducting rectangular solid, 358
Conduction current, 120, 124, 162
Conductivity, 37, 124, 674, 676
Conductor, 23
collisionless, 677
good, 147–148, 672
imperfect, 672–676
perfect, 676–678
Cones of silence, 633
Confocal resonator, 778
Conformal transformations, 36, 326–346
for Helmholtz equation, 343–346, 471
for Laplace equation, 331
Conservative property in electrostatics, 19
Continuity conditions at boundary, 143
Continuity of current, 120, 161, 174
Convection current, 120, 122, 124, 140, 162
Convergence factor, 562
Conversion between systems of units, 789
Convolution, 783
Coordinate systems, 791–796
Coplanar strip guide, 410–411
Correlation, 783
Coulomb (unit), 3
Coulomb, C.A., 2
Coulomb gauge, 167
Coulomb's law, 2–3
Coupled-mode theory, 481
Coupled-resonator filter, 545–546

Coupling:
to cavity resonators, 507–509, 564
to waveguides, 430–433
Critical angle, 307, 457
Critical coupling, 511, 522
Cross product of vectors, 70, 82
Curie temperature, 680
Curl:
of magnetic field, 86
of vector function, 82–84, 109
Current:
conduction, 120, 124, 162
continuity of, 120, 161, 174
convection, 120, 122, 124, 140, 162
defined for waveguide mode, 525
displacement, 121
element as radiator, 582–585
generator, 177
Curtis, H.E., 252
Curvilinear coordinates, 792
Cutoff, effective, 465
of nonuniform transmission lines, 263
of periodic systems, 481, 542–547
of waveguides, 398, 412, 418, 439, 443,
444–446
Cyclotron frequency, 705
Cylindrical coordinates, 791–795
Cylindrical harmonics, 360

Decibel, 242
Degenerate modes, 416, 494
Del, 29
Demagnetization factor, 728
Depletion region, 64
Depth of penetration, 151, 155, 282
Deschamps, G.A., 538
Desoer, C.A., 177
Detuned short, 511
Diamagnetic substances, 679
Dicke, R.H., 461, 525, 537, 553
Dielectric, 8, 280–283, 667–672
anisotropic, 689–703
artificial, 621–623
constant, 3
ellipsoid, 695
guide, circular cylindrical, 759–764
guide, curved, 344–346
guide, planar, 752–756
guide, rectangular, 756–759
guide, slab, 457–460, 477, 482
properties, 281, 667–672
radiators, 581, 761
resonators, 515–519, 522

Dielectric (*Continued*)
  wave reflections from, 283–293
  window, 314, 573
Difference equation, 321–325, 384, 385, 573
Diffraction, 612–623, 631–632, 781–783
  of gaussian beams, 769
  gratings, 301–302, 314, 631–632
  of lens systems, 781–783
Diffusion of charges in semiconductor, 47
Diffusion equation, 318
Digital signals, 220
Dipole:
  density, 679
  electric, 63, 64, 668
  magnetic, 95–96, 586
  moment, 63, 668
  radiator, 166, 582–586
Directional coupler, 551–553
Directive gain, 593
Directivity:
  antenna, 593
  aperture, 614
  of directional coupler, 553
Director, 636
Discontinuities in transmission systems, 444,
  567–571
  of transmission-line impedance, 224
Discontinuity capacitance, 568
Disk-loaded waveguide, 474
Dispersion, 128, 211, 253–254, 410, 446
  anomalous, 256
Displacement current, 111, 112, 119–123, 163
Displacement vector, 6
Distributed-circuit model, 196, 215–216
Divergence, 793
  of electrostatic field, 30–35
  of magnetic flux density, 96
  theorem, 33, 47, 64, 126
Dolph, C.L., 635
Domains, magnetic, 681
Dominant mode, 416, 524
Doppler shift, 708
Dot product of vectors, 13, 793
Double refraction, 697
Duality, 466–467, 483, 504, 518, 586
Dwight, H.B., 179
Dyadic, 689, 706

Earnshaw's theorem, 65
Eastwood, J.W., 321
Eddy currents, 683
Effective area, 653–654
Eikonal, 740, 784
Electric dipole, 25–26, 30

Electric dipole moment, 25
Electric dipole radiators, 582
Electric field, 4, 29, 61, 64, 124, 135
Electric flux, 53, 62
Electric flux density, 6
Electric induction, 6
Electric polarization, 8
Electric susceptibility, 9
Electrocardiography, 2
Electrodes, 27
Electroencephalography, 2
Electromagnetic horns, 617
Electromagnetic properties of materials, 666
Electromagnetic system of units, 789
Electromagnetic waves, 113
Electromagnets, 69
Electromotive force (emf), 114
Electron beam, 62, 65, 107, 507
Electron gun, 1
Electronic polarizability, 669–670
Electro-optic effects, 698–700
Electro-optic modulation, 699–701
Electrostatic equations, 1–67
Electrostatic equipotentials, 20
Electrostatic system of units (esu), 789
Elimination of wave reflections:
  from conducting surface, 287
  from dielectric interface, 292, 294
Elliott, R.S., 3, 591, 605, 621, 623, 647, 649,
  650
Ellipsometer, 316
Elliptic cylindrical conductor, 336, 386
Elliptic integrals, 190, 208, 408
Elliptic polarization, 278
Emde, F., 365
Emf method, 647
End-fire array, 629
Energy, 60
  conservative property of electrostatic fields,
  17
  density, 60
  of electrostatic system, 58, 67
  ot magnetostatic system, 104–105, 110
  of simple resonator, 491, 495
  relations for electromagnetic fields, 137–143
  relations for rays, 740
  of transmission line, 240, 267
  velocity, 402, 453, 455
Entire function, 558
Equipotential surfaces, 23, 29, 54, 63
Equivalent circuit:
  of N-port, 566–567
  of one-port, 558–565
  of receiving antenna, 658

for transverse electric wave, 454
for transverse magnetic wave, 453
Evanescent decay, 307, 445, 458
Excitation:
  of cavity modes, 507
  of waveguide by coaxial line, 431
  of waves in guides, 430–433
Exponential line, 262, 269
External inductance, 79, 200–201
  of circular loop, 190, 208
  of coaxial transmission line, 81, 207
  of parallel-wire transmission line, 184–185
External Q, 512
Extraordinary wave, 696–697

Fano, R.H., 296
Faraday, Michael, 68, 113, 714
Faraday rotation, 713–715, 719–724, 731
Faraday's law, 111, 113–119, 126, 143, 161,
    162, 165, 169–171, 178
  for moving system, 116–119
Far-zone fields, 585, 586–589
  of aperture, 612
Fay, C.E., 725
Feldman, C.B., 642
Ferrimagnetic materials, 680
Ferrite, 280, 680, 715–726, 731
  devices, 721–726
Ferroelectric materials, 685
Ferromagnetic domains, 682
Ferromagnetic materials, 680
Field:
  from ac current element, 159
  on axis of circular loop, 73
  in concentric spherical electrodes, 11
  of current loop (magnetic dipole), 94
  of finite line current, 74
  impedance, *see* Characteristic wave impedance
  in infinite solenoid, 75
  of line charge, 10
  maps, 56
  of ring of charge, 5
Field-displacement devices, 724
Filters:
  distributed-circuit type, 252–253, 267
  lumped-element, 175
  optical, 545–547
  periodic systems as, 478–481
  spatial, 783
  waveguide type, 530, 542–546, 574
Finite-difference solutions, 325
Finite element method, 321
Floquet's theorem, 481, 751
Flux, 6

density, 27, 62, 96
electric, 6–17
function, 16, 63, 329, 557
linkages, 187
magnetic, 79, 96
tubes, 17
Focal length, 734
Folded dipole, 605
Force, 3–4, 18, 70, 124
Foreshortened coaxial line resonator, 504–505
Foreshortened radial line resonator, 505
Foster, R.M., 556, 559
Foster's first canonical form, 558–560, 575
Foster's reactance theorem, 556
Foster's second canonical form, 561–562, 563
Fourier analysis, 127, 254, 321, 351–355
Fourier integral, 354–355, 388, 783
Fourier series, 351–353, 388
Fourier transforming properties of lenses, 781
Fox, A.G., 775
Fraunhofer diffraction, 609, 781
Free-space wavelength, 274
Frequency characteristics:
  of n ports, 566
  of one ports, 558
  of permittivity, 670
Frequency-independent antennas, 637–640
Fresnel diffraction, 609, 782
Fresnel equation of wave normals, 693, 729
Fresnel number, 775
Frey, J., 554
Friis, F.T., 583, 584, 655
Fringing fields, 27, 342, 389, 567

Gain-bandwidth product, 512
gaussian beam, 765–780, 786
  in graded-index medium, 765–770
  in homogeneous medium, 768–771
  in optical resonator, 774–780
  transformation by ray matrix, 771–774
gaussian cgs systems of units, 789
Gauss' law, 6–9, 13–15, 76, 89
  applications, 9–13, 28, 44–46, 63
  for boundary conditions, 38, 144
  for time-varying fields, 126
Generators, 69, 117
Geometrical optics, 733–752
  as limiting case of wave optics, 739–741
Ghose, R.N., 541
Ginzton, E.L., 538
Giorgi, G., 3
Gloge, D., 762, 763
Goell, J.E., 756, 758
Goodman, J.W., 781

Gordon, J.P., 775
Gorg, R., 410
Goubou, G., 477, 775
Graded-index fibers, 745, 759
Gradient, 28–30, 64, 793
Gradshteyn, I.S., 93, 615, 616
Graphical field mapping, 15, 53–58
Gravitational potential energy, 29
Gray, D.E., 672, 675
Green, E.I., 252
Green's function, 646
Gross, B., 668
Group:
  dispersion, 256, 448, 455, 762–763
  velocity, 254, 312, 399, 401, 424, 439
  velocity of spatial harmonics, 481
Guided waves, see Waveguide
Guide wavelength, 399, 419, 424
Gupta, K.C., 410
Gyrator, 721–722
Gyromagnetic ratio, 717
Gyroscope, 716
Gyrotropic media, 529, 710

Hahn, W.C., 571, 709
Half-wave:
  dielectric window, 292
  dipole, 589–591
Hall effect, 69
Hankel functions, 366, 501
Hansen, W.W., 558, 663
Harrington, R.F., 571
Haus, H.A., 685, 733
Helical sheet, 471–474
Helium-neon laser, 309
Helix, 471
Helmholtz, H., 117
Helmholtz coil, 107, 391
Helmholtz equation, 133, 136, 165, 318, 344, 393
Henry, J., 68
Hermite polynomials, 767, 771
Hertzian dipole, 582
Hertz vector potential, 167
Hockney, R.W., 321
Hogan, C.L., 721
Homogeneous medium, 8
Huygen's principle, 607
Hybrid mode, 395, 472
Hybrid networks, 553
Hyperbolic cylindrical conductors, 336
Hysteresis, 105, 681–682

Images, 48–52, 56, 596–597
  antennas, 596–597, 603

charges, in cylinder, 51
charges, in plane, 49–50
charges, in sphere, 51–52
multiple, 52
Impedance:
  of antennas, 592, 645–650
  coefficients, 530, 537
  functions, 555
  internal, see Internal impedance
  parameters, 528, 547
  of round wires, 180
  transformation by Smith chart, 233
  of transmission lines, 225, 229, 243, 250–251
  wave, see Characteristic wave impedance
Impedances, matching of, 220, 235–236, 263, 467, 483
Incidence of plane waves at arbitrary angles, 296–311
Inclined-plane transmission line, 467, 483
Incremental permeability, 684
Index ellipsoid, 690–696, 729
Index of wave normals, 690
Induced emf method, 205
Inductance, 79, 184–193
  as circuit element, 171
  distributed, 216
  from energy storage, 106, 110
  external, 79
  from flux linkages, 79
  internal, 79, 110, 152, 186
  mutual, 173, 189
  of practical coils, 191
Inductive diaphragm, 543, 569, 576
Inhomogeneous dielectric, 312
Integral equation for boundary-value problems, 571, 646
Interferometry, 701
Internal impedance, 151, 153, 181, 200, 208
International system of units, 3
Intrinsic impedance, 272, 281, 287
Ion beam, 62
Ionic polarizability, 669
Ionized gas, 674
Ionosphere, 165, 703, 727
Isawa, Y., 193
Isbell, D.E., 638
Isolator, 550, 722
Isotropic media, 8

Jackson, J.D., 158, 318, 671
Jahnke, E., 385
Jamieson, H.W., 571
Jasik, H., 607, 617
Jeffrey, A., 93
Jeffreys, B.S., 557

Jeffreys, H., 557
Jones, E.M.T., 542
Jordan, E.C., 598, 649
Junction parameters:
  by analysis, 567–572
  by measurement, 532–535, 538

Kamerlingh Onnes, H., 676
Kapany, N.S., 761
Karplus, W.J., 321
Kay, I.W., 741
Kerr effect, 698
Kerst, D.W., 114
King, D.D., 631
King, R.W.P., 647
Kino, G.S., 330
Kirchhoff's laws, 169, 216
  current law, 174
  voltage law, 169, 206
Kirstein, P.T., 330
Kittel, C., 668, 685, 681
Kline, M., 741
Klystron, 163, 507
Knopp, K., 558, 561
Kober, H., 332
Kock, W.E., 621, 623
Kogelnik, H., 753, 756, 769, 775
Kramers-Kronig relations, 557, 671
Kraus, J.D., 471, 595, 606
Kuh, E.S., 177

Laguerre polynomials, 767, 771
Landau, L.D., 690
Laplace's equation, 36, 65, 101, 216, 384, 567
  in circular cylindrical coordinates, 360–375
  coaxial cylinders, 42
  conformal transformations, 326–343
  numerical solution, 320–325
  product solutions, 346–383
  in rectangular coordinates, 346–360
  in spherical coordinates, 375–383
Laplacian, 36, 64
Laporte, E.A., 603
Laser, 166, 315
  resonator, 774–780
Lax, B., 715, 717
Leaky waveguide array, 581
Lee, S.H., 781
Legendre polynomials, 317, 376, 377
  associated, 499–500
Legendre's equation, 376
  associated, 499
Leibe, F.A., 252
Lenses, 581, 735–738

for directing radiation, 621
for Fourier transforming, 781–784
periodic systems of, 749–752
Levenson, M.D., 685
Li, K.K., 447
Li, T., 775
Lifshitz, E.M., 690
Light, velocity of, 132, 215, 271, back cover
Linear arrays, 627
Linear media, 8, 527, 667
Linear polarization, 276
Line charge, 16, 23
Line integral, 19
  of electric field, 18–20
  of magnetic field, 75–76
Lithium niobate, 699
Logarithmic transformation, 334, 386
London, F., 676
London gauge, 157
  penetration depth, 678
  theory, 677
Longitudinal modes, 780, 788
Longitudinal section waves, 449–451
Loop antenna, 581
  coupling in cylindrical cavity, 508, 564
Lorentz, H.A., 528
Lorentz gauge, 157, 167
Lorentz model of atom, 669, 685
Losch, F., 365
Loss-free networks, 548
Loss-free one ports, 556
Loss-free three ports, 550
Louisell, W.H., 481
Low-pass filter, 176, 541–544
Lumped-element circuits, 168–209
Lundstrom, O.C., 558
Luneberg lens, 622

Mach-Zehnder interferometer, 701
Maclane, S., 541
Magic T, 524, 553–554
Magnetic boundary, 516
Magnetic charge, 102, 110, 126, 608
Magnetic dipole, 96, 581, 586
Magnetic domains, 681
Magnetic energy, 103–107, 138
Magnetic field, 68–71
  between coaxial cylinders, 76
  of solenoid, 78
  inside uniform current, 77
Magnetic flux density, 70
Magnetic forces, 68
Magnetic materials, 679–684
Magnetic radiation vector, 609

Magnetic sphere in uniform field, 378
Magnetic susceptibility, 73
Magnetic vector potential, 91, 187
Magnetization, 72–73, 679–680, 717
Magnetized sphere, 102
Magnetron, 85
Mailloux, R.J., 630
Manley-Rowe relations, 688
Marcatili, E.A.J., 756
Marcuse, D., 460, 755, 781
Marcuvitz, N., 246, 461, 523, 534, 571
Matching of impedances, 220, 235–236, 263, 467, 483
Material:
    dispersion, 763
    parameters, 281, 283, 672, 675, 684
Matrix:
    inversion, 322
    permeability, 689, 714
    permittivity, 689
Matthei, G.L., 542
Maxwell, J.C., 71, 111, 137
Maxwell's equations, 31, 123–129
    complex form, 127–129, 164
    and continuity at boundary, 143–146
    differential form, 123–125
    large-scale form, 126
    and plane waves, 130–135
    potential formulation, 156–161
    and skin effect, 147–153
McLachlan, N.W., 179, 365
Mechtly, E.A., 789
Meissner-Ochsenfeld experiment, 677
Mermin, N.D., 668
Mesh relaxation, 323
Metal-semiconductor junction, 65
Meter-kilogram-second (mks) system, 789
Microstrip transmission lines, 407–410, 450, 543, 546
    as filters, 543, 545–546
Microwave bridge, 524
Microwave integrated circuit, 226
Microwave networks, 523–575
Microwaves lenses, 621
Midwinter, J.E., 685
Milford, F.J., 679, 680
Miller, S.E., 428
Mittag-Leffler theorem, 561
Modal dispersion, 763
Modes, 397
    in optical resonator, 776
    in rectangular resonator, 492–495
    in spherical resonator, 502–504

in step-index optical fiber, 761
in waveguides, 397
Modulated signals, 447
Modulated wave, 275
Modulation index, 700
Molecular polarizability, 668–669
Monopole, 604
Montgomery, C.G., 461, 525, 537, 553
Montgomery, D.B., 193
Morrish, A.H., 683
Morse, P.M., 501
Motional electric field, 117
Mottelay, P.F., 3
Mueller, G.E., 761
Multimode fiber, 786
Multiple coatings, 293, 309
Mutual couplings of antennas, 625, 650
Mutual inductance, 173, 187–190, 206, 208

Nagaoka formula, 191, 208
Neper, 242
Network formulation, 525–528
Neumann's form for mutual inductance, 189
Newton's law, 61
Nonlinear elements, 206
Nonlinear optics, 685–688
Nonlinear polarization, 686, 688
Nonreciprocal networks, 550, 709–726
North, D.O., 656
Norton theorem, 177
Numerical solution, 36, 322–325, 454
    of Laplace equation, 322
    of scalar Helmholtz equation, 324, 454

Oersted, H.C., 68, 69, 71
Ohmic conductors, 37
Ohm's law, 139, 148, 170
Open-circuited transmission line, 250, 563–564, 662
Optical fibers, 315, 759–768
Optical filter, 545–547
Optical indicatrix, 690
Optical information processing, 781
Optical resonators, 733, 768–780
Optical waveguides, 752–768
Ordinary wave, 696–697
Orthogonality properties:
    of Bessel functions, 371–373
    of Legendre functions, 378, 500
    of sinusoids, 351–352
Overcoupled cavity resonator, 511

Packard, R.F., 631

Papas, C.H., 343
Parabolic reflector, 581, 617, 735, 736, 746
Parallel conducting cylinders, 210, 338–339, 387
Parallel-plane:
  diode, 330
  transmission line, 79, 216, 246, 263, 396, 405
  waveguide, 397–407
Parallel-plate capacitor, 27
Paramagnetic substances, 679, 715
Parasitic element, 635
Paraxial approximation, 611, 746
Partial fraction expansion, 558
Periodic circuit, 522, 541–547
Periodic lens system, 749–752, 785
Periodic waveguiding systems, 477–481
Permanent dipoles, 670
Permanent magnetization, 100
Permeability, 71, 125
  initial, 683
  matrix for ferrites, 715
  maximum, 683
Permittivity, 62, 124, 668
  matrix, 705
  relative, 3
Perturbation of resonant cavities, 512–515
Phase constant, 133, 224, 248, 398, 403
Phase matching, 687
Phase scanning of arrays, 630
Phase velocity, 223, 253, 399, 413, 418, 424, 439
  for waves at oblique incidence, 299
Phasors, 127–129, 800–803
Pi equivalent circuit, 531
Pierce, J.R., 330, 471, 473, 775
Pierce gun, 330
Plane waves, 130–135, 270–315
  in anisotropic crystals, 691–698
  sources for radiation, 610
Plasma, 674
  Faraday rotation in, 713–715
  in magnetic field, 703–706
  moving, 707–709, 730
  TEM waves on, 709–713
Plasma frequency, 674, 676, 705
Plonsey, R., 679
pn junction, 44–46, 113
Pockels effect, 698
Poisson equation, 36, 44–46, 65, 97, 319, 321
Polarization, elliptic, 278
Polarization (material), 8, 31, 668–671
  nonlinear, 686
Polarization of electromagnetic waves, 165, 276–279, 418
  designation for waves at angle, 296
Polarization vector, 167
Polarizing (Brewster) angle, 308, 315
Polynomial formulation of arrays, 633
Portis, A.H., 157
Potential:
  electrostatic, 20–26, 29
  magnetic scalar, 98–99
  magnetic vector, 91–93, 96–97
  retarded, 156–161
Potential function, 156, 329, 557
Power:
  conservation, 552
  flow (Poynting's theorem), 137–143
  loss in plane conductor, 154, 313
  splitter, 266
  transfer, 441, 443
Poynting's theorem, 137–143
  for complex phasors, 141–143
Poynting vector, 138, 165
  complex, 142
  in conductor, 154
  in plane waves, 141, 166, 273, 312
  in radiating system, 585
P-polarization, 296
Principal coordinates, 690
Principal wave on antenna, 570
Probe excitation, 507, 522, 555
Product solutions, 346–384
Propagation, 131
  constant, 248, 397, 400
  time, 264
Pulse-forming line, 258, 269
  reflections, 218, 220, 260
Purcell, E.M., 525, 537, 553

Q (quality factor), 243–247, 492, 494, 496, 503, 509–512
  measurement, 509–512
  of resonant transmission lines, 243–247
  of resonator modes, 492, 494, 496, 503, 521
Quadrupole, 63, 659
Quarter-wave:
  coating for eliminating reflections, 293
  plate, 703, 730
  transmission line for impedance matching, 265
Quasistatic, 1, 2, 69, 162, 319, 567

Radiation:
  from aperture, 607
  efficiency, 595

Radiation (*Continued*)
  intensity, 588
  modes in dielectric guides, 460, 755
  patterns, 592
  resistance, 204, 205, 209, 595, 660
  from traveling wave, 598
  vector, 587
Radome, 314
Ramaswamy, V., 753
Ramo, S., 709
Ranganath, T.R., 701
Rat race, 554, 574
Ray matrix, 746–749, 785
  used with gaussian beams, 771–774
Ray optics, 733–752
Rays in inhomogeneous media, 741–745
Ray vector, 693
Reactive power, 142
Reactive wall, 474, 520
Receiving antenna, 199, 652–658
Reciprocal ellipsoid, 690
Reciprocity, 528–530, 537, 538, 541, 548, 573, 655–657
  for antennas, 655–657
  for networks, 529–548
  theorem, 528
Rectangular aperture, 613
Rectangular harmonic, 348
Rectangular resonator, 489–494
Rectangular waveguides, 411–422
Recurrence formulas for Bessel functions, 370
Reflection coefficient, 217, 249, 266, 289, 538
Reflection:
  of circularly polarized wave at oblique incidence, 305
  from good conductor, 290
  law of, 302
  losses, 547
  of plane waves from perfect conductors, 283
  of pulse, 219, 221
  with several dielectrics, 291
Reflectors for radiation, 581, 686
Refractive index, 274
  for silver and nickel, 675
Regular functions, 327
Reitz, J.R., 679, 680
Relaxation factor, 323
Remanence, 683
Residue, 559
Resistance element, 170
Resonance, 240
  isolator, 724, 731
Resonant cavities, *see* Cavity resonators
Resonant slot antenna, 619

Resonant transmission line, 238–240, 243–247, 487
Retarded potential, 156–161, 177
  for time-periodic case, 160
  use in circuit formulation, 198–206
Rhoderick, E.H., 87
Rhombic antenna, 603
Ridge waveguide, 469–470, 484
Rinehart, F.J., 622
Robbins, T.E., 571
Rose-Innes, A.C., 87
Rotation, 84
Rumsey, V.H., 637
Ryzhik, I.H., 93, 615, 616

Sakrison, D., 354
Saturation:
  induction, 683
  magnetization, 718
Saunders, W.K., 654
Savart, F., 68
Scalar magnetic potential, 98, 101, 110
Scalar product, 13, 18
Scattering matrix, 536–537, 548, 553, 573, 575
  measurement of coefficients, 538
Schawlow, A.L., 774
Scheingold, L.S., 538
Schelkunoff, S.A., 286, 501, 561, 562, 583, 584, 586, 633, 642, 645
Schneider, M.V., 409
Schottky barrier, 113
Schwarz transformation, 340–343, 387
Schwering, F., 775
Schwinger, J., 571
Sectoral electromagnetic horn, 464
Segerlind, L.J., 321
Seki, M., 715
Self-inductance, 184–187, 190–191
Semiconductor, 9, 39, 282, 672–674
  diode, 176
Separation of variables, *see* Product solutions
Serber, R., 114
Shielding, electrostatic, 195
Shorted transmission line, 285
Side lobes of antennas, 635
Siegman, A.E., 784
Siemens, 37
Silver, S., 145, 618, 653
Silvester, P., 324, 471
Single-mode fiber, 786
Singular points, 327
Sinusoidal wave, 132–134
  on ideal transmission lines, 223
S.I. system of units, 4, 789

Skin depth, 149–151

Skin effect, 147–153, 178–183, 186
  properties of metals, 150, 153
  in round wires, 178–183

Slater, J.C., 512

Slot antenna, 581, 614, 619–621

Slotline, 410–411

Slotted line, 229, 266

Slow-wave structures, 471–474

Smith, P.D., 644

Smith, P.H., 229

Smith chart, 229–238, 526
  for admittances, 235, 266
  for cavity coupling, 510
  for plane waves, 285, 291, 293

Smythe, W.R., 377

Snell's law, 303, 307

Snyder, A.W., 762

Solenoid, 75, 107, 208

Soohoo, R.F., 715

Sources for circuits, 176

Space-charge-limited diode, 64

Space-charge waves, 707–709

Space factor of array, 627

Spatial harmonics, 456, 477–481

Spherical aberration, 734–735

Spherical antennas, 640–643

Spherical capacitor, 28

Spherical harmonics, 375–381

Spherical magnetic medium, 378

Spherical mirror, 734, 735

Spherical mirror resonator, 749–752, 774–780

Spherical waves, 498–502, 521

Spiegel, M.R., 93, 179, 767

Spin, 72

S-polarization, 296

Stability:
  diagram for laser resonators, 779
  for periodic lens system, 751

Standing wave, 224–229, 238–240, 264,
    283–285, 298–299, 487–488
  in cavity resonators, 487–488, 496
  plane waves, incident at angle, 298–299
  plane waves, normal incidence, 283–285
  ratio, 226–229, 289–290
  transmission lines, 226–229, 238–240

Statcoulomb, 61

Stegun, I.A., 93, 365

Stein, S., 538

Step-index fiber, 759

Stiglitz, M.R., 516

Stix, T., 710

Stokes's theorem, 88–91, 109, 115, 117, 126,
    187

Stone, J.M., 294

Storer, J.E., 538

Stratton, J.A., 138, 145, 256, 498, 599, 610,
    642, 643, 689

Stripline, 265, 407–408, 450, 725

Stub tuner, 235

Stutzman, W.L., 602, 640

Summerfeld, A., 477

Superconducting solenoid, 104

Superconductor, 69, 87, 109, 110, 676–678

Supergain arrays, 633

Superposition, 319–320, 358–359

Surface:
  charge density, 66
  guiding, 474–477, 484
  integral, 13
  resistivity, 152

Susceptibility:
  electric, 9
  magnetic, 73

Tai, C.T., 117, 644

Tamir, T., 756

Tandem networks, 538, 748

Taylor, T.T., 663

Telegraphist's equations, 213

Television receiving antenna, 637

TEM waves, see Transverse electromagnetic
    (TEM) waves

Tensor, 689, 706

T equivalent circuit, 531

Terman, F.E., 192

TE waves, see Transverse electric (TE) waves

Thévenin theorem, 177, 572

Thiele, G.A., 602, 640

Thin-film transmission line, 226, 242

Thin lens, 736–738, 745, 746

Thomas, R.K., 631

Tien, P.K., 756

TM waves, see Transverse magnetic
    (TM) waves

Torque, 107, 109

Total reflection, 306–307, 344, 457

Townes, C.H., 774

Transfer matrix, 539

Transformer, 114
  audio, 151
  ideal, 532, 540

Transistor, 176

Transmission coefficient, 216–217, 224, 225,
    289
  of network, 535–537, 573

Transmission line, 55, 66, 69, 210–269
  analogy for plane waves, 285–288, 711

Transmission line (*Continued*)
 biconical, 467–469
 general formulas, summary of, 250
 losses, 241
 nonuniform, 261–263, 460
 parallel-plane, 396, 405
 radial, 460–465
 Smith chart for, 229–238
 specific lines, summary for, 252
 TEM waves on, 434–438
Transverse electric (TE) waves, 296, 299, 395,
   415, 424, 442
Transverse electromagnetic (TEM) waves, 216,
   264, 395–396, 434–438, 467–469, 494
Transverse magnetic (TM) waves, 296, 306,
   395, 412, 423, 438, 453, 498, 585
Transverse modes of optical resonator, 780
Traveling wave, 214, 223, 228, 272
 antenna, 581, 600
Tubes of flux, 15
Turner, C.W., 678
Two port, 530
Tyrrell, W.A., 761

Ueda, T., 409
Undercoupled resonator, 511
Uniaxial crystal, 691, 695–698
Uniform plane wave, 133, 135, 164, 271
Uniqueness, 47, 145, 527–528, 572
Unitary matrices, 550–552
Unit matrix, 549
Units, 3, 789–790. *See also* International
 system of units
Unloaded Q, 512
Unpolarized wave, 278
Unstable optical resonator, 751, 779
Uzdy, Z., 725

Vacuum diode, 122
Van Bladel, J., 518
Van Duzer, T., 678
V antenna, 600–603
 with standing waves, 661
Variational methods, 571
Vector:
 complex, 128
 differentiation, 794
 dot product, *see* Dot product of vectors
 magnetic potential, 91–93, 96, 101
 operator, 32
 product, 70. *See also* Cross product of
 vectors
 relations, 791, back cover

Velocity:
 of energy propagation, 256, 268
 group, *see* Group, velocity
 of light, 132, 215, 271, back cover
 modulation, 507
 phase, *see* Phase velocity
Vemuri, V., 321
Verdet constant, 715
Verweel, J., 684
Voltage:
 in circuits, 169–174
 generated by changing magnetic field, 114
 on transmission lines, 212–213, 434
 in waveguide modes, 525–526
Volume integral, 13, 64
Von Hippel, A.R., 668

Wait, J.R., 598
Wang, S., 701
Waters, W.E., 330
Watson, G.N., 365
Wave:
 equation, 131–136, 165
 impedance, *see* Characteristic wave
 impedance
 number, 133, 274
 velocity, 215
Waveguide:
 attenuators, 444
 discontinuities, 571
 modes, 397
Wavelength, 134, 224, 274
 guide, 399, 419
Waves:
 below cutoff, 444
 guided by conducting parallel plates, 396–407
 guided by dielectrics, 457–460, 752–764
 guided by hollow-pipe guides, 411–446
 guided by special types, 456–481
 in imperfect dielectrics and conductors,
   279–282
Weakly guiding fibers, 762–764
Weeks, W.L., 639, 661
Weissfloch, A., 534
Wheeler, H.A., 192, 409
Whinnery, J.R., 447, 571, 663, 785
Whittaker, E.T., 3, 71
Wire antenna, 589
Wold, E., 740
Wolf, E., 277, 609, 689, 695, 746
Wolff, E.A., 623
Woodyard, J.R., 633
Wooten, F., 676

Work integral, 18
Wyllie, G., 668

Yagi-Uda arrays, 635–637
Yamashita, E., 409
Yariv, A., 685, 688, 698, 733, 746
Yeh, P., 685, 733

Y junction, 550–574
Young, L., 542

Zenneck, J., 477
Zernike, F., 685
Zigzag rays, 482, 756
Zucker, F.J., 619, 647